STEPS FOR GRAPHING FUNCTIONS

STEP 1 Find the domain of f.
STEP 2 Locate the intercepts of f. (Skip the x-intercepts if they are too hard to find.)
STEP 3 Determine where the graph of f is increasing or decreasing.
A function f is increasing on an interval I if $f'(x) > 0$ on I.
A function f is decreasing on an interval I if $f'(x) < 0$ on I.
STEP 4 Find any local maxima or local minima of f by using the First Derivative Test or the Second Derivative Test.
If c is in the domain of a continuous function f and if f is decreasing for $x < c$ and is increasing for $x > c$, then at c there is a local minimum. $f(c)$ is the local minimum.
If c is in the domain of a continuous function f and if f is increasing for $x < c$ and is decreasing for $x > c$, then at c there is a local maximum. $f(c)$ is the local maximum.
If $f'(c) = 0$ and if $f''(c) > 0$, then at c there is a local minimum.
If $f'(c) = 0$ and if $f''(c) < 0$, then at c there is a local maximum.
STEP 5 Locate all points on the graph of f at which the tangent line is either horizontal or vertical.
STEP 6 Determine the end behavior and locate any asymptotes.
STEP 7 Locate the inflection points, if any, of the graph by determining the concavity of the graph.
If $f''(x) > 0$ on an interval I, the f is concave up on I.
If $f''(x) < 0$ on an interval I, the f is concave down on I.
Any point on the graph of f where the concavity changes is an inflection point.

INTEGRAL CALCULUS

Indefinite integral $\quad \int f(x)\, dx = F(x) + K \quad\quad F'(x) = f(x); K$ is a constant

Definite integral $\quad \int_a^b f(x)\, dx = F(b) - F(a) \quad\quad F'(x) = f(x)$

Integration by parts $\quad \int u\, dv = uv - \int v\, du$

Area A under the graph of f from a to b $\quad A = \int_a^b f(x)\, dx \quad\quad f(x) \geq 0$ on the interval $[a, b]$

PARTIAL DERIVATIVES

$$f_x(x, y) = \lim_{\Delta x \to 0} \frac{f(x + \Delta x, y) - f(x, y)}{\Delta x} \quad\quad f_y(x, y) = \lim_{\Delta y \to 0} \frac{f(x, y + \Delta y) - f(x, y)}{\Delta y}$$

eGrade Plus

www.wiley.com/college/sullivan
Based on the Activities You Do Every Day

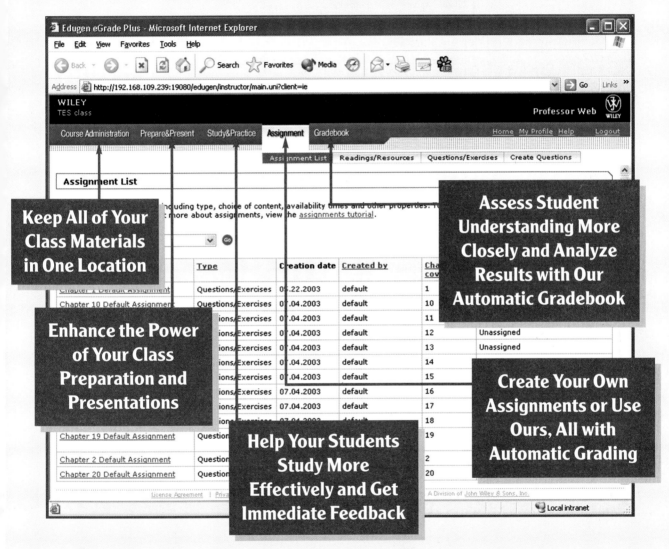

All the content and tools you need, all in one location, in an easy-to-use browser format. Choose the resources you need, or rely on the arrangement supplied by us.

Now, many of Wiley's textbooks are available with eGrade Plus, a powerful online tool that provides a completely integrated suite of teaching and learning resources in one easy-to-use website. eGrade Plus integrates Wiley's world-renowned content with media, including a multimedia version of the text. Upon adoption of eGrade Plus, you can begin to customize your course with the resources shown here.

See for yourself! Go to www.wiley.com/college/egradeplus for an online demonstration of this powerful new software.

Keep All of Your Class Materials in One Location

Course Administration tools allow you to manage your class and integrate your eGrade Plus resources with most Course Management Systems, allowing you to keep all of your class materials in one location.

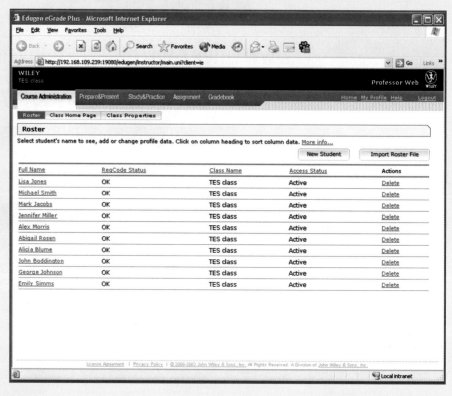

Enhance the Power of Your Class Preparation and Presentations

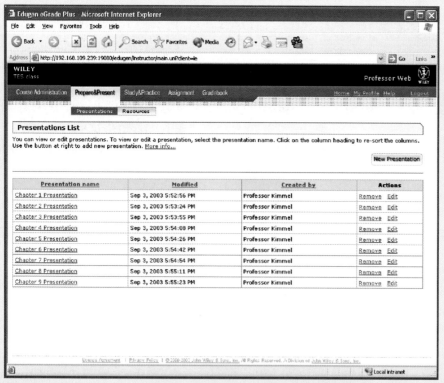

A Prepare and Present tool contains all of the Wiley-provided resources, making your preparation time more efficient. You may easily adapt, customize, and add to Wiley content to meet the needs of your course.

www.wiley.com/college/sullivan

Create Your Own Assignments or Use Ours, All with Automatic Grading

An **Assignment** area allows you to create **student homework** and **quizzes** by using **Wiley-provided question banks**, or by writing your own. You may also assign readings, activities and other work you want your students to complete. One of the most powerful features of eGrade Plus is that student assignments will be automatically graded and recorded in your gradebook. This will not only save you time but will provide your students with immediate feedback on their work.

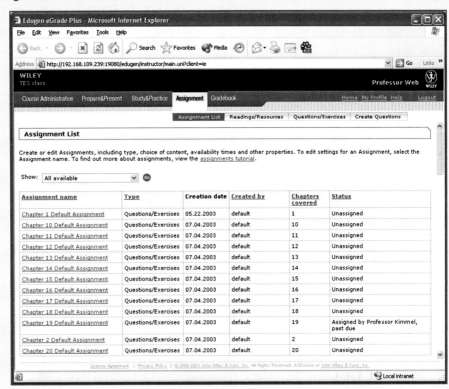

Assess Student Understanding More Closely

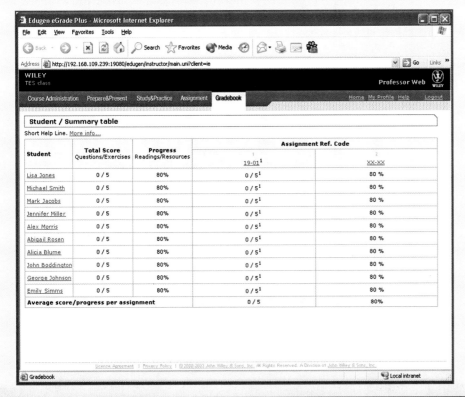

An **Instructor's Gradebook** will keep track of your students' progress and allow you to analyze individual and overall class results to determine their progress and level of understanding

www.wiley.com/college/sullivan

Students, eGrade Plus Allows You to:

Study More Effectively

Get Immediate Feedback When You Practice on Your Own

eGrade Plus problems link directly to relevant sections of the **electronic book content**, so that you can review the text while you study and complete homework online. Additional resources include **solutions manual material** and other problem-solving resources

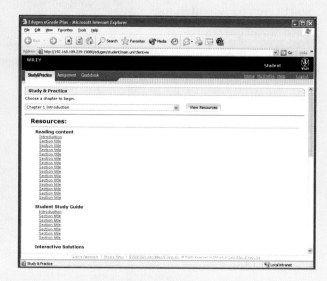

Complete Assignments / Get Help with Problem Solving

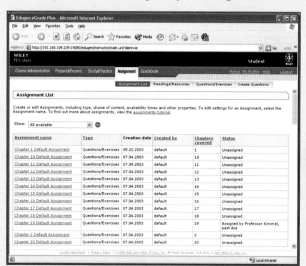

An **Assignment** area keeps all your assigned work in one location, making it easy for you to stay "on task." In addition, many homework problems contain a **link** to the relevant section of the **multimedia book**, providing you with a text explanation to help you conquer problem-solving obstacles as they arise. You will have access to a variety of resources for building your confidence and understanding.

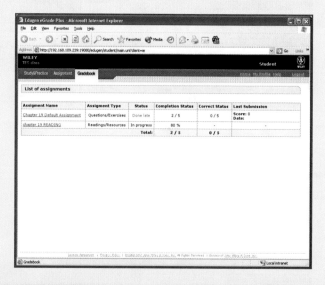

Keep Track of How You're Doing

A **Personal Gradebook** allows you to view your results from past assignments at any time.

www.wiley.com/college/sullivan

Brief Calculus
An Applied Approach

Eighth Edition

Michael Sullivan Chicago State University

JOHN WILEY & SONS, INC.

ACQUISITIONS EDITOR	Michael Boezi
ASSOCIATE PUBLISHER	Laurie Rosatone
FREELANCE DEVELOPMENTAL EDITOR	Anne Scanlan-Rohrer
EXECUTIVE MARKETING MANAGER	Julie Z. Lindstrom
SENIOR PRODUCTION EDITOR	Norine M. Pigliucci
SENIOR DESIGNER	Harry Nolan
COVER DESIGN	Howard Grossman
INTERIOR DESIGN	Jerry Wilke Design
ILLUSTRATION EDITOR	Sigmund Malinowski
ELECTRONIC ILLUSTRATIONS	Techsetters, Inc.
PHOTO EDITOR	Lisa Gee
ASSISTANT EDITOR	Jennifer Battista
EDITORIAL ASSISTANT	Kelly Boyle
PRODUCTION MANAGEMENT SERVICES	Suzanne Ingrao
COVER PHOTOS	(Peach) © Tim Turner/FoodPix/Getty Images (Apple) © Brian Hagiwara/FoodPix/Getty Images (Plum) © Richard Kolker/The Image Bank/Getty Images
INSET COVER PHOTOS	© Getty Images and Digital Vision

Excel is a trademark of Microsoft, Inc.

This book was set in Minion by Progressive Information Technologies and printed and bound by Von Hoffmann Corporation. The cover was printed by Von Hoffmann Corporation.

This book is printed on acid free paper. ∞

Copyright © 2005 John Wiley & Sons, Inc. All rights reserved.

No part of this publication may be reproduced, stored in a retrieval system or transmitted in any form or by any means, electronic, mechanical, photocopying, recording, scanning or otherwise, except as permitted under Sections 107 or 108 of the 1976 United States Copyright Act, without either the prior written permission of the Publisher, or authorization through payment of the appropriate per-copy fee to the Copyright Clearance Center, 222 Rosewood Drive, Danvers, MA 01923, (978) 750-8400, fax (978) 646-8600. Requests to the Publisher for permission should be addressed to the Permissions Department, John Wiley & Sons, Inc., 111 River Street, Hoboken, NJ 07030, (201) 748-6011, fax (201) 748-6008.

To order books or for customer service please, call 1(800)-CALL-WILEY (225-5945).

ISBN 0-471-45202-5

Printed in the United States of America

10 9 8 7 6 5 4 3 2 1

For Kathleen

About the Author

Michael Sullivan

is Professor Emeritus in the Department of Mathematics and Computer Science at Chicago State University where he taught for 35 years before retiring a few years ago. Dr. Sullivan is a member of the American Mathematical Society, the Mathematical Association of America, and the American Mathematical Association of Two Year Colleges. He is President of the Text and Academic Authors Association and represents that organization on the Authors Coalition of America. Mike has been writing textbooks in mathematics for over 30 years. He currently has 13 books in print: 3 texts with John Wiley & Sons and 10 with Prentice-Hall. Six of these titles are co-authored with his son, Michael Sullivan III.

Mike has four children: Kathleen, who teaches college mathematics; Michael, who teaches college mathematics, Dan, who is a Prentice-Hall sales representative, and Colleen, who teaches middle-school mathematics. Nine grandchildren round out the family.

Preface to the Instructor

The Eighth Edition

The Eighth Edition of *Brief Calculus* builds upon a solid foundation by integrating new features and techniques that further enhance student interest and involvement. The elements of previous editions that proved successful remain, while many changes have been made. Virtually every change is the result of thoughtful comments and suggestions from colleagues and students who have used previous editions. I am sincerely grateful for this feedback and have tried to incorporate changes that improve the flow and usability of the text.

New to the Eighth Edition

Chapter Projects
Each chapter begins with a student-oriented essay that previews a project involving the mathematics of the chapter. The Chapter Project appears at the end of the chapter and is designed for an individual or collaborative experience. Each project builds on the mathematics of the chapter but also requires students to stretch their understanding of the concepts.

A Look Back, a Look Forward
At the beginning of each chapter, a connection is made between previously studied material and material found in the current chapter.

Preparing for This Section
Most sections now open with a referenced list (by section and page number) of key items to review in preparation for the section ahead. This provides a just-in-time review for students.

Objectives
At the beginning of every section is a numbered list of objectives. The objective is cited at the appropriate place it is encountered in the section, and marked with a numbered icon .

Chapter Reviews
Each chapter concludes with a variety of features to synthesize the important ideas of the chapter.

Things to Know
A review of the important definitions, formulas, and equations from the chapter.

Objectives
The section objectives are listed again, with references to the review exercises that relate to them.

True/False Items; Fill in the Blanks
Short, quick-answer questions to test vocabulary and concepts.

Review Exercises
These now reflect the objectives of each section. Blue problem numbers can be used by the student as a Practice Test.

Mathematical Questions from Professional Exams
Where appropriate, questions from CPA, CMA, and Actuary Exams have been reproduced.

Exercises and Examples

Sourced Problems
Many new examples and exercises that contain sourced data or sourced facts have been added to each problem set.

Conceptual Problems
These new problems ask the student to verbalize or to write an answer to a problem that may have multiple solutions or require some research. These problems are clearly marked with an icon and green color to make them easy to identify.

Technology
As an optional feature, examples and exercises have been included that utilize a graphing utility. These examples and exercises are clearly identified using an icon and blue problem numbers. An Appendix provides explanations, examples, and exercises that relate to the optional use of a graphing utility. References to this appendix appear in the text.

Design
This edition has a fully integrated, pedagogically based, utilization of color in both the text and the illustrations. Just look at it to see the effect.

Using the Eighth Edition Effectively and Efficiently with Your Syllabus

The following chapter descriptions along with the Table of Contents will help you match the topics of this book with your syllabus. Please do not hesitate to contact me (through the publisher John Wiley & Sons) if you have any questions about the content and organization of this book.

Chapter 0 Review
This chapter consists of review material. It may be used as the first part of the course or as a just-in-time review when the content is required. Specific references to this chapter occur throughout the book to assist in the review process.

Chapter 1 Functions and Their Graphs
While this chapter reviews material studied in College Algebra or Precalculus, it also introduces many of the concepts needed later in calculus. For example, while the student is reviewing function notation, the difference quotient is introduced. When the graph of a function is discussed, the concepts of increasing and decreasing functions, local maximum and minimum, and average rate of change are introduced.

Chapter 2 Classes of Functions
Here quadratic, power, polynomial, rational, exponential, and logarithmic functions are introduced in the context of how they will be utilized in calculus. Section 2.6, Continuously Compounded Interest, may be omitted without a loss of continuity.

Chapter 3 The Limit of a Function
While intuitive in approach, the techniques introduced here will be used later in the discussion of the derivative and the integral of a function. Uses of limits as they apply to continuity, end behavior, and asymptotes are included.

Chapter 4 The Derivative of a Function
The derivative and its applications to geometry (tangent line), economics (marginal analysis), and physics (velocity and acceleration) are discussed in an organized, student-friendly manner. The subsection on Velocity and Acceleration may be omitted without a loss of continuity.

Chapter 5 Applications: Graphing Functions; Optimization
All the information required to obtain the graph of a function is given here using methodology from both algebra and calculus, with emphasis on the calculus. Additional applications are given to optimization (Section 5.4), elasticity of demand (Section 5.5), related rates (Section 5.6), and linear approximations using differentials (Section 5.7). These sections are independent of each other and may covered in any order or be omitted without any loss of continuity.

Chapter 6 The Integral of a Function and Applications
The indefinite integral and the definite integral are introduced here with various applications. Sections 6.5, 6.6, and 6.7 are independent of each other and may be covered in any order or be omitted without any loss of continuity.

Chapter 7 Other Applications and Extensions of the Integral
This chapter contains optional material. Section 7.3 depends on Section 7.1.

Chapter 8 Calculus of Functions of Two or More Variables
This chapter is optional. Sections 8.1–8.4 follow in sequence. Sections 8.4 and 8.5 are optional and may be covered in any order or omitted without loss of continuity.

Appendix: Graphing Utilities
This appendix provides an overview of some common uses of a graphing calculator in calculus.

Other Books in this Series

Also available are *Finite Mathematics: An Applied Approach, 9/e* (ISBN 0-471-45202-5), for a one-term course in finite mathematics and *Mathematics: An Applied Approach, 8/e* (ISBN 0-471-32784-0), for the combined course in finite mathematics and calculus.

The Faculty Resource Network

The *Faculty Resource Network* is a peer-to-peer network of academic faculty dedicated to the effective use of technology in the classroom. This group can help you apply innovative classroom techniques, implement specific software packages, and tailor the technology experience to the specific needs of each individual class. Ask your Wiley representative for more details.

Acknowledgments

There are many colleagues I would like to thank for their input, encouragement, patience, and support. They have my deepest thanks and appreciation. I apologize for any omissions.

Contributors
- Tim Comar of Benedictine University, Kathleen Miranda of SUNY Old Westbury, and Kurt Norlin of Laurel Technical Services for working on the answer sections.
- Thomas Polaski of Winthrop University for providing the chapter projects.
- Ken Brown of Southeastern Louisiana University, Mike Divinia of San Jose City College, and Henry Smith of River Parishes Community College for providing new applications and problems using real data.
- Kathleen Miranda for accuracy checking the text.

Reviewers

Stephanie L. Fitch	University of Missouri–Rolla
David Harpster	Minot State University
Richard Leedy	Polk Community College
Tsun-Zee Mai	University of Alabama
Bette Nelson	Alvin Community College

Krish Revuluri	Harper College
William H. Richardson, Jr.	Francis Marion University
Katie Stables	Western Washington University
Stephen J. Tillman	Wilkes University
Helene Tyler	Manhattan College
Cheryl Whitelaw	Southern Utah University
Yangbo Ye	The University of Iowa

Recognition and thanks are due particularly to the following individuals for their valuable assistance in the preparation of this edition:

- Michael Boezi, for his enthusiasm and support
- Kelly Boyle, for her skill at coordinating a complicated project
- Lisa Gee, for identifying representative photos
- Suzanne Ingrao, for her organizational skills as production manager
- Bonnie Lieberman, for taking a chance
- Julie Lindstrom, for her innovative marketing skills
- Sigmund Malinowski, for his dedication to quality illustrations
- Harry Nolan, for creating a beautifully designed book
- Kurt Norlin of Laurel Technical Services for checking all the answers and especially for his high regard for accuracy
- Norine Pigliucci, for making it all happen on time
- Laurie Rosatone and Bruce Spatz, for their forthright and candid viewpoints and strong encouragement
- Anne Scanlan-Rohrer, for her loyalty and professionalism as development editor

And, to Kathleen Miranda, who read page proofs to ensure accuracy, worked problems for the Solutions Manuals, offered many useful suggestions, and listened so often.

I also want to thank the Wiley sales staff for their continued support and confidence over the years.

Finally, I welcome comments and suggestions for improving this text. Please do not hesitate to contact me through the publisher, John Wiley & Sons.

Sincerely,

Michael Sullivan

Preface to the Student

As you begin this course in Brief Calculus, you may feel overwhelmed by the number of topics that the course contains. Some of these may be familiar to you, while others may be new to you. Either way, I have written this text with you, the student, in mind.

I have taught courses in Brief Calculus for over 30 years. I am also the father of four college graduates who, while in college, called home from time to time frustrated. I know what you are going through and have written a text that doesn't overwhelm or unnecessarily complicate.

This text was designed and written to help you master the terminology and basic concepts of Brief Calculus. Many learning aids are built into the format of the text to make your study of the material easier and more rewarding, helping you focus your efforts to get the most from the time and effort you invest.

How to Use this Book Effectively and Efficiently

First, and most important, this book is meant to be read! Please, read the material assigned to you. You will find that the text has additional explanations and examples that will help you.

Many sections begin with *Preparing for this Section*, a list of concepts that will be used in the section. Take the short amount of time required to refresh your memory. This will make the section easier to understand and will actually save you time and effort.

A list of *Objectives* is provided at the beginning of each section. Read them. They will help you recognize the important ideas and skills developed in the section.

After a concept has been introduced and an example given, you will see ↘ Now Work Problem xx. Go to the exercises at the end of the section, work the problem cited, and check your answer in the back of the book. If you get it right, you can be confident in continuing on in the section. If you don't get it right, go back over the explanations and examples to see what you might have missed. Then rework the problem. Ask for help if you miss it again.

If you follow these practices throughout the section, you will find that you have probably done many of your homework problems. In the exercises, every Now Work

Problem is in yellow with a pencil icon. All the odd-numbered problems have answers in the back of the book and worked-out solutions in the Student Solutions Manual. Be sure you have made an honest effort before looking at a worked-out solution.

At the end of each chapter, there is a Chapter Review. Use it to be sure you are completely familiar with the definitions, formulas, and equations listed under Things To Know. If you are unsure of an item here, use the page reference to go back and review it. Go through the Objectives and be sure you can answer 'Yes' to the question 'I should be able to . . . ' Review exercises that relate to each objective are listed to help you.

Lastly, do the problems in the Review Exercises identified with blue problem numbers. These are my suggestions for a Practice Test. Do some of the other problems in the review for more practice to prepare for your exam.

Please do not hesitate to contact me, through the publisher of this book, John Wiley and Sons, with any suggestions or comments that would improve this text. I look forward to hearing from you.

Best wishes.

Michael Sullivan

Supplements

The following ancillary materials are designed to support the text.

Student Solutions Manual
The Student Solutions Manual contains worked-out solutions to all of the odd-numbered problems. ISBN: 0-471-46644-1

Instructor's Solutions Manual
This manual contains worked-out solutions for all problems in the text.
ISBN: 0-471-46645-X

TI 83 Technology Resource Manual
This manual contains basic instructions for using technology with the text. Students get suggestions for using their calculators and a description of the steps used to solve particular problems from the text.
ISBN: 0-471-46642-5

Test Bank
The test bank contains a wide range of problems and their solutions, which are keyed to the text and exercise sets.
ISBN: 0-471-46643-3

Computerized Test Bank
Available in both IBM and Macintosh formats, the Computerized Test Bank allows instructors to create, customize, and print a test containing any combination of questions from the test bank. Instructors can also edit the questions or add their own.
ISBN: 0-471-46646-8

eGrade
eGrade is an online assessment system that contains a large bank of skill-building problems, homework problems, and solutions. Instructors can automate the process of assigning, delivering, grading, and routing all kinds of homework, quizzes and tests while providing students with immediate scoring and feedback on their work. Wiley eGrade "does the math". . . and much more. For more information, visit http://www.wiley.com/college/egrade or contact your Wiley representative.

Applications Index

Business and Economics
Advertising   95, 271, 279, 319, 448
Agriculture   118, 284, 334, 364, 437
Airfares   302
Airline performance   503
Architecture   75, 176, 233
Budget deficit   450
Cable installation   119
Capital value   490
Car costs and repairs   64, 94, 162, 214, 226, 299, 301, 317, 319, 437, 504
Cell phones   146, 257, 267, 415, 521
Checkbook balance   29
Cobb-Douglas model   530, 544–545, 554–555
College savings   227
Compound interest   223–225, 226–227, 232–234
Construction   76, 114, 135, 521, 545
Consumer's surplus   457–459. 462, 482
Cost   29, 92, 94, 114, 115, 119, 123, 136, 161, 234, 264, 267, 271, 280–281, 285, 291, 293, 294, 302, 318, 334, 344. 364, 375, 377, 386, 403, 404, 416–418, 428, 442, 446, 449, 480–481, 503, 544, 554
Cost of fencing   544
Cost of newspaper delivery   94
Demand   118, 161, 175, 284, 285, 302, 307, 332, 344, 396, 397–398, 405, 415, 416, 530, 555
Depreciation   201, 233
Diminishing returns   373
Drum   161
Economical speed   392
Economics   536
Economy   482
Elasticity of demand   393–396, 397–398, 415
Electricity rates   63, 94, 144
Employment   419, 437

Energy costs   144, 146
Inventory   417, 502, 555
Interest   225–227, 450
Investment   225, 226–227, 232–234
IRA   234
Life of a light bulb   502
Lorentz curve   482–483
Mail rates   146, 257
Manufacturing   29, 74, 123, 267, 402–403, 412
Marginal analysis   280, 390, 427–429, 480–481, 526
Marginal cost   526, 554
Marginal productivity   527, 529–530, 536
Market penetration   319
Maximum profit   416, 459, 460, 480, 534–535, 536, 541, 554
Maximum revenue   416, 462, 480
Minimizing cost   386, 418, 480, 542
Mining   459, 462
Oil drilling   459
Page design   118, 392
Price   171, 294, 320, 339, 397, 416, 475, 477, 482, 507
Producer's surplus   457–459, 462, 482
Production   92, 267, 271, 339, 363, 522, 526–527, 544, 554–555
Profit   115, 315, 319, 390, 393, 403, 404, 428, 430–431, 481, 507, 554
Quality control   409, 411, 412, 499, 508, 510
Real estate   63, 490
Revenue   115–116, 171, 175, 282, 285, 293, 306, 307, 318, 332, 344, 364, 397, 403, 404–405, 428, 437, 449, 482, 493, 554
Salaries   94, 319
Sales commissions   63
Sales forecasts   284, 319, 364, 373, 377, 380, 447–448, 450, 481
Sales of cars   380
Sales of tickets   284, 363, 397
Stocks   120, 190, 504
Supply   284

xvii

Applications Index

Taxes 63, 147, 271–272, 388, 392
Telephone boxes 391
Truck rental cost 94
Tuition costs 431
Value of a painting 301
Water bills 64

Life and Physical Sciences
Age of a fossil 476
Age of a tree 476
Air flow 389
Amino acids 308
Area 29, 75, 172, 175, 233, 385, 492, 537
Atmospheric pressure 200, 212, 318, 477
Average speed 492
Bacterial growth 302, 380, 473, 476, 482
Ball bearings 29, 408–409
Ballistics 327
Balloon travel 81
Balloon radius 400, 404, 411, 416
Bee and ant 411
Blood tests 344
Body temperature 29
Bomber 284
Building height 73, 412
Carbon dating 474–475, 482
Cell diameter 26
Celsius to Fahrenheit conversion 27
Chemistry 63, 177, 212, 364, 431, 476
Current in circuits 201
Dimensions of a body of water 76, 233, 401, 404
Dimensions of a box 384, 391, 544, 545, 555
Dimensions of a can 386, 391, 416
Dimensions of a cylinder 161, 391
Dimensions of a parcel of land 175, 233, 544
Dimensions of a rectangle 537
Distance between objects 76, 81
Drug concentrations 200, 208–209, 213, 214, 267, 302, 393, 417, 490, 536
Earth's surface area 26
Earthquakes 214
Elevation 135
Falling body 327, 343, 431
Gas pressure 334, 537
Glucose conversion 285
Growth of a tumor 416
Healing rates 200, 212
Heat index 521, 530
Humidity 262
Illumination 302
Instantaneous rate of change 285, 294
Loudness of sound 213, 233

Metal detector 536
Metal plate expansion 404, 411
Mountain heights 233
Optics 199, 233
Pendulum 412
Pollution 115, 267, 339, 405, 431, 435
Population growth 429, 441, 476, 482
Projectile motion 172, 175–176, 324
Radioactive decay 473–474, 476, 503
Rainfall 492–493
Rising object 302
Satellites 200
Size of a burn 416
Speed 392
Spread of a disease 94, 344, 380
Spread of a ripple 399, 416
Temperature changes 27, 95, 490–491
Transatlantic crossing 392
Travel time 392
Velocity 321–323, 327, 339
Volume of a cone 161
Volume of a cube 29
Volume of a cylinder 161
Volume of a rectangle 384
Volume of a reservoir 95, 431
Volume of a sphere 29, 412, 416
Voltage 29
Waste management 536
Water flow 401, 404, 431
Weber-Fechner law 318
Wind chill 147, 257
Windows 176, 392
Work output 294
Wound healing 417

Social Sciences
Age distribution 496
Arrival times 196–197, 200, 213, 508–509
Baseball 81, 325, 343, 530
Basketball 521, 530
College degrees 334
Crime rates 177, 234, 431
Critical thinking 177, 234
Death rates 493
Diversity 212
Enrollment trends 339
Gestation 503
Golf 135
Grades 63
Health care 177
Human body weight 504
Income distribution 176, 482–483

IQ scores **64, 521**
Learning curves **201, 213, 339**
Life expectancy **507**
Parcel regulations **536**
Population growth **234–235, 343, 431, 484, 492**

Population, United States **213**
SAT scores **118, 364**
Satisfaction and reward **303**
Spread of a rumor **200, 213, 380**
Waiting times **490, 498–499, 500, 503–503, 507–508**

Contents

CHAPTER 0 REVIEW 1
- **0.1** Real Numbers 2
- **0.2** Algebra Review 16
- **0.3** Polynomials and Rational Expressions 29
- **0.4** Solving Equations 40
- **0.5** Intervals; Solving Inequalities 51
- **0.6** nth Roots; Rational Exponents 64
- **0.7** Geometry Review 71
- **0.8** Rectangular Coordinates 76
- **0.9** Lines 81

CHAPTER 1 FUNCTIONS AND THEIR GRAPHS 97
- **1.1** Graphs of Equations 98
- **1.2** Functions 109
- **1.3** Graphs of Functions; Properties of Functions 120
- **1.4** Library of Functions; Piecewise-defined Functions 137
- **1.5** Graphing Techniques: Shifts and Reflections 148
- **Chapter Review** 156
- **Chapter Project: Evaluating Car Rental Costs** 162

CHAPTER 2 CLASSES OF FUNCTIONS 163
- **2.1** Quadratic Functions 164
- **2.2** Power Functions; Polynomial Functions; Rational Functions 178
- **2.3** Exponential Functions 186
- **2.4** Logarithmic Functions 202

xxi

2.5 Properties of Logarithms 214
2.6 Continuously Compounded Interest 223
Chapter Review 227
Chapter Project: Predicting Population Growth 234

CHAPTER 3 THE LIMIT OF A FUNCTION 236

3.1 Finding Limits using Tables and Graphs 237
3.2 Techniques for Finding Limits of Functions 243
3.3 One-sided Limits; Continuous Functions 251
3.4 Limits At Infinity; Infinite Limits; End-Behavior; Asymptotes 258
Chapter Review 267
Chapter Project: Tax Rates and Continuous Functions 271

CHAPTER 4 THE DERIVATIVE OF A FUNCTION 273

4.1 The Definition of a Derivative 274
4.2 The Derivative of a Power Function; Sum and Difference Formulas 286
4.3 Product and Quotient Formulas 295
4.4 The Power Rule 303
4.5 The Derivatives of the Exponential and Logarithmic Functions; the Chain Rule 308
4.6 Higher-Order Derivatives 320
4.7 Implicit Differentiation 328
4.8 The Derivative of $f(x) = x^{p/q}$ 334
Chapter Review 340
Chapter Project: Testing Blood Efficiently 344

CHAPTER 5 APPLICATIONS: GRAPHING FUNCTIONS; OPTIMIZATION 347

5.1 Horizontal and Vertical Tangent Lines; Continuity and Differentiability 348
5.2 Increasing and Decreasing Functions; the First Derivative Test 354
5.3 Concavity; the Second Derivative Test 365
5.4 Optimization 381
5.5 Elasticity of Demand 393
5.6 Related Rates 398
5.7 The Differential and Linear Approximations 405
Chapter Review 412
Chapter Project: Inventory Control 417

CHAPTER 6 THE INTEGRAL OF A FUNCTION AND APPLICATIONS 420

6.1 Antiderivatives; The Indefinite Integral; Marginal Analysis 421

6.2 Integration Using Substitution 432
6.3 Integration by Parts 437
6.4 The Definite Integral; Learning Curves; Sales over Time 442
6.5 Finding Areas; Consumer's Surplus, Producer's Surplus; Maximizing
Profit over Time 450
6.6 Approximating Definite Integrals Using Rectangles 463
6.7 Differential Equations... 470
Chapter Review 477
Chapter Project: Inequality of Income Distribution 482

CHAPTER 7 OTHER APPLICATIONS AND EXTENSIONS OF THE INTEGRAL 485

7.1 Improper Integrals 486
7.2 Average Value of a Function 490
7.3 Continuous Probability Functions 493
Chapter Review 505
Chapter Project: A Mathematical Model for Arrivals 509

CHAPTER 8 CALCULUS OF FUNCTIONS OF TWO OR MORE VARIABLES.. 511

8.1 Rectangular Coordinates in Space... 512
8.2 Functions and Their Graphs.. 516
8.3 Partial Derivatives 522
8.4 Local Maxima and Local Minima 531
8.5 Lagrange Multipliers 538
8.6 The Double Integral 547
Chapter Review 552
Chapter Project: Move Inventory Control with Calculus........................ 556

APPENDIX GRAPHING UTILITIES 558

A.1 The Viewing Rectangle 558
A.2 Using a Graphing Utility to Graph Equations 560
A.3 Square Screens... 564
A.4 Using a Graphing Utility To Locate Intercepts and Check for Symmetry 565
A.5 Using a Graphing Utility to Solve Equations 567

Answers to Odd-Numbered Problems   AN-1
Photo Credits   PC-1
Index   I-1

CHAPTER 0

Review*

OUTLINE

- **0.1** Real Numbers
- **0.2** Algebra Review
- **0.3** Polynomials and Rational Expressions
- **0.4** Solving Equations
- **0.5** Intervals; Solving Inequalities
- **0.6** nth Roots; Rational Exponents
- **0.7** Geometry Review
- **0.8** Rectangular Coordinates
- **0.9** Lines

This chapter consists of review material. It may be used as the first part of the course or later as a just-in-time review when the content is required. Specific references to this chapter occur throughout the book to assist in the review process.

*Sections 0.1–0.7 are based on material from *College Algebra*, 7th edition, by Michael Sullivan, used here with the permission of the author and Prentice-Hall, Inc.

0.1 Real Numbers

PREPARING FOR THIS BOOK *Before getting started, read the Preface to the Student on page xiv.*

OBJECTIVES
1. Classify numbers
2. Evaluate numerical expressions
3. Work with properties of real numbers

Sets

When we want to treat a collection of similar but distinct objects as a whole, we use the idea of a **set**. For example, the set of *digits* consists of the collection of numbers 0, 1, 2, 3, 4, 5, 6, 7, 8, and 9. If we use the symbol D to denote the set of digits, then we can write

$$D = \{0, 1, 2, 3, 4, 5, 6, 7, 8, 9\}$$

In this notation, the braces { } are used to enclose the objects, or **elements,** in the set. This method of denoting a set is called the **roster method.** A second way to denote a set is to use **set-builder notation,** where the set D of digits is written as

$$D = \{x \mid x \text{ is a digit}\}$$

read as "D is the set of all x such that x is a digit."

EXAMPLE 1 Using Set-Builder Notation and the Roster Method

(a) $E = \{x \mid x \text{ is an even digit}\} = \{0, 2, 4, 6, 8\}$
(b) $O = \{x \mid x \text{ is an odd digit}\} = \{1, 3, 5, 7, 9\}$

In listing the elements of a set, we do not list an element more than once because the elements of a set are distinct. Also, the order in which the elements are listed is not relevant. For example, {2, 3} and {3, 2} both represent the same set.

If every element of a set A is also an element of a set B, then we say that A **is a subset of** B. If two sets A and B have the same elements, then we say that A **equals** B. For example, {1, 2, 3} is a subset of {1, 2, 3, 4, 5}, and {1, 2, 3} equals {2, 3, 1}.

Finally, if a set has no elements, it is called the **empty set,** or the **null set,** and is denoted by the symbol $\varnothing$.

Classification of Numbers

It is helpful to classify the various kinds of numbers that we deal with as sets. The **counting numbers,** or **natural numbers,** are the numbers in the set $\{1, 2, 3, 4, \ldots\}$. (The three dots, called an **ellipsis,** indicate that the pattern continues indefinitely.) As their name implies, these numbers are often used to count things. For example, there are 26 letters in our alphabet; there are 100 cents in a dollar. The **whole numbers** are the numbers in the set $\{0, 1, 2, 3, \ldots\}$, that is, the counting numbers together with 0.

The **integers** are the numbers in the set $\{\ldots, -3, -2, -1, 0, 1, 2, 3, \ldots\}$.

These numbers are useful in many situations. For example, if your checking account has $10 in it and you write a check for $15, you can represent the current balance as $-\$5$.

Notice that the set of counting numbers is a subset of the set of whole numbers. Each time we expand a number system, such as from the whole numbers to the integers, we do so in order to be able to handle new, and usually more complicated, problems. The integers allow us to solve problems requiring both positive and negative counting numbers, such as profit/loss, height above/below sea level, temperature above/below 0°F, and so on.

But integers alone are not sufficient for *all* problems. For example, they do not answer the question "What part of a dollar is 38 cents?" To answer such a question, we enlarge our number system to include *rational numbers*. For example, $\frac{38}{100}$ answers the question "What part of a dollar is 38 cents?"

A **rational number** is a number that can be expressed as a quotient $\frac{a}{b}$ of two integers. The integer a is called the **numerator,** and the integer b, which cannot be 0, is called the **denominator.** The rational numbers are the numbers in the set $\{x \mid x = \frac{a}{b}, \text{ where } a \text{ and } b, b \neq 0, \text{ are integers}\}$.

Examples of rational numbers are $\frac{3}{4}, \frac{5}{2}, \frac{0}{4}, -\frac{2}{3},$ and $\frac{100}{3}$. Since $\frac{a}{1} = a$ for any integer a, it follows that the set of integers is a subset of the set of rational numbers.

Rational numbers may be represented as **decimals.** For example, the rational numbers $\frac{3}{4}, \frac{5}{2}, -\frac{2}{3},$ and $\frac{7}{66}$ may be represented as decimals by merely carrying out the indicated division:

$$\frac{3}{4} = 0.75 \qquad \frac{5}{2} = 2.5 \qquad -\frac{2}{3} = -0.666\ldots \qquad \frac{7}{66} = 0.1060606\ldots$$

Notice that the decimal representations of $\frac{3}{4}$ and $\frac{5}{2}$ terminate, or end. The decimal representations of $-\frac{2}{3}$ and $\frac{7}{66}$ do not terminate, but they do exhibit a pattern of repetition. For $-\frac{2}{3}$, the 6 repeats indefinitely; for $\frac{7}{66}$, the block 06 repeats indefinitely. It can be shown that every rational number may be represented by a decimal that either terminates or is nonterminating with a repeating block of digits, and vice versa.

On the other hand, there are decimals that do not fit into either of these categories. Such decimals represent **irrational numbers.** Every irrational number may be represented by a decimal that neither repeats nor terminates. In other words, irrational numbers cannot be written in the form $\frac{a}{b}$, where a and $b, b \neq 0$, are integers.

Irrational numbers occur naturally. For example, consider the isosceles right triangle whose legs are each of length 1. See Figure 1. The length of the hypotenuse is $\sqrt{2}$, an irrational number.

Also, the number that equals the ratio of the circumference C to the diameter d of any circle, denoted by the symbol π (the Greek letter pi), is an irrational number. See Figure 2.

FIGURE 1

FIGURE 2 $\pi = \frac{C}{d}$

Together, the rational numbers and irrational numbers form the set of **real numbers.**

Figure 3 shows the relationship of various types of numbers.

FIGURE 3

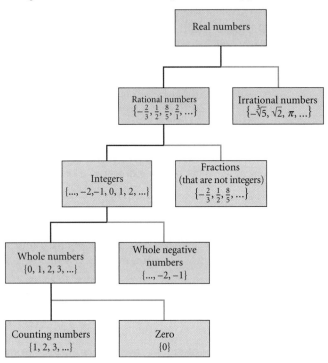

EXAMPLE 2   Classifying the Numbers in a Set

List the numbers in the set

$$\left\{-3, \frac{4}{3}, 0.12, \sqrt{2}, \pi, 2.151515 \ldots \text{ (where the block 15 repeats)}, 10\right\}$$

that are

(a) Natural numbers   (b) Integers   (c) Rational numbers
(d) Irrational numbers   (e) Real numbers

SOLUTION   (a) 10 is the only natural number.
(b) -3 and 10 are integers.
(c) $-3, \frac{4}{3}, 0.12, 2.151515 \ldots$, and 10 are rational numbers.
(d) $\sqrt{2}$ and π are irrational numbers.
(e) All the numbers listed are real numbers.

 NOW WORK PROBLEM 3.

Approximations

Every decimal may be represented by a real number (either rational or irrational), and every real number may be represented by a decimal.

The irrational numbers $\sqrt{2}$ and π have decimal representations that begin as follows:

$$\sqrt{2} = 1.414213\ldots \qquad \pi = 3.14159\ldots$$

In practice, decimals are generally represented by approximations. For example, using the symbol $\approx$ (read as "approximately equal to"), we can write

$$\sqrt{2} \approx 1.4142 \qquad \pi \approx 3.1416$$

In approximating decimals, we either *round off* or *truncate* to a given number of decimal places. The number of places establishes the location of the *final digit* in the decimal approximation.

Truncation: Drop all the digits that follow the specified final digit in the decimal.

Rounding: Identify the specified final digit in the decimal. If the next digit is 5 or more, add 1 to the final digit; if the next digit is 4 or less, leave the final digit as it is. Now truncate following the final digit.

EXAMPLE 3 Approximating a Decimal to Two Places

Approximate 20.98752 to two decimal places by

(a) Truncating   **(b)** Rounding

SOLUTION For 20.98752, the final digit is 8, since it is two decimal places from the decimal point.

(a) To truncate, we remove all digits following the final digit 8. The truncation of 20.98752 to two decimal places is 20.98.

(b) To round, we examine the digit following the final digit 8, which is 7. Since 7 is 5 or more, we add 1 to the final digit 8 and truncate. The rounded form of 20.98752 to two decimal places is 20.99.

EXAMPLE 4 Approximating a Decimal to Two and Four Places

Number	Rounded to Two Decimal Places	Rounded to Four Decimal Places	Truncated to Two Decimal Places	Truncated to Four Decimal Places
(a) 3.14159	3.14	3.1416	3.14	3.1415
(b) 0.056128	0.06	0.0561	0.05	0.0561
(c) 893.46125	893.46	893.4613	893.46	893.4612

NOW WORK PROBLEM 7.

Calculators

Calculators are finite machines. As a result, they are incapable of displaying decimals that contain a large number of digits. For example, some calculators are capable of displaying only eight digits. When a number requires more than eight digits, the calculator either truncates or rounds. To see how your calculator handles decimals, divide 2 by 3. How many digits do you see? Is the last digit a 6 or a 7? If it is a 6, your calculator truncates; if it is a 7, your calculator rounds.

There are different kinds of calculators. An **arithmetic** calculator can only add, subtract, multiply, and divide numbers; therefore, this type is not adequate for this course. **Scientific** calculators have all the capabilities of arithmetic calculators and also contain **function keys** labelled ln, log, sin, cos, tan, x^y, inv, and so on. **Graphing** calculators have all the capabilities of scientific calculators and contain a screen on which graphs can be displayed.

For those who have access to a graphing calculator, we have included comments, examples, and exercises marked with a ⌨, indicating that a graphing calculator is required. We have also included an Appendix that explains some of the capabilities of a graphing calculator. The ⌨ comments, examples, and exercises may be omitted without loss of continuity, if so desired.

Operations

In algebra, we use letters such as x, y, a, b, and c to represent numbers. The symbols used in algebra for the operations of addition, subtraction, multiplication, and division are $+$, $-$, $\cdot$, and $/$. The words used to describe the results of these operations are **sum, difference, product,** and **quotient.** Table 1 summarizes these ideas.

TABLE 1

Operation	Symbol	Words
Addition	$a + b$	Sum: a plus b
Subtraction	$a - b$	Difference: a minus b
Multiplication	$a \cdot b$, $(a) \cdot b$, $a \cdot (b)$, $(a) \cdot (b)$, ab, $(a)b$, $a(b)$, $(a)(b)$	Product: a times b
Division	a/b or $\dfrac{a}{b}$	Quotient: a divided by b

We generally avoid using the multiplication sign $\times$ and the division sign $\div$ so familiar in arithmetic. Notice also that when two expressions are placed next to each other without an operation symbol, as in ab, or in parentheses, as in $(a)(b)$, it is understood that the expressions, called **factors,** are to be multiplied.

We also prefer not to use mixed numbers. When mixed numbers are used, addition is understood; for example, $2\frac{3}{4}$ means $2 + \frac{3}{4}$. The use of a mixed number may be confusing because the absence of an operation symbol between two terms is generally taken to mean multiplication. The expression $2\frac{3}{4}$ is therefore written instead as 2.75 or as $\frac{11}{4}$.

The symbol $=$, called an **equal sign** and read as "equals" or "is," is used to express the idea that the number or expression on the left of the equal sign is equivalent to the number or expression on the right.

EXAMPLE 5 Writing Statements Using Symbols

(a) The sum of 2 and 7 equals 9. In symbols, this statement is written as $2 + 7 = 9$.
(b) The product of 3 and 5 is 15. In symbols, this statement is written as $3 \cdot 5 = 15$. ▶

NOW WORK PROBLEM 19.

Real Numbers 7

Evalute numerical **2**
expressions

Order of Operations

Consider the expression $2 + 3 \cdot 6$. It is not clear whether we should add 2 and 3 to get 5, and then multiply by 6 to get 30; or first multiply 3 and 6 to get 18, and then add 2 to get 20. To avoid this ambiguity, we have the following agreement.

> We agree that whenever the two operations of addition and multiplication separate three numbers, the multiplication operation will be performed first, followed by the addition operation.

For example, we find $2 + 3 \cdot 6$ as follows:
$$2 + 3 \cdot 6 = 2 + 18 = 20$$

EXAMPLE 6   Finding the Value of an Expression

Evaluate each expression.

(a) $3 + 4 \cdot 5$ (b) $8 \cdot 2 + 1$ (c) $2 + 2 \cdot 2$

SOLUTION

(a) $3 + 4 \cdot 5 = 3 + 20 = 23$
        ↑
      Multiply first

(b) $8 \cdot 2 + 1 = 16 + 1 = 17$
      ↑
      Multiply first

(c) $2 + 2 \cdot 2 = 2 + 4 = 6$

NOW WORK PROBLEM 31.

Look at Example 6, part (a). To first add 3 and 4 and then multiply the result by 5, we use parentheses and write $(3 + 4) \cdot 5$. Whenever parentheses appear in an expression, it means "perform the operations within the parentheses first!"

EXAMPLE 7   Finding the Value of an Expression

(a) $(5 + 3) \cdot 4 = 8 \cdot 4 = 32$
(b) $(4 + 5) \cdot (8 - 2) = 9 \cdot 6 = 54$

When we divide two expressions, as in
$$\frac{2 + 3}{4 + 8}$$

it is understood that the division bar acts like parentheses; that is,
$$\frac{2 + 3}{4 + 8} = \frac{(2 + 3)}{(4 + 8)}$$

The following list gives the rules for the order of operations.

Chapter 0 Review

> **Rules for the Order of Operations**
> 1. Begin with the innermost parentheses and work outward. Remember that in dividing two expressions the numerator and denominator are treated as if they were enclosed in parentheses.
> 2. Perform multiplications and divisions, working from left to right.
> 3. Perform additions and subtractions, working from left to right.

EXAMPLE 8 **Finding the Value of an Expression**

Evaluate each expression.

(a) $8 \cdot 2 + 3$ (b) $5 \cdot (3 + 4) + 2$

(c) $\dfrac{2 + 5}{2 + 4 \cdot 7}$ (d) $2 + [4 + 2 \cdot (10 + 6)]$

SOLUTION

(a) $8 \cdot 2 + 3 = 16 + 3 = 19$
   ↑
   Multiply first

(b) $5 \cdot (3 + 4) + 2 = 5 \cdot 7 + 2 = 35 + 2 = 37$
         ↑   ↑
   Parentheses first   Multiply before adding

(c) $\dfrac{2 + 5}{2 + 4 \cdot 7} = \dfrac{2 + 5}{2 + 28} = \dfrac{7}{30}$

(d) $2 + [4 + 2 \cdot (10 + 6)] = 2 + [4 + 2 \cdot (16)]$
    $= 2 + [4 + 32] = 2 + 36 = 38$

NOW WORK PROBLEMS 37 AND 45.

Properties of Real Numbers

Work with properties of real numbers ③ We have used the equal sign to mean that one expression is equivalent to another. Four important properties of equality are listed next. In this list, a, b, and c represent numbers.

> 1. The **reflexive property** states that a number always equals itself; that is, $a = a$.
> 2. The **symmetric property** states that if $a = b$ then $b = a$.
> 3. The **transitive property** states that if $a = b$ and $b = c$ then $a = c$.
> 4. The **principle of substitution** states that if $a = b$ then we may substitute b for a in any expression containing a.

Now, let's consider some other properties of real numbers. We begin with an example.

EXAMPLE 9 **Commutative Properties**

(a) $3 + 5 = 8$ (b) $2 \cdot 3 = 6$
 $5 + 3 = 8$ $3 \cdot 2 = 6$
 $3 + 5 = 5 + 3$ $2 \cdot 3 = 3 \cdot 2$

This example illustrates the **commutative property** of real numbers, which states that the order in which addition or multiplication takes place will not affect the final result.

Commutative Properties

$$a + b = b + a \qquad (1a)$$
$$a \cdot b = b \cdot a \qquad (1b)$$

Here, and in the properties listed next and on pages 10–13, a, b, and c represent real numbers.

EXAMPLE 10 Associative Properties

(a) $2 + (3 + 4) = 2 + 7 = 9$
$(2 + 3) + 4 = 5 + 4 = 9$
$2 + (3 + 4) = (2 + 3) + 4$

(b) $2 \cdot (3 \cdot 4) = 2 \cdot 12 = 24$
$(2 \cdot 3) \cdot 4 = 6 \cdot 4 = 24$
$2 \cdot (3 \cdot 4) = (2 \cdot 3) \cdot 4$

The way we add or multiply three real numbers will not affect the final result. So, expressions such as $2 + 3 + 4$ and $3 \cdot 4 \cdot 5$ present no ambiguity, even though addition and multiplication are performed on one pair of numbers at a time. This property is called the **associative property**.

Associative Properties

$$a + (b + c) = (a + b) + c = a + b + c \qquad (2a)$$
$$a \cdot (b \cdot c) = (a \cdot b) \cdot c = a \cdot b \cdot c \qquad (2b)$$

The next property is perhaps the most important.

Distributive Property

$$a \cdot (b + c) = a \cdot b + a \cdot c \qquad (3a)$$

The distributive property may be used in two different ways.

EXAMPLE 11 Distributive Property

(a) $2 \cdot (x + 3) = 2 \cdot x + 2 \cdot 3 = 2x + 6$ Use to remove parentheses.
(b) $3x + 5x = (3 + 5)x = 8x$ Use to combine two expressions.

NOW WORK PROBLEM 63.

The real numbers 0 and 1 have unique properties.

EXAMPLE 12 Identity Properties

(a) $4 + 0 = 0 + 4 = 4$ (b) $3 \cdot 1 = 1 \cdot 3 = 3$

The properties of 0 and 1 illustrated in Example 12 are called the **identity properties.**

Identity Properties

$$0 + a = a + 0 = a \tag{4a}$$
$$a \cdot 1 = 1 \cdot a = a \tag{4b}$$

We call 0 the **additive identity** and 1 the **multiplicative identity.**

For each real number a, there is a real number $-a$, called the **additive inverse** of a, having the following property:

Additive Inverse Property

$$a + (-a) = -a + a = 0 \tag{5a}$$

EXAMPLE 13 Finding an Additive Inverse

(a) The additive inverse of 6 is -6, because $6 + (-6) = 0$.
(b) The additive inverse of -8 is $-(-8) = 8$, because $-8 + 8 = 0$.

The additive inverse of a, that is, $-a$, is often called the *negative* of a or the *opposite* of a. The use of such terms can be dangerous, because they suggest that the additive inverse is a negative number, which it may not be. For example, the additive inverse of -3, namely $-(-3)$, equals 3, a positive number.

For each *nonzero* real number a, there is a real number $\dfrac{1}{a}$, called the **multiplicative inverse** of a, having the following property:

Multiplicative Inverse Property

$$a \cdot \frac{1}{a} = \frac{1}{a} \cdot a = 1 \quad \text{if } a \neq 0 \tag{5b}$$

The multiplicative inverse $\dfrac{1}{a}$ of a nonzero real number a is also referred to as the **reciprocal** of a.

EXAMPLE 14 Finding a Reciprocal

(a) The reciprocal of 6 is $\frac{1}{6}$, because $6 \cdot \frac{1}{6} = 1$.

(b) The reciprocal of -3 is $\frac{1}{-3}$, because $-3 \cdot \frac{1}{-3} = 1$.

(c) The reciprocal of $\frac{2}{3}$ is $\frac{3}{2}$, because $\frac{2}{3} \cdot \frac{3}{2} = 1$.

With these properties for adding and multiplying real numbers, we can now define the operations of subtraction and division as follows:

The **difference** $a - b$, also read "a less b" or "a minus b," is defined as

$$a - b = a + (-b) \tag{6}$$

To subtract b from a, add the opposite of b to a.

If b is a nonzero real number, the **quotient** $\frac{a}{b}$, also read as "a divided by b" or "the ratio of a to b," is defined as

$$\frac{a}{b} = a \cdot \frac{1}{b} \quad \text{if } b \neq 0 \tag{7}$$

EXAMPLE 15 Working with Differences and Quotients

(a) $8 - 5 = 8 + (-5) = 3$ (b) $4 - 9 = 4 + (-9) = -5$ (c) $\frac{5}{8} = 5 \cdot \frac{1}{8}$

For any number a, the product of a times 0 is always 0.

Multiplication by Zero

$$a \cdot 0 = 0 \tag{8}$$

For a nonzero number a, we have the following division properties.

Division Properties

$$\frac{0}{a} = 0 \qquad \frac{a}{a} = 1 \quad \text{if } a \neq 0 \tag{9}$$

NOTE: Division by 0 is *not defined.* One reason is to avoid the following difficulty: $\frac{2}{0} = x$ means to find x such that $0 \cdot x = 2$. But $0 \cdot x$ equals 0 for all x, so there is *no* number x such that $\frac{2}{0} = x$.

Rules of Signs

$$a(-b) = -(ab) \qquad (-a)b = -(ab) \qquad (-a)(-b) = ab$$
$$-(-a) = a \qquad \frac{a}{-b} = \frac{-a}{b} = -\frac{a}{b} \qquad \frac{-a}{-b} = \frac{a}{b}$$

(10)

EXAMPLE 16 Applying the Rules of Signs

(a) $2(-3) = -(2 \cdot 3) = -6$

(b) $(-3)(-5) = 3 \cdot 5 = 15$

(c) $\dfrac{3}{-2} = \dfrac{-3}{2} = -\dfrac{3}{2}$

(d) $\dfrac{-4}{-9} = \dfrac{4}{9}$

(e) $\dfrac{x}{-2} = \dfrac{1}{-2} \cdot x = -\dfrac{1}{2}x$

Cancellation Properties

$$ac = bc \text{ implies } a = b \quad \text{if } c \neq 0$$
$$\frac{ac}{bc} = \frac{a}{b} \qquad \text{if } b \neq 0, c \neq 0$$

(11)

EXAMPLE 17 Using the Cancellation Properties

(a) If $2x = 6$, then

$$2x = 6$$
$$2x = 2 \cdot 3 \quad \text{Factor 6.}$$
$$x = 3 \quad \text{Cancel the 2s.}$$

(b) $\dfrac{18}{12} = \dfrac{3 \cdot \cancel{6}}{2 \cdot \cancel{6}} = \dfrac{3}{2}$

↑ Cancel the 6s.

NOTE: We follow the common practice of using slash marks to indicate cancellations.

Zero-Product Property

$$\text{If } ab = 0, \text{ then } a = 0 \text{ or } b = 0, \text{ or both.} \tag{12}$$

EXAMPLE 18 Using the Zero-Product Property

If $2x = 0$, then either $2 = 0$ or $x = 0$. Since $2 \neq 0$, it follows that $x = 0$.

Arithmetic of Quotients

$$\frac{a}{b} + \frac{c}{d} = \frac{ad}{bd} + \frac{bc}{bd} = \frac{ad + bc}{bd} \qquad \text{if } b \neq 0, d \neq 0 \tag{13}$$

$$\frac{a}{b} \cdot \frac{c}{d} = \frac{ac}{bd} \qquad \text{if } b \neq 0, d \neq 0 \tag{14}$$

$$\frac{\frac{a}{b}}{\frac{c}{d}} = \frac{a}{b} \cdot \frac{d}{c} = \frac{ad}{bc} \qquad \text{if } b \neq 0, c \neq 0, d \neq 0 \tag{15}$$

EXAMPLE 19 Adding, Subtracting, Multiplying, and Dividing Quotients

(a) $\dfrac{2}{3} + \dfrac{5}{2} = \dfrac{2 \cdot 2}{3 \cdot 2} + \dfrac{3 \cdot 5}{3 \cdot 2} = \dfrac{2 \cdot 2 + 3 \cdot 5}{3 \cdot 2} = \dfrac{4 + 15}{6} = \dfrac{19}{6}$

　　　↑
　　By Equation (13)

(b) $\dfrac{3}{5} - \dfrac{2}{3} = \dfrac{3}{5} + \left(-\dfrac{2}{3}\right) = \dfrac{3}{5} + \dfrac{-2}{3}$

　　　　↑　　　　　　↑
　　By Equation (6)　By Equation (10)

$= \dfrac{3 \cdot 3 + 5 \cdot (-2)}{5 \cdot 3} = \dfrac{9 + (-10)}{15} = \dfrac{-1}{15} = -\dfrac{1}{15}$

　↑　　　　　　　↑
By Equation (13)　By Equation (10)

(c) $\dfrac{8}{3} \cdot \dfrac{15}{4} = \dfrac{8 \cdot 15}{3 \cdot 4} = \dfrac{2 \cdot \cancel{4} \cdot 3 \cdot 5}{3 \cdot \cancel{4} \cdot 1} = \dfrac{2 \cdot 5}{1} = 10$

　　　　↑　　　　↑　　　　↑
　By Equation (14)　Factor　By Equation (11)

NOTE: Slanting the cancellation marks in different directions for different factors, as shown here, is a good practice to follow, since it will help in checking for errors.

(d) $\dfrac{\frac{3}{5}}{\frac{7}{9}} = \dfrac{3}{5} \cdot \dfrac{9}{7} = \dfrac{3 \cdot 9}{5 \cdot 7} = \dfrac{27}{35}$

By Equation (15)   By Equation (14)

NOTE: In writing quotients, we shall follow the usual convention and write the quotient in lowest terms; that is, we write it so that any common factors of the numerator and the denominator have been removed using the cancellation properties, Equation (11). For example,

$$\frac{90}{24} = \frac{15 \cdot \cancel{6}}{4 \cdot \cancel{6}} = \frac{15}{4}$$

$$\frac{24x^2}{18x} = \frac{4 \cdot \cancel{6} \cdot x \cdot \cancel{x}}{3 \cdot \cancel{6} \cdot \cancel{x}} = \frac{4x}{3}, \quad x \neq 0$$

NOW WORK PROBLEMS 47, 51, AND 61.

Sometimes it is easier to add two fractions using *least common multiples* (LCM). The LCM of two numbers is the smallest number that each has as a common multiple.

EXAMPLE 20   Finding the Least Common Multiple of Two Numbers

Find the least common multiple of 15 and 12.

SOLUTION   To find the LCM of 15 and 12, we look at multiples of 15 and 12.

15, 30, 45, **60**, 75, 90, 105, **120**, . . .
12, 24, 36, 48, **60**, 72, 84, 96, 108, **120**, . . .

The *common* multiples are in blue. The *least* common multiple is 60.

EXAMPLE 21   Using the Least Common Multiple to Add Two Fractions

Find: $\dfrac{8}{15} + \dfrac{5}{12}$

SOLUTION   We use the LCM of the denominators of the fractions and rewrite each fraction using the LCM as a common denominator. The LCM of the denominators (12 and 15) is 60. Rewrite each fraction using 60 as the denominator.

$$\frac{8}{15} + \frac{5}{12} = \frac{8}{15} \cdot \frac{4}{4} + \frac{5}{12} \cdot \frac{5}{5} = \frac{32}{60} + \frac{25}{60} = \frac{32 + 25}{60} = \frac{57}{60}$$

NOW WORK PROBLEM 55.

Real Numbers

EXERCISE 0.1

In Problems 1–6, list the numbers in each set that are (a) natural numbers, (b) integers, (c) rational numbers, (d) irrational numbers, (e) real numbers.

1. $A = \left\{-6, \dfrac{1}{2}, -1.333\ldots \text{ (the 3s repeat)}, \pi, 2, 5\right\}$

2. $B = \left\{-\dfrac{5}{3}, 2.060606\ldots \text{ (the block 06 repeats)}, 1.25, 0, 1, \sqrt{5}\right\}$

3. $C = \left\{0, 1, \dfrac{1}{2}, \dfrac{1}{3}, \dfrac{1}{4}\right\}$

4. $D = \{-1, -1.1, -1.2, -1.3\}$

5. $E = \left\{\sqrt{2}, \pi, \sqrt{2}+1, \pi + \dfrac{1}{2}\right\}$

6. $F = \left\{-\sqrt{2}, \pi + \sqrt{2}, \dfrac{1}{2} + 10.3\right\}$

In Problems 7–18, approximate each number (a) rounded and (b) truncated to three decimal places.

7. 18.9526   **8.** 25.86134   **9.** 28.65319   **10.** 99.05249   **11.** 0.06291   **12.** 0.05388

13. 9.9985   **14.** 1.0006   **15.** $\dfrac{3}{7}$ **16.** $\dfrac{5}{9}$ **17.** $\dfrac{521}{15}$ **18.** $\dfrac{81}{5}$

In Problems 19–28, write each statement using symbols.

19. The sum of 3 and 2 equals 5.

20. The product of 5 and 2 equals 10.

21. The sum of x and 2 is the product of 3 and 4.

22. The sum of 3 and y is the sum of 2 and 2.

23. 3 times y is 1 plus 2.

24. 2 times x is 4 times 6.

25. x minus 2 equals 6.

26. 2 minus y equals 6.

27. x divided by 2 is 6.

28. 2 divided by x is 6.

In Problems 29–62, evaluate each expression.

29. $9 - 4 + 2$ **30.** $6 - 4 + 3$ **31.** $-6 + 4 \cdot 3$ **32.** $8 - 4 \cdot 2$

33. $4 + 5 - 8$ **34.** $8 - 3 - 4$ **35.** $4 + \dfrac{1}{3}$ **36.** $2 - \dfrac{1}{2}$

37. $6 - [3 \cdot 5 + 2 \cdot (3 - 2)]$ **38.** $2 \cdot [8 - 3(4 + 2)] - 3$ **39.** $2 \cdot (3 - 5) + 8 \cdot 2 - 1$ **40.** $1 - (4 \cdot 3 - 2 + 2)$

41. $10 - [6 - 2 \cdot 2 + (8 - 3)] \cdot 2$ **42.** $2 - 5 \cdot 4 - [6 \cdot (3 - 4)]$

43. $(5 - 3) \cdot \dfrac{1}{2}$ **44.** $(5 + 4) \cdot \dfrac{1}{3}$ **45.** $\dfrac{4 + 8}{5 - 3}$ **46.** $\dfrac{2 - 4}{5 - 3}$

47. $\dfrac{3}{5} \cdot \dfrac{10}{21}$ **48.** $\dfrac{5}{9} \cdot \dfrac{3}{10}$ **49.** $\dfrac{6}{25} \cdot \dfrac{10}{27}$ **50.** $\dfrac{21}{25} \cdot \dfrac{100}{3}$

51. $\dfrac{3}{4} + \dfrac{2}{5}$ **52.** $\dfrac{4}{3} + \dfrac{1}{2}$ **53.** $\dfrac{5}{6} + \dfrac{9}{5}$ **54.** $\dfrac{8}{9} + \dfrac{15}{2}$

55. $\dfrac{5}{18} + \dfrac{1}{12}$ **56.** $\dfrac{2}{15} + \dfrac{8}{9}$ **57.** $\dfrac{1}{30} - \dfrac{7}{18}$ **58.** $\dfrac{3}{14} - \dfrac{2}{21}$

59. $\dfrac{3}{20} - \dfrac{2}{15}$ **60.** $\dfrac{6}{35} - \dfrac{3}{14}$ **61.** $\dfrac{\frac{5}{18}}{\frac{11}{27}}$ **62.** $\dfrac{\frac{5}{21}}{\frac{2}{35}}$

In Problems 63–74, use the Distributive Property to remove the parentheses.

63. $6(x + 4)$ **64.** $4(2x - 1)$ **65.** $x(x - 4)$ **66.** $4x(x + 3)$

67. $(x + 2)(x + 4)$ **68.** $(x + 5)(x + 1)$ **69.** $(x - 2)(x + 1)$ **70.** $(x - 4)(x + 1)$

71. $(x - 8)(x - 2)$ **72.** $(x - 4)(x - 2)$ **73.** $(x + 2)(x - 2)$ **74.** $(x - 3)(x + 3)$

75. Explain to a friend how the Distributive Property is used to justify the fact that $2x + 3x = 5x$.

76. Explain to a friend why $2 + 3 \cdot 4 = 14$, whereas $(2 + 3) \cdot 4 = 20$.

77. Explain why $2(3 \cdot 4)$ is not equal to $(2 \cdot 3) \cdot (2 \cdot 4)$.

78. Explain why $\dfrac{4 + 3}{2 + 5}$ is not equal to $\dfrac{4}{2} + \dfrac{3}{5}$.

79. Is subtraction commutative? Support your conclusion with an example.

80. Is subtraction associative? Support your conclusion with an example.

81. Is division commutative? Support your conclusion with an example.

82. Is division associative? Support your conclusion with an example.

83. If $2 = x$, why does $x = 2$?

84. If $x = 5$, why does $x^2 + x = 30$?

85. Are there any real numbers that are both rational and irrational? Are there any real numbers that are neither? Explain your reasoning.

86. Explain why the sum of a rational number and an irrational number must he irrational.

87. What rational number does the repeating decimal 0.9999 . . . equal?

0.2 Algebra Review

OBJECTIVES
1. Graph inequalities
2. Find distance on the real number line
3. Evaluate algebraic expressions
4. Determine the domain of a variable
5. Use the laws of exponents
6. Evaluate square roots
7. Use a calculator to evaluate exponents
8. Use scientific notation

The Real Number Line

The real numbers can be represented by points on a line called the **real number line.** There is a one-to-one correspondence between real numbers and points on a line. That is, every real number corresponds to a point on the line, and each point on the line has a unique real number associated with it.

Pick a point on the line somewhere in the center, and label it O. This point, called the **origin,** corresponds to the real number 0. See Figure 4. The point 1 unit to the right of O corresponds to the number 1. The distance between 0 and 1 determines the **scale** of the number line. For example, the point associated with the number 2 is twice as far from O as 1 is. Notice that an arrowhead on the right end of the line indicates the direction in which the numbers increase. Figure 4 also shows the points associated with the irrational numbers $\sqrt{2}$ and π. Points to the left of the origin correspond to the real numbers -1, -2, and so on.

FIGURE 4

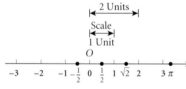

The real number associated with a point P is called the **coordinate** of P, and the line whose points have been assigned coordinates is called the **real number line.**

 NOW WORK PROBLEM 1.

FIGURE 5

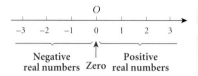

The real number line consists of three classes of real numbers, as shown in Figure 5:

1. The **negative real numbers** are the coordinates of points to the left of the origin O.
2. The real number **zero** is the coordinate of the origin O.
3. The **positive real numbers** are the coordinates of points to the right of the origin O.

Inequalities

FIGURE 6

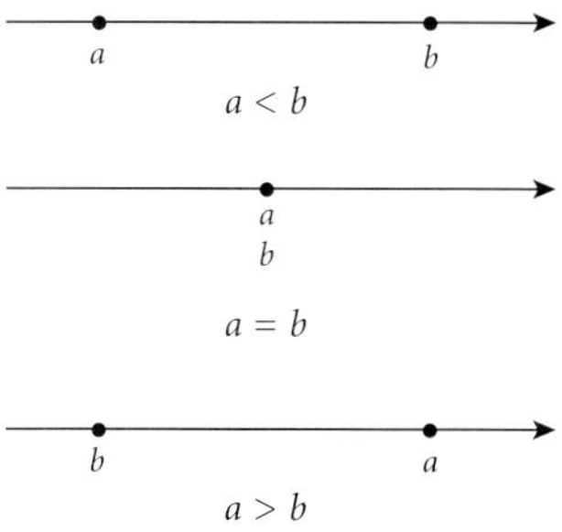

An important property of the real number line follows from the fact that, given two numbers (points) a and b, either a is to the left of b, a is at the same location as b, or a is to the right of b. See Figure 6.

If a is to the left of b, we say that "a is less than b" and write $a < b$. If a is to the right of b, we say that "a is greater than b" and write $a > b$. If a is at the same location as b, then $a = b$. If a is either less than or equal to b, we write $a \leq b$. Similarly, $a \geq b$ means that a is either greater than or equal to b. Collectively, the symbols $<, >, \leq$, and $\geq$ are called **inequality symbols.**

Note that $a < b$ and $b > a$ mean the same thing. It does not matter whether we write $2 < 3$ or $3 > 2$.

Furthermore, if $a < b$ or if $b > a$, then the difference $b - a$ is positive. Do you see why?

EXAMPLE 1   **Using Inequality Symbols**

(a) $3 < 7$ (b) $-8 > -16$ (c) $-6 < 0$
(d) $-8 < -4$ (e) $4 > -1$ (f) $8 > 0$

In Example 1(a), we conclude that $3 < 7$ either because 3 is to the left of 7 on the real number line or because the difference $7 - 3 = 4$ is a positive real number.

Similarly, we conclude in Example 1(b) that $-8 > -16$ either because -8 lies to the right of -16 on the real number line or because the difference $-8 - (-16) = -8 + 16 = 8$, is a positive real number.

Look again at Example 1. Note that the inequality symbol always points in the direction of the smaller number.

An **inequality** is a statement in which two expressions are related by an inequality symbol. The expressions are referred to as the **sides** of the inequality. Statements of the form $a < b$ or $b > a$ are called **strict inequalities,** whereas statements of the form $a \leq b$ or $b \geq a$ are called **nonstrict inequalities.**

Based on the discussion thus far, we conclude that

| $a > 0$ is equivalent to a is positive |
| $a < 0$ is equivalent to a is negative |

We sometimes read $a > 0$ by saying that "a is positive." If $a \geq 0$, then either $a > 0$ or $a = 0$, and we may read this as "a is nonnegative."

NOW WORK PROBLEMS 5 AND 15.

18 Chapter 0 Review

Graph inequalities 1   We shall find it useful in later work to graph inequalities on the real number line.

> **EXAMPLE 2**   **Graphing Inequalities**
>
> (a) On the real number line, graph all numbers x for which $x > 4$.
> (b) On the real number line, graph all numbers x for which $x \leq 5$.
>
> **SOLUTION**   (a) See Figure 7. Notice that we use a left parenthesis to indicate that the number 4 is *not* part of the graph.
> (b) See Figure 8. Notice that we use a right bracket to indicate that the number 5 is part of the graph.

FIGURE 7

FIGURE 8

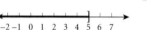

NOW WORK PROBLEM 21.

Absolute Value

The *absolute value* of a number a is the distance from 0 to a on the number line. For example, -4 is 4 units from 0; and 3 is 3 units from 0. See Figure 9. Thus, the absolute value of -4 is 4, and the absolute value of 3 is 3.

A more formal definition of absolute value is given next.

FIGURE 9

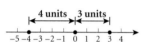

> The **absolute value** of a real number a, denoted by the symbol $|a|$, is defined by the rules
>
> $$|a| = a \text{ if } a \geq 0 \quad \text{and} \quad |a| = -a \text{ if } a < 0$$

For example, since $-4 < 0$, the second rule must be used to get $|-4| = -(-4) = 4$.

> **EXAMPLE 3**   **Computing Absolute Value**
>
> (a) $|8| = 8$ (b) $|0| = 0$ (c) $|-15| = -(-15) = 15$

Find distance on the real number line 2   Look again at Figure 9. The distance from -4 to 3 is 7 units. This distance is the difference $3 - (-4)$, obtained by subtracting the smaller coordinate from the larger. However, since $|3 - (-4)| = |7| = 7$ and $|-4 - 3| = |-7| = 7$, we can use absolute value to calculate the distance between two points without being concerned about which is smaller.

> If P and Q are two points on a real number line with coordinates a and b, respectively, the **distance between P and Q,** denoted by $d(P, Q)$, is
>
> $$d(P, Q) = |b - a|$$

Since $|b - a| = |a - b|$, it follows that $d(P, Q) = d(Q, P)$.

EXAMPLE 4 Finding Distance on a Number Line

Let P, Q, and R be points on a real number line with coordinates -5, 7, and -3, respectively. Find the distance

(a) between P and Q **(b)** between Q and R

SOLUTION We begin with Figure 10.

FIGURE 10

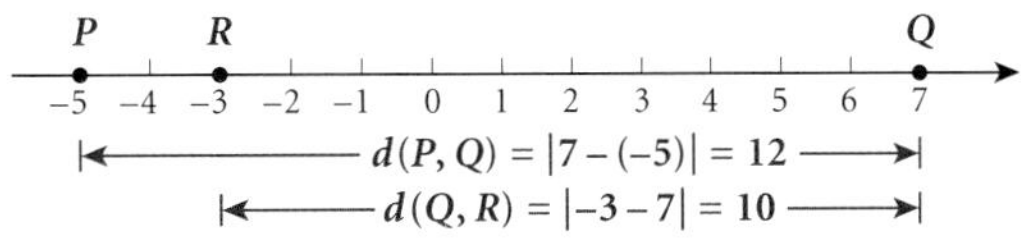

$$P \quad R \quad\quad\quad\quad\quad\quad Q$$
$$-5\ -4\ -3\ -2\ -1\ \ 0\ \ 1\ \ 2\ \ 3\ \ 4\ \ 5\ \ 6\ \ 7$$
$$\longleftarrow d(P, Q) = |7-(-5)| = 12 \longrightarrow$$
$$\longleftarrow d(Q, R) = |-3-7| = 10 \longrightarrow$$

(a) $d(P, Q) = |7 - (-5)| = |12| = 12$
(b) $d(Q, R) = |-3 - 7| = |-10| = 10$

NOW WORK PROBLEM 27.

Constants and Variables

As we said earlier, in algebra we use letters such as x, y, a, b, and c to represent numbers. If the letter used is to represent *any* number from a given set of numbers, it is called a **variable**. A **constant** is either a fixed number, such as 5 or $\sqrt{3}$, or a letter that represents a fixed (possibly unspecified) number.

Constants and variables are combined using the operations of addition, subtraction, multiplication, and division to form *algebraic expressions*. Examples of algebraic expressions include

$$x + 3 \quad\quad \frac{3}{1-t} \quad\quad 7x - 2y$$

Evaluate algebraic **3** To evaluate an algebraic expression, substitute for each variable its numerical value.
expressions

EXAMPLE 5 Evaluating an Algebraic Expression

Evaluate each expression if $x = 3$ and $y = -1$.

(a) $x + 3y$ **(b)** $5xy$ **(c)** $\dfrac{3y}{2-2x}$ **(d)** $|-4x + y|$

SOLUTION **(a)** Substitute 3 for x and -1 for y in the expression $x + 3y$.

$$x + 3y = 3 + 3(-1) = 3 + (-3) = 0$$
$$\uparrow$$
$$x = 3, y = -1$$

(b) If $x = 3$ and $y = -1$, then

$$5xy = 5(3)(-1) = -15$$

(c) If $x = 3$ and $y = -1$, then

$$\frac{3y}{2 - 2x} = \frac{3(-1)}{2 - 2(3)} = \frac{-3}{2 - 6} = \frac{-3}{-4} = \frac{3}{4}$$

(d) If $x = 3$ and $y = -1$, then

$$|-4x + y| = |-4(3) + (-1)| = |-12 + (-1)| = |-13| = 13$$

NOW WORK PROBLEMS 29 AND 37.

Determine the domain of a variable 4

In working with expressions or formulas involving variables, the variables may be allowed to take on values from only a certain set of numbers. For example, in the formula for the area A of a circle of radius r, $A = \pi r^2$, the variable r is necessarily restricted to the positive real numbers. In the expression $\frac{1}{x}$, the variable x cannot take on the value 0, since division by 0 is not defined.

The set of values that a variable may assume is called the **domain of the variable.**

EXAMPLE 6 **Finding the Domain of a Variable**

The domain of the variable x in the expression

$$\frac{5}{x - 2}$$

is $\{x \mid x \neq 2\}$, since, if $x = 2$, the denominator becomes 0, which is not defined.

EXAMPLE 7 **Circumference of a Circle**

In the formula for the circumference C of a circle of radius r,

$$C = 2\pi r$$

the domain of the variable r, representing the radius of the circle, is the set of positive real numbers. The domain of the variable C, representing the circumference of the circle, is also the set of positive real numbers.

In describing the domain of a variable, we may use either set notation or words, whichever is more convenient.

NOW WORK PROBLEM 47.

Exponents

Use the laws of exponents 5

Integer exponents provide a shorthand device for representing repeated multiplications of a real number. For example,

$$3^4 = 3 \cdot 3 \cdot 3 \cdot 3 = 81$$

Additionally, many formulas have exponents. For example,

- The formula for the horsepower rating H of an engine is

$$H = \frac{D^2 N}{2.5}$$

where D is the diameter of a cylinder and N is the number of cylinders.
- A formula for the resistance R of blood flowing in a blood vessel is

$$R = C \frac{L}{r^4}$$

where L is the length of the blood vessel, r is the radius, and C is a positive constant.

If a is a real number and n is a positive integer, then the symbol $\boldsymbol{a^n}$ represents the product of n factors of a. That is,

$$a^n = \underbrace{a \cdot a \cdot \ldots \cdot a}_{n \text{ factors}} \tag{1}$$

Here it is understood that $a^1 = a$.

Then $a^2 = a \cdot a$, $a^3 = a \cdot a \cdot a$, and so on. In the expression a^n, a is called the **base** and n is called the **exponent,** or **power.** We read a^n as "a raised to the power n" or as "a to the nth power." We usually read a^2 as "a squared" and a^3 as "a cubed."

In working with exponents, the operation of *raising to a power* is performed before any other operation. As examples,

$$4 \cdot 3^2 = 4 \cdot 9 = 36 \qquad 2^2 + 3^2 = 4 + 9 = 13$$
$$-2^4 = -16 \qquad 5 \cdot 3^2 + 2 \cdot 4 = 5 \cdot 9 + 2 \cdot 4 = 45 + 8 = 53$$

Parentheses are used to indicate operations to be performed first. For example,

$$(-2)^4 = (-2)(-2)(-2)(-2) = 16 \qquad (2 + 3)^2 = 5^2 = 25$$

If $a \neq 0$, we define

$$a^0 = 1 \quad \text{if } a \neq 0$$

If $a \neq 0$ and if n is a positive integer, then we define

$$a^{-n} = \frac{1}{a^n} \quad \text{if } a \neq 0$$

Whenever you encounter a negative exponent, think "reciprocal."

EXAMPLE 8 Evaluating Expressions Containing Negative Exponents

(a) $2^{-3} = \dfrac{1}{2^3} = \dfrac{1}{8}$ (b) $x^{-4} = \dfrac{1}{x^4}$ (c) $\left(\dfrac{1}{5}\right)^{-2} = \dfrac{1}{\left(\dfrac{1}{5}\right)^2} = \dfrac{1}{\dfrac{1}{25}} = 25$

NOW WORK PROBLEMS 65 AND 85.

The following properties, called the **Laws of Exponents,** can be proved using the preceding definitions. In the list, a and b are real numbers, and m and n are integers.

Laws of Exponents

$$a^m a^n = a^{m+n} \quad (a^m)^n = a^{mn} \quad (ab)^n = a^n b^n$$

$$\dfrac{a^m}{a^n} = a^{m-n} = \dfrac{1}{a^{n-m}}, \text{ if } a \neq 0 \quad \left(\dfrac{a}{b}\right)^n = \dfrac{a^n}{b^n}, \text{ if } b \neq 0$$

EXAMPLE 9 Using the Laws of Exponents

(a) $x^{-3} \cdot x^5 = x^{-3+5} = x^2, \quad x \neq 0$ (b) $(x^{-3})^2 = x^{-3 \cdot 2} = x^{-6} = \dfrac{1}{x^6}, \quad x \neq 0$

(c) $(2x)^3 = 2^3 \cdot x^3 = 8x^3$ (d) $\left(\dfrac{2}{3}\right)^4 = \dfrac{2^4}{3^4} = \dfrac{16}{81}$

(e) $\dfrac{x^{-2}}{x^{-5}} = x^{-2-(-5)} = x^3, \quad x \neq 0$

NOW WORK PROBLEM 67.

EXAMPLE 10 Using the Laws of Exponents

Write each expression so that all exponents are positive.

(a) $\dfrac{x^5 y^{-2}}{x^3 y}, \quad x \neq 0, \ y \neq 0$ (b) $\left(\dfrac{x^{-3}}{3y^{-1}}\right)^{-2}, \quad x \neq 0, \ y \neq 0$

SOLUTION (a) $\dfrac{x^5 y^{-2}}{x^3 y} = \dfrac{x^5}{x^3} \cdot \dfrac{y^{-2}}{y} = x^{5-3} \cdot y^{-2-1} = x^2 y^{-3} = x^2 \cdot \dfrac{1}{y^3} = \dfrac{x^2}{y^3}$

(b) $\left(\dfrac{x^{-3}}{3y^{-1}}\right)^{-2} = \dfrac{(x^{-3})^{-2}}{(3y^{-1})^{-2}} = \dfrac{x^6}{3^{-2}(y^{-1})^{-2}} = \dfrac{x^6}{\dfrac{1}{9}y^2} = \dfrac{9x^6}{y^2}$

NOW WORK PROBLEM 77.

Square Roots

Evaluate square roots 6

A real number is squared when it is raised to the power 2. The inverse of squaring is finding a **square root**. For example, since $6^2 = 36$ and $(-6)^2 = 36$, the numbers 6 and -6 are square roots of 36.

The symbol $\sqrt{}$, called a **radical sign,** is used to denote the **principal,** or nonnegative, square root. For example, $\sqrt{36} = 6$.

> In general, if a is a nonnegative real number, the nonnegative number b, such that $b^2 = a$ is the **principal square root** of a, and is denoted by $b = \sqrt{a}$.

The following comments are noteworthy:

1. Negative numbers do not have square roots (in the real number system), because the square of any real number is *nonnegative*. For example, $\sqrt{-4}$ is not a real number, because there is no real number whose square is -4.
2. The principal square root of 0 is 0, since $0^2 = 0$. That is, $\sqrt{0} = 0$.
3. The principal square root of a positive number is positive.
4. If $c \geq 0$, then $(\sqrt{c})^2 = c$. For example, $(\sqrt{2})^2 = 2$ and $(\sqrt{3})^2 = 3$.

EXAMPLE 11   **Evaluating Square Roots**

(a) $\sqrt{64} = 8$ (b) $\sqrt{\dfrac{1}{16}} = \dfrac{1}{4}$ (c) $(\sqrt{1.4})^2 = 1.4$

(d) $\sqrt{(-3)^2} = |-3| = 3$

Examples 11(a) and (b) are examples of square roots of perfect squares, since $64 = 8^2$ and $\dfrac{1}{16} = \left(\dfrac{1}{4}\right)^2$.

Notice the need for the absolute value in Example 11(d). Since $a^2 \geq 0$, the principal square root of a^2 is defined whether $a > 0$ or $a < 0$. However, since the principal square root is nonnegative, we need the absolute value to ensure the nonnegative result.

In general, we have

$$\sqrt{a^2} = |a| \tag{2}$$

EXAMPLE 12 Using Equation (2)

(a) $\sqrt{(2.3)^2} = |2.3| = 2.3$ (b) $\sqrt{(-2.3)^2} = |-2.3| = 2.3$ (c) $\sqrt{x^2} = |x|$

NOW WORK PROBLEM 73.

Calculator Use

Your calculator has either the caret key, , or the $\boxed{x^y}$ key, which is used for computations involving exponents.

EXAMPLE 13 Exponents on a Graphing Calculator

Evaluate: $(2.3)^5$

SOLUTION Figure 11 shows the result using a TI-83 graphing calculator.

FIGURE 11

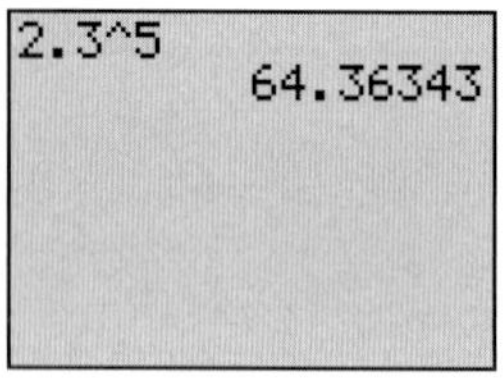

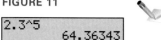

NOW WORK PROBLEM 103.

Scientific Notation

Measurements of physical quantities can range from very small to very large. For example, the mass of a proton is approximately 0.00000000000000000000000000167 kilogram and the mass of Earth is about 5,980,000,000,000,000,000,000,000 kilograms. These numbers obviously are tedious to write down and difficult to read, so we use exponents to rewrite each.

> When a number has been written as the product of a number x, where $1 \leq x < 10$, times a power of 10, it is said to be written in **scientific notation.**

In scientific notation,

$$\text{Mass of a proton} = 1.67 \times 10^{-27} \text{ kilogram}$$
$$\text{Mass of Earth} = 5.98 \times 10^{24} \text{ kilograms}$$

Converting a Decimal to Scientific Notation

To change a positive number into scientific notation:

1. Count the number N of places that the decimal point must be moved in order to arrive at a number x, where $1 \leq x < 10$.
2. If the original number is greater than or equal to 1, the scientific notation is $x \times 10^N$. If the original number is between 0 and 1, the scientific notation is $x \times 10^{-N}$.

EXAMPLE 14 Using Scientific Notation

Write each number in scientific notation.

(a) 9582   (b) 1.245   (c) 0.285   (d) 0.000561

SOLUTION (a) The decimal point in 9582 follows the 2. Thus, we count

$$9\underset{3}{\leftarrow}5\underset{2}{\leftarrow}8\underset{1}{\leftarrow}2.$$

stopping after three moves, because 9.582 is a number between 1 and 10. Since 9582 is greater than 1, we write

$$9582 = 9.582 \times 10^3$$

(b) The decimal point in 1.245 is between the 1 and 2. Since the number is already between 1 and 10, the scientific notation for it is $1.245 \times 10^0 = 1.245$.

(c) The decimal point in 0.285 is between the 0 and the 2. We count

$$0.\underset{1}{\overset{\rightarrow}{2}}85$$

stopping after one move, because 2.85 is a number between 1 and 10. Since 0.285 is between 0 and 1, we write

$$0.285 = 2.85 \times 10^{-1}$$

(d) The decimal point in 0.000561 is moved as follows:

$$0.\underset{1}{\overset{\rightarrow}{0}}\underset{2}{0}\underset{3}{0}\underset{4}{5}61$$

As a result,

$$0.000561 = 5.61 \times 10^{-4}$$

NOW WORK PROBLEM 109.

EXAMPLE 15 Changing from Scientific Notation to a Decimal

Write each number as a decimal.

(a) 2.1×10^4 (b) 3.26×10^{-5} (c) 1×10^{-2}

SOLUTION (a) $2.1 \times 10^4 = 2.\underset{1}{1}\underset{2}{0}\underset{3}{0}\underset{4}{0} \times 10^4 = 21{,}000$

(b) $3.26 \times 10^{-5} = 0\underset{5}{0}\underset{4}{0}\underset{3}{0}\underset{2}{0}\underset{1}{3}.26 \times 10^{-5} = 0.0000326$

(c) $1 \times 10^{-2} = 0\ \underset{2}{\overset{\uparrow}{0}}\ \underset{1}{\overset{\uparrow}{1}}\,.\ \times 10^{-2} = 0.01$

On a calculator, a number such as 3.615×10^{12} is usually displayed as $\boxed{3.615\text{E}12}$.

NOW WORK PROBLEM 117.

EXAMPLE 16 Using Scientific Notation

(a) The diameter of the smallest living cell is only about 0.00001 centimeter (cm). Express this number in scientific notation.

Source: Powers of Ten, Philip and Phylis Morrison.

(b) The surface area of Earth is about 1.97×10^8 square miles. Express the surface area as a whole number.

Source: 1998 Information Please Almanac.

SOLUTION **(a)** 0.00001 cm $= 1 \times 10^{-5}$ cm because the decimal point is moved five places and the number is less than 1.

(b) 1.97×10^8 square miles $= 197{,}000{,}000$ square miles.

EXERCISE 0.2

1. On the real number line, label the points with coordinates $0, 1, -1, \dfrac{5}{2}, -2.5, \dfrac{3}{4}$, and 0.25.

2. Repeat Problem 1 for the coordinates $0, -2, 2, -1.5, \dfrac{3}{2}, \dfrac{1}{3}$, and $\dfrac{2}{3}$.

In Problems 3–12, replace the question mark by $<$, $>$, or $=$, whichever is correct.

3. $\dfrac{1}{2}\ ?\ 0$
4. $5\ ?\ 6$
5. $-1\ ?\ -2$
6. $-3\ ?\ -\dfrac{5}{2}$
7. $\pi\ ?\ 3.14$
8. $\sqrt{2}\ ?\ 1.41$
9. $\dfrac{1}{2}\ ?\ 0.5$
10. $\dfrac{1}{3}\ ?\ 0.33$
11. $\dfrac{2}{3}\ ?\ 0.67$
12. $\dfrac{1}{4}\ ?\ 0.25$

In Problems 13–18, write each statement as an inequality.

13. x is positive
14. z is negative
15. x is less than 2
16. y is greater than -5
17. x is less than or equal to 1
18. x is greater than or equal to 2

In Problems 19–22, graph the numbers x on the real number line.

19. $x \geq -2$
20. $x < 4$
21. $x > -1$
22. $x \leq 7$

Algebra Review

In Problems 23–28, use the real number line below to compute each distance.

$$\begin{array}{cccccc} & A & & B & C & D & & E \\ \hline -4 & -3 & -2 & -1 & 0 & 1 & 2 & 3 & 4 & 5 & 6 \end{array}$$

23. $d(C, D)$ **24.** $d(C, A)$ **25.** $d(D, E)$ **26.** $d(C, E)$ **27.** $d(A, E)$ **28.** $d(D, B)$

In Problems 29–36, evaluate each expression if $x = -2$ and $y = 3$.

29. $x + 2y$ **30.** $3x + y$ **31.** $5xy + 2$ **32.** $-2x + xy$

33. $\dfrac{2x}{x - y}$ **34.** $\dfrac{x + y}{x - y}$ **35.** $\dfrac{3x + 2y}{2 + y}$ **36.** $\dfrac{2x - 3}{y}$

In Problems 37–46, find the value of each expression if $x = 3$ and $y = -2$.

37. $|x + y|$ **38.** $|x - y|$ **39.** $|x| + |y|$ **40.** $|x| - |y|$ **41.** $\dfrac{|x|}{x}$

42. $\dfrac{|y|}{y}$ **43.** $|4x - 5y|$ **44.** $|3x + 2y|$ **45.** $||4x| - |5y||$ **46.** $3|x| + 2|y|$

In Problems 47–54, determine which of the value(s) given below, if any, must be excluded from the domain of the variable in each expression:

(a) $x = 3$ (b) $x = 1$ (c) $x = 0$ (d) $x = -1$

47. $\dfrac{x^2 - 1}{x}$ **48.** $\dfrac{x^2 + 1}{x}$ **49.** $\dfrac{x}{x^2 - 9}$ **50.** $\dfrac{x}{x^2 + 9}$

51. $\dfrac{x^2}{x^2 + 1}$ **52.** $\dfrac{x^3}{x^2 - 1}$ **53.** $\dfrac{x^2 + 5x - 10}{x^3 - x}$ **54.** $\dfrac{-9x^2 - x + 1}{x^3 + x}$

In Problems 55–58, determine the domain of the variable x in each expression.

55. $\dfrac{4}{x - 5}$ **56.** $\dfrac{-6}{x + 4}$ **57.** $\dfrac{x}{x + 4}$ **58.** $\dfrac{x - 2}{x - 6}$

In Problems 59–62, use the formula $C = \dfrac{5}{9}(F - 32)$ for converting degrees Fahrenheit into degrees Celsius to find the Celsius measure of each Fahrenheit temperature.

59. $F = 32°$ **60.** $F = 212°$ **61.** $F = 77°$ **62.** $F = -4°$

In Problems 63–74, simplify each expression.

63. $(-4)^2$ **64.** -4^2 **65.** 4^{-2} **66.** -4^{-2} **67.** $3^{-6} \cdot 3^4$ **68.** $4^{-2} \cdot 4^3$

69. $(3^{-2})^{-1}$ **70.** $(2^{-1})^{-3}$ **71.** $\sqrt{25}$ **72.** $\sqrt{36}$ **73.** $\sqrt{(-4)^2}$ **74.** $\sqrt{(-3)^2}$

In Problems 75–84, simplify each expression. Express the answer so that all exponents are positive. Whenever an exponent is 0 or negative, we assume that the base is not 0.

75. $(8x^3)^2$ **76.** $(-4x^2)^{-1}$ **77.** $(x^2 y^{-1})^2$ **78.** $(x^{-1} y)^3$ **79.** $\dfrac{x^2 y^3}{xy^4}$

80. $\dfrac{x^{-2} y}{xy^2}$ **81.** $\dfrac{(-2)^3 x^4 (yz)^2}{3^2 xy^3 z}$ **82.** $\dfrac{4x^{-2}(yz)^{-1}}{2^3 x^4 y}$ **83.** $\left(\dfrac{3x^{-1}}{4y^{-1}}\right)^{-2}$ **84.** $\left(\dfrac{5x^{-2}}{6y^{-2}}\right)^{-3}$

Chapter 0 Review

In Problems 85–96, find the value of each expression if $x = 2$ and $y = -1$.

85. $2xy^{-1}$
86. $-3x^{-1}y$
87. $x^2 + y^2$
88. x^2y^2
89. $(xy)^2$
90. $(x + y)^2$
91. $\sqrt{x^2}$
92. $(\sqrt{x})^2$
93. $\sqrt{x^2 + y^2}$
94. $\sqrt{x^2} + \sqrt{y^2}$
95. x^y
96. y^x

97. Find the value of the expression $2x^3 - 3x^2 + 5x - 4$ if $x = 2$. What is the value if $x = 1$?

98. Find the value of the expression $4x^3 + 3x^2 - x + 2$ if $x = 1$. What is the value if $x = 2$?

99. What is the value of $\dfrac{(666)^4}{(222)^4}$?

100. What is the value of $(0.1)^3(20)^3$?

In Problems 101–108, use a calculator to evaluate each expression. Round your answer to three decimal places.

101. $(8.2)^6$
102. $(3.7)^5$
103. $(6.1)^{-3}$
104. $(2.2)^{-5}$
105. $(-2.8)^6$
106. $-(2.8)^6$
107. $(-8.11)^{-4}$
108. $-(8.11)^{-4}$

In Problems 109–116, write each number in scientific notation.

109. 454.2
110. 32.14
111. 0.013
112. 0.00421
113. 32,155
114. 21,210
115. 0.000423
116. 0.0514

In Problems 117–124, write each number as a decimal.

117. 6.15×10^4
118. 9.7×10^3
119. 1.214×10^{-3}
120. 9.88×10^{-4}
121. 1.1×10^8
122. 4.112×10^2
123. 8.1×10^{-2}
124. 6.453×10^{-1}

In Problems 125–134, express each statement as an equation involving the indicated variables. State the domain of each variable.

125. **Area of a Rectangle** The area A of a rectangle is its length l times its width w.

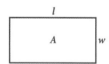

126. **Perimeter of a Rectangle** The perimeter P of a rectangle is twice the sum of its length l and its width w.

127. **Circumference of a Circle** The circumference C of a circle is π times its diameter d.

128. **Area of a Triangle** The area A of a triangle is one-half the base b times its height h.

129. **Area of an Equilateral Triangle** The area A of an equilateral triangle is $\dfrac{\sqrt{3}}{4}$ times the square of the length x of one side.

130. **Perimeter of an Equilateral Triangle** The perimeter P of an equilateral triangle is 3 times the length x of one side.

131. **Volume of a Sphere** The volume V of a sphere is $\frac{4}{3}$ times π times the cube of the radius r.

132. **Surface Area of a Sphere** The surface area S of a sphere is 4 times π times the square of the radius r.

133. **Volume of a Cube** The volume V of a cube is the cube of the length x of a side.

134. **Surface Area of a Cube** The surface area S of a cube is 6 times the square of the length x of a side.

135. **Manufacturing Cost** The weekly production cost C of manufacturing x watches is given by the formula $C = 4000 + 2x$, where the variable C is in dollars.
    (a) What is the cost of producing 1000 watches?
    (b) What is the cost of producing 2000 watches?

136. **Balancing a Checkbook** At the beginning of the month, Mike had a balance of $210 in his checking account. During the next month, he deposited $80, wrote a check for $120, made another deposit of $25, wrote two checks for $60 and $32, and was assessed a monthly service charge of $5. What was his balance at the end of the month?

137. **U.S. Voltage** In the United States, normal household voltage is 115 volts. It is acceptable for the actual voltage x to differ from normal by at most 5 volts. A formula that describes this is
$$|x - 115| \leq 5$$
(a) Show that a voltage of 113 volts is acceptable.
(b) Show that a voltage of 109 volts is not acceptable.

138. **Foreign Voltage** In other countries, normal household voltage is 220 volts. It is acceptable for the actual voltage x to differ from normal by at most 8 volts. A formula that describes this is
$$|x - 220| \leq 8$$
(a) Show that a voltage of 214 volts is acceptable.
(b) Show that a voltage of 209 volts is not acceptable.

139. **Making Precision Ball Bearings** The FireBall Company manufactures ball bearings for precision equipment. One of their products is a ball bearing with a stated radius of 3 centimeters (cm). Only ball bearings with a radius within 0.01 cm of this stated radius are acceptable. If x is the radius of a ball bearing, a formula describing this situation is
$$|x - 3| \leq 0.01$$
(a) Is a ball bearing of radius $x = 2.999$ acceptable?
(b) Is a ball bearing of radius $x = 2.89$ acceptable?

140. **Body Temperature** Normal human body temperature is 98.6°F. A temperature x that differs from normal by at least 1.5°F is considered unhealthy. A formula that describes this is
$$|x - 98.6| \geq 1.5$$
(a) Show that a temperature of 97°F is unhealthy.
(b) Show that a temperature of 100°F is not unhealthy.

141. Does $\frac{1}{3}$ equal 0.333? If not, which is larger? By how much?

142. Does $\frac{2}{3}$ equal 0.666? If not, which is larger? By how much?

143. Is there a positive real number "closest" to 0? Explain.

144. I'm thinking of a number! It lies between 1 and 10; its square is rational and lies between 1 and 10. The number is larger than π. Correct to two decimal places, name the number. Now think of your own number, describe it, and challenge a fellow student to name it.

145. Write a brief paragraph that illustrates the similarities and differences between "less than" ($<$) and "less than or equal to" ($\leq$).

0.3 Polynomials and Rational Expressions

OBJECTIVES
1. Recognize special products
2. Factor polynomials
3. Simplify rational expressions
4. Use the LCM to add rational expressions

As we said earlier, in algebra we use letters to represent real numbers. We shall use the letters at the end of the alphabet, such as x, y, and z, to represent variables and the letters at the beginning of the alphabet, such as a, b, and c, to represent constants. In the expressions $3x + 5$ and $ax + b$, it is understood that x is a variable and that a and b are constants, even though the constants a and b are unspecified. As you will find out, the context usually makes the intended meaning clear.

Now we introduce some basic vocabulary.

> A **monomial** in one variable is the product of a constant and a variable raised to a nonnegative integer power; that is, a monomial is of the form
>
> $$ax^k$$
>
> where a is a constant, x is a variable, and $k \geq 0$ is an integer. The constant a is called the **coefficient** of the monomial. If $a \neq 0$, then k is called the **degree** of the monomial.

Examples of monomials follow:

Monomial	Coefficient	Degree	
$6x^2$	6	2	
$-\sqrt{2}x^3$	$-\sqrt{2}$	3	
3	3	0	Since $3 = 3 \cdot 1 = 3x^0$
$-5x$	-5	1	Since $-5x = -5x^1$
x^4	1	4	Since $x^4 = 1 \cdot x^4$

Two monomials ax^k and bx^k with the same degree and the same variable are called **like terms**. Such monomials when added or subtracted can be combined into a single monomial by using the distributive property. For example,

$$2x^2 + 5x^2 = (2 + 5)x^2 = 7x^2 \quad \text{and} \quad 8x^3 - 5x^3 = (8 - 5)x^3 = 3x^3$$

The sum or difference of two monomials having different degrees is called a **binomial**. The sum or difference of three monomials with three different degrees is called a **trinomial**. For example,

$$x^2 - 2 \text{ is a binomial.}$$
$$x^3 - 3x + 5 \text{ is a trinomial.}$$
$$2x^2 + 5x^2 + 2 = 7x^2 + 2 \text{ is a binomial.}$$

> A **polynomial** in one variable is an algebraic expression of the form
>
> $$a_n x^n + a_{n-1} x^{n-1} + \cdots + a_1 x + a_0 \tag{1}$$
>
> where $a_n, a_{n-1}, \ldots, a_1, a_0$ are constants,* called the **coefficients** of the polynomial, $n \geq 0$ is an integer, and x is a variable. If $a_n \neq 0$, it is called the **leading coefficient**, and n is called the **degree** of the polynomial.

---

*The notation a_n is read as "a sub n." The number n is called a **subscript** and should not be confused with an exponent. We use subscripts in order to distinguish one constant from another when a large or undetermined number of constants is required.

The monomials that make up a polynomial are called its **terms.** If all the coefficients are 0, the polynomial is called the **zero polynomial,** which has no degree.

Polynomials are usually written in **standard form,** beginning with the nonzero term of highest degree and continuing with terms in descending order according to degree. If a power of x is missing, it is because its coefficient is zero. Examples of polynomials follow:

Polynomial	Coefficients	Degree
$3x^2 - 5 = 3x^2 + 0 \cdot x + (-5)$	$3, 0, -5$	2
$8 - 2x + x^2 = 1 \cdot x^2 - 2x + 8$	$1, -2, 8$	2
$5x + \sqrt{2} = 5x^1 + \sqrt{2}$	$5, \sqrt{2}$	1
$3 = 3 \cdot 1 = 3 \cdot x^0$	3	0
0	0	No degree

Although we have been using x to represent the variable, letters such as y or z are also commonly used.

$3x^4 - x^2 + 2$ is a polynomial (in x) of degree 4.
$9y^3 - 2y^2 + y - 3$ is a polynomial (in y) of degree 3.
$z^5 + \pi$ is a polynomial (in z) of degree 5.

Algebraic expressions such as

$$\frac{1}{x} \quad \text{and} \quad \frac{x^2 + 1}{x + 5}$$

are not polynomials. The first is not a polynomial because $\frac{1}{x} = x^{-1}$ has an exponent that is not a nonnegative integer. Although the second expression is the quotient of two polynomials, the polynomial in the denominator has degree greater than 0, so the expression cannot be a polynomial.

Recognize special products 1

Certain products, which we call **special products,** occur frequently in algebra. In the list that follows, $x, a, b, c,$ and d are real numbers.

Difference of Two Squares

$$\boxed{(x - a)(x + a) = x^2 - a^2} \tag{2}$$

Squares of Binomials, or Perfect Squares

$$\boxed{(x + a)^2 = x^2 + 2ax + a^2} \tag{3a}$$
$$\boxed{(x - a)^2 = x^2 - 2ax + a^2} \tag{3b}$$

Miscellaneous Trinomials

$$(x + a)(x + b) = x^2 + (a + b)x + ab \quad \text{(4a)}$$
$$(ax + b)(cx + d) = acx^2 + (ad + bc)x + bd \quad \text{(4b)}$$

Cubes of Binomials, or Perfect Cubes

$$(x + a)^3 = x^3 + 3ax^2 + 3a^2x + a^3 \quad \text{(5a)}$$
$$(x - a)^3 = x^3 - 3ax^2 + 3a^2x - a^3 \quad \text{(5b)}$$

Difference of Two Cubes

$$(x - a)(x^2 + ax + a^2) = x^3 - a^3 \quad \text{(6)}$$

Sum of Two Cubes

$$(x + a)(x^2 - ax + a^2) = x^3 + a^3 \quad \text{(7)}$$

The special product formulas in equations (2) through (7) are used often, and their patterns should be committed to memory. But if you forget one or are unsure of its form, you should be able to derive it as needed.

EXAMPLE 1 **Using Special Formulas**

(a) $(x - 4)(x + 4) = x^2 - 4^2 = x^2 - 16$
(b) $(2x + 5)(3x - 1) = 6x^2 - 2x + 15x - 5 = 6x^2 + 13x - 5$
(c) $(x - 2)^3 = x^3 - 3(2)x^2 + 3(2)^2x - (2)^3 = x^3 - 6x^2 + 12x - 8$

NOW WORK PROBLEM 3.

Factor polynomials 2 **Factoring**

Consider the following product:

$$(2x + 3)(x - 4) = 2x^2 - 5x - 12$$

The two polynomials on the left are called **factors** of the polynomial on the right. Expressing a given polynomial as a product of other polynomials, that is, finding the factors of a polynomial, is called **factoring**.

We shall restrict our discussion here to factoring polynomials in one variable into products of polynomials in one variable, where all coefficients are integers. We call this **factoring over the integers.**

Any polynomial can be written as the product of 1 times itself or as -1 times its additive inverse. If a polynomial cannot be written as the product of two other polynomials (excluding 1 and -1), then the polynomial is said to be **prime**. When a polynomial has been written as a product consisting only of prime factors, it is said to be **factored completely.** Examples of prime polynomials are

$$2, \quad 3, \quad 5, \quad x, \quad x+1, \quad x-1, \quad 3x+4$$

The first factor to look for in a factoring problem is a common monomial factor present in each term of the polynomial. If one is present, use the distributive property to factor it out. For example,

Polynomial	Common Monomial Factor	Remaining Factor	Factored Form
$2x + 4$	2	$x + 2$	$2x + 4 = 2(x + 2)$
$3x - 6$	3	$x - 2$	$3x - 6 = 3(x - 2)$
$2x^2 - 4x + 8$	2	$x^2 - 2x + 4$	$2x^2 - 4x + 8 = 2(x^2 - 2x + 4)$
$8x - 12$	4	$2x - 3$	$8x - 12 = 4(2x - 3)$
$x^2 + x$	x	$x + 1$	$x^2 + x = x(x + 1)$
$x^3 - 3x^2$	x^2	$x - 3$	$x^3 - 3x^2 = x^2(x - 3)$
$6x^2 + 9x$	$3x$	$2x + 3$	$6x^2 + 9x = 3x(2x + 3)$

The list of special products (2) through (7) given earlier provides a list of factoring formulas when the equations are read from right to left. For example, equation (2) states that if the polynomial is the difference of two squares, $x^2 - a^2$, it can be factored into $(x - a)(x + a)$. The following example illustrates several factoring techniques.

EXAMPLE 2 **Factoring Polynomials**

Factor completely each polynomial.

(a) $x^4 - 16$ (b) $x^3 - 1$ (c) $9x^2 - 6x + 1$

(d) $x^2 + 4x - 12$ (e) $3x^2 + 10x - 8$ (f) $x^3 - 4x^2 + 2x - 8$

SOLUTION (a) $x^4 - 16 = (x^2 - 4)(x^2 + 4) = (x - 2)(x + 2)(x^2 + 4)$
↑ ↑
Difference of squares   Difference of squares

(b) $x^3 - 1 = (x - 1)(x^2 + x + 1)$
  ↑
  Difference of cubes

(c) $9x^2 - 6x + 1 = (3x - 1)^2$
  ↑
  Perfect square

(d) $x^2 + 4x - 12 = (x + 6)(x - 2)$
  ↑
  The product of 6 and -2 is -12, and the sum of 6 and -2 is 4.

(e) $3x^2 + 10x - 8 = (3x - 2)(x + 4)$
  with $12x - 2x = 10x$, $3x^2$, -8

(f) $x^3 - 4x^2 + 2x - 8 = (x^3 - 4x^2) + (2x - 8)$
  ↑
  Regroup
  $= x^2(x - 4) + 2(x - 4) = (x^2 + 2)(x - 4)$
  ↑   ↑
  Distributive property   Distributive property

The technique used in Example 2(f) is called **factoring by grouping**.

 NOW WORK PROBLEMS 17, 33, AND 69.

Rational Expressions

If we form the quotient of two polynomials, the result is called a **rational expression**. Some examples of rational expressions are

(a) $\dfrac{x^3 + 1}{x}$ (b) $\dfrac{3x^3 + x - 2}{x^5 + 5}$ (c) $\dfrac{x}{x^2 - 1}$ (d) $\dfrac{xy^2}{(x - y)^2}$

Expressions (a), (b), and (c) are rational expressions in one variable, x, whereas (d) is a rational expression in two variables, x and y.

Rational expressions are described in the same manner as rational numbers. Thus, in expression (a), the polynomial $x^3 + 1$ is called the **numerator**, and x is called the **denominator**. When the numerator and denominator of a rational expression contain no common factors (except 1 and -1), we say that the rational expression is **reduced to lowest terms**, or **simplified**.

Simplify rational expressions   **3**

A rational expression is reduced to lowest terms by completely factoring the numerator and the denominator and canceling any common factors by using the cancellation property.

$$\dfrac{ac}{bc} = \dfrac{a}{b}, \quad b \neq 0, \quad c \neq 0$$

For example,

$$\dfrac{x^2 - 1}{x^2 - 2x - 3} = \dfrac{(x - 1)(x + 1)}{(x - 3)(x + 1)} = \dfrac{x - 1}{x - 3}$$

Polynomials and Rational Expressions   35

EXAMPLE 3   Simplifying Rational Expressions

Reduce each rational expression to lowest terms.

(a) $\dfrac{x^2 + 4x + 4}{x^2 + 3x + 2}$ (b) $\dfrac{x^3 - 8}{x^3 - 2x}$ (c) $\dfrac{8 - 2x}{x^2 - x - 12}$

SOLUTION

(a) $\dfrac{x^2 + 4x + 4}{x^2 + 3x + 2} = \dfrac{\cancel{(x+2)}(x+2)}{\cancel{(x+2)}(x+1)} = \dfrac{x+2}{x+1},\quad x \neq -2, -1$

(b) $\dfrac{x^3 - 8}{x^3 - 2x^2} = \dfrac{\cancel{(x-2)}(x^2 + 2x + 4)}{x^2\cancel{(x-2)}} = \dfrac{x^2 + 2x + 4}{x^2},\quad x \neq 0, 2$

(c) $\dfrac{8 - 2x}{x^2 - x - 12} = \dfrac{2(4 - x)}{(x - 4)(x + 3)} = \dfrac{2(-1)\cancel{(x-4)}}{\cancel{(x-4)}(x + 3)} = \dfrac{-2}{x + 3},\quad x \neq -3, 4$

NOW WORK PROBLEM 75.

The rules for multiplying and dividing rational expressions are the same as the rules for multiplying and dividing rational numbers.

$$\dfrac{a}{b} \cdot \dfrac{c}{d} = \dfrac{ac}{bd}, \quad \text{if } b \neq 0, d \neq 0 \tag{8}$$

$$\dfrac{\frac{a}{b}}{\frac{c}{d}} = \dfrac{a}{b} \cdot \dfrac{d}{c} = \dfrac{ad}{bc}, \quad \text{if } b \neq 0, c \neq 0, d \neq 0 \tag{9}$$

In using equations (8) and (9) with rational expressions, be sure first to factor each polynomial completely so that common factors can be canceled. We shall follow the practice of leaving our answers in factored form.

EXAMPLE 4   Finding Products and Quotients of Rational Expressions

Perform the indicated operation and simplify the result. Leave your answer in factored form.

(a) $\dfrac{x^2 - 2x + 1}{x^3 + x} \cdot \dfrac{4x^2 + 4}{x^2 + x - 2}$ (b) $\dfrac{\dfrac{x + 3}{x^2 - 4}}{\dfrac{x^2 - x - 12}{x^3 - 8}}$

SOLUTION (a) $\dfrac{x^2 - 2x + 1}{x^3 + x} \cdot \dfrac{4x^2 + 4}{x^2 + x - 2} = \dfrac{(x - 1)^2}{x(x^2 + 1)} \cdot \dfrac{4(x^2 + 1)}{(x + 2)(x - 1)}$

$= \dfrac{(x - 1)^2(4)\cancel{(x^2 + 1)}}{x\cancel{(x^2 + 1)}(x + 2)\cancel{(x - 1)}}$

$= \dfrac{4(x - 1)}{x(x + 2)},\quad x \neq -2, 0, 1$

(b) $\dfrac{\dfrac{x+3}{x^2-4}}{\dfrac{x^2-x-12}{x^3-8}} = \dfrac{x+3}{x^2-4} \cdot \dfrac{x^3-8}{x^2-x-12}$

$= \dfrac{x+3}{(x-2)(x+2)} \cdot \dfrac{(x-2)(x^2+2x+4)}{(x-4)(x+3)}$

$= \dfrac{\cancel{(x+3)}\cancel{(x-2)}(x^2+2x+4)}{\cancel{(x-2)}(x+2)(x-4)\cancel{(x+3)}}$

$= \dfrac{x^2+2x+4}{(x+2)(x-4)}, \quad x \neq -3, -2, 2, 4$

NOW WORK PROBLEM 53.

If the denominators of two rational expressions to be added (or subtracted) are equal, we add (or subtract) the numerators and keep the common denominator. That is, if $\dfrac{a}{b}$ and $\dfrac{c}{b}$ are two rational expressions, then

$$\dfrac{a}{b} + \dfrac{c}{b} = \dfrac{a+c}{b} \qquad \dfrac{a}{b} - \dfrac{c}{b} = \dfrac{a-c}{b}, \qquad \text{if } b \neq 0 \tag{10}$$

EXAMPLE 5 Finding the Sum of Two Rational Expressions

Perform the indicated operation and simplify the result. Leave your answer in factored form.

$$\dfrac{2x^2-4}{2x+5} + \dfrac{x+3}{2x+5}, \quad x \neq -\dfrac{5}{2}$$

SOLUTION

$\dfrac{2x^2-4}{2x+5} + \dfrac{x+3}{2x+5} = \dfrac{(2x^2-4)+(x+3)}{2x+5}$

$= \dfrac{2x^2+x-1}{2x+5} = \dfrac{(2x-1)(x+1)}{2x+5}$

If the denominators of two rational expressions to be added or subtracted are not equal, we can use the general formulas for adding and subtracting quotients.

$$\dfrac{a}{b} + \dfrac{c}{d} = \dfrac{a \cdot d}{b \cdot d} + \dfrac{b \cdot c}{b \cdot d} = \dfrac{ad+bc}{bd}, \qquad \text{if } b \neq 0, d \neq 0$$

$$\dfrac{a}{b} - \dfrac{c}{d} = \dfrac{a \cdot d}{b \cdot d} - \dfrac{b \cdot c}{b \cdot d} = \dfrac{ad-bc}{bd}, \qquad \text{if } b \neq 0, d \neq 0 \tag{11}$$

EXAMPLE 6 Finding the Difference of Two Rational Expressions

Perform the indicated operation and simplify the result. Leave your answer in factored form.

$$\frac{x^2}{x^2-4} - \frac{1}{x}, \quad x \neq -2, 0, 2$$

SOLUTION

$$\frac{x^2}{x^2-4} - \frac{1}{x} = \frac{x \cdot x^2}{x \cdot (x^2-4)} - \frac{1 \cdot (x^2-4)}{x \cdot (x^2-4)} = \frac{x^3 - (x^2-4)}{x \cdot (x^2-4)} = \frac{x^3 - x^2 + 4}{x(x-2)(x+2)}$$

Least Common Multiple (LCM)

If the denominators of two rational expressions to be added (or subtracted) have common factors, we usually do not use the general rules given by equation (11), since, in doing so, we make the problem more complicated than it needs to be. Instead, just as with fractions, we apply the **least common multiple (LCM) method** by using the polynomial of least degree that contains each denominator polynomial as a factor. Then we rewrite each rational expression using the LCM as the common denominator and use equation (10) to do the addition (or subtraction).

To find the least common multiple of two or more polynomials, first factor completely each polynomial. The LCM is the product of the different prime factors of each polynomial, each factor appearing the greatest number of times it occurs in each polynomial. The next example will give you the idea.

EXAMPLE 7 Finding the Least Common Multiple

Find the least common multiple of the following pair of polynomials:

$$x(x-1)^2(x+1) \quad \text{and} \quad 4(x-1)(x+1)^3$$

SOLUTION The polynomials are already factored completely as

$$x(x-1)^2(x+1) \quad \text{and} \quad 4(x-1)(x+1)^3$$

Start by writing the factors of the left-hand polynomial. (Alternatively, you could start with the one on the right.)

$$x(x-1)^2(x+1)$$

Now look at the right-hand polynomial. Its first factor, 4, does not appear in our list, so we insert it:

$$4x(x-1)^2(x+1)$$

The next factor, $x-1$, is already in our list, so no change is necessary. The final factor is $(x+1)^3$. Since our list has $x+1$ to the first power only, we replace $x+1$ in the list by $(x+1)^3$. The LCM is

$$4x(x-1)^2(x+1)^3$$

Notice that the LCM is, in fact, the polynomial of least degree that contains $x(x-1)^2(x+1)$ and $4(x-1)(x+1)^3$ as factors.

The next example illustrates how the LCM is used for adding and subtracting rational expressions.

EXAMPLE 8   Using the LCM to Add Rational Expressions

Perform the indicated operation and simplify the result. Leave your answer in factored form.

$$\frac{x}{x^2 + 3x + 2} + \frac{2x - 3}{x^2 - 1}, \qquad x \neq -2, -1, 1$$

SOLUTION   First, we find the LCM of the denominators.

$$x^2 + 3x + 2 = (x + 2)(x + 1)$$
$$x^2 - 1 = (x - 1)(x + 1)$$

The LCM is $(x + 2)(x + 1)(x - 1)$. Next, we rewrite each rational expression using the LCM as the denominator.

$$\frac{x}{x^2 + 3x + 2} = \frac{x}{(x + 2)(x + 1)} = \frac{x(x - 1)}{(x + 2)(x + 1)(x - 1)}$$

↑ Multiply numerator and denominator by $x - 1$ to get the LCM in the denominator.

$$\frac{2x - 3}{x^2 - 1} = \frac{2x - 3}{(x - 1)(x + 1)} = \frac{(2x - 3)(x + 2)}{(x - 1)(x + 1)(x + 2)}$$

↑ Multiply numerator and denominator by $x + 2$ to get the LCM in the denominator.

Now we can add using equation (10).

$$\frac{x}{x^2 + 3x + 2} + \frac{2x - 3}{x^2 - 1} = \frac{x(x - 1)}{(x + 2)(x + 1)(x - 1)} + \frac{(2x - 3)(x + 2)}{(x + 2)(x + 1)(x - 1)}$$

$$= \frac{(x^2 - x) + (2x^2 + x - 6)}{(x + 2)(x + 1)(x - 1)}$$

$$= \frac{3x^2 - 6}{(x + 2)(x + 1)(x - 1)} = \frac{3(x^2 - 2)}{(x + 2)(x + 1)(x - 1)}$$

If we had not used the LCM technique to add the quotients in Example 8, but decided instead to use the general rule of equation (11), we would have obtained a more complicated expression, as follows:

$$\frac{x}{x^2 + 3x + 2} + \frac{2x - 3}{x^2 - 1} = \frac{x(x^2 - 1) + (x^2 + 3x + 2)(2x - 3)}{(x^2 + 3x + 2)(x^2 - 1)}$$

$$= \frac{3x^3 + 3x^2 - 6x - 6}{(x^2 + 3x + 2)(x^2 - 1)} = \frac{3(x^3 + x^2 - 2x - 2)}{(x^2 + 3x + 2)(x^2 - 1)}$$

Now we are faced with a more complicated problem of expressing this quotient in lowest terms. It is always best to first look for common factors in the denominators of expressions to be added or subtracted and to use the LCM if any common factors are found.

NOW WORK PROBLEM 57.

EXERCISE 0.3

In Problems 1–10, perform the indicated operations. Express each answer as a polynomial written in standard form.

1. $(10x^5 - 8x^2) + (3x^3 - 2x^2 + 6)$
2. $3(x^2 - 3x + 1) + 2(3x^2 + x - 4)$
3. $(x + a)^2 - x^2$
4. $(x - a)^2 - x^2$
5. $(x + 8)(2x + 1)$
6. $(2x - 1)(x + 2)$
7. $(x^2 + x - 1)(x^2 - x + 1)$
8. $(x^2 + 2x + 1)(x^2 - 3x + 4)$
9. $(x + 1)^3 - (x - 1)^3$
10. $(x + 1)^3 - (x + 2)^3$

In Problems 11–52, factor completely each polynomial. If the polynomial cannot be factored, say it is prime.

11. $x^2 - 36$
12. $x^2 - 9$
13. $1 - 4x^2$
14. $1 - 9x^2$
15. $x^2 + 7x + 10$
16. $x^2 + 5x + 4$
17. $x^2 - 2x + 8$
18. $x^2 - 4x + 5$
19. $x^2 + 4x + 16$
20. $x^2 + 12x + 36$
21. $15 + 2x - x^2$
22. $14 + 6x - x^2$
23. $3x^2 - 12x - 36$
24. $x^3 + 8x^2 - 20x$
25. $y^4 + 11y^3 + 30y^2$
26. $3y^3 - 18y^2 - 48y$
27. $4x^2 + 12x + 9$
28. $9x^2 - 12x + 4$
29. $3x^2 + 4x + 1$
30. $4x^2 + 3x - 1$
31. $x^4 - 81$
32. $x^4 - 1$
33. $x^6 - 2x^3 + 1$
34. $x^6 + 2x^3 + 1$
35. $x^7 - x^5$
36. $x^8 - x^5$
37. $5 + 16x - 16x^2$
38. $5 + 11x - 16x^2$
39. $4y^2 - 16y + 15$
40. $9y^2 + 9y - 4$
41. $1 - 8x^2 - 9x^4$
42. $4 - 14x^2 - 8x^4$
43. $x(x + 3) - 6(x + 3)$
44. $5(3x - 7) + x(3x - 7)$
45. $(x + 2)^2 - 5(x + 2)$
46. $(x - 1)^2 - 2(x - 1)$
47. $6x(2 - x)^4 - 9x^2(2 - x)^3$
48. $6x(1 - x^2)^4 - 24x^3(1 - x^2)^3$
49. $x^3 + 2x^2 - x - 2$
50. $x^3 - 3x^2 - x + 3$
51. $x^4 - x^3 + x - 1$
52. $x^4 + x^3 + x + 1$

In Problems 53–64, perform the indicated operation and simplify the result. Leave your answer in factored form.

53. $\dfrac{3x - 6}{5x} \cdot \dfrac{x^2 - x - 6}{x^2 - 4}$
54. $\dfrac{9x^2 - 25}{2x - 2} \cdot \dfrac{1 - x^2}{6x - 10}$
55. $\dfrac{4x^2 - 1}{x^2 - 16} \cdot \dfrac{x^2 - 4x}{2x + 1}$
56. $\dfrac{12}{x^2 - x} \cdot \dfrac{x^2 - 1}{4x - 2}$
57. $\dfrac{x}{x^2 - 7x + 6} - \dfrac{x}{x^2 - 2x - 24}$
58. $\dfrac{x}{x - 3} - \dfrac{x + 1}{x^2 + 5x - 24}$
59. $\dfrac{4}{x^2 - 4} - \dfrac{2}{x^2 + x - 6}$
60. $\dfrac{3}{x - 1} - \dfrac{x - 4}{x^2 - 2x + 1}$
61. $\dfrac{1}{x} - \dfrac{2}{x^2 + x} + \dfrac{3}{x^3 - x^2}$
62. $\dfrac{x}{(x - 1)^2} + \dfrac{2}{x} - \dfrac{x + 1}{x^3 - x^2}$
63. $\dfrac{1}{h}\left(\dfrac{1}{x + h} - \dfrac{1}{x}\right)$
64. $\dfrac{1}{h}\left[\dfrac{1}{(x + h)^2} - \dfrac{1}{x^2}\right]$

In Problems 65–74, expressions that occur in calculus are given. Factor completely each expression.

65. $2(3x + 4)^2 + (2x + 3) \cdot 2(3x + 4) \cdot 3$
66. $5(2x + 1)^2 + (5x - 6) \cdot 2(2x + 1) \cdot 2$
67. $2x(2x + 5) + x^2 \cdot 2$
68. $3x^2(8x - 3) + x^3 \cdot 8$
69. $2(x + 3)(x - 2)^3 + (x + 3)^2 \cdot 3(x - 2)^2$
70. $4(x + 5)^3(x - 1)^2 + (x + 5)^4 \cdot 2(x - 1)$
71. $(4x - 3)^2 + x \cdot 2(4x - 3) \cdot 4$
72. $3x^2(3x + 4)^2 + x^3 \cdot 2(3x + 4) \cdot 3$
73. $2(3x - 5) \cdot 3(2x + 1)^3 + (3x - 5)^2 \cdot 3(2x + 1)^2 \cdot 2$
74. $3(4x + 5)^2 \cdot 4(5x + 1)^2 + (4x + 5)^3 \cdot 2(5x + 1) \cdot 5$

In Problems 75–82, expressions that occur in calculus are given. Reduce each expression to lowest terms.

75. $\dfrac{(2x + 3) \cdot 3 - (3x - 5) \cdot 2}{(3x - 5)^2}$

76. $\dfrac{(4x + 1) \cdot 5 - (5x - 2) \cdot 4}{(5x - 2)^2}$

77. $\dfrac{x \cdot 2x - (x^2 + 1) \cdot 1}{(x^2 + 1)^2}$

78. $\dfrac{x \cdot 2x - (x^2 - 4) \cdot 1}{(x^2 - 4)^2}$

79. $\dfrac{(3x + 1) \cdot 2x - x^2 \cdot 3}{(3x + 1)^2}$

80. $\dfrac{(2x - 5) \cdot 3x^2 - x^3 \cdot 2}{(2x - 5)^2}$

81. $\dfrac{(x^2 + 1) \cdot 3 - (3x + 4) \cdot 2x}{(x^2 + 1)^2}$

82. $\dfrac{(x^2 + 9) \cdot 2 - (2x - 5) \cdot 2x}{(x^2 + 9)^2}$

0.4 Solving Equations

PREPARING FOR THIS SECTION *Before getting started, review the following:*

> Factoring Polynomials (Section 0.3, pp. 31–34)
> Zero-Product Property (Section 0.2, p. 13)
> Square Roots (Section 0.2, pp. 23–24)
> Absolute Value (Section 0.2, pp. 18–19)

OBJECTIVES
1. Solve equations
2. Solve quadratic equations by factoring
3. Know how to complete the square
4. Solve a quadratic equation by completing the square
5. Solve a quadratic equation using the quadratic formula

Solve equations **1** An **equation in one variable** is a statement in which two expressions, at least one containing the variable, are equal. The expressions are called the **sides** of the equation. Since an equation is a statement, it may be true or false, depending on the value of the variable. Unless otherwise restricted, the admissible values of the variable are those in the domain of the variable. Those admissible values of the variable, if any, that result in a true statement are called **solutions**, or **roots**, of the equation. To **solve an equation** means to find all the solutions of the equation.

For example, the following are all equations in one variable, x:

$$x + 5 = 9 \qquad x^2 + 5x = 2x - 2 \qquad \dfrac{x^2 - 4}{x + 1} = 0 \qquad \sqrt{x^2 + 9} = 5$$

The first of these statements, $x + 5 = 9$, is true when $x = 4$ and false for any other choice of x. So, 4 is a solution of the equation $x + 5 = 9$. We also say that 4 **satisfies** the equation $x + 5 = 9$, because, when we substitute 4 for x, a true statement results.

Sometimes an equation will have more than one solution. For example, the equation

$$\dfrac{x^2 - 4}{x + 1} = 0$$

has $x = -2$ and $x = 2$ as solutions.

Usually, we will write the solution of an equation in set notation. This set is called the **solution set** of the equation. For example, the solution set of the equation $x^2 - 9 = 0$ is $\{-3, 3\}$.

Some equations have no real solution. For example, $x^2 + 9 = 5$ has no real solution, because there is no real number whose square when added to 9 equals 5.

An equation that is satisfied for every choice of the variable for which both sides are defined is called an **identity**. For example, the equation

$$3x + 5 = x + 3 + 2x + 2$$

is an identity, because this statement is true for any real number x.

Two or more equations that have precisely the same solution set are called **equivalent equations**.

For example, all the following equations are equivalent, because each has only the solution $x = 5$:

$$2x + 3 = 13$$
$$2x = 10$$
$$x = 5$$

These three equations illustrate one method for solving many types of equations: Replace the original equation by an equivalent equation, and continue until an equation with an obvious solution, such as $x = 5$, is reached. The question, though, is "How do I obtain an equivalent equation?" In general, there are five ways to do so.

> **Procedures That Result in Equivalent Equations**
>
> 1. Interchange the two sides of the equation:
>
>   Replace $\quad 3 = x \quad$ by $\quad x = 3$
>
> 2. Simplify the sides of the equation by combining like terms, eliminating parentheses, and so on:
>
>   Replace $\quad (x + 2) + 6 = 2x + (x + 1)$
>   by $\quad x + 8 = 3x + 1$
>
> 3. Add or subtract the same expression on both sides of the equation:
>
>   Replace $\quad 3x - 5 = 4$
>   by $\quad (3x - 5) + 5 = 4 + 5$
>
> 4. Multiply or divide both sides of the equation by the same nonzero expression:
>
>   Replace $\quad \dfrac{3x}{x - 1} = \dfrac{6}{x - 1}, \quad x \neq 1$
>   by $\quad \dfrac{3x}{x - 1} \cdot (x - 1) = \dfrac{6}{x - 1} \cdot (x - 1)$
>
> 5. If one side of the equation is 0 and the other side can be factored, then we may use the Zero-Product Property* and set each factor equal to 0:
>
>   Replace $\quad x(x - 3) = 0$
>   by $\quad x = 0 \quad$ or $\quad x - 3 = 0$

WARNING: Squaring both sides of an equation does not necessarily lead to an equivalent equation. ▶

*The Zero-Product Property says that if $ab = 0$ then $a = 0$ or $b = 0$ or both equal 0.

Whenever it is possible to solve an equation in your head, do so. For example:

The solution of $2x = 8$ is $x = 4$.
The solution of $3x - 15 = 0$ is $x = 5$.

Often, though, some rearrangement is necessary.

EXAMPLE 1   Solving an Equation

Solve the equation: $3x - 5 = 4$

SOLUTION We replace the original equation by a succession of equivalent equations.

$$3x - 5 = 4$$
$$(3x - 5) + 5 = 4 + 5 \quad \text{Add 5 to both sides.}$$
$$3x = 9 \quad \text{Simplify.}$$
$$\frac{3x}{3} = \frac{9}{3} \quad \text{Divide both sides by 3.}$$
$$x = 3 \quad \text{Simplify.}$$

The last equation, $x = 3$, has the single solution 3. All these equations are equivalent, so 3 is the only solution of the original equation, $3x - 5 = 4$.

Check: It is a good practice to check the solution by substituting 3 for x in the original equation.

$$3x - 5 = 4$$
$$3(3) - 5 \stackrel{?}{=} 4$$
$$9 - 5 \stackrel{?}{=} 4$$
$$4 = 4$$

The solution checks.

NOW WORK PROBLEMS 15 AND 21.

In the next examples, we use the Zero-Product Property.

EXAMPLE 2   Solving Equations by Factoring

Solve the equations: **(a)** $x^2 = 4x$ **(b)** $x^3 - x^2 - 4x + 4 = 0$

SOLUTION   **(a)** We begin by collecting all terms on one side. This results in 0 on one side and an expression to be factored on the other.

$$x^2 = 4x$$
$$x^2 - 4x = 0$$
$$x(x - 4) = 0 \quad \text{Factor.}$$
$$x = 0 \quad \text{or} \quad x - 4 = 0 \quad \text{Apply the Zero-Product Property.}$$
$$x = 4$$

The solution set is $\{0, 4\}$.

Check: $x = 0$: $0^2 = 4 \cdot 0$ So 0 is a solution.
$x = 4$: $4^2 = 4 \cdot 4$ So 4 is a solution.

(b) We group the terms of $x^3 - x^2 - 4x + 4 = 0$ as follows:
$$(x^3 - x^2) - (4x - 4) = 0$$
Factor x^2 from the first grouping and 4 from the second.
$$x^2(x - 1) - 4(x - 1) = 0$$
This reveals the common factor $(x - 1)$, so we have
$$(x^2 - 4)(x - 1) = 0$$

$(x - 2)(x + 2)(x - 1) = 0$		Factor again.
$x - 2 = 0 \quad$ or $\quad x + 2 = 0 \quad$ or $\quad x - 1 = 0$		Set each factor equal to 0.
$x = 2 \qquad\qquad\qquad x = -2 \qquad\qquad\qquad x = 1$		Solve.

The solution set is $\{-2, 1, 2\}$.

Check: $x = -2$: $(-2)^3 - (-2)^2 - 4(-2) + 4 = -8 - 4 + 8 + 4 = 0$ -2 is a solution.
$\qquad\quad x = 1$: $\;\;1^3 - 1^2 - 4(1) + 4 = 1 - 1 - 4 + 4 = 0$ $\qquad\qquad$ 1 is a solution.
$\qquad\quad x = 2$: $\;\;2^3 - 2^2 - 4(2) + 4 = 8 - 4 - 8 + 4 = 0$ $\qquad\qquad$ 2 is a solution.

NOW WORK PROBLEM 25.

There are two points whose distance from the origin is 5 units, -5 and 5, so the equation $|x| = 5$ will have the solution set $\{-5, 5\}$.

EXAMPLE 3 **Solving an Equation Involving Absolute Value**

Solve the equation: $\;|x + 4| = 13$

SOLUTION There are two possibilities:
$$x + 4 = 13 \quad \text{or} \quad x + 4 = -13$$
$$x = 9 \qquad\qquad\qquad x = -17$$

The solution set is $\{-17, 9\}$.

WARNING: Since the absolute value of any real number is nonnegative, equations such as $|x| = -2$ have no solution.

NOW WORK PROBLEM 37.

Quadratic Equations

A **quadratic equation** is an equation equivalent to one written in the **standard form** $ax^2 + bx + c = 0$, where a, b, and c are real numbers and $a \neq 0$.

When a quadratic equation is written in the standard form, $ax^2 + bx + c = 0$, it may be possible to factor the expression on the left side as the product of two first-degree polynomials.

2 **EXAMPLE 4** **Solving a Quadratic Equation by Factoring**

Solve the equation: $\;2x^2 = x + 3$

SOLUTION We put the equation in standard form by adding $-x - 3$ to both sides.
$$2x^2 = x + 3$$
$$2x^2 - x - 3 = 0 \qquad \text{Add } -x - 3 \text{ to both sides.}$$

The left side may now be factored as
$$(2x - 3)(x + 1) = 0$$
so that
$$2x - 3 = 0 \quad \text{or} \quad x + 1 = 0$$
$$x = \frac{3}{2} \quad \text{or} \quad x = -1$$

The solution set is $\left\{-1, \frac{3}{2}\right\}$.

When the left side factors into two linear equations with the same solution, the quadratic equation is said to have a **repeated solution**. We also call this solution a **root of multiplicity 2**, or a **double root**.

EXAMPLE 5 **Solving a Quadratic Equation by Factoring**

Solve the equation: $9x^2 - 6x + 1 = 0$

SOLUTION This equation is already in standard form, and the left side can be factored.
$$9x^2 - 6x + 1 = 0$$
$$(3x - 1)(3x - 1) = 0$$
so
$$x = \frac{1}{3} \quad \text{or} \quad x = \frac{1}{3}$$

This equation has only the repeated solution $\frac{1}{3}$.

NOW WORK PROBLEM 55.

The Square Root Method

Suppose that we wish to solve the quadratic equation
$$x^2 = p \tag{1}$$
where p is a nonnegative number. We proceed as in the earlier examples.

$$x^2 - p = 0 \quad \text{Put in standard form.}$$
$$(x - \sqrt{p})(x + \sqrt{p}) = 0 \quad \text{Factor (over the real numbers).}$$
$$x = \sqrt{p} \quad \text{or} \quad x = -\sqrt{p} \quad \text{Solve.}$$

We have the following result:

$$\boxed{\text{If } x^2 = p \text{ and } p \geq 0, \text{ then } x = \sqrt{p} \text{ or } x = -\sqrt{p}.} \tag{2}$$

When statement (2) is used, it is called the **Square Root Method.** In statement (2), note that if $p > 0$ the equation $x^2 = p$ has two solutions, $x = \sqrt{p}$ and $x = -\sqrt{p}$. We usually abbreviate these solutions as $x = \pm\sqrt{p}$, read as "x equals plus or minus the square root of p."

For example, the two solutions of the equation

$$x^2 = 4$$

are

$$x = \pm\sqrt{4} \quad \text{Use the Square Root Method.}$$

and, since $\sqrt{4} = 2$, we have

$$x = \pm 2$$

The solution set is $\{-2, 2\}$.

NOW WORK PROBLEM 69.

Completing The Square

3 Know how to complete the square

We now introduce the method of **completing the square.** The idea behind this method is to *adjust* the left side of a quadratic equation, $ax^2 + bx + c = 0$, so that it becomes a perfect square, that is, the square of a first-degree polynomial. For example, $x^2 + 6x + 9$ and $x^2 - 4x + 4$ are perfect squares because

$$x^2 + 6x + 9 = (x + 3)^2 \quad \text{and} \quad x^2 - 4x + 4 = (x - 2)^2$$

How do we adjust the left side? We do it by adding the appropriate number to the left side to create a perfect square. For example, to make $x^2 + 6x$ a perfect square, we add 9.

Let's look at several examples of completing the square when the coefficient of x^2 is 1:

Start	Add	Result
$x^2 + 4x$	4	$x^2 + 4x + 4 = (x + 2)^2$
$x^2 + 12x$	36	$x^2 + 12x + 36 = (x + 6)^2$
$x^2 - 6x$	9	$x^2 - 6x + 9 = (x - 3)^2$
$x^2 + x$	$\dfrac{1}{4}$	$x^2 + x + \dfrac{1}{4} = \left(x + \dfrac{1}{2}\right)^2$

Do you see the pattern? Provided that the coefficient of x^2 is 1, we complete the square by adding the square of $\frac{1}{2}$ of the coefficient of x.

Procedure for completing a square

Start	Add	Result
$x^2 + mx$	$\left(\dfrac{m}{2}\right)^2$	$x^2 + mx + \left(\dfrac{m}{2}\right)^2 = \left(x + \dfrac{m}{2}\right)^2$

NOW WORK PROBLEM 73.

EXAMPLE 6 Solving a Quadratic Equation by Completing the Square

Solve by completing the square: $2x^2 - 12x - 5 = 0$

SOLUTION First, we rewrite the equation as follows:
$$2x^2 - 12x - 5 = 0$$
$$2x^2 - 12x = 5$$

Next, we divide both sides by 2 so that the coefficient of x^2 is 1. (This enables us to complete the square at the next step.)

$$x^2 - 6x = \frac{5}{2}$$

Now complete the square by adding 9 to both sides.

$$x^2 - 6x + 9 = \frac{5}{2} + 9$$

$$(x - 3)^2 = \frac{23}{2}$$

$$x - 3 = \pm\sqrt{\frac{23}{2}} \quad \text{Use the Square Root Method.}$$

$$x - 3 = \pm\frac{\sqrt{46}}{2} \qquad \sqrt{\frac{23}{2}} = \frac{\sqrt{23}}{\sqrt{2}} \cdot \frac{\sqrt{2}}{\sqrt{2}} = \frac{\sqrt{46}}{2}$$

$$x = 3 \pm \frac{\sqrt{46}}{2}$$

The solution set is $\left\{3 - \frac{\sqrt{46}}{2}, 3 + \frac{\sqrt{46}}{2}\right\}$

NOTE: If we wanted an approximation, say rounded to two decimal places, of these solutions, we would use a calculator to get $\{-0.39, 6.39\}$.

NOW WORK PROBLEM 79.

Solve a quadratic equation using the quadratic formula

5 The Quadratic Formula

We can use the method of completing the square to obtain a general formula for solving the quadratic equation.

$$ax^2 + bx + c = 0, \quad a \neq 0$$

NOTE: There is no loss in generality to assume that $a > 0$, since if $a < 0$ we can multiply by -1 to obtain an equivalent equation with a positive leading coefficient.

As in Example 6, we rearrange the terms as

$$ax^2 + bx = -c, \quad a > 0$$

Since $a > 0$, we can divide both sides by a to get

$$x^2 + \frac{b}{a}x = -\frac{c}{a}$$

Now the coefficient of x^2 is 1. To complete the square on the left side, add the square of $\frac{1}{2}$ of the coefficient of x; that is, add

$$\left(\frac{1}{2} \cdot \frac{b}{a}\right)^2 = \frac{b^2}{4a^2}$$

to both sides. Then

$$x^2 + \frac{b}{a}x + \frac{b^2}{4a^2} = \frac{b^2}{4a^2} - \frac{c}{a}$$

$$\left(x + \frac{b}{2a}\right)^2 = \frac{b^2 - 4ac}{4a^2} \qquad \frac{b^2}{4a^2} - \frac{c}{a} = \frac{b^2}{4a^2} - \frac{4ac}{4a^2} = \frac{b^2 - 4ac}{4a^2} \qquad (3)$$

Provided that $b^2 - 4ac \geq 0$, we now can use the Square Root Method to get

$$x + \frac{b}{2a} = \pm\sqrt{\frac{b^2 - 4ac}{4a^2}} \qquad \text{Square Root Method.}$$

$$x + \frac{b}{2a} = \frac{\pm\sqrt{b^2 - 4ac}}{2a} \qquad \text{The square root of a quotient equals the quotient of the square roots. Also, } \sqrt{4a^2} = 2a \text{ since } a > 0.$$

$$x = -\frac{b}{2a} \pm \frac{\sqrt{b^2 - 4ac}}{2a} \qquad \text{Add } -\frac{b}{2a} \text{ to both sides.}$$

$$x = \frac{-b \pm \sqrt{b^2 - 4ac}}{2a} \qquad \text{Combine the quotients on the right.}$$

What if $b^2 - 4ac$ is negative? Then equation (3) states that the left expression (a real number squared) equals the right expression (a negative number). Since this occurrence is impossible for real numbers, we conclude that if $b^2 - 4ac < 0$ the quadratic equation has no *real* solution.

We now state the *quadratic formula*.

Quadratic Formula

Consider the quadratic equation

$$\boxed{ax^2 + bx + c = 0, \qquad a \neq 0}$$

If $b^2 - 4ac < 0$, this equation has no real solution.
If $b^2 - 4ac \geq 0$, the real solution(s) of this equation is (are) given by the **quadratic formula.**

$$\boxed{x = \frac{-b \pm \sqrt{b^2 - 4ac}}{2a}} \qquad (4)$$

The quantity $b^2 - 4ac$ is called the **discriminant** of the quadratic equation, because its value tells us whether the equation has real solutions. In fact, it also tells us how many solutions to expect.

> **Discriminant of a Quadratic Equation**
>
> For a quadratic equation $ax^2 + bx + c = 0$:
>
> 1. If $b^2 - 4ac > 0$, there are two unequal real solutions.
> 2. If $b^2 - 4ac = 0$, there is a repeated solution, a root of multiplicity 2.
> 3. If $b^2 - 4ac < 0$, there is no real solution.

When asked to find the real solutions, if any, of a quadratic equation, always evaluate the discriminant first to see how many real solutions there are.

EXAMPLE 7 Solving a Quadratic Equation Using the Quadratic Formula

Use the quadratic formula to find the real solutions, if any, of the equation

$$3x^2 - 5x + 1 = 0$$

SOLUTION The equation is in standard form, so we compare it to $ax^2 + bx + c = 0$ to find a, b, and c.

$$3x^2 - 5x + 1 = 0$$
$$ax^2 + bx + c = 0 \quad a = 3, b = -5, c = 1$$

With $a = 3$, $b = -5$, and $c = 1$, we evaluate the discriminant $b^2 - 4ac$.

$$b^2 - 4ac = (-5)^2 - 4(3)(1) = 25 - 12 = 13$$

Since $b^2 - 4ac > 0$, there are two real solutions, which can be found using the quadratic formula.

$$x = \frac{-b \pm \sqrt{b^2 - 4ac}}{2a} = \frac{-(-5) \pm \sqrt{13}}{2(3)} = \frac{5 \pm \sqrt{13}}{6}$$

The solution set is $\left\{ \dfrac{5 - \sqrt{13}}{6}, \dfrac{5 + \sqrt{13}}{6} \right\}$.

EXAMPLE 8 Solving a Quadratic Equation Using the Quadratic Formula

Use the quadratic formula to find the real solutions, if any, of the equation

$$3x^2 + 2 = 4x$$

SOLUTION The equation, as given, is not in standard form.

$$3x^2 + 2 = 4x$$
$$3x^2 - 4x + 2 = 0 \quad \text{Put in standard form.}$$
$$ax^2 + bx + c = 0 \quad \text{Compare to standard form.}$$

Solving Equations

With $a = 3$, $b = -4$, and $c = 2$, we find
$$b^2 - 4ac = (-4)^2 - 4(3)(2) = 16 - 24 = -8$$
Since $b^2 - 4ac < 0$, the equation has no real solution.

NOW WORK PROBLEMS 85 AND 91.

SUMMARY

Procedure for Solving a Quadratic Equation

To solve a quadratic equation, first put it in standard form:
$$ax^2 + bx + c = 0$$
Then:

Step 1: Identify a, b, and c.
Step 2: Evaluate the discriminant, $b^2 - 4ac$.
Step 3: (a) If the discriminant is negative, the equation has no real solution.
(b) If the discriminant is zero, the equation has one real solution, a repeated root.
(c) If the discriminant is positive, the equation has two distinct real solutions.

For conditions (b) and (c), if you can easily spot factors, use the factoring method to solve the equation. Otherwise, use the quadratic formula or the method of completing the square.

EXERCISE 0.4

In Problems 1–66, solve each equation.

1. $3x = 21$
2. $3x = -24$
3. $5x + 15 = 0$
4. $3x + 18 = 0$
5. $2x - 3 = 5$
6. $3x + 4 = -8$
7. $\frac{1}{3}x = \frac{5}{12}$
8. $\frac{2}{3}x = \frac{9}{2}$
9. $6 - x = 2x + 9$
10. $3 - 2x = 2 - x$
11. $2(3 + 2x) = 3(x - 4)$
12. $3(2 - x) = 2x - 1$
13. $8x - (2x + 1) = 3x - 10$
14. $5 - (2x - 1) = 10$
15. $\frac{1}{2}x - 4 = \frac{3}{4}x$
16. $1 - \frac{1}{2}x = 5$
17. $0.9t = 0.4 + 0.1t$
18. $0.9t = 1 + t$
19. $\frac{2}{y} + \frac{4}{y} = 3$
20. $\frac{4}{y} - 5 = \frac{5}{2y}$
21. $(x + 7)(x - 1) = (x + 1)^2$
22. $(x + 2)(x - 3) = (x - 3)^2$
23. $z(z^2 + 1) = 3 + z^3$
24. $w(4 - w^2) = 8 - w^3$
25. $x^2 = 9x$
26. $x^3 = x^2$
27. $t^3 - 9t^2 = 0$
28. $4z^3 - 8z^2 = 0$
29. $\frac{3}{2x - 3} = \frac{2}{x + 5}$
30. $\frac{-2}{x + 4} = \frac{-3}{x + 1}$
31. $(x + 2)(3x) = (x + 2)(6)$
32. $(x - 5)(2x) = (x - 5)(4)$
33. $\frac{2}{x - 2} = \frac{3}{x + 5} + \frac{10}{(x + 5)(x - 2)}$
34. $\frac{1}{2x + 3} + \frac{1}{x - 1} = \frac{1}{(2x + 3)(x - 1)}$
35. $|2x| = 6$
36. $|3x| = 12$
37. $|2x + 3| = 5$
38. $|3x - 1| = 2$
39. $|1 - 4t| = 5$
40. $|1 - 2z| = 3$
41. $|-2x| = 8$
42. $|-x| = 1$
43. $|-2|x = 4$
44. $|3|x = 9$
45. $|x - 2| = -\frac{1}{2}$
46. $|2 - x| = -1$

47. $|x^2 - 4| = 0$
48. $|x^2 - 9| = 0$
49. $|x^2 - 2x| = 3$
50. $|x^2 + x| = 12$
51. $|x^2 + x - 1| = 1$
52. $|x^2 + 3x - 2| = 2$
53. $x^2 = 4x$
54. $x^2 = -8x$
55. $z^2 + 4z - 12 = 0$
56. $v^2 + 7v + 12 = 0$
57. $2x^2 - 5x - 3 = 0$
58. $3x^2 + 5x + 2 = 0$
59. $x(x - 7) + 12 = 0$
60. $x(x + 1) = 12$
61. $4x^2 + 9 = 12x$
62. $25x^2 + 16 = 40x$
63. $6x - 5 = \dfrac{6}{x}$
64. $x + \dfrac{12}{x} = 7$
65. $\dfrac{4(x - 2)}{x - 3} + \dfrac{3}{x} = \dfrac{-3}{x(x - 3)}$
66. $\dfrac{5}{x + 4} = 4 + \dfrac{3}{x - 2}$

In Problems 67–72, solve each equation by the Square Root Method.

67. $x^2 = 25$
68. $x^2 = 36$
69. $(x - 1)^2 = 4$
70. $(x + 2)^2 = 1$
71. $(2x + 3)^2 = 9$
72. $(3x - 2)^2 = 4$

In Problems 73–78, what number should be added to complete the square of each expression?

73. $x^2 + 8x$
74. $x^2 - 4x$
75. $x^2 + \dfrac{1}{2}x$
76. $x^2 - \dfrac{1}{3}x$
77. $x^2 - \dfrac{2}{3}x$
78. $x^2 - \dfrac{2}{5}x$

In Problems 79–84, solve each equation by completing the square.

79. $x^2 + 4x = 21$
80. $x^2 - 6x = 13$
81. $x^2 - \dfrac{1}{2}x - \dfrac{3}{16} = 0$
82. $x^2 + \dfrac{2}{3}x - \dfrac{1}{3} = 0$
83. $3x^2 + x - \dfrac{1}{2} = 0$
84. $2x^2 - 3x - 1 = 0$

In Problems 85–96, find the real solutions, if any, of each equation. Use the quadratic formula.

85. $x^2 - 4x + 2 = 0$
86. $x^2 + 4x + 2 = 0$
87. $x^2 - 5x - 1 = 0$
88. $x^2 + 5x + 3 = 0$
89. $2x^2 - 5x + 3 = 0$
90. $2x^2 + 5x + 3 = 0$
91. $4y^2 - y + 2 = 0$
92. $4t^2 + t + 1 = 0$
93. $4x^2 = 1 - 2x$
94. $2x^2 = 1 - 2x$
95. $x^2 + \sqrt{3}x - 3 = 0$
96. $x^2 + \sqrt{2}x - 2 = 0$

In Problems 97–102, use the discriminant to determine whether each quadratic equation has two unequal real solutions, a repeated real solution, or no real solution, without solving the equation.

97. $x^2 - 5x + 7 = 0$
98. $x^2 + 5x + 7 = 0$
99. $9x^2 - 30x + 25 = 0$
100. $25x^2 - 20x + 4 = 0$
101. $3x^2 + 5x - 8 = 0$
102. $2x^2 - 3x - 4 = 0$

In Problems 103–108, solve each equation. The letters a, b, and c are constants.

103. $ax - b = c, \quad a \neq 0$
104. $1 - ax = b, \quad a \neq 0$
105. $\dfrac{x}{a} + \dfrac{x}{b} = c, \quad a \neq 0, b \neq 0, a \neq -b$
106. $\dfrac{a}{x} + \dfrac{b}{x} = c, \quad c \neq 0$
107. $\dfrac{1}{x - a} + \dfrac{1}{x + a} = \dfrac{2}{x - 1}, \quad a \neq 1$
108. $\dfrac{b + c}{x + a} = \dfrac{b - c}{x - a}, \quad c \neq 0, a \neq 0$

Problems 109–114 list some formulas that occur in applications. Solve each formula for the indicated variable.

109. Electricity $\dfrac{1}{R} = \dfrac{1}{R_1} + \dfrac{1}{R_2}$ for R

110. Finance $A = P(1 + rt)$ for r

111. Mechanics $F = \dfrac{mv^2}{R}$ for R

112. Chemistry $PV = nRT$ for T

113. Mathematics $S = \dfrac{a}{1 - r}$ for r

114. Mechanics $v = -gt + v_0$ for t

---

115. Show that the sum of the roots of a quadratic equation is $-\dfrac{b}{a}$.

116. Show that the product of the roots of a quadratic equation is $\dfrac{c}{a}$.

117. Find k such that the equation $kx^2 + x + k = 0$ has a repeated real solution.

118. Find k such that the equation $x^2 - kx + 4 = 0$ has a repeated real solution.

119. Show that the real solutions of the equation $ax^2 + bx + c = 0$ are the negatives of the real solutions of the equation $ax^2 - bx + c = 0$. Assume that $b^2 - 4ac \geq 0$.

120. Show that the real solutions of the equation $ax^2 + bx + c = 0$ are the reciprocals of the real solutions of the equation $cx^2 + bx + a = 0$. Assume that $b^2 - 4ac \geq 0$.

121. Which of the following pairs of equations are equivalent? Explain.

(a) $x^2 = 9$; $x = 3$
(b) $x = \sqrt{9}$; $x = 3$
(c) $(x - 1)(x - 2) = (x - 1)^2$; $x - 2 = x - 1$

122. The equation
$$\dfrac{5}{x + 3} + 3 = \dfrac{8 + x}{x + 3}$$
has no solution, yet when we go through the process of solving it we obtain $x = -3$. Write a brief paragraph to explain what causes this to happen.

123. Make up an equation that has no solution and give it to a fellow student to solve. Ask the fellow student to write a critique of your equation.

124. Describe three ways you might solve a quadratic equation. State your preferred method; explain why you chose it.

125. Explain the benefits of evaluating the discriminant of a quadratic equation before attempting to solve it.

126. Make up three quadratic equations: one having two distinct solutions, one having no real solution, and one having exactly one real solution.

127. The word *quadratic* seems to imply four (*quad*), yet a quadratic equation is an equation that involves a polynomial of degree 2. Investigate the origin of the term *quadratic* as it is used in the expression *quadratic equation*. Write a brief essay on your findings.

0.5 Intervals; Solving Inequalities

PREPARING FOR THIS SECTION *Before getting started, review the following:*

> Real Number Line, Inequalities, Absolute Value (Section 0.2, pp. 16–19)

OBJECTIVES
1. Use interval notation
2. Use properties of inequalities
3. Solve inequalities
4. Solve combined inequalities
5. Solve polynomial and rational inequalities

Suppose that a and b are two real numbers and $a < b$. We shall use the notation $a < x < b$ to mean that x is a number *between* a and b. The expression $a < x < b$ is equivalent to the two inequalities $a < x$ and $x < b$. Similarly, the expression $a \leq x \leq b$ is equivalent to the two inequalities $a \leq x$ and $x \leq b$. The remaining two possibilities, $a \leq x < b$ and $a < x \leq b$, are defined similarly.

Although it is acceptable to write $3 \geq x \geq 2$, it is preferable to reverse the inequality symbols and write instead $2 \leq x \leq 3$ so that, as you read from left to right, the values go from smaller to larger.

A statement such as $2 \leq x \leq 1$ is false because there is no number x for which $2 \leq x$ and $x \leq 1$. Finally, we never mix inequality symbols, as in $2 \leq x \geq 3$.

Intervals

Use internal notation 1 Let a and b represent two real numbers with $a < b$.

> A **closed interval**, denoted by $[a, b]$, consists of all real numbers x for which $a \leq x \leq b$.
>
> An **open interval**, denoted by (a, b), consists of all real numbers x for which $a < x < b$.
>
> The **half-open**, or **half-closed**, **intervals** are $(a, b]$, consisting of all real numbers x for which $a < x \leq b$, and $[a, b)$, consisting of all real numbers x for which $a \leq x < b$.

In each of these definitions, a is called the **left endpoint** and b the **right endpoint** of the interval.

The symbol ∞ (read as "infinity") is not a real number, but a notational device used to indicate unboundedness in the positive direction. The symbol $-\infty$ (read as "negative infinity") also is not a real number, but a notational device used to indicate unboundedness in the negative direction. Using the symbols ∞ and $-\infty$, we can define five other kinds of intervals:

$[a, \infty)$	consists of all real numbers x for which $x \geq a$ $(a \leq x < \infty)$
(a, ∞)	consists of all real numbers x for which $x > a$ $(a < x < \infty)$
$(-\infty, a]$	consists of all real numbers x for which $x \leq a$ $(-\infty < x \leq a)$
$(-\infty, a)$	consists of all real numbers x for which $x < a$ $(-\infty < x < a)$
$(-\infty, \infty)$	consists of all real numbers x $(-\infty < x < \infty)$

Note that ∞ and $-\infty$ are never included as endpoints, since neither is a real number.

Table 2 summarizes interval notation, corresponding inequality notation, and their graphs.

TABLE 2

Interval	Inequality	Graph
The open interval (a, b)	$a < x < b$	
The closed interval $[a, b]$	$a \leq x \leq b$	
The half-open interval $[a, b)$	$a \leq x < b$	
The half-open interval $(a, b]$	$a < x \leq b$	
The interval $[a, \infty)$	$x \geq a$	
The interval (a, ∞)	$x > a$	
The interval $(-\infty, a]$	$x \leq a$	
The interval $(-\infty, a)$	$x < a$	
The interval $(-\infty, \infty)$	All real numbers	

EXAMPLE 1 Writing Inequalities Using Interval Notation

Write each inequality using interval notation.

(a) $1 \leq x \leq 3$ (b) $-4 < x < 0$ (c) $x > 5$ (d) $x \leq 1$

SOLUTION (a) $1 \leq x \leq 3$ describes all numbers x between 1 and 3, inclusive. In interval notation, we write $[1, 3]$.
(b) In interval notation, $-4 < x < 0$ is written $(-4, 0)$.
(c) $x > 5$ consists of all numbers x greater than 5. In interval notation, we write $(5, \infty)$.
(d) In interval notation, $x \leq 1$ is written $(-\infty, 1]$.

EXAMPLE 2 Writing Intervals Using Inequality Notation

Write each interval as an inequality involving x.

(a) $[1, 4)$ (b) $(2, \infty)$ (c) $[2, 3]$ (d) $(-\infty, -3]$

SOLUTION (a) $[1, 4)$ consists of all numbers x for which $1 \leq x < 4$.
(b) $(2, \infty)$ consists of all numbers x for which $x > 2$ ($2 < x < \infty$).
(c) $[2, 3]$ consists of all numbers x for which $2 \leq x \leq 3$.
(d) $(-\infty, -3]$ consists of all numbers x for which $x \leq -3$ ($-\infty < x \leq -3$).

NOW WORK PROBLEMS 1, 7, AND 15.

Properties of Inequalities

Use properties of inequalities ②

The product of two positive real numbers is positive, the product of two negative real numbers is positive, and the product of 0 and 0 is 0. For any real number a, the value of a^2 is 0 or positive; that is, a^2 is nonnegative. This is called the **nonnegative property.**

For any real number a, we have the following:

Nonnegative Property

$$a^2 \geq 0 \tag{1}$$

If we add the same number to both sides of an inequality, we obtain an equivalent inequality. For example, since $3 < 5$, then $3 + 4 < 5 + 4$ or $7 < 9$. This is called the **addition property** of inequalities.

Addition Property of Inequalities

$$\text{If } a < b, \text{ then } a + c < b + c. \tag{2a}$$
$$\text{If } a > b, \text{ then } a + c > b + c. \tag{2b}$$

The addition property states that the sense, or direction, of an inequality remains unchanged if the same number is added to each side. Now let's see what happens if we multiply each side of an inequality by a non-zero number.

Begin with $3 < 7$ and multiply each side by 2. The numbers 6 and 14 that result yield the inequality $6 < 14$.

Begin with $9 > 2$ and multiply each side by -4. The numbers -36 and -8 that result yield the inequality $-36 < -8$.

Note that the effect of multiplying both sides of $9 > 2$ by the negative number -4 is that the direction of the inequality symbol is reversed. We are led to the following general **multiplication properties** for inequalities:

Multiplication Properties for Inequalities

$$\text{If } a < b \text{ and if } c > 0, \text{ then } ac < bc.$$
$$\text{If } a < b \text{ and if } c < 0, \text{ then } ac > bc. \quad (3a)$$
$$\text{If } a > b \text{ and if } c > 0, \text{ then } ac > bc.$$
$$\text{If } a > b \text{ and if } c < 0, \text{ then } ac < bc. \quad (3b)$$

The multiplication properties state that the sense, or direction, of an inequality *remains the same* if each side is multiplied by a *positive* real number, whereas the direction is *reversed* if each side is multiplied by a *negative* real number.

NOW WORK PROBLEMS 29 AND 35.

The **reciprocal property** states that the reciprocal of a positive real number is positive and that the reciprocal of a negative real number is negative.

Reciprocal Property for Inequalities

$$\text{If } a > 0, \text{ then } \frac{1}{a} > 0. \quad (4a)$$

$$\text{If } a < 0, \text{ then } \frac{1}{a} < 0. \quad (4b)$$

Solving Inequalities

An **inequality in one variable** is a statement involving two expressions, at least one containing the variable, separated by one of the inequality symbols $<$, $\leq$, $>$, or $\geq$. To **solve an inequality** means to find all values of the variable for which the statement is true. These values are called **solutions** of the inequality.

For example, the following are all inequalities involving one variable, x:

$$x + 5 < 8 \qquad 2x - 3 \geq 4 \qquad x^2 - 1 \leq 3 \qquad \frac{x+1}{x-2} > 0$$

Two inequalities having exactly the same solution set are called **equivalent inequalities**. As with equations, one method for solving an inequality is to replace it by a series of equivalent inequalities until an inequality with an obvious solution, such as $x < 3$, is obtained. We obtain equivalent inequalities by applying some of the same properties as those used to find equivalent equations. The addition property and the multiplication properties form the basis for the following procedures.

> **Procedures That Leave the Inequality Symbol Unchanged**
>
> **1.** Simplify both sides of the inequality by combining like terms and eliminating parentheses:
>
> $$\text{Replace} \quad (x + 2) + 6 > 2x + 5(x + 1)$$
> $$\text{by} \quad x + 8 > 7x + 5$$
>
> **2.** Add or subtract the same expression on both sides of the inequality:
>
> $$\text{Replace} \quad 3x - 5 < 4$$
> $$\text{by} \quad (3x - 5) + 5 < 4 + 5$$
>
> **3.** Multiply or divide both sides of the inequality by the same positive expression:
>
> $$\text{Replace} \quad 4x > 16 \quad \text{by} \quad \frac{4x}{4} > \frac{16}{4}$$
>
> **Procedures That Reverse the Sense or Direction of the Inequality Symbol**
>
> **1.** Interchange the two sides of the inequality:
>
> $$\text{Replace} \quad 3 < x \quad \text{by} \quad x > 3$$
>
> **2.** Multiply or divide both sides of the inequality by the same *negative* expression.
>
> $$\text{Replace} \quad -2x > 6 \quad \text{by} \quad \frac{-2x}{-2} < \frac{6}{-2}$$

As the examples that follow illustrate, we solve inequalities using many of the same steps that we would use to solve equations. In writing the solution of an inequality, we may use either set notation or interval notation, whichever is more convenient.

EXAMPLE 3 Solving an Inequality

Solve the inequality: $4x + 7 \geq 2x - 3$
Graph the solution set.

SOLUTION

$$4x + 7 \geq 2x - 3$$
$$4x + 7 - 7 \geq 2x - 3 - 7 \qquad \text{Subtract 7 from both sides.}$$
$$4x \geq 2x - 10 \qquad \text{Simplify.}$$
$$4x - 2x \geq 2x - 10 - 2x \qquad \text{Subtract } 2x \text{ from both sides.}$$
$$2x \geq -10 \qquad \text{Simplify.}$$
$$\frac{2x}{2} \geq \frac{-10}{2} \qquad \text{Divide both sides by 2. (The sense of the inequality symbol is unchanged.)}$$
$$x \geq -5 \qquad \text{Simplify.}$$

FIGURE 12

The solution set is $\{x \mid x \geq -5\}$ or, using interval notation, all numbers in the interval $[-5, \infty)$. See Figure 12 for the graph.

NOW WORK PROBLEM 43.

4 EXAMPLE 4   Solving Combined Inequalities

Solve the inequality: $-5 < 3x - 2 < 1$
Graph the solution set.

SOLUTION Recall that the inequality

$$-5 < 3x - 2 < 1$$

is equivalent to the two inequalities

$$-5 < 3x - 2 \quad \text{and} \quad 3x - 2 < 1$$

We will solve each of these inequalities separately.

$-5 < 3x - 2$		$3x - 2 < 1$
$-5 + 2 < 3x - 2 + 2$	Add 2 to both sides.	$3x - 2 + 2 < 1 + 2$
$-3 < 3x$	Simplify.	$3x < 3$
$\dfrac{-3}{3} < \dfrac{3x}{3}$	Divide both sides by 3.	$\dfrac{3x}{3} < \dfrac{3}{3}$
$-1 < x$	Simplify.	$x < 1$

The solution set of the original pair of inequalities consists of all x for which

$$-1 < x \quad \text{and} \quad x < 1$$

FIGURE 13

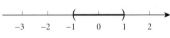

This may be written more compactly as $\{x \mid -1 < x < 1\}$. In interval notation, the solution is $(-1, 1)$. See Figure 13 for the graph.

NOW WORK PROBLEM 63.

EXAMPLE 5   Using the Reciprocal Property to Solve an Inequality

Solve the inequality: $(4x - 1)^{-1} > 0$
Graph the solution set.

SOLUTION Since $(4x - 1)^{-1} = \dfrac{1}{4x - 1}$ and since the Reciprocal Property states that when $\dfrac{1}{a} > 0$ then $a > 0$, we have

$$(4x - 1)^{-1} > 0$$

$$\dfrac{1}{4x - 1} > 0$$

$$4x - 1 > 0 \quad \text{Reciprocal Property.}$$

$$4x > 1$$

$$x > \dfrac{1}{4}$$

FIGURE 14

The solution set is $\left\{x \mid x > \dfrac{1}{4}\right\}$; that is, all x in the interval $\left(\dfrac{1}{4}, \infty\right)$. Figure 14 illustrates the graph.

NOW WORK PROBLEM 67.

Intervals; Solving Inequalities

Polynomial and Rational Inequalities

Solve polynomial and rational inequalities 5

The next four examples deal with polynomial and rational inequalities which are important for solving certain types of problems in calculus.

EXAMPLE 6 Solving a Quadratic Inequality

Solve the inequality $x^2 + x - 12 > 0$, and graph the solution set.

SOLUTION We factor the left side, obtaining

$$x^2 + x - 12 > 0$$
$$(x + 4)(x - 3) > 0$$

We then construct a graph that uses the solutions to the equation

$$x^2 + x - 12 = (x + 4)(x - 3) = 0$$

namely, $x = -4$ and $x = 3$. These numbers separate the real number line into three parts:

$$-\infty < x < -4 \qquad -4 < x < 3 \qquad 3 < x < \infty$$

or, in interval notation, into $(-\infty, -4)$, $(-4, 3)$, and $(3, \infty)$. See Figure 15(a).

Now if $x < -4$, then $x + 4 < 0$. We indicate this fact about the expression $x + 4$ by placing minus signs $(- - -)$ to the left of -4. If $x > -4$, then $x + 4 > 0$. We indicate this fact about $x + 4$ by placing plus signs $(+ + +)$ to the right of -4.

Similarly, if $x < 3$, then $x - 3 < 0$. We indicate this fact about $x - 3$ by placing minus signs to the left of 3. If $x > 3$, then $x - 3 > 0$. We indicate this fact about $x - 3$ by placing plus signs to the right of 3. See Figure 15 (b).

FIGURE 15

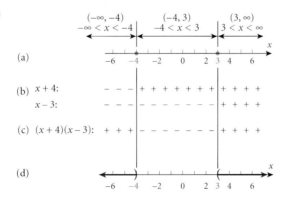

Now we prepare Figure 15(c). Since we know that the expressions $x + 4$ and $x - 3$ are both negative for $x < -4$, it follows that their product will be positive for $x < -4$. Since we know that $x + 4$ is positive and $x - 3$ is negative for $-4 < x < 3$, it follows that their product is negative for $-4 < x < 3$. Finally, since both expressions are positive for $x > 3$, their product is positive for $x > 3$.

We conclude that the product $(x + 4)(x - 3) = x^2 + x - 12$ is positive when $x < -4$ or when $x > 3$. The solution set is $\{x \mid x < -4 \text{ or } x > 3\}$. In interval notation the solution consists of the intervals $(-\infty, -4)$ or $(3, \infty)$. See Figure 15(d). ▶

The preceding discussion demonstrates that the sign of each factor of the expression and, consequently, the sign of the expression itself, is the same on each interval that the real number line was divided into. An alternative, and simpler, approach to obtaining

Figure 15(c) would be to select a **test number** in each interval and use it to evaluate the expression to see if it is positive or negative. You may choose any number in the interval as a test number. See Figure 16.

FIGURE 16

(a)

(b) Value of $x^2 + x - 12$:

(c) Sign of $x^2 + x - 12$:

(d)

In Figure 16(a) the test numbers we selected, $-5, 1, 4$, have been circled. We evaluate the expression $x^2 + x - 12$ at each test number.

For $x = -5$: $(-5)^2 + (-5) - 12 = 8$, a positive number.
For $x = 1$: $1^2 + 1 - 12 = -10$, a negative number.
Fox $x = 4$: $4^2 + 4 - 12 = 8$, a positive number.

See Figure 16(b) and 16(c).

The rest of Figure 16 is obtained as before.

We shall employ the method of using a test number to solve inequalities. Here is another example showing all the details.

EXAMPLE 7 Solving a Quadratic Inequality

Solve the inequality $x^2 \leq 4x + 12$, and graph the solution set.

SOLUTION First, we rearrange the inequality so that 0 is on the right side:

$$x^2 \leq 4x + 12$$
$$x^2 - 4x - 12 \leq 0 \quad \text{Subtract } 4x \text{ and } 12 \text{ from each side.}$$

This inequality is equivalent to the one we seek to solve.

Next, we set the left side equal to 0 and solve the resulting equation:

$$x^2 - 4x - 12 = 0$$
$$(x + 2)(x - 6) = 0 \quad \text{Factor.}$$
$$x + 2 = 0 \quad \text{or} \quad x - 6 = 0 \quad \text{Zero-Product Property.}$$
$$x = -2 \quad \text{or} \quad x = 6$$

The solutions of the equation are -2 and 6, and they separate the real number line into three parts:

$$-\infty < x < -2 \qquad -2 < x < 6 \qquad 6 < x < \infty$$

See Figure 17.

In each part, select a test number. We will choose $-3, 1,$ and 8, which are circled. See Figure 17(a).

Next we evaluate the expression $x^2 - 4x - 12$ at each test number.

For $x = -3$: $(-3)^2 - 4(-3) - 12 = 9$, a positive number.
For $x = 1$: $1^2 - 4(1) - 12 = -15$, a negative number.
For $x = 8$: $8^2 - 4(8) - 12 = 20$, a positive number.

See Figure 17(b) and 17(c).

The expression $x^2 - 4x - 12 < 0$ for $-2 < x < 6$. However, because the inequality we wish to solve is nonstrict, numbers x that satisfy the equation $x^2 - 4x + 12 = 0$ are also solutions of the inequality $x^2 - 4x - 12 \leq 0$. We include -2 and 6, and the solution set of the given inequality is $\{x| -2 \leq x \leq 6\}$; that is, all x in the interval $[-2, 6]$. See Figure 17(d).

FIGURE 17

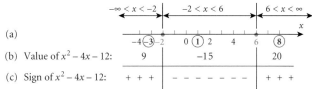

(a)
(b) Value of $x^2 - 4x - 12$:
(c) Sign of $x^2 - 4x - 12$:
(d)

 NOW WORK PROBLEM 75.

EXAMPLE 8 Solving a Polynomial Inequality

Solve the inequality $x^4 < x$, and graph the solution set.

SOLUTION We rewrite the inequality so that 0 is on the right side:

$$x^4 < x$$
$$x^4 - x < 0 \quad \text{Subtract } x \text{ from both sides.}$$

This inequality is equivalent to the one we wish to solve.
We proceed to solve the equation $x^4 - x = 0$ using factoring.

$$x^4 - x = 0$$
$$x(x^3 - 1) = 0 \quad \text{Factor out } x.$$
$$x(x - 1)(x^2 + x + 1) = 0 \quad \text{Difference of two cubes.}$$
$$x = 0 \quad \text{or} \quad x - 1 = 0 \quad \text{or} \quad x^2 + x + 1 = 0 \quad \text{Zero-Product Property.}$$

The solutions are 0 and 1, since the equation $x^2 + x + 1 = 0$ has no real solution.
Next we use 0 and 1 to separate the real number line into three parts:

$$-\infty < x < 0 \qquad 0 < x < 1 \qquad 1 < x < \infty$$

In each part, select a test number. We will choose $-1, \dfrac{1}{2}$, and 2. See Figure 18.

FIGURE 18

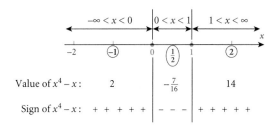

We evaluate the expression $x^4 - x$ at each test number.

For $x = -1$: $(-1)^4 - (-1) = 2$, a positive number.

For $x = \dfrac{1}{2}$: $\left(\dfrac{1}{2}\right)^4 - \left(\dfrac{1}{2}\right) = -\dfrac{7}{16}$, a negative number.

For $x = 2$: $2^4 - 2 = 14$, a positive number.

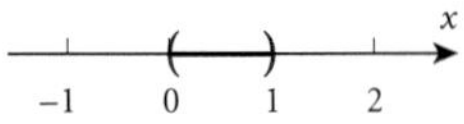

The expression $x^4 - x < 0$ for $0 < x < 1$. The solution set is $\{x \mid 0 < x < 1\}$; that is, all x in the interval $(0, 1)$. See Figure 19.

FIGURE 19

We have been solving inequalities by rearranging the inequality so that 0 is on the right side, setting the left side equal to 0, and solving the resulting equation. The solutions are then used to separate the real number line into intervals. But what if the resulting equation has no real solution? In this case we rely on the following result.

Theorem

If a polynomial equation has no real solutions, the polynomial is either always positive or always negative.

For example, the equation

$$x^2 + 5x + 8 = 0$$

has no real solutions. (Do you see why? Its discriminant, $b^2 - 4ac = 25 - 32 = -7$, is negative.) The value of $x^2 + 5x + 8$ is therefore always positive or always negative. To see which is true, we test its value at some number (0 is the easiest). Because $0^2 + 5(0) + 8 = 8$ is positive, we conclude that $x^2 + 5x + 8 > 0$ for all x.

NOW WORK PROBLEM 79.

Next we solve a rational inequality.

EXAMPLE 9 **Solving a Rational Inequality**

Solve the inequality $\dfrac{4x + 5}{x + 2} \geq 3$, and graph the solution set.

SOLUTION We first note that the domain of the variable consists of all real numbers except -2. We rearrange terms so that 0 is on the right side:

$$\dfrac{4x + 5}{x + 2} \geq 3$$

$$\dfrac{4x + 5}{x + 2} - 3 \geq 0 \quad \text{Subtract 3 from both sides.}$$

$$\dfrac{4x + 5 - 3(x + 2)}{x + 2} \geq 0 \quad \text{Rewrite using } x + 2 \text{ as the denominator.}$$

$$\dfrac{x - 1}{x + 2} \geq 0 \quad \text{Simplify.}$$

This inequality is equivalent to the one we wish to solve.

For rational expressions we set both the numerator and the denominator equal to 0 to determine the numbers to use to separate the number line. For this example we use -2 and 1 to divide the number line into three parts:

$$-\infty < x < -2 \qquad -2 < x < 1 \qquad 1 < x < \infty$$

Construct Figure 20, using $-3, 0,$ and 2 as test numbers.

FIGURE 20

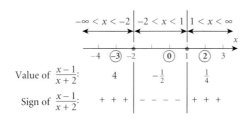

We conclude that $\dfrac{x-1}{x+2} > 0$ for $x < -2$ or for $x > 1$.

However, we want to know where the expression $\dfrac{x-1}{x+2}$ is positive or 0. Since $\dfrac{x-1}{x+2} = 0$ only if $x = 1$, we conclude that the solution set is $\{x \mid x < -2 \text{ or } x \geq 1\}$; that is, all x in the intervals $(-\infty, -2)$ or $[1, \infty)$. See Figure 21.

FIGURE 21

In Example 9 you may wonder why we did not first multiply both sides of the inequality by $x + 2$ to clear the denominator. The reason is that we do not know whether $x + 2$ is positive or negative and, as a result, we do not know whether to reverse the sense of the inequality symbol after multiplying by $x + 2$. However, there is nothing to prevent us from multiplying both sides by $(x + 2)^2$, which is always positive, since $x \neq -2$. (Do you see why?)

$$\dfrac{4x+5}{x+2} \geq 3 \qquad\qquad x \neq -2$$

$$\dfrac{4x+5}{x+2}(x+2)^2 \geq 3(x+2)^2$$

$$(4x+5)(x+2) \geq 3(x^2+4x+4)$$

$$4x^2 + 13x + 10 \geq 3x^2 + 12x + 12$$

$$x^2 + x - 2 \geq 0$$

$$(x+2)(x-1) \geq 0 \qquad\qquad x \neq -2$$

This last expression leads to the same solution set obtained in Example 9.

 NOW WORK PROBLEM 91.

EXERCISE 0.5

In Problems 1–6, express the graph shown in color using interval notation. Also express each as an inequality involving x.

1.

2.

3.

4. (number line: closed at -1, arrow left through -2)   **5.** (number line: closed at -1, open at 3)   **6.** (number line: open at -1, open at 3)

In Problems 7–14, write each inequality using interval notation, and graph each inequality using the real number line.

7. $0 \le x \le 4$ **8.** $-1 < x < 5$ **9.** $4 \le x < 6$ **10.** $-2 < x < 0$

11. $x \ge 4$ **12.** $x \le 5$ **13.** $x < -4$ **14.** $x > 1$

In Problems 15–22, write each interval as an inequality involving x, and graph each inequality using the real number line.

15. $[2, 5]$ **16.** $(1, 2)$ **17.** $(-3, -2)$ **18.** $[0, 1)$

19. $[4, \infty)$ **20.** $(-\infty, 2]$ **21.** $(-\infty, -3)$ **22.** $(-8, \infty)$

In Problems 23–28, an inequality is given. Write the inequality obtained by:
(a) Adding 3 to each side of the given inequality.
(b) Subtracting 5 from each side of the given inequality.
(c) Multiplying each side of the given inequality by 3.
(d) Multiplying each side of the given inequality by -2.

23. $3 < 5$ **24.** $2 > 1$ **25.** $4 > -3$

26. $-3 > -5$ **27.** $2x + 1 < 2$ **28.** $1 - 2x > 5$

In Problems 29–42, fill in the blank with the correct inequality symbol.

29. If $x < 5$, then $x - 5$ _____ 0.   **30.** If $x < -4$, then $x + 4$ _____ 0.

31. If $x > -4$, then $x + 4$ _____ 0.   **32.** If $x > 6$, then $x - 6$ _____ 0.

33. If $x \ge -4$, then $3x$ _____ -12.   **34.** If $x \le 3$, then $2x$ _____ 6.

35. If $x > 6$, then $-2x$ _____ -12.   **36.** If $x > -2$, then $-4x$ _____ 8.

37. If $x \ge 5$, then $-4x$ _____ -20.   **38.** If $x \le -4$, then $-3x$ _____ 12.

39. If $2x > 6$, then x _____ 3.   **40.** If $3x \le 12$, then x _____ 4.

41. If $-\dfrac{1}{2}x \le 3$, then x _____ -6.   **42.** If $-\dfrac{1}{4}x > 1$, then x _____ -4.

In Problems 43–100, solve each inequality. Express your answer using set notation or interval notation. Graph the solution set.

43. $x + 1 < 5$ **44.** $x - 6 < 1$ **45.** $1 - 2x \le 3$

46. $2 - 3x \le 5$ **47.** $3x - 7 > 2$ **48.** $2x + 5 > 1$

49. $3x - 1 \ge 3 + x$ **50.** $2x - 2 \ge 3 + x$ **51.** $-2(x + 3) < 8$

52. $-3(1 - x) < 12$ **53.** $4 - 3(1 - x) \le 3$ **54.** $8 - 4(2 - x) \le -2x$

55. $\dfrac{1}{2}(x - 4) > x + 8$ **56.** $3x + 4 > \dfrac{1}{3}(x - 2)$ **57.** $\dfrac{x}{2} \ge 1 - \dfrac{x}{4}$

58. $\dfrac{x}{3} \ge 2 + \dfrac{x}{6}$ **59.** $0 \le 2x - 6 \le 4$ **60.** $4 \le 2x + 2 \le 10$

61. $-5 \le 4 - 3x \le 2$ **62.** $-3 \le 3 - 2x \le 9$ **63.** $-3 < \dfrac{2x - 1}{4} < 0$

64. $0 < \dfrac{3x+2}{2} < 4$

65. $1 < 1 - \dfrac{1}{2}x < 4$

66. $0 < 1 - \dfrac{1}{3}x < 1$

67. $(4x+2)^{-1} < 0$

68. $(2x-1)^{-1} > 0$

69. $0 < \dfrac{2}{x} < \dfrac{3}{5}$

70. $0 < \dfrac{4}{x} < \dfrac{2}{3}$

71. $(x-3)(x+1) < 0$

72. $(x-1)(x+2) < 0$

73. $-x^2 + 9 > 0$

74. $-x^2 + 1 > 0$

75. $x^2 + x > 12$

76. $x^2 + 7x < -12$

77. $x(x-7) > -12$

78. $x(x+1) > 12$

79. $4x^2 + 9 < 6x$

80. $25x^2 + 16 < 40x$

81. $(x-1)(x^2+x+1) > 0$

82. $(x+2)(x^2-x+1) > 0$

83. $(x-1)(x-2)(x-3) < 0$

84. $(x+1)(x+2)(x+3) < 0$

85. $-x^3 + 2x^2 + 8x < 0$

86. $-x^3 - 2x^2 + 8x < 0$

87. $x^3 > x$

88. $x^3 < 4x$

89. $x^3 > x^2$

90. $x^3 < 3x^2$

91. $\dfrac{x+1}{1-x} < 0$

92. $\dfrac{3-x}{x+1} < 0$

93. $\dfrac{(x-1)(x+1)}{x} < 0$

94. $\dfrac{(x-3)(x+2)}{x-1} < 0$

95. $\dfrac{x-2}{x^2-1} \geq 0$

96. $\dfrac{x+5}{x^2-4} \geq 0$

97. $\dfrac{x+4}{x-2} \leq 1$

98. $\dfrac{x+2}{x-4} \geq 1$

99. $\dfrac{2x+5}{x+1} > \dfrac{x+1}{x-1}$

100. $\dfrac{1}{x+2} > \dfrac{3}{x+1}$

---

101. Computing Grades In your Economics 101 class, you have scores of 68, 82, 87, and 89 on the first four of five tests. To get a grade of B, the average of the first five test scores must be greater than or equal to 80 and less than 90. Solve an inequality to find the range of the score that you need on the last test to get a B.

102. General Chemistry For a certain ideal gas, the volume V (in cubic centimeters) equals 20 times the temperature T (in degrees Celsius). If the temperature varies from 80° to 120°C inclusive, what is the corresponding range of the volume of the gas?

103. Real Estate A real estate agent agrees to sell a large apartment complex according to the following commission schedule: $45,000 plus 25% of the selling price in excess of $900,000. Assuming that the complex will sell at some price between $900,000 and $1,100,000 inclusive, over what range does the agent's commission vary? How does the commission vary as a percent of selling price?

104. Sales Commission A used car salesperson is paid a commission of $25 plus 40% of the selling price in excess of owner's cost. The owner claims that used cars typically sell for at least owner's cost plus $70 and at most owner's cost plus $300. For each sale made, over what range can the salesperson expect the commission to vary?

105. Federal Tax Withholding The percentage method of withholding for federal income tax (2003) states that a single person whose weekly wages, after subtracting withholding allowances, are over $592, but not over $1317, shall have $74.35 plus 25% of the excess over $592 withheld. Over what range does the amount withheld vary if the weekly wages vary from $600 to $800 inclusive?

Source: Internal Revenue Service, 2003.

106. Federal Tax Withholding Rework Problem 105 if the weekly wages vary from $800 to $1000 inclusive.

107. Electricity Rates Commonwealth Edison Company's charge for electricity in May 2003 is 8.275¢ per kilowatt-hour. In addition, each monthly bill contains a customer charge of $7.58. If last summer's bills ranged from a low of $63.47 to a high of $214.53, over what range did usage vary (in kilowatt-hours)?

Source: Commonwealth Edison Co., Chicago, Illinois, 2003.

108. Water Bills The Village of Oak Lawn charges homeowners $27.18 per quarter-year plus $1.90 per 1000 gallons for water usage in excess of 12,000 gallons. In 2003, one homeowner's quarterly bill ranged from a high of $76.52 to a low of $34.78. Over what range did water usage vary?

Source: Village of Oak Lawn, Illinois, 2003.

109. Markup of a New Car The sticker price of a new car ranges from 12% to 18% higher than the dealer's cost. If the sticker price is $8800, over what range will the dealer's cost vary?

110. IQ Tests A standard intelligence test has an average score of 100. According to statistical theory, of the people who take the test, the 2.5% with the highest scores will have scores of more than 1.96σ above the average, where σ (sigma, a number called the **standard deviation**) depends on the nature of the test. If $\sigma = 12$ for this test and there is (in principle) no upper limit to the score possible on the test, write the interval of possible test scores of the people in the top 2.5%.

111. Make up an inequality that has no solution. Make up one that has exactly one solution.

112. The inequality $x^2 + 1 < -5$ has no solution. Explain why.

113. Do you prefer to use inequality notation or interval notation to express the solution to an inequality? Give your reasons. Are there particular circumstances when you prefer one to the other? Cite examples.

114. How would you explain to a fellow student the underlying reason for the multiplication properties for inequalities (page 54); that is, the sense or direction of an inequality remains the same if each side is multiplied by a positive real number, whereas the direction is reversed if each side is multiplied by a negative real number.

0.6 *n*th Roots; Rational Exponents

PREPARING FOR THIS SECTION *Before getting started, review the following:*

> Exponents, Square Roots (Section 0.2, pp. 20–24)

OBJECTIVES
1. Work with *n*th roots
2. Simplify radicals
3. Rationalize denominators
4. Solve radical equations
5. Simplify expressions with rational exponents

*n*th ROOTS

The **principal *n*th root of a number *a*,** symbolized by $\sqrt[n]{a}$, where $n \geq 2$ is an integer, is defined as follows:

$$\sqrt[n]{a} = b \quad \text{means} \quad a = b^n$$

where $a \geq 0$ and $b \geq 0$ if $n \geq 2$ is even, and a, b are any real numbers if $n \geq 3$ is odd.

Notice that if a is negative and n is even then $\sqrt[n]{a}$ is not defined. When it is defined, the principal *n*th root of a number is unique.

1 Work with *n*th roots   The symbol $\sqrt[n]{a}$ for the principal *n*th root of a is sometimes called a **radical;** the integer n is called the **index,** and a is called the **radicand.** If the index of a radical is 2, we call $\sqrt[n]{a}$ the **square root** of a and omit the index 2 by simply writing $\sqrt{a}$. If the index is 3, we call $\sqrt[3]{a}$ the **cube root** of a.

EXAMPLE 1 Evaluating Principal nth Roots

(a) $\sqrt[3]{8} = \sqrt[3]{2^3} = 2$

(b) $\sqrt[3]{-64} = \sqrt[3]{(-4)^3} = -4$

(c) $\sqrt[4]{\dfrac{1}{16}} = \sqrt[4]{\left(\dfrac{1}{2}\right)^4} = \dfrac{1}{2}$

(d) $\sqrt[6]{(-2)^6} = |-2| = 2$

These are examples of **perfect roots,** since each simplifies to a rational number. Notice the absolute value in Example 1(d). If n is even, the principal nth root must be nonnegative.

In general, if $n \geq 2$ is a positive integer and a is a real number, we have

$$\sqrt[n]{a^n} = a, \quad \text{if } n \geq 3 \text{ is odd} \tag{1a}$$
$$\sqrt[n]{a^n} = |a|, \quad \text{if } n \geq 2 \text{ is even} \tag{1b}$$

NOW WORK PROBLEM 1.

Properties of Radicals

Let $n \geq 2$ and $m \geq 2$ denote positive integers, and let a and b represent real numbers. Assuming that all radicals are defined, we have the following properties:

$$\sqrt[n]{ab} = \sqrt[n]{a}\,\sqrt[n]{b} \tag{2a}$$
$$\sqrt[n]{\dfrac{a}{b}} = \dfrac{\sqrt[n]{a}}{\sqrt[n]{b}} \tag{2b}$$
$$\sqrt[n]{a^m} = (\sqrt[n]{a})^m \tag{2c}$$

When used in reference to radicals, the direction to "simplify" will mean to remove from the radicals any perfect roots that occur as factors. Let's look at some examples of how the preceding rules are applied to simplify radicals.

EXAMPLE 2 Simplifying Radicals

(a) $\sqrt{32} = \sqrt{16 \cdot 2} = \sqrt{16} \cdot \sqrt{2} = 4\sqrt{2}$
     ↑   ↑
  16 is a perfect square. (2a)

(b) $\sqrt[3]{16} = \sqrt[3]{8 \cdot 2} = \sqrt[3]{8} \cdot \sqrt[3]{2} = 2\sqrt[3]{2}$
     ↑   ↑
  8 is a perfect cube.   (2a)

(c) $\sqrt[3]{-16x^4} = \sqrt[3]{-8 \cdot 2 \cdot x^3 \cdot x} = \sqrt[3]{(-8x^3)(2x)}$
$\phantom{(c) \sqrt[3]{-16x^4}} \uparrow \phantom{= \sqrt[3]{-8 \cdot 2 \cdot x^3 \cdot x} = }\uparrow$
$\phantom{(c) \sqrt[3]{-16x^4}}$ Factor perfect Combine perfect cubes.
$\phantom{(c) \sqrt[3]{-16x^4}}$ cubes inside radical.

$ = \sqrt[3]{(-2x)^3 \cdot 2x} = \sqrt[3]{(-2x)^3} \cdot \sqrt[3]{2x}$
$\phantom{(c) = \sqrt[3]{(-2x)^3 \cdot 2x}} \uparrow$
$ = -2x \cdot \sqrt[3]{2x} \text{(2a)}$

 NOW WORK PROBLEM 7.

EXAMPLE 3 **Combining Like Radicals**

(a) $-8\sqrt{12} + \sqrt{3} = -8\sqrt{4 \cdot 3} + \sqrt{3} = -8 \cdot \sqrt{4}\sqrt{3} + \sqrt{3} = -8 \cdot 2\sqrt{3} + \sqrt{3}$
$\phantom{\text{(a)} -8\sqrt{12} + \sqrt{3}} = -16\sqrt{3} + \sqrt{3} = -15\sqrt{3}$

(b) $\sqrt[3]{8x^4} + \sqrt[3]{-x} + 4\sqrt[3]{27x} = \sqrt[3]{2^3 x^3 x} + \sqrt[3]{-1 \cdot x} + 4\sqrt[3]{3^3 x}$
$\phantom{\text{(b)} \sqrt[3]{8x^4} + \sqrt[3]{-x} + 4\sqrt[3]{27x}} = \sqrt[3]{(2x)^3} \cdot \sqrt[3]{x} + \sqrt[3]{-1} \cdot \sqrt[3]{x} + 4\sqrt[3]{3^3} \cdot \sqrt[3]{x}$
$\phantom{\text{(b)} \sqrt[3]{8x^4} + \sqrt[3]{-x} + 4\sqrt[3]{27x}} = 2x \cdot \sqrt[3]{x} - 1 \cdot \sqrt[3]{x} + 12 \cdot \sqrt[3]{x}$
$\phantom{\text{(b)} \sqrt[3]{8x^4} + \sqrt[3]{-x} + 4\sqrt[3]{27x}} = (2x + 11)\sqrt[3]{x}$

 NOW WORK PROBLEM 25.

Rationalizing

When radicals occur in quotients, it is customary to rewrite the quotient so that the denominator contains no radicals. This process is referred to as **rationalizing the denominator.**

The idea is to multiply by an appropriate expression so that the new denominator contains no radicals. For example:

If Denominator Contains the Factor	Multiply By	To Obtain Denominator Free of Radicals
$\sqrt{3}$	$\sqrt{3}$	$(\sqrt{3})^2 = 3$
$\sqrt{3} + 1$	$\sqrt{3} - 1$	$(\sqrt{3})^2 - 1^2 = 3 - 1 = 2$
$\sqrt{2} - 3$	$\sqrt{2} + 3$	$(\sqrt{2})^2 - 3^2 = 2 - 9 = -7$
$\sqrt{5} - \sqrt{3}$	$\sqrt{5} + \sqrt{3}$	$(\sqrt{5})^2 - (\sqrt{3})^2 = 5 - 3 = 2$
$\sqrt[3]{4}$	$\sqrt[3]{2}$	$\sqrt[3]{4} \cdot \sqrt[3]{2} = \sqrt[3]{8} = 2$

In rationalizing the denominator of a quotient, be sure to multiply both the numerator and the denominator by the same expression.

3 **EXAMPLE 4** **Rationalizing Denominators**

Rationalize the denominator of each expression.

(a) $\dfrac{4}{\sqrt{2}}$ **(b)** $\dfrac{\sqrt{3}}{\sqrt[3]{2}}$ **(c)** $\dfrac{\sqrt{x} - 2}{\sqrt{x} + 2}, x \geq 0$

SOLUTION **(a)** $\dfrac{4}{\sqrt{2}} = \dfrac{4}{\sqrt{2}} \cdot \dfrac{\sqrt{2}}{\sqrt{2}} = \dfrac{4\sqrt{2}}{(\sqrt{2})^2} = \dfrac{4\sqrt{2}}{2} = 2\sqrt{2}$
$\phantom{\text{SOLUTION (a)} \dfrac{4}{\sqrt{2}} =} \uparrow$
$\phantom{\text{SOLUTION (a)} \dfrac{4}{\sqrt{2}}}$ Multiply by $\dfrac{\sqrt{2}}{\sqrt{2}}$.

(b) $\dfrac{\sqrt{3}}{\sqrt[3]{2}} = \dfrac{\sqrt{3}}{\sqrt[3]{2}} \cdot \dfrac{\sqrt[3]{4}}{\sqrt[3]{4}} = \dfrac{\sqrt{3}\sqrt[3]{4}}{\sqrt[3]{8}} = \dfrac{\sqrt{3}\sqrt[3]{4}}{2}$

Multiply by $\dfrac{\sqrt[3]{4}}{\sqrt[3]{4}}$.

(c) $\dfrac{\sqrt{x} - 2}{\sqrt{x} + 2} = \dfrac{\sqrt{x} - 2}{\sqrt{x} + 2} \cdot \dfrac{\sqrt{x} - 2}{\sqrt{x} - 2} = \dfrac{(\sqrt{x} - 2)^2}{(\sqrt{x})^2 - 2^2}$

$= \dfrac{(\sqrt{x})^2 - 4\sqrt{x} + 4}{x - 4} = \dfrac{x - 4\sqrt{x} + 4}{x - 4}$

NOW WORK PROBLEM 33.

Equations Containing Radicals

When the variable in an equation occurs in a square root, cube root, and so on, that is, when it occurs under a radical, the equation is called a **radical equation.** Sometimes a suitable operation will change a radical equation to one that is linear or quadratic. The most commonly used procedure is to isolate the most complicated radical on one side of the equation and then eliminate it by raising each side to a power equal to the index of the radical. Care must be taken, because extraneous solutions may result. Thus, when working with radical equations, we always check apparent solutions. Let's look at an example.

4

EXAMPLE 5   **Solving Radical Equations**

Solve the equation: $\sqrt[3]{2x - 4} - 2 = 0$

SOLUTION   The equation contains a radical whose index is 3. We isolate it on the left side.

$\sqrt[3]{2x - 4} - 2 = 0$
$\sqrt[3]{2x - 4} = 2$ Add 2 to both sides.

Now raise each side to the third power (since the index of the radical is 3) and solve.

$(\sqrt[3]{2x - 4})^3 = 2^3$ Raise each side to the 3rd power.
$2x - 4 = 8$ Simplify.
$2x = 12$ Solve for x.
$x = 6$

Check: $x = 6$: $\sqrt[3]{2(6) - 4} - 2 = \sqrt[3]{12 - 4} - 2 = \sqrt[3]{8} - 2 = 2 - 2 = 0$.

The solution is $x = 6$.

EXAMPLE 6   **Solving Radical Equations**

Solve the equation: $\sqrt{3x + 4} = x$

SOLUTION   The index of a square root is 2, so we square both sides

$\sqrt{3x + 4} = x$
$(\sqrt{3x + 4})^2 = x^2$ Square both sides.
$3x + 4 = x^2$ Simplify.
$x^2 - 3x - 4 = 0$ Place in standard form.

$$(x+1)(x-4) = 0 \quad \text{Factor.}$$
$$x+1 = 0 \quad \text{or} \quad x-4 = 0 \quad \text{Zero-Product Property.}$$
$$x = -1 \quad \text{or} \quad x = 4$$

There are two apparent solutions that need to be checked.

Check: $x = -1$: $\sqrt{3x+4} = \sqrt{3(-1)+4} = \sqrt{1} = 1 \neq -1$ $x = -1$ is extraneous

$x = 4$: $\sqrt{3x-4} = \sqrt{3(4)+4} = \sqrt{16} = 4$

The only solution is $x = 4$.

NOW WORK PROBLEM 41.

Rational Exponents

Radicals are used to define rational exponents.

If a is a real number and $n \geq 2$ is an integer, then

$$a^{1/n} = \sqrt[n]{a} \tag{3}$$

provided that $\sqrt[n]{a}$ exists.

Note that if n is even and $a < 0$, then $\sqrt[n]{a}$ and $a^{1/n}$ do not exist.

EXAMPLE 7   **Using Equation (3)**

(a) $4^{1/2} = \sqrt{4} = 2$ (b) $(-27)^{1/3} = \sqrt[3]{-27} = -3$
(c) $8^{1/2} = \sqrt{8} = 2\sqrt{2}$ (d) $16^{1/3} = \sqrt[3]{16} = 2\sqrt[3]{2}$

If a is a real number and m and n are integers containing no common factors with $n \geq 2$, then

$$a^{m/n} = \sqrt[n]{a^m} = (\sqrt[n]{a})^m \tag{4}$$

provided that $\sqrt[n]{a}$ exists.

We have two comments about equation (4):

1. The exponent m/n must be in lowest terms and n must be positive.
2. In simplifying $a^{m/n}$, either $\sqrt[n]{a^m}$ or $(\sqrt[n]{a})^m$ may be used. Generally, taking the root first, as in $(\sqrt[n]{a})^m$, is easier.

EXAMPLE 8   **Using Equation (4)**

(a) $4^{3/2} = (\sqrt{4})^3 = 2^3 = 8$ (b) $(-8)^{4/3} = (\sqrt[3]{-8})^4 = (-2)^4 = 16$
(c) $(32)^{-2/5} = (\sqrt[5]{32})^{-2} = 2^{-2} = \dfrac{1}{4}$

NOW WORK PROBLEM 45.

nth Roots; Rational Exponents

It can be shown that the laws of exponents hold for rational exponents. We use the laws of exponent in the next example.

EXAMPLE 9 Simplifying Expressions with Rational Exponents

Simplify each expression. Express your answer so that only positive exponents occur. Assume that the variables are positive.

(a) $\left(\dfrac{2x^{1/3}}{y^{2/3}}\right)^{-3}$ (b) $(x^{2/3}y)(x^{-2}y)^{1/2}$

SOLUTION (a) $\left(\dfrac{2x^{1/3}}{y^{2/3}}\right)^{-3} = \left(\dfrac{y^{2/3}}{2x^{1/3}}\right)^{3} = \dfrac{(y^{2/3})^3}{(2x^{1/3})^3} = \dfrac{y^2}{2^3(x^{1/3})^3} = \dfrac{y^2}{8x}$

(b) $(x^{2/3}y)(x^{-2}y)^{1/2} = (x^{2/3}y)[(x^{-2})^{1/2}y^{1/2}]$

$= x^{2/3}yx^{-1}y^{1/2} = (x^{2/3}x^{-1})(y \cdot y^{1/2})$

$= x^{-1/3}y^{3/2} = \dfrac{y^{3/2}}{x^{1/3}}$

NOW WORK PROBLEM 61.

The next two examples illustrate some algebra that you will need to know for certain calculus problems.

EXAMPLE 10 Writing an Expression as a Single Quotient

Write the following expression as a single quotient in which only positive exponents appear.

$$(x^2 + 1)^{1/2} + x \cdot \dfrac{1}{2}(x^2 + 1)^{-1/2} \cdot 2x$$

SOLUTION

$(x^2 + 1)^{1/2} + x \cdot \dfrac{1}{2}(x^2 + 1)^{-1/2} \cdot 2x = (x^2 + 1)^{1/2} + \dfrac{x^2}{(x^2 + 1)^{1/2}}$

$= \dfrac{(x^2 + 1)^{1/2}(x^2 + 1)^{1/2} + x^2}{(x^2 + 1)^{1/2}}$

$= \dfrac{(x^2 + 1) + x^2}{(x^2 + 1)^{1/2}}$

$= \dfrac{2x^2 + 1}{(x^2 + 1)^{1/2}}$

NOW WORK PROBLEM 65.

EXAMPLE 11 Factoring an Expression Containing Rational Exponents

Factor: $4x^{1/3}(2x + 1) + 2x^{4/3}$

SOLUTION We begin by looking for factors that are common to the two terms. Notice that 2 and $x^{1/3}$ are common factors. Then,

$$4x^{1/3}(2x + 1) + 2x^{4/3} = 2x^{1/3}[2(2x + 1) + x]$$
$$= 2x^{1/3}(5x + 2)$$

NOW WORK PROBLEM 79.

EXERCISE 0.6

In Problems 1–28, simplify each expression. Assume that all variables are positive when they appear.

1. $\sqrt[3]{27}$
2. $\sqrt[4]{16}$
3. $\sqrt[3]{-8}$
4. $\sqrt[3]{-1}$
5. $\sqrt{8}$
6. $\sqrt[3]{54}$
7. $\sqrt[3]{-8x^4}$
8. $\sqrt[4]{48x^5}$
9. $\sqrt[4]{x^{12}y^8}$
10. $\sqrt[5]{x^{10}y^5}$
11. $\sqrt[4]{\dfrac{x^9 y^7}{xy^3}}$
12. $\sqrt[3]{\dfrac{3xy^2}{81x^4 y^2}}$
13. $\sqrt{36x}$
14. $\sqrt{9x^5}$
15. $\sqrt{3x^2}\sqrt{12x}$
16. $\sqrt{5x}\sqrt{20x^3}$
17. $(\sqrt{5}\sqrt[3]{9})^2$
18. $(\sqrt[3]{3}\sqrt{10})^4$
19. $(3\sqrt{6})(2\sqrt{2})$
20. $(5\sqrt{8})(-3\sqrt{3})$
21. $(\sqrt{3}+3)(\sqrt{3}-1)$
22. $(\sqrt{5}-2)(\sqrt{5}+3)$
23. $(\sqrt{x}-1)^2$
24. $(\sqrt{x}+\sqrt{5})^2$
25. $3\sqrt{2}-4\sqrt{8}$
26. $\sqrt[3]{-x^4}+\sqrt[3]{8x}$
27. $\sqrt[3]{16x^4}-\sqrt[3]{2x}$
28. $\sqrt[4]{32x}+\sqrt[4]{2x^5}$

In Problems 29–40, rationalize the denominator of each expression. Assume that all variables are positive when they appear.

29. $\dfrac{1}{\sqrt{2}}$
30. $\dfrac{6}{\sqrt[3]{4}}$
31. $\dfrac{-\sqrt{3}}{\sqrt{5}}$
32. $\dfrac{-\sqrt[3]{3}}{\sqrt[3]{8}}$
33. $\dfrac{\sqrt{3}}{5-\sqrt{2}}$
34. $\dfrac{\sqrt{2}}{\sqrt{7}+2}$
35. $\dfrac{2-\sqrt{5}}{2+3\sqrt{5}}$
36. $\dfrac{\sqrt{3}-1}{2\sqrt{3}+3}$
37. $\dfrac{5}{\sqrt[3]{2}}$
38. $\dfrac{-2}{\sqrt[3]{9}}$
39. $\dfrac{\sqrt{x+h}-\sqrt{x}}{\sqrt{x+h}+\sqrt{x}}$
40. $\dfrac{\sqrt{x+h}+\sqrt{x-h}}{\sqrt{x+h}-\sqrt{x-h}}$

In Problems 41–44, solve each equation.

41. $\sqrt[3]{2t-1}=2$
42. $\sqrt[3]{3t+1}=-2$
43. $\sqrt{15-2x}=x$
44. $\sqrt{12-x}=x$

In Problems 45–56, simplify each expression.

45. $8^{2/3}$
46. $4^{3/2}$
47. $(-27)^{1/3}$
48. $16^{3/4}$
49. $16^{3/2}$
50. $64^{3/2}$
51. $9^{-3/2}$
52. $25^{-5/2}$
53. $\left(\dfrac{9}{8}\right)^{3/2}$
54. $\left(\dfrac{27}{8}\right)^{2/3}$
55. $\left(\dfrac{8}{9}\right)^{-3/2}$
56. $\left(\dfrac{8}{27}\right)^{-2/3}$

In Problems 57–64 simplify each expression. Express your answer so that only positive exponents occur. Assume that the variables are positive.

57. $x^{3/4} x^{1/3} x^{-1/2}$
58. $x^{2/3} x^{1/2} x^{-1/4}$
59. $(x^3 y^6)^{1/3}$
60. $(x^4 y^8)^{3/4}$
61. $(x^2 y)^{1/3}(xy^2)^{2/3}$
62. $(xy)^{1/4}(x^2 y^2)^{1/2}$
63. $(16x^2 y^{-1/3})^{3/4}$
64. $(4x^{-1} y^{1/3})^{3/2}$

In Problems 65–78, expressions that occur in calculus are given. Write each expression as a single quotient in which only positive exponents and/or radicals appear.

65. $\dfrac{x}{(1+x)^{1/2}}+2(1+x)^{1/2}, \quad x>-1$
66. $\dfrac{1+x}{2x^{1/2}}+x^{1/2}, \quad x>0$
67. $2x(x^2+1)^{1/2}+x^2 \cdot \dfrac{1}{2}(x^2+1)^{-1/2}\cdot 2x$
68. $(x+1)^{1/3}+x \cdot \dfrac{1}{3}(x+1)^{-2/3}, \quad x \neq -1$

69. $\sqrt{4x+3} \cdot \dfrac{1}{2\sqrt{x-5}} + \sqrt{x-5} \cdot \dfrac{1}{5\sqrt{4x+3}}, \quad x > 5$

70. $\dfrac{\sqrt[3]{8x+1}}{3\sqrt[3]{(x-2)^2}} + \dfrac{\sqrt[3]{x-2}}{24\sqrt[3]{(8x+1)^2}}, \quad x \neq 2, x \neq -\dfrac{1}{8}$

71. $\dfrac{\sqrt{1+x} - x \cdot \dfrac{1}{2\sqrt{1+x}}}{1+x}, \quad x > -1$

72. $\dfrac{\sqrt{x^2+1} - x \cdot \dfrac{2x}{2\sqrt{x^2+1}}}{x^2+1}$

73. $\dfrac{(x+4)^{1/2} - 2x(x+4)^{-1/2}}{x+4}, \quad x > -4$

74. $\dfrac{(9-x^2)^{1/2} + x^2(9-x^2)^{-1/2}}{9-x^2}, \quad -3 < x < 3$

75. $\dfrac{\dfrac{x^2}{(x^2-1)^{1/2}} - (x^2-1)^{1/2}}{x^2}, \quad x < -1 \text{ or } x > 1$

76. $\dfrac{(x^2+4)^{1/2} - x^2(x^2+4)^{-1/2}}{x^2+4}$

77. $\dfrac{\dfrac{1+x^2}{2\sqrt{x}} - 2x\sqrt{x}}{(1+x^2)^2}, \quad x > 0$

78. $\dfrac{2x(1-x^2)^{1/3} + \dfrac{2}{3}x^3(1-x^2)^{-2/3}}{(1-x^2)^{2/3}}, \quad x \neq -1, x \neq 1$

In Problems 79–90, expressions that occur in calculus are given. Factor each expression. Express your answer so that only positive exponents occur.

79. $(x+1)^{3/2} + x \cdot \dfrac{3}{2}(x+1)^{1/2}, \quad x \geq -1$

80. $(x^2+4)^{4/3} + x \cdot \dfrac{4}{3}(x^2+4)^{1/3} \cdot 2x$

81. $6x^{1/2}(x^2+x) - 8x^{3/2} - 8x^{1/2}, \quad x \geq 0$

82. $6x^{1/2}(2x+3) + x^{3/2} \cdot 8, \quad x \geq 0$

83. $3(x^2+4)^{4/3} + x \cdot 4(x^2+4)^{1/3} \cdot 2x$

84. $2x(3x+4)^{4/3} + x^2 \cdot 4(3x+4)^{1/3}$

85. $4(3x+5)^{1/3}(2x+3)^{3/2} + 3(3x+5)^{4/3}(2x+3)^{1/2}, \quad x \geq -\dfrac{3}{2}$

86. $6(6x+1)^{1/3}(4x-3)^{3/2} + 6(6x+1)^{4/3}(4x-3)^{1/2}, \quad x \geq \dfrac{3}{4}$

87. $3x^{-1/2} + \dfrac{3}{2}x^{1/2}, \quad x > 0$

88. $8x^{1/3} - 4x^{-2/3}, \quad x \neq 0$

89. $x\left(\dfrac{1}{2}\right)(8-x^2)^{-1/2}(-2x) + (8-x^2)^{1/2}$

90. $2x(1-x^2)^{3/2} + x^2\left(\dfrac{3}{2}\right)(1-x^2)^{1/2}(-2x)$

0.7 Geometry Review

OBJECTIVES
1. Use the Pythagorean Theorem and its converse
2. Know geometry formulas

In this section we review some topics studied in geometry that we shall need for calculus.

Pythagorean Theorem

Use the Pythagorean Theorem and its converse 1 The *Pythagorean Theorem* is a statement about *right triangles*. A **right triangle** is one that contains a **right angle,** that is, an angle of 90°. The side of the triangle opposite the 90° angle is called the **hypotenuse;** the remaining two sides are called **legs.** In Figure 22 we have used c to represent the length of the hypotenuse and a and b to represent the lengths of the legs. Notice the use of the symbol ⌐ to show the 90° angle. We now state the Pythagorean Theorem.

FIGURE 22

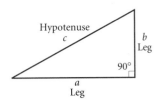

Pythagorean Theorem

In a right triangle, the square of the length of the hypotenuse is equal to the sum of the squares of the lengths of the legs. That is, in the right triangle shown in Figure 22,

$$c^2 = a^2 + b^2 \qquad (1)$$

EXAMPLE 1 **Finding the Hypotenuse of a Right Triangle**

In a right triangle, one leg is of length 4 and the other is of length 3. What is the length of the hypotenuse?

SOLUTION Since the triangle is a right triangle, we use the Pythagorean Theorem with $a = 4$ and $b = 3$ to find the length c of the hypotenuse. From equation (1), we have

$$c^2 = a^2 + b^2$$
$$c^2 = 4^2 + 3^2 = 16 + 9 = 25$$
$$c = \sqrt{25} = 5$$

 NOW WORK PROBLEM 3.

The converse of the Pythagorean Theorem is also true.

Converse of the Pythagorean Theorem

In a triangle, if the square of the length of one side equals the sum of the squares of the lengths of the other two sides, then the triangle is a right triangle. The 90° angle is opposite the longest side.

EXAMPLE 2 **Verifying That a Triangle Is a Right Triangle**

FIGURE 23

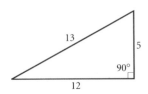

Show that a triangle whose sides are of lengths 5, 12, and 13 is a right triangle. Identify the hypotenuse.

SOLUTION We square the lengths of the sides.

$$5^2 = 25, \qquad 12^2 = 144, \qquad 13^2 = 169$$

Notice that the sum of the first two squares (25 and 144) equals the third square (169). So the triangle is a right triangle. The longest side, 13, is the hypotenuse. See Figure 23.

 NOW WORK PROBLEM 11.

EXAMPLE 3 **Applying the Pythagorean Theorem**

The tallest inhabited building in the world is the Sears Tower in Chicago. If the observation tower is 1450 feet above ground level, how far can a person standing in the observation tower see (with the aid of a telescope)? Use 3960 miles for the radius of Earth. See Figure 24.

FIGURE 24

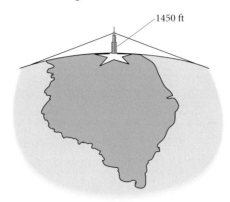

Source: Council on Tall Buildings and Urban Habitat (1997): Sears Tower No. 1 for tallest roof (1450 ft) and tallest occupied floor (1431 ft).

FIGURE 25

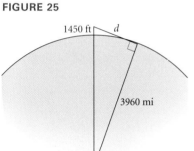

SOLUTION From the center of Earth, draw two radii: one through the Sears Tower and the other to the farthest point a person can see from the tower. See Figure 25. Apply the Pythagorean Theorem to the right triangle.

Since 1 mile = 5280 feet, 1450 feet = $\dfrac{1450}{5280}$ miles. Then we have

$$d^2 + (3960)^2 = \left(3960 + \dfrac{1450}{5280}\right)^2$$

$$d^2 = \left(3960 + \dfrac{1450}{5280}\right)^2 - (3960)^2 \approx 2175.08$$

$$d \approx 46.64$$

A person can see about 47 miles from the observation tower.

 NOW WORK PROBLEM 37.

Geometry Formulas

Know geometry formulas 2 Certain formulas from geometry are useful in solving calculus problems. We list some of these formulas next.

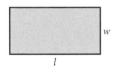

For a rectangle of length l and width w,

$$\boxed{\text{Area} = lw \qquad \text{Perimeter} = 2l + 2w}$$

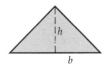

For a triangle with base b and altitude h,

$$\boxed{\text{Area} = \dfrac{1}{2}bh}$$

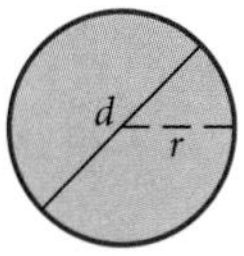

For a circle of radius r (diameter $d = 2r$),

$$\text{Area} = \pi r^2 \qquad \text{Circumference} = 2\pi r = \pi d$$

For a closed rectangular box of length l, width w, and height h,

$$\text{Volume} = lwh \qquad \text{Surface area} = 2lw + 2lh + 2wh$$

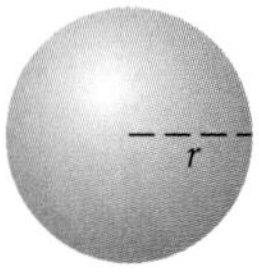

For a sphere of radius r,

$$\text{Volume} = \frac{4}{3}\pi r^3 \qquad \text{Surface area} = 4\pi r^2$$

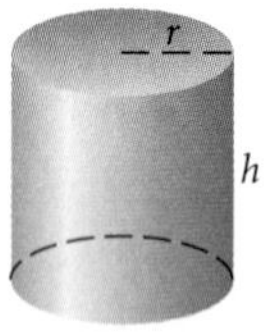

For a right circular cylinder of height h and radius r,

$$\text{Volume} = \pi r^2 h \qquad \text{Surface area} = 2\pi r^2 + 2\pi r h$$

NOW WORK PROBLEM 19.

EXAMPLE 4 Using Geometry Formulas

A Christmas tree ornament is in the shape of a semicircle on top of a triangle. How many square centimeters (cm) of copper are required to make the ornament if the height of the triangle is 6 cm and the base is 4 cm?

FIGURE 26

SOLUTION See Figure 26. The amount of copper required equals the shaded area. This area is the sum of the area of the triangle and the semicircle. The triangle has height $h = 6$ and base $b = 4$. The semicircle has diameter $d = 4$, so its radius is $r = 2$.

$$\text{Area} = \text{Area of triangle} + \text{Area of semicircle}$$

$$= \frac{1}{2}bh + \frac{1}{2}\pi r^2 = \frac{1}{2}(4)(6) + \frac{1}{2}\pi \cdot 2^2 \qquad b = 4; h = 6; r = 2.$$

$$= 12 + 2\pi \approx 18.28 \text{ cm}^2$$

About 18.28 cm² of copper are required.

NOW WORK PROBLEM 33.

EXERCISE 0.7

In Problems 1–6, the lengths of the legs of a right triangle are given. Find the hypotenuse.

1. $a = 5, \quad b = 12$ **2.** $a = 6, \quad b = 8$ **3.** $a = 10, \quad b = 24$

4. $a = 4, \quad b = 3$ **5.** $a = 7, \quad b = 24$ **6.** $a = 14, \quad b = 48$

Geometry Review 75

In Problems 7–14, the lengths of the sides of a triangle are given. Determine which are right triangles. For those that are, identify the hypotenuse.

7. 3, 4, 5
8. 6, 8, 10
9. 4, 5, 6
10. 2, 2, 3
11. 7, 24, 25
12. 10, 24, 26
13. 6, 4, 3
14. 5, 4, 7

15. Find the area A of a rectangle with length 4 inches and width 2 inches.
16. Find the area A of a rectangle with length 9 centimeters and width 4 centimeters.
17. Find the area A of a triangle with height 4 inches and base 2 inches.
18. Find the area A of a triangle with height 9 centimeters and base 4 centimeters.
19. Find the area A and circumference C of a circle of radius 5 meters.
20. Find the area A and circumference C of a circle of radius 2 feet.
21. Find the volume V and surface area S of a rectangular box with length 8 feet, width 4 feet, and height 7 feet.
22. Find the volume V and surface area S of a rectangular box with length 9 inches, width 4 inches, and height 8 inches.
23. Find the volume V and surface area S of a sphere of radius 4 centimeters.
24. Find the volume V and surface area S of a sphere of radius 3 feet.
25. Find the volume V and surface area S of a right circular cylinder with radius 9 inches and height 8 inches.
26. Find the volume V and surface area S of a right circular cylinder with radius 8 inches and height 9 inches.

In Problems 27–30, find the area of the shaded region.

27.
28.
29.
30.

31. How many feet does a wheel with a diameter of 16 inches travel after four revolutions?
32. How many revolutions will a circular disk with a diameter of 4 feet have completed after it has rolled 20 feet?
33. In the figure shown, *ABCD* is a square, with each side of length 6 feet. The width of the border (shaded portion) between the outer square *EFGH* and *ABCD* is 2 feet. Find the area of the border.

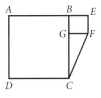

35. **Architecture** A Norman window consists of a rectangle surmounted by a semicircle. Find the area of the Norman window shown in the illustration. How much wood frame is needed to enclose the window?

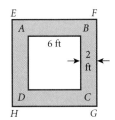

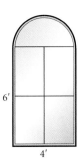

34. Refer to the figure above Problem 35. Square *ABCD* has an area of 100 square feet; square *BEFG* has an area of 16 square feet. What is the area of the triangle *CGF*?

76 Chapter 0 Review

36. **Construction** A circular swimming pool, 20 feet in diameter, is enclosed by a wooden deck that is 3 feet wide. What is the area of the deck? How much fence is required to enclose the deck?

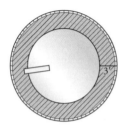

In Problems 37–39, use the facts that the radius of Earth is 3960 miles and 1 mile = 5280 feet.

37. **How Far Can You See?** The conning tower of the U.S.S. *Silversides*, a World War II submarine now permanently stationed in Muskegon, Michigan, is approximately 20 feet above sea level. How far can you see from the conning tower?

38. **How Far Can You See?** A person who is 6 feet tall is standing on the beach in Fort Lauderdale, Florida, and looks out onto the Atlantic Ocean. Suddenly, a ship appears on the horizon. How far is the ship from shore?

39. **How Far Can You See?** The deck of a destroyer is 100 feet above sea level. How far can a person see from the deck? How far can a person see from the bridge, which is 150 feet above sea level?

40. Suppose that m and n are positive integers with $m > n$. If $a = m^2 - n^2$, $b = 2mn$, and $c = m^2 + n^2$, show that a, b, and c are the lengths of the sides of a right triangle. (This formula can be used to find the sides of a right triangle that are integers, such as 3, 4, 5; 5, 12, 13; and so on. Such triplets of integers are called **Pythagorean triples.**)

41. You have 1000 feet of flexible pool siding and wish to construct a swimming pool. Experiment with rectangular-shaped pools with perimeters of 1000 feet. How do their areas vary? What is the shape of the rectangle with the largest area? Now compute the area enclosed by a circular pool with a perimeter (circumference) of 1000 feet. What would be your choice of shape for the pool? If rectangular, what is your preference for dimensions? Justify your choice. If your only consideration is to have a pool that encloses the most area, what shape should you use?

0.8 Rectangular Coordinates

PREPARING FOR THIS SECTION *Before getting started, review the following:*

> Algebra Review (Section 0.2, pp. 16–24)

> Geometry Review (Section 0.7, pp. 71–74)

OBJECTIVE 1 Use the Distance Formula

We locate a point on the real number line by assigning it a single real number, called the *coordinate of the point*. For work in a two-dimensional plane, we locate points by using two numbers.

We begin with two real number lines located in the same plane: one horizontal and the other vertical. We call the horizontal line the ***x*-axis**; the vertical line the ***y*-axis**; and the point of intersection the **origin** O. We assign coordinates to every point on these

Rectangular Coordinates 77

FIGURE 27

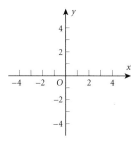

number lines as shown in Figure 27, using a convenient scale. In mathematics, we usually use the same scale on each axis; in applications, a different scale is often used on each axis.

The origin O has a value of 0 on both the x-axis and the y-axis. We follow the usual convention that points on the x-axis to the right of O are associated with positive real numbers, and those to the left of O are associated with negative real numbers. Points on the y-axis above O are associated with positive real numbers, and those below O are associated with negative real numbers. In Figure 1, the x-axis and y-axis are labeled as x and y, respectively, and we have used an arrow at the end of each axis to denote the positive direction.

The coordinate system described here is called a **rectangular** or **Cartesian* coordinate system.** The plane formed by the x-axis and y-axis is sometimes called the **xy-plane,** and the x-axis and y-axis are referred to as the **coordinate axes.**

Any point P in the xy-plane can then be located by using an **ordered pair** (x, y) of real numbers. Let x denote the signed distance of P from the y-axis (*signed* in the sense that, if P is to the right of the y-axis, then $x > 0$, and if P is to the left of the y-axis, then $x < 0$); and let y denote the signed distance of P from the x-axis. The ordered pair (x, y), also called the **coordinates** of P, then gives us enough information to locate the point P in the plane.

FIGURE 28

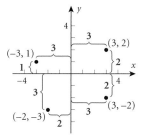

For example, to locate the point whose coordinates are $(-3, 1)$, go 3 units along the x-axis to the left of O and then go straight up 1 unit. We **plot** this point by placing a dot at this location. See Figure 28, in which the points with coordinates $(-3, 1)$, $(-2, -3)$, $(3, -2)$, and $(3, 2)$ are plotted.

The origin has coordinates $(0, 0)$. Any point on the x-axis has coordinates of the form $(x, 0)$, and any point on the y-axis has coordinates of the form $(0, y)$.

If (x, y) are the coordinates of a point P, then x is called the **x-coordinate,** or **abscissa,** of P, and y is called the **y-coordinate,** or **ordinate,** of P. We identify the point P by its coordinates (x, y) by writing $P = (x, y)$, referring to it as "the point (x, y)," rather than "the point whose coordinates are (x, y)."

FIGURE 29

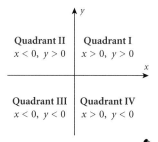

The coordinate axes divide the xy-plane into four sections, called **quadrants,** as shown in Figure 29. In quadrant I, both the x-coordinate and the y-coordinate of all points are positive; in quadrant II, x is negative and y is positive; in quadrant III, both x and y are negative; and in quadrant IV, x is positive and y is negative. Points on the coordinate axes belong to no quadrant.

NOW WORK PROBLEM 1.

FIGURE 30

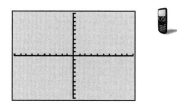

COMMENT: On a graphing calculator, you can set the scale on each axis. Once this has been done, you obtain the **viewing rectangle.** See Figure 30 for a typical viewing rectangle. You should now read Section A.1, *The Viewing Rectangle,* in the Appendix.

Use the distance formula

Distance between Points

If the same units of measurement, such as inches or centimeters, are used for both the x-axis and the y-axis, then all distances in the xy-plane can be measured using this unit of measurement.

*Named after René Descartes (1596–1650), a French mathematician, philosopher, and theologian.

EXAMPLE 1 Finding the Distance between Two Points

Find the distance d between the points $(1, 3)$ and $(5, 6)$.

SOLUTION First we plot the points $(1, 3)$ and $(5, 6)$ as shown in Figure 31(a). Then we draw a horizontal line from $(1, 3)$ to $(5, 3)$ and a vertical line from $(5, 3)$ to $(5, 6)$, forming a right triangle, as in Figure 31(b). One leg of the triangle is of length 4 and the other is of length 3. By the Pythagorean Theorem, the square of the distance d that we seek is

$$d^2 = 4^2 + 3^2 = 16 + 9 = 25$$
$$d = 5$$

FIGURE 31

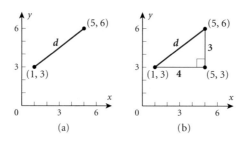

The **distance formula** provides a straightforward method for computing the distance between two points.

Distance Formula

The distance between two points $P_1 = (x_1, y_1)$ and $P_2 = (x_2, y_2)$, denoted by $d(P_1, P_2)$, is

$$d(P_1, P_2) = \sqrt{(x_2 - x_1)^2 + (y_2 - y_1)^2} \qquad (1)$$

That is, to compute the distance between two points, find the difference of the x-coordinates, square it, and add this to the square of the difference of the y-coordinates. The square root of this sum is the distance. See Figure 32.

FIGURE 32

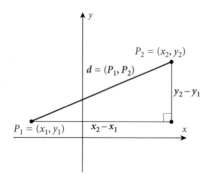

Rectangular Coordinates 79

EXAMPLE 2 Finding the Distance between Two Points

Find the distance d between the points $(-4, 5)$ and $(3, 2)$.

SOLUTION Using the distance formula (1), the solution is obtained as follows:

$$d = \sqrt{[3 - (-4)]^2 + (2 - 5)^2} = \sqrt{7^2 + (-3)^2}$$
$$= \sqrt{49 + 9} = \sqrt{58} \approx 7.62$$

NOW WORK PROBLEMS 5 AND 9.

The distance between two points $P_1 = (x_1, y_1)$ and $P_2 = (x_2, y_2)$ is never a negative number. Furthermore, the distance between two points is 0 only when the points are identical, that is, when $x_1 = x_2$ and $y_1 = y_2$. Also, because $(x_2 - x_1)^2 = (x_1 - x_2)^2$ and $(y_2 - y_1)^2 = (y_1 - y_2)^2$, it makes no difference whether the distance is computed from P_1 to P_2 or from P_2 to P_1; that is, $d(P_1, P_2) = d(P_2, P_1)$.

Rectangular coordinates enable us to translate geometry problems into algebra problems, and vice versa. The next example shows how algebra (the distance formula) can be used to solve geometry problems.

EXAMPLE 3 Using Algebra to Solve Geometry Problems

Consider the three points $A = (-2, 1)$, $B = (2, 3)$, and $C = (3, 1)$.

(a) Plot each point and form the triangle ABC.
(b) Find the length of each side of the triangle.
(c) Verify that the triangle is a right triangle.
(d) Find the area of the triangle.

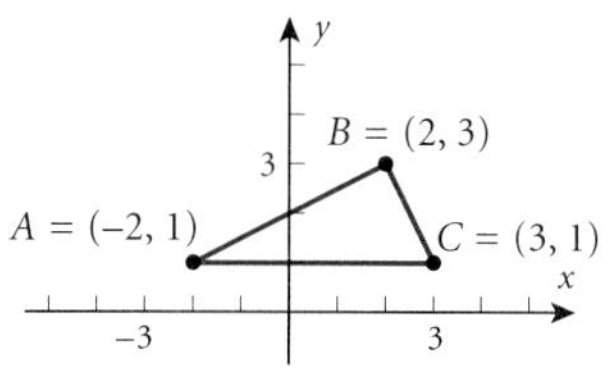

SOLUTION **(a)** Points A, B, and C and triangle ABC are plotted in Figure 33.

(b) $d(A, B) = \sqrt{[2 - (-2)]^2 + (3 - 1)^2} = \sqrt{16 + 4} = \sqrt{20} = 2\sqrt{5}$
  $d(B, C) = \sqrt{(3 - 2)^2 + (1 - 3)^2} = \sqrt{1 + 4} = \sqrt{5}$
  $d(A, C) = \sqrt{[3 - (-2)]^2 + (1 - 1)^2} = \sqrt{25 + 0} = 5$

FIGURE 33

(c) To show that the triangle is a right triangle, we need to show that the sum of the squares of the lengths of two of the sides equals the square of the length of the third side. (Why is this sufficient?) Looking at Figure 33, it seems reasonable to conjecture that the right angle is at vertex B. To verify, we check to see whether

$$[d(A, B)]^2 + [d(B, C)]^2 = [d(A, C)]^2$$

We find that

$$[d(A, B)]^2 + [d(B, C)]^2 = (2\sqrt{5})^2 + (\sqrt{5})^2$$
$$= 20 + 5 = 25 = [d(A, C)]^2$$

so it follows from the converse of the Pythagorean Theorem that triangle ABC is a right triangle.

(d) Because the right angle is at B, the sides AB and BC form the base and altitude of the triangle. Its area is therefore

$$\text{Area} = \frac{1}{2}(\text{Base})(\text{Altitude}) = \frac{1}{2}(2\sqrt{5})(\sqrt{5}) = 5 \text{ square units}$$

NOW WORK PROBLEM 19.

EXERCISE 0.8

In Problems 1 and 2, plot each point in the xy-plane. Tell in which quadrant or on what coordinate axis each point lies.

1. (a) $A = (-3, 2)$ (d) $D = (6, 5)$
   (b) $B = (6, 0)$ (e) $E = (0, -3)$
   (c) $C = (-2, -2)$ (f) $F = (6, -3)$

2. (a) $A = (1, 4)$ (d) $D = (4, 1)$
   (b) $B = (-3, -4)$ (e) $E = (0, 1)$
   (c) $C = (-3, 4)$ (f) $F = (-3, 0)$

3. Plot the points $(2, 0), (2, -3), (2, 4), (2, 1)$, and $(2, -1)$. Describe the set of all points of the form $(2, y)$, where y is a real number.

4. Plot the points $(0, 3), (1, 3), (-2, 3), (5, 3)$, and $(-4, 3)$. Describe the set of all points of the form $(x, 3)$ where x is a real number.

In Problems 5–18, find the distance $d(P_1, P_2)$ between the points P_1 and P_2.

5.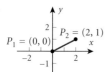
6.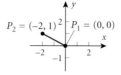
7.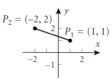
8.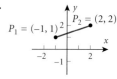

9. $P_1 = (3, -4); P_2 = (5, 4)$
10. $P_1 = (-1, 0); P_2 = (2, 4)$
11. $P_1 = (-3, 2); P_2 = (6, 0)$
12. $P_1 = (2, -3); P_2 = (4, 2)$
13. $P_1 = (4, -3); P_2 = (6, 4)$
14. $P_1 = (-4, -3); P_2 = (6, 2)$
15. $P_1 = (-0.2, 0.3); P_2 = (2.3, 1.1)$
16. $P_1 = (1.2, 2.3); P_2 = (-0.3, 1.1)$
17. $P_1 = (a, b); P_2 = (0, 0)$
18. $P_1 = (a, a); P_2 = (0, 0)$

In Problems 19–24, plot each point and form the triangle ABC. Verify that the triangle is a right triangle. Find its area.

19. $A = (-2, 5); B = (1, 3); C = (-1, 0)$
20. $A = (-2, 5); B = (12, 3); C = (10, -11)$
21. $A = (-5, 3); B = (6, 0); C = (5, 5)$
22. $A = (-6, 3); B = (3, -5); C = (-1, 5)$
23. $A = (4, -3); B = (0, -3); C = (4, 2)$
24. $A = (4, -3); B = (4, 1); C = (2, 1)$

25. Find all points having an x-coordinate of 2 whose distance from the point $(-2, -1)$ is 5.

26. Find all points having a y-coordinate of -3 whose distance from the point $(1, 2)$ is 13.

27. Find all points on the x-axis that are 5 units from the point $(4, -3)$.

28. Find all points on the y-axis that are 5 units from the point $(4, 4)$.

In Problems 29–32, find the length of the line segment. Assume that the endpoints of each line segment have integer coordinates.

29.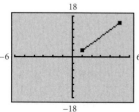
30.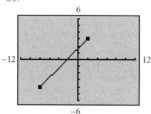
31.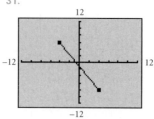
32.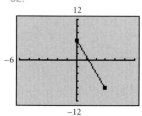

33. **Baseball** A major league baseball "diamond" is actually a square, 90 feet on a side (see the figure). What is the distance directly from home plate to second base (the diagonal of the square)?

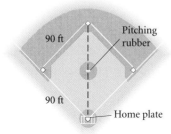

34. **Little League Baseball** The layout of a Little League playing field is a square, 60 feet on a side. How far is it directly from home plate to second base (the diagonal of the square)?

   Source: Little League Baseball, Official Regulations and Playing Rules, 2003.

35. **Baseball** Refer to Problem 33. Overlay a rectangular coordinate system on a major league baseball diamond so that the origin is at home plate, the positive *x*-axis lies in the direction from home plate to first base, and the positive *y*-axis lies in the direction from home plate to third base.

   (a) What are the coordinates of first base, second base, and third base? Use feet as the unit of measurement.
   (b) If the right fielder is located at (310, 15), how far is it from the right fielder to second base?
   (c) If the center fielder is located at (300, 300), how far is it from the center fielder to third base?

36. **Little League Baseball** Refer to Problem 34. Overlay a rectangular coordinate system on a Little League baseball diamond so that the origin is at home plate, the positive *x*-axis lies in the direction from home plate to first base, and the positive *y*-axis lies in the direction from home plate to third base.

   (a) What are the coordinates of first base, second base, and third base? Use feet as the unit of measurement.
   (b) If the right fielder is located at (180, 20), how far is it from the right fielder to second base?
   (c) If the center fielder is located at (220, 220), how far is it from the center fielder to third base?

37. A Dodge Intrepid and a Mack truck leave an intersection at the same time. The Intrepid heads east at an average speed of 30 miles per hour, while the truck heads south at an average speed of 40 miles per hour. Find an expression for their distance apart d (in miles) at the end of t hours.

38. A hot-air balloon, headed due east at an average speed of 15 miles per hour and at a constant altitude of 100 feet, passes over an intersection (see the figure). Find an expression for the distance d (measured in feet) from the balloon to the intersection t seconds later.

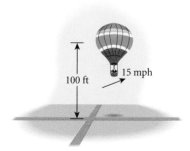

0.9 Lines

PREPARING FOR THIS SECTION   *Before getting started, review the following:*

> Algebra Review (Section 0.2, pp. 19–24)

OBJECTIVES
1. Graph linear equations
2. Find the equation of a vertical line
3. Calculate and interpret the slope of a line
4. Graph a line given a point on the line and the slope
5. Use the point–slope form of a line
6. Find the equation of a horizontal line
7. Find the equation of a line given two points
8. Use the slope–intercept form of a line

Graphs of Linear Equations in Two Variables

A linear equation in two variables is an equation of the form

$$Ax + By = C \quad (1)$$

where A and B are not both zero.

Examples of linear equations are

$3x - 5y - 6 = 0$ This equation can be written as
$\quad 3x - 5y = 6 \quad A = 3, B = -5, C = 6$

$-3x = 2y - 1$ This equation can be written as
$\quad -3x - 2y = -1 \quad A = -3, B = -2, C = -1$
or as
$\quad 3x + 2y = 1 \quad A = 3, B = 2, C = 1$

$y = \frac{3}{4}x - 5$ Here we can write
$\quad -\frac{3}{4}x + y = -5 \quad A = -\frac{3}{4}, B = 1, C = -5$
or
$\quad 3x - 4y = 20 \quad A = 3, B = -4, C = 20$

$y = -5$ Here we can write
$\quad 0 \cdot x + y = -5 \quad A = 0, B = 1, C = -5$

$x = 4$ Here we can write
$\quad x + 0 \cdot y = 4 \quad A = 1, B = 0, C = 4$

The **graph** of an equation is the set of all points (x, y) whose coordinates satisfy the equation. For example, $(0, 4)$ is a point on the graph of the equation $3x + 4y = 16$, because when we substitute 0 for x and 4 for y in the equation, we get

$$3 \cdot 0 + 4 \cdot 4 = 16 \quad 3x + 4y = 16, x = 0, y = 4$$

which is a true statement.

It can be shown that if A, B, and C are real numbers, with A and B not both zero, then the graph of the equation

$$Ax + By = C$$

is a **line**. This is the reason we call it a **linear equation**.

Conversely, any line is the graph of an equation of the form $Ax + By = C$.

Since any line can be written as an equation in the form $Ax + By = C$, we call this form the **general equation** of a line.

Graph linear equations **1** Given a linear equation, we can obtain its graph by plotting two points that satisfy its equation and connecting them with a line. The easiest two points to plot are the *intercepts*. For example, the line shown in Figure 34 has the intercepts $(0, -4)$ and $(3, 0)$.

Intercepts

The points at which the graph of a linear equation crosses the axes are called **intercepts**. The **x-intercept** is the point at which the graph crosses the x-axis; the **y-intercept** is the point at which the graph crosses the y-axis.

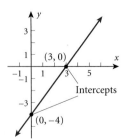

FIGURE 34

> **Steps for Finding the Intercepts of a Linear Equation**
>
> To find the intercepts of a linear equation $Ax + By = C$, with $A \neq 0$ or $B \neq 0$, follow these steps.
>
> **STEP 1:** Let $y = 0$ and solve for x. This determines the x-intercept of the line.
> **STEP 2:** Let $x = 0$ and solve for y. This determines the y-intercept of the line.

EXAMPLE 1 **Finding the Intercepts of a Linear Equation**

Find the intercepts of the equation $2x + 3y = 6$. Graph the equation.

SOLUTION **Step 1** To find the x-intercept, we need to find the number x for which $y = 0$. We let $y = 0$ in the equation and proceed to solve for x:

$$2x + 3y = 6$$
$$2x + 3(0) = 6 \quad \text{$y = 0$}$$
$$2x = 6 \quad \text{Simplify.}$$
$$x = 3 \quad \text{Solve for x.}$$

The x-intercept is $(3, 0)$.

Step 2 To find the y-intercept, we let $x = 0$ in the equation and solve for y:

$$2x + 3y = 6$$
$$2(0) + 3y = 6 \quad \text{$x = 0$}$$
$$3y = 6 \quad \text{Simplify.}$$
$$y = 2 \quad \text{Solve for y.}$$

The y-intercept is $(0, 2)$.

Since the equation is a linear equation, its graph is a line. We use the two intercepts $(3, 0)$ and $(0, 2)$ to graph it. See Figure 35.

FIGURE 35

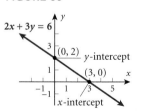

EXAMPLE 2 **Graphing a Linear Equation**

Graph the equation: $y = 2x + 5$

SOLUTION This equation can be written as

$$-2x + y = 5$$

This is a linear equation, so its graph is a line. The intercepts are $(0, 5)$ and $\left(-\frac{5}{2}, 0\right)$, which you should verify. For reassurance we'll find a third point. Arbitrarily, we let $x = 10$. Then $y = 2x + 5 = 2(10) + 5 = 25$, so $(10, 25)$ is a point on the graph. See Figure 36.

x	y
0	5
$-\frac{5}{2}$	0
10	25

FIGURE 36

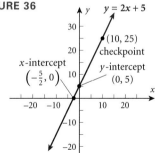

NOW WORK PROBLEM 1.

When a line passes through the origin, it has only one intercept. To graph such lines, we need to locate an additional point on the graph.

EXAMPLE 3 **Graphing a Linear Equation**

Graph the equation: $-x + 2y = 0$

SOLUTION This is a linear equation, so its graph is a line. The only intercept is (0, 0). To locate another point on the graph, let $x = 4$. (This choice is arbitrary; any choice of x other than 0 could also be used). Then,

$$-4 + 2y = 0 \quad -x + 2y = 0, x = 4$$
$$2y = 4$$
$$y = 2$$

So, $y = 2$ when $x = 4$ and (4, 2) is a point on the graph. See Figure 37.

Next we discuss linear equations whose graphs are vertical lines.

FIGURE 37

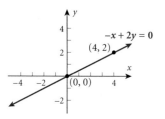

EXAMPLE 4 **Graphing a Linear Equation (a Vertical Line)**

Graph the equation: $x = 3$

SOLUTION We are looking for all points (x, y) in the plane for which $x = 3$. Since $x = 3$, no matter what y-coordinate is used, the corresponding x-coordinate always equals 3. Consequently, the graph of the equation $x = 3$ is a vertical line with x-intercept (3, 0) as shown in Figure 38.

As suggested by Example 4, we have the following result:

FIGURE 38

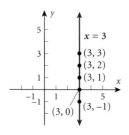

> **Equation of a Vertical Line**
>
> A vertical line is given by an equation of the form
>
> $$x = a$$
>
> where $(a, 0)$ is the x-intercept.

2 **EXAMPLE 5** **Finding the Equation of a Vertical Line**

Find an equation for the vertical line containing the point $(-1, 6)$.

SOLUTION The x-coordinate of any point on a vertical line is always the same. Since $(-1, 6)$ is a point on the vertical line, its equation is $x = -1$.

NOW WORK PROBLEM 5(a).

Lines 85

Calculate and interpret the slope of a line 3

Slope of a Line

An important characteristic of a line, called its *slope*, is best defined by using rectangular coordinates.

> **Slope of a Line**
>
> Let $P = (x_1, y_1)$ and $Q = (x_2, y_2)$ be two distinct points. If $x_1 \neq x_2$, the **slope** m of the nonvertical line L containing P and Q is defined by the formula
>
> $$m = \frac{y_2 - y_1}{x_2 - x_1} \qquad x_1 \neq x_2 \qquad (2)$$
>
> If $x_1 = x_2$, L is a vertical line and the slope m of L is **undefined** (since this results in division by 0).

Figure 39(a) provides an illustration of the slope of a nonvertical line; Figure 39(b) illustrates a vertical line.

FIGURE 39

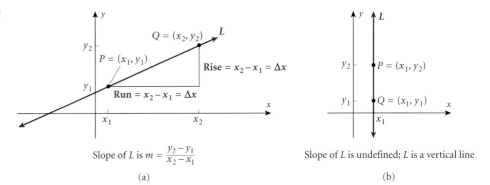

Slope of L is $m = \frac{y_2 - y_1}{x_2 - x_1}$

(a)

Slope of L is undefined; L is a vertical line

(b)

As Figure 39(a) illustrates, the slope m of a nonvertical line may be given as

$$m = \frac{y_2 - y_1}{x_2 - x_1} = \frac{\text{Rise}}{\text{Run}} = \frac{\text{Change in } y}{\text{Change in } x}$$

The change in y is usually denoted by Δy, read "delta y," and the change in x is denoted by Δx.

The slope m of a nonvertical line L measures the amount y changes, Δy, as x changes from x_1 to x_2, Δx. This is called the **average rate of change of y with respect to x**. Then, the slope m is

$$m = \frac{\Delta y}{\Delta x} = \text{average rate of change of } y \text{ with respect to } x$$

EXAMPLE 6 Finding and Interpreting the Slope of a Line

The slope m of the line containing the points $(3, -2)$ and $(1, 5)$ is

$$m = \frac{\Delta y}{\Delta x} = \frac{5 - (-2)}{1 - 3} = \frac{7}{-2} = \frac{-7}{2} = -\frac{7}{2}$$

We interpret the slope to mean that for every 2-unit change in x, y will change by −7 units. That is, if x increases by 2 units, then y decreases by 7 units. The average rate of change of y with respect to x is $\dfrac{-7}{2} = -\dfrac{7}{2}$.

NOW WORK PROBLEMS 9 AND 13.

Two comments about computing the slope of a nonvertical line may prove helpful:

1. Any two distinct points on the line can be used to compute the slope of the line. (See Figure 40 for justification.)
2. The slope of a line may be computed from $P = (x_1, y_1)$ to $Q = (x_2, y_2)$ or from Q to P because

$$\frac{y_2 - y_1}{x_2 - x_1} = \frac{y_1 - y_2}{x_1 - x_2}$$

FIGURE 40
Triangles ABC and PQR are similar (they have equal angles). So ratios of corresponding sides are proportional. Then:
Slope using P and Q = $\dfrac{y_2 - y_1}{x_2 - x_1}$ = $\dfrac{d(B, C)}{d(A, C)}$ = Slope using A and B
where $d(B,C)$ denotes the distance from B to C and $d(A, C)$ denotes the distance from A to C.

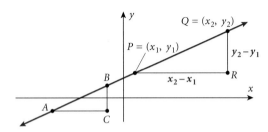

To get a better idea of the meaning of the slope m of a line L, consider the following example.

EXAMPLE 7 Finding the Slopes of Various Lines Containing the Same Point (2, 3)

Compute the slopes of the line L_1, L_2, L_3, and L_4 containing the following pairs of points. Graph all four lines on the same set of coordinate axes.

$$L_1: \quad P = (2, 3) \quad Q_1 = (-1, -2)$$
$$L_2: \quad P = (2, 3) \quad Q_2 = (3, -1)$$
$$L_3: \quad P = (2, 3) \quad Q_3 = (5, 3)$$
$$L_4: \quad P = (2, 3) \quad Q_4 = (2, 5)$$

FIGURE 41

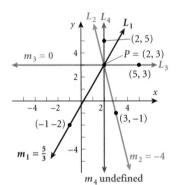

SOLUTION Let m_1, m_2, m_3, and m_4 denote the slopes of the lines L_1, L_2, L_3, and L_4, respectively. Then

$m_1 = \dfrac{-2 - 3}{-1 - 2} = \dfrac{-5}{-3} = \dfrac{5}{3}$ A rise of 5 divided by a run of 3

$m_2 = \dfrac{-1 - 3}{3 - 2} = \dfrac{-4}{1} = -4$ A rise of −4 divided by a run of 1

$m_3 = \dfrac{3 - 3}{5 - 2} = \dfrac{0}{3} = 0$ A rise of 0 divided by a run of 3

m_4 is undefined   The x coordinates of P and Q_4 are equal ($x_1 = x_2 = 2$).

The graphs of these lines are given in Figure 41.

Lines 87

As Figure 41 illustrates,

1. When the slope m of a line is positive, the line slants upward from left to right (L_1).
2. When the slope m is negative, the line slants downward from left to right (L_2).
3. When the slope m is 0, the line is horizontal (L_3).
4. When the slope m is undefined, the line is vertical (L_4).

COMMENT: Now read Section A.3, *Square Screens,* in the Appendix.

SEEING THE CONCEPT: On the same square screen, graph the following equations:

$Y_1 = 0$ Slope of line is 0.
$Y_2 = \dfrac{1}{4}x$ Slope of line is $\dfrac{1}{4}$.
$Y_3 = \dfrac{1}{2}x$ Slope of line is $\dfrac{1}{2}$.
$Y_4 = x$ Slope of line is 1.
$Y_5 = 2x$ Slope of line is 2.
$Y_6 = 6x$ Slope of line is 6.

FIGURE 42

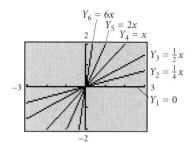

See Figure 42.

SEEING THE CONCEPT: On the same square screen, graph the following equations:

$Y_1 = 0$ Slope of line is 0.
$Y_2 = -\dfrac{1}{4}x$ Slope of line is $-\dfrac{1}{4}$.
$Y_3 = -\dfrac{1}{2}x$ Slope of line is $-\dfrac{1}{2}$.
$Y_4 = -x$ Slope of line is -1.
$Y_5 = -2x$ Slope of line is -2.
$Y_6 = -6x$ Slope of line is -6.

FIGURE 43

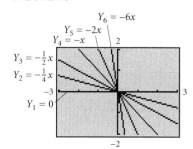

See Figure 43.

Figures 42 and 43 illustrate that the closer the line is to the vertical position, the greater the magnitude of the slope.

The next example illustrates how the slope of a line can be used to graph the line.

EXAMPLE 8   **Graphing a Line When its Slope and a Point Are Given**

Draw a graph of the line that contains the point (3, 2) and has a slope of

(a) $\dfrac{3}{4}$ (b) $-\dfrac{4}{5}$

SOLUTION   (a) Slope = $\dfrac{\text{rise}}{\text{run}}$. The fact that the slope is $\dfrac{3}{4}$ means that for every horizontal movement (run) of 4 units to the right, there will be a vertical movement (rise) of 3 units. If we start at the given point (3, 2) and move 4 units to the right and 3 units up, we reach the point (7, 5). By drawing the line through this point and the point (3, 2), we have the graph. See Figure 44(a).

(b) The fact that the slope is $-\dfrac{4}{5} = \dfrac{-4}{5}$ means that for every horizontal movement of 5 units to the right, there will be a corresponding vertical movement of -4 units (a downward movement). If we start at the given point (3, 2) and move 5 units to the right and then 4 units down, we arrive at the point $(8, -2)$. By drawing the line through these points, we have the graph. See Figure 44(b).

Alternatively, we can set $-\dfrac{4}{5} = \dfrac{4}{-5}$ so that for every horizontal movement of -5 units (a movement to the left), there will be a corresponding vertical movement of 4 units (upward). This approach brings us to the point $(-2, 6)$, which is also on the graph shown in Figure 44(b).

FIGURE 44

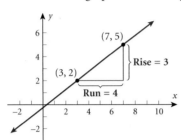

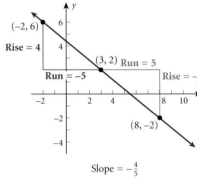

(a)   (b)

 NOW WORK PROBLEM 21.

Other Forms of the Equation of a Line

FIGURE 45

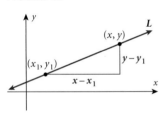

Let L be a nonvertical line with slope m and containing the point (x_1, y_1). See Figure 45. Since any two distinct points on L can be used to compute its slope, for any other point (x, y) on L, we have

$$m = \dfrac{y - y_1}{x - x_1} \quad \text{or} \quad y - y_1 = m(x - x_1)$$

Point–Slope Form of an Equation of a Line

An equation of a nonvertical line with slope m that contains the point (x_1, y_1) is

$$\boxed{y - y_1 = m(x - x_1)} \tag{3}$$

5   **EXAMPLE 9**   **Using the Point–Slope Form of a Line**

FIGURE 46

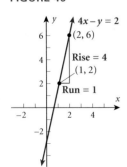

An equation of the line with slope 4 and containing the point (1, 2) can be found by using the point–slope form with $m = 4$, $x_1 = 1$, and $y_1 = 2$:

$$y - y_1 = m(x - x_1) \quad \text{Point–slope form.}$$
$$y - 2 = 4(x - 1) \quad m = 4, x_1 = 1, y_1 = 2$$
$$y - 2 = 4x - 4$$
$$4x - y = 2 \quad \text{General equation.}$$

See Figure 46.

 NOW WORK PROBLEM 33.

Lines   89

EXAMPLE 10 Finding the Equation of a Horizontal Line

Find an equation of the horizontal line containing the point (3, 2).

SOLUTION The slope of a horizontal line is 0. To get an equation, we use the point–slope form with $m = 0$, $x_1 = 3$, and $y_1 = 2$:

$$y - y_1 = m(x - x_1) \quad \text{Point–slope form.}$$
$$y - 2 = 0 \cdot (x - 3) \quad m = 0, x_1 = 3, y_1 = 2$$
$$y - 2 = 0$$
$$y = 2$$

See Figure 47 for the graph.

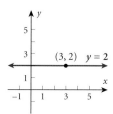

FIGURE 47

As suggested by Example 10, we have the following result:

Equation of a Horizontal Line

A horizontal line is given by an equation of the form

$$y = b$$

where $(0, b)$ is the y-intercept.

 NOW WORK PROBLEM 5(b).

EXAMPLE 11 Finding the Equation of a Line Given Two Points

Find an equation of the line containing the points (2, 3) and (−4, 5). Graph the line.

SOLUTION Since two points are given, we first compute the slope of the line:

$$m = \frac{5 - 3}{-4 - 2} = \frac{2}{-6} = \frac{1}{-3} = -\frac{1}{3}$$

We use the point (2, 3) and the fact that the slope $m = -\dfrac{1}{3}$ to get the point–slope form of the equation of the line:

$$y - 3 = -\frac{1}{3}(x - 2)$$

See Figure 48 for the graph.

FIGURE 48

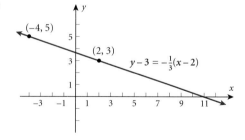

 NOW WORK PROBLEM 37.

90 Chapter 0 Review

In the solution to Example 11 we could have used the point $(-4, 5)$ instead of the point $(2, 3)$. The equation that results, although it looks different, is equivalent to the equation we obtained in the example. (Try it for yourself.)

The general form of the equation of the line in Example 11 can be obtained by multiplying both sides of the point–slope equation by 3 and collecting terms:

$$y - 3 = -\frac{1}{3}(x - 2) \quad \text{Point–slope equation.}$$

$$3(y - 3) = 3\left(-\frac{1}{3}\right)(x - 2) \quad \text{Multiply by 3.}$$

$$3y - 9 = -1(x - 2) \quad \text{Simplify.}$$

$$3y - 9 = -x + 2 \quad \text{Simplify.}$$

$$x + 3y = 11 \quad \text{General equation.}$$

This is the general form of the equation of the line.

Use the slope–intercept form of a line   **8**   Another useful equation of a line is obtained when the slope m and y-intercept $(0, b)$ are known. In this case we know both the slope m of the line and a point $(0, b)$ on the line. Then we can use the point–slope form, Equation (3), to obtain the following equation:

$$y - y_1 = m(x - x_1) \quad \text{Point–slope form.}$$

$$y - b = m(x - 0) \quad x_1 = 0, y_1 = b$$

$$y = mx + b \quad \text{Simplify and solve for } y.$$

Slope–Intercept Form of an Equation of a Line

An equation of a line L with slope m and y-intercept $(0, b)$ is

$$\boxed{y = mx + b} \tag{4}$$

SEEING THE CONCEPT: To see the role that the slope m plays in the equation $y = mx + b$, graph the following lines on the same square screen.

$$Y_1 = 2$$
$$Y_2 = x + 2$$
$$Y_3 = -x + 2$$
$$Y_4 = 3x + 2$$
$$Y_5 = -3x + 2$$

See Figure 49. What do you conclude about the lines $y = mx + 2$?

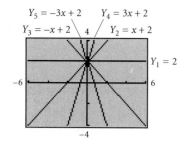

FIGURE 49

SEEING THE CONCEPT: To see the role of the *y*-intercept *b* in the equation $y = mx + b$, graph the following lines on the same square screen.

$$Y_1 = 2x$$
$$Y_2 = 2x + 1$$
$$Y_3 = 2x - 1$$
$$Y_4 = 2x + 4$$
$$Y_5 = 2x - 4$$

FIGURE 50

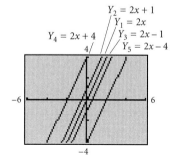

See Figure 50. What do you conclude about the lines $y = 2x + b$?

When an equation of a line is written in slope–intercept form, it is easy to find the slope *m* and *y*-intercept $(0, b)$ of the line. For example, suppose the equation of the line is

$$y = -2x + 3$$

Compare it to $y = mx + b$:

$$y = -2x + 3$$
$$\uparrow\uparrow$$
$$y = mx + b$$

The slope of this line is -2 and its *y*-intercept is $(0, 3)$.

Let's look at another example.

EXAMPLE 12 **Finding the Slope and *y*-Intercept of a Line**

Find the slope *m* and *y*-intercept $(0, b)$ of the line $2x + 4y = 8$. Graph the line.

SOLUTION To obtain the slope and *y*-intercept, we transform the equation into its slope–intercept form. To do this, we need to solve for *y*.

$$2x + 4y = 8$$
$$4y = -2x + 8$$
$$y = -\frac{1}{2}x + 2$$

The coefficient of *x*, $-\frac{1}{2}$, is the slope, and the *y*-intercept is $(0, 2)$.

We can graph the line in either of two ways:

1. Use the fact that the *y*-intercept is $(0, 2)$ and the slope is $-\frac{1}{2}$. Then, starting at the point $(0, 2)$, go to the right 2 units and then down 1 unit to the point $(2, 1)$. Plot these points and draw the line containing them. See Figure 51.

FIGURE 51

2. Locate the intercepts. The *y*-intercept is $(0, 2)$. To obtain the *x*-intercept, we let $y = 0$ and solve for *x*. When $y = 0$, we have

$$2x + 4 \cdot 0 = 8$$
$$2x = 8$$
$$x = 4$$

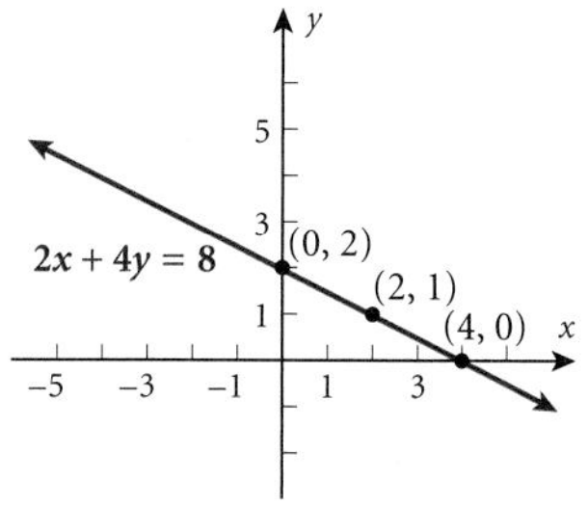

The intercepts are $(4, 0)$ and $(0, 2)$. Plot these points and draw the line containing them. See Figure 51.

NOTE: The second method, locating the intercepts, only produces one point when the line passes through the origin. In this case some other point on the line must be found in order to graph the line. Refer back to Example 3.

NOW WORK PROBLEM 55.

EXAMPLE 13 Daily Cost of Production

A certain factory has daily fixed overhead expenses of $2000, while each item produced costs $100. Find an equation that relates the daily cost C to the number x of items produced each day.

SOLUTION The fixed overhead expense of $2000 represents the fixed cost, the cost incurred no matter how many items are produced. Since each item produced costs $100, the variable cost of producing x items is $100x$. Then the total daily cost C of production is

$$C = 100x + 2000$$

The graph of this equation is given by the line in Figure 52. Notice that the fixed cost $2000 is represented by the y-intercept, while the $100 cost of producing each item is the slope. Also notice that, for convenience, a different scale is used on each axis. ▶

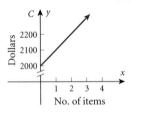

FIGURE 52

SUMMARY The graph of a linear equation, $Ax + By = C$, where A and B are not both zero, is a line. In this form it is referred to as the general equation of a line.

1. Given the general equation of a line, information can be found about the line:
   (a) Let $x = 0$ and solve for y to find the y-intercept.
   (b) Let $y = 0$ and solve for x to find the x-intercept.
   (c) Place the equation in slope–intercept form $y = mx + b$ to find the slope m and y-intercept $(0, b)$.
2. Given information about a line, an equation of the line can be found. The form of the equation to use depends on the given information. See the table below:

Given	Use	Equation
Point (x_1, y_1), slope m	Point–slope form	$y - y_1 = m(x - x_1)$
Two points (x_1, y_1), (x_2, y_2)	If $x_1 = x_2$, the line is vertical	$x = x_1$
	If $x_1 \neq x_2$, find the slope m: $$m = \frac{y_2 - y_1}{x_2 - x_1}$$ Then use the point–slope form	$y - y_1 = m(x - x_1)$
Slope m, y-intercept $(0, b)$	Slope–intercept form	$y = mx + b$

EXERCISE 0.9

In Problems 1–4, use the given equations to fill in the missing values in each table. Use these points to graph each equation.

1. $y = 2x + 4$

x	0	2	-2	4	-4
y		0			

2. $y = -3x + 6$

x	0	2	-2	4	-4
y		0			

Lines 93

3. $2x - y = 6$

x	0		2	-2	4	-4
y		0				

4. $x + 2y = 8$

x	0		2	-2	4	-4
y		0				

In Problems 5–8: (a) find the equation of the vertical line containing the given point;
(b) find the equation of the horizontal line containing the given point.

5. $(2, -3)$ **6.** $(5, 4)$ **7.** $(-4, 1)$ **8.** $(-6, -3)$

In Problems 9–12, find the slope of the line. Give an interpretation of the slope.

9.   **10.**   **11.** 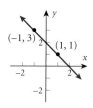   **12.**

In Problems 13–20, plot each pair of points and find the slope of the line containing them. Interpret the slope and graph the line.

13. $(2, 3); (1, 0)$ **14.** $(1, 2); (3, 4)$ **15.** $(-2, 3); (2, 1)$ **16.** $(-1, 1); (2, 3)$

17. $(-3, -1); (2, -1)$ **18.** $(4, 2); (-5, 2)$ **19.** $(-1, 2); (-1, -2)$ **20.** $(2, 0); (2, 2)$

In Problems 21–28, graph the line containing the point P and having slope m.

21. $P = (1, 2); m = 2$ **22.** $P = (2, 1); m = 3$ **23.** $P = (2, 4); m = -\dfrac{3}{4}$ **24.** $P = (1, 3); m = -\dfrac{2}{3}$

25. $P = (-1, 3); m = 0$ **26.** $P = (2, -4); m = 0$ **27.** $P = (0, 3)$; slope undefined   **28.** $P = (-2, 0)$; slope undefined

In Problems 29–32, write the equation of each line in the form $Ax + By = C$.

29.   **30.**   **31.**   **32.**

In Problems 33–50, write the equation of each line in the form $Ax + By = C$.

33. Slope = 2; containing the point $(-4, 1)$ **34.** Slope = 3; containing the point $(-3, 4)$

35. Slope = $-\dfrac{2}{3}$; containing the point $(1, -1)$ **36.** Slope = $\dfrac{1}{2}$; containing the point $(3, 1)$

37. Containing the points $(1, 3)$ and $(-1, 2)$ **38.** Containing the points $(-3, 4)$ and $(2, 5)$

39. Slope = -2; y-intercept = $(0, 3)$ **40.** Slope = -3; y-intercept = $(0, -2)$

41. Slope = 3; x-intercept = $(-4, 0)$ **42.** Slope = -4; x-intercept = $(2, 0)$

43. Slope = $\dfrac{4}{5}$; containing the point $(0, 0)$ **44.** Slope = $\dfrac{7}{3}$; containing the point $(0, 0)$

45. x-intercept = (2, 0); y-intercept = (0, −1)

46. x-intercept = (−4, 0); y-intercept = (0, 4)

47. Slope undefined; containing the point (1, 4)

48. Slope undefined; containing the point (2, 1)

49. Slope = 0; containing the point (1, 4)

50. Slope = 0; containing the point (2, 1)

In Problems 51–66, find the slope and y-intercept of each line. Graph the line.

51. $y = 2x + 3$

52. $y = -3x + 4$

53. $\frac{1}{2}y = x - 1$

54. $\frac{1}{3}x + y = 2$

55. $2x - 3y = 6$

56. $3x + 2y = 6$

57. $x + y = 1$

58. $x - y = 2$

59. $x = -4$

60. $y = -1$

61. $y = 5$

62. $x = 2$

63. $y - x = 0$

64. $x + y = 0$

65. $2y - 3x = 0$

66. $3x + 2y = 0$

---

67. Find the equation of the horizontal line containing the point (−1, −3).

68. Find the equation of the vertical line containing the point (−2, 5).

69. Cost of Operating a Car According to the American Automobile Association (AAA), the average cost of operating a standard-sized car, including gasoline, oil, tires, and maintenance increased to $0.122 per mile in 2000. Write an equation that relates the average cost C of operating a standard-sized car and the number x of miles it is driven.

Source: AAA Traveler Magazine.

70. Cost of Renting a Truck The cost of renting a truck is $280 per week plus a charge of $0.30 per mile driven. Write an equation that relates the cost C for a weekly rental in which the truck is driven x miles.

71. Electricity Rates in Illinois Commonwealth Edison Company supplies electricity to residential customers for a monthly customer charge of $7.58 plus 8.275 cents per kilowatt-hour for up to 400 kilowatt-hours.

(a) Write an equation that relates the monthly charge C, in dollars, to the number x of kilowatt-hours used in a month, $0 \leq x \leq 400$.
(b) Graph this equation.
(c) What is the monthly charge for using 100 kilowatt-hours?
(d) What is the monthly charge for using 300 kilowatt-hours?
(e) Interpret the slope of the line.

Source: Commonwealth Edison Company, December 2003.

72. Electricity Rates in Florida Florida Power & Light Company supplies electricity to residential customers for a monthly customer charge of $5.25 plus 6.787 cents per kilowatt-hour for up to 750 kilowatt-hours.

(a) Write an equation that relates the monthly charge C, in dollars, to the number x of kilowatt-hours used in a month, $0 \leq x \leq 750$.
(b) Graph this equation.
(c) What is the monthly charge for using 200 kilowatt-hours?
(d) What is the monthly charge for using 500 kilowatt-hours?
(e) Interpret the slope of the line.

Source: Florida Power & Light Company, January 2003.

73. Weight–Height Relation in the U.S. Army Assume the recommended weight w of females aged 17–20 years in the U.S. Army is linearly related to their height h. If an Army female who is 67 inches tall should weigh 139 pounds and if an Army female who is 70 inches tall should weigh 151 pounds, find an equation that expresses weight in terms of height.

Source: http://www.nutribase.com/nutrition-fwchartf.htm.

74. Wages of a Car Salesperson Dan receives $375 per week for selling new and used cars at a car dealership in Omaha, Nebraska. In addition, he receives 5% of the profit on any sales he generates. Write an equation that relates Dan's weekly salary S when he has sales that generate a profit of x dollars.

75. Cost of Sunday Home Delivery The cost to the *Chicago Tribune* for Sunday home delivery is approximately $0.53 per newspaper with fixed costs of $1,070,000. Write an equation that relates the cost C and the number x of copies delivered.

Source: Chicago Tribune, 2002.

76. Disease Propagation Research indicates that in a controlled environment, the number of diseased mice will increase linearly each day after one of the mice in the cage is infected with a particular type of disease-causing germ. There were 8 diseased mice 4 days after the first exposure and 14 diseased mice after 6 days. Write an equation that will give the number of diseased mice after any given number of days. If there were 40 mice in the cage, how long will it take until they are all infected?

77. Temperature Conversion The relationship between Celsius (°C) and Fahrenheit (°F) degrees for measuring temperature is linear. Find an equation relating °C and °F if 0°C corresponds to 32°F and 100°C corresponds to 212°F. Use the equation to find the Celsius measure of 68°F.

78. Temperature Conversion The Kelvin (K) scale for measuring temperature is obtained by adding 273 to the Celsius temperature.

(a) Write an equation relating K and °C.
(b) Write an equation relating K and °F (see Problem 77).

79. Water Preservation At Harlan County Dam in Nebraska, the U.S. Bureau of Reclamation reports that the storage content of the reservoir decreased from 162,400 acre-feet (52.5 billion gallons of water) on November 8, 2002 to 161,200 acre-feet (52.5 billion gallons of water) on December 8, 2002. Suppose that the rate of loss of water remains constant.

(a) Write an equation that relates the amount A of water, in billions of gallons, to the time t, in days. Use $t = 1$ for November 1, $t = 2$ for November 2, and so on.
(b) How much water was in the reservoir on November 20 ($t = 20$)?
(c) Interpret the slope.
(d) How much water will be in the reservoir on December 31, 2002 ($t = 61$)?
(e) When will the reservoir be empty?
(f) Comment on your answer to part (e).

Source: U.S. Bureau of Reclamation.

80. Product Promotion A cereal company finds that the number of people who will buy one of its products the first month it is introduced is linearly related to the amount of money it spends on advertising. If it spends $400,000 on advertising, then 100,000 boxes of cereal will be sold, and if it spends $600,000, then 140,000 boxes will be sold.

(a) Write an equation describing the relation between the amount spent on advertising and the number of boxes sold.
(b) How much advertising is needed to sell 200,000 boxes of cereal?
(c) Interpret the slope.

In Problems 81–88, use a graphing utility to graph each linear equation. Be sure to use a viewing rectangle that shows the intercepts. Then locate each intercept rounded to two decimal places.

81. $1.2x + 0.8y = 2$

82. $-1.3x + 2.7y = 8$

83. $21x - 15y = 53$

84. $5x - 3y = 82$

85. $\dfrac{4}{17}x + \dfrac{6}{23}y = \dfrac{2}{3}$

86. $\dfrac{9}{14}x - \dfrac{3}{8}y = \dfrac{2}{7}$

87. $\pi x - \sqrt{3}\, y = \sqrt{6}$

88. $x + \pi y = \sqrt{15}$

In Problems 89–92, match each graph with the correct equation:

(a) $y = x$; (b) $y = 2x$; (c) $y = \dfrac{x}{2}$; (d) $y = 4x$.

89.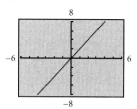

90.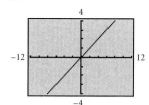

91.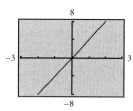

92.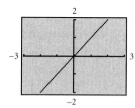

In Problems 93–96, write an equation of each line. Express your answer using either the general form or the slope-intercept form of the equation of a line, whichever you prefer.

93.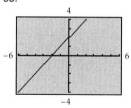

94.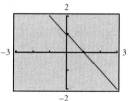

95.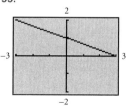

96.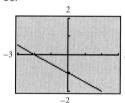

97. Which of the following equations might have the graph shown. (More than one answer is possible.)

    (a) $2x + 3y = 6$
    (b) $-2x + 3y = 6$
    (c) $3x - 4y = -12$
    (d) $x - y = 1$
    (e) $x - y = -1$
    (f) $y = 3x - 5$
    (g) $y = 2x + 3$
    (h) $y = -3x + 3$

98. Which of the following equations might have the graph shown. (More than one answer is possible.)

    (a) $2x + 3y = 6$
    (b) $2x - 3y = 6$
    (c) $3x + 4y = 12$
    (d) $x - y = 1$
    (e) $x - y = -1$
    (f) $y = -2x + 1$
    (g) $y = -\dfrac{1}{2}x + 10$
    (h) $y = x + 4$

99. Write the general equation of the x-axis.

100. Write the general equation of the y-axis.

101. Which form of the equation of a line do you prefer to use? Justify your position with an example that shows that your choice is better than another. Have reasons.

102. Can every line be written in slope–intercept form? Explain.

103. Does every line have two distinct intercepts? Explain. Are there lines that have no intercepts? Explain.

104. What can you say about two lines that have equal slopes and equal y-intercepts?

105. What can you say about two lines with the same x-intercept and the same y-intercept? Assume that the x-intercept is not $(0, 0)$.

106. If two lines have the same slope, but different x-intercepts, can they have the same y-intercept?

107. If two lines have the same y-intercept, but different slopes, can they have the same x-intercept? What is the only way that this can happen?

108. The accepted symbol used to denote the slope of a line is the letter m. Investigate the origin of this symbolism. Begin by consulting a French dictionary and looking up the French word *monter*. Write a brief essay on your findings.

CHAPTER 1

Functions and Their Graphs

OUTLINE

1.1 Graphs of Equations
1.2 Functions
1.3 Graphs of Functions; Properties of Functions
1.4 Library of Functions; Piecewise-defined Functions
1.5 Graphing Techniques: Shifts and Reflections
- **Chapter Review**
- **Chapter Project**
- **Mathematical Questions from Professional Exams**

On the way home from college, you and a friend decide to stop off in Charlotte, North Carolina. Because you have only one full day to see the sights, you decide that renting a car is the best way to see the most. But which car rental company should you use? Naturally, the cheapest! But what is the cheapest? Is it the one with unlimited mileage or the one with a better daily rate and a mileage charge? The mathematics of this chapter provides the background for solving this problem. The Chapter Project at the end of the chapter will help you understand how to make the best decision.

A LOOK BACK, A LOOK FORWARD

In Chapter 0 we studied rectangular coordinates and used them to graph linear equations of the form $Ax + By = C$, where either $A \neq 0$ or $B \neq 0$. We continue the study of graphing equations in two variables here. In particular, we look at a special type of equation involving two variables, called a *function*. We will define what a function is, how to graph functions, what properties functions have, and develop a "library" of functions.

The word function apparently was introduced by René Descartes in 1637. For him, a function simply meant any positive integral power of a variable x. Gottfried Wilhelm Leibniz (1646–1716), who always emphasized the geometric side of mathematics, used the word function to denote any quantity associated with a curve, such as the coordinates of a point on the curve. Leonhard Euler (1707–1783) employed the word to mean any equation or formula involving variables and constants. His idea of a function is similar to the one most often seen in courses that precede calculus. Later, the use of functions in investigating heat flow equations led to a very broad definition, due to Lejeune Dirichlet (1805–1859), which describes a function as a correspondence between two sets. It is his definition that we use here.

1.1 Graphs of Equations

PREPARING FOR THIS SECTION   *Before getting started, review the following:*

> Evaluating Algebraic Expressions (Chapter 0, Section 0.2, pp. 19–20)
> Rectangular coordinates (Chapter 0, Section 0.8, pp. 76–77)
> Solving equations (Chapter 0, Section 0.4, pp. 40–49)
> Lines (Chapter 0, Section 0.9, pp. 81–92)

OBJECTIVES
1. Graph equations by plotting points
2. Find intercepts from a graph
3. Find intercepts from an equation
4. Test an equation for symmetry with respect to the (a) *x*-axis, (b) *y*-axis, and (c) origin

An **equation in two variables,** say x and y, is a statement in which two expressions involving x and y are equal. The expressions are called the **sides** of the equation. Since an equation is a statement, it may be true or false, depending on the value of the variables. Any values of x and y that result in a true statement are said to **satisfy** the equation.

For example, the following are all equations in two variables x and y:

$$x^2 + y^2 = 5 \qquad 2x^2 - y = 6 \qquad y = 2x + 5 \qquad x^2 = y^3$$

The first of these, $x^2 + y^2 = 5$, is satisfied for $x = 1$, $y = 2$, since $1^2 + 2^2 = 1 + 4 = 5$. Other choices of x and y also satisfy this equation. It is not satisfied for $x = 2$ and $y = 3$, since $2^2 + 3^2 = 4 + 9 = 13 \neq 5$.

The **graph of an equation** in two variables x and y consists of the set of points in the xy-plane whose coordinates (x, y) satisfy the equation.

For example, as we learned in Chapter 0, the graph of any equation of the form $Ax + By = C$, where either $A \neq 0$ or $B \neq 0$, is a line.

Graphs play an important role in helping us to visualize the relationships that exist between two variable quantities. Figure 1 shows the monthly closing prices of Intel stock from August 31, 2002, through August 31, 2003. For example, the closing price on May 31, 2003, was about $21 per share.

FIGURE 1 Monthly closing prices of Intel stock, 8/31/02 to 8/31/03.

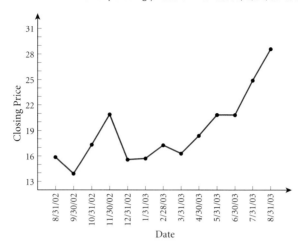

EXAMPLE 1 Determining Whether a Point Is on the Graph of an Equation

Determine if the following points are on the graph of the equation $2x - y = 6$.

(a) $(2, 3)$ **(b)** $(2, -2)$

SOLUTION **(a)** For the point $(2, 3)$, we check to see if $x = 2, y = 3$ satisfy the equation $2x - y = 6$.

$$2x - y = 2(2) - 3 = 4 - 3 = 1 \neq 6$$

The equation is not satisfied, so the point $(2, 3)$ is not on the graph.

(b) For the point $(2, -2)$, we have

$$2x - y = 2(2) - (-2) = 4 + 2 = 6$$

The equation is satisfied, so the point $(2, -2)$ is on the graph. ▸

NOW WORK PROBLEM 23.

EXAMPLE 2 Graphing an Equation by Plotting Points

Graph the equation: $y = x^2$

SOLUTION Table 1 on page 100 provides several points on the graph. In Figure 2 we plot these points and connect them with a smooth curve to obtain the graph (a *parabola*). ▸

The graph of the equation shown in Figure 2 does not show all points. For example, the point $(5, 25)$ is a part of the graph of $y = x^2$, but it is not shown. Since the graph of $y = x^2$ could be extended out as far as we please, we use arrows to indicate that the pattern shown continues. It is important when illustrating a graph to present enough of the graph so that any viewer of the illustration will "see" the rest of it as an obvious continuation of what is actually there. This is referred to as a **complete graph.**

One way to obtain a complete graph of an equation is to plot a sufficient number of points on the graph until a pattern becomes evident. Then these points are connected

TABLE 1

x	y = x²	(x, y)
−4	16	(−4, 16)
−3	9	(−3, 9)
−2	4	(−2, 4)
−1	1	(−1, 1)
0	0	(0, 0)
1	1	(1, 1)
2	4	(2, 4)
3	9	(3, 9)
4	16	(4, 16)

FIGURE 2

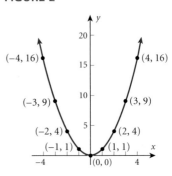

with a smooth curve following the suggested pattern. But how many points are sufficient? Sometimes knowledge about the equation tells us. For example, we learned in Chapter 0 that if an equation is of the form $y = mx + b$, then its graph is a line. In this case, only two points are needed to obtain the graph.

One purpose of this book is to investigate the properties of equations in order to decide whether a graph is complete. At first we shall graph equations by plotting a sufficient number of points. Shortly, we shall investigate various techniques that will enable us to graph an equation without plotting so many points. Other times we shall graph equations based solely on properties of the equation.

 COMMENT: Another way to obtain the graph of an equation is to use a graphing utility. Read Section A.2, *Using a Graphing Utility to Graph Equations,* in the Appendix. ▶

EXAMPLE 3 **Graphing an Equation by Plotting Points**

Graph the equation: $y = x^3$

SOLUTION We set up Table 2, listing several points on the graph. Figure 3 illustrates some of these points and the graph of $y = x^3$.

TABLE 2

x	y = x³	(x, y)
−3	−27	(−3, −27)
−2	−8	(−2, −8)
−1	−1	(−1, −1)
0	0	(0, 0)
1	1	(1, 1)
2	8	(2, 8)
3	27	(3, 27)

FIGURE 3

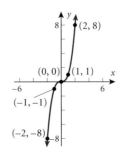

▶

EXAMPLE 4 Graphing an Equation by Plotting Points

Graph the equation: $x = y^2$

SOLUTION We set up Table 3, listing several points on the graph. In this case, because of the form of the equation, we assign some numbers to y and find corresponding values of x. Figure 4 illustrates some of these points and the graph of $x = y^2$.

TABLE 3

y	$x = y^2$	(x, y)
-3	9	$(9, -3)$
-2	4	$(4, -2)$
-1	1	$(1, -1)$
0	0	$(0, 0)$
1	1	$(1, 1)$
2	4	$(4, 2)$
3	9	$(9, 3)$
4	16	$(16, 4)$

FIGURE 4

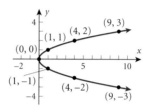

If we restrict y so that $y \geq 0$, the equation $x = y^2$, $y \geq 0$, may be written equivalently as $y = \sqrt{x}$. The portion of the graph of $x = y^2$ in quadrant I is therefore the graph of $y = \sqrt{x}$. See Figure 5.

FIGURE 5

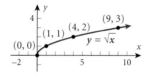

COMMENT: To see the graph of the equation $x = y^2$ on a graphing calculator, you will need to graph two equations: $Y_1 = \sqrt{x}$ and $Y_2 = -\sqrt{x}$. See Figure 6. We discuss why a little later in this chapter.

FIGURE 6

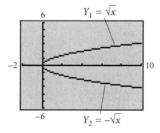

NOW WORK PROBLEM 45.

We said earlier that we would discuss techniques that reduce the number of points required to graph an equation. Two such techniques involve finding *intercepts* and checking for *symmetry*.

2 Intercepts

Find intercepts from a graph

The points, if any, at which a graph crosses or touches the coordinate axes are called the **intercepts.** See Figure 7. The point at which the graph crosses or touches the *x*-axis is an ***x*-intercept,** and the point at which the graph crosses or touches the *y*-axis is a ***y*-intercept.**

FIGURE 7

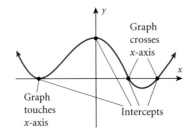

EXAMPLE 5 Finding Intercepts from a Graph

FIGURE 8

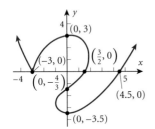

Find the intercepts of the graph in Figure 8. What are its *x*-intercepts? What are its *y*-intercepts?

SOLUTION The intercepts of the graph are the points

$$(-3, 0), \quad (0, 3), \quad \left(\frac{3}{2}, 0\right), \quad \left(0, -\frac{4}{3}\right), \quad (0, -3.5), \quad (4.5, 0)$$

The *x*-intercepts are $(-3, 0)$, $\left(\frac{3}{2}, 0\right)$, and $(4.5, 0)$. The *y*-intercepts are $(0, 3)$, $\left(0, -\frac{4}{3}\right)$, and $(0, -3.5)$. ◼

NOW WORK PROBLEM 11(a).

Find intercepts from an equation 3 The intercepts of the graph of an equation can be found from the equation by using the fact that points on the *x*-axis have *y*-coordinates equal to 0, and points on the *y*-axis have *x*-coordinates equal to 0.

> **Procedure for Finding Intercepts**
> 1. To find the *x*-intercept(s), if any, of the graph of an equation, let $y = 0$ in the equation and solve for *x*.
> 2. To find the *y*-intercept(s), if any, of the graph of an equation, let $x = 0$ in the equation and solve for *y*.

Because the *x*-intercepts of the graph of an equation are those *x*-values for which $y = 0$, they are also called the **zeros** (or **roots**) of the equation.

EXAMPLE 6 Finding Intercepts from an Equation

Find the x-intercept(s) and the y-intercept(s), if any, of the graph of $y = x^2 - 4$. Graph $y = x^2 - 4$.

SOLUTION To find the x-intercept(s), we let $y = 0$ and obtain the equation

$$x^2 - 4 = 0$$
$$(x + 2)(x - 2) = 0 \quad \text{Factor.}$$
$$x + 2 = 0 \quad \text{or} \quad x - 2 = 0 \quad \text{Zero-Product Property.}$$
$$x = -2 \quad \text{or} \quad x = 2$$

The equation has the solution set $\{-2, 2\}$. The x-intercepts are $(-2, 0)$ and $(2, 0)$.

To find the y-intercept(s), we let $x = 0$ and obtain the equation

$$y = -4$$

The y-intercept is $(0, -4)$.

Since $x^2 \geq 0$ for all x, we deduce from the equation $y = x^2 - 4$ that $y \geq -4$ for all x. This information, the intercepts, and the points from Table 4, enable us to graph $y = x^2 - 4$. See Figure 9.

TABLE 4

x	$y = x^2 - 4$	(x, y)
-3	5	$(-3, 5)$
-1	-3	$(-1, -3)$
1	-3	$(1, -3)$
3	5	$(3, 5)$

FIGURE 9

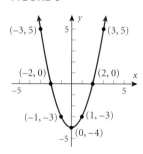

 NOW WORK PROBLEM 33 (List the Intercepts).

 COMMENT: For many equations, finding intercepts may not be so easy. In such cases, a graphing utility can be used. Read Section A.4, *Using a Graphing Utility to Locate Intercepts and Check for Symmetry* in the Appendix, to find out how a graphing utility locates intercepts.

Symmetry

We have just seen the role that intercepts play in obtaining key points on the graph of an equation. Another helpful tool for graphing equations involves *symmetry*, particularly symmetry with respect to the x-axis, the y-axis, and the origin.

FIGURE 10 Symmetry with respect to the x-axis

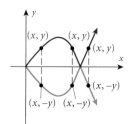

A graph is said to be **symmetric with respect to the x-axis** if, for every point (x, y) on the graph, the point $(x, -y)$ is also on the graph.

Figure 10 illustrates the definition. Notice that, when a graph is symmetric with respect to the x-axis, the part of the graph above the x-axis is a reflection or mirror image of the part below it, and vice versa.

EXAMPLE 7 Points Symmetric with Respect to the x-Axis

If a graph is symmetric with respect to the x-axis and the point (3, 2) is on the graph, then the point (3, −2) is also on the graph.

FIGURE 11 Symmetry with respect to the y-axis

A graph is said to be **symmetric with respect to the y-axis** if, for every point (x, y) on the graph, the point $(-x, y)$ is also on the graph.

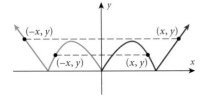

Figure 11 illustrates the definition. Notice that, when a graph is symmetric with respect to the y-axis, the part of the graph to the right of the y-axis is a reflection of the part to the left of it, and vice versa.

EXAMPLE 8 Points Symmetric with Respect to the y-Axis

If a graph is symmetric with respect to the y-axis and the point (5, 8) is on the graph, then the point (−5, 8) is also on the graph.

FIGURE 12 Symmetry with respect to the origin

A graph is said to be **symmetric with respect to the origin** if, for every point (x, y) on the graph, the point $(-x, -y)$ is also on the graph.

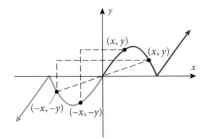

Figure 12 illustrates the definition. Notice that symmetry with respect to the origin may be viewed in two ways:

1. As a reflection about the y-axis, followed by a reflection about the x-axis.
2. As a projection along a line through the origin so that the distances from the origin are equal.

EXAMPLE 9 Points Symmetric with Respect to the Origin

If a graph is symmetric with respect to the origin and the point (4, 2) is on the graph, then the point (−4, −2) is also on the graph.

NOW WORK PROBLEMS 1 AND 11(b).

Test an equation for symmetry with respect to the (a) x-axis, (b) y-axis, and (c) origin

When the graph of an equation is symmetric with respect to a coordinate axis or the origin, the number of points that you need to plot in order to see the pattern is reduced. For example, if the graph of an equation is symmetric with respect to the y-axis, then, once points to the right of the y-axis are plotted, an equal number of points on the graph can be obtained by reflecting them about the y-axis. Because of this, before we graph an equation, we first want to determine whether it has any symmetry. The following tests are used for this purpose.

Tests for Symmetry

To test the graph of an equation for symmetry with respect to the

x-Axis Replace y by $-y$ in the equation. If an equivalent equation results, the graph of the equation is symmetric with respect to the x-axis.
y-Axis Replace x by $-x$ in the equation. If an equivalent equation results, the graph of the equation is symmetric with respect to the y-axis.
Origin Replace x by $-x$ and y by $-y$ in the equation. If an equivalent equation results, the graph of the equation is symmetric with respect to the origin.

Let's look at an equation that we have already graphed to see how these tests are used.

EXAMPLE 10 Testing an Equation for Symmetry ($x = y^2$)

(a) To test the graph of the equation $x = y^2$ for symmetry with respect to the x-axis, we replace y by $-y$ in the equation, as follows:

$$x = y^2 \quad \text{Original equation.}$$
$$x = (-y)^2 \quad \text{Replace } y \text{ by } -y.$$
$$x = y^2 \quad \text{Simplify.}$$

When we replace y by $-y$, the result is the same equation. The graph is symmetric with respect to the x-axis.

(b) To test the graph of the equation $x = y^2$ for symmetry with respect to the y-axis, we replace x by $-x$ in the equation:

$$x = y^2 \quad \text{Original equation.}$$
$$-x = y^2 \quad \text{Replace } x \text{ by } -x.$$

Because we arrive at the equation $-x = y^2$, which is not equivalent to the original equation, we conclude that the graph is not symmetric with respect to the y-axis.

(c) To test for symmetry with respect to the origin, we replace x by $-x$ and y by $-y$:

$$x = y^2 \quad \text{Original equation.}$$
$$-x = (-y)^2 \quad \text{Replace } x \text{ by } -x \text{ and } y \text{ by } -y.$$
$$-x = y^2 \quad \text{Simplify.}$$

The resulting equation, $-x = y^2$, is not equivalent to the original equation. We conclude that the graph is not symmetric with respect to the origin.

Figure 13(a) on page 106 illustrates the graph of $x = y^2$. In forming a table of points on the graph of $x = y^2$, we can restrict ourselves to points whose y-coordinates are positive. Once these are plotted and connected, a reflection about the x-axis (because of the symmetry) provides the rest of the graph.

Figures 13(b) and (c) illustrate two other equations, $y = x^2$ and $y = x^3$, that we graphed earlier. Test each of these equations for symmetry to verify the conclusions stated in Figures 13(b) and (c). Notice how the existence of symmetry reduces the number of points that we need to plot.

FIGURE 13

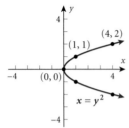
(a) Symmetry with respect to the x-axis

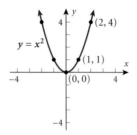
(b) Symmetry with respect to the y-axis

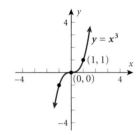
(c) Symmetry with respect to the origin

 NOW WORK PROBLEM 33 (Test for Symmetry).

EXAMPLE 11 Graphing the Equation $y = \dfrac{1}{x}$

Graph the equation: $y = \dfrac{1}{x}$

Find any intercepts and check for symmetry first.

SOLUTION We check for intercepts first. If we let $x = 0$, we obtain a 0 denominator, which is not defined. We conclude that there is no y-intercept. If we let $y = 0$, we get the equation $\dfrac{1}{x} = 0$, which has no solution. We conclude that there is no x-intercept. The graph of $y = \dfrac{1}{x}$ does not cross or touch the coordinate axes.

Next we check for symmetry:

x-Axis Replacing y by $-y$ yields $-y = \dfrac{1}{x}$, which is not equivalent to $y = \dfrac{1}{x}$.

y-Axis Replacing x by $-x$ yields $y = \dfrac{1}{-x}$, which is not equivalent to $y = \dfrac{1}{x}$.

Origin Replacing x by $-x$ and y by $-y$ yields $-y = \dfrac{1}{-x} = -\dfrac{1}{x}$, which is equivalent to $y = \dfrac{1}{x}$.

The graph is symmetric with respect to the origin.

Finally, we set up Table 5, listing several points on the graph. Because of the symmetry with respect to the origin, we use only positive values of x.

From Table 5 we infer that if x is a large and positive number then $y = \dfrac{1}{x}$ is a positive number close to 0. We also infer that if x is a positive number close to 0 then $y = \dfrac{1}{x}$ is a large and positive number. Armed with this information, we can graph the equation. Figure 14 illustrates some of these points and the graph of $y = \dfrac{1}{x}$. Observe how the absence of intercepts and the existence of symmetry with respect to the origin were utilized.

TABLE 5

x	$y = \dfrac{1}{x}$	(x, y)
$\dfrac{1}{10}$	10	$\left(\dfrac{1}{10}, 10\right)$
$\dfrac{1}{3}$	3	$\left(\dfrac{1}{3}, 3\right)$
$\dfrac{1}{2}$	2	$\left(\dfrac{1}{2}, 2\right)$
1	1	$(1, 1)$
2	$\dfrac{1}{2}$	$\left(2, \dfrac{1}{2}\right)$
3	$\dfrac{1}{3}$	$\left(3, \dfrac{1}{3}\right)$
10	$\dfrac{1}{10}$	$\left(10, \dfrac{1}{10}\right)$

FIGURE 14

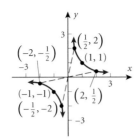

COMMENT: Look at Figure 14. The line $y = 0$ (the x-axis) is called a horizontal *asymptote* of the graph. The line $x = 0$ (the y-axis) is called a vertical *asymptote* of the graph. We will discuss asymptotes in more detail in Chapter 3.

COMMENT: Refer to Example 3 in Section A.4 of the Appendix for the graph of $y = \dfrac{1}{x}$ using a graphing utility.

EXERCISE 1.1

In Problems 1–10, plot each point. Then plot the point that is symmetric to it with respect to (a) the x-axis; (b) the y-axis; (c) the origin.

1. $(3, 4)$
2. $(5, 3)$
3. $(-2, 1)$
4. $(4, -2)$
5. $(1, 1)$
6. $(-1, -1)$
7. $(-3, -4)$
8. $(4, 0)$
9. $(0, -3)$
10. $(-3, 0)$

In Problems 11–22, the graph of an equation is given.
(a) List the intercepts of the graph.
(b) Based on the graph, tell whether the graph is symmetric with respect to the x-axis, the y-axis, and/or the origin.

11.
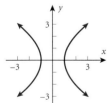

12.
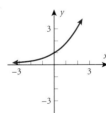

13.
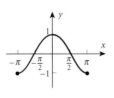

14.
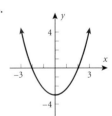

15.
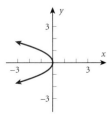

16.
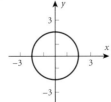

17.
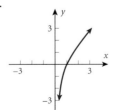

18.
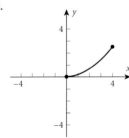

108 Chapter 1 Functions and Their Graphs

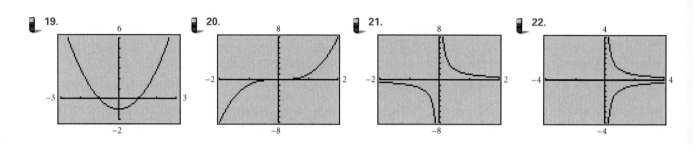

In Problems 23–28, determine whether the given points are on the graph of the equation.

23. Equation: $y = x^4 - \sqrt{x}$
    Points: $(0, 0); (1, 1); (-1, 0)$

24. Equation: $y = x^3 - 2\sqrt{x}$
    Points: $(0, 0); (1, 1); (1, -1)$

25. Equation: $y^2 = x^2 + 9$
    Points: $(0, 3); (3, 0); (-3, 0)$

26. Equation: $y^3 = x + 1$
    Points: $(1, 2); (0, 1); (-1, 0)$

27. Equation: $x^2 + y^2 = 4$
    Points: $(0, 2); (-2, 2); (\sqrt{2}, \sqrt{2})$

28. Equation: $x^2 + 4y^2 = 4$
    Points: $(0, 1); (2, 0); \left(2, \dfrac{1}{2}\right)$

In Problems 29–44, list the intercepts and test for symmetry.

29. $x^2 = y$

30. $y^2 = x$

31. $y = 3x$

32. $y = -5x$

33. $x^2 + y - 9 = 0$

34. $y^2 - x - 4 = 0$

35. $9x^2 + 4y^2 = 36$

36. $4x^2 + y^2 = 4$

37. $y = x^3 - 27$

38. $y = x^4 - 1$

39. $y = x^2 - 3x - 4$

40. $y = x^2 + 4$

41. $y = \dfrac{3x}{x^2 + 9}$

42. $y = \dfrac{x^2 - 4}{2x^4}$

43. $y = |x|$

44. $y = \sqrt{x}$

In Problems 45–48, graph each equation by plotting points.

45. $y = x^3 - 1$

46. $x = y^2 + 1$

47. $y = 2\sqrt{x}$

48. $y = x^2 + 2$

49. If $(a, 2)$ is a point on the graph of $y = 3x + 5$, what is a?

50. If $(2, b)$ is a point on the graph of $y = x^2 + 4x$, what is b?

51. If (a, b) is a point on the graph of $2x + 3y = 6$, write an equation that relates a to b.

52. If $(2, 0)$ and $(0, 5)$ are points on the graph of $y = mx + b$, what are m and b?

In Problem 53, you may use a graphing utility, but it is not required.

53. (a) Graph $y = \sqrt{x^2}$, $y = x$, $y = |x|$, and $y = (\sqrt{x})^2$, noting which graphs are the same.
    (b) Explain why the graphs of $y = \sqrt{x^2}$ and $y = |x|$ are the same.
    (c) Explain why the graphs of $y = x$ and $y = (\sqrt{x})^2$ are not the same.
    (d) Explain why the graphs of $y = \sqrt{x^2}$ and $y = x$ are not the same.

54. Make up an equation with the intercepts $(2, 0)$, $(4, 0)$, and $(0, 1)$. Compare your equation with a friend's equation. Comment on any similarities.

55. An equation is being tested for symmetry with respect to the x-axis, the y-axis, and the origin. Explain why, if two of these symmetries are present, the remaining one must also be present.

56. Draw a graph that contains the points $(-2, -1), (0, 1), (1, 3)$, and $(3, 5)$. Compare your graph with those of other students. Are most of the graphs almost straight lines? How many are "curved"? Discuss the various ways that these points might be connected.

1.2 Functions

PREPARING FOR THIS SECTION *Before getting started, review the following:*

> Intervals (Chapter 0, Section 0.5, pp. 51–53)
> Evaluating Algebraic Expressions; Domain of a Variable (Chapter 0, Section 0.2, pp. 19–20)
> Square Root Method (Chapter 0, Section 0.4, pp. 44–45)
> Solving Inequalities (Chapter 0, Section 0.5, pp. 53–61)

OBJECTIVES
1. Find the value of a function
2. Find the difference quotient of a function
3. Find the domain of a function
4. Solve applied problems involving functions

In many applications a correspondence (such as an equation) exists between two variables. For example, the relation between the revenue R resulting from the sale of x items selling for $10 each may be expressed by the equation $R = 10x$. If we know how many items have been sold, then we can calculate the revenue by using the equation $R = 10x$. This equation is an example of a *function*.

As another example, suppose that an icicle falls off a building from a height of 64 feet above the ground. According to a law of physics, the distance s (in feet) of the icicle from the ground after t seconds is given (approximately) by the formula $s = 64 - 16t^2$. When $t = 0$ seconds, the icicle is $s = 64$ feet above the ground. After 1 second, the icicle is $s = 64 - 16(1)^2 = 48$ feet above the ground. After 2 seconds, the icicle strikes the ground. The formula $s = 64 - 16t^2$ provides a way of finding the distance s for any time t $(0 \leq t \leq 2)$. There is a correspondence between each time t in the interval $0 \leq t \leq 2$ and the distance s. We say that the distance s is a function of the time t because:

1. There is a correspondence between the set of times and the set of distances.
2. There is exactly one distance s obtained for any time t in the interval $0 \leq t \leq 2$.

Let's now look at the definition of a function.

> **Definition of Function**
>
> Let X and Y be two nonempty sets of real numbers. A **function** from X into Y is a correspondence that associates with each number in X exactly one number in Y.

FIGURE 15

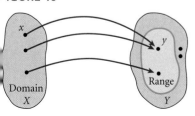

The set X is called the **domain** of the function. For each number x in X, the corresponding number y in Y is called the **value** of the function at x, or the **image** of x. The set of all images of the numbers in the domain is called the **range** of the function. See Figure 15.

Since there may be some numbers in Y that are not the image of any x in X, it follows that the range of a function is a subset of Y, as shown in Figure 15.

EXAMPLE 1 Example of a Function

Consider the function defined by the equation

$$y = 2x - 5, \quad 1 \leq x \leq 6$$

Notice that for each number x there corresponds exactly one number y. For example, if $x = 1$, then $y = 2(1) - 5 = -3$. If $x = 3$, then $y = 2(3) - 5 = 1$. For this reason, the equation is a function. Since we restrict the numbers x to the real numbers between 1 and 6, inclusive, the domain of the function is $\{x | 1 \leq x \leq 6\}$. The function specifies that in order to get the image of x we multiply x by 2 and then subtract 5 from this product. ▶

Find the value of a function **1** ### Function Notation

Functions are often denoted by letters such as f, F, g, G, and others. If f is a function, then for each number x in its domain the corresponding image in the range is designated by the symbol $f(x)$, read as "f of x" or as "f at x." We refer to $f(x)$ as the **value of f at the number x;** $f(x)$ is the number that results when x is given and the function f is applied; $f(x)$ does *not* mean "f times x." For example, the function given in Example 1 may be written as $y = f(x) = 2x - 5$, $1 \leq x \leq 6$. Then $f(1) = -3$.

Figure 16 illustrates some other functions. Notice that in every function illustrated, for each x in the domain there is one value in the range.

FIGURE 16

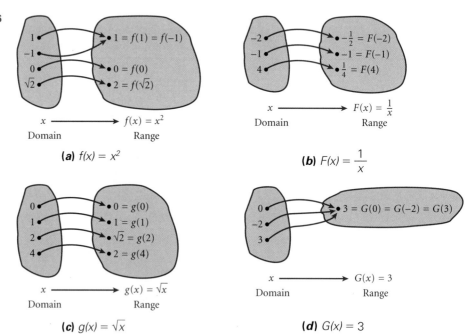

(a) $f(x) = x^2$

(b) $F(x) = \dfrac{1}{x}$

(c) $g(x) = \sqrt{x}$

(d) $G(x) = 3$

FIGURE 17

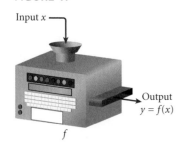

Sometimes it is helpful to think of a function f as a machine that receives as input a number from the domain, manipulates it, and outputs the value. See Figure 17.

The restrictions on this input/output machine are as follows:

1. It only accepts numbers from the domain of the function.
2. For each input, there is exactly one output (which may be repeated for different inputs).

For a function $y = f(x)$, the variable x is called the **independent variable,** because it can be assigned any of the permissible numbers from the domain. The variable y is called the **dependent variable,** because its value depends on x.

Any symbol can be used to represent the independent and dependent variables. For example, if f is the *cube function*, then f can be given by $f(x) = x^3$ or $f(t) = t^3$ or $f(z) = z^3$. All three functions are the same. Each tells us to cube the independent variable. In practice, the symbols used for the independent and dependent variables are based on common usage, such as using C for cost in business.

The independent variable is also called the **argument** of the function. Thinking of the independent variable as an argument can sometimes make it easier to find the value of a function. For example, if f is the function defined by $f(x) = x^3$, then f tells us to cube the argument. For example, $f(2)$ means to cube 2, $f(a)$ means to cube the number a, and $f(x + h)$ means to cube the quantity $x + h$.

EXAMPLE 2   Finding Values of a Function

For the function f defined by $f(x) = 2x^2 - 3x$, evaluate

(a) $f(3)$ (b) $f(x) + f(3)$ (c) $f(-x)$

(d) $-f(x)$ (e) $f(x + 3)$ (f) $\dfrac{f(x + h) - f(x)}{h}$, $h \neq 0$

SOLUTION (a) We substitute 3 for x in the equation for f to get
$$f(3) = 2(3)^2 - 3(3) = 18 - 9 = 9$$

(b) $f(x) + f(3) = (2x^2 - 3x) + 9 = 2x^2 - 3x + 9$

(c) We substitute $-x$ for x in the equation for f.
$$f(-x) = 2(-x)^2 - 3(-x) = 2x^2 + 3x$$

(d) $-f(x) = -(2x^2 - 3x) = -2x^2 + 3x$

(e) We substitute $x + 3$ for x in the equation for f.
$$\begin{aligned} f(x + 3) &= 2(x + 3)^2 - 3(x + 3) \quad \text{Notice the use of parentheses here.} \\ &= 2(x^2 + 6x + 9) - 3x - 9 \\ &= 2x^2 + 12x + 18 - 3x - 9 \\ &= 2x^2 + 9x + 9 \end{aligned}$$

(f) $\dfrac{f(x + h) - f(x)}{h} = \dfrac{[2(x + h)^2 - 3(x + h)] - [2x^2 - 3x]}{h}$
$\qquad\qquad\qquad\qquad\quad \uparrow$
$\qquad\qquad\qquad f(x + h) = 2(x + h)^2 - 3(x + h)$

$= \dfrac{2(x^2 + 2xh + h^2) - 3x - 3h - 2x^2 + 3x}{h}$ Simplify.

$= \dfrac{2x^2 + 4xh + 2h^2 - 3h - 2x^2}{h}$

$= \dfrac{4xh + 2h^2 - 3h}{h}$

$= \dfrac{h(4x + 2h - 3)}{h}$ Factor out h.

$= 4x + 2h - 3$ Cancel h.

Notice in this example that $f(x + 3) \neq f(x) + f(3)$ and $f(-x) \neq -f(x)$.

NOW WORK PROBLEM 1.

112 Chapter 1 Functions and Their Graphs

The expression in part (f) of Example 2 is called the *difference quotient* of f, an important expression in calculus.

> **Difference Quotient**
>
> The **difference quotient** of a function $y = f(x)$ is:
>
> $$\frac{f(x+h) - f(x)}{h} \quad h \neq 0$$

EXAMPLE 3 Finding Difference Quotients

Find the difference quotient of

(a) $f(x) = b$ (b) $f(x) = mx + b$ (c) $f(x) = x^2$

where m and b are constants.

SOLUTION (a) If $f(x) = b$, then $f(x + h) = b$ and the difference quotient of f is

$$\frac{f(x+h) - f(x)}{h} = \frac{b - b}{h} = \frac{0}{h} = 0$$

(b) If $f(x) = mx + b$, then $f(x + h) = m(x + h) + b = mx + mh + b$ and the difference quotient of f is

$$\frac{f(x+h) - f(x)}{h} = \frac{mx + mh + b - (mx + b)}{h} = \frac{mh}{h} = m$$

(c) If $f(x) = x^2$, then $f(x + h) = (x + h)^2 = x^2 + 2xh + h^2$ and the difference quotient of f is

$$\frac{f(x+h) - f(x)}{h} = \frac{x^2 + 2xh + h^2 - x^2}{h} = \frac{(2x + h)h}{h} = 2x + h \quad \blacktriangleright$$

NOW WORK PROBLEM 9.

Most calculators have special keys that enable you to find the value of certain commonly used functions. For example, you should be able to find the square function $f(x) = x^2$, the square root function $f(x) = \sqrt{x}$, the reciprocal function $f(x) = \frac{1}{x} = x^{-1}$, and many others that will be discussed later in the book (such as ln x and log x). Verify the results of Example 6, which follows, on your calculator.

EXAMPLE 4 Finding Values of a Function on a Calculator

(a) $f(x) = x^2$; $f(1.234) = 1.522756$

(b) $f(x) = \dfrac{1}{x}$; $F(1.234) = 0.8103727715$

(c) $g(x) = \sqrt{x}$; $g(1.234) = 1.110855526$

Functions 113

 COMMENT: Graphing calculators can be used to evaluate any function that you wish. Figure 18 shows the result obtained in Example 2(a) on a TI-83 graphing calculator with the function to be evaluated, $f(x) = 2x^2 - 3x$, in Y_1.*

FIGURE 18

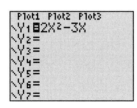 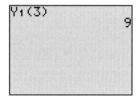

Implicit Form of a Function

In general, when a function f is defined by an equation in x and y, we say that the function f is given **implicitly**. If it is possible to solve the equation for y in terms of x, then we write $y = f(x)$ and say that the function is given **explicitly**. For example,

Implicit Form	*Explicit Form*
$3x + y = 5$	$y = f(x) = -3x + 5$
$x^2 - y = 6$	$y = f(x) = x^2 - 6$
$xy = 4$	$y = f(x) = \dfrac{4}{x}$

Not all equations in x and y define a function $y = f(x)$. If an equation is solved for y and two or more values of y can be obtained for a given x, then the equation does not define a function.

EXAMPLE 5   Determining Whether an Equation Is a Function

Determine if the equation $x^2 + y^2 = 1$ is a function.

SOLUTION   To determine whether the equation $x^2 + y^2 = 1$ is a function, we need to solve the equation for y.

$$x^2 + y^2 = 1$$
$$y^2 = 1 - x^2$$
$$y = \pm\sqrt{1 - x^2} \quad \text{Square Root Method.}$$

For values of x between -1 and 1, two values of y result. This means that the equation $x^2 + y^2 = 1$ does not define a function.

NOW WORK PROBLEM 23.

 COMMENT: The explicit form of a function is the form required by a graphing calculator. Now do you see why it is necessary to graph some equations in two "pieces"?

We list next a summary of some important facts to remember about a function f.

---

*Consult your owner's manual for the required keystrokes.

SUMMARY IMPORTANT FACTS ABOUT FUNCTIONS

(a) To each x in the domain of f, there is exactly one image $f(x)$ in the range; however, a number in the range can result from more than one x in the domain.
(b) f is the symbol that we use to denote the function. It is symbolic of the equation that we use to get from an x in the domain to $f(x)$ in the range.
(c) If $y = f(x)$, then the function f is given explicitly; x is called the independent variable, or argument, of f and y is called the dependent variable or the value of f at x.

Find the domain of a function

3 Domain of a Function

Often the domain of a function f is not specified; instead, only the equation defining the function is given. In such cases, we agree that the domain of f is the largest set of real numbers for which the value $f(x)$ is a real number. The domain of a function f is the same as the domain of the variable x in the expression $f(x)$.

EXAMPLE 6 **Finding the Domain of a Function**

Find the domain of each of the following functions:

(a) $f(x) = x^2 + 5x$ (b) $g(x) = \dfrac{3x}{x^2 - 4}$ (c) $h(t) = \sqrt{4 - 3t}$

SOLUTION (a) The function f tells us to square a number and then add five times the number. Since these operations can be performed on any real number, we conclude that the domain of f is all real numbers.

(b) The function g tells us to divide $3x$ by $x^2 - 4$. Since division by 0 is not defined, the denominator $x^2 - 4$ can never be 0, so x can never equal -2 or 2. The domain of the function g is $\{x \mid x \neq -2, x \neq 2\}$.

(c) The function h tells us to take the square root of $4 - 3t$. But only nonnegative numbers have real square roots, so the expression under the square root must be nonnegative. This requires that

$$4 - 3t \geq 0$$
$$-3t \geq -4$$
$$t \leq \frac{4}{3}$$

The domain of h is $\left\{t \mid t \leq \dfrac{4}{3}\right\}$ or the interval $\left(-\infty, \dfrac{4}{3}\right]$.

NOW WORK PROBLEM 33.

If x is in the domain of a function f, we shall say that **f is defined at x,** or $f(x)$ **exists.** If x is not in the domain of f, we say that **f is not defined at x,** or $f(x)$ **does not exist.**

For example, if $f(x) = \dfrac{x}{x^2 - 1}$, then $f(0)$ exists, but $f(1)$ and $f(-1)$ do not exist. (Do you see why?)

We have not said much about finding the range of a function. The reason is that when a function is defined by an equation it is often difficult to find the range. Therefore, we shall usually be content to find just the domain of a function when only

Solve applied problems involving functions

Applications

When we use functions in applications, the domain may be restricted by physical or geometric considerations. For example, the domain of the function f defined by $f(x) = x^2$ is the set of all real numbers. However, if f is used to obtain the area of a square when the length x of a side is known, then we must restrict the domain of f to the positive real numbers, since the length of a side can never be 0 or negative.

EXAMPLE 7 Constructing a Cost Function

The cost per square foot to build a house is $110.

(a) Express the cost C as a function of x, the number of square feet.
(b) What is the cost to build a 2000-square-foot house?

SOLUTION (a) The cost C of building a house containing x square feet is $110x$ dollars. A function expressing this relationship is

$$C(x) = 110x$$

where x is the independent variable and C is the dependent variable. In this setting the domain is $\{x|x > 0\}$ since a house cannot have 0 or negative square feet.

(b) The cost to build a 2000-square-foot house is

$$C(2000) = 110(2000) = \$220{,}000$$

NOW WORK PROBLEM 49.

Observe in the solution to Example 7 that we used the symbol C in two ways: it is used to name the function, and it is used to symbolize the dependent variable. This double use is common in applications and should not cause any difficulty.

EXAMPLE 8 Determining the Cost of Removing Pollutants

The cost of eliminating a large part of the pollutants from the atmosphere (or from water) is relatively cheap. However, removing the last traces of pollutants results in a significant increase in cost. A typical relationship between the cost C, in millions of dollars, for removal and the percent x of pollutant removed is given by the function

$$C(x) = \frac{3x}{105 - x}$$

Since x is a percentage, the domain of C consists of all real numbers x for which $0 \leq x \leq 100$. The cost of removing 0% of the pollutant is

$$C(0) = 0$$

The cost of removing 50% of the pollutant is

$$C(50) = \frac{150}{55} = 2.727 \text{ million dollars}$$

The costs of removing 60% and 70% are

$$C(60) = \frac{180}{45} = 4 \text{ million dollars}$$

and

$$C(70) = \frac{210}{35} = 6 \text{ million dollars}$$

Observe that the cost of removing an additional 10% of the pollutant after 50% had been removed is $1,273,000, while the cost of removing an additional 10% after 60% is removed is $2,000,000.

REVENUE, COST, AND PROFIT FUNCTIONS **Revenue** is the amount of money derived from the sale of a product and equals the price of the product times the quantity of the product that is actually sold. But the price and the quantity sold are not independent. As the price falls, the demand for the product increases; and when the price rises, the demand decreases.

The equation that relates the price p of a quantity bought and the amount x of a quantity demanded is called the **demand equation.** If in this equation we solve for p, we have

$$p = d(x)$$

The function d is called the **price function** and $d(x)$ is the price per unit when x units are demanded. If x is the number of units sold and $d(x)$ is the price for each unit, the **revenue function** $R(x)$ is defined as

$$R(x) = xp = x\, d(x)$$

If we denote the **cost function** by $C(x)$, then the **profit function** $P(x)$ is defined as

$$P(x) = R(x) - C(x)$$

EXAMPLE 9 Constructing a Revenue Function

No matter how much wheat a farmer can grow, it can be sold at $4 per bushel. Find the price function. What is the revenue function?

SOLUTION Since the price per bushel is fixed at $4 per bushel, the price function is

$$p = \$4$$

The revenue function is

$$R(x) = xp = 4x$$

EXAMPLE 10 Constructing a Revenue Function

The manager of a toy store has observed that each week 1000 toy trucks are sold at a price of $5 per truck. When there is a special sale, the trucks sell for $4 each and 1200 per week are sold. Assuming a linear price function, construct the price function. What is the revenue function?

SOLUTION Let p be the price of each truck and let x be the number sold. If the price function $p = d(x)$ is linear, then we know that $(x_1, p_1) = (1000, 5)$ and $(x_2, p_2) = (1200, 4)$ are two points on the line $p = d(x)$. The slope of the line is

$$\frac{p_2 - p_1}{x_2 - x_1} = \frac{4 - 5}{1200 - 1000} = \frac{-1}{200} = -\frac{1}{200}$$

Use the point–slope form of the equation of a line:

$$p - p_1 = m(x - x_1) \qquad \text{Point–slope form.}$$

$$p - 5 = -\frac{1}{200}(x - 1000) \qquad m = -\frac{1}{200}; p_1 = 5; x_1 = 1000$$

$$p = -\frac{1}{200}x + 10$$

The price function is

$$p = d(x) = -\frac{1}{200}x + 10$$

The revenue function is

$$R(x) = xp = x\left(-\frac{1}{200}x + 10\right) = -\frac{1}{200}x^2 + 10x \qquad \blacktriangleright$$

The price function obtained in Example 10 is not meant to reflect extreme situations. For example, we do not expect to sell $x = 0$ trucks nor do we expect to sell too many trucks in excess of 1500, since even during a special sale only 1200 are sold. The price function does represent the relationship between price and quantity in a certain range—in this case, perhaps $500 < x < 1500$.

NOW WORK PROBLEM 51.

SUMMARY We list here some of the important vocabulary introduced in this section, with a brief description of each term.

Function A relation between two sets of real numbers so that each number x in the first set, the domain, has corresponding to it exactly one number y in the second set.
The range is the set of y values of the function for the x values in the domain.
A function f may be defined implicitly by an equation involving x and y or explicitly by writing $y = f(x)$.

Unspecified domain If a function f is defined by an equation and no domain is specified, then the domain will be taken to be the largest set of real numbers for which the equation defines a real number.

Function notation $y = f(x)$
f is a symbol for the function.
x is the independent variable or argument.
y is the dependent variable.
$f(x)$ is the value of the function at x, or the image of x.

EXERCISE 1.2

In Problems 1–8, find the following values for each function:

(a) $f(0)$ (b) $f(1)$ (c) $f(-1)$ (d) $f(-x)$ (e) $-f(x)$ (f) $f(x+1)$ (g) $f(2x)$ (h) $f(x+h)$

1. $f(x) = 3x^2 + 2x - 4$
2. $f(x) = -2x^2 + x - 1$
3. $f(x) = \dfrac{x}{x^2 + 1}$
4. $f(x) = \dfrac{x^2 - 1}{x + 4}$

5. $f(x) = |x| + 4$
6. $f(x) = \sqrt{x^2 + x}$
7. $f(x) = \dfrac{2x + 1}{3x - 5}$
8. $f(x) = 1 - \dfrac{1}{(x + 2)^2}$

In Problems 9–16, find the difference quotient of f, that is, find $\dfrac{f(x+h) - f(x)}{h}$, $h \neq 0$, *for each function. Be sure to simplify.*

9. $f(x) = 4x + 3$
10. $f(x) = -3x + 1$
11. $f(x) = x^2 - x + 4$
12. $f(x) = x^2 + 5x - 1$

13. $f(x) = x^3$
14. $f(x) = x^3 - 2x$
15. $f(x) = x^4$
16. $f(x) = \dfrac{1}{x}$

In Problems 17–28, determine whether the equation is a function.

17. $y = x^2$
18. $y = x^3$
19. $y = \dfrac{1}{x}$
20. $y = |x|$

21. $y^2 = 4 - x^2$
22. $y = \pm\sqrt{1 - 2x}$
23. $x = y^2$
24. $x + y^2 = 1$

25. $y = 2x^2 - 3x + 4$
26. $y = \dfrac{3x - 1}{x + 2}$
27. $2x^2 + 3y^2 = 1$
28. $x^2 - 4y^2 = 1$

In Problems 29–42, find the domain of each function.

29. $f(x) = -5x + 4$
30. $f(x) = x^2 + 2$
31. $f(x) = \dfrac{x}{x^2 + 1}$
32. $f(x) = \dfrac{x^2}{x^2 + 1}$

33. $g(x) = \dfrac{x}{x^2 - 16}$
34. $h(x) = \dfrac{2x}{x^2 - 4}$
35. $F(x) = \dfrac{x - 2}{x^3 + x}$
36. $G(x) = \dfrac{x + 4}{x^3 - 4x}$

37. $h(x) = \sqrt{3x - 12}$
38. $G(x) = \sqrt{1 - x}$
39. $f(x) = \dfrac{4}{\sqrt{x - 9}}$
40. $f(x) = \dfrac{x}{\sqrt{x - 4}}$

41. $p(x) = \sqrt{\dfrac{2}{x - 1}}$
42. $q(x) = \sqrt{-x - 2}$

---

43. If $f(x) = 2x^3 + Ax^2 + 4x - 5$ and $f(2) = 5$, what is the value of A?

44. If $f(x) = 3x^2 - Bx + 4$ and $f(-1) = 12$, what is the value of B?

45. If $f(x) = \dfrac{3x + 8}{2x - A}$ and $f(0) = 2$, what is the value of A? Where is f undefined?

46. If $f(x) = \dfrac{2x - B}{3x + 4}$ and $f(2) = \dfrac{1}{2}$, what is the value of B? Where is f undefined?

47. If $f(x) = \dfrac{2x - A}{x - 3}$ and $f(4) = 0$, what is the value of A? Where is f not defined?

48. If $f(x) = \dfrac{x - B}{x - A}$, $f(2) = 0$, and $f(1)$ is undefined, what are the values of A and B?

49. **Constructing Functions** Express the gross salary G of a person who earns $10 per hour as a function of the number x of hours worked.

50. Constructing Functions Tiffany, a commissioned salesperson, earns $100 base pay plus $10 per item sold. Express her gross salary G as a function of the number x of items sold.

51. Demand Equation The price p and the quantity x sold of a certain product obey the demand equation

$$p = -\frac{1}{5}x + 100 \qquad 0 \leq x \leq 500$$

Express the revenue $R = xp$ as a function of x.

52. Demand Equation The price p and the quantity x sold of a certain product obey the demand equation

$$p = -\frac{1}{4}x + 100 \qquad 0 \leq x \leq 400$$

Express the revenue $R = xp$ as a function of x.

53. Demand Equation The price p and the quantity x sold of a certain product obey the demand equation

$$x = -20p + 100 \qquad 0 \leq p \leq 5$$

Express the revenue $R = xp$ as a function of x.

54. Demand Equation The price p and the quantity x sold of a certain product obey the demand equation

$$x = -5p + 500 \qquad 0 \leq p \leq 100$$

Express the revenue $R = xp$ as a function of x.

55. Wheat Production The amount of wheat planted annually in the United States is given in the table.

Year	1990	1991	1992	1993	1994	1995
Thousand Acres	77,041	69,881	72,219	72,168	70,349	69,031

Year	1996	1997	1998	1999	2000
Thousand Acres	75,105	70,412	65,821	62,714	62,629

If 1990 is taken as year 0, the number (in thousands) of acres of wheat planted in the United States can be approximated by $A(t) = -119t^2 + 113t + 73{,}367$. If this function remains valid, project the number (in thousands) of acres of wheat that will be planted in 2010.

Source: U.S. Department of Agriculture.

56. Wheat Production Suppose the number (in thousands) of acres of wheat in 2010 is 28,027, which is consistent with the current trend. If, at that time, there is a movement to increase the number of acres of wheat planted annually, and the equation selected to achieve this is $A(t) = 28{,}027 + 200\sqrt{t}$, how much wheat would be planted in 2020? Assume $t = 0$ for the year 2010.

57. SAT Scores The data for the mathematics scores on the SAT can be approximated by $S(t) = -0.04t^3 + 0.43t^2 + 0.24t + 506$, where t is the number of years since 1994. If the trend continues, what would be the expected score in 2010?

58. Page Design A page with dimensions of 11 inches by 7 inches has a border of uniform width x surrounding the printed matter of the page, as shown in the figure. Write a formula for the area A of the printed part of the page as a function of the width x of the border. Give the domain and range of A.

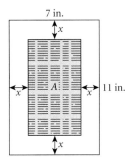

59. Cost of Flying An airplane crosses the Atlantic Ocean (3000 miles) with an airspeed of 500 miles per hour. The cost C (in dollars) per passenger is

$$C(x) = 100 + \frac{x}{10} + \frac{36{,}000}{x}$$

where x is the ground speed (airspeed ± wind).

(a) What is the cost per passenger for quiescent (no-wind) conditions?

(b) What is the cost per passenger with a head wind of 50 miles per hour?

(c) What is the cost per passenger with a tail wind of 100 miles per hour?

(d) What is the cost per passenger with a head wind of 100 miles per hour?

60. Cable Installation A cable TV company is asked to provide service to a customer whose house is located 2 miles from the road along which the cable is buried. The nearest connection box for the cable is located 5 miles down the road.

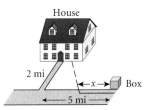

(a) If the installation cost is $100 per mile along the road and $140 per mile off the road, express the total cost C of installation as a function of the distance x (in miles) from the connection box to the point where the cable installation turns off the road.

(b) What is the domain of C?

(c) Compute the cost for $x = 1$, $x = 2$, $x = 3$, and $x = 4$.

61. Some functions f have the property that $f(a + b) = f(a) + f(b)$ for all real numbers a and b. Which of the following functions have this property?
(a) $h(x) = 2x$
(b) $g(x) = x^2$
(c) $F(x) = 5x - 2$
(d) $G(x) = \dfrac{1}{x}$

62. Are the functions $f(x) = x - 1$ and $g(x) = \dfrac{x^2 - 1}{x + 1}$ the same? Explain.

63. Investigate when, historically, the use of the function notation $y = f(x)$ first appeared.

1.3 Graphs of Functions; Properties of Functions

PREPARING FOR THIS SECTION *Before getting started, review the following:*

> Rectangular Coordinates (Chapter 0, Section 0.8, pp. 76–77)
> Intervals (Chapter 0, Section 0.5, pp. 51–53)
> Slope of a Line (Chapter 0, Section 0.9, pp. 85–87)
> Point–Slope Equation of a Line (Chapter 0, Section 0.9, pp. 87–89)

OBJECTIVES
1. Identify the graph of a function
2. Obtain information from or about the graph of a function
3. Determine even and odd functions from a graph
4. Identify even and odd functions from the equation
5. Use a graph to determine where a function is increasing, is decreasing, or is constant
6. Use a graph to locate local maxima and minima
7. Use a graphing utility to approximate local maxima and minima and to determine where a function is increasing or decreasing
8. Find the average rate of change of a function

TABLE 6

Date	Closing Price ($)
8/31/02	15.86
9/30/02	13.89
10/31/02	17.30
11/30/02	20.88
12/31/02	15.57
1/31/03	15.70
2/28/03	17.26
3/31/03	16.28
4/30/03	18.37
5/31/03	20.82
6/30/03	20.81
7/31/03	24.89
8/31/03	28.59

In applications, a graph often demonstrates more clearly the relationship between two variables than, say, an equation or table would. For example, Table 6 shows the price per share of Intel stock at the end of each month from 8/31/02 through 8/31/03. If we plot these data using the date as the x-coordinate and the price as the y-coordinate and then connect the points, we obtain Figure 19.

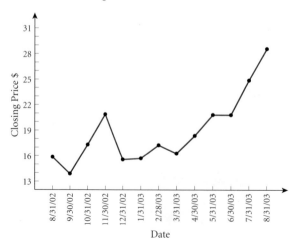

FIGURE 19 Monthly closing prices of Intel stock 8/31/02 through 8/31/03.

We can see from the graph that the price of the stock was rising rapidly from 9/30/02 through 11/30/02 and was falling slightly from 8/31/02 through 9/30/02. The graph also shows that the lowest price occurred at the end of September, 2002 whereas the highest occurred at the end of August, 2003. Equations and tables, on the other hand, usually require some calculations and interpretation before this kind of information can be "seen."

Look again at Figure 19. The graph shows that for each date on the horizontal axis there is only one price on the vertical axis. Thus, the graph represents a function, although the exact rule for getting from date to price is not given.

When a function is defined by an equation in x and y, the **graph of the function** is the graph of the equation, that is, the set of points (x, y) in the xy-plane that satisfies the equation.

For example, the graph of the function $f(x) = mx + b$ is a line with slope m and y-intercept $(0, b)$. Because of this, functions of the form $f(x) = mx + b$ are called **linear functions**.

COMMENT: When we select a viewing rectangle to graph a function, the values of Xmin, Xmax give the domain that we wish to view, while Ymin, Ymax give the range that we wish to view. These settings usually do not represent the actual domain and range of the function.

Identify the graph of a function

Not every collection of points in the xy-plane represents the graph of a function. Remember, for a function, each number x in the domain has exactly one image y in the range. This means that the graph of a function cannot contain two points with the same x-coordinate and different y-coordinates. Therefore, the graph of a function must satisfy the following **vertical-line test**.

Vertical-Line Test

A set of points in the xy-plane is the graph of a function if and only if every vertical line intersects the graph in at most one point.

In other words, if any vertical line intersects a graph at more than one point, the graph is not the graph of a function.

EXAMPLE 1 **Identifying the Graph of a Function**

Which of the graphs in Figure 20 are graphs of functions?

FIGURE 20

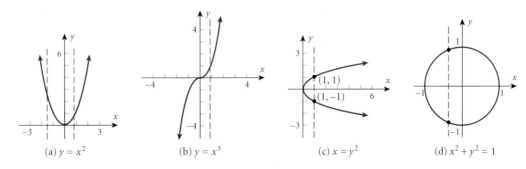

(a) $y = x^2$ (b) $y = x^3$ (c) $x = y^2$ (d) $x^2 + y^2 = 1$

SOLUTION The graphs in Figures 20(a) and 20(b) are graphs of functions, because every vertical line intersects each graph in at most one point. The graphs in Figures 20(c) and 20(d) are not graphs of functions, because there is a vertical line that intersects each graph in more than one point.

NOW WORK PROBLEM 5.

Obtain information from or about the graph of a function

2 If (x, y) is a point on the graph of a function f, then y is the value of f at x; that is, $y = f(x)$. The next example illustrates how to obtain information about a function if its graph is given.

FIGURE 21

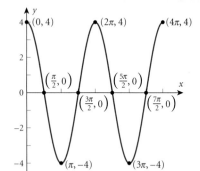

EXAMPLE 2 Obtaining Information from the Graph of a Function

Let f be the function whose graph is given in Figure 21.

(a) What is $f(0)$, $f\left(\dfrac{3\pi}{2}\right)$, and $f(3\pi)$?

(b) What is the domain of f?

(c) What is the range of f?

(d) List the intercepts. (Recall that these are the points, if any, where the graph crosses or touches the coordinate axes.)

(e) How often does the line $y = 2$ intersect the graph?

(f) For what values of x does $f(x) = -4$?

(g) For what values of x is $f(x) > 0$?

SOLUTION (a) Since $(0, 4)$ is on the graph of f, the y-coordinate 4 is the value of f at the x-coordinate 0; that is, $f(0) = 4$. In a similar way, we find that when $x = \dfrac{3\pi}{2}$ then $y = 0$, so $f\left(\dfrac{3\pi}{2}\right) = 0$. When $x = 3\pi$, then $y = -4$, so $f(3\pi) = -4$.

(b) To determine the domain of f, we notice that the points on the graph of f will have x-coordinates between 0 and 4π, inclusive; and for each number x between 0 and 4π there is a point $(x, f(x))$ on the graph. The domain of f is $\{x | 0 \le x \le 4\pi\}$ or the interval $[0, 4\pi]$.

(c) The points on the graph all have y-coordinates between -4 and 4, inclusive; and for each such number y there is at least one number x in the domain. The range of f is $\{y | -4 \le y \le 4\}$ or the interval $[-4, 4]$.

(d) The intercepts are $(0, 4)$, $\left(\dfrac{\pi}{2}, 0\right)$, $\left(\dfrac{3\pi}{2}, 0\right)$, $\left(\dfrac{5\pi}{2}, 0\right)$, and $\left(\dfrac{7\pi}{2}, 0\right)$.

(e) Draw the horizontal line $y = 2$ on the graph in Figure 21. Then we find that it intersects the graph four times.

(f) Since $(\pi, -4)$ and $(3\pi, -4)$ are the only points on the graph for which $y = f(x) = -4$, we have $f(x) = -4$ when $x = \pi$ and $x = 3\pi$.

(g) To determine where $f(x) > 0$, we look at Figure 21 and determine the x-values for which the y-coordinate is positive. This occurs on the intervals $\left[0, \dfrac{\pi}{2}\right)$, $\left(\dfrac{3\pi}{2}, \dfrac{5\pi}{2}\right)$, and $\left(\dfrac{7\pi}{2}, 4\pi\right]$. Using inequality notation, $f(x) > 0$ for $0 \le x < \dfrac{\pi}{2}$, $\dfrac{3\pi}{2} < x < \dfrac{5\pi}{2}$, and $\dfrac{7\pi}{2} < x \le 4\pi$.

Graphs of Functions; Properties of Functions 123

When the graph of a function is given, its domain may be viewed as the shadow created by the graph on the x-axis by vertical beams of light. Its range can be viewed as the shadow created by the graph on the y-axis by horizontal beams of light. Try this technique with the graph given in Figure 21.

NOW WORK PROBLEMS 3, 13, and 31(a) and (b).

EXAMPLE 3   Obtaining Information about the Graph of a Function

Consider the function: $f(x) = \dfrac{x}{x + 2}$

(a) Is the point $\left(1, \dfrac{1}{2}\right)$ on the graph of f?

(b) If $x = -1$, what is $f(-1)$? What point is on the graph of f?

(c) If $f(x) = 2$, what is x? What point is on the graph of f?

SOLUTION (a) When $x = 1$, then

$$f(x) = \dfrac{x}{x + 2}$$

$$f(1) = \dfrac{1}{1 + 2} = \dfrac{1}{3}$$

The point $\left(1, \dfrac{1}{3}\right)$ is on the graph of f; the point $\left(1, \dfrac{1}{2}\right)$ is not.

(b) If $x = -1$, then

$$f(x) = \dfrac{x}{x + 2}$$

$$f(-1) = \dfrac{-1}{-1 + 2} = -1$$

The point $(-1, -1)$ is on the graph of f.

(c) If $f(x) = 2$, then

$$f(x) = 2$$

$$\dfrac{x}{x + 2} = 2$$

$x = 2(x + 2)$ Multiply both sides by $x + 2$.

$x = 2x + 4$ Remove parentheses.

$x = -4$ Solve for x.

If $f(x) = 2$, then $x = -4$. The point $(-4, 2)$ is on the graph of f.

NOW WORK PROBLEM 17.

EXAMPLE 4   Average Cost Function

The average cost $\overline{C}$ of manufacturing x computers per day is given by the function

$$\overline{C}(x) = 0.56x^2 - 34.39x + 1212.57 + \dfrac{20{,}000}{x}$$

Determine the average cost of manufacturing the following:

(a) 30 computers in a day.
(b) 40 computers in a day.
(c) 50 computers in a day.
(d) Graph the function $\overline{C} = \overline{C}(x), 0 < x \leq 80$.
(e) Create a TABLE with TblStart = 1 and ΔTbl = 1.* Which value of x minimizes the average cost?

SOLUTION (a) The average cost of manufacturing $x = 30$ computers is

$$\overline{C}(30) = 0.56(30)^2 - 34.39(30) + 1212.57 + \frac{20{,}000}{30} = \$1351.54$$

(b) The average cost of manufacturing $x = 40$ computers is

$$\overline{C}(40) = 0.56(40)^2 - 34.39(40) + 1212.57 + \frac{20{,}000}{40} = \$1232.97$$

(c) The average cost of manufacturing $x = 50$ computers is

$$\overline{C}(50) = 0.56(50)^2 - 34.39(50) + 1212.57 + \frac{20{,}000}{50} = \$1293.07$$

(d) See Figure 22 for the graph of $\overline{C} = \overline{C}(x)$.

FIGURE 22

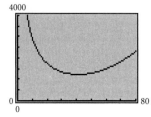

(e) With the function $\overline{C} = \overline{C}(x)$ in Y_1, we create Table 7. We scroll down until we find a value of x for which Y_1 is smallest. Table 8 shows that manufacturing $x = 41$ computers minimizes the average cost at \$1231.75 per computer.

TABLE 7

X	Y1
1	21179
2	11146
3	7781.1
4	6084
5	5054.6
6	4359.7
7	3856.4

Y1■.56X²-34.39X...

TABLE 8

X	Y1
38	1240.7
39	1235.9
40	1233
41	1231.7
42	1232.2
43	1234.4
44	1238.1

Y1=1231.74487805

 NOW WORK PROBLEM 79.

It is easiest to obtain the graph of a function $y = f(x)$ by knowing certain properties that the function has and the impact of these properties on the way that the graph will look. We describe next some properties of functions that we will use in subsequent chapters.

We begin with intercepts and symmetry.

*Consult your user's manual for the keystrokes to use.

Intercepts

If $x = 0$ is in the domain of a function $y = f(x)$, then the y-intercept of the graph of f is obtained by finding the value of f at 0, which is $f(0)$. The x-intercepts of the graph of f, if there are any, are obtained by finding the solutions of the equation $f(x) = 0$.

The x-intercepts of the graph of a function f are called the **zeros of f**.

Even and Odd Functions

Determine even and odd functions from a graph 3

The words *even* and *odd*, when applied to a function f, describe the symmetry that exists for the graph of the function.

A function f is even if and only if whenever the point (x, y) is on the graph of f then the point $(-x, y)$ is also on the graph. Using function notation, we define an even function as follows:

> A function f is **even** if, for every number x in its domain, the number $-x$ is also in the domain and
> $$f(-x) = f(x)$$

A function f is odd if and only if whenever the point (x, y) is on the graph of f then the point $(-x, -y)$ is also on the graph. Using function notation, we define an odd function as follows:

> A function f is **odd** if, for every number x in its domain, the number $-x$ is also in the domain and
> $$f(-x) = -f(x)$$

Refer to Section 1.1, where the tests for symmetry are explained. The following results are then evident.

Theorem

A function is even if and only if its graph is symmetric with respect to the y-axis. A function is odd if and only if its graph is symmetric with respect to the origin.

EXAMPLE 5 Determining Even and Odd Functions from the Graph

Determine whether each graph given in Figure 23 is the graph of an even function, an odd function, or a function that is neither even nor odd.

FIGURE 23

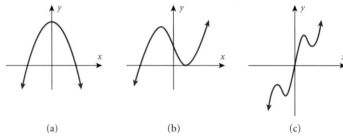

(a)   (b)   (c)

SOLUTION   The graph in Figure 23(a) is that of an even function, because the graph is symmetric with respect to the y-axis. The function whose graph is given in Figure 23(b) is neither even nor odd, because the graph is neither symmetric with respect to the y-axis nor symmetric with respect to the origin. The function whose graph is given in Figure 23(c) is odd, because its graph is symmetric with respect to the origin.

NOW WORK PROBLEM 31(d).

Identify even and odd functions from the equation   4   In the next example, we use algebraic techniques to verify whether a given function is even, odd, or neither.

EXAMPLE 6 Identifying Even and Odd Functions Algebraically

Determine whether each of the following functions is even, odd, or neither. Then determine whether the graph is symmetric with respect to the y-axis or with respect to the origin.

(a) $f(x) = x^2 - 5$
(b) $g(x) = x^3 - 1$
(c) $h(x) = 5x^3 - x$
(d) $F(x) = |x|$

SOLUTION   (a) To determine whether f is even, odd, or neither, we replace x by $-x$ in $f(x) = x^2 - 5$. Then

$$f(-x) = (-x)^2 - 5 = x^2 - 5 = f(x)$$

Since $f(-x) = f(x)$, we conclude that f is an even function, and the graph is symmetric with respect to the y-axis.

(b) We replace x by $-x$ in $g(x) = x^3 - 1$. Then

$$g(-x) = (-x)^3 - 1 = -x^3 - 1$$

Since $g(-x) \neq g(x)$ and $g(-x) \neq -g(x) = -(x^3 - 1) = -x^3 + 1$, we conclude that g is neither even nor odd. The graph is not symmetric with respect to the y-axis nor is it symmetric with respect to the origin.

(c) We replace x by $-x$ in $h(x) = 5x^3 - x$. Then

$$h(-x) = 5(-x)^3 - (-x) = -5x^3 + x = -(5x^3 - x) = -h(x)$$

Since $h(-x) = -h(x)$, h is an odd function, and the graph of h is symmetric with respect to the origin.

(d) We replace x by $-x$ in $F(x) = |x|$. Then

$$F(-x) = |-x| = |-1| \cdot |x| = |x| = F(x)$$

Since $F(-x) = F(x)$, F is an even function, and the graph of F is symmetric with respect to the y-axis.

NOW WORK PROBLEM 57.

Increasing and Decreasing Functions

Use a graph to determine where a function is increasing, is decreasing, or is constant

5 Consider the graph given in Figure 24. If you look from left to right along the graph of the function, you will notice that parts of the graph are rising, parts are falling, and parts are horizontal. In such cases, the function is described as *increasing*, *decreasing*, or *constant*, respectively.

FIGURE 24

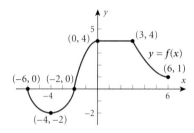

EXAMPLE 7 Determining Where a Function Is Increasing, Decreasing, or Constant from Its Graph

Where is the function in Figure 24 increasing? Where is it decreasing? Where is it constant?

SOLUTION To answer the question of where a function is increasing, where is it decreasing, and where it is constant, we use strict inequalities involving the independent variable x, or we use open intervals of x-coordinates. The graph in Figure 24 is rising (increasing) from the point $(-4, -2)$ to the point $(0, 4)$, so we conclude that it is increasing on the open interval $(-4, 0)$ or for $-4 < x < 0$. The graph is falling (decreasing) from the point $(-6, 0)$ to the point $(-4, -2)$ and from the point $(3, 4)$ to the point $(6, 1)$. We conclude that the graph is decreasing on the open intervals $(-6, -4)$ and $(3, 6)$ or for $-6 < x < -4$ and $3 < x < 6$. The graph is constant on the open interval $(0, 3)$ or for $0 < x < 3$.

More precise definitions follow:

A function f is **increasing** on an open interval I if, for any choice of x_1 and x_2 in I, with $x_1 < x_2$, we have $f(x_1) < f(x_2)$.

A function f is **decreasing** on an open interval I if, for any choice of x_1 and x_2 in I, with $x_1 < x_2$, we have $f(x_1) > f(x_2)$.

A function f is **constant** on an open interval I if, for all choices of x in I, the values $f(x)$ are equal.

Figure 25 illustrates the definitions. The graph of an increasing function goes up from left to right, the graph of a decreasing function goes down from left to right, and the graph of a constant function remains at a fixed height.

FIGURE 25

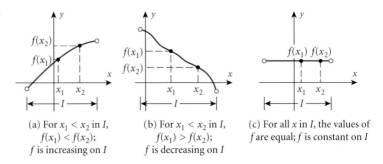

(a) For $x_1 < x_2$ in I, $f(x_1) < f(x_2)$; f is increasing on I

(b) For $x_1 < x_2$ in I, $f(x_1) > f(x_2)$; f is decreasing on I

(c) For all x in I, the values of f are equal; f is constant on I

In Chapter 5 we develop a method for determining where a function is increasing or decreasing using calculus.

NOW WORK PROBLEMS 21, 23, 25, and 31(c).

Local Maximum; Local Minimum

Use a graph to locate local maxima and minima

6 When the graph of a function is increasing to the left of $x = c$ and decreasing to the right of $x = c$, then at c the value of f is largest. This value is called a *local maximum* of f. See Figure 26(a).

FIGURE 26

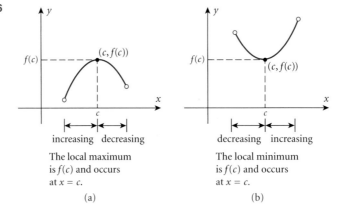

(a) The local maximum is $f(c)$ and occurs at $x = c$.

(b) The local minimum is $f(c)$ and occurs at $x = c$.

When the graph of a function is decreasing to the left of $x = c$ and is increasing to the right of $x = c$, then at c the value of f is the smallest. This value is called a *local minimum* of f. See Figure 26(b).

> A function f has a **local maximum at c** if there is an open interval I containing c so that, for all $x \neq c$ in I, $f(x) < f(c)$. We call $f(c)$ a **local maximum of f**.
>
> A function f has a **local minimum at c** if there is an open interval I containing c so that, for all $x \neq c$ in I, $f(x) > f(c)$. We call $f(c)$ a **local minimum of f**.

Graphs of Functions; Properties of Functions   129

If f has a local maximum at c, then the value of f at c is greater than the values of f near c. If f has a local minimum at c, then the value of f at c is less than the values of f near c. The word *local* is used to suggest that it is only near c that the value $f(c)$ is largest or smallest.

EXAMPLE 8   **Finding Local Maxima and Local Minima from the Graph of a Function and Determining Where the Function Is Increasing, Decreasing, or Constant**

FIGURE 27

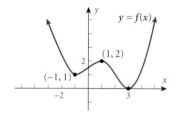

Figure 27 shows the graph of a function f.

(a) At what number(s), if any, does f have a local maximum?
(b) What are the local maxima?
(c) At what number(s), if any, does f have a local minimum?
(d) What are the local minima?
(e) List the intervals on which f is increasing.
(f) List the intervals on which f is decreasing.

SOLUTION (a) f has a local maximum at 1, since for all x close to 1, $x \neq 1$, we have $f(x) < f(1)$.
(b) The local maximum is $f(1) = 2$.
(c) f has local minima at -1 and at 3.
(d) The local minima are $f(-1) = 1$ and $f(3) = 0$.
(e) The function whose graph is given in Figure 27 is increasing on the interval $(-1, 1)$. It is also increasing for all values of x greater than 3. That is, the function is increasing on the intervals $(-1, 1)$ and $(3, \infty)$ or for $-1 < x < 1$ and $x > 3$.
(f) The function is decreasing for all values of x less than -1. It is also decreasing on the interval $(1, 3)$. That is, the function is decreasing on the intervals $(-\infty, -1)$ and $(1, 3)$ or for $x < -1$ and $1 < x < 3$.

  NOW WORK PROBLEMS 27 AND 29.

In Chapter 5, we use calculus to determine the local maxima and the local minima of a function.

A graphing utility may be used to approximate these values by using the MAXIMUM and MINIMUM features.*

  EXAMPLE 9   **Using a Graphing Utility to Approximate Local Maxima and Minima and to Determine Where a Function Is Increasing or Decreasing**

(a) Use a graphing utility to graph $f(x) = 6x^3 - 12x + 5$ for $-2 < x < 2$. Approximate where f has a local maximum and where f has a local minimum.
(b) Determine where f is increasing and where it is decreasing.

SOLUTION (a) Graphing utilities have a feature that finds the maximum or minimum point of a graph within a given interval. Graph the function f for $-2 < x < 2$. Using MAXIMUM, we find that the local maximum is 11.53 and it occurs at $x = -0.82$, rounded to two decimal places. See Figure 28(a). Using MINIMUM, we find that the local minimum is -1.53 and it occurs at $x = 0.82$, rounded to two decimal places. See Figure 28(b).

*Consult your owner's manual for the appropriate keystrokes.

FIGURE 28

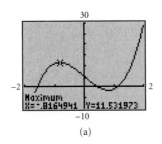
(a)

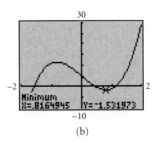
(b)

(b) Looking at Figures 28(a) and (b), we see that the graph of f is increasing from $x = -2$ to $x = -0.82$ and from $x = 0.82$ to $x = 2$, so f is increasing on the intervals $(-2, -0.82)$ and $(0.82, 2)$ or for $-2 < x < -0.82$ and $0.82 < x < 2$. The graph is decreasing from $x = -0.82$ to $x = 0.82$, so f is decreasing on the interval $(-0.82, 0.82)$ or for $-0.82 < x < 0.82$. ◀

  NOW WORK PROBLEM 69.

Average Rate of Change

Find the average rate of change of a function ⑧ Often we are interested in the rate at which functions change. To find the average rate of change of a function between any two points on its graph, we calculate the slope of the line containing the two points.

> If c is in the domain of a function $y = f(x)$, the **average rate of change of f** from c to x is defined as
>
> $$\text{Average rate of change} = \frac{\Delta y}{\Delta x} = \frac{f(x) - f(c)}{x - c}, \quad x \neq c \qquad (1)$$

Recall that the symbol Δy in (1) is the "change in y," and Δx is the "change in x." The average rate of change of f is the change in y divided by the change in x.

EXAMPLE 10 **Finding the Average Rate of Change**

Find the average rate of change of $f(x) = 3x^2$:

(a) From 1 to 3 (b) From 1 to 5 (c) From 1 to 7

SOLUTION (a) The average rate of change of $f(x) = 3x^2$ from 1 to 3 is

$$\frac{\Delta y}{\Delta x} = \frac{f(3) - f(1)}{3 - 1} = \frac{27 - 3}{3 - 1} = \frac{24}{2} = 12$$

(b) The average rate of change of $f(x) = 3x^2$ from 1 to 5 is

$$\frac{\Delta y}{\Delta x} = \frac{f(5) - f(1)}{5 - 1} = \frac{75 - 3}{5 - 1} = \frac{72}{4} = 18$$

(c) The average rate of change of $f(x) = 3x^2$ from 1 to 7 is

$$\frac{\Delta y}{\Delta x} = \frac{f(7) - f(1)}{7 - 1} = \frac{147 - 3}{7 - 1} = \frac{144}{6} = 24$$ ▶

The average rate of change of a function has an important geometric interpretation. Look at the graph of $y = f(x)$ in Figure 29. We have labeled two points on the graph: $(c, f(c))$ and $(x, f(x))$. The line containing these two points is called a **secant line;** its slope is

$$m_{\sec} = \frac{f(x) - f(c)}{x - c}$$

FIGURE 29

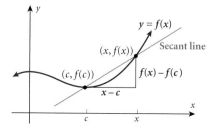

Slope of the Secant Line

The average rate of change of a function equals the slope of the secant line containing two points on its graph.

EXAMPLE 11   **Finding the Average Rate of Change of a Function**

(a) Find the average rate of change of $f(x) = 2x^2 - 3x$ from 1 to x.
(b) Use this result to find the slope of the secant line containing $(1, f(1))$ and $(2, f(2))$.
(c) Find an equation of this secant line.

SOLUTION (a) The average rate of change of f from 1 to x is

$$\frac{\Delta y}{\Delta x} = \frac{f(x) - f(1)}{x - 1} \qquad x \neq 1$$

$$= \frac{2x^2 - 3x - (-1)}{x - 1} \qquad f(x) = 2x^2 - 3x; f(1) = 2 \cdot 1^2 - 3(1) = -1$$

$$= \frac{2x^2 - 3x + 1}{x - 1} \qquad \text{Simplify.}$$

$$= \frac{(2x - 1)(x - 1)}{x - 1} \qquad \text{Factor the numerator.}$$

$$= 2x - 1 \qquad x \neq 1; \text{cancel } x - 1.$$

(b) The slope of the secant line containing $(1, f(1))$ and $(2, f(2))$ is the average rate of change of f from 1 to 2. Using $x = 2$ in part (a), we obtain $m_{\sec} = 2(2) - 1 = 3$.

(c) Use the point–slope form to find the equation of the secant line.

$$y - y_1 = m_{\sec}(x - x_1) \qquad \text{Point–slope form of the secant line.}$$
$$y + 1 = 3(x - 1) \qquad x_1 = 1, y_1 = f(1) = -1; m_{\sec} = 3$$
$$y + 1 = 3x - 3$$
$$y = 3x - 4 \qquad \text{Slope–intercept form of the secant line.}$$

 NOW WORK PROBLEM 49.

EXERCISE 1.3

In Problems 1–12, determine whether the graph is that of a function by using the vertical-line test. If it is, use the graph to find:

(a) Its domain and range; (b) The intercepts, if any; (c) Any symmetry with respect to the x-axis, the y-axis, or the origin.

1.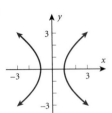
2.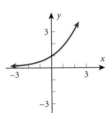
3.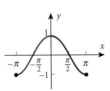
4.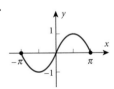
5.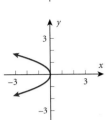
6.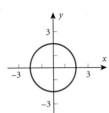
7.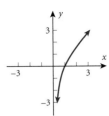
8.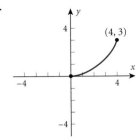
9.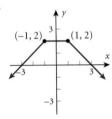
10.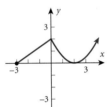
11.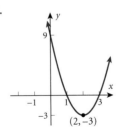
12.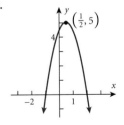

13. Use the graph of the function f given below to answer parts (a)–(n).

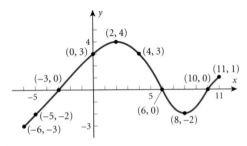

(a) Find $f(0)$ and $f(-6)$.
(b) Find $f(6)$ and $f(11)$.
(c) Is $f(3)$ positive or negative?
(d) Is $f(-4)$ positive or negative?
(e) For what numbers x is $f(x) = 0$?
(f) For what numbers x is $f(x) > 0$?
(g) What is the domain of f?
(h) What is the range of f?
(i) What are the x-intercepts?
(j) What is the y-intercept?
(k) How often does the line $y = \dfrac{1}{2}$ intersect the graph?
(l) How often does the line $x = 5$ intersect the graph?
(m) For what values of x does $f(x) = 3$?
(n) For what values of x does $f(x) = -2$?

14. Use the graph of the function f given below to answer parts (a)–(n).

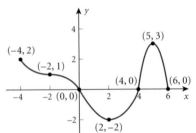

(a) Find $f(0)$ and $f(6)$.
(b) Find $f(2)$ and $f(-2)$.
(c) Is $f(3)$ positive or negative?
(d) Is $f(-1)$ positive or negative?

(e) For what numbers is $f(x) = 0$?
(f) For what numbers is $f(x) < 0$?
(g) What is the domain of f?
(h) What is the range of f?
(i) What are the x-intercepts?
(j) What is the y-intercept?
(k) How often does the line $y = -1$ intersect the graph?
(l) How often does the line $x = 1$ intersect the graph?
(m) For what value of x does $f(x) = 3$?
(n) For what value of x does $f(x) = -2$?

In Problems 15–20, answer the questions about the given function.

15. $f(x) = 2x^2 - x - 1$

(a) Is the point $(-1, 2)$ on the graph of f?
(b) If $x = -2$, what is $f(x)$? What point is on the graph of f?
(c) If $f(x) = -1$, what is x? What points(s) are on the graph of f?
(d) What is the domain of f?
(e) List the x-intercepts, if any, of the graph of f.
(f) List the y-intercept, if there is one, of the graph of f.

16. $f(x) = -3x^2 + 5x$

(a) Is the point $(-1, 2)$ on the graph of f?
(b) If $x = -2$, what is $f(x)$? What point is on the graph of f?
(c) If $f(x) = -2$, what is x? What point(s) are on the graph of f?
(d) What is the domain of f?
(e) List the x-intercepts, if any, of the graph of f.
(f) List the y-intercept, if there is one, of the graph of f.

17. $f(x) = \dfrac{x + 2}{x - 6}$

(a) Is the point $(3, 14)$ on the graph of f?
(b) If $x = 4$, what is $f(x)$? What point is on the graph of f?
(c) If $f(x) = 2$, what is x? What point(s) are on the graph of f?
(d) What is the domain of f?
(e) List the x-intercepts, if any, of the graph of f.
(f) List the y-intercept, if there is one, of the graph of f.

18. $f(x) = \dfrac{x^2 + 2}{x + 4}$

(a) Is the point $\left(1, \dfrac{3}{5}\right)$ on the graph of f?
(b) If $x = 0$, what is $f(x)$? What point is on the graph of f?
(c) If $f(x) = \dfrac{1}{2}$, what is x? What point(s) are on the graph of f?
(d) What is the domain of f?
(e) List the x-intercepts, if any, of the graph of f.
(f) List the y-intercept, if there is one, of the graph of f.

19. $f(x) = \dfrac{2x^2}{x^4 + 1}$

(a) Is the point $(-1, 1)$ on the graph of f?
(b) If $x = 2$, what is $f(x)$? What point is on the graph of f?
(c) If $f(x) = 1$, what is x? What points(s) are on the graph of f?
(d) What is the domain of f?
(e) List the x-intercepts, if any, of the graph of f.
(f) List the y-intercept, if there is one, of the graph of f.

20. $f(x) = \dfrac{2x}{x - 2}$

(a) Is the point $\left(\dfrac{1}{2}, -\dfrac{2}{3}\right)$ on the graph of f?
(b) If $x = 4$, what is $f(x)$? What point is on the graph of f?
(c) If $f(x) = 1$, what is x? What point(s) are on the graph of f?
(d) What is the domain of f?
(e) List the x-intercepts, if any, of the graph of f.
(f) List the y-intercept, if there is one, of the graph of f.

In Problems 21–30, use the graph of the function f given below.

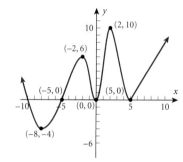

21. Is f increasing on the interval $(-8, -2)$?

22. Is f decreasing on the interval $(-8, -4)$?

23. Is f increasing on the interval $(2, 10)$?

24. Is f decreasing on the interval $(2, 5)$?

25. List the interval(s) on which f is increasing.

26. List the interval(s) on which f is decreasing.

27. Is there a local maximum at 2? If yes, what is it?

28. Is there a local maximum at 5? If yes, what is it?

29. List the numbers at which f has a local maximum. What are these local maxima?

30. List the numbers at which f has a local minimum. What are these local minima?

In Problems 31–38, the graph of a function is given. Use the graph to find:

(a) The intercepts, if any
(b) Its domain and range
(c) The intervals on which it is increasing, decreasing, or constant
(d) Whether it is even, odd, or neither

31.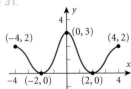
32.
33.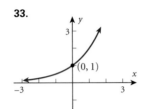
34.

35.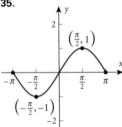
36.
37.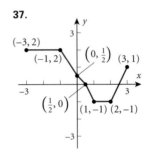
38.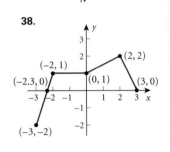

In Problems 39–42, the graph of a function f is given. Use the graph to find:

(a) The numbers, if any, at which f has a local maximum. What are these local maxima?
(b) The numbers, if any, at which f has a local minimum. What are these local minima?

39.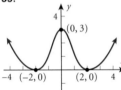
40.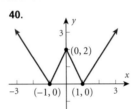
41.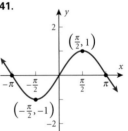
42.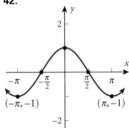

43. Find the average rate of change of $f(x) = -2x^2 + 4$
  (a) from 0 to 2   (b) from 1 to 3   (c) from 1 to 4

44. Find the average rate of change of $f(x) = -x^3 + 1$
  (a) from 0 to 2   (b) from 1 to 3   (c) from -1 to 1

In Problems 45–56, (a) for each function find the average rate of change of f from 1 to x:

$$\frac{f(x) - f(1)}{x - 1}, \quad x \neq 1$$

(b) Use the result from part (a) to compute the average rate of change from $x = 1$ to $x = 2$. Be sure to simplify.
(c) Find an equation of the secant line containing $(1, f(1))$ and $(2, f(2))$.

45. $f(x) = 5x$
46. $f(x) = -4x$
47. $f(x) = 1 - 3x$
48. $f(x) = x^2 + 1$

49. $f(x) = x^2 - 2x$
50. $f(x) = x - 2x^2$
51. $f(x) = x^3 - x$
52. $f(x) = x^3 + x$

53. $f(x) = \dfrac{2}{x + 1}$
54. $f(x) = \dfrac{1}{x^2}$
55. $f(x) = \sqrt{x}$
56. $f(x) = \sqrt{x + 3}$

Problems 57–68, determine algebraically whether each function is even, odd, or neither.

57. $f(x) = 4x^3$
58. $f(x) = 2x^4 - x^2$
59. $g(x) = -3x^2 - 5$
60. $h(x) = 3x^3 + 5$

61. $F(x) = \sqrt[3]{x}$
62. $G(x) = \sqrt{x}$
63. $f(x) = x + |x|$
64. $f(x) = \sqrt[3]{2x^2 + 1}$

65. $g(x) = \dfrac{1}{x^2}$
66. $h(x) = \dfrac{x}{x^2 - 1}$
67. $h(x) = \dfrac{-x^3}{3x^2 - 9}$
68. $F(x) = \dfrac{2x^2}{x^3 + 1}$

Graphs of Functions; Properties of Functions   135

In Problems 69–76, use a graphing utility to graph each function over the indicated interval and approximate any local maxima and local minima. Determine where the function is increasing and where it is decreasing. Round answers to two decimal places.

69. $f(x) = x^3 - 3x + 2$; $(-2, 2)$

70. $f(x) = x^3 - 3x^2 + 5$; $(-1, 3)$

71. $f(x) = x^5 - x^3$; $(-2, 2)$

72. $f(x) = x^4 - x^2$; $(-2, 2)$

73. $f(x) = -0.2x^3 - 0.6x^2 + 4x - 6$; $(-6, 4)$

74. $f(x) = -0.4x^3 + 0.6x^2 + 3x - 2$; $(-4, 5)$

75. $f(x) = 0.25x^4 + 0.3x^3 - 0.9x^2 + 3$; $(-3, 2)$

76. $f(x) = -0.4x^4 - 0.5x^3 + 0.8x^2 - 2$; $(-3, 2)$

77. For the function $f(x) = x^2$, compute each average rate of change:
   (a) from 0 to 1
   (b) from 0 to 0.5
   (c) from 0 to 0.1
   (d) from 0 to 0.01
   (e) from 0 to 0.001
   (f) Graph each of the secant lines. Set the viewing rectangle to: Xmin $= -0.2$, Xmax $= 1.2$, Xscl $= 0.1$, Ymin $= -0.2$, Ymax $= 1.2$, Yscl $= 0.1$.
   (g) What do you think is happening to the secant lines?
   (h) What is happening to the slopes of the secant lines? Is there some number they are getting closer to? What is that number?

78. For the function $f(x) = x^2$, compute each average rate of change:
   (a) from 1 to 2
   (b) from 1 to 1.5
   (c) from 1 to 1.1
   (d) from 1 to 1.01
   (e) from 1 to 1.001
   (f) Graph each of the secant lines. Set the viewing rectangle to: Xmin $= -0.5$, Xmax $= 2.5$, Xscl $= 0.1$, Ymin $= -1$, Ymax $= 4$, Yscl $= 0.1$.
   (g) What do you think is happening to the secant lines?
   (h) What is happening to the slopes of the secant lines? Is there some number they are getting closer to? What is that number?

79. **Motion of a Golf Ball** A golf ball is hit with an initial velocity of 130 feet per second at an inclination of 45° to the horizontal. In physics, it is established that the height h of the golf ball is given by the function

$$h(x) = \frac{-32x^2}{130^2} + x$$

where x is the horizontal distance that the golf ball has traveled.

(a) Determine the height of the golf ball after it has traveled 100 feet.
(b) What is the height after it has traveled 300 feet?
(c) What is the height after it has traveled 500 feet?
(d) How far was the golf ball hit?
(e) Use a graphing utility to graph the function $h = h(x)$.
(f) Use a graphing utility to determine the distance that the ball has traveled when the height of the ball is 90 feet.
(g) Create a TABLE with TblStart $= 0$ and ΔTbl $= 25$.
(h) To the nearest 25 feet, how far does the ball travel before it reaches a maximum height? What is the maximum height?
(i) Adjust the value of ΔTbl until you determine the distance, to within 1 foot, that the ball travels before it reaches a maximum height.

80. **Effect of Elevation on Weight** If an object weighs m pounds at sea level, then its weight W (in pounds) at a height of h miles above sea level is given approximately by

$$W(h) = m\left(\frac{4000}{4000 + h}\right)^2$$

(a) If Amy weighs 120 pounds at sea level, how much will she weigh on Pike's Peak, which is 14,110 feet above sea level? (1 mile = 5280 feet)
(b) Use a graphing utility to graph the function $W = W(h)$. Use $m = 120$ pounds.
(c) Create a Table with TblStart $= 0$ and ΔTbl $= 0.5$ to see how the weight W varies as h changes from 0 to 5 miles.
(d) At what height will Amy weigh 119.95 pounds?
(e) Does your answer to part (d) seem reasonable?

81. **Constructing an Open Box** An open box with a square base is to be made from a square piece of cardboard 24 inches on a side by cutting out a square from each corner and turning up the sides (see the figure).

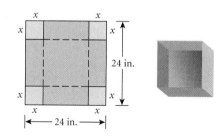

(a) Express the volume V of the box as a function of the length x of the side of the square cut from each corner.
(b) What is the volume if a 3-inch square is cut out?
(c) What is the volume if a 10-inch square is cut out?
(d) Graph $V = V(x)$. For what value of x is V largest?

82. **Constructing an Open Box** A open box with a square base is required to have a volume of 10 cubic feet.

    (a) Express the amount A of material used to make such a box as a function of the length x of a side of the square base.
    (b) How much material is required for a base 1 foot by 1 foot?
    (c) How much material is required for a base 2 feet by 2 feet?
    (d) Graph $A = A(x)$. For what value of x is A smallest?

83. **Minimum Average Cost** The average cost of producing x riding lawn mowers is given by

$$\overline{C}(x) = 0.3x^2 + 21x - 251 + \frac{2500}{x}$$

    (a) Use a graphing utility to graph $\overline{C}$.
    (b) Determine the number of riding lawn mowers to produce in order to minimize average cost.
    (c) What is the minimum average cost?

84. Match each function with the graph that best describes the situation. Discuss the reason for your choice.

    (a) The cost of building a house as a function of its square footage
    (b) The height of an egg dropped from a 300-foot building as a function of time
    (c) The height of a human as a function of time
    (d) The demand for Big Macs as a function of price
    (e) The height of a child on a swing as a function of time

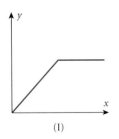
(I)

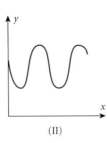
(II)

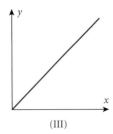
(III)

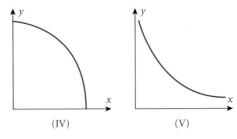
(IV)   (V)

85. Match each function with the graph that best describes the situation. Discuss the reason for your choice.

    (a) The temperature of a bowl of soup as a function of time
    (b) The number of hours of daylight per day over a two-year period
    (c) The population of Florida as a function of time
    (d) The distance of a car traveling at a constant velocity as a function of time
    (e) The height of a golf ball hit with a 7-iron as a function of time

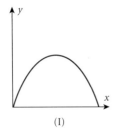
(I)

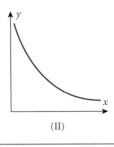
(II)

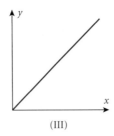
(III)

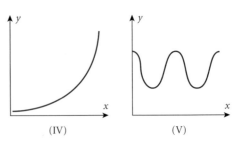
(IV)   (V)

86. Draw the graph of a function that has the following characteristics:

    Domain: all real numbers
    Range: all real numbers
    Intercepts: $(0, -3)$ and $(2, 0)$

    A local maximum of -2 is at -1; a local minimum of -6 is at 2. Compare your graph with others. Comment on any differences.

87. Redo Problem 86 with the following additional information:

    Increasing on $(-\infty, -1), (2, \infty)$
    Decreasing on $(-1, 2)$

    Again compare your graph with others and comment on any differences.

88. How many x-intercepts can a function defined on an interval have if it is increasing on that interval? Explain.

89. Suppose that a friend of yours does not understand the idea of increasing and decreasing functions. Provide an explanation complete with graphs that clarifies the idea.

90. Can a function be both even and odd? Explain.

91. Describe how you would proceed to find the domain and range of a function if you were given its graph. How would your strategy change if, instead, you were given the equation defining the function?

92. Is a graph that consists of a single point the graph of a function? If so, can you write the equation of such a function?

93. Define some functions that pass through $(0, 0)$ and $(1, 1)$ and are increasing for $x \geq 0$. Begin your list with $y = \sqrt{x}$, $y = x$, and $y = x^2$. Can you propose a general result about such functions?

1.4 Library of Functions; Piecewise-defined Functions

PREPARING FOR THIS SECTION *Before getting started, review the following:*

> Square Roots (Chapter 0, Section 0.2, pp. 23–24)
> nth Roots (Chapter 0, Section 0.6, pp. 64–65)
> Absolute Value (Chapter 0, Section 0.2, p. 18)

OBJECTIVES **1** Graph the functions listed in the library of functions
2 Graph piecewise-defined functions

FIGURE 30

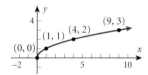

We now introduce a few more functions to add to our list of important functions. We begin with the *square root function*.

In Section 1.1 we graphed the equation $x = y^2$. If we solve the equation for y and restrict y so that $y \geq 0$, the equation $x = y^2$, $y \geq 0$, can be written as $y = f(x) = \sqrt{x}$. Figure 30 shows a graph of $f(x) = \sqrt{x}$.

Based on the graph of $f(x) = \sqrt{x}$, we have the following properties:

Properties of $f(x) = \sqrt{x}$

1. The domain and range of $f(x) = \sqrt{x}$ are the set of nonnegative real numbers.
2. The x-intercept of the graph of $f(x) = \sqrt{x}$ is $(0, 0)$. The y-intercept of the graph of $f(x) = \sqrt{x}$ is also $(0, 0)$.
3. The function is neither even nor odd.
4. It is increasing on the interval $(0, \infty)$.

EXAMPLE 1 **Graphing the Cube Root Function**

(a) Determine whether $f(x) = \sqrt[3]{x}$ is even, odd, or neither. State whether the graph of f is symmetric with respect to the y-axis or symmetric with respect to the origin.
(b) Determine the intercepts, if any, of the graph of $f(x) = \sqrt[3]{x}$.
(c) Graph $f(x) = \sqrt[3]{x}$.

SOLUTION (a) Because

$$f(-x) = \sqrt[3]{-x} = -\sqrt[3]{x} = -f(x)$$

the function is odd. The graph of f is symmetric with respect to the origin.

(b) Since $f(0) = \sqrt[3]{0} = 0$, the y-intercept is $(0, 0)$. The x-intercept is found by solving the equation $f(x) = 0$.

$$f(x) = 0$$
$$\sqrt[3]{x} = 0 \quad f(x) = \sqrt[3]{x}$$
$$x = 0 \quad \text{Cube both sides of the equation.}$$

The x-intercept is also $(0, 0)$.

(c) We use the function to form Table 9 and obtain some points on the graph. Because of the symmetry with respect to the origin, we only need to find points (x, y) for which $x \geq 0$. Figure 31 shows the graph of $f(x) = \sqrt[3]{x}$.

TABLE 9

x	$y = \sqrt[3]{x}$	(x, y)
0	0	$(0, 0)$
$\frac{1}{8}$	$\frac{1}{2}$	$\left(\frac{1}{8}, \frac{1}{2}\right)$
1	1	$(1, 1)$
2	$\sqrt[3]{2} \approx 1.26$	$(2, \sqrt[3]{2})$
8	2	$(8, 2)$

FIGURE 31

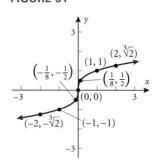

From the results of Example 1 and Figure 31, we have the following properties of the cube root function.

Properties of $f(x) = \sqrt[3]{x}$

1. The domain and range of $f(x) = \sqrt[3]{x}$ are the set of real numbers.
2. The x-intercept of the graph of $f(x) = \sqrt[3]{x}$ is $(0, 0)$. The y-intercept of the graph of $f(x) = \sqrt[3]{x}$ is also $(0, 0)$.
3. The function is odd.
4. It is increasing on the interval $(-\infty, \infty)$.

EXAMPLE 2 **Graphing the Absolute Value Function**

(a) Determine whether $f(x) = |x|$ is even, odd, or neither. State whether the graph of f is symmetric with respect to the y-axis or symmetric with respect to the origin.
(b) Determine the intercepts, if any, of the graph of $f(x) = |x|$.
(c) Graph $f(x) = |x|$.

SOLUTION (a) Because

$$f(-x) = |-x| = |x| = f(x)$$

the function is even. The graph of f is symmetric with respect to the y-axis.

(b) Since $f(0) = |0| = 0$, the y-intercept is $(0, 0)$. The x-intercept is found by solving the equation $f(x) = |x| = 0$. So the x-intercept is also $(0, 0)$.

(c) We use the function to form Table 10 and obtain some points on the graph. Because of the symmetry with respect to the y-axis, we only need to find points (x, y) for which $x \geq 0$. Figure 32 shows the graph of $f(x) = |x|$.

TABLE 10

x	$y = |x|$	(x, y)
0	0	$(0, 0)$
1	1	$(1, 1)$
2	2	$(2, 2)$
3	3	$(3, 3)$

FIGURE 32

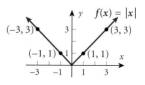

From the results of Example 2 and Figure 32, we have the following properties of the absolute value function.

Properties of $f(x) = |x|$

1. The domain of $f(x) = |x|$ is the set of real numbers; the range is the set of nonnegative real numbers.
2. The x-intercept of the graph of $f(x) = |x|$ is $(0, 0)$. The y-intercept of the graph of $f(x) = |x|$ is also $(0, 0)$.
3. The function is even.
4. It is decreasing on the interval $(-\infty, 0)$. It is increasing on the interval $(0, \infty)$.

SEEING THE CONCEPT: Graph $y = |x|$ on a square screen and compare what you see with Figure 32. Note that some graphing calculators use the symbols abs(x) for absolute value. If your utility has no built-in absolute value function, you can still graph $y = |x|$ by using the fact that $|x| = \sqrt{x^2}$.

Library of Functions

Graph the functions listed in the library of functions

We now provide a summary of the key functions that we have encountered. In going through this list, pay special attention to the properties of each function, particularly to the shape of each graph. Knowing these graphs will lay the foundation for later graphing techniques.

Linear Function

$$f(x) = mx + b \qquad m \text{ and } b \text{ are real numbers}$$

FIGURE 33 Linear function

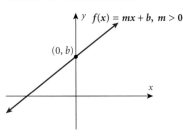

FIGURE 34 Constant function

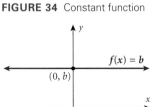

See Figure 33.

The domain of a **linear function** is the set of all real numbers. The graph of this function is a nonvertical line with slope m and y-intercept $(0, b)$. A linear function is increasing if $m > 0$, decreasing if $m < 0$, and constant if $m = 0$.

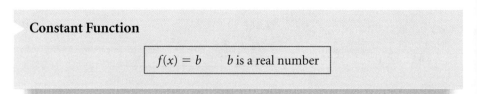

Constant Function

$$f(x) = b \quad b \text{ is a real number}$$

See Figure 34.

A **constant function** is a special linear function ($m = 0$). Its domain is the set of all real numbers; its range is the set consisting of a single number b. Its graph is a horizontal line whose y-intercept is $(0, b)$. The constant function is an even function whose graph is constant over its domain.

Identity Function

$$f(x) = x$$

FIGURE 35 Identity function

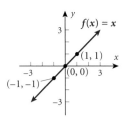

See Figure 35.

The **identity function** is also a special linear function. Its domain and range are the set of all real numbers. Its graph is a line whose slope is $m = 1$ and whose only intercept is $(0,0)$. The line consists of all points for which the x-coordinate equals the y-coordinate. The identity function is an odd function that is increasing over its domain. Note that the graph bisects quadrants I and III.

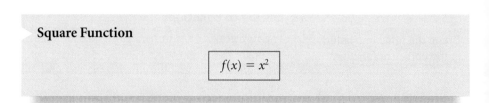

Square Function

$$f(x) = x^2$$

FIGURE 36 Square function

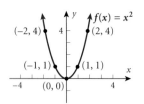

See Figure 36.

The domain of the **square function** f is the set of all real numbers; its range is the set of nonnegative real numbers. The graph of this function is a parabola whose only intercept is $(0, 0)$. The square function is an even function that is decreasing on the interval $(-\infty, 0)$ and increasing on the interval $(0, \infty)$.

Library of Functions; Piecewise-defined Functions   141

FIGURE 37 Cube function

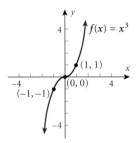

Cube Function

$$f(x) = x^3$$

See Figure 37.
The domain and range of the **cube function** are the set of all real numbers. The only intercept of the graph is $(0, 0)$. The cube function is odd and is increasing on the interval $(-\infty, \infty)$.

Square Root Function

$$f(x) = \sqrt{x}$$

FIGURE 38
Square root function

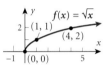

See Figure 38.
The domain and range of the **square root function** are the set of nonnegative real numbers. The only intercept of the graph is $(0, 0)$. The square root function is neither even nor odd and is increasing on the interval $(0, \infty)$.

FIGURE 39 Cube root function

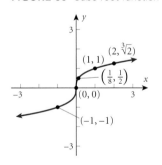

Cube Root Function

$$f(x) = \sqrt[3]{x}$$

See Figure 39.
The domain and the range of the **cube root function** is the set of all real numbers. The intercept of the graph is at $(0, 0)$. The cube root function is an odd function that is increasing on the interval $(-\infty, \infty)$.

Reciprocal Function

$$f(x) = \frac{1}{x}$$

FIGURE 40
Reciprocal function

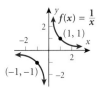

Refer to Example 11, p. 106 for a discussion of the equation $y = \dfrac{1}{x}$. See Figure 40.

The domain and range of the **reciprocal function** are the set of all nonzero real numbers. The graph has no intercepts. The reciprocal function is decreasing on the intervals $(-\infty, 0)$ and $(0, \infty)$ and is an odd function.

Absolute Value Function

$$f(x) = |x|$$

FIGURE 41 Absolute value function

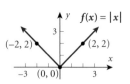

See Figure 41.

The domain of the **absolute value function** is the set of all real numbers; its range is the set of nonnegative real numbers. The intercept of the graph is at (0, 0). If $x \geq 0$, then $f(x) = x$, and the graph of f is part of the line $y = x$; if $x < 0$, then $f(x) = -x$, and the graph of f is part of the line $y = -x$. The absolute value function is an even function; it is decreasing on the interval $(-\infty, 0)$ and is increasing on the interval $(0, \infty)$.

The notation int(x) stands for the largest integer less than or equal to x. For example,

$$\text{int}(1) = 1 \quad \text{int}(2.5) = 2 \quad \text{int}(\tfrac{1}{2}) = 0 \quad \text{int}(-\tfrac{3}{4}) = -1 \quad \text{int}(\pi) = 3$$

This type of correspondence occurs frequently enough in mathematics that we give it a name.

Greatest Integer Function

$$f(x) = \text{int}(x) = \text{Greatest integer less than or equal to } x$$

NOTE: Some books use the notation $f(x) = [x]$ instead of int(x).

We obtain the graph of $f(x) = \text{int}(x)$ by plotting several points. See Table 11. For values of x, $-1 \leq x < 0$, the value of $f(x) = \text{int}(x)$ is -1; for values of x, $0 \leq x < 1$, the value of f is 0. See Figure 42 for the graph.

TABLE 11

x	$y = \text{int}(x)$	(x, y)
-1	-1	$(-1, -1)$
$-\tfrac{1}{2}$	-1	$\left(-\tfrac{1}{2}, -1\right)$
$-\tfrac{1}{4}$	-1	$\left(-\tfrac{1}{4}, -1\right)$
0	0	$(0, 0)$
$\tfrac{1}{4}$	0	$\left(\tfrac{1}{4}, 0\right)$
$\tfrac{1}{2}$	0	$\left(\tfrac{1}{2}, 0\right)$
$\tfrac{3}{4}$	0	$\left(\tfrac{3}{4}, 0\right)$

FIGURE 42 Greatest integer function

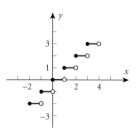

The domain of the **greatest integer function** is the set of all real numbers; its range is the set of integers. The y-intercept of the graph is (0, 0). The x-intercepts lie in the interval $[0, 1)$. The greatest integer function is neither even nor odd. It is constant on every interval of the form $[k, k + 1)$, for k an integer. In Figure 42, we use a solid dot to indicate, for example, that at $x = 1$ the value of f is $f(1) = 1$; we use an open circle to illustrate that the function does not assume the value of 0 at $x = 1$.

From the graph of the greatest integer function, we can see why it is also called a **step function**. At $x = 0$, $x = \pm 1$, $x = \pm 2$, and so on, this function exhibits what is called a *discontinuity*; that is, at integer values, the graph suddenly "steps" from one value to another without taking on any of the intermediate values. For example, to the immediate left of $x = 3$, the y-coordinates are 2, and to the immediate right of $x = 3$, the y-coordinates are 3.

COMMENT: When graphing a function, you can choose either the **connected mode**, in which points plotted on the screen are connected, making the graph appear without any breaks, or the **dot mode**, in which only the points plotted appear. When graphing the greatest integer function with a graphing utility, it is necessary to be in the **dot mode**. This is to prevent the utility from "connecting the dots" when $f(x)$ changes from one integer value to the next. See Figure 43.

FIGURE 43

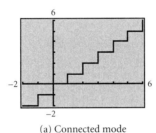
(a) Connected mode

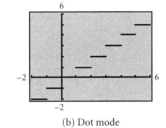
(b) Dot mode

The functions that we have discussed so far are basic. Whenever you encounter one of them, you should see a mental picture of its graph. For example, if you encounter the function $f(x) = x^2$, you should see in your mind's eye a picture like Figure 36.

NOW WORK PROBLEMS 1–8.

Piecewise-defined Functions

2 Graph piecewise-defined functions

Sometimes a function is defined differently on different parts of its domain. For example, the absolute value function $f(x) = |x|$ is actually defined by two equations: $f(x) = x$ if $x \geq 0$ and $f(x) = -x$ if $x < 0$. For convenience, we generally combine these equations into one expression as

$$f(x) = |x| = \begin{cases} x & \text{if } x \geq 0 \\ -x & \text{if } x < 0 \end{cases}$$

When functions are defined by more than one equation, they are called **piecewise-defined** functions.

Let's look at another example of a piecewise-defined function.

EXAMPLE 3 **Analyzing a Piecewise-Defined Function**

The function f is defined as

$$f(x) = \begin{cases} -x + 1 & \text{if } -1 \leq x < 1 \\ 2 & \text{if } x = 1 \\ x^2 & \text{if } x > 1 \end{cases}$$

(a) Find $f(0), f(1)$, and $f(2)$. (b) Determine the domain of f.
(c) Graph f. (d) Use the graph to find the range of f.

SOLUTION (a) To find $f(0)$, we observe that when $x = 0$ the equation for f is given by $f(x) = -x + 1$. So we have

$$f(0) = -0 + 1 = 1$$

When $x = 1$, the equation for f is $f(x) = 2$. So

$$f(1) = 2$$

When $x = 2$, the equation for f is $f(x) = x^2$. So

$$f(2) = 2^2 = 4$$

FIGURE 44

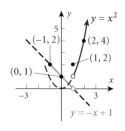

(b) To find the domain of f, we look at its definition. We conclude that the domain of f is $\{x | x \geq -1\}$, or the interval $[-1, \infty)$.

(c) To graph f, we graph "each piece." First we graph the line $y = -x + 1$ and keep only the part for which $-1 \leq x < 1$. Then we plot the point $(1, 2)$ because, when $x = 1$, $f(x) = 2$. Finally, we graph the parabola $y = x^2$ and keep only the part for which $x > 1$. See Figure 44.

(d) From the graph, we conclude that the range of f is $\{y | y > 0\}$, or the interval $(0, \infty)$.

 NOW WORK PROBLEM 21.

EXAMPLE 4 **Cost of Electricity**

In May 2003, Commonwealth Edison Company supplied electricity to residences for a monthly customer charge of $7.58 plus 8.275¢ per kilowatt-hour (kWhr) for the first 400 kWhr supplied in the month and 6.574¢ per kWhr for all usage over 400 kWhr in the month.

(a) What is the charge for using 300 kWhr in a month?
(b) What is the charge for using 700 kWhr in a month?
(c) If C is the monthly charge for x kWhr, express C as a function of x.

Source: Commonwealth Edison Co., Chicago, Illinois, 2003.

SOLUTION (a) For 300 kWhr, the charge is $7.58 plus 8.275¢ = $0.08275 per kWhr. That is,

$$\text{Charge} = \$7.58 + \$0.08275(300) = \$32.41$$

(b) For 700 kWhr, the charge is $7.58 plus 8.275¢ per kWhr for the first 400 kWhr plus 6.574¢ per kWhr for the 300 kWhr in excess of 400. That is,

$$\text{Charge} = \$7.58 + \$0.08275(400) + \$0.06574(300) = \$60.40$$

(c) If $0 \leq x \leq 400$, the monthly charge C (in dollars) can be found by multiplying x times $0.08275 and adding the monthly customer charge of $7.58. So, if $0 \leq x \leq 400$, then $C(x) = 0.08275x + 7.58$. For $x > 400$, the charge is $0.08275(400) + 7.58 + 0.06574(x - 400)$, since $x - 400$ equals the usage in excess of 400 kWhr, which costs $0.06574 per kWhr. That is, if $x > 400$, then

$$C(x) = 0.08275(400) + 7.58 + 0.06574(x - 400)$$
$$= 40.68 + 0.06574(x - 400)$$
$$= 0.06574x + 14.38$$

Library of Functions; Piecewise-defined Functions   145

FIGURE 45

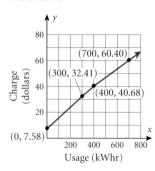

The rule for computing C follows two equations:

$$C(x) = \begin{cases} 0.08275x + 7.58 & \text{if } 0 \leq x \leq 400 \\ 0.06574x + 14.38 & \text{if } x > 400 \end{cases}$$

See Figure 45 for the graph.

EXERCISE 1.4

In Problems 1–8, match each graph to the function whose graph most resembles the one given.

A. Constant function B. Linear function C. Square function
D. Cube function E. Square root function F. Reciprocal function
G. Absolute value function H. Cube root function

1. 2. 3. 4.

5. 6. 7. 8.

In Problems 9–16, sketch the graph of each function. Be sure to label at least three points on the graph.

9. $f(x) = x$ **10.** $f(x) = x^2$ **11.** $f(x) = x^3$ **12.** $f(x) = \sqrt{x}$

13. $f(x) = \dfrac{1}{x}$ **14.** $f(x) = |x|$ **15.** $f(x) = \sqrt[3]{x}$ **16.** $f(x) = 3$

17. If $f(x) = \begin{cases} x^2 & \text{if } x < 0 \\ 2 & \text{if } x = 0 \\ 2x + 1 & \text{if } x > 0 \end{cases}$
find: (a) $f(-2)$ (b) $f(0)$ (c) $f(2)$

18. If $f(x) = \begin{cases} x^3 & \text{if } x < 0 \\ 3x + 2 & \text{if } x \geq 0 \end{cases}$
find: (a) $f(-1)$ (b) $f(0)$ (c) $f(1)$

19. If $f(x) = \text{int}(2x)$, find: (a) $f(1.2)$ (b) $f(1.6)$ (c) $f(-1.8)$

20. If $f(x) = \text{int}\left(\dfrac{x}{2}\right)$, find: (a) $f(1.2)$ (b) $f(1.6)$ (c) $f(-1.8)$

In Problems 21–32:
(a) Find the domain of each function. (b) Locate any intercepts.
(c) Graph each function (d) Based on the graph, find the range.

21. $f(x) = \begin{cases} 2x & \text{if } x \neq 0 \\ 1 & \text{if } x = 0 \end{cases}$

22. $f(x) = \begin{cases} 3x & \text{if } x \neq 0 \\ 4 & \text{if } x = 0 \end{cases}$

23. $f(x) = \begin{cases} -2x + 3 & x < 1 \\ 3x - 2 & x \geq 1 \end{cases}$

146 Chapter 1 Functions and Their Graphs

24. $f(x) = \begin{cases} x + 3 & x < -2 \\ -2x - 3 & x \geq -2 \end{cases}$

25. $f(x) = \begin{cases} x + 3 & -2 \leq x < 1 \\ 5 & x = 1 \\ -x + 2 & x > 1 \end{cases}$

26. $f(x) = \begin{cases} 2x + 5 & -3 \leq x < 0 \\ -3 & x = 0 \\ -5x & x > 0 \end{cases}$

27. $f(x) = \begin{cases} 1 + x & \text{if } x < 0 \\ x^2 & \text{if } x \geq 0 \end{cases}$

28. $f(x) = \begin{cases} \dfrac{1}{x} & \text{if } x < 0 \\ \sqrt[3]{x} & \text{if } x \geq 0 \end{cases}$

29. $f(x) = \begin{cases} |x| & \text{if } -2 \leq x < 0 \\ 1 & \text{if } x = 0 \\ x^3 & \text{if } x > 0 \end{cases}$

30. $f(x) = \begin{cases} 3 + x & \text{if } -3 \leq x < 0 \\ 3 & \text{if } x = 0 \\ \sqrt{x} & \text{if } x > 0 \end{cases}$

31. $f(x) = 2 \text{ int } (x)$

32. $f(x) = \text{int } (2x)$

In Problems 33–36, the graph of a piecewise-defined function is given. Write a definition for each function.

33.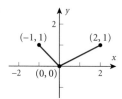

34.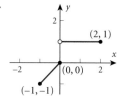

35.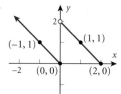

36.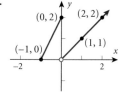

---

37. Cell Phone Service Sprint PCS offers a monthly cellular phone plan for $39.99. It includes 350 anytime minutes plus $0.25 per minute for additional minutes. The following function is used to compute the monthly cost for a subscriber

$$C(x) = \begin{cases} 39.99 & \text{if } 0 < x \leq 350 \\ 0.25x - 47.51 & \text{if } x > 350 \end{cases}$$

where x is the number of anytime minutes used. Compute the monthly cost of the cellular phone for the following anytime minutes:

(a) 200 (b) 365 (c) 351

Source: Sprint PCS.

38. First-class Letter According to the U.S. Postal Service, first-class mail is used for personal and business correspondence. Any mailable item may be sent as first-class mail. It includes postcards, letters, large envelopes, and small packages. The maximum weight is 13 ounces. The following function is used to compute the cost of mailing a first-class item.

$$C(x) = \begin{cases} 0.37 & \text{if } 0 < x \leq 1 \\ 0.23 \text{ int}(x) + 0.37 & \text{if } 1 < x \leq 13 \end{cases}$$

where x is the weight of the item in ounces. Compute the cost of mailing the following items first-class:

(a) A letter weighing 4.3 ounces
(b) A postcard weighing 0.4 ounces
(c) A package weighing 12.2 ounces

Source: United States Postal Service.

39. Cost of Natural Gas In May 2003, the People Gas Company had the following rate schedule for natural gas usage in single-family residences:

Monthly service charge	$9.45
Per therm service charge	
1st 50 therms	$0.36375/therm
Over 50 therms	$0.11445/therm
Gas charge	0.6338/therm

(a) What is the charge for using 50 therms in a month?
(b) What is the charge for using 500 therms in a month?
(c) Construct a function that relates the monthly charge C for x therms of gas.
(d) Graph this function.

Source: The Peoples Gas Company, Chicago, Illinois, 2003.

40. Cost of Natural Gas In May 2003, Nicor Gas had the following rate schedule for natural gas usage in single-family residences:

Monthly customer charge	$6.45
Distribution charge	
1st 20 therms	$0.2012/therm
Next 30 therms	$0.1117/therm
Over 50 therms	$0.0374/therm
Gas supply charge	$0.7268/therm

(a) What is the charge for using 40 therms in a month?
(b) What is the charge for using 202 therms in a month?
(c) Construct a function that gives the monthly charge C for x therms of gas.
(d) Graph this function.

Source: Nicor Gas, Aurora, Illinois, 2003.

41. Wind Chill The wind chill factor represents the equivalent air temperature at a standard wind speed that would produce the same heat loss as the given temperature and wind speed. One formula for computing the equivalent temperature is

$$W = \begin{cases} t & 0 \leq v < 1.79 \\ 33 - \dfrac{(10.45 + 10\sqrt{v} - v)(33 - t)}{22.04} & 1.79 \leq v \leq 20 \\ 33 - 1.5958(33 - t) & v > 20 \end{cases}$$

where v represents the wind speed (in meters per second) and t represents the air temperature (°C). Compute the wind chill for the following:

(a) An air temperature of 10°C and a wind speed of 1 meter per second (m/sec).
(b) An air temperature of 10°C and a wind speed of 5 m/sec.
(c) An air temperature of 10°C and a wind speed of 15 m/sec.
(d) An air temperature of 10°C and a wind speed of 25 m/sec.
(e) Explain the physical meaning of the equation corresponding to $0 \leq v < 1.79$.
(f) Explain the physical meaning of the equation corresponding to $v > 20$.

42. Wind Chill Redo Problem 41(a)–(d) for an air temperature of $-10°C$.

43. Federal Income Tax Two 2003 Tax Rate Schedules are given in the accompanying tables. If x equals taxable income and y equals the tax due, construct a function $y = f(x)$ for Schedule X.

Revised 2003 Tax Rate Schedules

| | If TAXABLE INCOME | | The TAX is | | |
| | | Then | | | |
	Is Over	But Not Over	This Amount	Plus This %	Of the Excess Over
SCHEDULE X—					
Single	$0	$7,000	$0.00	10%	$0.00
	$7,000	$28,400	$700.00	15%	$7,000
	$28,400	$68,800	$3,910.00	25%	$28,400
	$68,800	$143,500	$14,010.00	28%	$68,800
	$143,500	$311,950	$34,926.00	33%	$143,500
	$311,950	—	$90,514.50	35%	$311,950

| | If TAXABLE INCOME | | The TAX is | | |
| | | Then | | | |
	Is Over	But Not Over	This Amount	Plus This %	Of the Excess Over
SCHEDULE Y-1—					
Married Filing Jointly or Qualifying Widow(er)	$0	$14,000	$0.00	10%	$0.00
	$14,000	$56,800	$1,400.00	15%	$14,000
	$56,800	$114,650	$7,820.00	25%	$56,800
	$114,650	$174,700	$22,282.50	28%	$114,650
	$174,700	$311,950	$39,096.50	33%	$174,700
	$311,950	—	$84,389.00	35%	$311,950

Source: Internal Revenue Service.

44. Federal Income Tax Refer to the revised 2003 tax rate schedules. If x equals the taxable income and y equals the tax due, construct a function $y = f(x)$ for Schedule Y-1.

45. Exploration Graph $y = x^2$. Then on the same screen graph $y = x^2 + 2$, followed by $y = x^2 + 4$, followed by $y = x^2 - 2$. What pattern do you observe? Can you predict the graph of $y = x^2 - 4$? Of $y = x^2 + 5$?

46. Exploration Graph $y = x^2$. Then on the same screen graph $y = (x - 2)^2$, followed by $y = (x - 4)^2$, followed by $y = (x + 2)^2$. What pattern do you observe? Can you predict the graph of $y = (x + 4)^2$? Of $y = (x - 5)^2$?

47. Exploration Graph $y = x^2$. Then on the same screen graph $y = -x^2$. What pattern do you observe? Now try $y = |x|$ and $y = -|x|$. What do you conclude?

48. Exploration Graph $y = \sqrt{x}$. Then on the same screen graph $y = \sqrt{-x}$. What pattern do you observe? Now try $y = 2x + 1$ and $y = 2(-x) + 1$. What do you conclude?

49. Exploration Graph $y = x^3$. Then on the same screen graph $y = (x - 1)^3 + 2$. Could you have predicted the result?

50. Exploration Graph $y = x^2$, $y = x^4$, and $y = x^6$ on the same screen. What do you notice is the same about each graph? What do you notice that is different?

51. Exploration Graph $y = x^3$, $y = x^5$, and $y = x^7$ on the same screen. What do you notice is the same about each graph? What do you notice that is different? What is its y-intercept, if any? What are its x-intercepts, if any? Is it even, odd, or neither? How would you describe its graph?

52. Consider the equation $y = \begin{cases} 1 & \text{if } x \text{ is rational} \\ 0 & \text{if } x \text{ is irrational} \end{cases}$

Is this a function? What is its domain? What is its range?

1.5 GRAPHING TECHNIQUES: Shifts and Reflections

OBJECTIVES
1. Graph functions using horizontal and vertical shifts
2. Graph functions using reflections about the x-axis or y-axis

At this stage, if you were asked to graph any of the functions defined by $y = x$, $y = x^2$, $y = x^3$, $y = \sqrt{x}$, $y = \sqrt[3]{x}$, $y = |x|$, or $y = \dfrac{1}{x}$, your response should be, "Yes, I recognize these functions and know the general shapes of their graphs." (If this is not your answer, review the previous section, Figures 35 through 41.)

Sometimes we are asked to graph a function that is "almost" like one that we already know how to graph. In this section, we look at some of these functions and develop techniques for graphing them. Collectively, these techniques are referred to as **transformations**.

Graph functions using horizontal and vertical shifts

1 Vertical Shifts

EXAMPLE 1 Vertical Shift Up

Use the graph of $f(x) = x^2$ to obtain the graph of $g(x) = x^2 + 3$.

SOLUTION We begin by obtaining some points on the graphs of f and g. For example, when $x = 0$, then $y = f(0) = 0$ and $y = g(0) = 3$. When $x = 1$, then $y = f(1) = 1$ and $y = g(1) = 4$. Table 12 lists these and a few other points on each graph. We conclude that the graph of g is identical to that of f, except that it is shifted vertically up 3 units. See Figure 46.

TABLE 12

x	$y = f(x) = x^2$	$y = g(x) = x^2 + 3$
-2	4	7
-1	1	4
0	0	3
1	1	4
2	4	7

FIGURE 46

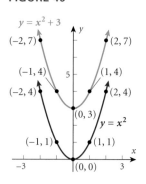

Graphing Techniques: Shifts and Reflections   149

SEEING THE CONCEPT: On the same screen, graph each of the following functions:

$$Y_1 = x^2$$
$$Y_2 = x^2 + 1$$
$$Y_3 = x^2 + 2$$
$$Y_4 = x^2 - 1$$
$$Y_5 = x^2 - 2$$

FIGURE 47

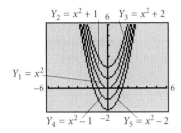

Figure 47 illustrates the graphs. You should have observed a general pattern. With $Y_1 = x^2$ on the screen, the graph of $Y_2 = x^2 + 1$ is identical to that of $Y_1 = x^2$, except that it is shifted vertically up 1 unit. Similarly, $Y_3 = x^2 + 2$ is identical to that of $Y_1 = x^2$, except that it is shifted vertically up 2 units. The graph of $Y_4 = x^2 - 1$ is identical to that of $Y_1 = x^2$, except that it is shifted vertically down 1 unit.

We are led to the following conclusion:

> If a real number k is added to the right side of a function $y = f(x)$, the graph of the new function $y = f(x) + k$ is the graph of f **shifted vertically up** (if $k > 0$) or **down** (if $k < 0$).

Let's look at another example.

EXAMPLE 2   **Vertical Shift Down**

Use the graph of $f(x) = x^2$ to obtain the graph of $g(x) = x^2 - 4$.

SOLUTION   Table 13 lists some points on the graphs of f and g. Notice that each y-coordinate of g is 4 units less than the corresponding y-coordinate of f. The graph of g is identical to that of f, except that it is shifted down 4 units. See Figure 48.

TABLE 13

x	$y = f(x)$ $= x^2$	$y = g(x)$ $= x^2 - 4$
-2	4	0
-1	1	-3
0	0	-4
1	1	-3
2	4	0

FIGURE 48

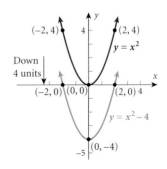

  NOW WORK PROBLEMS 11 AND 21.

Horizontal Shifts

EXAMPLE 3 **Horizontal Shift to the Right**

Use the graph of $f(x) = x^2$ to obtain the graph of $g(x) = (x - 2)^2$.

SOLUTION The function $g(x) = (x - 2)^2$ is basically a square function. Table 14 lists some points on the graphs of f and g. Note that when $f(x) = 0$ then $x = 0$, and when $g(x) = 0$, then $x = 2$. Also, when $f(x) = 4$, then $x = -2$ or 2, and when $g(x) = 4$, then $x = 0$ or 4. We conclude that the graph of g is identical to that of f, except that it is shifted 2 units to the right. See Figure 49.

TABLE 14

x	$y = f(x)$ $= x^2$	$y = g(x)$ $= (x - 2)^2$
-2	4	16
0	0	4
2	4	0
4	16	4

FIGURE 49

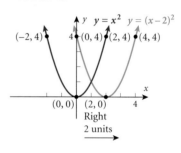

 SEEING THE CONCEPT: On the same screen, graph each of the following functions:

$$Y_1 = x^2$$
$$Y_2 = (x - 1)^2$$
$$Y_3 = (x - 3)^2$$
$$Y_4 = (x + 2)^2$$

FIGURE 50

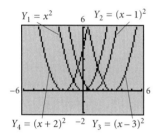

Figure 50 illustrates the graphs.

You should have observed the following pattern. With the graph of $Y_1 = x^2$ on the screen, the graph of $Y_2 = (x - 1)^2$ is identical to that of $Y = x^2$, except that it is shifted horizontally to the right 1 unit. Similarly, the graph of $Y_3 = (x - 3)^2$ is identical to that of $Y_1 = x^2$, except that it is shifted horizontally to the right 3 units. Finally, the graph of $Y_4 = (x + 2)^2$ is identical to that of $Y_1 = x^2$, except that it is shifted horizontally to the left 2 units.

We are led to the following conclusion.

> If the argument x of a function f is replaced by $x - h$, h a real number, the graph of the new function $y = f(x - h)$ is the graph of f **shifted horizontally left** (if $h < 0$) or **right** (if $h > 0$).

EXAMPLE 4 **Horizontal Shift to the Left**

Use the graph of $f(x) = x^2$ to obtain the graph of $g(x) = (x + 4)^2$.

SOLUTION The function $g(x) = (x + 4)^2$ is basically a square function. Its graph is the same as that of f, except that it is shifted horizontally 4 units to the left. (Do you see why? $(x + 4)^2 = [x - (-4)]^2$) See Figure 51.

Graphing Techniques: Shifts and Reflections 151

FIGURE 51

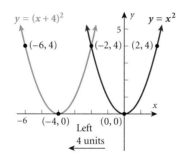

 NOW WORK PROBLEMS 9 AND 25.

Vertical and horizontal shifts are sometimes combined.

EXAMPLE 5 **Combining Vertical and Horizontal Shifts**

Graph the function: $f(x) = (x + 3)^2 - 5$

SOLUTION We graph f in steps. First, we note that the rule for f is basically a square function, so we begin with the graph of $y = x^2$ as shown in Figure 52(a). Next, to get the graph of $y = (x + 3)^2$, we shift the graph of $y = x^2$ horizontally 3 units to the left. See Figure 52(b). Finally, to get the graph of $y = (x + 3)^2 - 5$, we shift the graph of $y = (x + 3)^2$ vertically down 5 units. See Figure 52(c). Note the points plotted on each graph. Using key points can be helpful in keeping track of the transformation that has taken place.

FIGURE 52

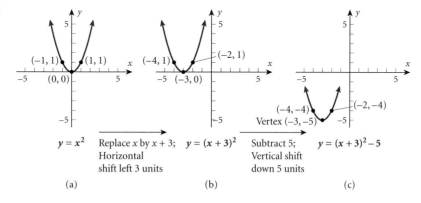

CHECK: Graph $Y_1 = f(x) = (x + 3)^2 - 5$ and compare the graph to Figure 52(c).

In Example 5, if the vertical shift had been done first, followed by the horizontal shift, the final graph would have been the same. Try it for yourself.

 NOW WORK PROBLEM 27.

Graph functions using reflections about the x-axis or y-axis

2 Reflections about the x-Axis and the y-Axis

EXAMPLE 6 Reflection about the x-Axis

Graph the function: $f(x) = -x^2$

SOLUTION We begin with the graph of $y = x^2$, as shown in Figure 53. For each point (x, y) on the graph of $y = x^2$, the point $(x, -y)$ is on the graph of $y = -x^2$, as indicated in Table 15. We can draw the graph of $y = -x^2$ by reflecting the graph of $y = x^2$ about the x-axis. See Figure 53.

TABLE 15

x	$y = x^2$	$y = -x^2$
-2	4	-4
-1	1	-1
0	0	0
1	1	-1
2	4	-4

FIGURE 53

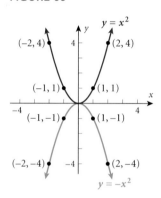

When the right side of the function $y = f(x)$ is multiplied by -1, the graph of the new function $y = -f(x)$ is the **reflection about the x-axis** of the graph of the function $y = f(x)$.

  NOW WORK PROBLEM 29.

EXAMPLE 7 Reflection about the y-Axis

Graph the function: $f(x) = \sqrt{-x}$

SOLUTION First, notice that the domain of f consists of all real numbers x for which $-x \geq 0$ or, equivalently, $x \leq 0$. To get the graph of $f(x) = \sqrt{-x}$, we begin with the graph of $y = \sqrt{x}$. For each point (x, y) on the graph of $y = \sqrt{x}$, the point $(-x, y)$ is on the graph of $y = \sqrt{-x}$. We obtain the graph of $y = \sqrt{-x}$ by reflecting the graph of $y = \sqrt{x}$ about the y-axis. See Figure 54.

FIGURE 54

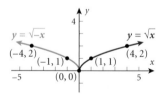

When the graph of the function $y = f(x)$ is known, the graph of the new function $y = f(-x)$ is the **reflection about the y-axis** of the graph of the function $y = f(x)$.

  NOW WORK PROBLEMS 13 AND 37.

SUMMARY: GRAPHING TECHNIQUES

Table 16 summarizes the graphing procedures that we have just discussed.

TABLE 16

To Graph:	Draw the Graph of f and:	Functional Change to f(x)
Vertical shifts		
$y = f(x) + k, \quad k > 0$	Raise the graph of f by k units.	Add k to $f(x)$.
$y = f(x) - k, \quad k > 0$	Lower the graph of f by k units.	Subtract k from $f(x)$.
Horizontal shifts		
$y = f(x + h), \quad h > 0$	Shift the graph of f to the left h units.	Replace x by $x + h$.
$y = f(x - h), \quad h > 0$	Shift the graph of f to the right h units.	Replace x by $x - h$.
Reflection about the x-axis		
$y = -f(x)$	Reflect the graph of f about the x-axis.	Multiply $f(x)$ by -1.
Reflection about the y-axis		
$y = f(-x)$	Reflect the graph of f about the y-axis.	Replace x by $-x$.

The examples that follow combine some of the procedures outlined in this section to get the required graph.

EXAMPLE 8 Determining the Function Obtained from a Series of Transformations

Find the function that is finally graphed after the following three transformations are applied to the graph of $y = |x|$.

1. Shift left 2 units. 2. Shift up 3 units. 3. Reflect about the y-axis.

SOLUTION
1. Shift left 2 units: Replace x by $x + 2$. $y = |x + 2|$
2. Shift up 3 units: Add 3. $y = |x + 2| + 3$
3. Reflect about the y-axis: Replace x by $-x$. $y = |-x + 2| + 3$

NOW WORK PROBLEM 15.

EXAMPLE 9 Combining Graphing Procedures

Graph the function: $f(x) = \sqrt{1 - x} + 2$

SOLUTION We use the following steps to get the graph of $y = \sqrt{1 - x} + 2$:

STEP 1: $y = \sqrt{x}$ Square root function.

STEP 2: $y = \sqrt{x + 1}$ Replace x by $x + 1$; horizontal shift left 1 unit.

STEP 3: $y = \sqrt{-x + 1} = \sqrt{1 - x}$ Replace x by $-x$; reflect about y-axis.

STEP 4: $y = \sqrt{1 - x} + 2$ Add 2; vertical shift up 2 units.

See Figure 55.

FIGURE 55

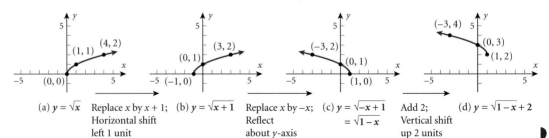

(a) $y = \sqrt{x}$ Replace x by $x + 1$;   (b) $y = \sqrt{x + 1}$ Replace x by $-x$;   (c) $y = \sqrt{-x + 1}$ Add 2;   (d) $y = \sqrt{1 - x} + 2$
              Horizontal shift   Reflect $= \sqrt{1 - x}$ Vertical shift
              left 1 unit   about y-axis   up 2 units

EXERCISE 1.5

In Problems 1–8, match each graph to one of the following functions.

A. $y = x^2 + 2$ B. $y = -x^2 + 2$ C. $y = |x| + 2$ D. $y = -|x| + 2$

E. $y = (x - 2)^2$ F. $y = -(x + 2)^2$ G. $y = |x - 2|$ H. $y = -|x + 2|$

1. 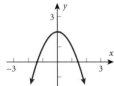   **2.** 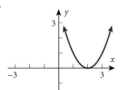   **3.** 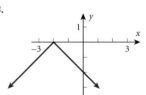   **4.**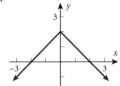

5. 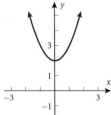   **6.** 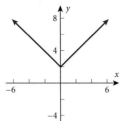   **7.** 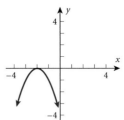   **8.**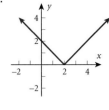

In Problems 9–14, write the function whose graph is the graph of $y = x^3$, but is:

 9. Shifted to the right 4 units   10. Shifted to the left 4 units   11. Shifted up 4 units

12. Shifted down 4 units   13. Reflected about the y-axis   14. Reflected about the x-axis

In Problems 15–18, find the function that is finally graphed after the following transformations are applied to the graph of $y = \sqrt{x}$.

 15. (1) Shift up 2 units   16. (1) Reflect about the x-axis
      (2) Reflect about the x-axis   (2) Shift right 3 units
      (3) Reflect about the y-axis   (3) Shift down 2 units

17. (1) Reflect about the *x*-axis
(2) Shift up 2 units
(3) Shift left 3 units

18. (1) Shift up 2 units
(2) Reflect about the *y*-axis
(3) Shift left 3 units

19. If (3, 0) is a point on the graph of $y = f(x)$, which of the following must be on the graph of $y = -f(x)$?
(a) (0, 3)   (b) (0, −3)   (c) (3, 0)   (d) (−3, 0)

20. If (3, 0) is a point on the graph of $y = f(x)$, which of the following must be on the graph of $y = f(-x)$?
(a) (0, 3)   (b) (0, −3)   (c) (3, 0)   (d) (−3, 0)

In Problems 21–38, graph each function using the techniques of shifting and/or reflecting. Start with the graph of the basic function (for example, $y = x^2$) and show all stages.

21. $f(x) = x^2 - 1$

22. $f(x) = x^2 + 4$

23. $g(x) = x^3 + 1$

24. $g(x) = x^3 - 1$

25. $h(x) = \sqrt{x - 2}$

26. $h(x) = \sqrt{x + 1}$

27. $f(x) = (x - 1)^3 + 2$

28. $f(x) = (x + 2)^3 - 3$

29. $f(x) = -\sqrt[3]{x}$

30. $f(x) = -\sqrt{x}$

31. $g(x) = |-x|$

32. $g(x) = \sqrt[3]{-x}$

33. $h(x) = -x^3 + 2$

34. $h(x) = \dfrac{1}{-x} + 2$

35. $g(x) = \sqrt{x - 2} + 1$

36. $g(x) = |x + 1| - 3$

37. $h(x) = \sqrt{-x} - 2$

38. $f(x) = -(x + 1)^3 - 1$

In Problems 39–40, the graph of a function f is illustrated. Use the graph of f as the first step toward graphing each of the following functions:

(a) $F(x) = f(x) + 3$ (b) $G(x) = f(x + 2)$ (c) $P(x) = -f(x)$ (d) $H(x) = f(x + 1) - 2$ (e) $g(x) = f(-x)$

39.

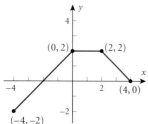

40.

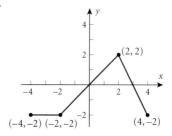

41. Exploration
(a) Use a graphing utility to graph $y = x + 1$ and $y = |x + 1|$.
(b) Graph $y = 4 - x^2$ and $y = |4 - x^2|$.
(c) Graph $y = x^3 + x$ and $y = |x^3 + x|$.
(d) What do you conclude about the relationship between the graphs of $y = f(x)$ and $y = |f(x)|$?

42. Exploration
(a) Use a graphing utility to graph $y = x + 1$ and $y = |x| + 1$.
(b) Graph $y = 4 - x^2$ and $y = 4 - |x|^2$.
(c) Graph $y = x^3 + x$ and $y = |x|^3 + |x|$.
(d) What do you conclude about the relationship between the graphs of $y = f(x)$ and $y = f(|x|)$?

43. The graph of a function *f* is illustrated in the figure.
(a) Draw the graph of $y = |f(x)|$.
(b) Draw the graph of $y = f(|x|)$.

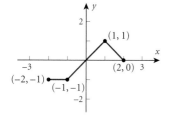

44. The graph of a function *f* is illustrated in the figure.
(a) Draw the graph of $y = |f(x)|$.
(b) Draw the graph of $y = f(|x|)$.

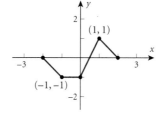

Chapter 1 Review

OBJECTIVES

Section	You should be able to	Review Exercises
1.1	1 Graph equations by plotting points	1
	2 Find intercepts from a graph	2, 27(b), 28(c), 29(f), 30(f)
	3 Find intercepts from an equation	3–10, 53(b)–56(b)
	4 Test an equation for symmetry with respect to the (a) x-axis, (b) y-axis, and (c) origin	3–10
1.2	1 Find the value of a function	11–16, 27(c), 28(b), 69, 70
	2 Find the difference quotient of a function	25, 26
	3 Find the domain of a function	17–24, 53(a)–56(a)
	4 Solve applied problems involving functions	71–78
1.3	1 Identify the graph of a function	49
	2 Obtain information from or about the graph of a function	27–30
	3 Determine even and odd functions from a graph	29(e), 30(e)
	4 Determine even and odd functions from the equation	31–38
	5 Use a graph to determine where a function is increasing, is decreasing, or is constant	29(b), 30(b)
	6 Use a graph to locate local maxima and minima	29(c), 30(c)
	7 Use a graphing utility to approximate local maxima and minima and to determine where a function is increasing or decreasing	39–42
	8 Find the average rate of change of a function	43–48
1.4	1 Graph the functions listed in the library of functions	50–52
	2 Graph piecewise-defined functions	53(c)–56(c)
1.5	1 Graph functions using horizontal and vertical shifts	57–60, 63, 64, 65(c), (d), 66(c), (d)
	2 Graph functions using reflections about the x-axis or y-axis	61, 62, 65(a), (b), 66(a), (b)

THINGS TO KNOW

Function (p. 109)   A relation between two sets of real numbers so that each number x in the first set, the domain, has corresponding to it exactly one number y in the second set. The range is the set of y values of the function for the x values in the domain.

x is the independent variable; y is the dependent variable.

A function f may be defined implicitly by an equation involving x and y or explicitly by writing $y = f(x)$.

Function notation (p. 109) $y = f(x)$

f is a symbol for the function.

x is the argument, or independent variable.

y is the dependent variable.

Domain (p. 113)   If unspecified, the domain of a function f is the largest set of real numbers for which $f(x)$ is a real number.

Difference quotient of f (p. 111) $\dfrac{f(x+h)-f(x)}{h}, \quad h \neq 0$

Vertical-line test (p. 121) A set of points in the plane is the graph of a function if and only if every vertical line intersects the graph in at most one point.

Even function f (p. 125) $f(-x) = f(x)$ for every x in the domain ($-x$ must also be in the domain).

Odd function f (p. 125) $f(-x) = -f(x)$ for every x in the domain ($-x$ must also be in the domain).

Increasing function (p. 127) A function f is increasing on an open interval I if, for any choice of x_1 and x_2 in I, with $x_1 < x_2$, we have $f(x_1) < f(x_2)$.

Decreasing function (p. 127) A function f is decreasing on an open interval I if, for any choice of x_1 and x_2 in I, with $x_1 < x_2$, we have $f(x_1) > f(x_2)$.

Constant function (p. 127) A function f is constant on an interval I if, for all choices of x in I, the values of $f(x)$ are equal.

Local maximum (p. 128) A function f has a local maximum at c if there is an open interval I containing c so that, for all $x \neq c$ in I, $f(x) < f(c)$.

Local minimum (p. 128) A function f has a local minimum at c if there is an open interval I containing c so that, for all $x \neq c$ in I, $f(x) > f(c)$.

Average rate of change of a function (p. 130) The average rate of change of f from c to x is

$$\dfrac{\Delta y}{\Delta x} = \dfrac{f(x) - f(c)}{x - c}, \quad x \neq c$$

LIBRARY OF FUNCTIONS

Constant function (p. 140)
$f(x) = b$
Graph is a horizontal line with y-intercept $(0, b)$.

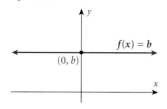

Linear function (p. 140)
$f(x) = mx + b$
Graph is a line with slope m and y-intercept $(0, b)$.

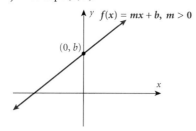

Identity function (p. 140)
$f(x) = x$
Graph is a line with slope 1 and y-intercept $(0, 0)$.

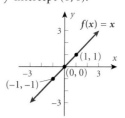

Square function (p. 140)
$f(x) = x^2$
Graph is a parabola with intercept at $(0, 0)$.

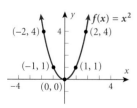

Cube function (p. 141)
$f(x) = x^3$

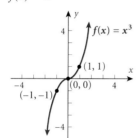

Square root function (p. 141)
$f(x) = \sqrt{x}$

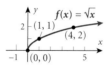

Cube root function (p. 141)
$f(x) = \sqrt[3]{x}$

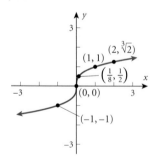

Reciprocal function (p. 141)
$f(x) = \dfrac{1}{x}$

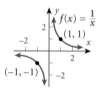

Absolute value function (p. 142)
$f(x) = |x|$

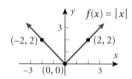

Greatest integer function (p. 142)
$f(x) = \text{int}(x)$

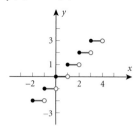

TRUE–FALSE ITEMS

T F **1.** The domain of the function $f(x) = \sqrt{x}$ is the set of all real numbers.

T F **2.** For any function f, it follows that
$$f(x + h) = f(x) + f(h).$$

T F **3.** A function can have more than one y-intercept.

T F **4.** The graph of a function $y = f(x)$ always crosses the y-axis.

T F **5.** The y-intercept of the graph of the function $y = f(x)$, whose domain is all real numbers, is $(0, f(0))$.

T F **6.** To obtain the graph of $y = f(x + 2) - 3$, shift the graph of $y = f(x)$ horizontally to the right 2 units and vertically down 3 units.

FILL IN THE BLANKS

1. If f is a function defined by the equation $y = f(x)$, then x is called the _____ variable and y is the _____ variable.

2. A set of points in the xy-plane is the graph of a function if and only if every _____ line intersects the graph in at most one point.

3. If the point $(5, -3)$ is a point on the graph of f, then $f(\_\_\_\_) = \_\_\_\_$.

4. If the point $(-1, 2)$ is on the graph of $f(x) = ax^2 + 4$, then $a = \_\_\_\_$.

5. Suppose that the x-intercepts of the graph of $y = f(x)$ are $(-2, 0)$, $(1, 0)$, and $(5, 0)$. The x-intercepts of $y = f(x + 3)$ are _____, _____, and _____.

REVIEW EXERCISES Blue problem numbers indicate the author's suggestion for a practice test.

1. Graph $y = x^2 + 4$ by plotting points.

2. List the intercepts of the graph shown.

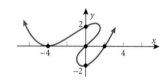

Chapter 1 Review 159

In Problems 3–10, list the intercepts and test for symmetry with respect to the x-axis, the y-axis, and the origin.

3. $2x = 3y^2$
4. $y = 5x$
5. $x^2 + 4y^2 = 16$
6. $9x^2 - y^2 = 9$
7. $y = x^4 + 2x^2 + 1$
8. $y = x^3 - x$
9. $x^2 + x + y^2 + 2y = 0$
10. $x^2 + 4x + y^2 - 2y = 0$

In Problems 11–16, find the following for each function:

(a) $f(2)$ (b) $f(-2)$ (c) $f(-x)$ (d) $-f(x)$ (e) $f(x-2)$ (f) $f(2x)$

11. $f(x) = \dfrac{3x}{x^2 - 1}$
12. $f(x) = \dfrac{x^2}{x+1}$
13. $f(x) = \sqrt{x^2 - 4}$
14. $f(x) = |x^2 - 4|$
15. $f(x) = \dfrac{x^2 - 4}{x^2}$
16. $f(x) = \dfrac{x^3}{x^2 - 9}$

In Problems 17–24, find the domain of each function.

17. $f(x) = \dfrac{x}{x^2 - 9}$
18. $f(x) = \dfrac{3x^2}{x-2}$
19. $f(x) = \sqrt{2-x}$
20. $f(x) = \sqrt{x+2}$
21. $h(x) = \dfrac{\sqrt{x}}{|x|}$
22. $g(x) = \dfrac{|x|}{x}$
23. $f(x) = \dfrac{x}{x^2 + 2x - 3}$
24. $F(x) = \dfrac{1}{x^2 - 3x - 4}$

In Problems 25–26, find the difference quotient of each function f, that is, find $\dfrac{f(x+h) - f(x)}{h}$, $h \ne 0$.

25. $f(x) = -2x^2 + x + 1$
26. $f(x) = 3x^2 - 2x + 4$

27. Using the graph of the function f shown below,

(a) Find the domain and range of f.
(b) List the intercepts.
(c) Find $f(-2)$.
(d) For what values of x does $f(x) = -3$?
(e) Solve $f(x) > 0$.

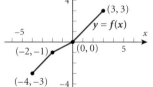

28. Using the graph of the function g shown below,

(a) Find the domain and range of g.
(b) Find $g(-1)$.
(c) List the intercepts of g.
(d) For what value of x does $g(x) = -3$?
(e) Solve $g(x) > 0$.

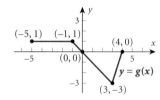

In Problems 29 and 30, use the graph of the function f to find:

(a) The domain and the range of f.
(b) The intervals on which f is increasing, decreasing, or constant.
(c) The local minima and local maxima.
(d) Whether the graph is symmetric with respect to the x axis, the y-axis, or the origin.
(e) Whether the function is even, odd, or neither.
(f) The intercepts, if any.

29.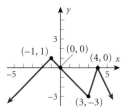

30.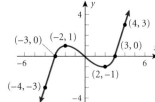

In Problems 31–38, determine (algebraically) whether the given function is even, odd, or neither.

31. $f(x) = x^3 - 4x$
32. $g(x) = \dfrac{4 + x^2}{1 + x^4}$
33. $h(x) = \dfrac{1}{x^4} + \dfrac{1}{x^2} + 1$
34. $F(x) = \sqrt{1 - x^3}$
35. $G(x) = 1 - x + x^3$
36. $H(x) = 1 + x + x^2$
37. $f(x) = \dfrac{x}{1 + x^2}$
38. $g(x) = \dfrac{1 + x^2}{x^3}$

In Problems 39–42, use a graphing utility to graph each function over the indicated interval. Approximate any local maxima and local minima. Determine where the function is increasing and where it is decreasing.

39. $f(x) = 2x^3 - 5x + 1 \quad (-3, 3)$
40. $f(x) = -x^3 + 3x - 5 \quad (-3, 3)$
41. $f(x) = 2x^4 - 5x^3 + 2x + 1 \quad (-2, 3)$
42. $f(x) = -x^4 + 3x^3 - 4x + 3 \quad (-2, 3)$

In Problems 43–44, find the average rate of change of f.
(a) from 1 to 2   (b) from 0 to 1   (c) from 2 to 4

43. $f(x) = 8x^2 - x$
44. $f(x) = 2x^3 + x$

In Problems 45–48, find the average rate of change from 2 to x for each function f. Be sure to simplify.

45. $f(x) = 2 - 5x$
46. $f(x) = 2x^2 + 7$
47. $f(x) = 3x - 4x^2$
48. $f(x) = x^2 - 3x + 2$

49. Which of the following are graphs of functions.

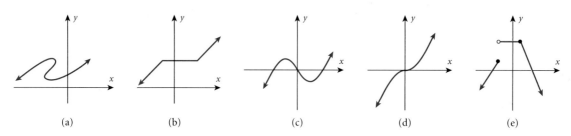

(a) (b) (c) (d) (e)

In Problems 50–52, sketch the graph of each function. Be sure to label at least three points.

50. $f(x) = |x|$
51. $f(x) = \sqrt[3]{x}$
52. $f(x) = \sqrt{x}$

In Problems 53–56:
(a) Find the domain of each function.
(b) Locate any intercepts.
(c) Graph each function.
(d) Based on the graph, find the range.

53. $f(x) = \begin{cases} 3x & -2 < x \le 1 \\ x + 1 & x > 1 \end{cases}$
54. $f(x) = \begin{cases} x - 1 & -3 < x < 0 \\ 3x - 1 & x \ge 0 \end{cases}$
55. $f(x) = \begin{cases} x & -4 \le x < 0 \\ 1 & x = 0 \\ 3x & x > 0 \end{cases}$
56. $f(x) = \begin{cases} x^2 & -2 \le x \le 2 \\ 2x - 1 & x > 2 \end{cases}$

In Problems 57–64, graph each function using shifting and/or reflections. Identify any intercepts of the graph. State the domain and, based on the graph, find the range.

57. $F(x) = |x| - 4$
58. $f(x) = |x| + 4$
59. $h(x) = \sqrt{x - 1}$
60. $h(x) = \sqrt{x} - 1$
61. $f(x) = \sqrt{1 - x}$
62. $f(x) = -\sqrt{x + 3}$
63. $h(x) = (x - 1)^2 + 2$
64. $h(x) = (x + 2)^2 - 3$

65. For the graph of the function f shown below, draw the graph of:
(a) $y = f(-x)$ (b) $y = -f(x)$
(c) $y = f(x + 2)$ (d) $y = f(x) + 2$

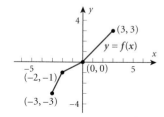

66. For the graph of the function g shown below, draw the graph of:
(a) $y = g(-x)$ (b) $y = -g(x)$
(c) $y = g(x + 2)$ (d) $y = g(x) + 2$

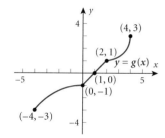

67. Given that f is a linear function, $f(4) = -5$ and $f(0) = 3$, write the equation that defines f.

68. Given that g is a linear function with slope $= -4$ and $g(-2) = 2$, write the equation that defines g.

69. A function f is defined by
$$f(x) = \frac{Ax + 5}{6x - 2}$$
If $f(1) = 4$, find A.

70. A function g is defined by
$$g(x) = \frac{A}{x} + \frac{8}{x^2}$$
If $g(-1) = 0$, find A.

71. Volume of a Cylinder The volume V of a right circular cylinder of height h and radius r is $V = \pi r^2 h$. If the height is twice the radius, express the volume V as a function of r.

72. Volume of a Cone The volume V of a right circular cone is $V = \frac{1}{3}\pi r^2 h$. If the height is twice the radius, express the volume V as a function of r.

73. Demand Equation The price p and the quantity x sold of a certain product obey the demand equation
$$p = -\frac{1}{6}x + 100, \quad 0 \le x \le 600$$
(a) Express the revenue R as a function of x. (Remember, $R = xp$.)
(b) What is the revenue if 200 units are sold?

74. Demand Equation The price p and the quantity x sold of a certain product obey the demand equation
$$p = -\frac{1}{3}x + 100, \quad 0 \le x \le 300$$
(a) Express the revenue R as a function of x.
(b) What is the revenue if 100 units are sold?

75. Demand Equation The price p and the quantity x sold of a certain product obey the demand equation
$$x = -5p + 100, \quad 0 \le p \le 20$$
(a) Express the revenue R as a function of x.
(b) What is the revenue if 15 units are sold?

76. Demand Equation The price p and the quantity x sold of a certain product obey the demand equation
$$x = -20p + 500, \quad 0 \le p \le 25$$
(a) Express the revenue R as a function of x.
(b) What is the revenue if 20 units are sold?

77. Cost of a Drum A drum in the shape of a right circular cylinder is required to have a volume of 500 cubic centimeters. The top and bottom are made of material that costs 6¢ per square centimeter; the sides are made of material that costs 4¢ per square centimeter.

(a) Express the total cost C of the material as a function of the radius r of the cylinder. [**Hint:** The volume V of a right circular cylinder of height h and radius r is $V = \pi r^2 h$.]
(b) What is the cost if the radius is 4 cm?
(c) What is the cost if the radius is 8 cm?
(d) Graph $C = C(r)$. For what value of r is the cost C least?

78. Material Needed to Make a Drum A steel drum in the shape of a right circular cylinder is required to have a volume of 100 cubic feet.

(a) Express the amount A of material required to make the drum as a function of the radius r of the cylinder.

[**Hint:** The surface area S of a right circular cylinder of height h and radius r is $S = 2\pi r^2 + 2\pi rh$.]

(b) How much material is required if the drum is of radius 3 feet?
(c) Of radius 4 feet?
(d) Of radius 5 feet?
(e) Graph $A = A(r)$. For what value of r is A smallest?

Chapter 1 Project

For the one-day sightseeing trip to Charlotte, North Carolina, you and your friend decide to rent a mid-size car and, naturally, you want to do this as cheaply as possible.

You begin by contacting two well-known car rental companies: Avis and Enterprise. Avis offers a mid-size car for $64.99 per day with unlimited mileage, so the number of miles you actually drive the car will not matter. Enterprise, on the other hand, offers a mid-size car for $45.87 per day with 150 free miles, but will charge $0.25 per mile for each mile in excess of 150 miles. Enterprise is the better deal as long as you drive less than 150 miles, but at what point, if any, will Avis be better? We'll use piecewise-defined functions to arrive at the answer.

1. Let x denote the number of miles the rental car is driven. Find the function $A = A(x)$ that gives the cost of driving the Avis car x miles. What kind of function will this be?

2. Find the function $E = E(x)$ that gives the cost of driving the Enterprise car x miles. Remember that the rule for computing this cost changes when x exceeds 150 miles, so a piecewise function is required.

3. Graph the functions $A = A(x)$ and $E = E(x)$ on the same set of axes. At what number of miles does the Avis rental car become a better choice?

4. In an effort to find an even better deal, you contact SaveALot Car Rental. They offer a mid-size car for $36.99 per day with 100 free miles, but each mile in excess of 100 will cost $0.30. Find the function $S = S(x)$ that gives the cost of driving the SaveALot car x miles. Graph this function along with $A = A(x)$ and $E = E(x)$.

5. Determine the number of miles driven and the companies that minimize the cost of the car rental.

6. In one last attempt to save money, you contact USave Car Rental and are offered a mid-size car for $35.99 per day with 50 free miles, but each mile in excess of 50 will cost $0.35. Find the function $U = U(x)$ that gives the cost of driving the USave car x miles. Graph this function along with the other three.

7. Determine the mileage and companies that minimize the cost of renting.

8. Comment on which car rental company you would use. Be sure to provide reasons.

MATHEMATICAL QUESTIONS FROM PROFESSIONAL EXAMS*

1. **Actuary Exam—Part I** What is the range of the function $f(x) = x^2$ with domain $(-1, 2]$?

   (a) $1 < y \leq 4$ (b) $1 < y < 4$
   (c) $0 < y \leq 4$ (d) $0 \leq y < 4$
   (e) $1 \leq y \leq 4$

2. **Actuary Exam—Part I** If $f(x + 1) = x^3 + 6x^2 + x + 3$ for all real x, then $f(4) = ?$

   (a) 22   (b) 42   (c) 57   (d) 87   (e) 116

3. **Actuary Exam—Part I** What is the largest possible subset of $\mathbb{R}$ (real numbers) that can be a domain for the function $f(x) = \sqrt{x^3 - x}$?

   (a) $[1, \infty)$
   (b) $\{0\} \cup [1, \infty)$
   (c) $[-1, 0] \cup [1, \infty)$
   (d) $(-\infty, -1] \cup [1, \infty)$
   (e) $(-\infty, -1] \cup \{0\} \cup [1, \infty)$

*Copyright © 1998, 1999 by the Society of Actuaries. Reprinted with permission.

CHAPTER 2

Classes of Functions

OUTLINE

2.1 Quadratic Functions

2.2 Power Functions; Polynomial Functions; Rational Functions

2.3 Exponential Functions

2.4 Logarithmic Functions

2.5 Properties of Logarithms

2.6 Continuously Compounded Interest

- **Chapter Review**
- **Chapter Project**
- **Mathematical Questions from Professional Exams**

Almost everyday you can pick up a newspaper and read about some prediction. Headlines appear like 'World population to increase 46% by 2050'. Or 'Projected growth in population of North America will be 41.8% over the next 50 years. Of course, predictions like these are important for proper planning and allocation of future resources. But where did these numbers come from? The Chapter Project provides one method of making predictions about future populations.

163

Chapter 2 Classes of Functions

A LOOK BACK, A LOOK FORWARD

In Chapter 1, we began our discussion of functions. We defined domain and range and independent and dependent variables; we found the value of a function and graphed functions. We continued our study of functions by listing the properties that a function might have, like being even or odd, and we created a library of functions, naming key functions and listing their properties, including their graphs.

In this chapter, we look at four general classes of functions—polynomial functions, rational functions, exponential functions, and logarithmic functions—and examine their properties. Polynomial functions are arguably the simplest expressions in algebra. For this reason, they are often used to approximate other, more complicated functions. Rational functions are simply ratios of polynomial functions.

We begin with a discussion of quadratic functions, a type of polynomial function.

2.1 Quadratic Functions

PREPARING FOR THIS SECTION *Before getting started, review the following:*

> Completing the Square (Chapter 0, Section 0.4, pp. 45–46)

> Square Root Method (Chapter 0, Section 0.4, pp. 44–45)

> Quadratic Equations (Chapter 0, Section 0.4, pp. 43–49)

> Geometry Review (Chapter 0, Section 0.7, pp. 71–74)

OBJECTIVES
1. Locate the vertex and axis of symmetry of a quadratic function
2. Graph quadratic functions
3. Find the maximum or the minimum value of a quadratic function
4. Use the maximum or the minimum value of a quadratic function to solve applied problems

A *quadratic function* is a function that is defined by a second-degree polynomial in one variable.

A **quadratic function** is a function of the form

$$f(x) = ax^2 + bx + c \quad (1)$$

where a, b, and c are real numbers and $a \neq 0$. The domain of a quadratic function is the set of all real numbers.

Many applications require a knowledge of the properties of the graph of a quadratic function. For example, suppose that Texas Instruments collects the data shown in Table 1 that relate the number of calculators sold at the price p per calculator. Since the price of a product determines the quantity that will be purchased, we treat price as the independent variable.

TABLE 1

Price per Calculator, p (Dollars)	Number of Calculators, x
60	11,100
65	10,115
70	9,652
75	8,731
80	8,087
85	7,205
90	6,439

A linear relationship between the number of calculators and the price p per calculator may be given by the equation

$$x = 21{,}000 - 150p$$

Then the revenue R derived from selling x calculators at the price p per calculator is

$$R = xp$$
$$R(p) = (21{,}000 - 150p)p$$
$$= -150p^2 + 21{,}000p$$

So the revenue R is a quadratic function of the price p. Figure 1 illustrates the graph of this revenue function, whose domain is $0 \leq p \leq 140$, since both x and p must be nonnegative. Later in this section we shall determine the price p that maximizes revenue.

A second situation in which a quadratic function appears involves the motion of a projectile. Based on Newton's second law of motion (force equals mass times acceleration, $F = ma$), it can be shown that, ignoring air resistance, the path of a projectile propelled upward at an inclination to the horizontal is the graph of a quadratic function. See Figure 2 for an illustration. Later in this section we shall analyze the path of a projectile.

FIGURE 1 Graph of a revenue function: $R = -150p^2 + 21{,}000p$

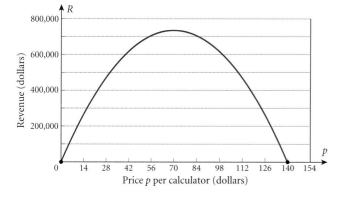

FIGURE 2 Path of a cannonball

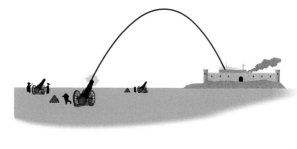

Graphing Quadratic Functions

We know how to graph the quadratic function $f(x) = x^2$. Figure 3 on page 166 shows the graph of three functions of the form $f(x) = ax^2$, $a > 0$, for $a = 1$, $a = \frac{1}{2}$, and $a = 3$. Notice that the larger the value of a, the "narrower" the graph, and the smaller the value of a, the "wider" the graph.

Figure 4 shows the graphs of $f(x) = ax^2$ for $a < 0$. Notice that these graphs are reflections about the x-axis of the graphs in Figure 3. Based on the results of these two figures, we can draw some general conclusions about the graph of $f(x) = ax^2$. First, as $|a|$ increases, the graph becomes *narrower*, and as $|a|$ gets closer to zero, the graph gets *wider*. Second, if a is positive, then the graph opens *up*, and if a is negative, the graph opens *down*.

The graphs in Figures 3 and 4 are typical of the graphs of all quadratic functions, which we call **parabolas**. Refer to Figure 5, where two parabolas are pictured. The one

FIGURE 3

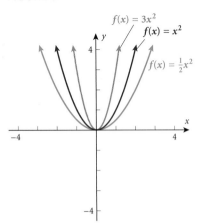

FIGURE 4

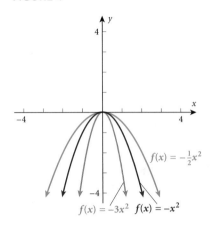

FIGURE 5 Graphs of a quadratic function, $f(x) = ax^2 + bx + c$, $a \neq 0$

(a) Opens up $a > 0$

(b) Opens down $a < 0$

on the left **opens up** and has a lowest point; the one on the right **opens down** and has a highest point. The lowest or highest point of a parabola is called the **vertex.** The vertical line passing through the vertex in each parabola in Figure 5 is called the **axis of symmetry** (sometimes abbreviated to **axis**) of the parabola. Because the parabola is symmetric about its axis, the axis of symmetry of a parabola can be used to find additional points on the parabola.

The parabolas shown in Figure 5 are the graphs of a quadratic function $f(x) = ax^2 + bx + c$, $a \neq 0$. Notice that the coordinate axes are not included in the figure. Depending on the values of a, b, and c, the axes could be anywhere. The important fact is that the shape of the graph of a quadratic function will look like one of the parabolas in Figure 5.

A key element in graphing a quadratic function is locating the vertex. To find a formula, we begin with a quadratic function $f(x) = ax^2 + bx + c$, $a \neq 0$, and complete the square in x.

$$f(x) = ax^2 + bx + c \qquad a \neq 0$$

$$= a\left(x^2 + \frac{b}{a}x\right) + c \qquad \text{Factor out } a \text{ from } ax^2 + bx$$

$$= a\left(x^2 + \frac{b}{a}x + \frac{b^2}{4a^2}\right) + c - a\left(\frac{b^2}{4a^2}\right) \qquad \text{Complete the square by adding and subtracting } a\frac{b^2}{4a^2}. \text{ Look closely at this step!}$$

$$= a\left(x + \frac{b}{2a}\right)^2 + c - \frac{b^2}{4a} \qquad \text{Factor the perfect square.}$$

$$f(x) = a\left(x + \frac{b}{2a}\right)^2 + \frac{4ac - b^2}{4a} \qquad c - \frac{b^2}{4a} = c \cdot \frac{4a}{4a} - \frac{b^2}{4a} = \frac{4ac - b^2}{4a} \quad (2)$$

Suppose that $a > 0$. If $x = -\frac{b}{2a}$, then the term $a\left(x + \frac{b}{2a}\right)^2 = a\left(-\frac{b}{2a} + \frac{b}{2a}\right)^2 = a \cdot 0 = 0$. For any other value of x, the term $a\left(x + \frac{b}{2a}\right)^2$ will be positive. [Do you see why? $\left(x + \frac{b}{2a}\right)^2$ is positive for $x \neq -\frac{b}{2a}$ because it is a nonzero quantity squared, a is positive, and the product of two positive quantities is positive.] Because $a\left(x + \frac{b}{2a}\right)^2$ is zero if $x = -\dfrac{b}{2a}$ and is positive if $x \neq -\dfrac{b}{2a}$, the value of the function f given in (2) will be smallest when $x = -\dfrac{b}{2a}$. That is, if $a > 0$, the parabola opens up, the vertex is at $\left(-\dfrac{b}{2a}, f\left(-\dfrac{b}{2a}\right)\right)$, and the vertex is a minimum point.

Similarly, if $a < 0$, the term $a\left(x + \dfrac{b}{2a}\right)^2$ is zero for $x = -\dfrac{b}{2a}$ and is negative if $x \neq -\dfrac{b}{2a}$. In this case the largest value of $f(x)$ occurs when $x = -\dfrac{b}{2a}$. That is, if $a < 0$, the parabola opens down, the vertex is at $\left(-\dfrac{b}{2a}, f\left(-\dfrac{b}{2a}\right)\right)$, and the vertex is a maximum point.

We summarize these remarks as follows:

Properties of the Graph of a Quadratic Function

$$f(x) = ax^2 + bx + c, \quad a \neq 0$$

Vertex $= \left(-\dfrac{b}{2a}, f\left(-\dfrac{b}{2a}\right)\right)$ Axis of symmetry: the line $x = -\dfrac{b}{2a}$

Parabola opens up if $a > 0$; the vertex is a minimum point.
Parabola opens down if $a < 0$; the vertex is a maximum point.

1 EXAMPLE 1 **Locating the Vertex and Axis of Symmetry**

Locate the vertex and axis of symmetry of the parabola defined by $f(x) = -3x^2 + 6x + 1$. Does it open up or down?

SOLUTION For this quadratic function, $a = -3$, $b = 6$, and $c = 1$. The x-coordinate of the vertex is

$$-\frac{b}{2a} = -\frac{6}{-6} = 1$$

The y-coordinate of the vertex is

$$f\left(-\frac{b}{2a}\right) = f(1) = -3 + 6 + 1 = 4$$

The vertex is located at the point $(1, 4)$. The axis of symmetry is the line $x = 1$. Because $a = -3 < 0$, the parabola opens down. ▶

Graph quadratic functions 2   The facts that we gathered in Example 1, together with the location of the intercepts, usually provide enough information to graph $f(x) = ax^2 + bx + c, a \neq 0$.

The y-intercept is found by finding the value of f at $x = 0$; that is, by finding $f(0) = c$.
The x-intercepts, if there are any, are found by solving the equation

$$f(x) = ax^2 + bx + c = 0$$

This equation has two, one, or no real solutions, depending on whether the discriminant $b^2 - 4ac$ is positive, 0, or negative. Depending on the value of the discriminant, the graph of f has x-intercepts as follows:

The x-Intercepts of a Quadratic Function

1. If the discriminant $b^2 - 4ac > 0$, the graph of $f(x) = ax^2 + bx + c$ has two distinct x-intercepts and so will cross the x-axis in two places.
2. If the discriminant $b^2 - 4ac = 0$, the graph of $f(x) = ax^2 + bx + c$ has one x-intercept and touches the x-axis at its vertex.
3. If the discriminant $b^2 - 4ac < 0$, the graph of $f(x) = ax^2 + bx + c$ has no x-intercept and so will not cross or touch the x-axis.

Figure 6 illustrates these possibilities for parabolas that open up.

FIGURE 6
$f(x) = ax^2 + bx + c, a > 0$

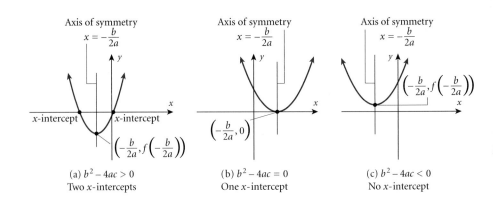

(a) $b^2 - 4ac > 0$
Two x-intercepts

(b) $b^2 - 4ac = 0$
One x-intercept

(c) $b^2 - 4ac < 0$
No x-intercept

EXAMPLE 2 Graphing a Quadratic Function Using Its Vertex, Axis, and Intercepts

Use the information from Example 1 and the locations of the intercepts to graph $f(x) = -3x^2 + 6x + 1$.

SOLUTION In Example 1, we found the vertex to be at $(1, 4)$ and the axis of symmetry to be $x = 1$. The y-intercept is found by letting $x = 0$. Since $f(0) = 1$, the y-intercept is $(0, 1)$. The x-intercepts are found by solving the equation $f(x) = 0$. This results in the equation

$$-3x^2 + 6x + 1 = 0 \quad a = -3, b = 6, c = 1$$

The discriminant $b^2 - 4ac = (6)^2 - 4(-3)(1) = 36 + 12 = 48 > 0$, so the equation has two real solutions and the graph has two x-intercepts. Using the quadratic formula, we find that

$$x = \frac{-b + \sqrt{b^2 - 4ac}}{2a} = \frac{-6 + \sqrt{48}}{-6} = \frac{-6 + 4\sqrt{3}}{-6} \approx -0.15$$

and

$$x = \frac{-b - \sqrt{b^2 - 4ac}}{2a} = \frac{-6 - \sqrt{48}}{-6} = \frac{-6 - 4\sqrt{3}}{-6} \approx 2.15$$

The x-intercepts are approximately $(-0.15, 0)$ and $(2.15, 0)$.

The graph is illustrated in Figure 7. Notice how we used the y-intercept $(0, 1)$ and the axis of symmetry, $x = 1$, to obtain the additional point $(2, 1)$ on the graph.

FIGURE 7

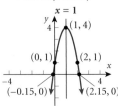

Quadratic Functions 169

 CHECK: Graph $f(x) = -3x^2 + 6x + 1$. Use ROOT or ZERO to locate the two x-intercepts and use MAXIMUM to locate the vertex. ▶

 NOW WORK PROBLEM 9.

If the graph of a quadratic function has only one x-intercept or none, it is usually necessary to plot an additional point to obtain the graph.

EXAMPLE 3 **Graphing a Quadratic Function Using Its Vertex, Axis, and Intercepts**

Graph $f(x) = x^2 - 6x + 9$ by determining whether the graph opens up or down. Find its vertex, axis of symmetry, y-intercept, and x-intercepts, if any.

FIGURE 8

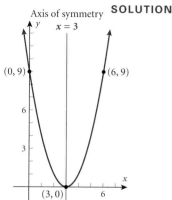

SOLUTION For $f(x) = x^2 - 6x + 9$, we have $a = 1$, $b = -6$, and $c = 9$. Since $a = 1 > 0$, the parabola opens up. The x-coordinate of the vertex is

$$-\frac{b}{2a} = -\frac{-6}{2(1)} = 3$$

The y-coordinate of the vertex is

$$f(3) = (3)^2 - 6(3) + 9 = 0$$

So the vertex is at $(3, 0)$. The axis of symmetry is the line $x = 3$. Since $f(0) = 9$, the y-intercept is $(0, 9)$. Since the vertex $(3, 0)$ lies on the x-axis, the graph touches the x-axis at the x-intercept. By using the axis of symmetry and the y-intercept $(0, 9)$, we can locate the additional point $(6, 9)$ on the graph. See Figure 8. ▶

 NOW WORK PROBLEM 17.

EXAMPLE 4 **Graphing a Quadratic Function Using Its Vertex, Axis, and Intercepts**

Graph $f(x) = 2x^2 + x + 1$ by determining whether the graph opens up or down. Find its vertex, axis of symmetry, y-intercept, and x-intercepts, if any.

SOLUTION For $f(x) = 2x^2 + x + 1$, we have $a = 2$, $b = 1$, and $c = 1$. Since $a = 2 > 0$, the parabola opens up. The x-coordinate of the vertex is

$$-\frac{b}{2a} = -\frac{1}{4}$$

FIGURE 9

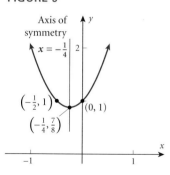

The y-coordinate of the vertex is

$$k = f\left(-\frac{1}{4}\right) = 2\left(\frac{1}{16}\right) + \left(-\frac{1}{4}\right) + 1 = \frac{7}{8}$$

So the vertex is at $\left(-\frac{1}{4}, \frac{7}{8}\right)$. The axis of symmetry is the line $x = -\frac{1}{4}$. Since $f(0) = 1$, the y-intercept is $(0, 1)$. The x-intercept(s), if any, obey the equation $2x^2 + x + 1 = 0$. Since the discriminant $b^2 - 4ac = (1)^2 - 4(2)(1) = -7 < 0$, this equation has no real solutions, and therefore the graph has no x-intercepts. We use the point (0, 1) and the axis of symmetry $x = -\dfrac{1}{4}$ to locate the additional point $\left(-\dfrac{1}{2}, 1\right)$ on the graph. See Figure 9.

NOW WORK PROBLEM 21.

SUMMARY

Steps for Graphing a Quadratic Function $f(x) = ax^2 + bx + c, \quad a \neq 0$

STEP 1: Determine the vertex, $\left(-\dfrac{b}{2a}, f\left(-\dfrac{b}{2a}\right)\right)$.

STEP 2: Determine the axis of symmetry, $x = -\dfrac{b}{2a}$.

STEP 3: Determine the y-intercept by finding $f(0) = c$.

STEP 4: Evaluate the discriminant $b^2 - 4ac$.
  (a) If $b^2 - 4ac > 0$, then the graph of the quadratic function has two x-intercepts, which are found by solving the equation $ax^2 + bx + c = 0$.
  (b) If $b^2 - 4ac = 0$, the vertex is the x-intercept.
  (c) If $b^2 - 4ac < 0$, there are no x-intercepts.

STEP 5: Determine an additional point if $b^2 - 4ac \leq 0$ by using the y-intercept and the axis of symmetry.

STEP 6: Plot the points and draw the graph.

Quadratic Models

When a mathematical model leads to a quadratic function, the properties of this quadratic function can provide important information about the model. For example, for a quadratic revenue function, we can find the maximum revenue; for a quadratic cost function, we can find the minimum cost.

To see why, recall that the graph of a quadratic function $f(x) = ax^2 + bx + c$ is a parabola with vertex at $\left(-\dfrac{b}{2a}, f\left(-\dfrac{b}{2a}\right)\right)$. This vertex is the highest point on the graph if $a < 0$ and the lowest point on the graph if $a > 0$. If the vertex is the highest point, $(a < 0)$, then $f\left(-\dfrac{b}{2a}\right)$ is the **maximum value** of f. If the vertex is the lowest point, $(a > 0)$, then $f\left(-\dfrac{b}{2a}\right)$ is the **minimum value** of f.

This property of the graph of a quadratic function enables us to answer questions involving optimization (finding maximum or minimum values) in models involving quadratic functions.

EXAMPLE 5 Finding the Maximum or Minimum Value of a Quadratic Function

Determine whether the quadratic function

$$f(x) = x^2 - 4x + 7$$

has a maximum or minimum value. Then find the maximum or minimum value.

Quadratic Functions 171

SOLUTION We compare $f(x) = x^2 - 4x + 7$ to $f(x) = ax^2 + bx + c$. We conclude that $a = 1$, $b = -4$, and $c = 7$. Since $a > 0$, the graph of f opens up, so the vertex is a minimum point. The minimum value occurs at

$$x = -\frac{b}{2a} \underset{\substack{\uparrow \\ a=1, b=-4}}{=} -\frac{-4}{2(1)} = \frac{4}{2} = 2$$

The minimum value is

$$f\left(-\frac{b}{2a}\right) = f(2) = 2^2 - 4(2) + 7 = 4 - 8 + 7 = 3$$

NOW WORK PROBLEM 29.

4 EXAMPLE 6 Maximizing Revenue

The marketing department at Texas Instruments has found that, when certain calculators are sold at a price of p dollars per unit, the revenue R (in dollars) as a function of the price p is

$$R(p) = -150p^2 + 21,000p$$

What unit price should be established in order to maximize revenue? If this price is charged, what is the maximum revenue?

SOLUTION The revenue R is

$$R(p) = -150p^2 + 21,000p \qquad R(p) = ap^2 + bp + c$$

The function R is a quadratic function with $a = -150$, $b = 21,000$, and $c = 0$. Because $a < 0$, the vertex is the highest point of the parabola. The revenue R is therefore a maximum when the price p is

$$p = -\frac{b}{2a} = -\frac{21,000}{2(-150)} = \frac{-21,000}{-300} = \$70.00$$

The maximum revenue R is

$$R(70) = -150(70)^2 + 21,000(70) = \$735,000$$

See Figure 10 for an illustration.

FIGURE 10

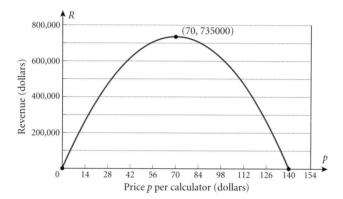

NOW WORK PROBLEM 37.

EXAMPLE 7 Maximizing the Area Enclosed by a Fence

A farmer has 2000 yards of fence to enclose a rectangular field. What is the largest area that can be enclosed?

SOLUTION Figure 11 illustrates the situation. The available fence represents the perimeter of the rectangle. If ℓ is the length and w is the width, then

$$\text{perimeter} = 2\ell + 2w$$
$$2\ell + 2w = 2000 \qquad (3)$$

FIGURE 11

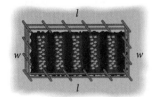

The area A of the rectangle is

$$A = \ell w$$

To express A in terms of a single variable, we solve equation (3) for w and substitute the result in $A = \ell w$. Then A involves only the variable ℓ. [You could also solve equation (3) for ℓ and express A in terms of w alone. Try it!]

$$2\ell + 2w = 2000 \qquad \text{Equation (3).}$$
$$2w = 2000 - 2\ell \qquad \text{Solve for } w.$$
$$w = \frac{2000 - 2\ell}{2} = 1000 - \ell$$

Then the area A is

$$A = \ell w = \ell(1000 - \ell) = -\ell^2 + 1000\ell$$

Now, A is a quadratic function of ℓ.

$$A(\ell) = -\ell^2 + 1000\ell \qquad a = -1, b = 1000, c = 0$$

Since $a < 0$, the vertex is a maximum point on the graph of A. The maximim value occurs at

$$\ell = -\frac{b}{2a} = -\frac{1000}{2(-1)} = 500$$

The maximum value of A is

$$A\left(-\frac{b}{2a}\right) = A(500) = -500^2 + 1000(500)$$
$$= -250,000 + 500,000 = 250,000$$

The largest area that can be enclosed by 2000 yards of fence in the shape of a rectangle is 250,000 square yards.

FIGURE 12

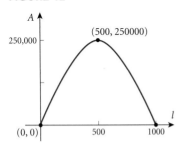

Figure 12 shows the graph of $A(\ell) = -\ell^2 + 1000\ell$.

NOW WORK PROBLEM 43.

EXAMPLE 8 Analyzing the Motion of a Projectile

A projectile is fired from a cliff 500 feet above the water at an inclination of 45° to the horizontal, with a muzzle velocity of 400 feet per second. In physics, it is established

FIGURE 13

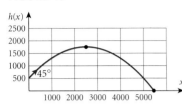

that the height h of the projectile above the water is given by

$$h(x) = \frac{-32x^2}{(400)^2} + x + 500$$

where x is the horizontal distance of the projectile from the base of the cliff. See Figure 13.

(a) Find the maximum height of the projectile.
(b) How far from the base of the cliff will the projectile strike the water?

SOLUTION **(a)** The height of the projectile is given by a quadratic function.

$$h(x) = \frac{-32x^2}{(400)^2} + x + 500 = \frac{-1}{5000}x^2 + x + 500 \quad a = \frac{-1}{5000}, \; b = 1, \; c = 500$$

We are looking for the maximum value of h. Since the maximum value is obtained at the vertex, we compute

$$x = -\frac{b}{2a} = -\frac{1}{2\left(\frac{-1}{5000}\right)} = \frac{5000}{2} = 2500$$

The maximum height of the projectile is

$$h(2500) = \frac{-1}{5000}(2500)^2 + 2500 + 500$$
$$= -1250 + 2500 + 500 = 1750 \text{ ft}$$

(b) The projectile will strike the water when the height is zero. To find the distance x traveled, we need to solve the equation

$$h(x) = \frac{-1}{5000}x^2 + x + 500 = 0$$

We use the quadratic formula with

$$b^2 - 4ac = 1 - 4\left(\frac{-1}{5000}\right)(500) = 1.4$$

$$x = \frac{-1 \pm \sqrt{1.4}}{2\left(\frac{-1}{5000}\right)} \approx \begin{cases} -458 \\ 5458 \end{cases}$$

We discard the negative solution and find that the projectile will strike the water at a distance of about 5458 feet from the base of the cliff. ▶

SEEING THE CONCEPT: Graph

$$h(x) = \frac{-1}{5000}x^2 + x + 500, \quad 0 \leq x \leq 5500$$

Use MAXIMUM to find the maximum height of the projectile, and use ROOT or ZERO to find the distance from the base of the cliff to where the projectile strikes the water. Compare your results with those obtained in Example 8. TRACE the path of the projectile. How far from the base of the cliff is the projectile when its height is 1000 ft? 1500 ft? ▶

NOW WORK PROBLEM 47.

Chapter 2 Classes of Functions

EXERCISE 2.1

In Problems 1–8, match each graph to one the following functions without using a graphing utility.

1. $f(x) = x^2 - 1$
2. $f(x) = -x^2 - 1$
3. $f(x) = x^2 - 2x + 1$
4. $f(x) = x^2 + 2x + 1$
5. $f(x) = x^2 - 2x + 2$
6. $f(x) = x^2 + 2x$
7. $f(x) = x^2 - 2x$
8. $f(x) = x^2 + 2x + 2$

(A)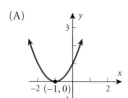

(B)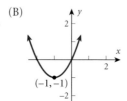

(C)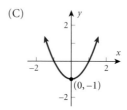

(D)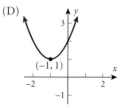

(E)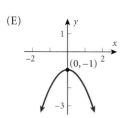

(F)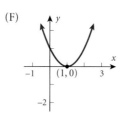

(G)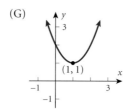

(H)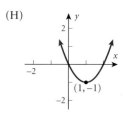

In Problems 9–26, graph each quadratic function by determining whether its graph opens up or down and by finding its vertex, axis of symmetry, y-intercept, and x-intercepts, if any. Determine the domain and the range of the function. Determine where the function is increasing and where it is decreasing.

9. $f(x) = x^2 + 2x$
10. $f(x) = x^2 - 4x$
11. $f(x) = -x^2 - 6x$
12. $f(x) = -x^2 + 4x$
13. $f(x) = 2x^2 - 8x$
14. $f(x) = 3x^2 + 18x$
15. $f(x) = x^2 + 2x - 8$
16. $f(x) = x^2 - 2x - 3$
17. $f(x) = x^2 + 2x + 1$
18. $f(x) = x^2 + 6x + 9$
19. $f(x) = 2x^2 - x + 2$
20. $f(x) = 4x^2 - 2x + 1$
21. $f(x) = -2x^2 + 2x - 3$
22. $f(x) = -3x^2 + 3x - 2$
23. $f(x) = 3x^2 + 6x + 2$
24. $f(x) = 2x^2 + 5x + 3$
25. $f(x) = -4x^2 - 6x + 2$
26. $f(x) = 3x^2 - 8x + 2$

In Problems 27–34, determine, without graphing, whether the given quadratic function has a maximum value or a minimum value and then find the value.

27. $f(x) = 2x^2 + 12x$
28. $f(x) = -2x^2 + 12x$
29. $f(x) = 2x^2 + 12x - 3$
30. $f(x) = 4x^2 - 8x + 3$
31. $f(x) = -x^2 + 10x - 4$
32. $f(x) = -2x^2 + 8x + 3$
33. $f(x) = -3x^2 + 12x + 1$
34. $f(x) = 4x^2 - 4x$

Answer Problems 35 and 36 using the following discussion: A quadratic function of the form $f(x) = ax^2 + bx + c$ with $b^2 - 4ac > 0$ may also be written in the form $f(x) = a(x - r_1)(x - r_2)$, where r_1 and r_2 are the x-intercepts of the graph of the quadratic function.

35. (a) Find a quadratic function whose x-intercepts are -3 and 1 with $a = 1$; $a = 2$; $a = -2$; $a = 5$.
    (b) How does the value of a affect the intercepts?
    (c) How does the value of a affect the axis of symmetry?
    (d) How does the value of a affect the vertex?
    (e) Compare the x-coordinate of the vertex with the midpoint of the x-intercepts. What might you conclude?

36. (a) Find a quadratic function whose x-intercepts are -5 and 3 with $a = 1$; $a = 2$; $a = -2$; $a = 5$.
    (b) How does the value of a affect the intercepts?
    (c) How does the value of a affect the axis of symmetry?
    (d) How does the value of a affect the vertex?
    (e) Compare the x-coordinate of the vertex with the midpoint of the x-intercepts. What might you conclude?

37. **Maximizing Revenue** Suppose that the manufacturer of a gas clothes dryer has found that, when the unit price is p dollars, the revenue R (in dollars) is

$$R(p) = -4p^2 + 4000p$$

What unit price should be established for the dryer to maximize revenue? What is the maximum revenue?

38. **Maximizing Revenue** The John Deere company has found that the revenue from sales of heavy-duty tractors is a function of the unit price p that it charges. If the revenue R is

$$R(p) = -\frac{1}{2}p^2 + 1900p$$

what unit price p should be charged to maximize revenue? What is the maximum revenue?

39. **Demand Equation** The price p and the quantity x sold of a certain product obey the demand equation

$$p = -\frac{1}{6}x + 100, \quad 0 \le x \le 600$$

(a) Express the revenue R as a function of x. (Remember, $R = xp$.)
(b) What is the revenue if 200 units are sold?
(c) What quantity x maximizes revenue? What is the maximum revenue?
(d) What price should the company charge to maximize revenue?

40. **Demand Equation** The price p and the quantity x sold of a certain product obey the demand equation

$$p = -\frac{1}{3}x + 100, \quad 0 \le x \le 300$$

(a) Express the revenue R as a function of x.
(b) What is the revenue if 100 units are sold?
(c) What quantity x maximizes revenue? What is the maximum revenue?
(d) What price should the company charge to maximize revenue?

41. **Demand Equation** The price p and the quantity x sold of a certain product obey the demand equation

$$x = -5p + 100, \quad 0 \le p \le 20$$

(a) Express the revenue R as a function of x.
(b) What is the revenue if 15 units are sold?
(c) What quantity x maximizes revenue? What is the maximum revenue?
(d) What price should the company charge to maximize revenue?

42. **Demand Equation** The price p and the quantity x sold of a certain product obey the demand equation

$$x = -20p + 500, \quad 0 \le p \le 25$$

(a) Express the revenue R as a function of x.
(b) What is the revenue if 20 units are sold?

(c) What quantity x maximizes revenue? What is the maximum revenue?
(d) What price should the company charge to maximize revenue?

43. **Enclosing a Rectangular Field** David has available 400 yards of fencing and wishes to enclose a rectangular area.

(a) Express the area A of the rectangle as a function of the width x of the rectangle.
(b) For what value of x is the area largest?
(c) What is the maximum area?

44. **Enclosing a Rectangular Field** Beth has 3000 feet of fencing available to enclose a rectangular field.

(a) Express the area A of the rectangle as a function of x, where x is the length of the rectangle.
(b) For what value of x is the area largest?
(c) What is the maximum area?

45. **Enclosing the Most Area with a Fence** A farmer with 4000 meters of fencing wants to enclose a rectangular plot that borders on a river. If the farmer does not fence the side along the river, what is the largest area that can be enclosed? (See the figure.)

46. **Enclosing the Most Area with a Fence** A farmer with 2000 meters of fencing wants to enclose a rectangular plot that borders on a straight highway. If the farmer does not fence the side along the highway, what is the largest area that can be enclosed?

47. **Analyzing the Motion of a Projectile** A projectile is fired from a cliff 200 feet above the water at an inclination of 45° to the horizontal, with a muzzle velocity of 50 feet per second. The height h of the projectile above the water is given by

$$h(x) = \frac{-32x^2}{(50)^2} + x + 200$$

where x is the horizontal distance of the projectile from the base of the cliff.

(a) How far from the base of the cliff is the height of the projectile a maximum?
(b) Find the maximum height of the projectile.

(c) How far from the base of the cliff will the projectile strike the water?
(d) Using a graphing utility, graph the function h, $0 \leq x \leq 200$.
(e) When the height of the projectile is 100 feet above the water, how far is it from the cliff?

48. **Analyzing the Motion of a Projectile** A projectile is fired at an inclination of 45° to the horizontal, with a muzzle velocity of 100 feet per second. The height h of the projectile is given by

$$h(x) = \frac{-32x^2}{(100)^2} + x$$

where x is the horizontal distance of the projectile from the firing point.

(a) How far from the firing point is the height of the projectile a maximum?
(b) Find the maximum height of the projectile.
(c) How far from the firing point will the projectile strike the ground?
(d) Using a graphing utility, graph the function h, $0 \leq x \leq 350$.
(e) When the height of the projectile is 50 feet above the ground, how far has it traveled horizontally?

49. **Constructing Rain Gutters** A rain gutter is to be made of aluminum sheets that are 12 inches wide by turning up the edges 90°. What depth will provide maximum cross-sectional area, allowing the most water to flow?

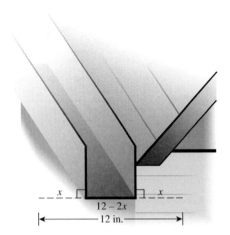

50. **Norman Windows** A Norman window has the shape of a rectangle surmounted by a semicircle of diameter equal to the width of the rectangle (see the figure). If the perimeter of the window is 20 feet, what dimensions will admit the most light (maximize the area)?
[**Hint:** Circumference of a circle = $2\pi r$; area of a circle = πr^2, where r is the radius of the circle.]

51. **Constructing a Stadium** A track and field playing area is in the shape of a rectangle with semicircles at each end (see the figure). The inside perimeter of the track is to be 400 meters. What should the dimensions of the rectangle be so that the area of the rectangle is a maximum?

52. **Architecture** A special window has the shape of a rectangle surmounted by an equilateral triangle (see the figure). If the perimeter of the window is 16 feet, what dimensions will admit the most light?
[**Hint:** Area of an equilateral triangle = $\frac{\sqrt{3}}{4}x^2$, where x is the length of a side of the triangle.]

53. **Hunting** The function $H(x) = -1.01x^2 + 114.3x + 451.0$ models the number of individuals who engage in hunting activities whose annual income is x thousand dollars.

(a) What is the income level for which there are the most hunters? Approximately how many hunters earn this amount?
(b) Using a graphing utility, graph $H = H(x)$. Are the number of hunters increasing or decreasing for individuals earning between $20,000 and $40,000?

Source: National Sporting Goods Association.

54. Advanced Degrees The function
$$P(x) = -0.008x^2 + 0.868x - 11.884$$
models the percentage of the U.S. population whose age is given by x that have earned an advanced degree (more than a bachelor's degree) in March 2000.

(a) What is the age for which the highest percentage of Americans have earned an advanced degree? What is the highest percentage?

(b) Using a graphing utility, graph $P = P(x)$. Is the percentage of Americans that have earned an advanced degree increasing or decreasing for individuals between the ages of 40 and 50?

Source: U.S. Census Bureau.

55. Male Murder Victims The function
$$M(x) = 0.76x^2 - 107.00x + 3854.18$$
models the number of male murder victims who are x years of age ($20 \leq x < 90$).

(a) Use the model to approximate the number of male murder victims who are $x = 23$ years of age.

(b) At what age is the number of male murder victims 1456?

(c) Using a graphing utility, graph $M = M(x)$.

(d) Based on the graph obtained in part (c), describe what happens to the number of male murder victims as age increases.

Source: Federal Bureau of Investigation.

56. Health Care Expenditures The function
$$H(x) = 0.004x^2 - 0.197x + 5.406$$
models the percentage of total income that an individual that is x years of age spends on health care.

(a) Use the model to approximate the percentage of total income an individual 45 years of age spends on health care.

(b) At what age is the percentage of income spent on health care 10%?

(c) Using a graphing utility, graph $H = H(x)$.

(d) Based on the graph obtained in part (c), describe what happens to the percentage of income spent on health care as individuals age.

Source: Bureau of Labor Statistics.

57. Chemical Reactions A self-catalytic chemical reaction results in the formation of a compound that causes the formation ratio to increase. If the reaction rate V is given by
$$V(x) = kx(a - x), \quad 0 \leq x \leq a$$
where k is a positive constant, a is the initial amount of the compound, and x is the variable amount of the compound, for what value of x is the reaction rate a maximum?

58. Calculus: Simpson's Rule The figure shows the graph of $y = ax^2 + bx + c$. Suppose that the points $(-h, y_0)$, $(0, y_1)$, and (h, y_2) are on the graph. It can be shown that the area enclosed by the parabola, the x-axis, and the lines $x = -h$ and $x = h$ is
$$\text{Area} = \frac{h}{3}(2ah^2 + 6c)$$
Show that this area may also be given by
$$\text{Area} = \frac{h}{3}(y_0 + 4y_1 + y_2)$$

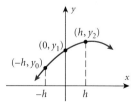

59. Use the result obtained in Problem 58 to find the area enclosed by $f(x) = -5x^2 + 8$, the x-axis, and the lines $x = -1$ and $x = 1$.

60. Use the result obtained in Problem 58 to find the area enclosed by $f(x) = 2x^2 + 8$, the x-axis, and the lines $x = -2$ and $x = 2$.

61. Use the result obtained in Problem 58 to find the area enclosed by $f(x) = x^2 + 3x + 5$, the x-axis, and the lines $x = -4$ and $x = 4$.

62. Use the result obtained in Problem 58 to find the area enclosed by $f(x) = -x^2 + x + 4$, the x-axis, and the lines $x = -1$ and $x = 1$.

63. A rectangle has one vertex on the line $y = 10 - x$, $x > 0$, another at the origin, one on the positive x-axis, and one on the positive y-axis. Find the largest area A that can be enclosed by the rectangle.

64. Let $f(x) = ax^2 + bx + c$, where a, b, and c are odd integers. If x is an integer, show that $f(x)$ must be an odd integer. [**Hint:** x is either an even integer or an odd integer.]

65. Make up a quadratic function that opens down and has only one x-intercept. Compare yours with others in the class. What are the similarities? What are the differences?

66. On one set of coordinate axes, graph the family of parabolas $f(x) = x^2 + 2x + c$ for $c = -3$, $c = 0$, and $c = 1$. Describe the characteristics of a member of this family.

67. On one set of coordinate axes, graph the family of parabolas $f(x) = x^2 + bx + 1$ for $b = -4$, $b = 0$, and $b = 4$. Describe the general characteristics of this family.

68. State the circumstances under which the graph of a quadratic function $f(x) = ax^2 + bx + c$ has no x-intercepts.

69. Why does the graph of a quadratic function open up if $a > 0$ and down if $a < 0$?

70. Refer to Example 6 on page 171. Notice that if the price charged for the calculators is $0 or $140 the revenue is $0. It is easy to explain why revenue would be $0 if the priced charged is $0, but how can revenue be $0 if the price charged is $140?

178   Chapter 2   Classes of Functions

2.2 Power Functions; Polynomial Functions; Rational Functions

PREPARING FOR THIS SECTION   *Before getting started, review the following:*

> Polynomials (Chapter 0, Section 0.3, pp. 29–34)
> Graphing Techniques (Chapter 1, Section 1.5, pp. 148–154)
> Domain of a Function (Chapter 1, Section 1.2, pp. 114–115)

OBJECTIVES
1. Know the properties of power functions
2. Graph functions using shifts and/or reflections
3. Identify polynomial functions and their degree
4. Find the end behavior of a polynomial function
5. Find the domain of a rational function

We begin by discussing *power functions*, a special kind of polynomial.

Power Functions

A **power function of degree n** is a function of the form

$$f(x) = ax^n \tag{1}$$

where a is a real number, $a \neq 0$, and $n > 0$ is an integer.

In other words, a power function is a function that is defined by a single monomial.

The graph of a power function of degree 1, $f(x) = ax$, is a straight line, with slope a, that passes through the origin. The graph of a power function of degree 2, $f(x) = ax^2$, is a parabola, with vertex at the origin, that opens up if $a > 0$ and down if $a < 0$.

Know the properties of $\boxed{1}$
power functions

We begin with power functions of even degree of the form $f(x) = x^n$, $n \geq 2$ and n even. The domain of f is the set of all real numbers, and the range is the set of nonnegative real numbers. Such a power function is an even function (do you see why?), so its graph is symmetric with respect to the y-axis. Its graph always contains the origin and the points $(-1, 1)$ and $(1, 1)$.

If $n = 2$, the graph is the familiar parabola $y = x^2$ that opens up, with vertex at the origin. If $n \geq 4$, the graph of $f(x) = x^n$, n even, will be closer to the x-axis than the parabola $y = x^2$, if $-1 < x < 1$, and farther from the x-axis than the parabola $y = x^2$, if $x < -1$ or if $x > 1$. Figure 14(a) illustrates this conclusion. Figure 14(b) shows the graphs of $y = x^4$ and $y = x^8$ for comparison.

From Figure 14, we can see that as n increases the graph of $f(x) = x^n$, $n \geq 2$ and n even, tends to flatten out near the origin and to increase very rapidly when x is far from 0. For large n, it may appear that the graph coincides with the x-axis near the origin, but it does not; the graph actually touches the x-axis only at the origin (see Table 2). Also, for large n, it may appear that for $x < -1$ or for $x > 1$ the graph is vertical, but it is not; it is only increasing very rapidly in these intervals. If the graphs were enlarged many times, these distinctions would be clear.

FIGURE 14

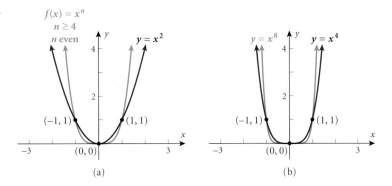

$f(x) = x^n$
$n \geq 4$
n even

(a)

(b)

TABLE 2

	$x = 0.1$	$x = 0.3$	$x = 0.5$
$f(x) = x^8$	10^{-8}	0.0000656	0.0039063
$f(x) = x^{20}$	10^{-20}	$3.487 \cdot 10^{-11}$	0.000001
$f(x) = x^{40}$	10^{-40}	$1.216 \cdot 10^{-21}$	$9.095 \cdot 10^{-13}$

SEEING THE CONCEPT: Graph $Y_1 = x^4$, $Y_2 = x^8$, and $Y_3 = x^{12}$ using the viewing rectangle $-2 \leq x \leq 2$, $-4 \leq y \leq 16$. Then graph each again using the viewing rectangle $-1 \leq x \leq 1$, $0 \leq y \leq 1$. See Figure 15. TRACE along one of the graphs to confirm that for x close to 0 the graph is above the x-axis and that for $x > 0$ the graph is increasing.

FIGURE 15

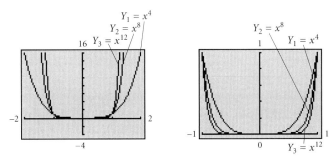

Properties of Power Functions, $f(x) = x^n$, n Is an Even Integer

1. The domain is the set of all real numbers. The range is the set of nonnegative real numbers.
2. The graph always contains the points $(0, 0)$, $(1, 1)$, and $(-1, 1)$.
3. The graph is symmetric with respect to the y-axis; the function is even.
4. As the exponent n increases in magnitude, the graph becomes more vertical when $x < -1$ or $x > 1$; but for x near the origin, the graph tends to flatten out and lie closer to the x-axis.

FIGURE 16

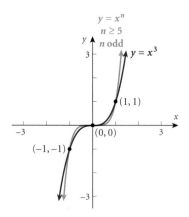

Now we consider power functions of odd degree of the form $f(x) = x^n$, $n \geq 3$ and n odd. The domain and range of f are the set of real numbers. Such a power function is an odd function (do you see why?), so its graph is symmetric with respect to the origin. Its graph always contains the origin and the points $(-1, -1)$ and $(1, 1)$.

The graph of $f(x) = x^n$ when $n = 3$ has been shown several times and is repeated in Figure 16. If $n \geq 5$, the graph of $f(x) = x^n$, n odd, will be closer to the x-axis than that of $y = x^3$, if $-1 < x < 1$, and farther from the x-axis than that of $y = x^3$, if $x < -1$ or if $x > 1$. Figure 16 also illustrates this conclusion.

Figure 17 shows the graph of $y = x^5$ and the graph of $y = x^9$ for further comparison.

It appears that each graph coincides with the x-axis near the origin, but it does not; each graph actually touches the x-axis only at the origin. Also, it appears that as x increases the graph becomes vertical, but it does not; each graph is increasing very rapidly.

SEEING THE CONCEPT: Graph $Y_1 = x^3$, $Y_2 = x^7$, and $Y_3 = x^{11}$ using the viewing rectangle $-2 \leq x \leq 2$, $-16 \leq y \leq 16$. Then graph each again using the viewing rectangle $-1 \leq x \leq 1$, $-1 \leq y \leq 1$. See Figure 18. TRACE along one of the graphs to confirm that the graph is increasing and only touches the x-axis at the origin.

FIGURE 17

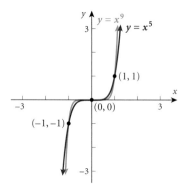

FIGURE 18

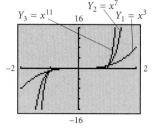 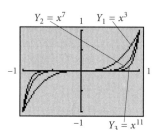

To summarize:

Properties of Power Functions, $f(x) = x^n$, n Is an Odd Integer

1. The domain and range are the set of all real numbers.
2. The graph always contains the points $(0, 0)$, $(1, 1)$, and $(-1, -1)$.
3. The graph is symmetric with respect to the origin; the function is odd.
4. As the exponent n increases in magnitude, the graph becomes more vertical when $x < -1$ or $x > 1$; but for x near the origin, the graph tends to flatten out and lie closer to the x-axis.

---

2 EXAMPLE 1 Graphing Functions Using Shifts and/or Reflections

Graph the function $f(x) = -x^3 + 1$.

SOLUTION We begin with the power function $y = x^3$. See Figure 19 for the steps.

FIGURE 19

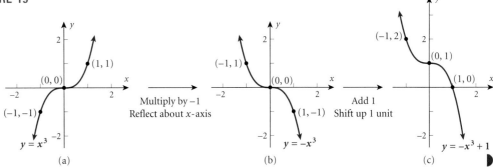

  NOW WORK PROBLEMS 1 AND 5.

Polynomial Functions

Polynomial functions are among the simplest expressions in algebra. They are easy to evaluate: only addition and repeated multiplication are required. Because of this, they are often used to approximate other, more complicated functions. In this section, we investigate characteristics of this important class of function.

> A **polynomial function** is a function of the form
> $$f(x) = a_n x^n + a_{n-1} x^{n-1} + \cdots + a_1 x + a_0 \tag{2}$$
> where $a_n, a_{n-1}, \ldots, a_1, a_0$ are real numbers and n is a nonnegative integer. The domain of a polynomial function consists of all real numbers.

A polynomial function is a function whose rule is given by a polynomial in one variable. The **degree** of a polynomial function is the degree of the polynomial in one variable, that is, the largest power of x that appears.

EXAMPLE 2 Identifying Polynomial Functions and Their Degree

Determine which of the following are polynomial functions. For those that are, state the degree; for those that are not, tell why not.

(a) $f(x) = 2 - 3x^4$ **(b)** $g(x) = \sqrt{x}$ **(c)** $h(x) = \dfrac{x^2 - 2}{x^3 - 1}$

(d) $F(x) = 0$ **(e)** $G(x) = 8$ **(f)** $H(x) = -2x^3(x - 1)^2$

SOLUTION **(a)** f is a polynomial function of degree 4.
(b) g is not a polynomial function. The variable x is raised to the $\tfrac{1}{2}$ power, which is not a nonnegative integer.
(c) h is not a polynomial function. It is the ratio of two polynomials, and the polynomial in the denominator is of positive degree.

182 Chapter 2 Classes of Functions

(d) F is the zero polynomial function; it is not assigned a degree.
(e) G is a nonzero constant function, a polynomial function of degree 0 since $G(x) = 8 = 8x^0$.
(f) $H(x) = -2x^3(x - 1)^2 = -2x^3(x^2 - 2x + 1) = -2x^5 + 4x^4 - 2x^3$. So H is a polynomial function of degree 5. Do you see how to find the degree of H without multiplying out? ▸

 NOW WORK PROBLEMS 11 AND 15.

We have already discussed in detail polynomial functions of degrees 0, 1, and 2. See Table 3 for a summary of the properties of the graphs of these polynomial functions.

TABLE 3

Degree	Form	Name	Graph
No degree	$f(x) = 0$	Zero function	The x-axis
0	$f(x) = a_0, \; a_0 \neq 0$	Constant function	Horizontal line with y-intercept $(0, a_0)$
1	$f(x) = a_1 x + a_0, \; a_1 \neq 0$	Linear function	Nonvertical, nonhorizontal line with slope a_1 and y-intercept $(0, a_0)$
2	$f(x) = a_2 x^2 + a_1 x + a_0, \; a_2 \neq 0$	Quadratic function	Parabola: Graph opens up if $a_2 > 0$; graph opens down if $a_2 < 0$. The y intercept is $(0, a_0)$.

One of the objectives of this book is to analyze the graph of a polynomial function. You will learn that the graph of every polynomial function is both smooth and continuous. By *smooth*, we mean that the graph contains no sharp corners or cusps; by *continuous*, we mean that the graph has no gaps or holes and can be drawn without lifting pencil from paper. Later we use calculus to define these concepts more carefully. See Figures 20(a) and (b).

FIGURE 20

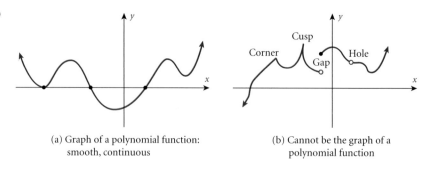

(a) Graph of a polynomial function: smooth, continuous

(b) Cannot be the graph of a polynomial function

Figure 21 shows the graph of a polynomial function with four x-intercepts. Notice that at the x-intercepts the graph must either cross the x-axis or touch the x-axis. Consequently, between consecutive x-intercepts the graph is either above the x-axis or below the x-axis. Notice also that the graph has two local maxima and two local minima. In Chapter 5 we will use calculus to locate the local maxima and minima of polynomial functions so that we can draw a complete graph.

FIGURE 21

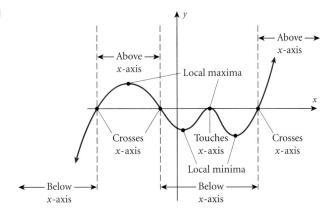

The behavior of the graph of a function for large values of x, either positive or negative, is referred to as its **end behavior.**

For polynomial functions, we have this important result.

End Behavior

For large values of x, either positive or negative, the graph of the polynomial

$$f(x) = a_n x^n + a_{n-1} x^{n-1} + \cdots + a_1 x + a_0$$

resembles the graph of the power function

$$y = a_n x^n$$

EXAMPLE 3   **Finding the End Behavior of a Polynomial Function**

(a) For large values of x, the graph of the polynomial function
$$f(x) = 3x^5 - 4x^4 + 8x - 4$$
resembles that of the power function $y = 3x^5$.

(b) For large values of x, the graph of the polynomial function
$$f(x) = -4x^7 + 2x^5 - 4x^2 + 2x - 10$$
resembles that of the power function $y = -4x^7$.

NOW WORK PROBLEM 25.

Look back at Figures 14 and 16. Based on the above theorem and the previous discussion on power functions, the end behavior of a polynomial can only be of four types. See Figure 22.

FIGURE 22 End behavior

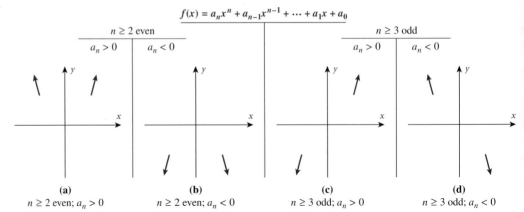

For example, consider the polynomial function $f(x) = -2x^4 + x^3 + 4x^2 - 7x + 1$. The graph of f will resemble the graph of the power function $y = -2x^4$ for large values of x, either positive or negative. The graph of f will look like Figure 22(b) for large values of x.

Rational Functions

Ratios of integers are called *rational numbers*. Similarly, ratios of polynomial functions are called *rational functions*.

> A **rational function** is a function of the form
> $$R(x) = \frac{p(x)}{q(x)}$$
> where p and q are polynomial functions and q is not the zero polynomial. The domain of a rational function consists of all real numbers except those for which the denominator q is 0.

EXAMPLE 4 Finding the Domain of a Rational Function

(a) The domain of $R(x) = \dfrac{2x^2 - 4}{x + 5}$ consists of all real numbers x except -5, that is, $\{x \mid x \neq -5\}$.

(b) The domain of $R(x) = \dfrac{1}{x^2 - 4}$ consists of all real numbers x except -2 and 2, that is, $\{x \mid x \neq -2, x \neq 2\}$.

(c) The domain of $R(x) = \dfrac{x^3}{x^2 + 1}$ consists of all real numbers.

(d) The domain of $R(x) = \dfrac{-x^2 + 2}{3}$ consists of all real numbers.

(e) The domain of $R(x) = \dfrac{x^2 - 1}{x - 1}$ consists of all real numbers x except 1, that is, $\{x \mid x \neq 1\}$.

It is important to observe that the functions

$$R(x) = \frac{x^2 - 1}{x - 1} \quad \text{and} \quad f(x) = x + 1$$

are not equal, since the domain of R is $\{x \mid x \neq 1\}$ and the domain of f is all real numbers.

If $R(x) = \dfrac{p(x)}{q(x)}$ is a rational function and if p and q have no common factors, then the rational function R is said to be in **lowest terms**. For a rational function $R(x) = \dfrac{p(x)}{q(x)}$ in lowest terms, the zeros, if any, of the numerator are the x-intercepts of the graph of R and so will play a major role in the graph of R. The zeros of the denominator of R [that is, the numbers x, if any, for which $q(x) = 0$], although not in the domain of R, also play a major role in the graph of R. We will discuss this role in Chapter 3.

NOW WORK PROBLEM 31.

EXERCISE 2.2

1. Name three points on the graph of the power function $f(x) = x^5$.

2. Name three points on the graph of the power function $f(x) = x^6$.

3. The graph of the power function $f(x) = x^5$ is symmetric with respect to the _____.

4. The graph of the power function $f(x) = x^6$ is symmetric with respect to the _____.

In Problems 5–10, graph each function using shifts and/or reflections. Be sure to label at least three points.

5. $f(x) = x^6 + 2$ 6. $f(x) = x^5 - 3$ 7. $f(x) = -x^5 + 2$ 8. $f(x) = -x^4 + 9$ 9. $f(x) = (x-2)^4$ 10. $f(x) = (x+3)^5$

In Problems 11–22, determine which functions are polynomial functions. For those that are, state the degree. For those that are not, tell why not.

11. $f(x) = 4x + x^3$ 12. $f(x) = 5x^2 + 4x^4$ 13. $g(x) = \dfrac{1-x^2}{2}$ 14. $h(x) = 3 - \dfrac{1}{2}x$

15. $f(x) = 1 - \dfrac{1}{x}$ 16. $f(x) = x(x-1)$ 17. $g(x) = x^{3/2} - x^2 + 2$ 18. $h(x) = \sqrt{x}(\sqrt{x} - 1)$

19. $F(x) = 5x^4 - \pi x^3 + \dfrac{1}{2}$ 20. $F(x) = \dfrac{x^2 - 5}{x^3}$ 21. $G(x) = 2(x-1)^2(x^2+1)$ 22. $G(x) = -3x^2(x+2)^3$

In Problems 23–28, find the power function that the graph of f resembles for large values of x. That is, find the end behavior of each polynomial function.

23. $f(x) = 3x^4 - 2x^2 + 1$ 24. $f(x) = 4x^5 - 6x^3 - x$ 25. $f(x) = -2x^5 + 8x^4$

26. $f(x) = -3x^4 - 5x + 1$ 27. $f(x) = 5(x+1)^2(x-2)$ 28. $f(x) = 6x(x^2 + 4)^2$

In Problems 29–40, find the domain of each rational function.

29. $R(x) = \dfrac{4x}{x-3}$ 30. $R(x) = \dfrac{5x^2}{3+x}$ 31. $H(x) = \dfrac{-4x^2}{(x-2)(x+4)}$ 32. $G(x) = \dfrac{6}{(x+3)(4-x)}$

33. $F(x) = \dfrac{3x(x-1)}{2x^2 - 5x - 3}$ 34. $Q(x) = \dfrac{-x(1-x)}{3x^2 + 5x - 2}$ 35. $R(x) = \dfrac{x}{x^3 - 8}$ 36. $R(x) = \dfrac{x}{x^4 - 1}$

37. $H(x) = \dfrac{3x^2 + x}{x^2 + 4}$ 38. $G(x) = \dfrac{x-3}{x^4 + 1}$ 39. $R(x) = \dfrac{3(x^2 - x - 6)}{4(x^2 - 9)}$ 40. $F(x) = \dfrac{-2(x^2 - 4)}{3(x^2 + 4x + 4)}$

41. Union Membership The percentage of the labor force who are union members is given below for 1930–2000.

Year	1930	1940	1950	1960	1970	1980	1990	2000
Percentage	11.6	26.9	31.5	31.4	27.3	21.9	16.1	13.2

This data can be modeled by the polynomial function

$$u(t) = 11.93 + 1.9t - 0.052t^2 + 0.00037t^3$$

where t is the number of years since 1930.

(a) Use $u(t)$ to find the percentage of union membership in 2000.

(b) Find $u(75)$. Write a sentence explaining what it means.

Source: Bureau of Labor Statistics, U.S. Department of Labor.

2.3 Exponential Functions

PREPARING FOR THIS SECTION   *Before getting started, review the following:*

> Exponents (Chapter 0, Section 0.2, pp. 20–24 and Section 0.6, pp. 64–69)
> Slope of a Line (Chapter 0, Section 0.9, pp. 81–87)
> Graphing Techniques (Chapter 1, Section 1.5, pp. 148–154)
> Solving Equations (Chapter 0, Section 0.4, pp. 40–45)

OBJECTIVES
1. Evaluate exponents
2. Graph exponential functions
3. Define the number e
4. Solve exponential equations

Evaluate exponents   **1**   In Chapter 0, Section 0.6, we give a definition for raising a real number a to a rational power. Based on that discussion, we gave meaning to expressions of the form

$$a^r$$

where the base a is a positive real number and the exponent r is a rational number.

But what is the meaning of a^x, where the base a is a positive real number and the exponent x is an irrational number? Although a rigorous definition requires advanced methods, the basis for the definition is easy to follow: Select a rational number r that is formed by truncating (removing) all but a finite number of digits from the irrational number x. Then it is reasonable to expect that

$$a^x \approx a^r$$

For example, take the irrational number $\pi = 3.14159\ldots$. Then, an approximation to a^π is

$$a^\pi \approx a^{3.14}$$

where the digits after the hundredths position have been removed from the value for π. A better approximation would be

$$a^\pi \approx a^{3.14159}$$

where the digits after the hundred-thousandths position have been removed. Continuing in this way, we can obtain approximations to a^π to any desired degree of accuracy.

Most calculators have an $\boxed{x^y}$ key or a caret key $\boxed{\wedge}$ for working with exponents. To evaluate expressions of the form a^x, enter the base a, then press the $\boxed{x^y}$ key (or the $\boxed{\wedge}$ key), enter the exponent x, and press $\boxed{=}$ (or $\boxed{\text{enter}}$).

EXAMPLE 1 Using a Calculator to Evaluate Powers of 2

Using a calculator, evaluate:

(a) $2^{1.4}$ (b) $2^{1.41}$ (c) $2^{1.414}$ (d) $2^{1.4142}$ (e) $2^{\sqrt{2}}$

SOLUTION
(a) $2^{1.4} \approx 2.639015822$
(b) $2^{1.41} \approx 2.657371628$
(c) $2^{1.414} \approx 2.66474965$
(d) $2^{1.4142} \approx 2.665119089$
(e) $2^{\sqrt{2}} \approx 2.665144143$

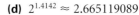

NOW WORK PROBLEM 1.

It can be shown that the Laws of Exponents hold for real exponents.

Laws of Exponents

If s, t, a, and b are real numbers, with $a > 0$ and $b > 0$, then

$$a^s \cdot a^t = a^{s+t} \qquad (a^s)^t = a^{st} \qquad (ab)^s = a^s \cdot b^s$$
$$1^s = 1 \qquad a^{-s} = \frac{1}{a^s} = \left(\frac{1}{a}\right)^s \qquad a^0 = 1 \qquad (1)$$

We are now ready for the following definition:

An **exponential function** is a function of the form

$$f(x) = a^x$$

where a is a positive real number ($a > 0$) and $a \neq 1$. The domain of f is the set of all real numbers.

We exclude the base $a = 1$ because this function is simply the constant function $f(x) = 1^x = 1$. We also need to exclude the bases that are negative, because, otherwise, we would have to exclude many values of x from the domain, such as $x = \frac{1}{2}$ and $x = \frac{3}{4}$. [Recall that $(-2)^{1/2} = \sqrt{-2}$, $(-3)^{3/4} = \sqrt[4]{(-3)^3} = \sqrt[4]{-27}$, and so on, are not defined in the system of real numbers.]

COMMENT: It is important to distinguish a power function $g(x) = x^n$, $n \geq 2$ an integer, from an exponential function $f(x) = a^x$, $a > 0$, $a \neq 1$, a real. In a power function, the base is a variable and the exponent is a constant. In an exponential function, the base is a constant and the exponent is a variable.

Some examples of exponential functions are

$$f(x) = 2^x, \quad F(x) = \left(\frac{1}{3}\right)^x$$

Notice that in each example, the base is a constant and the exponent is a variable.

You may wonder what role the base a plays in the exponential function $f(x) = a^x$. We use the following Exploration to find out.

EXPLORATION

(a) Evaluate $f(x) = 2^x$ at $x = -2, -1, 0, 1, 2,$ and 3.
(b) Evaluate $g(x) = 3x + 2$ at $x = -2, -1, 0, 1, 2,$ and 3.
(c) Comment on the pattern that exists in the values of f and g.

Result

(a) Table 4 shows the values of $f(x) = 2^x$ for $x = -2, -1, 0, 1, 2,$ and 3.
(b) Table 5 shows the values of $g(x) = 3x + 2$ for $x = -2, -1, 0, 1, 2,$ and 3.

TABLE 4

x	$f(x) = 2^x$
-2	$f(-2) = 2^{-2} = \frac{1}{2^2} = \frac{1}{4}$
-1	$\frac{1}{2}$
0	1
1	2
2	4
3	8

TABLE 5

x	$g(x) = 3x + 2$
-2	$g(-2) = 3(-2) + 2 = -4$
-1	-1
0	2
1	5
2	8
3	11

(c) In Table 4 we notice that each value of the exponential function $f(x) = a^x = 2^x$ could be found by multiplying the previous value of the function by the base, $a = 2$. For example,

$$f(-1) = 2 \cdot f(-2) = 2 \cdot \frac{1}{4} = \frac{1}{2}, \quad f(0) = 2 \cdot f(-1) = 2 \cdot \frac{1}{2} = 1, \quad f(1) = 2 \cdot f(0) = 2 \cdot 1 = 2$$

and so on.

Put another way, we see that the ratio of consecutive outputs is constant for unit increases in the inputs. The constant equals the value of the base of the exponential function a. For example, for the function $f(x) = 2^x$, we notice that

$$\frac{f(-1)}{f(-2)} = \frac{\frac{1}{2}}{\frac{1}{4}} = 2, \quad \frac{f(1)}{f(0)} = \frac{2}{1} = 2, \quad \frac{f(x+1)}{f(x)} = \frac{2^{x+1}}{2^x} = 2$$

and so on.

From Table 5 we see that ratios of consecutive outputs of $g(x) = 3x + 2$ are not constant. For example,

$$\frac{g(-1)}{g(-2)} = \frac{-1}{-4} = \frac{1}{4} \neq \frac{g(1)}{g(0)} = \frac{5}{2}$$

Instead, because $g(x) = 3x + 2$ is a linear function, for unit increases in the input, the outputs increase by a fixed amount equal to the value of the slope, 3.

The conclusions reached in the Exploration lead to the following theorem.

For an exponential function $f(x) = a^x$, $a > 0$, $a \neq 1$, if x is any real number, then

$$\boxed{\frac{f(x + 1)}{f(x)} = a}$$

Proof

$$\frac{f(x + 1)}{f(x)} = \frac{a^{x+1}}{a^x} = a^{x+1-x} = a^1 = a \quad \blacksquare$$

NOW WORK PROBLEM 11.

Graphs of Exponential Functions

First, we graph the exponential function $f(x) = 2^x$.

EXAMPLE 2 **Graphing an Exponential Function**

Graph the exponential function: $f(x) = 2^x$

SOLUTION The domain of $f(x) = 2^x$ consists of all real numbers. We begin by locating some points on the graph of $f(x) = 2^x$, as listed in Table 6 on page 190.

Since $2^x > 0$ for all x, the range of f is the interval $(0, \infty)$. From this, we conclude that the graph has no x-intercepts, and, in fact, the graph will lie above the x-axis. As Table 6 indicates, the y-intercept is 1. Table 6 also indicates that as x becomes unbounded in the negative direction, the value of $f(x) = 2^x$ get closer and closer to 0. This means that the line $y = 0$ (the x-axis) is a horizontal asymptote to the graph as x becomes unbounded in the negative direction.* This gives us the end behavior of the graph for x large and negative.

---

* A horizontal asymptote is a line that the graph of a function gets closer and closer to, as x becomes unbounded in the positive (or negative) direction. We discuss asymptotes in detail in Chapter 3.

190 Chapter 2 Classes of Functions

TABLE 6

x	$f(x) = 2^x$
-10	$2^{-10} \approx 0.00098$
-3	$2^{-3} = \dfrac{1}{8}$
-2	$2^{-2} = \dfrac{1}{4}$
-1	$2^{-1} = \dfrac{1}{2}$
0	$2^0 = 1$
1	$2^1 = 2$
2	$2^2 = 4$
3	$2^3 = 8$
10	$2^{10} = 1024$

To determine the end behavior for x large and positive, look again at Table 6. As x becomes unbounded in the positive direction, $f(x) = 2^x$ grows very quickly, causing the graph of $f(x) = 2^x$ to rise very rapidly. It is apparent that f is an increasing function.

Using all this information, we plot some of the points from Table 6 and connect them with a smooth, continuous curve, as shown in Figure 23.

FIGURE 23

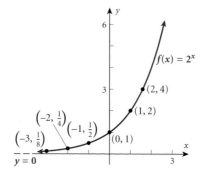

As we shall see, graphs that look like the one in Figure 23 occur very frequently in a variety of situations. For example, look at the graph in Figure 24, which illustrates the closing price of a share of Harley Davidson stock. Investors might conclude from this graph that the price of Harley Davidson is *behaving exponentially*; that is, the graph exhibits rapid, or exponential, growth.

FIGURE 24

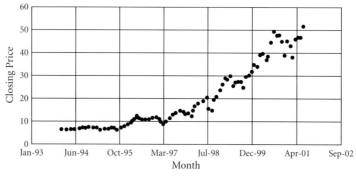

FIGURE 25

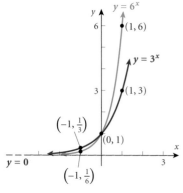

We shall have more to say about situations that lead to exponential growth later. For now, we continue to seek properties of exponential functions.

The graph of $f(x) = 2^x$ in Figure 23 is typical of all exponential functions that have a base larger than 1. Such functions are increasing functions. Their graphs lie above the x-axis, pass through the point $(0, 1)$, and thereafter rise rapidly as x becomes unbounded in the positive direction. As x becomes unbounded in the negative direction, the line $y = 0$ (the x-axis) is a horizontal asymptote. Finally, the graphs are smooth and continuous, with no corners or gaps.

Figure 25 illustrates the graphs of two more exponential functions whose bases are larger than 1. Notice that for the larger base the graph is steeper when $x > 0$ and is closer to the x-axis when $x < 0$.

 SEEING THE CONCEPT: Graph $y = 2^x$ and compare what you see to Figure 23. Clear the screen and graph $y = 3^x$ and $y = 6^x$ and compare what you see to Figure 25. Clear the screen and graph $y = 10^x$ and $y = 100^x$. What viewing rectangle seems to work best?

Exponential Functions

The following list summarizes the information that we have about $f(x) = a^x$, $a > 1$.

Properties of the Exponential Function $f(x) = a^x$, $a > 1$

1. The domain is the set of all real numbers; the range is the set of positive real numbers.
2. There is no x-intercept; the y-intercept is $(0, 1)$.
3. The line $y = 0$ (the x-axis) is a horizontal asymptote as x becomes unbounded in the negative direction.
4. $f(x) = a^x$, $a > 1$, is an increasing function.
5. The graph of f contains the points $(0, 1)$, $(1, a)$, and $\left(-1, \dfrac{1}{a}\right)$.
6. The graph of f is smooth and continuous, with no corners or gaps. See Figure 26.

FIGURE 26

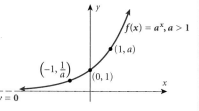

Now we consider $f(x) = a^x$ when $0 < a < 1$.

EXAMPLE 3 Graphing an Exponential Function

Graph the exponential function: $f(x) = \left(\dfrac{1}{2}\right)^x$

SOLUTION The domain of $f(x) = \left(\dfrac{1}{2}\right)^x$ consists of all real numbers. As before, we locate some points on the graph by creating Table 7. Since $\left(\dfrac{1}{2}\right)^x > 0$ for all x, the range of f is the interval $(0, \infty)$. The graph lies above the x-axis and so has no x-intercepts. The y-intercept is $(0, 1)$. As x becomes unbounded in the negative direction, $f(x) = \left(\dfrac{1}{2}\right)^x$ grows very quickly. As x becomes unbounded in the positive direction, the values of $f(x)$ approach 0. The line $y = 0$ (the x-axis) is a horizontal asymptote as x becomes unbounded in the positive direction. It is apparent that f is a decreasing function. Figure 27 illustrates the graph.

TABLE 7

x	$f(x) = \left(\dfrac{1}{2}\right)^x$
-10	$\left(\dfrac{1}{2}\right)^{-10} = 1024$
-3	$\left(\dfrac{1}{2}\right)^{-3} = 8$
-2	$\left(\dfrac{1}{2}\right)^{-2} = 4$
-1	$\left(\dfrac{1}{2}\right)^{-1} = 2$
0	$\left(\dfrac{1}{2}\right)^{0} = 1$
1	$\left(\dfrac{1}{2}\right)^{1} = \dfrac{1}{2}$
2	$\left(\dfrac{1}{2}\right)^{2} = \dfrac{1}{4}$
3	$\left(\dfrac{1}{2}\right)^{3} = \dfrac{1}{8}$
10	$\left(\dfrac{1}{2}\right)^{10} \approx 0.00098$

FIGURE 27

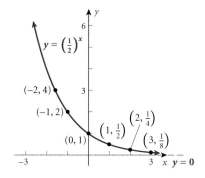

We could have obtained the graph of $y = \left(\frac{1}{2}\right)^x$ from the graph of $y = 2^x$ using a reflection. If $f(x) = 2^x$, then $f(-x) = 2^{-x} = \frac{1}{2^x} = \left(\frac{1}{2}\right)^x$. The graph of $y = \left(\frac{1}{2}\right)^x = 2^{-x}$ is a reflection about the y-axis of the graph of $y = 2^x$. See Figures 28(a) and (b).

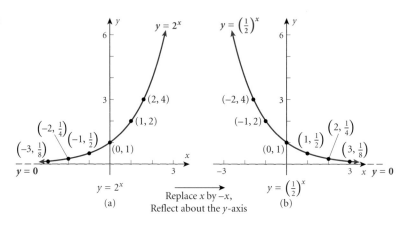

FIGURE 28

SEEING THE CONCEPT: Using a graphing utility, simultaneously graph

(a) $Y_1 = 3^x$, $Y_2 = \left(\frac{1}{3}\right)^x$ (b) $Y_1 = 6^x$, $Y_2 = \left(\frac{1}{6}\right)^x$

Conclude that the graph of $Y_2 = \left(\frac{1}{a}\right)^x$, for $a > 0$, is the reflection about the y-axis of the graph of $Y_1 = a^x$.

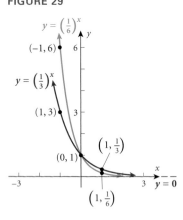

FIGURE 29

The graph of $f(x) = \left(\frac{1}{2}\right)^x$ in Figure 28(b) is typical of all exponential functions that have a base between 0 and 1. Such functions are decreasing. Their graphs lie above the x-axis and pass through the point (0,1). The graphs rise rapidly as x becomes unbounded in the negative direction. As x becomes unbounded in the positive direction, the x-axis is a horizontal asymptote. Finally, the graphs are smooth and continuous with no corners or gaps.

Figure 29 illustrates the graphs of two more exponential functions whose bases are between 0 and 1. Notice that the choice of a base closer to 0 results in a graph that is steeper when $x < 0$ and closer to the x-axis when $x > 0$.

SEEING THE CONCEPT: Graph $Y = \left(\frac{1}{2}\right)^x$ and compare what you see to Figure 28(b).

Clear the screen and graph $Y_1 = \left(\frac{1}{3}\right)^x$ and $Y_2 = \left(\frac{1}{6}\right)^x$ and compare what you see to Figure 29. Clear the screen and graph $Y_1 = \left(\frac{1}{10}\right)^x$ and $Y_2 = \left(\frac{1}{100}\right)^x$. What viewing rectangle seems to work best?

The following list summarizes the information that we have about the function $f(x) = a^x, 0 < a < 1$.

Properties of the Graph of an Exponential Function
$f(x) = a^x, 0 < a < 1$

1. The domain is the set of all real numbers; the range is the set of positive real numbers.
2. There is no x-intercept; the y-intercept is $(0,1)$.
3. The line $y = 0$ (the x-axis) is a horizontal asymptote as x become unbounded in the positive direction.
4. $f(x) = a^x, 0 < a < 1$, is a decreasing function.
5. The graph of f contains the points $(0, 1)$, $(1, a)$, and $\left(-1, \dfrac{1}{a}\right)$.
6. The graph of f is smooth and continuous, with no corners or gaps. See Figure 30.

FIGURE 30

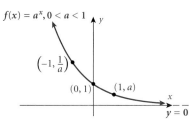

$f(x) = a^x, 0 < a < 1$

EXAMPLE 4 Graphing Exponential Functions Using Shifts and/or Reflections

Graph $f(x) = 2^{-x} - 3$ and determine the domain, range, and horizontal asymptote of f.

SOLUTION We begin with the graph of $y = 2^x$. Figure 31 shows the various steps.

FIGURE 31

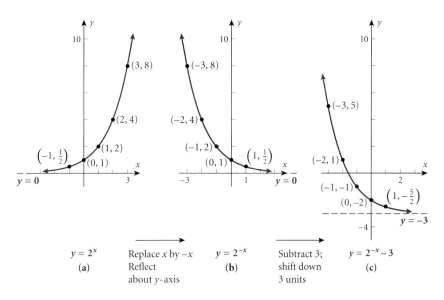

As Figure 31(c) illustrates, the domain of $f(x) = 2^{-x} - 3$ is the interval $(-\infty, \infty)$ and the range is the interval $(-3, \infty)$. The horizontal asymptote of f is the line $y = -3$.

NOW WORK PROBLEM 27.

The Base e

Define the number e Many applied problems require the use of an exponential function whose base is a certain irrational number, symbolized by the letter e.

Let's look now at one way of arriving at this important number e.

The **number e** is defined as the number that the expression

$$\left(1 + \frac{1}{n}\right)^n \qquad (2)$$

approaches as n becomes unbounded in the positive direction. In calculus, this is expressed using limit notation as

$$e = \lim_{n \to \infty}\left(1 + \frac{1}{n}\right)^n$$

Table 8 illustrates what happens to the defining expression (2) as n becomes unbounded in the positive direction. The last number in the last column in the table gives e correct to nine decimal places and is the same as the entry given for e on your calculator (if expressed correctly to nine decimal places).

TABLE 8

n	$\dfrac{1}{n}$	$1 + \dfrac{1}{n}$	$\left(1 + \dfrac{1}{n}\right)^n$
1	1	2	2
2	0.5	1.5	2.25
5	0.2	1.2	2.48832
10	0.1	1.1	2.59374246
100	0.01	1.01	2.704813829
1,000	0.001	1.001	2.716923932
10,000	0.0001	1.0001	2.718145927
100,000	0.00001	1.00001	2.718268237
1,000,000	0.000001	1.000001	2.718280469
1,000,000,000	10^{-9}	$1 + 10^{-9}$	2.718281827

TABLE 9

x	e^x
-2	0.14
-1	0.37
0	1
1	2.72
2	7.39

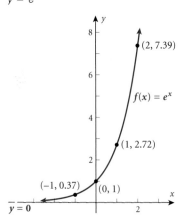

FIGURE 32
$y = e^x$

The exponential function $f(x) = e^x$, whose base is the number e, occurs with such frequency in applications that it is usually referred to as *the* exponential function. Indeed, most calculators have the key $\boxed{e^x}$ or $\boxed{\exp(x)}$, which may be used to evaluate the exponential function for a given value of x. (Consult your owner's manual if there is no such key.)

Now use your calculator to approximate e^x for $x = -2$, $x = -1$, $x = 0$, $x = 1$, and $x = 2$, as we have done to create Table 9.

The graph of the exponential function $f(x) = e^x$ is given in Figure 32. Since $2 < e < 3$, the graph of $y = e^x$ lies between the graphs of $y = 2^x$ and $y = 3^x$. Do you see why? (Refer to Figures 23 and 25.)

SEEING THE CONCEPT: Graph $Y_1 = e^x$ and compare what you see to Figure 32. Use eVALUEate or TABLE to verify the entries in Table 9. Now graph $Y_2 = 2^x$ and $Y_3 = 3^x$

on the same screen as $Y_1 = e^x$. Notice that the graph of $Y_1 = e^x$ lies between these two graphs.

EXAMPLE 5   Graphing Exponential Functions Using Shifts and/or Reflections

Graph $f(x) = -e^x + 1$ and determine the domain, range, and horizontal asymptote of f.

SOLUTION   We begin with the graph of $y = e^x$. Figure 33 shows the various steps.

FIGURE 33

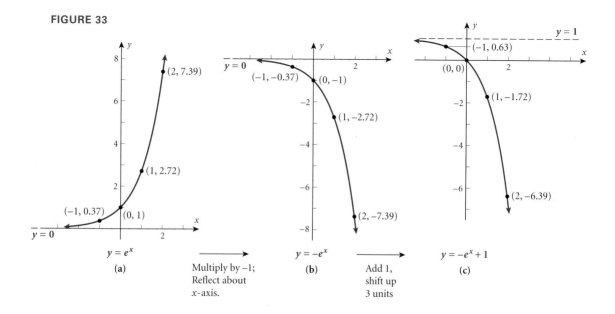

As Figure 33(c) illustrates, the domain of $f(x) = -e^x + 1$ is the interval $(-\infty, \infty)$ and the range is the interval $(-\infty, 1)$. The horizontal asymptote is the line $y = 1$.

 NOW WORK PROBLEM 31.

Exponential Equations

Equations that involve terms of the form a^x, $a > 0$, $a \neq 1$, are often referred to as **exponential equations.** Such equations can sometimes be solved by appropriately applying the Laws of Exponents and statement (3) below.

$$\text{If } a^u = a^v, \text{ then } u = v \tag{3}$$

To use property (3), each side of the equality must be written with the same base.

EXAMPLE 6 Solving an Exponential Equation

Solve: $3^{x+1} = 81$

SOLUTION Since $81 = 3^4$, we can write the equation as

$$3^{x+1} = 81$$
$$3^{x+1} = 3^4$$

Now we have the same base, 3, on each side, so we can apply property (3) to obtain

$$x + 1 = 4$$
$$x = 3$$

 NOW WORK PROBLEM 35.

EXAMPLE 7 Solving an Exponential Equation

Solve: $e^{-x^2} = (e^x)^2 \cdot \dfrac{1}{e^3}$

SOLUTION We use Laws of Exponents first to get the base e on the right side.

$$(e^x)^2 \cdot \dfrac{1}{e^3} = e^{2x} \cdot e^{-3} = e^{2x-3}$$

As a result,

$$e^{-x^2} = e^{2x-3}$$
$$-x^2 = 2x - 3 \quad \text{Apply Property (3).}$$
$$x^2 + 2x - 3 = 0 \quad \text{Place the quadratic equation in standard form.}$$
$$(x + 3)(x - 1) = 0 \quad \text{Factor.}$$
$$x = -3 \quad \text{or} \quad x = 1 \quad \text{Use the Zero-Product Property.}$$

The solution set is $\{-3, 1\}$.

Application

Many applications involve the exponential function. Let's look at one.

EXAMPLE 8 Exponential Probability

Between 9:00 PM and 10:00 PM cars arrive at Burger King's drive-thru at the rate of 12 cars per hour (0.2 car per minute). The following formula from probability can be used to determine the probability that a car will arrive within t minutes of 9:00 PM.

$$F(t) = 1 - e^{-0.2t}$$

(a) Determine the probability that a car will arrive within 5 minutes of 9 PM (that is, before 9:05 PM).

(b) Determine the probability that a car will arrive within 30 minutes of 9 PM (before 9:30 PM).

(c) What value does F approach as t becomes unbounded in the positive direction?

(d) Graph $F(t) = 1 - e^{-0.2t}$, $t > 0$. Use eVALUEate or TABLE to compare the values of F at $t = 5$ [part (a)] and at $t = 30$ [part (b)].

(e) Within how many minutes of 9 PM will the probability of a car arriving equal 50%? [**Hint:** Use TRACE or TABLE].

SOLUTION (a) The probability that a car will arrive within 5 minutes is found by evaluating $F(t)$ at $t = 5$.
$$F(5) = 1 - e^{-0.2(5)} \approx 0.63212$$
↑
Use a calculator

We conclude that there is a 63% probability that a car will arrive within 5 minutes.

(b) The probability that a car will arrive within 30 minutes is found by evaluating $F(t)$ at $t = 30$.
$$F(30) = 1 - e^{-0.2(30)} \approx 0.9975$$
↑
Use a calculator

There is a 99.75% probability that a car will arrive within 30 minutes.

(c) As time passes, the probability that a car will arrive increases. The value that F approaches can be found by letting t become unbounded in the positive direction. Since $e^{-0.2t} = \dfrac{1}{e^{0.2t}}$, it follows that $e^{-0.2t}$ approaches 0 as t becomes unbounded in the positive direction. Thus, F approaches 1 as t becomes unbounded in the positive direction.

(d) See Figure 34 for the graph of F.

(e) Within 3.5 minutes of 9 PM, the probability of a car arriving equals 50%.

FIGURE 34

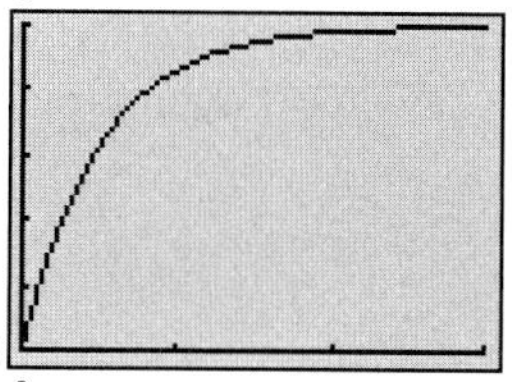

NOW WORK PROBLEM 63.

SUMMARY **Properties of the Exponential Function**

$f(x) = a^x$, $a > 1$ Domain: the interval $(-\infty, \infty)$; Range: the interval $(0, \infty)$; x-intercept: none; y-intercept: $(0, 1)$ horizontal asymptote: the line $y = 0$ (the x-axis), as x becomes unbounded in the negative direction increasing; smooth; continuous See Figure 26 for a typical graph.

$f(x) = a^x$, $0 < a < 1$ Domain: the interval $(-\infty, \infty)$; Range: the interval $(0, \infty)$; x-intercept: none; y-intercept: $(0, 1)$ horizontal asymptote: the line $y = 0$ (the x-axis), as x becomes unbounded in the positive direction decreasing; smooth; continuous See Figure 30 for a typical graph.

If $a^u = a^v$, then $u = v$.

EXERCISE 2.3

In Problems 1–10, approximate each number using a calculator. Express your answer rounded to four decimal places.

1. (a) $3^{2.2}$ (b) $3^{2.23}$ (c) $3^{2.236}$ (d) $3^{\sqrt{5}}$ **2.** (a) $5^{1.7}$ (b) $5^{1.73}$ (c) $5^{1.732}$ (d) $5^{\sqrt{3}}$

3. (a) $2^{3.14}$ (b) $2^{3.141}$ (c) $2^{3.1415}$ (d) 2^{π} **4.** (a) $2^{2.7}$ (b) $2^{2.71}$ (c) $2^{2.718}$ (d) 2^{e}

5. (a) $3.1^{2.7}$ (b) $3.14^{2.71}$ (c) $3.141^{2.718}$ (d) π^e 6. (a) $2.7^{3.1}$ (b) $2.71^{3.14}$ (c) $2.718^{3.141}$ (d) e^π

7. $e^{1.2}$ 8. $e^{-1.3}$ 9. $e^{-0.85}$ 10. $e^{2.1}$

In Problems 11–18, determine whether the given function is exponential or not. For those that are exponential functions, identify the value of a. [**Hint:** *Look at the ratio of consecutive values.*]

11.

x	$f(x)$
-1	3
0	6
1	12
2	18
3	30

12.

x	$g(x)$
-1	2
0	5
1	8
2	11
3	14

13.

x	$H(x)$
-1	$\frac{1}{4}$
0	1
1	4
2	16
3	64

14.

x	$F(x)$
-1	$\frac{2}{3}$
0	1
1	$\frac{3}{2}$
2	$\frac{9}{4}$
3	$\frac{27}{8}$

15.

x	$f(x)$
-1	$\frac{3}{2}$
0	3
1	6
2	12
3	24

16.

x	$g(x)$
-1	6
0	1
1	0
2	3
3	10

17.

x	$H(x)$
-1	2
0	4
1	6
2	8
3	10

18.

x	$F(x)$
-1	$\frac{1}{2}$
0	$\frac{1}{4}$
1	$\frac{1}{8}$
2	$\frac{1}{16}$
3	$\frac{1}{32}$

In Problems 19–26, the graph of an exponential function is given. Match each graph to one of the following functions.

A. $y = 3^x$ B. $y = 3^{-x}$ C. $y = -3^x$ D. $y = -3^{-x}$

E. $y = 3^x - 1$ F. $y = 3^{x-1}$ G. $y = 3^{1-x}$ H. $y = 1 - 3^x$

19.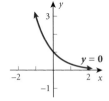

20.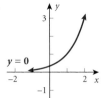

21.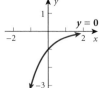

22.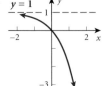

23.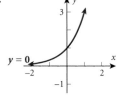

24.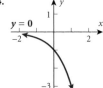

25.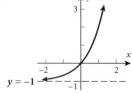

26.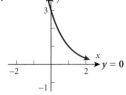

In Problems 27–34, use shifts and/or reflections to graph each function. Determine the domain, range, and horizontal asymptote of each function.

27. $f(x) = 2^x + 1$
28. $f(x) = 2^{x+2}$
29. $f(x) = 3^{-x} - 2$
30. $f(x) = -3^x + 1$
31. $f(x) = e^{-x}$
32. $f(x) = -e^x$
33. $f(x) = e^{x-2} - 1$
34. $f(x) = -e^x - 1$

In Problems 35–48, solve each equation.

35. $2^{2x+1} = 4$
36. $5^{1-2x} = \dfrac{1}{5}$
37. $3^{x^3} = 9^x$
38. $4^{x^2} = 2^x$
39. $8^{x^2-2x} = \dfrac{1}{2}$
40. $9^{-x} = \dfrac{1}{3}$
41. $2^x \cdot 8^{-x} = 4^x$
42. $\left(\dfrac{1}{2}\right)^{1-x} = 4$
43. $\left(\dfrac{1}{5}\right)^{2-x} = 25$
44. $4^x - 2^x = 0$
45. $4^x = 8$
46. $9^{2x} = 27$
47. $e^{x^2} = (e^{3x}) \cdot \dfrac{1}{e^2}$
48. $(e^4)^x \cdot e^{x^2} = e^{12}$

49. If $4^x = 7$, what does 4^{-2x} equal?
50. If $2^x = 3$, what does 4^{-x} equal?
51. If $3^{-x} = 2$, what does 3^{2x} equal?
52. If $5^{-x} = 3$, what does 5^{3x} equal?

In Problems 53–56, determine the exponential function whose graph is given.

53.
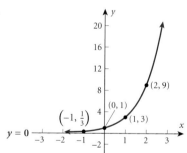

54.
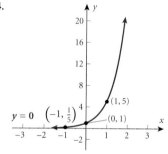

55.
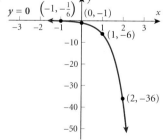

56.
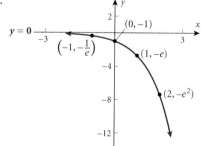

57. Optics If a single pane of glass obliterates 3% of the light passing through it, then the percent p of light that passes through n successive panes is given approximately by the function

$$p(n) = 100e^{-0.03n}$$

(a) What percent of light will pass through 10 panes?
(b) What percent of light will pass through 25 panes?

58. Atmospheric Pressure The atmospheric pressure p on a balloon or airplane decreases with increasing height. This pressure, measured in millimeters of mercury, is related to the number of kilometers h above sea level by the function

$$p(h) = 760e^{-0.145h}$$

(a) Find the atmospheric pressure at a height of 2 kilometers (over 1 mile).
(b) What is it at a height of 10 kilometers (over 30,000 feet)?

59. Space Satellites The number of watts w provided by a space satellite's power supply after a period of d days is given by the function

$$w(d) = 50e^{-0.004d}$$

(a) How much power will be available after 30 days?
(b) How much power will be available after 1 year (365 days)?

60. Healing of Wounds The normal healing of wounds can be modeled by an exponential function. If A_0 represents the original area of the wound and if A equals the area of the wound after n days, then the function

$$A(n) = A_0 e^{-0.35n}$$

describes the area of the wound on the nth day following an injury when no infection is present to retard the healing. Suppose that a wound initially had an area of 100 square millimeters.

(a) If healing is taking place, how large will the area of the wound be after 3 days?
(b) How large will it be after 10 days?

61. Drug Medication The function

$$D(h) = 5e^{-0.4h}$$

can be used to find the number of milligrams D of a certain drug that is in a patient's bloodstream h hours after the drug has been administered. How many milligrams will be present after 1 hour? After 6 hours?

62. Spreading of Rumors A model for the number of people N in a college community who have heard a certain rumor is

$$N = P(1 - e^{-0.15d})$$

where P is the total population of the community and d is the number of days that have elapsed since the rumor began. In a community of 1000 students, how many students will have heard the rumor after 3 days?

63. Exponential Probability Between 12:00 PM and 1:00 PM, cars arrive at Citibank's drive-thru at the rate of 6 cars per hour (0.1 car per minute). The following formula from probability can be used to determine the probability that a car will arrive within t minutes of 12:00 PM:

$$F(t) = 1 - e^{-0.1t}$$

(a) Determine the probability that a car will arrive within 10 minutes of 12:00 PM (that is, before 12:10 PM).
(b) Determine the probability that a car will arrive within 40 minutes of 12:00 PM (before 12:40 PM).
(c) What value does F approach as t becomes unbounded in the positive direction?
(d) Graph F using your graphing utility.
(e) Using TRACE, determine how many minutes are needed for the probability to reach 50%.

64. Exponential Probability Between 5:00 PM and 6:00 PM, cars arrive at Jiffy Lube at the rate of 9 cars per hour (0.15 car per minute). The following formula from probability can be used to determine the probability that a car will arrive within t minutes of 5:00 PM:

$$F(t) = 1 - e^{-0.15t}$$

(a) Determine the probability that a car will arrive within 15 minutes of 5:00 PM (that is, before 5:15 PM).
(b) Determine the probability that a car will arrive within 30 minutes of 5:00 PM (before 5:30 PM).
(c) What value does F approach as t becomes unbounded in the positive direction?
(d) Graph F using your graphing utility.
(e) Using TRACE, determine how many minutes are needed for the probability to reach 60%.

65. Poisson Probability Between 5:00 PM and 6:00 PM, cars arrive at McDonald's drive-thru at the rate of 20 cars per hour. The following formula from probability can be used to determine the probability that x cars will arrive between 5:00 PM and 6:00 PM.

$$P(x) = \frac{20^x e^{-20}}{x!}$$

where

$$x! = x \cdot (x-1) \cdot (x-2) \cdots \cdots 3 \cdot 2 \cdot 1$$

(a) Determine the probability that $x = 15$ cars will arrive between 5:00 PM and 6:00 PM.
(b) Determine the probability that $x = 20$ cars will arrive between 5:00 PM and 6:00 PM.

66. Poisson Probability People enter a line for the *Demon Roller Coaster* at the rate of 4 per minute. The following formula from probability can be used to determine the probability that x people will arrive within the next minute.

$$P(x) = \frac{4^x e^{-4}}{x!}$$

where

$$x! = x \cdot (x-1) \cdot (x-2) \cdots \cdots 3 \cdot 2 \cdot 1$$

(a) Determine the probability that $x = 5$ people will arrive within the next minute.
(b) Determine the probability that $x = 8$ people will arrive within the next minute.

67. Depreciation The price p of a Honda Civic DX Sedan that is x years old is given by

$$p(x) = 16{,}630(0.90)^x$$

(a) How much does a 3-year-old Civic DX Sedan cost?
(b) How much does a 9-year-old Civic DX Sedan cost?

68. Learning Curve Suppose that a student has 500 vocabulary words to learn. If the student learns 15 words after 5 minutes, the function

$$L(t) = 500(1 - e^{-0.0061t})$$

approximates the number of words L that the student will learn after t minutes.

(a) How many words will the student learn after 30 minutes?
(b) How many words will the student learn after 60 minutes?

69. Alternating Current in a RL Circuit The equation governing the amount of current I (in amperes) after time t (in seconds) in a single RL circuit consisting of a resistance R (in ohms), an inductance L (in henrys), and an electromotive force E (in volts) is

$$I = \frac{E}{R}[1 - e^{-(R/L)t}]$$

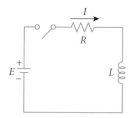

(a) If $E = 120$ volts, $R = 10$ ohms, and $L = 5$ henrys, how much current I_1 is flowing after 0.3 second? After 0.5 second? After 1 second?
(b) Graph the function $I = I_1(t)$, measuring I along the y-axis and t along the x-axis.
(c) What is the maximum current?
(d) If $E = 120$ volts, $R = 5$ ohms, and $L = 10$ henrys, how much current I_2 is flowing after 0.3 second? After 0.5 second? After 1 second?
(e) Graph the function $I = I_2(t)$ on the same screen as $I_1(t)$.
(f) What is the maximum current?

70. Alternating Current in a RC Circuit The equation governing the amount of current I (in amperes) after time t (in microseconds) in a single RC circuit consisting of a resistance R (in ohms), a capacitance C (in microfarads), and an electromotive force E (in volts) is

$$I = \frac{E}{R} e^{-t/(RC)}$$

(a) If $E = 120$ volts, $R = 2000$ ohms, and $C = 1.0$ microfarad, how much current I_1 is flowing initially ($t = 0$)? After 1000 microseconds? After 3000 microseconds?
(b) Graph the function $I = I_1(t)$, measuring I along the y-axis and t along the x-axis.
(c) What is the maximum current?
(d) If $E = 120$ volts, $R = 1000$ ohms, and $C = 2.0$ microfarads, how much current I_2 is flowing initially? After 1000 microseconds? After 3000 microseconds?
(e) Graph the function $I = I_2(t)$ on the same screen as $I_1(t)$.
(f) What is the maximum current?

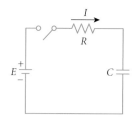

71. Another Formula for e Use a calculator to compute the values of

$$2 + \frac{1}{2!} + \frac{1}{3!} + \cdots + \frac{1}{n!}$$

for $n = 4, 6, 8,$ and 10. Compare each result with e.
[**Hint:** $1! = 1, 2! = 2 \cdot 1, 3! = 3 \cdot 2 \cdot 1$,
$n! = n(n-1) \cdots \cdots (3)(2)(1)$]

72. Another Formula for e Use a calculator to compute the first five values of the expression. The first one is $2 + 1 = 3$; the second one is $2 + \dfrac{1}{1+1} = 2.5$. Compare the values to e.

$$2 + \cfrac{1}{1 + \cfrac{1}{2 + \cfrac{2}{3 + \cfrac{3}{4 + 4}}}}$$

etc.

73. Difference Quotient If $f(x) = a^x$, show that

$$\frac{f(x+h) - f(x)}{h} = a^x\left(\frac{a^h - 1}{h}\right)$$

74. If $f(x) = a^x$, show that $f(A + B) = f(A) \cdot f(B)$.

75. If $f(x) = a^x$, show that $f(-x) = \dfrac{1}{f(x)}$.

76. If $f(x) = a^x$, show that $f(\alpha x) = [f(x)]^\alpha$.

77. Relative Humidity The relative humidity is the ratio (expressed as a percent) of the amount of water vapor in the air to the maximum amount that it can hold at a specific temperature. The relative humidity, R, is found using the following formula:

$$R = 10^{\frac{4221}{T+459.4} - \frac{4221}{D+459.4} + 2}$$

where T is the air temperature (in °F) and D is the dew point temperature (in °F).

(a) Determine the relative humidity if the air temperature is 50° Fahrenheit and the dew point temperature is 41° Fahrenheit.
(b) Determine the relative humidity if the air temperature is 68° Fahrenheit and the dew point temperature is 59° Fahrenheit.
(c) What is the relative humidity if the air temperature and the dew point temperature are the same?

78. Historical Problem Pierre de Fermat (1601–1665) conjectured that the function

$$f(x) = 2^{(2^x)} + 1$$

for $x = 1, 2, 3, \ldots$, would always have a value equal to a prime number. But Leonhard Euler (1707–1783) showed that this formula fails for $x = 5$. Use a calculator to determine the prime numbers produced by f for $x = 1, 2, 3, 4$. Then show that $f(5) = 641 \times 6{,}700{,}417$, which is not prime.

Problems 79 and 80 provide definitions for two other functions.

79. The **hyperbolic sine function**, designated by sinh x, is defined as

$$\sinh x = \frac{1}{2}(e^x - e^{-x})$$

(a) Show that $f(x) = \sinh x$ is an odd function.
(b) Graph $f(x) = \sinh x$ using a graphing utility.

80. The **hyperbolic cosine function**, designated by cosh x, is defined as

$$\cosh x = \frac{1}{2}(e^x + e^{-x})$$

(a) Show that $f(x) = \cosh x$ is an even function.
(b) Graph $f(x) = \cosh x$ using a graphing utility.
(c) Refer to Problem 79. Show that, for every x,

$$(\cosh x)^2 - (\sinh x)^2 = 1.$$

81. The bacteria in a 4-liter container double every minute. After 60 minutes the container is full. How long did it take to fill half the container? Explain your reasoning.

82. Explain in your own words what the number e is. Provide at least two applications that require the use of this number.

83. Do you think that there is a power function that increases more rapidly than an exponential function whose base is greater than 1? Explain.

84. As the base a of an exponential function $f(x) = a^x$, $a > 1$, increases, what happens to the behavior of its graph for $x > 0$? What happens to the behavior of the graph for $x < 0$?

85. The graphs of $y = a^{-x}$ and $y = \left(\dfrac{1}{a}\right)^x$ are identical. Why?

2.4 Logarithmic Functions

PREPARING FOR THIS SECTION *Before getting started, review the following:*

> Solving Inequalities (Chapter 0, Section 0.5, pp. 54–61)
> Graphing Techniques (Chapter 1, Section 1.5, pp. 148–154)
> Vertical Line Test (Chapter 1, Section 1.3, p. 121)

OBJECTIVES
1. Change exponential expressions to logarithmic expressions
2. Change logarithmic expressions to exponential expressions
3. Evaluate logarithmic functions
4. Find the domain of a logarithmic function
5. Graph logarithmic functions
6. Solve logarithmic equations

Logarithmic Functions

We begin with the exponential function

$$y = 3^x$$

If we interchange the variables x and y, we obtain the equation

$$x = 3^y$$

FIGURE 35

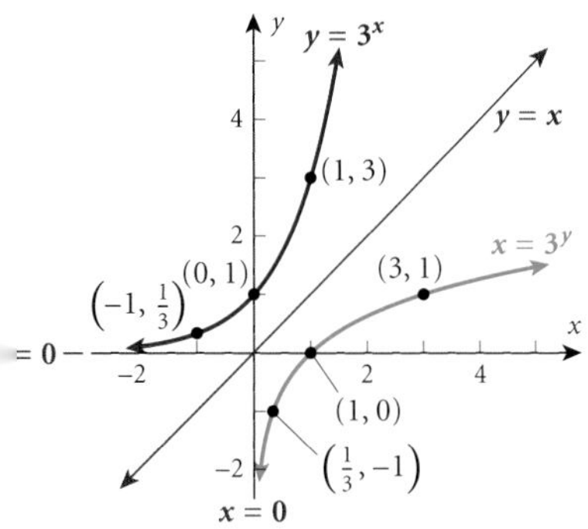

Let's compare the graphs of these two equations. For example, the point $(1, 3)$ is on the graph of $y = 3^x$ and the point $(3, 1)$ is on the graph of $x = 3^y$. Also, the point $(0, 1)$ is on the graph of $y = 3^x$ and the point $(1, 0)$ is on the graph of $x = 3^y$. In general, if the point (a, b) is on the graph of $y = 3^x$, then the point (b, a) will be on the graph of $x = 3^y$. See Figure 35.

Notice in Figure 35 that we show the line $y = x$. You should see that the graphs of $y = 3^x$ and $x = 3^y$ are symmetric with respect to the line $y = x$, a fact we shall not prove. This means we could have obtained the graph of $x = a^y$, $a > 0$, $a \neq 1$, by reflecting the graph of $y = a^x$ about the line $y = x$.

Look again at Figure 35. We see from the graph of the equation $x = 3^y$ that it is the graph of a function. (Do you see why? Apply the Vertical Line Test.) We call this function a *logarithmic function*. The general definition is given next.

> The **logarithmic function to the base a,** where $a > 0$ and $a \neq 1$, is denoted by $y = \log_a x$ (read as "y is the logarithm to the base a of x") and is defined by
>
> $$\boxed{y = \log_a x \quad \text{if and only if} \quad x = a^y}$$
>
> The domain of the logarithmic function $y = \log_a x$ is $x > 0$.

A *logarithm* is merely a name for a certain exponent.

EXAMPLE 1 Relating Logarithms to Exponents

(a) If $y = \log_3 x$, then $x = 3^y$. For example, $2 = \log_3 9$ is equivalent to $9 = 3^2$.

(b) If $y = \log_5 x$, then $x = 5^y$. For example, $-1 = \log_5\left(\dfrac{1}{5}\right)$ is equivalent to $\dfrac{1}{5} = 5^{-1}$.

1 EXAMPLE 2 Changing Exponential Expressions to Logarithmic Expressions

Change each exponential expression to an equivalent expression involving a logarithm.

(a) $1.2^3 = m$ (b) $e^b = 9$ (c) $a^4 = 24$

SOLUTION We use the fact that $y = \log_a x$ and $x = a^y$, $a > 0$, $a \neq 1$, are equivalent.

(a) If $1.2^3 = m$, then $3 = \log_{1.2} m$.   (b) If $e^b = 9$, then $b = \log_e 9$.

(c) If $a^4 = 24$, then $4 = \log_a 24$.

NOW WORK PROBLEM 1.

EXAMPLE 3 Changing Logarithmic Expressions to Exponential Expressions

Change each logarithmic expression to an equivalent expression involving an exponent.

(a) $\log_a 4 = 5$ (b) $\log_e b = -3$ (c) $\log_3 5 = c$

SOLUTION (a) If $\log_a 4 = 5$, then $a^5 = 4$.   (b) If $\log_e b = -3$, then $e^{-3} = b$.

(c) If $\log_3 5 = c$, then $3^c = 5$.

NOW WORK PROBLEM 13.

To find the exact value of a logarithm, we write the logarithm in exponential notation and use the fact that if $a^u = a^v$ then $u = v$.

EXAMPLE 4 Finding the Exact Value of a Logarithmic Expression

Find the exact value of

(a) $\log_2 16$ (b) $\log_3 \dfrac{1}{27}$

SOLUTION (a) $y = \log_2 16$

$2^y = 16$ Change to exponential form.

$2^y = 2^4$ $16 = 2^4$

$y = 4$ Equate exponents.

Therefore, $\log_2 16 = 4$.

(b) $y = \log_3 \dfrac{1}{27}$

$3^y = \dfrac{1}{27}$ Change to exponential form.

$3^y = 3^{-3}$ $\dfrac{1}{27} = \dfrac{1}{3^3} = 3^{-3}$

$y = -3$ Equate exponents.

Therefore, $\log_3 \dfrac{1}{27} = -3$.

NOW WORK PROBLEM 25.

The domain of a logarithmic function consists of the *positive* real numbers, so the argument of a logarithmic function must be greater than zero.

EXAMPLE 5 Finding the Domain of a Logarithmic Function

Find the domain of each logarithmic function

(a) $F(x) = \log_2(x - 5)$ (b) $g(x) = \log_5\left(\dfrac{1+x}{1-x}\right)$ (c) $h(x) = \log_{1/2}|x|$

SOLUTION (a) The domain of F consists of all x for which $x - 5 > 0$, that is, all $x > 5$, or using interval notation, $(5, \infty)$.

(b) The domain of g is restricted to
$$\frac{1 + x}{1 - x} > 0$$

Solving this inequality, we find that the domain of g consists of all x between -1 and 1, that is, $-1 < x < 1$, or using interval notation, $(-1, 1)$.

(c) Since $|x| > 0$, provided that $x \neq 0$, the domain of h consists of all nonzero real numbers, or using interval notation, $(-\infty, 0)$ or $(0, \infty)$. ◀

NOW WORK PROBLEMS 37 AND 39.

Graph logarithmic functions **5** **Graphs of Logarithmic Functions**

As we said earlier, we can obtain the graph of $x = a^y$, $a > 0$, $a \neq 1$, or equivalently the graph of $y = \log_a x$, by reflecting the graph of $y = a^x$ about the line $y = x$. For example, to graph $y = \log_2 x$, graph $y = 2^x$ and reflect it about the line $y = x$. See Figure 36. To graph $y = \log_{1/3} x$, graph $y = \left(\frac{1}{3}\right)^x$ and reflect it about the line $y = x$. See Figure 37.

FIGURE 36

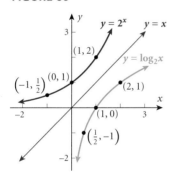

FIGURE 37

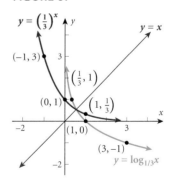

The graph of the logarithmic function $y = \log_a x$ is the reflection about the line $y = x$ of the graph of the exponential function $y = a^x$, as shown in Figures 38 and 39.

FIGURE 38

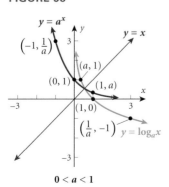

$0 < a < 1$

FIGURE 39

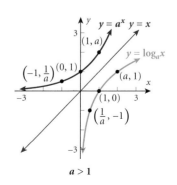

$a > 1$

NOW WORK PROBLEM 51.

The graphs in Figures 38 and 39 lead to the following result.

> **Properties of the Graph of a Logarithmic Function $f(x) = \log_a x$**
> 1. The domain is the set of positive real numbers; the range is all real numbers.
> 2. The x-intercept of the graph is $(1, 0)$ There is no y-intercept.
> 3. The line $x = 0$ (the y-axis) is a vertical asymptote of the graph.
> 4. A logarithmic function is decreasing if $0 < a < 1$ and is increasing if $a > 1$.
> 5. The graph of f contains the points $(1, 0)$, $(a, 1)$, and $\left(\dfrac{1}{a}, -1\right)$.
> 6. The graph is smooth and continuous, with no corners or gaps.

If the base of a logarithmic function is the number e, then we have the **natural logarithm function**. This function occurs so frequently in applications that it is given a special symbol, **ln** (from the Latin, *logarithmus naturalis*). That is,

$$y = \log_e x = \ln x \quad \text{if and only if} \quad x = e^y \tag{1}$$

We can obtain the graph of $y = \ln x$ by reflecting the graph of $y = e^x$ about the line $y = x$. See Figure 40.

FIGURE 40

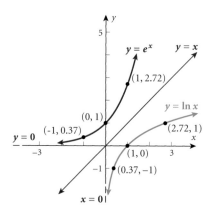

TABLE 10

x	$\ln x$
0.5	-0.69
2	0.69
3	1.10

Using a calculator with an $\boxed{\ln}$ key, we can obtain other points on the graph of $f(x) = \ln x$. See Table 10.

SEEING THE CONCEPT: Graph $Y_1 = e^x$ and $Y_2 = \ln x$ on the same square screen. Use eVALUEate to verify the points on the graph given in Figure 41. Do you see the symmetry of the two graphs with respect to the line $y = x$?

EXAMPLE 6 **Graphing Logarithmic Functions Using Shifts and/or Reflections**

Graph $f(x) = -\ln(x + 2)$ by starting with the graph of $y = \ln x$. Determine the domain, range, and vertical asymptote of f.

SOLUTION The domain of f consists of all x for which

$$x + 2 > 0 \quad \text{or} \quad x > -2$$

so the domain is $\{x \mid x > -2\}$.

To obtain the graph of $y = -\ln(x + 2)$, we use the steps illustrated in Figure 41.

FIGURE 41

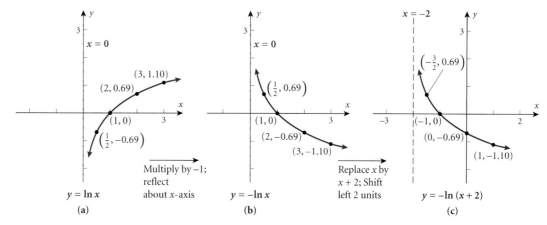

The range of $f(x) = -\ln(x + 2)$ is the interval $(-\infty, \infty)$, and the vertical asymptote is $x = -2$. [Do you see why? The original asymptote ($x = 0$) is shifted to the left 2 units.]

NOW WORK PROBLEM 63.

If the base of a logarithmic function is the number 10, then we have the **common logarithm function.** If the base a of the logarithmic function is not indicated, it is understood to be 10. Thus,

$$y = \log x \quad \text{if and only if} \quad x = 10^y$$

We can obtain the graph of $y = \log x$ by reflecting the graph of $y = 10^x$ about the line $y = x$. See Figure 42.

FIGURE 42

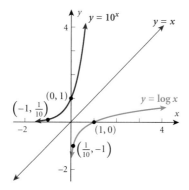

Logarithmic Equations

Equations that contain logarithms are called **logarithmic equations.** Care must be taken when solving logarithmic equations. Be sure to check each apparent solution in the original equation and discard any that are extraneous. In the expression $\log_a M$, remember that a and M are positive and $a \neq 1$.

Some logarithmic equations can be solved by changing from a logarithmic expression to an exponential expression.

EXAMPLE 7 Solving a Logarithmic Equation

Solve: **(a)** $\log_3(4x - 7) = 2$ **(b)** $\log_x 64 = 2$

SOLUTION **(a)** We can obtain an exact solution by changing the logarithm to exponential form

$$\log_3(4x - 7) = 2$$
$$4x - 7 = 3^2 \quad \text{Change to exponential form.}$$
$$4x - 7 = 9$$
$$4x = 16$$
$$x = 4$$

CHECK: $\log_3(4x - 7) = \log_3(16 - 7) = \log_3 9 = 2 \quad 3^2 = 9$

(b) We can obtain an exact solution by changing the logarithm to exponential form.

$$\log_x 64 = 2$$
$$x^2 = 64 \quad \text{Change to exponential form.}$$
$$x = \pm\sqrt{64} = \pm 8$$

The base of a logarithm is always positive. As a result, we discard -8; the only solution is 8.

CHECK: $\log_8 64 = 2 \quad 8^2 = 64.$

EXAMPLE 8 Using Logarithms to Solve Exponential Equations

Solve: $e^{2x} = 5$

SOLUTION We can obtain an exact solution by changing the exponential equation to logarithmic form.

$$e^{2x} = 5$$
$$\ln 5 = 2x \quad \text{Change to a logarithmic expression.}$$
$$x = \frac{\ln 5}{2} \quad \text{Exact solution.}$$
$$\approx 0.805 \quad \text{Approximate solution.}$$

NOW WORK PROBLEMS 71 AND 83.

EXAMPLE 9 Alcohol and Driving

The concentration of alcohol in a person's blood is measurable. Recent medical research suggests that the risk R (given as a percent) of having an accident while driving a car can be modeled by the equation

$$R = 6e^{kx}$$

where x is the variable concentration of alcohol in the blood and k is a constant.

(a) Suppose that a concentration of alcohol in the blood of 0.04 results in a 10% risk ($R = 10$) of an accident. Find the constant k in the equation.

(b) Using this value of k, what is the risk if the concentration is 0.17?
(c) Using the same value of k, what concentration of alcohol corresponds to a risk of 100%?
(d) If the law asserts that anyone with a risk of having an accident of 20% or more should not have driving privileges, at what concentration of alcohol in the blood should a driver be arrested and charged with a DUI (Driving Under the Influence)?

SOLUTION

(a) For a concentration of alcohol in the blood of 0.04 and a risk of 10%, we let $x = 0.04$ and $R = 10$ in the equation and solve for k.

$$R = 6e^{kx}$$
$$10 = 6e^{k(0.04)} \qquad R = 10; x = 0.04.$$
$$\frac{10}{6} = e^{0.04k} \qquad \text{Divide both sides by 6.}$$
$$0.04k = \ln\frac{10}{6} = 0.5108256 \qquad \text{Change to a logarithmic expression.}$$
$$k = 12.77 \qquad \text{Solve for } k.$$

(b) Using $k = 12.77$ and $x = 0.17$ in the equation, we find the risk R to be

$$R = 6e^{kx} = 6e^{(12.77)(0.17)} = 52.6$$

For a concentration of alcohol in the blood of 0.17, the risk of an accident is about 52.6%.

(c) Using $k = 12.77$ and $R = 100$ in the equation, we find the concentration x of alcohol in the blood to be

$$R = 6e^{kx}$$
$$100 = 6e^{12.77x} \qquad R = 100; k = 12.77.$$
$$\frac{100}{6} = e^{12.77x} \qquad \text{Divide both sides by 6.}$$
$$12.77x = \ln\frac{100}{6} = 2.8134 \qquad \text{Change to a logarithmic expression.}$$
$$x = 0.22 \qquad \text{Solve for } x.$$

For a concentration of alcohol in the blood of 0.22, the risk of an accident is 100%.

(d) Using $k = 12.77$ and $R = 20$ in the equation, we find the concentration x of alcohol in the blood to be

$$R = 6e^{kx}$$
$$20 = 6e^{12.77x} \qquad R = 20; k = 12.77$$
$$\frac{20}{6} = e^{12.77x}$$
$$12.77x = \ln\frac{20}{6} = 1.204$$
$$x = 0.094$$

A driver with a concentration of alcohol in the blood of 0.094 or more (9.4%) should be arrested and charged with DUI.

[**NOTE:** Most states use 0.08 or 0.10 as the blood alcohol content at which a DUI citation is given.]

SUMMARY Properties of the Logarithmic Function

$f(x) = \log_a x, \ a > 1$
$(y = \log_a x \text{ means } x = a^y)$
Domain: the interval $(0, \infty)$; Range: the interval $(-\infty, \infty)$;
x-intercept: $(1, 0)$; y-intercept: none;
vertical asymptote: $x = 0$ (y-axis); increasing.
See Figure 43 for a typical graph.

$f(x) = \log_a x, \ 0 < a < 1$
$(y = \log_a x \text{ means } x = a^y)$
Domain: the interval $(0, \infty)$; Range: the interval $(-\infty, \infty)$;
x-intercept: $(1, 0)$; y-intercept: none;
vertical asymptote: $x = 0$ (y-axis); decreasing.
See Figure 44 for a typical graph.

FIGURE 43

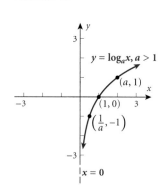

FIGURE 44

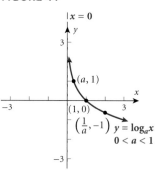

EXERCISE 2.4

In Problems 1–12, change each exponential expression to an equivalent expression involving a logarithm.

1. $9 = 3^2$
2. $16 = 4^2$
3. $a^2 = 1.6$
4. $a^3 = 2.1$
5. $1.1^2 = M$
6. $2.2^3 = N$
7. $2^x = 7.2$
8. $3^x = 4.6$
9. $x^{\sqrt{2}} = \pi$
10. $x^\pi = e$
11. $e^x = 8$
12. $e^{2.2} = M$

In Problems 13–24, change each logarithmic expression to an equivalent expression involving an exponent.

13. $\log_2 8 = 3$
14. $\log_3 \left(\dfrac{1}{9}\right) = -2$
15. $\log_a 3 = 6$
16. $\log_b 4 = 2$
17. $\log_3 2 = x$
18. $\log_2 6 = x$
19. $\log_2 M = 1.3$
20. $\log_3 N = 2.1$
21. $\log_{\sqrt{2}} \pi = x$
22. $\log_\pi x = \dfrac{1}{2}$
23. $\ln 4 = x$
24. $\ln x = 4$

In Problems 25–36, find the exact value of each logarithm without using a calculator.

25. $\log_2 1$
26. $\log_8 8$
27. $\log_5 25$
28. $\log_3 \left(\dfrac{1}{9}\right)$

29. $\log_{1/2} 16$ **30.** $\log_{1/3} 9$ **31.** $\log_{10} \sqrt{10}$ **32.** $\log_5 \sqrt[3]{25}$

33. $\log_{\sqrt{2}} 4$ **34.** $\log_{\sqrt{3}} 9$ **35.** $\ln \sqrt{e}$ **36.** $\ln e^3$

In Problems 37–44, find the domain of each function.

37. $f(x) = \ln(x - 3)$ **38.** $g(x) = \ln(x - 1)$ **39.** $F(x) = \log_2 x^2$ **40.** $H(x) = \log_5 x^3$

41. $f(x) = 3 - 2\log_4 \dfrac{x}{2}$ **42.** $g(x) = 8 + 5\ln(2x)$ **43.** $f(x) = \sqrt{\ln x}$ **44.** $g(x) = \dfrac{1}{\ln x}$

In Problems 45–48, use a calculator to evaluate each expression. Round your answer to three decimal places.

45. $\ln \dfrac{5}{3}$ **46.** $\dfrac{\ln 5}{3}$ **47.** $\dfrac{\ln \dfrac{10}{3}}{0.04}$ **48.** $\dfrac{\ln \dfrac{2}{3}}{-0.1}$

49. Find a so that the graph of $f(x) = \log_a x$ contains the point $(2, 2)$.

50. Find a so that the graph of $f(x) = \log_a x$ contains the point $\left(\dfrac{1}{2}, -4\right)$.

In Problems 51–54, graph each logarithmic function.

51. $y = \log_3 x$ **52.** $y = \log_{1/3} x$ **53.** $y = \log_{1/5} x$ **54.** $y = \log_5 x$

In Problems 55–62, the graph of a logarithmic function is given. Match each graph to one of the following functions:

A. $y = \log_3 x$ **B.** $y = \log_3(-x)$ **C.** $y = -\log_3 x$ **D.** $y = -\log_3(-x)$

E. $y = \log_3 x - 1$ **F.** $y = \log_3(x - 1)$ **G.** $y = \log_3(1 - x)$ **H.** $y = 1 - \log_3 x$

55. 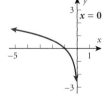   **56.** 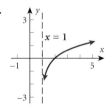   **57.** 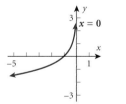   **58.**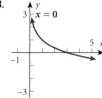

59. 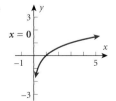   **60.** 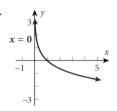   **61.** 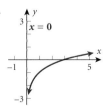   **62.**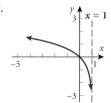

In Problems 63–70, use shifts and/or reflections to graph each function. Determine the domain, range, and vertical asymptote of each function.

63. $f(x) = \ln(x + 4)$ **64.** $f(x) = \ln(x - 3)$ **65.** $f(x) = 2 + \ln x$ **66.** $f(x) = -\ln(-x)$

67. $f(x) = \log(x - 4)$ **68.** $f(x) = \log(x + 5)$ **69.** $h(x) = \log x + 2$ **70.** $g(x) = \log x - 5$

In Problems 71–90, solve each equation.

71. $\log_3 x = 2$ **72.** $\log_5 x = 3$ **73.** $\log_2(2x+1) = 3$ **74.** $\log_3(3x-2) = 2$

75. $\log_x 4 = 2$ **76.** $\log_x\left(\dfrac{1}{8}\right) = 3$ **77.** $\ln e^x = 5$ **78.** $\ln e^{-2x} = 8$

79. $\log_4 64 = x$ **80.** $\log_5 625 = x$ **81.** $\log_3 243 = 2x+1$ **82.** $\log_6 36 = 5x+3$

83. $e^{3x} = 10$ **84.** $e^{-2x} = \dfrac{1}{3}$ **85.** $e^{2x+5} = 8$ **86.** $e^{-2x+1} = 13$

87. $\log_3(x^2+1) = 2$ **88.** $\log_5(x^2+x+4) = 2$ **89.** $\log_2 8^x = -3$ **90.** $\log_3 3^x = -1$

---

91. Chemistry The pH of a chemical solution is given by the formula

$$\text{pH} = -\log_{10}[\text{H}^+]$$

where $[\text{H}^+]$ is the concentration of hydrogen ions in moles per liter. Values of pH range from 0 (acidic) to 14 (alkaline). Distilled water has a pH of 7.

(a) What is the pH of a solution for which $[\text{H}^+]$ is 0.1?
(b) What is the pH of a solution for which $[\text{H}^+]$ is 0.01?
(c) What is the pH of a solution for which $[\text{H}^+]$ is 0.001?
(d) What happens to pH as the hydrogen ion concentration decreases?
(e) Determine the hydrogen ion concentration of an orange (pH = 3.5).
(f) Determine the hydrogen ion concentration of human blood (pH = 7.4).

92. Diversity Index Shannon's diversity index is a measure of the diversity of a population. The diversity index is given by the formula

$$H = -(p_1 \log p_1 + p_2 \log p_2 + \cdots + p_n \log p_n)$$

where p_1 is the proportion of the population that is species 1, p_2 is the proportion of the population that is species 2, and so on.

The distribution of race in the United States in 2000 was as follows:

Race	Proportion
American Indian or Native Alaskan	0.014
Asian	0.041
Black or African American	0.128
Hispanic	0.124
Native Hawaiian or Pacific Islander	0.003
White	0.690

Source: U.S. Census Bureau.

(a) Compute the diversity index of the United States in 2000.
(b) The largest value of the diversity index is given by $H_{\max} = \log(S)$, where S is the number of categories of race. Compute $H_{\max}$.
(c) The evenness ratio is given by $E_H = \dfrac{H}{H_{\max}}$, where $0 \le E_H \le 1$. If $E_H = 1$, there is complete evenness. Compute the evenness ratio for the United States.
(d) Obtain the distribution of race for the United States in 1990 from the Census Bureau. Compute Shannon's diversity index. Is the United States becoming more diverse? Why?

93. Atmospheric Pressure The atmospheric pressure p on a balloon or an aircraft decreases with increasing height. This pressure, measured in millimeters of mercury, is related to the height h (in kilometers) above sea level by the formula.

$$p = 760e^{-0.145h}$$

(a) Find the height of an aircraft if the atmospheric pressure is 320 millimeters of mercury.
(b) Find the height of a mountain if the atmospheric pressure is 667 millimeters of mercury.

94. Healing of Wounds The normal healing of wounds can be modeled by an exponential function. If A_0 represents the original area of the wound and if A equals the area of the wound after n days, then the formula

$$A = A_0 e^{-0.35n}$$

describes the area of the wound on the nth day following an injury when no infection is present to retard the healing. Suppose that a wound initially had an area of 100 square millimeters.

(a) If healing is taking place, how many days should pass before the wound is one-half its original size?
(b) How long before the wound is 10% of its original size?

95. Exponential Probability Between 12:00 PM and 1:00 PM, cars arrive at Citibank's drive-thru at the rate of 6 cars per hour (0.1 car per minute). The following formula from statistics can be used to determine the probability that a car will arrive within t minutes of 12:00 PM.

$$F(t) = 1 - e^{-0.1t}$$

(a) Determine how many minutes are needed for the probability to reach 50%.
(b) Determine how many minutes are needed for the probability to reach 80%.
(c) Is it possible for the probability to equal 100%? Explain.

96. Exponential Probability Between 5:00 PM and 6:00 PM, cars arrive at Jiffy Lube at the rate of 9 cars per hour (0.15 car per minute). The following formula from statistics can be used to determine the probability that a car will arrive within t minutes of 5:00 PM.

$$F(t) = 1 - e^{-0.15t}$$

(a) Determine how many minutes are needed for the probability to reach 50%.
(b) Determine how many minutes are needed for the probability to reach 80%.

97. Drug Medication The formula

$$D = 5e^{-0.4h}$$

can be used to find the number of milligrams D of a certain drug that is in a patient's bloodstream h hours after the drug has been administered. When the number of milligrams reaches 2, the drug is to be administered again. What is the time between injections?

98. Spreading of Rumors A model for the number of people N in a college community who have heard a certain rumor is

$$N = P(1 - e^{-0.15d})$$

where P is the total population of the community and d is the number of days that have elapsed since the rumor began. In a community of 1000 students, how many days will elapse before 450 students have heard the rumor?

99. Current in an RL Circuit The equation governing the amount of current I (in amperes) after time t (in seconds) in a simple RL circuit consisting of a resistance R (in ohms), an inductance L (in henrys), and an electromotive force E (in volts) is

$$I = \frac{E}{R}[1 - e^{-(R/L)t}]$$

If $E = 12$ volts, $R = 10$ ohms, and $L = 5$ henrys, how long does it take to obtain a current of 0.5 ampere? Of 1.0 ampere?

100. Learning Curve Psychologists sometimes use the function

$$L(t) = A(1 - e^{-kt})$$

to measure the amount L learned at time t. The number A represents the amount to be learned, and the number k measures the rate of learning. Suppose that a student has an amount A of 200 vocabulary words to learn. A psychologist determines that the student learned 20 vocabulary words after 5 minutes.

(a) Determine the rate of learning k.
(b) Approximately how many words will the student have learned after 10 minutes?
(c) After 15 minutes?
(d) How long does it take for the student to learn 180 words?

101. U.S. Population According to the U.S. Census Bureau, the population of the United States is projected to be 298,710,000 on January 1, 2010. Suppose this projection is correct, but after January 1, 2010, the population grows according to $P(t) = 298{,}710{,}000 + 10{,}000{,}000 \log t$. Project the population, to the nearest thousand, on January 1, 2020.
[**Hint:** $t = 1$ corresponds to the year 2010.]

102. U.S. Population A reasonable projection for the population of the United States on January 1, 2025, is 336,566,000.

(a) If 2025 is taken as year 1, and the formula for year 2025 onward that represents a new trend in the population growth is $P(t) = 336{,}566{,}000 + 8{,}000{,}000 \log t$, what would be the population in 2045?
(b) Is this higher or lower than the current U.S. Census Bureau's estimate of 363,077,000?

Loudness of Sound Problems 103–106 use the following discussion: The **loudness** $L(x)$, measured in decibels, of a sound of intensity x, measured in watts per square meter, is defined as $L(x) = 10 \log \dfrac{x}{I_0}$, where $I_0 = 10^{-12}$ watt per square meter is the least intense sound that a human ear can detect.

Determine the loudness, in decibels, of each of the following sounds.

103. Normal conversation: intensity of $x = 10^{-7}$ watt per square meter.

104. Heavy city traffic: intensity of $x = 10^{-3}$ watt per square meter.

105. Amplified rock music: intensity of 10^{-1} watt per square meter.

106. Diesel truck traveling 40 miles per hour 50 feet away: intensity 10 times that of a passenger car traveling 50 miles per hour 50 feet away whose loudness is 70 decibels.

Problems 107 and 108 use the following discussion: The **Richter scale** is one way of converting seismographic readings into numbers that provide an easy reference for measuring the magnitude M of an earthquake. All earthquakes are compared to a **zero-level earthquake** whose seismographic reading measures 0.001 millimeter at a distance of 100 kilometers from the epicenter. An earthquake whose seismographic reading measures x millimeters has **magnitude** M(x) given by

$$M(x) = \log\left(\frac{x}{x_0}\right)$$

where $x_0 = 10^{-3}$ is the reading of a zero-level earthquake the same distance from its epicenter. Determine the magnitude of the following earthquakes.

107. Magnitude of an Earthquake Mexico City in 1985: seismographic reading of 125,892 millimeters 100 kilometers from the center.

108. Magnitude of an Earthquake San Francisco in 1906: seismographic reading of 7943 millimeters 100 kilometers from the center.

109. Alcohol and Driving The concentration of alcohol in a person's blood is measurable. Suppose that the risk R (given as a percent) of having an accident while driving a car can be modeled by the equation

$$R = 3e^{kx}$$

where x is the variable concentration of alcohol in the blood and k is a constant.

(a) Suppose that a concentration of alcohol in the blood of 0.06 results in a 10% risk ($R = 10$) of an accident. Find the constant k in the equation.
(b) Using this value of k, what is the risk if the concentration is 0.17?
(c) Using the same value of k, what concentration of alcohol corresponds to a risk of 100%?
(d) If the law asserts that anyone with a risk of having an accident of 15% or more should not have driving privileges, at what concentration of alcohol in the blood should a driver be arrested and charged with a DUI?
(e) Compare this situation with that of Example 9. If you were a lawmaker, which situation would you support? Give your reasons.

110. Is there any function of the form $y = x^\alpha$, $0 < \alpha < 1$, that increases more slowly than a logarithmic function whose base is greater than 1? Explain.

111. In the definition of the logarithmic function, the base a is not allowed to equal 1. Why?

112. Critical Thinking In buying a new car, one consideration might be how well the price of the car holds up over time. Different makes of cars have different depreciation rates. One way to compute a depreciation rate for a car is given here. Suppose that the current prices of a certain Mercedes automobile are as follows:

	Age in Years				
New	1	2	3	4	5
$38,000	$36,600	$32,400	$28,750	$25,400	$21,200

(a) Use the formula New = Old(e^{Rt}) to find R, the annual depreciation rate, for a specific time t.
(b) When might be the best time to trade in the car?
(c) Consult the NADA ("blue") book and compare two like models that you are interested in. Which has the better depreciation rate?

2.5 Properties of Logarithms

OBJECTIVES
1. Work with the properties of logarithms
2. Write a logarithmic expression as a sum or difference of logarithms
3. Write a logarithmic expression as a single logarithm
4. Evaluate logarithms whose base is neither 10 nor e

Work with the properties of logarithms **1** Logarithms have some very useful properties that can be derived directly from the definition and the laws of exponents.

EXAMPLE 1 Establishing Properties of Logarithms

(a) Show that $\log_a 1 = 0$.   (b) Show that $\log_a a = 1$.

SOLUTION (a) This fact was established when we graphed $y = \log_a x$ (see Figure 25). To show the result algebraically, let $y = \log_a 1$. Then

$$y = \log_a 1$$
$$a^y = 1 \qquad \text{Change to an exponent.}$$
$$a^y = a^0 \qquad a^0 = 1$$
$$y = 0 \qquad \text{Solve for } y.$$
$$\log_a 1 = 0 \qquad y = \log_a 1$$

(b) Let $y = \log_a a$. Then

$$y = \log_a a$$
$$a^y = a \qquad \text{Change to an exponent.}$$
$$a^y = a^1 \qquad a^1 = a$$
$$y = 1 \qquad \text{Solve for } y.$$
$$\log_a a = 1 \qquad y = \log_a a$$

To summarize:

$$\boxed{\log_a 1 = 0 \quad \log_a a = 1}$$

Properties of Logarithms

In the properties given next, M and a are positive real numbers, with $a \neq 1$, and r is any real number.

The number $\log_a M$ is the exponent to which a must be raised to obtain M. That is,

$$\boxed{a^{\log_a M} = M} \tag{1}$$

The logarithm to the base a of a raised to a power equals that power. That is,

$$\boxed{\log_a a^r = r} \tag{2}$$

The proof uses the fact that $x = a^y$ and $y = \log_a x$ are equivalent.

PROOF: Since $x = a^y$ and $y = \log_a x$ are equivalent, we have

$$a^{\log_a x} = a^y = x$$

Now let $x = M$ to obtain equation (1).

216 Chapter 2 Classes of Functions

To prove (2), we use the fact that $x = a^y$ and $y = \log_a x$ are equivalent. Then,

$$\log_a a^y = \log_a x = y$$

Now let $y = r$ to obtain equation (2). ■

EXAMPLE 2   **Using Properties (1) and (2)**

(a) $2^{\log_2 \pi} = \pi$ (b) $\log_{0.2} 0.2^{-\sqrt{2}} = -\sqrt{2}$ (c) $\ln e^{kt} = kt$ ■

NOW WORK PROBLEM 3.

Other useful properties of logarithms are given below.

Properties of Logarithms

In the following properties M, N, and a are positive real numbers, with $a \neq 1$, and r is any real number.

The Log of a Product Equals the Sum of the Logs

$$\log_a(MN) = \log_a M + \log_a N \quad (3)$$

The Log of a Quotient Equals the Difference of the Logs

$$\log_a\left(\frac{M}{N}\right) = \log_a M - \log_a N \quad (4)$$

The Log of a Power Equals the Product of the Power and the Log

$$\log_a M^r = r \log_a M \quad (5)$$

We shall derive properties (3) and (5) and will leave the derivation of property (4) as an exercise (see Problem 95).

PROOF OF PROPERTY (3)   Let $A = \log_a M$ and let $B = \log_a N$. These expressions are equivalent to the exponential expressions

$$a^A = M \quad \text{and} \quad a^B = N$$

Now

$$\log_a(MN) = \log_a(a^A a^B) = \log_a a^{A+B} \quad \text{Law of Exponents.}$$
$$= A + B \quad\quad\quad\quad\quad\quad \text{Property (2) of logarithms.}$$
$$= \log_a M + \log_a N$$

PROOF OF PROPERTY (5) Let $A = \log_a M$. This expression is equivalent to
$$a^A = M$$
Now
$$\begin{aligned}\log_a M^r &= \log_a(a^A)^r = \log_a a^{rA} & \text{Law of Exponents.} \\ &= rA & \text{Property (2) of logarithms.} \\ &= r \log_a M\end{aligned}$$

NOW WORK PROBLEM 7.

Logarithms can be used to transform products into sums, quotients into differences, and powers into factors. Such transformations prove useful in certain types of calculus problems.

EXAMPLE 3 Writing a Logarithmic Expression as a Sum of Logarithms

Write $\log_a(x\sqrt{x^2 + 1})$, $x > 0$, as a sum of logarithms. Express all powers as factors.

SOLUTION
$$\begin{aligned}\log_a(x\sqrt{x^2 + 1}) &= \log_a x + \log_a \sqrt{x^2 + 1} & \text{Property (3).} \\ &= \log_a x + \log_a(x^2 + 1)^{1/2} \\ &= \log_a x + \frac{1}{2}\log_a(x^2 + 1) & \text{Property (5).}\end{aligned}$$

EXAMPLE 4 Writing a Logarithmic Expression as a Difference of Logarithms

Write
$$\ln \frac{x^2}{(x - 1)^3}, \quad x > 1$$
as a difference of logarithms. Express all powers as factors.

SOLUTION
$$\ln \frac{x^2}{(x - 1)^3} = \underset{\text{Property (4)}}{\ln x^2 - \ln(x - 1)^3} = \underset{\text{Property (5)}}{2 \ln x - 3 \ln(x - 1)}$$

EXAMPLE 5 Writing a Logarithmic Expression as a Sum and Difference of Logarithms

Write
$$\log_a \frac{\sqrt{x^2 + 1}}{x^3(x + 1)^4}, \quad x > 0$$
as a sum and difference of logarithms. Express all powers as factors.

SOLUTION
$$\begin{aligned}\log_a \frac{\sqrt{x^2 + 1}}{x^3(x + 1)^4} &= \log_a \sqrt{x^2 + 1} - \log_a[x^3(x + 1)^4] & \text{Property (4).} \\ &= \log_a \sqrt{x^2 + 1} - [\log_a x^3 + \log_a(x + 1)^4] & \text{Property (3).} \\ &= \log_a(x^2 + 1)^{1/2} - \log_a x^3 - \log_a(x + 1)^4 \\ &= \frac{1}{2}\log_a(x^2 + 1) - 3\log_a x - 4\log_a(x + 1) & \text{Property (5).}\end{aligned}$$

CAUTION: In using properties (3) through (5), be careful about the values that the variable may assume. For example, the domain of the variable for $\log_a x$ is $x > 0$ and for $\log_a(x - 1)$ it is $x > 1$. If we add these functions, the domain of the sum function is $x > 1$. That is, the equality

$$\log_a x + \log_a(x - 1) = \log_a[x(x - 1)]$$

is true only for $x > 1$.

NOW WORK PROBLEM 39.

Another use of properties (3) through (5) is to write sums and/or differences of logarithms with the same base as a single logarithm.

EXAMPLE 6 Writing Expressions as a Single Logarithm

Write each of the following as a single logarithm.

(a) $\log_a 7 + 4\log_a 3$ (b) $\dfrac{2}{3}\ln 8 - \ln(3^4 - 8)$

(c) $\log_a x + \log_a 9 + \log_a(x^2 + 1) - \log_a 5$

SOLUTION (a) $\log_a 7 + 4\log_a 3 = \log_a 7 + \log_a 3^4$ Property (5).

$\qquad\qquad\qquad\qquad\quad = \log_a 7 + \log_a 81$

$\qquad\qquad\qquad\qquad\quad = \log_a(7 \cdot 81)$ Property (3).

$\qquad\qquad\qquad\qquad\quad = \log_a 567$

(b) $\dfrac{2}{3}\ln 8 - \ln(3^4 - 8) = \ln 8^{2/3} - \ln(81 - 8)$ Property (5).

$\qquad\qquad\qquad\qquad\quad = \ln 4 - \ln 73$

$\qquad\qquad\qquad\qquad\quad = \ln\left(\dfrac{4}{73}\right)$ Property (4).

(c) $\log_a x + \log_a 9 + \log_a(x^2 + 1) - \log_a 5 = \log_a(9x) + \log_a(x^2 + 1) - \log_a 5$

$\qquad\qquad\qquad\qquad\qquad\qquad\qquad\qquad\quad = \log_a[9x(x^2 + 1)] - \log_a 5$

$\qquad\qquad\qquad\qquad\qquad\qquad\qquad\qquad\quad = \log_a\left[\dfrac{9x(x^2 + 1)}{5}\right]$

WARNING: A common error made by some students is to express the logarithm of a sum as the sum of logarithms.

$\qquad\qquad\log_a(M + N)$ is not equal to $\log_a M + \log_a N$

Correct statement $\log_a(MN) = \log_a M + \log_a N$ Property (3).

Another common error is to express the difference of logarithms as the quotient of logarithms.

$\qquad\qquad\log_a M - \log_a N$ is not equal to $\dfrac{\log_a M}{\log_a N}$

Correct statement $\log_a M - \log_a N = \log_a\left(\dfrac{M}{N}\right)$ Property (4).

A third common error is to express a logarithm raised to a power as the product of the power times the logarithm.

$(\log_a M)^r$ is not equal to $r \log_a M$

Correct statement $\log_a M^r = r \log_a M$ Property (5).

NOW WORK PROBLEM 45.

Two other properties of logarithms that we need to know are given next.

Properties of Logarithms

In the following properties, M, N, and a are positive real numbers, with $a \neq 1$.

$$\text{If } M = N, \text{ then } \log_a M = \log_a N. \qquad (6)$$
$$\text{If } \log_a M = \log_a N, \text{ then } M = N. \qquad (7)$$

When property (6) is used, we start with the equation $M = N$ and say "take the logarithm of both sides" to obtain $\log_a M = \log_a N$.

Using a Calculator to Evaluate Logarithms with Bases Other Than 10 or e

Logarithms to the base 10, common logarithms, were used to facilitate arithmetic computations before the widespread use of calculators. Natural logarithms, that is, logarithms whose base is the number e, remain very important because they arise frequently in the study of natural phenomena.

Common logarithms are usually abbreviated by writing **log,** with the base understood to be 10, just as natural logarithms are abbreviated by **ln,** with the base understood to be e.

Most calculators have both $\boxed{\log}$ and $\boxed{\ln}$ keys to calculate the common logarithm and natural logarithm of a number. Let's look at an example to see how to approximate logarithms having a base other than 10 or e.

EXAMPLE 7 Approximating Logarithms Whose Base Is Neither 10 Nor e

Approximate $\log_2 7$. Round the answer to four decimal places.

SOLUTION Let $y = \log_2 7$. Then

$2^y = 7$ Change to an exponential expression.

$\ln 2^y = \ln 7$ Property (6).

$y \ln 2 = \ln 7$ Property (5).

$y = \dfrac{\ln 7}{\ln 2}$ Exact solution.

$y \approx 2.8074$ Approximate solution rounded to four decimal places.

Example 7 shows how to approximate a logarithm whose base is 2 by changing to logarithms involving the base e. In general, we use the **Change-of-Base Formula.**

Change-of-Base Formula

If $a \neq 1$, $b \neq 1$, and M are positive real numbers, then

$$\log_a M = \frac{\log_b M}{\log_b a} \qquad (8)$$

PROOF We derive this formula as follows: Let $y = \log_a M$. Then

$$a^y = M \qquad \text{Change to an exponential expression.}$$
$$\log_b a^y = \log_b M \qquad \text{Property (6).}$$
$$y \log_b a = \log_b M \qquad \text{Property (5).}$$
$$y = \frac{\log_b M}{\log_b a} \qquad \text{Solve for } y.$$
$$\log_a M = \frac{\log_b M}{\log_b a} \qquad y = \log_a M$$

▶

Since calculators have keys only for $\boxed{\log}$ and $\boxed{\ln}$, in practice the Change-of-Base Formula uses either $b = 10$ or $b = e$. That is,

$$\log_a M = \frac{\log M}{\log a} \quad \text{and} \quad \log_a M = \frac{\ln M}{\ln a} \qquad (9)$$

EXAMPLE 8 Using the Change-of-Base Formula

Approximate: **(a)** $\log_5 89$ **(b)** $\log_{\sqrt{2}} \sqrt{5}$

Round answers to four decimal places.

SOLUTION **(a)** $\log_5 89 = \dfrac{\log 89}{\log 5} \approx \dfrac{1.949390007}{0.6989700043} \approx 2.7889$

or

$$\log_5 89 = \frac{\ln 89}{\ln 5} \approx \frac{4.48863637}{1.609437912} \approx 2.7889$$

(b) $\log_{\sqrt{2}} \sqrt{5} = \dfrac{\log \sqrt{5}}{\log \sqrt{2}} \approx 2.3219$

or

$$\log_{\sqrt{2}} \sqrt{5} = \frac{\ln \sqrt{5}}{\ln \sqrt{2}} \approx 2.3219$$

▶

Properties of Logarithms 221

COMMENT: To graph logarithmic functions when the base is different from e or 10 requires the Change-of-Base Formula. For example, to graph $y = \log_2 x$, we would instead graph $y = \dfrac{\ln x}{\ln 2}$. Try it.

NOW WORK PROBLEMS 11 AND 59.

SUMMARY **Properties of Logarithms**

In the summary that follows, $a > 0$, $a \neq 1$, and $b > 0$, $b \neq 1$; also, $M > 0$ and $N > 0$.

Definition $\quad y = \log_a x$ means $x = a^y$

Properties of logarithms $\quad \log_a 1 = 0 \quad \log_a a = 1 \quad a^{\log_a M} = M \quad \log_a a^r = r$

$\log_a(MN) = \log_a M + \log_a N$

$\log_a\left(\dfrac{M}{N}\right) = \log_a M - \log_a N$

$\log_a M^r = r \log_a M$

If $M = N$, then $\log_a M = \log_a N$.

If $\log_a M = \log_a N$, then $M = N$.

Change-of-Base Formula $\quad \log_a M = \dfrac{\log_b M}{\log_b a}$

EXERCISE 2.5

In Problems 1–10, use properties of logarithms to find the exact value of each expression. Do not use a calculator.

1. $\log_3 3^{71}$ **2.** $\log_2 2^{-13}$ **3.** $\ln e^{-4}$ **4.** $\ln e^{\sqrt{2}}$

5. $2^{\log_2 7}$ **6.** $e^{\ln 8}$ **7.** $\log_8 2 + \log_8 4$ **8.** $\log_6 9 + \log_6 4$

9. $\log_6 18 - \log_6 3$ **10.** $\log_8 16 - \log_8 2$ **11.** $\log_2 6 \cdot \log_6 4$ **12.** $\log_3 8 \cdot \log_8 9$

13. $3^{\log_3 5 - \log_3 4}$ **14.** $5^{\log_5 6 + \log_5 7}$ **15.** $e^{\log_{e^2} 16}$ **16.** $e^{\log_{e^2} 9}$

In Problems 17–24, suppose that $\ln 2 = a$ and $\ln 3 = b$. Use properties of logarithms to write each logarithm in terms of a and b.

17. $\ln 6$ **18.** $\ln \dfrac{2}{3}$ **19.** $\ln 1.5$ **20.** $\ln 0.5$

21. $\ln 8$ **22.** $\ln 27$ **23.** $\ln \sqrt[5]{6}$ **24.** $\ln \sqrt[4]{\dfrac{2}{3}}$

In Problems 25–44, write each expression as a sum and/or difference of logarithms. Express powers as factors.

25. $\log_5(25x)$ **26.** $\log_3 \dfrac{x}{9}$ **27.** $\log_2 z^3$ **28.** $\log_7(x^5)$

29. $\ln(ex)$ **30.** $\ln \dfrac{e}{x}$ **31.** $\ln(xe^x)$ **32.** $\ln \dfrac{x}{e^x}$

33. $\log_a(u^2 v^3)$, $u > 0, v > 0$

34. $\log_2\left(\dfrac{a}{b^2}\right)$, $a > 0, b > 0$

35. $\ln(x^2\sqrt{1-x})$, $0 < x < 1$

36. $\ln(x\sqrt{1+x^2})$, $x > 0$

37. $\log_2\left(\dfrac{x^3}{x-3}\right)$, $x > 3$

38. $\log_5\left(\dfrac{\sqrt[3]{x^2+1}}{x^2-1}\right)$, $x > 1$

39. $\log\left[\dfrac{x(x+2)}{(x+3)^2}\right]$, $x > 0$

40. $\log\left[\dfrac{x^3\sqrt{x+1}}{(x-2)^2}\right]$, $x > 2$

41. $\ln\left[\dfrac{x^2-x-2}{(x+4)^2}\right]^{1/3}$, $x > 2$

42. $\ln\left[\dfrac{(x-4)^2}{x^2-1}\right]^{2/3}$, $x > 4$

43. $\ln\dfrac{5x\sqrt{1+3x}}{(x-4)^3}$, $x > 4$

44. $\ln\left[\dfrac{5x^2\sqrt[3]{1-x}}{4(x+1)^2}\right]$, $0 < x < 1$

In Problems 45–58, write each expression as a single logarithm.

45. $3\log_5 u + 4\log_5 v$

46. $2\log_3 u - \log_3 v$

47. $\log_3 \sqrt{x} - \log_3 x^3$

48. $\log_2\left(\dfrac{1}{x}\right) + \log_2\left(\dfrac{1}{x^2}\right)$

49. $\log_4(x^2-1) - 5\log_4(x+1)$

50. $\log(x^2+3x+2) - 2\log(x+1)$

51. $\ln\left(\dfrac{x}{x-1}\right) + \ln\left(\dfrac{x+1}{x}\right) - \ln(x^2-1)$

52. $\log\left(\dfrac{x^2+2x-3}{x^2-4}\right) - \log\left(\dfrac{x^2+7x+6}{x+2}\right)$

53. $8\log_2\sqrt{3x-2} - \log_2\left(\dfrac{4}{x}\right) + \log_2 4$

54. $21\log_3 \sqrt[3]{x} + \log_3(9x^2) - \log_3 9$

55. $2\log_a(5x^3) - \dfrac{1}{2}\log_a(2x+3)$

56. $\dfrac{1}{3}\log(x^3+1) + \dfrac{1}{2}\log(x^2+1)$

57. $2\log_2(x+1) - \log_2(x+3) - \log_2(x-1)$

58. $3\log_5(3x+1) - 2\log_5(2x-1) - \log_5 x$

In Problems 59–66, use the Change-of-Base Formula and a calculator to evaluate each logarithm. Round your answer to three decimal places.

59. $\log_3 21$

60. $\log_5 18$

61. $\log_{1/3} 71$

62. $\log_{1/2} 15$

63. $\log_{\sqrt{2}} 7$

64. $\log_{\sqrt{5}} 8$

65. $\log_\pi e$

66. $\log_\pi \sqrt{2}$

In Problems 67–72, graph each function using a graphing utility and the Change-of-Base Formula.

67. $y = \log_4 x$

68. $y = \log_5 x$

69. $y = \log_2(x+2)$

70. $y = \log_4(x-3)$

71. $y = \log_{x-1}(x+1)$

72. $y = \log_{x+2}(x-2)$

In Problems 73–82, express y as a function of x. The constant C is a positive number.

73. $\ln y = \ln x + \ln C$

74. $\ln y = \ln(x + C)$

75. $\ln y = \ln x + \ln(x+1) + \ln C$

76. $\ln y = 2\ln x - \ln(x+1) + \ln C$

77. $\ln y = 3x + \ln C$

78. $\ln y = -2x + \ln C$

79. $\ln(y-3) = -4x + \ln C$

80. $\ln(y+4) = 5x + \ln C$

81. $3\ln y = \dfrac{1}{2}\ln(2x+1) - \dfrac{1}{3}\ln(x+4) + \ln C$

82. $2\ln y = -\dfrac{1}{2}\ln x + \dfrac{1}{3}\ln(x^2+1) + \ln C$

83. Find the value of $\log_2 3 \cdot \log_3 4 \cdot \log_4 5 \cdot \log_5 6 \cdot \log_6 7 \cdot \log_7 8$.

84. Find the value of $\log_2 4 \cdot \log_4 6 \cdot \log_6 8$.

85. Find the value of $\log_2 3 \cdot \log_3 4 \cdots \log_n(n+1) \cdot \log_{n+1} 2$.
86. Find the value of $\log_2 2 \cdot \log_2 4 \cdots \log_2 2^n$.
87. Show that $\log_a(x + \sqrt{x^2 - 1}) + \log_a(x - \sqrt{x^2 - 1}) = 0$.
88. Show that $\log_a(\sqrt{x} + \sqrt{x-1}) + \log_a(\sqrt{x} - \sqrt{x-1}) = 0$.
89. Show that $\ln(1 + e^{2x}) = 2x + \ln(1 + e^{-2x})$.
90. **Difference Quotient** If $f(x) = \log_a x$, show that
$$\frac{f(x+h) - f(x)}{h} = \log_a\left(1 + \frac{h}{x}\right)^{1/h}, \quad h \neq 0.$$
91. If $f(x) = \log_a x$, show that $-f(x) = \log_{1/a} x$.
92. If $f(x) = \log_a x$, show that $f(AB) = f(A) + f(B)$.
93. If $f(x) = \log_a x$, show that $f\left(\dfrac{1}{x}\right) = -f(x)$.
94. If $f(x) = \log_a x$, show that $f(x^\alpha) = \alpha f(x)$.
95. Show that $\log_a\left(\dfrac{M}{N}\right) = \log_a M - \log_a N$, where a, M, and N are positive real numbers, with $a \neq 1$.
96. Show that $\log_a\left(\dfrac{1}{N}\right) = -\log_a N$, where a and N are positive real numbers, with $a \neq 1$.
97. **Graph** $Y_1 = \log(x^2)$ and $Y_2 = 2\log(x)$ on your graphing utility. Are they equivalent? What might account for any differences in the two functions?

2.6 Continuously Compounded Interest

OBJECTIVES
1. Use the compound interest formula
2. Find the present value of a dollar amount
3. Find the time required to double an investment
4. Find the rate of interest needed to double an investment

Suppose a principal P is to be invested at an annual rate of interest r, which is compounded n times per year. The interest earned on a principal P at each compounding period is then $P \cdot \dfrac{r}{n}$. The amount A after 1 year with

1 compoundings per year (annually)
$$A = P + P \cdot \left(\frac{r}{1}\right) = P \cdot \left(1 + \frac{r}{1}\right)$$

2 compoundings per year (semiannually)
$$A = P \cdot \left(1 + \frac{r}{2}\right) + P \cdot \left(1 + \frac{r}{2}\right)\left(\frac{r}{2}\right)$$
$$= P \cdot \left(1 + \frac{r}{2}\right)\left(1 + \frac{r}{2}\right) = P \cdot \left(1 + \frac{r}{2}\right)^2$$

4 compoundings per year (quarterly)
$$A = P \cdot \left(1 + \frac{r}{4}\right)^3 + P \cdot \left(1 + \frac{r}{4}\right)^3\left(\frac{r}{4}\right)$$
$$= P \cdot \left(1 + \frac{r}{4}\right)^3\left(1 + \frac{r}{4}\right) = P \cdot \left(1 + \frac{r}{4}\right)^4$$

⋮

n compoundings per year per year
$$A = P \cdot \left(1 + \frac{r}{n}\right)^n$$

What happens to the amount A after 1 year if the number of times, n, that the interest is compounded per year gets larger and larger? The answer turns out to involve the number e.

Rewrite the expression for A as follows:

$$A = P \cdot \left(1 + \frac{r}{n}\right)^n = P \cdot \left[\left(1 + \frac{r}{n}\right)^{n/r}\right]^r$$

To simplify the calculation, let

$$k = \frac{n}{r} \quad \text{so} \quad \frac{1}{k} = \frac{r}{n}$$

We substitute to get

$$A = P \cdot \left[\left(1 + \frac{r}{n}\right)^{n/r}\right]^r = P \cdot \left[\left(1 + \frac{1}{k}\right)^k\right]^r$$

As n gets larger and larger, so does k and, since $\left(1 + \frac{1}{k}\right)^k$ approaches e as k becomes unbounded in the positive direction, it follows that

$$P \cdot \left[\left(1 + \frac{1}{k}\right)^k\right]^r \to P \cdot [(e)]^r = Pe^r$$

That is, no matter how often the interest is compounded during the year, the amount after 1 year has the definite ceiling Pe^r. When interest is compounded so that the amount after 1 year is Pe^r, we say that the interest is **compounded continuously.**

For example, the amount A due to investing $1000 for 1 year at an annual rate of 10% compounded continuously is

$$A = 1000e^{0.1} = \$1105.17$$

The formula $A = Pe^r$ gives the amount A after 1 year resulting from investing a principal P at the annual rate of interest r compounded continuously.

Compound Interest Formula – Continuous Compounding

The amount A due to investing a principal P for a period of t years at the annual rate of interest r compounded continuously is

$$\boxed{A = Pe^{rt}} \tag{1}$$

EXAMPLE 1 Using the Compound Interest Formula

If $1000 is invested at 10% compounded continuously, how much is in the account

(a) after 3 years   (b) after 5 years

SOLUTION (a) If $1000 is invested at 10% compounded continuously, the amount A after 3 years is

$$A = Pe^{rt} = 1000e^{(0.1)(3)} = 1000e^{0.3} = \$1349.86$$

(b) After 5 years the amount A is

$$A = 1000e^{(0.1)(5)} = \$1648.72$$

NOW WORK PROBLEM 1.

Continuously Compounded Interest

The Compound Interest Formula states that a principal P earning an annual rate of interest r compounded continuously will, after t years, be worth the amount A, where

$$A = Pe^{rt}$$

If we solve for P, we obtain

$$\boxed{P = \frac{A}{e^{rt}} = Ae^{-rt}} \tag{2}$$

In this formula P is called the **present value** of the amount A. In other words, P is the amount that must be invested now in order to accumulate the amount A in t years.

2 EXAMPLE 2 Computing the Present Value of $10,000

How much money should be invested now at 8% per annum compounded continuously, so that after 2 years the amount will be $10,000?

SOLUTION In this problem we want to find the principal P needed now to get the amount $A = \$10,000$ after $t = 2$ years. That is, we want to find the present value of $10,000. We use formula (2) with $r = .08$:

$$P = Ae^{-rt} = \$10,000e^{-.08(2)} = \$8521.44$$

If you invest $8521.44 now at 8% per annum compounded continuously, you will have $10,000 after 2 years. ▸

NOW WORK PROBLEM 5.

3 EXAMPLE 3 Finding the Time Required to Double an Investment

Find the time required to double an investment if the rate of interest is 5% compounded continuously.

SOLUTION If P is the principal invested, it will double when the amount $A = 2P$. We use the Compound Interest Formula (1).

$$A = Pe^{rt} \quad \text{Formula (1).}$$
$$2P = Pe^{0.05t} \quad A = 2P; r = 0.05$$
$$2 = e^{0.05t} \quad \text{Cancel the } P\text{'s.}$$
$$0.05t = \ln 2 \quad \text{Change to a logarithmic expression.}$$
$$t = 13.863 \quad \text{Solve for } t.$$

It will take almost 14 years to double an investment at 5% compounded continuously. ▸

NOW WORK PROBLEM 15.

EXAMPLE 4 Finding the Rate of Interest to Double an Investment

Find the rate of interest required to double an investment in 8 years if the interest is compounded continuously.

SOLUTION If P is the principle invested, it will double when the amount $A = 2P$. We use the Compound Interest Formula (1).

$$A = Pe^{rt} \quad \text{Formula (1).}$$
$$2P = Pe^{8r} \quad A = 2P; t = 8.$$
$$2 = e^{8r} \quad \text{Cancel the } P\text{'s.}$$
$$8r = \ln 2 \quad \text{Change to a logarithmic expression.}$$
$$r = 0.0866 \quad \text{Solve for } r.$$

It will require an interest rate of about 8.7% compounded continuously to double an investment in 5 years.

NOW WORK PROBLEM 13.

EXERCISE 2.6

In Problems 1–4, find the amount if:

1. $1000 is invested at 4% compounded continuously for 3 years.
2. $100 is invested at 6% compounded continuously for $1\frac{1}{2}$ years.
3. $500 is invested at 5% compounded continuously for 3 years.
4. $200 is invested at 10% compounded continuously for 10 years.

In Problems 5–8, find the principal needed now to get each amount.

5. To get $100 in 6 months at 4% compounded continuously.
6. To get $500 in 1 year at 6% compounded continuously.
7. To get $500 in 1 year at 7% compounded continuously.
8. To get $800 in 2 years at 5% compounded continuously.
9. If $1000 is invested at 2% compounded continuously, what is the amount after 1 year? How much interest is earned?
10. If $2000 is invested at 5% compounded continuously, what is the amount after 5 years? How much interest is earned?
11. If a bank pays 3% compounded continuously, how much should be deposited now to have $5000
    (a) 4 years later? (b) 8 years later?
12. If a bank pays 2% compounded continuously, how much should be deposited now to have $10,000
    (a) 5 years later? (b) 10 years later?
13. What annual rate of interest compounded continuously is required to double an investment in 3 years?
14. What annual rate of interest compounded continuously is required to double an investment in 10 years?
15. Approximately how long will it take to triple an investment at 10% compounded continuously?
16. Approximately how long will it take to triple an investment at 9% compounded continuously?
17. What principal is needed now to get $1000 in 1 year at 9% compounded continuously? How much should be invested to get $1000 in 2 years?
18. **Buying a Car** Laura wishes to have $8000 available to buy a car in 3 years. How much should she invest in a savings account now so that she will have enough if the bank pays 8% interest compounded continuously?

19. Down Payment on a House Tami and Todd will need $40,000 for a down payment on a house in 4 years. How much should they invest in a savings account now so that they will be able to do this? The bank pays 3% compounded continuously.

20. Saving for College A newborn child receives a $3000 gift toward a college education. How much will the $3000 be worth in 17 years if it is invested at 10% compounded continuously?

21. What annual rate of interest compounded continuously is required to triple an investment in 5 years?

22. What annual rate of interest compounded continuously is required to triple an investment in 7 years?

Problems 23 and 24 require the following discussion: In asking for the time required to double an investment, we use formula (1) with $A = 2P$.

$$A = Pe^{rt} \qquad \text{Formula (1).}$$
$$2P = Pe^{rt} \qquad A = 2P$$
$$2 = e^{rt} \qquad \text{Cancel the } P\text{'s.}$$
$$\ln 2 = rt \qquad \text{Change to a logarithm.}$$
$$t = \frac{\ln 2}{r} = \frac{0.6931471806}{r} \qquad \text{Solve for } t.$$

23. The Rule of 70 This rule uses the approximation $\ln 2 = 0.70$ to find t. Compare the approximation given by the rule of 70 to the actual solution if

(a) $r = 1\%$ (b) $r = 5\%$ (c) $r = 10\%$

24. The Rule of 72 This rule uses the approximation $\ln 2 = 0.72$. Compare the approximation given by the rule of 72 to the actual solution if

(a) $r = 1\%$ (b) $r = 8\%$ (c) $r = 12\%$

Chapter 2 Review

OBJECTIVES

Section	You should be able to	Review Exercises
2.1	1 Locate the vertex and axis of symmetry of a quadratic function	1–10
	2 Graph quadratic functions	1–10
	3 Find the maximum or the minimum value of a quadratic function	11–16
	4 Use the maximum or the minimum value of a quadratic function to solve applied problems	91–93, 99, 100
2.2	1 Know the properties of power functions	17, 18
	2 Graph functions using shifts and/or reflections	19–22
	3 Identify polynomial functions and their degree	23–26
	4 Find the end behavior of a polynomial function	27, 28
	5 Find the domain of a rational function	29–32

Section		Objective	Example(s) and Review Exercises
2.3	1	Evaluate exponential functions	33 (a), (c), 34 (a), (c), 94(a)
	2	Graph exponential functions	67–70
	3	Define the number e	44, 47, 69, 70
	4	Solve exponential equations	73–76, 79, 80, 82
2.4	1	Change exponential expressions to logarithmic expressions	35, 36
	2	Change logarithmic expressions to exponential expressions	37, 38
	3	Evaluate logarithmic functions	33 (b), (d), 34 (b), (d), 43–48, 89, 90, 95 (a), (b), 96
	4	Find the domain of a logarithmic function	39–42
	5	Graph logarithmic functions	71, 72
	6	Solve logarithmic equations	77, 78, 81
2.5	1	Work with the properties of logarithms	43–52
	2	Write a logarithmic expression as a sum or difference of logarithms	53-58
	3	Write a logarithmic expression as a single logarithm	59–64
	4	Evaluate logarithms whose base is neither 10 nor e	65–66
2.6	1	Use the Compound Interest Formula	83, 84, 97
	2	Find the present value of a dollar amount	85, 86
	3	Find the time required to double an investment	87
	4	Find the rate of interest needed to double an investment	88

THINGS TO KNOW

Quadratic function (pp. 167 and 168)

$f(x) = ax^2 + bx + c, \quad a \neq 0$

Graph is a parabola that opens up if $a > 0$ and opens down if $a < 0$.

Vertex: $\left(-\dfrac{b}{2a}, f\left(-\dfrac{b}{2a}\right)\right)$

Axis of symmetry: $x = -\dfrac{b}{2a}$

y-intercept: Found by evaluating $f(0) = c$.
x-intercept(s): Found by finding the real solutions, if any, of the equation $ax^2 + bx + c = 0$.

Polynomial function (p. 181)

$f(x) = a_n x^n + a_{n-1} x^{n-1} + \cdots + a_1 x + a_0, \quad a_n \neq 0, n \geq 0$ an integer
Domain: all real numbers

Rational function (p. 184)

$R(x) = \dfrac{p(x)}{q(x)}$

p, q are polynomial functions. Domain: $\{x \mid q(x) \neq 0\}$

Exponential Function $f(x) = a^x$, $a > 1$ Domain: the interval $(-\infty, \infty)$;
(pp. 191 and 193)   Range: the interval $(0, \infty)$;
x-intercept: none; y-intercept: $(0, 1)$
horizontal asymptote: the line
$y = 0$ (the x-axis), as x becomes
unbounded in the negative
direction; increasing; smooth;
continuous

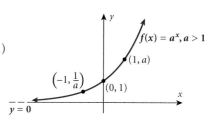

$f(x) = a^x$, $0 < a < 1$ Domain: the interval $(-\infty, \infty)$;
Range: the interval $(0, \infty)$;
x-intercept: none;
y-intercept: $(0, 1)$ horizontal
asymptote: the line $y = 0$
(the x-axis), as x becomes
unbounded in the positive
direction; decreasing; smooth;
continuous

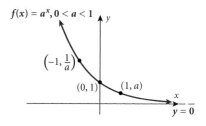

Number e (p. 194)   Value approached by the expression $\left(1 + \dfrac{1}{n}\right)^n$ as n becomes unbounded in the positive direction

That is, $\displaystyle\lim_{n \to \infty} \left(1 + \dfrac{1}{n}\right)^n = e$.

Property of exponents   If $a^u = a^v$, then $u = v$.
(p. 195)

Logarithmic Functions $f(x) = \log_a x$, $a > 1$ Domain: the interval $(0, \infty)$;
(p. 206) $(y = \log_a x$ means $x = a^y)$ Range: the interval $(-\infty, \infty)$;
x-intercept: $(1, 0)$;
y-intercept: none; vertical
asymptote: the line $x = 0$
(the y-axis); increasing; smooth;
continuous

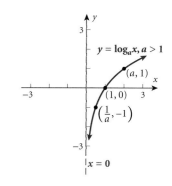

$f(x) = \log_a x$, $0 < a < 1$ Domain: the interval $(0, \infty)$;
$(y = \log_a x$ means $x = a^y)$ Range: the interval $(-\infty, \infty)$;
x-intercept: $(1, 0)$;
y-intercept: none;
vertical asymptote: the line
$x = 0$ (the y-axis);
decreasing; smooth; continuous

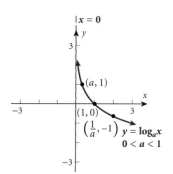

Natural Logarithm $y = \ln x$ means $x = e^y$.
(p. 206)

Properties of logarithms
(pp. 215, 216, and 219)

$\log_a 1 = 0 \quad \log_a a = 1 \quad a^{\log_a M} = M \quad \log_a a^r = r$

$\log_a(MN) = \log_a M + \log_a N \quad \log_a\left(\dfrac{M}{N}\right) = \log_a M - \log_a N$

$\log_a M^r = r \log_a M$

If $M = N$, then $\log_a M = \log_a N$.

If $\log_a M = \log_a N$, then $M = N$.

Change-of-Base Formula (p. 220)

$\log_a M = \dfrac{\log_b M}{\log_b a}$

Compound Interest Formula
(continuously compounded interest)
(p. 224)

$A = Pe^{rt} \qquad A =$ amount; $P =$ principal; $r =$ rate of interest
$t =$ time (in years)

TRUE–FALSE ITEMS

T F **1.** The graph of $f(x) = 2x^2 + 3x - 4$ opens up.
T F **2.** The x-coordinate of the vertex of $f(x) = -x^2 + 4x + 5$ is $f(2)$.
T F **3.** If the discriminant $b^2 - 4ac = 0$, the graph of $f(x) = ax^2 + bx + c$, $a \neq 0$, will touch the x-axis at its vertex.
T F **4.** The graph of an exponential function $f(x) = a^x$, $0 < a < 1$, is decreasing.
T F **5.** The graphs of $y = 3^x$ and $y = \left(\dfrac{1}{3}\right)^x$ are identical.

T F **6.** The range of the exponential function $f(x) = a^x$, $a > 0$, $a \neq 1$, is the set of all real numbers.
T F **7.** If $y = \log_a x$, then $y = a^x$.
T F **8.** The graph of every logarithmic function $f(x) = \log_a x$, $a > 0$, $a \neq 1$, will contain the points $(1, 0)$, $(a, 1)$ and $\left(\dfrac{1}{a}, -1\right)$.
T F **9.** $\log_a(MN) = \log_a(M + N)$, $M > 0$, $N > 0$, $a > 0$

FILL IN THE BLANKS

1. The graph of a quadratic function is called a _____.

2. The vertical line passing through the vertex of a parabola is called the _____.

3. The x-coordinate of the vertex of $f(x) = ax^2 + bx + c$, $a \neq 0$, is _____.

4. The graph of every exponential function $f(x) = a^x$, $a > 0$, $a \neq 1$, passes through three points: _____, _____, and _____.

5. If the graph of the exponential function $f(x) = a^x$, $a > 0$, $a \neq 1$, is decreasing, then a must be less than _____.

6. If $3^x = 3^4$, then $x =$ _____.

7. The domain of the logarithmic function $f(x) = \log_a x$ is _____.

8. The graph of every logarithmic function $f(x) = \log_a x$, $a > 0$, $a \neq 1$, passes through three points: _____, _____, and _____.

9. If the graph of a logarithmic function $f(x) = \log_a x$, $a > 0$, $a \neq 1$ is increasing, then its base must be larger than _____.

10. $\log_a M^r =$ _____, $a > 0$, $M > 0$, $r > 0$.

REVIEW EXERCISES Blue problem numbers indicate the author's suggestions for a practice test.

In Problems 1–10, graph each quadratic function by determining whether its graph opens up or down and by finding its vertex, axis of symmetry, y-intercept, and x-intercepts, if any.

1. $f(x) = (x - 2)^2 + 2$
2. $f(x) = (x + 1)^2 - 4$
3. $f(x) = \dfrac{1}{4}x^2 - 16$
4. $f(x) = -\dfrac{1}{2}x^2 + 2$

5. $f(x) = -4x^2 + 4x$

6. $f(x) = 9x^2 - 6x + 3$

7. $f(x) = \frac{9}{2}x^2 + 3x + 1$

8. $f(x) = -x^2 + x + \frac{1}{2}$

9. $f(x) = 3x^2 + 4x - 1$

10. $f(x) = -2x^2 - x + 4$

In Problems 11–16, determine whether the given quadratic function has maximum value or a minimum value, and then find the value.

11. $f(x) = 3x^2 - 6x + 4$

12. $f(x) = 2x^2 + 8x + 5$

13. $f(x) = -x^2 + 8x - 4$

14. $f(x) = -x^2 - 10x - 3$

15. $f(x) = -3x^2 + 12x + 4$

16. $f(x) = -2x^2 + 4$

17. Name three points the graph of the power function $f(x) = x^6$ contains.

18. The graph of the power function $f(x) = x^5$ is symmetric with respect to the _____.

In Problems 19–22, graph each function using shifts and/or reflections. Be sure to label at least three points.

19. $f(x) = x^4 + 2$

20. $f(x) = (x - 3)^3$

21. $f(x) = -x^5 + 1$

22. $f(x) = -x^4 - 2$

In Problems 23–26, determine which functions are polynomial functions. For those that are, state the degree. For those that are not, tell why not.

23. $f(x) = 4x^5 - 3x^2 + 5x - 2$

24. $f(x) = \frac{3x^5}{2x + 1}$

25. $f(x) = 3x^2 + 5x^{1/2} - 1$

26. $f(x) = 3$

In Problems 27–28, find the power function that the graph of f resembles for large values of x. That is, find the end behavior of each polynomial function.

27. $f(x) = -2x^4 + 3x^3 - 6x + 4$

28. $f(x) = 5x^6 - 7x^4 + 8x - 10$

In Problems 29–32, find the domain of each rational function.

29. $R(x) = \dfrac{x + 2}{x^2 - 9}$

30. $R(x) = \dfrac{x^2 + 4}{x - 2}$

31. $R(x) = \dfrac{x^2 + 3x + 2}{(x + 2)^2}$

32. $R(x) = \dfrac{x^3}{x^3 - 1}$

In Problems 33 and 34, suppose that $f(x) = 3^x$ and $g(x) = \log_3 x$.

33. Evaluate the following: (a) $f(4)$ (b) $g(9)$ (c) $f(-2)$ (d) $g\left(\dfrac{1}{27}\right)$

34. Evaluate the following: (a) $f(1)$ (b) $g(81)$ (c) $f(-4)$ (d) $g\left(\dfrac{1}{243}\right)$

In Problems 35 and 36, convert each exponential expression to an equivalent expression involving a logarithm. In Problems 37 and 38, convert each logarithmic expression to an equivalent expression involving an exponent.

35. $5^2 = z$

36. $a^5 = m$

37. $\log_5 u = 13$

38. $\log_a 4 = 3$

In Problems 39–42, find the domain of each logarithmic function.

39. $f(x) = \log(3x - 2)$

40. $F(x) = \log_5(2x + 1)$

41. $H(x) = \log_2(-3x + 2)$

42. $F(x) = \ln(-2x - 9)$

232 Chapter 2 Classes of Functions

In Problems 43–52, evaluate each expression. Do not use a calculator

43. $\log_2\left(\dfrac{1}{8}\right)$ 44. $\ln e$ 45. $\log_3 81$ 46. $\log 10$ 47. $\ln e^2$

48. $\log_3\left(\dfrac{1}{3}\right)$ 49. $\ln e^{\sqrt{2}}$ 50. $e^{\ln 0.1}$ 51. $2^{\log_2 0.4}$ 52. $\log_2 2^{\sqrt{3}}$

In Problems 53–58, write each expression as the sum and/or difference of logarithms. Express powers as factors.

53. $\log_3\left(\dfrac{uv^2}{w}\right)$, $u > 0, v > 0, w > 0$ 54. $\log_2(a^2\sqrt{b})^4$, $a > 0, b > 0$ 55. $\log(x^2\sqrt{x^3 + 1})$, $x > 0$

56. $\log_5\left(\dfrac{x^2 + 2x + 1}{x^2}\right)$, $x > 0$ 57. $\ln\left(\dfrac{x\sqrt[3]{x^2+1}}{x-3}\right)$, $x > 3$ 58. $\ln\left(\dfrac{2x+3}{x^2-3x+2}\right)^2$, $x > 2$

In Problems 59–64, write each expression as a single logarithm.

59. $3\log_4 x^2 + \dfrac{1}{2}\log_4 \sqrt{x}$

60. $-2\log_3\left(\dfrac{1}{3}\right) + \dfrac{1}{3}\log_3 \sqrt{x}$

61. $\ln\left(\dfrac{x-1}{x}\right) + \ln\left(\dfrac{x}{x+1}\right) - \ln(x^2 - 1)$

62. $\log(x^2 - 9) - \log(x^2 + 7x + 12)$

63. $2\log 2 + 3\log x - \dfrac{1}{2}[\log(x+3) + \log(x-2)]$

64. $\dfrac{1}{2}\ln(x^2+1) - 4\ln\dfrac{1}{2} - \dfrac{1}{2}[\ln(x-4) + \ln x]$

In Problems 65 and 66, use the Change-of-Base Formula and a calculator to evaluate each logarithm. Round your answer to three decimal places.

65. $\log_4 19$ 66. $\log_2 21$

In Problems 67–72, use shifts and/or reflections to graph each function. Determine the domain, range, and any asymptotes.

67. $f(x) = 2^{x-3}$ 68. $f(x) = -2^x + 3$ 69. $f(x) = 1 - e^x$

70. $f(x) = 3 - e^{-x}$ 71. $f(x) = 3 + \ln x$ 72. $f(x) = 4 - \ln(-x)$

In Problems 73–82, solve each equation.

73. $4^{1-2x} = 2$ 74. $8^{6+3x} = 4$ 75. $3^{x^2+x} = \sqrt{3}$ 76. $4^{x-x^2} = \dfrac{1}{2}$ 77. $\log_x 64 = -3$

78. $\log_{\sqrt{2}} x = -6$ 79. $9^{2x} = 27^{3x-4}$ 80. $25^{2x} = 5^{x^2-12}$ 81. $\log_3(x-2) = 2$ 82. $2^{x+1} \cdot 8^{-x} = 4$

83. Find the amount of an investment of $100 after 2 years and 3 months at 10% compounded continuously.

84. Mike places $200 in a savings account that pays 4% per annum compounded continuously. How much is in his account after 9 months?

85. A bank pays 4% per annum compounded continuously. How much should I invest now so that 2 years from now I will have $1000 in the account?

86. **Saving for a Bicycle** Katy wants to buy a bicycle that costs $75 and will purchase it in 6 months. How much should she put in her savings account now if she can get 10% per annum compounded continuously?

87. **Doubling Money** Marcia has $220,000 saved for her retirement. How long will it take for the investment to double in value if it earns 6% compounded continuously?

88. **Doubling Money** What annual rate of interest is required to double an investment in 4 years?

In Problems 89 and 90, use the following result: If x is the atmospheric pressure (measured in millimeters of mercury), then the formula for the altitude h(x) (measured in meters above sea level) is

$$h(x) = (30T + 8000) \log\left(\frac{P_0}{x}\right)$$

where T is the temperature (in degrees Celsius) and P_0 is the atmospheric pressure at sea level, which is approximately 760 millimeters of mercury.

89. **Finding the Altitude of an Airplane** At what height is a Piper Cub whose instruments record an outside temperature of 0°C and a barometric pressure of 300 millimeters of mercury?

90. **Finding the Height of a Mountain** How high is a mountain if instruments placed on its peak record a temperature of 5°C and a barometric pressure of 500 millimeters of mercury?

91. **Landscaping** A landscape engineer has 200 feet of border to enclose a rectangular pond. What dimensions will result in the largest pond?

92. **Enclosing the Most Area with a Fence** A farmer with 10,000 meters of fencing wants to enclose a rectangular field and then divide it into two plots with a fence parallel to one of the sides (see the figure). What is the largest area that can be enclosed?

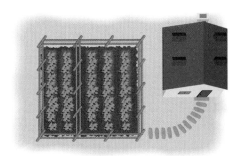

93. **Architecture** A special window in the shape of a rectangle with semicircles at each end is to be constructed so that the outside dimensions are 100 feet in length. See the illustration. Find the dimensions of the rectangle that maximizes its area.

94. **Amplifying Sound** An amplifier's power output P (in watts) is related to its decibel voltage gain d by the formula $P = 25e^{0.1d}$.

    (a) Find the power output for a decibel voltage gain of 4 decibels.
    (b) For a power output of 50 watts, what is the decibel voltage gain?

95. **Limiting Magnitude of a Telescope** A telescope is limited in its usefulness by the brightness of the star it is aimed at and by the diameter of its lens. One measure of a star's brightness is its *magnitude*; the dimmer the star, the larger its magnitude. A formula for the limiting magnitude L of a telescope, that is, the magnitude of the dimmest star that it can be used to view, is given by

    $$L = 9 + 5.1 \log d$$

    where d is the diameter (in inches) of the lens.

    (a) What is the limiting magnitude of a 3.5-inch telescope?
    (b) What diameter is required to view a star of magnitude 14?

96. **Salvage Value** The number of years n for a piece of machinery to depreciate to a known salvage value can be found using the formula

    $$n = \frac{\log s - \log i}{\log(1 - d)}$$

    where s is the salvage value of the machinery, i is its initial value, and d is the annual rate of depreciation.

    (a) How many years will it take for a piece of machinery to decline in value from $90,000 to $10,000 if the annual rate of depreciation is 0.20 (20%)?

(b) How many years will it take for a piece of machinery to lose half of its value if the annual rate of depreciation is 15%?

97. **Funding an IRA** First Colonial Bankshares Corporation advertised the following IRA investment plans.

TARGET IRA PLANS For each $5000 Maturity Value Desired	
Deposit	For a Term of:
$620.17	20 Years
$1045.02	15 Years
$1760.92	10 Years
$2967.26	5 Years

(a) Assuming continuous compounding, what was the annual rate of interest that they offered?

(b) First Colonial Bankshares claims that $4000 invested today will have a value of over $32,000 in 20 years. Use the answer found in part (a) to find the actual value of $4000 in 20 years. Assume continuous compounding.

98. Find the point on the line $y = x$ that is closest to the point $(3, 1)$.

[**Hint:** Find the minimum value of the function $f(x) = d^2$, where d is the distance from $(3, 1)$ to a point on the line.]

99. **Minimizing Marginal Cost** The marginal cost of a product can be thought of as the cost of producing one additional unit of output. For example, if the marginal cost of producing the 50th product is $6.20, then it cost $6.20 to increase production from 49 to 50 units of output. Callaway Golf Company has determined that the marginal cost C of manufacturing x Big Bertha golf clubs may be expressed by the quadratic function

$$C(x) = 4.9x^2 - 617.4x + 19{,}600$$

(a) How many clubs should be manufactured to minimize the marginal cost?

(b) At this level of production, what is the marginal cost?

100. **Violent Crimes** The function $V(t) = -10.0t^2 + 39.2t + 1862.6$ models the number V (in thousands) of violent crimes committed in the United States t years after 1990 based on data obtained from the Federal Bureau of Investigation. So $t = 0$ represents 1990, $t = 1$ represents 1991, and so on.

(a) Determine the year in which the most violent crimes were committed.

(b) Approximately how many violent crimes were committed during this year?

(c) Using a graphing utility, graph $V = V(t)$. Were the number of violent crimes increasing or decreasing during the years 1994 to 1998?

Chapter 2 Project

The table below lists historical data on the population of Houston, Texas from 1850 through 2000. Use the information provided to do the following problems.

Year	Population (in thousands)
2000	1953.6
1990	1630.6
1980	1595.1
1970	1232.8
1960	938.2
1950	596.2
1940	384.5
1930	292.4
1920	138.3
1910	78.8
1900	44.6
1890	27.6
1880	16.5
1870	9.4
1860	4.8
1850	2.4

Source: U.S. Census.

1. Plot the data measuring population (in thousands) on the y-axis and time (in years) on the x-axis. Use $t = 0$ for 1850, $t = 10$ for 1860, and so on.

2. Assume the relationship between population P and time t is exponential. That is, assume $P = P_0 a^t$, where $P_0 = P(0)$. Then find a as follows:
   Form a table that lists the ratios of the populations for consecutive years. For example, from 1850 to 1860 the ratio is $\frac{4.8}{2.4} = 2$. From 1860 to 1870 the ratio is $\frac{9.4}{4.8} = 1.96$. Now find the average of all these ratios. Use this number for the base a of the function $P = P_0 a^t$. (Refer to the theorem on page 189 to review the basis for this).

3. (a) According to the result found above, what is the population of Houston in 2000?
   (b) How close is this to the actual population as given in the table?
   (c) Can you provide an explanation for any differences?
   (d) Use the exponential growth function to predict the population in 2010.
   (e) What is the predicted population of Houston in 2050?

4. Write the exponential function found in Problem 2 in the form $P = P_0 e^{kt}$. Here k is the growth rate of the population. [Hint: $a^t = e^{kt} = (e^k)^t$. Now solve for k.]

5. What is the growth rate of the population of Houston?

6.2–6.5 Repeat Problems 2–5 using the data from the table, but beginning with 1900 instead of 1850.

7. Compare the results of Problems 3 and 6.3. What differences do you get in the exponential growth function and the prediction of future population of Houston? Explain any differences. Which predictions seem more reasonable?

8.2–8.5 Repeat Problems 2–5 using the data from the table, but beginning with 1950 instead of 1850.

9. Compare the results of Problems 3, 6.3, and 8.3. What differences do you get in the exponential growth function and the prediction of future population of Houston? Explain any differences. Which predictions seem more reasonable?

10. Use a graphing utility to find the exponential curve of best fit using the data from the table beginning with 1950 and ending with 2000.

11. Work Problems 3, 4, and 5 using the result found in Problem 10. Compare the results with those of Problems 8.3, 8.4, and 8.5. Explain any differences.

MATHEMATICAL QUESTIONS FROM PROFESSIONAL EXAMS*

1. **Actuary Exam—Part I** If $\log_6 2 = b$, which of the following is equal to $\log_6(4 \cdot 27)$?
   (a) $3 - b$ (b) $2 + b$ (c) $5b$ (d) $3b$ (e) $2b + \log_6 27$

2. **Actuary Exam—Part II** If b and c are real numbers and there are two different real numbers x such that $(e^x)^2 + be^x + c = 0$, which of the following must be true?
   I. $b^2 - 4c > 0$ II. $b < 0$ III. $c > 0$
   (a) None (b) I only (c) I and II only
   (d) I and III only (e) I, II, and III

3. **Actuary Exam—Part I** If $y = \frac{e^x - e^{-x}}{2}$, then find x in terms of y.
   (a) $x = \frac{e^y - e^{-y}}{2}$ (b) $x = \ln y$
   (c) $x = \ln(y + \sqrt{y^2 + 1})$ (d) $x = \ln 2y$
   (e) $x = \ln(y + y^2)$

4. **Actuary Exam—Part I** If $(\log_x x)(\log_5 x) = 3$ then $x = ?$
   (a) 3 (b) 5 (c) 25 (d) 75 (e) 125

5. **Actuary Exam—Part I** $(\log_a b)(\log_b a) = ?$
   (a) 0 (b) 1 (c) $\log_a b$ (d) $(\log_a b)^2$ (e) $\frac{1}{\log_b a}$

6. **Actuary Exam—Part I** The product $(\log_3 2)(\log_2 9)$ is equal to:
   (a) 2 (b) 3 (c) 9 (d) $\sqrt[3]{9}$ (e) $\sqrt{2}$

7. **Actuary Exam—Part I** For what real value of $x > 1$ is $e^{2\ln(x-1)} = 4$?
   (a) 3 (b) 9 (c) 17 (d) $1 + 4e^{-2}$ (e) $\frac{2 + e^4}{2}$

*Copyright © 1998, 1999 by the Society of Actuaries, Inc. Reprinted with permission.

CHAPTER 3

The Limit of a Function

OUTLINE

3.1 Finding Limits Using Tables and Graphs

3.2 Techniques for Finding Limits of Functions

3.3 One-sided Limits; Continuous Functions

3.4 Limits at Infinity; Infinite Limits; End Behavior; Asymptotes

- **Chapter Review**
- **Chapter Project**
- **Mathematical Questions from Professional Exams**

It's almost April 15th, time to start thinking about taxes. As usual, a headache starts, especially when you look at the tax rate schedules the IRS supplies (see page 272). You begin to wonder what tax bracket you are in. You see that the tax rate can be as low as 10% and as high as 35%. If your taxable income is less than $28,400 the tax rate is 15%, but if your taxable income is just over $28,400, the tax rate jumps from 15% to 25%. That's a big difference. Does that mean the amount you pay in taxes will also have a big jump? No, that wouldn't make any sense. But how can you have a jump in the tax rate, but not in the amount you pay? The Chapter Project will provide an answer and the discussion in this chapter will help explain.

A LOOK BACK, A LOOK FORWARD

In Chapter 1 we defined a function and many of the properties that functions can have. In Chapter 2 we discussed classes of functions and properties that the classes have. With this as background we are ready to study the *limit of a function*. This concept is the bridge that takes us from the mathematics of algebra and geometry to the mathematics of calculus.

Calculus actually consists of two parts: the *differential calculus*, which we discuss in Chapters 4 and 5 and the *integral calculus*, discussed in Chapters 6 and 7. In Chapter 8 we study the calculus of functions of two or more variables.

In differential calculus we introduce another property of functions, namely the *derivative of a function*. We shall find that the derivative opens up a way for doing many applied problems in business, economics, and social sciences. Many of these applications involve an analysis of the graph of a function.

3.1 Finding Limits Using Tables and Graphs

PREPARING FOR THIS SECTION *Before getting started, review the following:*

> Evaluating Functions (Chapter 1, Section 1.2, pp. 108–112)
> Piecewise-defined Functions (Chapter 1, Section 1.4, pp. 143–145)
> Library of Functions (Chapter 1, Section 1.4, pp. 137–143)

OBJECTIVES
1. Find a limit using a table
2. Find a limit using a graph

The idea of the limit of a function is what connects algebra and geometry to calculus. In working with the limit of a function, we encounter notation of the form

$$\lim_{x \to c} f(x) = N$$

This is read as "the limit of $f(x)$ as x approaches c equals the number N." Here f is a function defined on some open interval containing the number c; f need not be defined at c, however.

We may describe the meaning of $\lim_{x \to c} f(x) = N$ as follows:

> For all values of x approximately equal to c, with $x \neq c$, the corresponding value $f(x)$ is approximately equal to N.

Another description of $\lim_{x \to c} f(x) = N$ is

> As x gets closer to c, but remains unequal to c, the corresponding value of $f(x)$ gets closer to N.

Tables generated with the help of a calculator are useful for finding limits.

EXAMPLE 1 **Finding a Limit Using a Table**

Find: $\lim_{x \to 3} (5x^2)$

SOLUTION Here $f(x) = 5x^2$ and $c = 3$. We choose values of x close to 3, arbitrarily starting with 2.99. Then we select additional numbers that get closer to 3, but remain less than 3. Next we choose values of x greater than 3, starting with 3.01, that get closer to 3. Finally, we evaluate f at each choice to obtain Table 1.

TABLE 1

x	2.99	2.999	2.9999 →	← 3.0001	3.001	3.01
$f(x) = 5x^2$	44.701	44.97	44.997 →	← 45.003	45.030	45.301

From Table 1, we infer that as x gets closer to 3 the value of $f(x) = 5x^2$ gets closer to 45. That is,

$$\lim_{x \to 3} (5x^2) = 45$$

TABLE 2

When choosing the values of x in a table, the number to start with and the subsequent entries are arbitrary. However, the entries should be chosen so that the table makes it clear what the corresponding values of f are getting close to.

COMMENT: A graphing utility with a TABLE feature can be used to generate the entries. Table 2 shows the result using a TI-83 Plus.

NOW WORK PROBLEMS 1 AND 9.

EXAMPLE 2 **Finding a Limit Using a Table**

Find: **(a)** $\lim_{x \to 2} \dfrac{x^2 - 4}{x - 2}$ **(b)** $\lim_{x \to 2} (x + 2)$

SOLUTION **(a)** Here $f(x) = \dfrac{x^2 - 4}{x - 2}$ and $c = 2$. Notice that the domain of f is $\{x \mid x \neq 2\}$, so f is not defined at 2. We proceed to choose values of x close to 2 and evaluate f at each choice, as shown in Table 3.

TABLE 3

x	1.99	1.999	1.9999 →	← 2.0001	2.001	2.01
$f(x) = \dfrac{x^2 - 4}{x - 2}$	3.99	3.999	3.9999 →	← 4.0001	4.001	4.01

We infer that as x gets closer to 2 the value of $f(x) = \dfrac{x^2 - 4}{x - 2}$ gets closer to 4. That is,

$$\lim_{x \to 2} \frac{x^2 - 4}{x - 2} = 4$$

(b) Here $g(x) = x + 2$ and $c = 2$. The domain of g is all real numbers. As before, we choose values of x close to 2, and evaluate the function f at each choice. See Table 4.

TABLE 4

x	1.99	1.999	1.9999 →	← 2.0001	2.001	2.01
$f(x) = x + 2$	3.99	3.999	3.9999 →	← 4.0001	4.001	4.01

We infer that as x gets closer to 2 the value of $g(x)$ gets closer to 4. That is,

$$\lim_{x \to 2} (x + 2) = 4$$

CHECK: Use a graphing utility with a TABLE feature to verify the results obtained in Example 2.

The conclusion that $\lim_{x \to 2}(x + 2) = 4$ could have been obtained without the use of Table 4; as x gets closer to 2, it follows that $x + 2$ will get closer to $2 + 2 = 4$.

Also, for part (a), you are right if you make the observation that, since $x \neq 2$, then

$$f(x) = \frac{x^2 - 4}{x - 2} = \frac{(x - 2)(x + 2)}{x - 2} = x + 2, \quad x \neq 2$$

Now it is easy to conclude that

$$\lim_{x \to 2} \frac{x^2 - 4}{x - 2} = \lim_{x \to 2}(x + 2) = 4$$

The graph of a function f can also be of help in finding limits. See Figure 1.

FIGURE 1

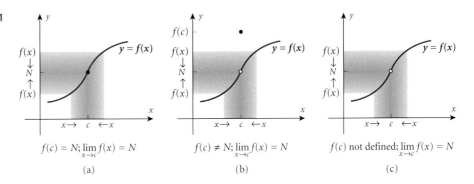

In each graph, notice that as x gets closer to c, the value of f gets closer to the number N. We conclude that

$$\lim_{x \to c} f(x) = N$$

This is the conclusion regardless of the value of f at c. In Figure 1(a), $f(c) = N$, and in Figure 1(b), $f(c) \neq N$. Figure 1(c) illustrates that $\lim_{x \to c} f(x) = N$, even if f is not defined at c.

EXAMPLE 3 Finding a Limit by Graphing

Find: $\lim_{x \to 2} f(x)$ if $f(x) = \begin{cases} 3x - 2 & \text{if } x \neq 2 \\ 3 & \text{if } x = 2 \end{cases}$

SOLUTION The function f is a piecewise-defined function. Its graph is shown in Figure 2. We conclude from the graph that $\lim_{x \to 2} f(x) = 4$.

Notice in Example 3 that the value of f at 2, that is, $f(2) = 3$, plays no role in the conclusion that $\lim_{x \to 2} f(x) = 4$. In fact, even if f were undefined at 2, it would still happen that $\lim_{x \to 2} f(x) = 4$.

NOW WORK PROBLEMS 17 AND 23.

Sometimes there is no *single* number that the value of f gets closer to as x gets closer to c. In this case, we say that f **has no limit as x approaches c** or that $\lim_{x \to c} f(x)$ **does not exist.**

FIGURE 2

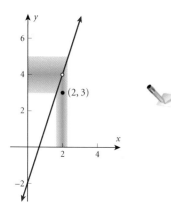

EXAMPLE 4 A Function That Has No Limit at 0

FIGURE 3

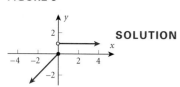

Find: $\lim_{x \to 0} f(x)$ if $f(x) = \begin{cases} x & \text{if } x \leq 0 \\ 1 & \text{if } x > 0 \end{cases}$

SOLUTION See Figure 3. As x gets closer to 0, but remains negative, the value of f also gets closer to 0. As x gets closer to 0, but remains positive, the value of f always equals 1. Since there is no single number that the values of f are close to when x is close to 0, we conclude that $\lim_{x \to 0} f(x)$ does not exist.

NOW WORK PROBLEM 35.

EXAMPLE 5 Using a Graphing Utility to Find a Limit

Find: $\lim_{x \to 2} \dfrac{x^3 - 2x^2 + 4x - 8}{x^4 - 2x^3 + x - 2}$

SOLUTION We create Table 5. From the table we conclude that

$$\lim_{x \to 2} \dfrac{x^3 - 2x^2 + 4x - 8}{x^4 - 2x^3 + x - 2} = 0.889$$

rounded to three decimal places.

TABLE 5

NOW WORK PROBLEM 41.

In the next section, we will see how to obtain exact solutions to limits like the one in Example 5.

EXERCISE 3.1

In Problems 1–8, complete each table and evaluate the indicated limit.

1.
x	0.9	0.99	0.999
$f(x) = 2x$			
x	1.1	1.01	1.001
$f(x) = 2x$			

$\lim_{x \to 1} f(x) = $ _____

2.
x	1.9	1.99	1.999
$f(x) = x + 3$			
x	2.1	2.01	2.001
$f(x) = x + 3$			

$\lim_{x \to 2} f(x) = $ _____

3.
x	-0.1	-0.01	-0.001
$f(x) = x^2 + 2$			
x	0.1	0.01	0.001
$f(x) = x^2 + 2$			

$\lim_{x \to 0} f(x) = $ _____

4.
x	-1.1	-1.01	-1.001
$f(x) = x + 2$			
x	-0.9	-0.99	-0.999
$f(x) = x^2 - 2$			

$\lim_{x \to -1} f(x) = $ _____

5.
x	-1.9	-1.99	-1.999
$f(x) = \dfrac{x^2 - 4}{x + 2}$			
x	-2.1	-2.01	-2.001
$f(x) = \dfrac{x^2 - 4}{x + 2}$			

$\lim_{x \to -2} f(x) = $ _____

6.
x	-1.1	-1.01	-1.001
$f(x) = \dfrac{x^2 - 1}{x + 1}$			
x	-0.9	-0.99	-0.999
$f(x) = \dfrac{x^2 - 1}{x + 1}$			

$\lim_{x \to -1} f(x) = $ _____

7.

x	-1.1	-1.01	-1.001
$f(x) = \dfrac{x^3 + 1}{x + 1}$			
x	-0.9	-0.99	-0.999
$f(x) = \dfrac{x^3 + 1}{x + 1}$			

$\lim\limits_{x \to -1} f(x) = $ _____

8.

x	2.9	2.99	2.999
$f(x) = \dfrac{x^3 - 27}{x - 3}$			
x	3.1	3.01	3.001
$f(x) = \dfrac{x^3 - 27}{x - 3}$			

$\lim\limits_{x \to 3} f(x) = $ _____

In Problems 9–16, use a table to find the indicated limit.

9. $\lim\limits_{x \to 2}(4x^3)$

10. $\lim\limits_{x \to 3}(2x^2 + 1)$

11. $\lim\limits_{x \to 0} \dfrac{x + 1}{x^2 + 1}$

12. $\lim\limits_{x \to 0} \dfrac{2 - x}{x^2 + 4}$

13. $\lim\limits_{x \to 4} \dfrac{x^2 - 4x}{x - 4}$

14. $\lim\limits_{x \to 3} \dfrac{x^2 - 9}{x^2 - 3x}$

15. $\lim\limits_{x \to 0}(e^x + 1)$

16. $\lim\limits_{x \to 0} \dfrac{e^x - e^{-x}}{2}$

In Problems 17–22, use the graph shown to determine if the limit exists. If it does, find it.

17.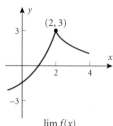

$\lim\limits_{x \to 2} f(x)$

18.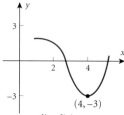

$\lim\limits_{x \to 4} f(x)$

19.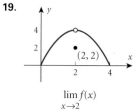

$\lim\limits_{x \to 2} f(x)$

20.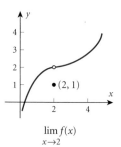

$\lim\limits_{x \to 2} f(x)$

21.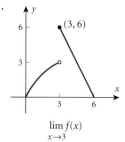

$\lim\limits_{x \to 3} f(x)$

22.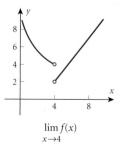

$\lim\limits_{x \to 4} f(x)$

In Problems 23–40, graph each function. Use the graph to find the indicated limit, if it exists.

23. $\lim\limits_{x \to 4} f(x), \quad f(x) = 3x + 1$

24. $\lim\limits_{x \to -1} f(x), \quad f(x) = 2x - 1$

25. $\lim\limits_{x \to 2} f(x), \quad f(x) = 1 - x^2$

26. $\lim\limits_{x \to -1} f(x), \quad f(x) = x^3 - 1$

27. $\lim\limits_{x \to -3} f(x), \quad f(x) = |x| - 2$

28. $\lim\limits_{x \to 4} f(x), \quad f(x) = \sqrt{x + 5}$

29. $\lim\limits_{x \to 0} f(x), \quad f(x) = e^x$

30. $\lim\limits_{x \to 1} f(x), \quad f(x) = \ln x$

31. $\lim\limits_{x \to -1} f(x), \quad f(x) = \dfrac{1}{x}$

32. $\lim\limits_{x \to 8} f(x), \quad f(x) = \sqrt[3]{x}$

33. $\lim\limits_{x \to 0} f(x), \quad f(x) = \begin{cases} x^2 & x < 0 \\ 2x & x \geq 0 \end{cases}$

34. $\lim\limits_{x \to 0} f(x), \quad f(x) = \begin{cases} x - 1 & x < 0 \\ 3x - 1 & x \geq 0 \end{cases}$

35. $\lim\limits_{x \to 1} f(x), \quad f(x) = \begin{cases} 3x & x \leq 1 \\ x + 1 & x > 1 \end{cases}$

36. $\lim\limits_{x \to 2} f(x), \quad f(x) = \begin{cases} x^2 & x \leq 2 \\ 2x - 1 & x > 2 \end{cases}$

37. $\lim\limits_{x \to 0} f(x), \quad f(x) = \begin{cases} x & x < 0 \\ 1 & x = 0 \\ 3x & x > 0 \end{cases}$

38. $\lim_{x \to 0} f(x)$, $f(x) = \begin{cases} 1 & x < 0 \\ -1 & x > 0 \end{cases}$

39. $\lim_{x \to 0} f(x)$, $f(x) = \begin{cases} e^x - 1 & x \leq 0 \\ x^2 & x > 0 \end{cases}$

40. $\lim_{x \to 0} f(x)$, $f(x) = \begin{cases} e^x & x \leq 0 \\ 1 - x & x > 0 \end{cases}$

In Problems 41–46, use a graphing utility to find the indicated limit. Round answers to two decimal places.

41. $\lim_{x \to 1} \dfrac{x^3 - x^2 + x - 1}{x^4 - x^3 + 2x - 2}$

42. $\lim_{x \to -1} \dfrac{x^3 + x^2 + 3x + 3}{x^4 + x^3 + 2x + 2}$

43. $\lim_{x \to 2} \dfrac{x^3 - 2x^2 + 4x - 8}{x^2 + x - 6}$

44. $\lim_{x \to 1} \dfrac{x^3 - x^2 + 3x - 3}{x^2 + 3x - 4}$

45. $\lim_{x \to -1} \dfrac{x^3 + 2x^2 + x}{x^4 + x^3 + 2x + 2}$

46. $\lim_{x \to 3} \dfrac{x^3 - 3x^2 + 4x - 12}{x^4 - 3x^3 + x - 3}$

3.2 Techniques for Finding Limits of Functions

PREPARING FOR THIS SECTION *Before getting started, review the following:*

> Power Functions (Chapter 2, Section 2.2, pp. 178–181)
> Polynomial Functions (Chapter 2, Section 2.2, pp. 181–184)
> Rational Functions (Chapter 2, Section 2.2, pp. 184–185)
> Library of Functions (Chapter 1, Section 1.4, pp. 137–143)
> Average Rate of Change (Chapter 1, Section 1.3, pp. 130–131)

OBJECTIVES
1. Find the limit of a sum, a difference, and a product
2. Find the limit of a polynomial function
3. Find the limit of a function involving a power or a root
4. Find the limit of a quotient
5. Find the limit of an average rate of change

We can find the limit of most functions by developing two formulas involving limits and by using properties of limits.

Two Formulas: $\lim_{x \to c} b$ **and** $\lim_{x \to c} x$

Limit of the Constant Function

For the constant function $f(x) = b$,

$$\lim_{x \to c} f(x) = \lim_{x \to c} b = b \qquad (1)$$

where c is any number.

Limit of the Identity Function

For the identity function $f(x) = x$,

$$\lim_{x \to c} f(x) = \lim_{x \to c} x = c \qquad (2)$$

where c is any number.

Chapter 3 The Limit of a Function

We use graphs to establish formulas (1) and (2). Since the graph of a constant function is a horizontal line, it follows that, no matter how close x is to c, the corresponding value of $f(x)$ equals b. That is, $\lim_{x \to c} b = b$. See Figure 4.

The identity function is $f(x) = x$. For any choice of c, as x gets closer to c, the corresponding value of $f(x)$ is just as close to c. That is, $\lim_{x \to c} x = c$. See Figure 5.

FIGURE 4

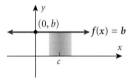

FIGURE 5

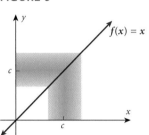

EXAMPLE 1 Using Formulas (1) and (2)

(a) $\lim_{x \to 3} 5 = 5$ (b) $\lim_{x \to 3} x = 3$ (c) $\lim_{x \to 0} (-8) = -8$ (d) $\lim_{x \to -1/2} x = -\dfrac{1}{2}$

NOW WORK PROBLEM 1.

Formulas (1) and (2), when used with the algebraic properties that follow, enable us to evaluate limits of more complicated functions.

Algebraic Properties of Limits

Find the limit of a sum, a difference, and a product   **1**   In the following properties, we assume that f and g are two functions for which both $\lim_{x \to c} f(x)$ and $\lim_{x \to c} g(x)$ exist.

Limit of a Sum

$$\lim_{x \to c}[f(x) + g(x)] = \lim_{x \to c} f(x) + \lim_{x \to c} g(x) \qquad (3)$$

In words, the limit of the sum of two functions equals the sum of their limits.

EXAMPLE 2 Finding the Limit of a Sum

Find: $\lim_{x \to -3} (x + 4)$

SOLUTION   The limit we seek is the sum of the two functions $f(x) = x$ and $g(x) = 4$. From Formulas (1) and (2), we know that

$$\lim_{x \to -3} f(x) = \lim_{x \to -3} x = -3 \quad \text{and} \quad \lim_{x \to -3} g(x) = \lim_{x \to -3} 4 = 4$$

From Formula (3), it follows that

$$\lim_{x \to -3} (x + 4) = \lim_{x \to -3} x + \lim_{x \to -3} 4 = -3 + 4 = 1$$

Limit of a Difference

$$\lim_{x \to c} [f(x) - g(x)] = \lim_{x \to c} f(x) - \lim_{x \to c} g(x) \qquad (4)$$

In words, the limit of the difference of two functions equals the difference of their limits.

EXAMPLE 3 **Finding the Limit of a Difference**

Find: $\lim_{x \to 4} (6 - x)$

SOLUTION The limit we seek is the difference of the two functions $f(x) = 6$ and $g(x) = x$. From Formulas (1) and (2), we know that

$$\lim_{x \to 4} f(x) = \lim_{x \to 4} 6 = 6 \quad \text{and} \quad \lim_{x \to 4} g(x) = \lim_{x \to 4} x = 4$$

From Formula (4), it follows that

$$\lim_{x \to 4} (6 - x) = \lim_{x \to 4} 6 - \lim_{x \to 4} x = 6 - 4 = 2$$

Limit of a Product

$$\lim_{x \to c} [f(x) \cdot g(x)] = \left[\lim_{x \to c} f(x) \right] \left[\lim_{x \to c} g(x) \right] \qquad (5)$$

In words, the limit of the product of two functions equals the product of their limits.

EXAMPLE 4 **Finding the Limit of a Product**

Find: $\lim_{x \to -5} (-4x)$

SOLUTION The limit we seek is the product of the two functions $f(x) = -4$ and $g(x) = x$. From Formulas (1) and (2), we know that

$$\lim_{x \to -5} f(x) = \lim_{x \to -5} (-4) = -4 \quad \text{and} \quad \lim_{x \to -5} g(x) = \lim_{x \to -5} x = -5$$

From Formula (5), it follows that

$$\lim_{x \to -5} (-4x) = \left[\lim_{x \to -5} (-4) \right] \left[\lim_{x \to -5} x \right] = (-4)(-5) = 20$$

EXAMPLE 5   Finding Limits Using Algebraic Properties

Find: **(a)** $\lim_{x \to -2} (3x - 5)$ **(b)** $\lim_{x \to 2} (5x^2)$

SOLUTION   **(a)** $\lim_{x \to -2} (3x - 5) = \lim_{x \to -2} (3x) - \lim_{x \to -2} 5 = \left[\lim_{x \to -2} 3 \right] \left[\lim_{x \to -2} x \right] - \lim_{x \to -2} 5$

$$= (3)(-2) - 5 = -6 - 5 = -11$$

(b) $\lim_{x \to 2} (5x^2) = \left[\lim_{x \to 2} 5 \right] \left[\lim_{x \to 2} x^2 \right] = 5 \cdot \lim_{x \to 2} (x \cdot x) = 5 \cdot \left[\lim_{x \to 2} x \right] \left[\lim_{x \to 2} x \right]$

$$= 5 \cdot 2 \cdot 2 = 20$$

NOW WORK PROBLEM 5.

Notice in the solution to part (b) of Example 5 that $\lim_{x \to 2} (5x^2) = 5 \cdot 2^2$.

Limit of a Power Function

If $n \geq 1$ is a positive integer and a is a constant, then

$$\lim_{x \to c} (ax^n) = ac^n \qquad (6)$$

for any number c. That is, if $f(x) = ax^n$, then

$$\lim_{x \to c} f(x) = f(c) \qquad (7)$$

Proof $\lim_{x \to c} (ax^n) = \left[\lim_{x \to c} a \right] \left[\lim_{x \to c} x^n \right] = a \left[\lim_{x \to c} \underbrace{(x \cdot x \cdot x \cdot \ldots \cdot x)}_{n \text{ factors}} \right]$

$$= a \underbrace{\left[\lim_{x \to c} x \right] \left[\lim_{x \to c} x \right] \left[\lim_{x \to c} x \right] \ldots \left[\lim_{x \to c} x \right]}_{n \text{ factors}}$$

$$= a \cdot \underbrace{c \cdot c \cdot c \cdot \ldots \cdot c}_{n \text{ factors}} = ac^n \quad \blacksquare$$

EXAMPLE 6   Finding the Limit of a Power Function

Find: $\lim_{x \to 2} (-4x^3)$

SOLUTION $\lim_{x \to 2} (-4x^3) = -4 \cdot 2^3 = -4 \cdot 8 = -32$

Find the limit of a polynomial function 2   Since a polynomial is a sum of power functions, we can use Formula (6) and repeated use of Formula (3) to obtain the following result:

Limit of a Polynomial Function

If P is a polynomial function, then

$$\lim_{x \to c} P(x) = P(c) \qquad (8)$$

for any number c.

Proof   If P is a polynomial function, that is, if

$$P(x) = a_n x^n + a_{n-1} x^{n-1} + \cdots + a_1 x + a_0$$

then

$$\lim_{x \to c} P(x) = \lim_{x \to c} [a_n x^n + a_{n-1} x^{n-1} + \cdots + a_1 x + a_0]$$

$$= \lim_{x \to c}(a_n x^n) + \lim_{x \to c}(a_{n-1} x^{n-1}) + \cdots + \lim_{x \to c}(a_1 x) + \lim_{x \to c} a_0 \qquad \text{Formula (3)}$$

$$= a_n c^n + a_{n-1} c^{n-1} + \cdots + a_1 c + a_0 \qquad \text{Formula (6)}$$

$$= P(c) \qquad \blacksquare$$

Formula (8) states that to find the limit of a polynomial as x approaches c, all we need to do is to evaluate the polynomial at c.

EXAMPLE 7   Finding the Limit of a Polynomial Function

Find: $\lim\limits_{x \to 2} [5x^4 - 6x^3 + 3x^2 + 4x - 2]$

SOLUTION $\lim\limits_{x \to 2} [5x^4 - 6x^3 + 3x^2 + 4x - 2] = 5 \cdot 2^4 - 6 \cdot 2^3 + 3 \cdot 2^2 + 4 \cdot 2 - 2$

$$= 5 \cdot 16 - 6 \cdot 8 + 3 \cdot 4 + 8 - 2$$

$$= 80 - 48 + 12 + 6 = 50$$

NOW WORK PROBLEM 7.

Limit of a Power or Root

If $\lim\limits_{x \to c} f(x)$ exists and if $n \geq 2$ is a positive integer, then

$$\lim_{x \to c} [f(x)]^n = \left[\lim_{x \to c} f(x)\right]^n \qquad (9)$$

and

$$\lim_{x \to c} \sqrt[n]{f(x)} = \sqrt[n]{\lim_{x \to c} f(x)} \qquad (10)$$

In Formula (10), we require that both $\sqrt[n]{f(x)}$ and $\sqrt[n]{\lim\limits_{x \to c} f(x)}$ be defined.

EXAMPLE 8 Finding the Limit of a Function Involving a Power or a Root

Find: **(a)** $\lim_{x \to 1}(3x - 5)^4$ **(b)** $\lim_{x \to 0}\sqrt{5x^2 + 8}$ **(c)** $\lim_{x \to -1}(5x^3 - x + 3)^{4/3}$

SOLUTION **(a)** $\lim_{x \to 1}(3x - 5)^4 = \left[\lim_{x \to 1}(3x - 5)\right]^4 = (-2)^4 = 16$

(b) $\lim_{x \to 0}\sqrt{5x^2 + 8} = \sqrt{\lim_{x \to 0}(5x^2 + 8)} = \sqrt{8} = 2\sqrt{2}$

(c) $\lim_{x \to -1}(5x^3 - x + 3)^{4/3} = \sqrt[3]{\lim_{x \to -1}(5x^3 - x + 3)^4}$
$= \sqrt[3]{\left[\lim_{x \to -1}(5x^3 - x + 3)\right]^4} = \sqrt[3]{(-1)^4} = \sqrt[3]{1} = 1$

NOW WORK PROBLEM 17.

Limit of a Quotient

$$\lim_{x \to c}\left[\frac{f(x)}{g(x)}\right] = \frac{\lim_{x \to c} f(x)}{\lim_{x \to c} g(x)} \tag{11}$$

provided that $\lim_{x \to c} g(x) \neq 0$. In words, the limit of the quotient of two functions equals the quotient of their limits, provided that the limit of the denominator is not zero.

Since a rational function is a quotient of polynomials, we can use Formulas (8) and (11) to establish the following result.

Limit of a Rational Function

If R is a rational function and if c is in the domain of R, then

$$\lim_{x \to c} R(x) = R(c) \tag{12}$$

Proof Suppose $R(x) = \dfrac{p(x)}{q(x)}$, where p and q are polynomial functions. If c is in the domain of R, then $q(c) \neq 0$. By Formula (8), $\lim_{x \to c} p(x) = p(c)$ and $\lim_{x \to c} q(x) = q(c)$. Since $q(c) \neq 0$, by Formula (11) we have

$$\lim_{x \to 0} R(x) = \lim_{x \to c}\frac{p(x)}{q(x)} = \frac{\lim_{x \to c} p(x)}{\lim_{x \to c} q(x)} = \frac{p(c)}{q(c)} = R(c)$$
$$\qquad\qquad\qquad\quad \uparrow \qquad\qquad \uparrow$$
$$\qquad\qquad\qquad \text{Formula (11)} \quad \text{Formula (8)}$$

EXAMPLE 9 Finding the Limit of a Rational Function

Find: $\lim_{x \to 1} \dfrac{5x^3 - x + 2}{3x + 4}$

SOLUTION The limit we seek is the limit of a rational function whose domain is $\left\{x \mid x \neq -\dfrac{4}{3}\right\}$. Since 1 is in the domain, we use Formula (12).

$$\lim_{x \to 1} \frac{5x^3 - x + 2}{3x + 4} = \frac{5 \cdot 1^3 - 1 + 2}{3 \cdot 1 + 4} = \frac{6}{7}$$

NOW WORK PROBLEM 15.

When the limit of the denominator of a quotient is zero, Formula (11) cannot be used. In such cases, other strategies need to be used. Let's look at an example.

EXAMPLE 10 Finding the Limit of a Quotient

Find: $\lim_{x \to 3} \dfrac{x^2 - x - 6}{x^2 - 9}$

SOLUTION The domain of the rational function $R(x) = \dfrac{x^2 - x - 6}{x^2 - 9}$ is $\{x \mid x \neq -3, x \neq 3\}$. Since 3 is not in the domain, we cannot use Formula (12). Also, the limit of the denominator equals zero, so Formula (11) cannot be used. Instead, we notice that the expression can be factored as

$$\frac{x^2 - x - 6}{x^2 - 9} = \frac{(x - 3)(x + 2)}{(x - 3)(x + 3)}$$

When we compute a limit as x approaches 3, we are interested in the values of the function when x is close to 3, but unequal to 3. Since $x \neq 3$, we can cancel the $(x - 3)$'s. Formula (11) can then be used.

$$\lim_{x \to 3} \frac{x^2 - x - 6}{x^2 - 9} = \lim_{x \to 3} \frac{\cancel{(x - 3)}(x + 2)}{\cancel{(x - 3)}(x + 3)} = \frac{\lim_{x \to 3}(x + 2)}{\lim_{x \to 3}(x + 3)} = \frac{5}{6}$$

Now let's work Example 5 of Section 3.1.

EXAMPLE 11 Finding Limits Using Algebraic Properties

Find: $\lim_{x \to 2} \dfrac{x^3 - 2x^2 + 4x - 8}{x^4 - 2x^3 + x - 2}$

SOLUTION The limit of the denominator is zero, so Formula (11) cannot be used. We factor the expression.

$$\frac{x^3 - 2x^2 + 4x - 8}{x^4 - 2x^3 + x - 2} \underset{\text{Factor by grouping}}{=} \frac{x^2(x - 2) + 4(x - 2)}{x^3(x - 2) + 1(x - 2)} \underset{\text{Factor}}{=} \frac{(x^2 + 4)(x - 2)}{(x^3 + 1)(x - 2)}$$

Then,

250 Chapter 3 The Limit of a Function

$$\lim_{x\to 2}\frac{x^3-2x^2+4x-8}{x^4-2x^3+x-2}=\lim_{x\to 2}\frac{(x^2+4)(x-2)}{(x^3+1)(x-2)}=\frac{8}{9}$$

which is exact.

Compare the exact solution above with the approximate solution found in Example 5 of Section 3.1.

EXAMPLE 12 Finding the Limit of an Average Rate of Change

Find the limit as x approaches 2 of the average rate of change of the function

$$f(x)=x^2+3x$$

from 2 to x.

SOLUTION The average rate of change of f from 2 to x is

$$\frac{\Delta y}{\Delta x}=\frac{f(x)-f(2)}{x-2}=\frac{x^2+3x-10}{x-2}=\frac{(x+5)(x-2)}{x-2}$$

↑ Factor the numerator

The limit as x approaches 2 of the average rate of change is

$$\lim_{x\to 2}\frac{f(x)-f(2)}{x-2}=\lim_{x\to 2}\frac{x^2+3x-10}{x-2}=\lim_{x\to 2}\frac{(x+5)(x-2)}{x-2}=7$$

NOW WORK PROBLEM 35.

SUMMARY To find exact values for $\lim_{x\to c} f(x)$, try the following:

1. If f is a polynomial Function or if f is a rational function and c is in the domain, then $\lim_{x\to c} f(x)=f(c)$ [Formula (8) or Formula (12)].
2. If f is a polynomial raised to a power or is the root of a polynomial, use Formulas (8) and (9) with Formula (7).
3. If f is a quotient and the limit of the denominator is not zero, use the fact that the limit of a quotient is the quotient of the limits.
4. If f is a quotient and the limit of the denominator is zero, use other techniques, such as factoring.

EXERCISE 3.2

In Problems 1–32, find each limit.

1. $\lim_{x\to 1} 5$
2. $\lim_{x\to 1} (-3)$
3. $\lim_{x\to 4} x$
4. $\lim_{x\to -3} x$
5. $\lim_{x\to 2} (3x+2)$
6. $\lim_{x\to 3} (2-5x)$
7. $\lim_{x\to -1} (3x^2-5x)$
8. $\lim_{x\to 2} (8x^2-4)$
9. $\lim_{x\to 1}(5x^4-3x^2+6x-9)$
10. $\lim_{x\to -1} (8x^5-7x^3+8x^2+x-4)$
11. $\lim_{x\to 1}(x^2+1)^3$
12. $\lim_{x\to 2}(3x-4)^2$
13. $\lim_{x\to 1}\sqrt{5x+4}$
14. $\lim_{x\to 0}\sqrt{1-2x}$
15. $\lim_{x\to 0}\frac{x^2-4}{x^2+4}$
16. $\lim_{x\to 2}\frac{3x+4}{x^2+x}$
17. $\lim_{x\to 2}(3x-2)^{5/2}$
18. $\lim_{x\to -1}(2x+1)^{5/3}$

19. $\lim_{x \to 2} \dfrac{x^2 - 4}{x^2 - 2x}$

20. $\lim_{x \to -1} \dfrac{x^2 + x}{x^2 - 1}$

21. $\lim_{x \to -3} \dfrac{x^2 - x - 12}{x^2 - 9}$

22. $\lim_{x \to -3} \dfrac{x^2 + x - 6}{x^2 + 2x - 3}$

23. $\lim_{x \to 1} \dfrac{x^3 - 1}{x - 1}$

24. $\lim_{x \to 1} \dfrac{x^4 - 1}{x - 1}$

25. $\lim_{x \to -1} \dfrac{(x+1)^2}{x^2 - 1}$

26. $\lim_{x \to 2} \dfrac{x^3 - 8}{x^2 - 4}$

27. $\lim_{x \to 1} \dfrac{x^3 - x^2 + x - 1}{x^4 - x^3 + 2x - 2}$

28. $\lim_{x \to -1} \dfrac{x^3 + x^2 + 3x + 3}{x^4 + x^3 + 2x + 2}$

29. $\lim_{x \to 2} \dfrac{x^3 - 2x^2 + 4x - 8}{x^2 + x - 6}$

30. $\lim_{x \to 1} \dfrac{x^3 - x^2 + 3x - 3}{x^2 + 3x - 4}$

31. $\lim_{x \to -1} \dfrac{x^3 + 2x^2 + x}{x^4 + x^3 + 2x + 2}$

32. $\lim_{x \to 3} \dfrac{x^3 - 3x^2 + 4x - 12}{x^4 - 3x^3 + x - 3}$

In Problems 33–44, find the limit as x approaches c of the average rate of change of each function from c to x.

33. $c = 2;\ f(x) = 5x - 3$
34. $c = -2;\ f(x) = 4 - 3x$
35. $c = 3;\ f(x) = x^2$
36. $c = 3;\ f(x) = x^3$
37. $c = -1;\ f(x) = x^2 + 2x$
38. $c = -1;\ f(x) = 2x^2 - 3x$
39. $c = 0;\ f(x) = 3x^3 - 2x^2 + 4$
40. $c = 0;\ f(x) = 4x^3 - 5x + 8$
41. $c = 1;\ f(x) = \dfrac{1}{x}$
42. $c = 1;\ f(x) = \dfrac{1}{x^2}$
43. $c = 4;\ f(x) = \sqrt{x}$
44. $c = 1;\ f(x) = \sqrt{x}$

In Problems 45–52, assume that $\lim_{x \to c} f(x) = 5$ and $\lim_{x \to c} g(x) = 2$ to find each limit.

45. $\lim_{x \to c} [2f(x)]$
46. $\lim_{x \to c} [f(x) - g(x)]$
47. $\lim_{x \to c} [g(x)^3]$
48. $\lim_{x \to c} \dfrac{f(x)}{g(x)}$
49. $\lim_{x \to c} \dfrac{4}{f(x)}$
50. $\lim_{x \to c} \dfrac{3}{g(x)}$
51. $\lim_{x \to c} [4f(x) - 5g(x)]$
52. $\lim_{x \to c} [8f(x) \cdot g(x)]$

3.3 One-Sided Limits; Continuous Functions

PREPARING FOR THIS SECTION *Before getting started, review the following:*

> Piecewise–defined Functions (Chapter 1, Section 1.4, pp. 143–145)
> Library of Functions (Chapter 1, Section 1.4, pp. 137–143)
> Properties of the Logarithmic Function (Chapter 2, Section 2.4, p. 206)
> Domain of Rational Functions (Chapter 2, Section 2.2, p. 184)
> Properties of the Exponential Function (Chapter 2, Section 2.3, pp. 191–193)

OBJECTIVES 1 Find the one-sided limits of a function
2 Determine whether a function is continuous

Find the one-sided limits of a function

1 Earlier we described $\lim_{x \to c} f(x) = N$ by saying that as x gets closer to c, but remains unequal to c, the corresponding value of $f(x)$ gets closer to N. Whether we use a numerical argument or the graph of the function f, the variable x can get closer to c in only two ways: either by approaching c from the left, through numbers less than c, or by approaching c from the right, through numbers greater than c.

If we only approach c from one side, we have a **one-sided limit**. The notation

$$\lim_{x \to c^-} f(x) = L$$

sometimes called the **left limit,** read as "the limit of $f(x)$ as x approaches c from the left equals L," may be described by the following statement:

> As x gets closer to c, but remains less than c, the corresponding value of $f(x)$ gets closer to L.

The notation $x \to c^-$ is used to remind us that x is less than c.
The notation

$$\lim_{x \to c^+} f(x) = R$$

sometimes called the **right limit,** read as "the limit of $f(x)$ as x approaches c from the right equals R," may be described by the following statement:

> As x gets closer to c, but remains greater than c, the corresponding value of $f(x)$ gets closer to R.

The notation $x \to c^+$ is used to remind us that x is greater than c.
Figure 6 illustrates left and right limits.

FIGURE 6

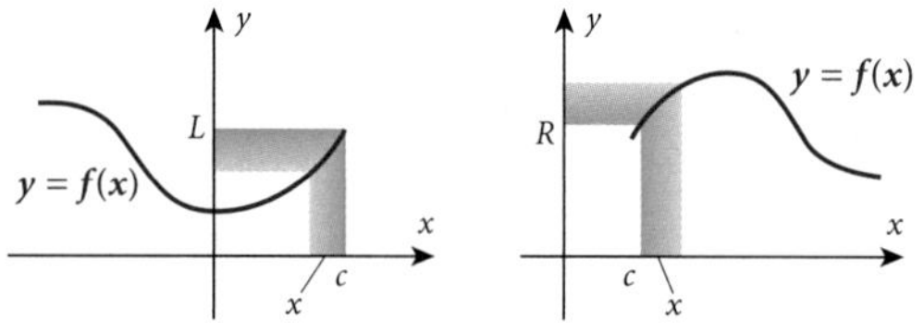

(a) $x < c$, $\lim_{x \to c^-} f(x) = L$

(b) $x > c$, $\lim_{x \to c^+} f(x) = R$

The left and right limits can be used to determine whether $\lim_{x \to c} f(x)$ exists. See Figure 7.

FIGURE 7

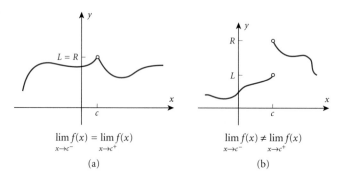

(a) $\lim_{x \to c^-} f(x) = \lim_{x \to c^+} f(x)$

(b) $\lim_{x \to c^-} f(x) \neq \lim_{x \to c^+} f(x)$

As Figure 7(a) illustrates, $\lim_{x \to c} f(x)$ exists and equals the common value of the left limit and the right limit ($L = R$). In Figure 7(b), we see that $\lim_{x \to c} f(x)$ does not exist and that $L \neq R$. This leads us to the following result:

Suppose that $\lim_{x \to c^-} f(x) = L$ and $\lim_{x \to c^+} f(x) = R$. Then $\lim_{x \to c} f(x)$ exists if and only if $L = R$. Furthermore, if $L = R$, then $\lim_{x \to c} f(x) = L (=R)$.

Collectively, the left and right limits of a function are called **one-sided limits** of the function.

EXAMPLE 1 Finding One-Sided Limits of a Function

For the function

$$f(x) = \begin{cases} 2x - 1 & \text{if } x < 2 \\ 1 & \text{if } x = 2 \\ x - 2 & \text{if } x > 2 \end{cases}$$

find: **(a)** $\lim_{x \to 2^-} f(x)$ **(b)** $\lim_{x \to 2^+} f(x)$ **(c)** $\lim_{x \to 2} f(x)$

SOLUTION Figure 8 shows the graph of f.

FIGURE 8

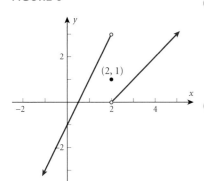

(a) To find $\lim_{x \to 2^-} f(x)$, we look at the values of f when x is close to 2, but less than 2. Since $f(x) = 2x - 1$ for such numbers, we conclude that

$$\lim_{x \to 2^-} f(x) = \lim_{x \to 2^-} (2x - 1) = 3$$

(b) To find $\lim_{x \to 2^+} f(x)$, we look at the values of f when x is close to 2, but greater than 2. Since $f(x) = x - 2$ for such numbers, we conclude that

$$\lim_{x \to 2^+} f(x) = \lim_{x \to 2^+} (x - 2) = 0$$

(c) Since the left and right limits are unequal, $\lim_{x \to 2} f(x)$ does not exist. ▶

NOW WORK PROBLEMS 9 AND 23.

Continuous Functions

Determine whether a **2** function is continuous

We have observed that the value of a function f at c, namely, $f(c)$, plays no role in determining the one-sided limits of f at c. What is the role of the value of a function at c and its one-sided limits at c? Let's look at some of the possibilities. See Figure 9.

FIGURE 9

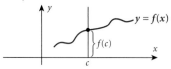
$\lim\limits_{x\to c^-} f(x) = \lim\limits_{x\to c^+} f(x)$, so $\lim\limits_{x\to c} f(x)$ exists;
$\lim\limits_{x\to c} f(x) = f(c)$
(a)

$\lim\limits_{x\to c^-} f(x) = \lim\limits_{x\to c^+} f(x)$, so $\lim\limits_{x\to c} f(x)$ exists;
$\lim\limits_{x\to c} f(x) \neq f(c)$
(b)

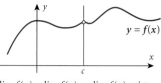
$\lim\limits_{x\to c^-} f(x) = \lim\limits_{x\to c^+} f(x)$, so $\lim\limits_{x\to c} f(x)$ exists;
$f(c)$ is not defined
(c)

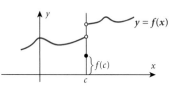
$\lim\limits_{x\to c^-} f(x) \neq \lim\limits_{x\to c^+} f(x)$, so $\lim\limits_{x\to c} f(x)$ does not exist;
$f(c)$ is defined
(d)

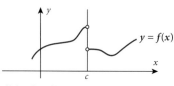
$\lim\limits_{x\to c^-} f(x) \neq \lim\limits_{x\to c^+} f(x)$, so $\lim\limits_{x\to c} f(x)$ does not exist;
$f(c)$ is not defined
(e)

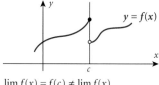
$\lim\limits_{x\to c^-} f(x) = f(c) \neq \lim\limits_{x\to c^+} f(x)$
so $\lim\limits_{x\to c} f(x)$ does not exist
$f(c)$ is defined
(f)

Much earlier in this book, we said that a function f was *continuous* if its graph could be drawn without lifting pencil from paper. In looking at Figure 9, the only graph that has this characteristic is the graph in Figure 9(a), for which the one-sided limits at c each exist and are equal to the value of f at c. This leads us to the following definition:

A function f is **continuous** at c if:

1. f is defined at c; that is, c is in the domain of f so that $f(c)$ equals a number.
2. $\lim\limits_{x\to c^-} f(x) = f(c)$
3. $\lim\limits_{x\to c^+} f(x) = f(c)$

In other words, a function f is continuous at c if

$$\lim\limits_{x\to c} f(x) = f(c)$$

If f is not continuous at c, we say that f is **discontinuous at c**. Each of the functions whose graphs appear in Figures 9(b) to 9(f) is discontinuous at c.

 NOW WORK PROBLEM 15.

Look again at Formula (8) on page 247. Based on (8), we conclude that a polynomial function is continuous at every number. Look at Formula (12). We conclude that a rational function is continuous at every number, except numbers at which it is not defined.

As we mentioned in Chapter 2, pp. 191, 192, 193, and 206, the exponential and logarithmic functions are continuous at every number in their domain. Look at the graphs of the square root function, the absolute value function, and the greatest integer function on pages 141–143. We see that the square root function and absolute value function are continuous at every number in their domain. The function $f(x) = \text{int}(x)$ is continuous except for $x =$ an integer, where a jump occurs in the graph.

Piecewise-defined functions require special attention.

EXAMPLE 2 **Determining Where a Piecewise-Defined Function Is Continuous**

Determine the numbers at which the following function is continuous.

$$f(x) = \begin{cases} x^2 & \text{if } x \leq 0 \\ x + 1 & \text{if } 0 < x < 2 \\ 5 - x & \text{if } 2 \leq x \leq 5 \end{cases}$$

SOLUTION The "pieces" of f, that is, $y = x^2$, $y = x + 1$, and $y = 5 - x$, are each continuous for every number since they are polynomials. In other words, when we graph the pieces, we will not lift our pencil. When we graph the function f, however, we have to be careful, because the pieces change at $x = 0$ and at $x = 2$. So the numbers we need to investigate further are $x = 0$ and $x = 2$.

For $x = 0$: $f(0) = 0^2 = 0$

$$\lim_{x \to 0^-} f(x) = \lim_{x \to 0^-} x^2 = 0$$

$$\lim_{x \to 0^+} f(x) = \lim_{x \to 0^+} (x + 1) = 1$$

FIGURE 10

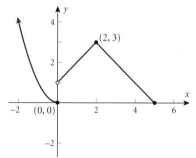

Since $\lim_{x \to 0^+} f(x) \neq f(0)$, we conclude that f is not continuous at $x = 0$.

For $x = 2$: $f(2) = 5 - 2 = 3$

$$\lim_{x \to 2^-} f(x) = \lim_{x \to 2^-} (x + 1) = 3$$

$$\lim_{x \to 2^+} f(x) = \lim_{x \to 2^+} (5 - x) = 3$$

We conclude that f is continuous at $x = 2$.

The function f is continuous for all x, except $x = 0$. The graph of f, given in Figure 10, demonstrates this conclusion.

NOW WORK PROBLEMS 41 AND 49.

SUMMARY **Continuity Properties**

Function	Domain	Property
Polynomial function	All real numbers	Continuous at every number in the domain
Rational function $R(x) = \dfrac{p(x)}{q(x)}$ p, q are polynomials	$\{x \mid q(x) \neq 0\}$	Continuous at every number in the domain
Exponential function	All real numbers	Continuous at every number in the domain
Logarithmic function	Positive real numbers	Continuous at every number in the domain

EXERCISE 3.3

In Problems 1–20, use the accompanying graph of y = f(x).

1. What is the domain of f?
2. What is the range of f?
3. Find the x-intercept(s), if any, of f.
4. Find the y-intercept(s), if any, of f.
5. Find $f(-8)$ and $f(-4)$.
6. Find $f(2)$ and $f(6)$.
7. Find $\lim_{x \to -6^-} f(x)$.
8. Find $\lim_{x \to -6^+} f(x)$.

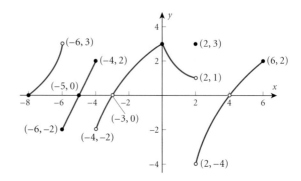

9. Find $\lim_{x \to -4^-} f(x)$.
10. Find $\lim_{x \to -4^+} f(x)$.
11. Find $\lim_{x \to 2^-} f(x)$.
12. Find $\lim_{x \to 2^+} f(x)$.
13. Does $\lim_{x \to 4} f(x)$ exist? If it does, what is it?
14. Does $\lim_{x \to 0} f(x)$ exist? If it does, what is it?
15. Is f continuous at -6?
16. Is f continuous at -4?
17. Is f continuous at 0?
18. Is f continuous at 2?
19. Is f continuous at 4?
20. Is f continuous at 5?

In Problems 21–32, find the one-sided limit.

21. $\lim_{x \to 1^+} (2x + 3)$
22. $\lim_{x \to 2^-} (4 - 2x)$
23. $\lim_{x \to 1^-} (2x^3 + 5x)$
24. $\lim_{x \to -2^+} (3x^2 - 8)$
25. $\lim_{x \to 0^-} e^x$
26. $\lim_{x \to 0^+} e^x$
27. $\lim_{x \to 2^+} \dfrac{x^2 - 4}{x - 2}$
28. $\lim_{x \to 1^-} \dfrac{x^3 - x}{x - 1}$
29. $\lim_{x \to -1^-} \dfrac{x^2 - 1}{x^3 + 1}$
30. $\lim_{x \to 0^+} \dfrac{x^3 - x^2}{x^4 + x^2}$
31. $\lim_{x \to -2^+} \dfrac{x^2 + x - 2}{x^2 + 2x}$
32. $\lim_{x \to -4^-} \dfrac{x^2 + x - 12}{x^2 + 4x}$

In Problems 33–48, determine whether f is continuous at c. Justify your answer.

33. $f(x) = x^3 - 3x^2 + 2x - 6 \quad c = 2$
34. $f(x) = 3x^2 - 6x + 5 \quad c = -3$
35. $f(x) = \dfrac{x^2 + 5}{x - 6} \quad c = 3$
36. $f(x) = \dfrac{x^3 - 8}{x^2 + 4} \quad c = 2$
37. $f(x) = \dfrac{x + 3}{x - 3} \quad c = 3$
38. $f(x) = \dfrac{x - 6}{x + 6} \quad c = -6$
39. $f(x) = \dfrac{x^3 + 3x}{x^2 - 3x} \quad c = 0$
40. $f(x) = \dfrac{x^2 - 6x}{x^2 + 6x} \quad c = 0$
41. $f(x) = \begin{cases} \dfrac{x^3 + 3x}{x^2 - 3x} & \text{if } x \neq 0 \\ 1 & \text{if } x = 0 \end{cases} \quad c = 0$
42. $f(x) = \begin{cases} \dfrac{x^2 - 6x}{x^2 + 6x} & \text{if } x \neq 0 \\ -2 & \text{if } x = 0 \end{cases} \quad c = 0$

One-Sided Limits; Continuous Functions   257

43. $f(x) = \begin{cases} \dfrac{x^3 + 3x}{x^2 - 3x} & \text{if } x \neq 0 \\ -1 & \text{if } x = 0 \end{cases}$ $c = 0$

44. $f(x) = \begin{cases} \dfrac{x^2 - 6x}{x^2 + 6x} & \text{if } x \neq 0 \\ -1 & \text{if } x = 0 \end{cases}$ $c = 0$

45. $f(x) = \begin{cases} \dfrac{x^3 - 1}{x^2 - 1} & \text{if } x < 1 \\ 2 & \text{if } x = 1 \\ \dfrac{3}{x + 1} & \text{if } x > 1 \end{cases}$ $c = 1$

46. $f(x) = \begin{cases} \dfrac{x^2 - 2x}{x - 2} & \text{if } x < 2 \\ 2 & \text{if } x = 2 \\ \dfrac{x - 4}{x - 1} & \text{if } x > 2 \end{cases}$ $c = 2$

47. $f(x) = \begin{cases} 2e^x & \text{if } x < 0 \\ 2 & \text{if } x = 0 \\ \dfrac{x^3 + 2x^2}{x^2} & \text{if } x > 0 \end{cases}$ $c = 0$

48. $f(x) = \begin{cases} 3e^{-x} & \text{if } x < 0 \\ 3 & \text{if } x = 0 \\ \dfrac{x^3 + 3x^2}{x^2} & \text{if } x > 0 \end{cases}$ $c = 0$

In Problems 49–62, find the numbers at which f is continuous. At which numbers is f discontinuous?

49. $f(x) = 2x + 3$
50. $f(x) = 4 - 3x$
51. $f(x) = 3x^2 + x$
52. $f(x) = -3x^3 + 7$

53. $f(x) = 4 \ln x$
54. $f(x) = -2 \ln(x - 3)$
55. $f(x) = 3e^x$
56. $f(x) = 4e^{-x}$

57. $f(x) = \dfrac{2x + 5}{x^2 - 4}$
58. $f(x) = \dfrac{x^2 - 4}{x^2 - 9}$
59. $f(x) = \dfrac{x - 3}{\ln x}$
60. $f(x) = \dfrac{\ln x}{x - 3}$

61. $f(x) = \begin{cases} 3x + 1 & \text{if } x \leq 0 \\ -x^2 & \text{if } 0 < x \leq 2 \\ \dfrac{1}{2}x - 5 & \text{if } x > 2 \end{cases}$

62. $f(x) = \begin{cases} -1 & \text{if } x < -2 \\ x + 1 & \text{if } -2 \leq x \leq 1 \\ x^2 + x + 1 & \text{if } x > 1 \end{cases}$

63. **Cell Phone Service** Sprint PCS offers a monthly cellular phone plan for $39.99. It includes 350 anytime minutes plus $0.25 per minute for additional minutes. The following function is used to compute the monthly cost for a subscriber

$$C(x) = \begin{cases} 39.99 & \text{if } 0 < x \leq 350 \\ 0.25x - 47.51 & \text{if } x > 350 \end{cases}$$

where x is the number of anytime minutes used.

(a) Find $\lim_{x \to 350^-} C(x)$
(b) Find $\lim_{x \to 350^+} C(x)$
(c) Is C continuous at 350?
(d) Give an explanation for your answer in part (c).

64. **First-class Letter** According to the U.S. Postal Service, first-class mail is used for personal and business correspondence. Any mailable item may be sent as first-class mail. It includes postcards, letters, large envelopes, and small packages. The maximum weight is 13 ounces. The following function is used to compute the cost of mailing an item first-class.

$$C(x) = \begin{cases} 0.37 & \text{if } 0 < x \leq 1 \\ 0.23 \, \text{int}(x) + 0.37 & \text{if } 1 < x \leq 13 \end{cases}$$

where x is the weight of the package in ounces.

(a) Find $\lim_{x \to 1^-} C(x)$
(b) Find $\lim_{x \to 1^+} C(x)$
(c) Is C continuous at 1?
(d) Give an explanation for your answer in part (c).

65. **Wind Chill** The wind chill factor represents the equivalent air temperature at a standard wind speed that would produce the same heat loss as the given temperature and wind speed. One formula for computing the equivalent temperature is

$$W = \begin{cases} t & 0 \leq v < 1.79 \\ 33 - \dfrac{(10.45 + 10\sqrt{v} - v)(33 - t)}{22.04} & 1.79 \leq v \leq 20 \\ 33 - 1.5958(33 - t) & v > 20 \end{cases}$$

where v represents the wind speed (in meters per second) and t represents the air temperature (in °C). Suppose $t = 10°C$.

(a) Write the function $W = W(v)$
(b) Find $\lim_{v \to 0^+} W(v)$
(c) Find $\lim_{v \to 1.79^-} W(v)$
(d) Find $\lim_{v \to 1.79^+} W(v)$
(e) Find $W(1.79)$.
(f) Is W continuous at $v = 1.79$?
(g) Round the answers obtained in parts (c), (d), and (e) to two decimal places. Now is W continuous at $v = 1.79$?

(h) Comment on your answers to parts (f) and (g).
(i) Find $\lim_{v \to 20^-} W(v)$
(j) Find $\lim_{v \to 20^+} W(v)$
(k) Find $W(20)$.
(l) Is W continuous at $v = 20$?
(m) Round the answers obtained in parts (i), (j) and (k) to two decimal places. Now is W continuous at $v = 20$?
(n) Comment on your answers to parts (l) and (m).

66. Rework Problem 65 where $t = 0°$ C.

3.4 Limits at Infinity; Infinite Limits; End Behavior; Asymptotes

PREPARING FOR THIS SECTION *Before getting started, review the following:*

> Rational Functions (Chapter 2, Section 2.2, pp. 184–185)

> End Behavior of a Polynomial Function (Chapter 2, Section 2.2, pp. 183–184)

> Graph of $f(x) = \dfrac{1}{x}$ (Chapter 1, Section 1.1, p. 106)

OBJECTIVES
1. Find limits at infinity
2. Find infinite limits
3. Find horizontal asymptotes
4. Find vertical asymptotes
5. Analyze the graph of a rational function at points of discontinuity

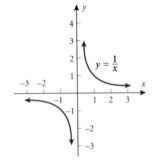

FIGURE 11

In Section 3.2 we described $\lim_{x \to c} f(x) = N$ by saying that the value of f can be made as close as we please to N by choosing numbers x sufficiently close to c. It was understood that N and c were real numbers. In this section we extend the language of limits to allow c to be ∞ or $-\infty$ (*limits at infinity*) and to allow N to be ∞ or $-\infty$ (*infinite limits*).* These limits, it turns out, are useful for finding the end behavior and locating *asymptotes* which help in obtaining the graph of certain functions.

We begin with limits at infinity.

Limits at Infinity

Let's look again at the graph of the function $f(x) = \dfrac{1}{x}$, whose domain is $\{x \mid x \neq 0\}$, that was discussed in Section 1.1, p. 106. See Figure 11.

Table 6 illustrates that the values of f can be made as close as we please to 0 as x becomes unbounded in the positive direction.

TABLE 6

x	1	10	100	1000	10,000	100,000
$f(x) = \dfrac{1}{x}$	1	0.1	0.01	0.001	0.0001	0.00001

*Remember that the symbols ∞ (infinity) and $-\infty$ (minus infinity) are not numbers. Infinity expresses the idea of unboundedness in the positive direction; minus infinity expresses the idea of unboundedness in the negative direction.

This conclusion is expressed by saying that $f(x) = \dfrac{1}{x}$ has the limit 0 as x approaches ∞ and is symbolized by writing

$$\lim_{x \to \infty} \frac{1}{x} = 0 \tag{1}$$

In the same way, we can write

$$\lim_{x \to -\infty} \frac{1}{x} = 0 \tag{2}$$

to indicate that $\dfrac{1}{x}$ can be made as close as we please to 0 as x becomes unbounded in the negative direction. We summarize statements (1) and (2) by saying that $f(x) = \dfrac{1}{x}$ has **limits at infinity.**

Recall that as x becomes unbounded in the positive direction or unbounded in the negative direction, the graph of a polynomial function

$$f(x) = a_n x^n + a_{n-1} x^{n-1} + \ldots + a_1 x + a_0 \qquad a_n \neq 0,$$

behaves the same as the graph of $y = a_n x^n$. In other words, as $x \to -\infty$ or as $x \to \infty$, we can replace $a_n x^n + a_{n-1} x^{n-1} + \ldots + a_1 x + a_0$ by $a_n x^n$. We use this fact to find limits of rational functions at infinity.

1 EXAMPLE 1 Finding Limits at Infinity

Find: **(a)** $\displaystyle\lim_{x \to \infty} \frac{3x - 2}{4x - 1}$ **(b)** $\displaystyle\lim_{x \to \infty} \frac{5x^2 - 3x + 2}{x^3 + 5}$

SOLUTION **(a)** $\displaystyle\lim_{x \to \infty} \frac{3x - 2}{4x - 1} = \lim_{x \to \infty} \frac{3x}{4x}$ As $x \to \infty$, $3x - 2 = 3x$ and $4x - 1 = 4x$

$$= \lim_{x \to \infty} \frac{3}{4} = \frac{3}{4}$$

(b) We follow the same procedure as in part (a):

$$\lim_{x \to \infty} \frac{5x^2 - 3x + 2}{x^3 + 5} = \lim_{x \to \infty} \frac{5x^2}{x^3} = 5 \lim_{x \to \infty} \frac{1}{x} = 0$$

$$\uparrow$$
$$\lim_{x \to \infty} \frac{1}{x} = 0$$

NOW WORK PROBLEM 1.

Infinite Limits

Again we use the function $f(x) = \dfrac{1}{x}$, whose graph is given in Figure 11, to introduce the idea of **infinite limits.** Table 7 gives values of f for selected numbers x that are close to 0 and positive:

TABLE 7

x	1	0.1	0.01	0.001	0.0001	0.00001
$f(x) = \dfrac{1}{x}$	1	10	100	1000	10,000	100,000

We see that as x gets closer to 0 from the right, the value of $f(x) = \dfrac{1}{x}$ is becoming unbounded in the positive direction. We express this fact by writing

$$\lim_{x \to 0^+} \frac{1}{x} = \infty \tag{3}$$

Similarly, we use the notation

$$\lim_{x \to 0^-} \frac{1}{x} = -\infty \qquad (4)$$

to indicate that as x gets closer to 0, and is negative, the values of $\frac{1}{x}$ are becoming unbounded in the negative direction. We summarize (3) and (4) by saying that $f(x) = \frac{1}{x}$ has **one-sided infinite limits** at 0.

EXAMPLE 2 Finding Infinite Limits

Find: $\lim\limits_{x \to 4^+} \dfrac{2-x}{x-4}$

SOLUTION As x gets closer to 4, $x > 4$, then $2 - x$ gets closer to -2 and $\dfrac{1}{x-4}$ is positive and unbounded. As a result, as $x \to 4^+$, the expression $\dfrac{2-x}{x-4}$ is negative and unbounded. That is,

$$\lim_{x \to 4^+} \frac{2-x}{x-4} = -\infty \qquad \blacktriangleright$$

NOW WORK PROBLEM 13.

We now apply the ideas of limits at infinity and infinite limits to the problem of finding end behavior and locating horizontal asymptotes.

End Behavior; Horizontal Asymptotes

The limit at infinity of a function provides information about the end behavior of the graph. This limit can be infinite, indicating that the graph is becoming unbounded as $x \to -\infty$ or as $x \to \infty$. When this limit is a number, a horizontal asymptote describes the end behavior of the graph.

For example, if $\lim\limits_{x \to \infty} f(x) = N$, it means that as x becomes unbounded in the positive direction, the value of f can be made as close as we please to N. That is, the graph of $y = f(x)$ for x sufficiently positive is as close as we please to the horizontal line $y = N$. Similarly, $\lim\limits_{x \to -\infty} f(x) = M$ means that the graph of $y = f(x)$ for x sufficiently negative is as close as we please to the horizontal line $y = M$. These lines are called **horizontal asymptotes** of the graph of f. See Figure 12.

FIGURE 12

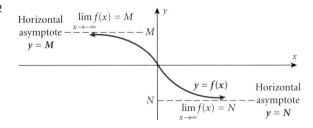

EXAMPLE 3 Finding Horizontal Asymptotes

Find the horizontal asymptotes, if any, of the graph of

$$f(x) = \frac{4x^2}{x^2 + 2}$$

SOLUTION To find any horizontal asymptotes, we need to examine two limits: $\lim\limits_{x \to \infty} f(x)$ and $\lim\limits_{x \to -\infty} f(x)$.

$$\lim_{x \to \infty} f(x) = \lim_{x \to \infty} \frac{4x^2}{x^2 + 2} = \lim_{x \to \infty} \frac{4x^2}{x^2} = 4$$

Limits at Infinity; Infinite Limits; End Behavior; Asymptotes

We conclude that the line $y = 4$ is a horizontal asymptote of the graph when x is sufficiently positive.

$$\lim_{x \to -\infty} f(x) = \lim_{x \to -\infty} \frac{4x^2}{x^2 + 2} = \lim_{x \to -\infty} \frac{4x^2}{x^2} = 4$$

We conclude that the line $y = 4$ is a horizontal asymptote of the graph when x is sufficiently negative.

These conclusions also explain the end behavior of the graph.

Vertical Asymptotes

Infinite limits are used to find vertical asymptotes. Figure 13 illustrates some of the possibilities that can occur when a function has an infinite limit.

FIGURE 13

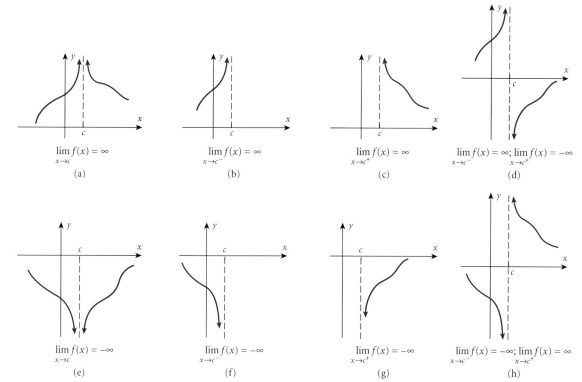

Whenever

$$\lim_{x \to c^-} f(x) = \infty \ (\text{or} -\infty) \quad \text{or} \quad \lim_{x \to c^+} f(x) = \infty \ (\text{or} -\infty)$$

we call the line $x = c$ a **vertical asymptote** of the graph of f.

For rational functions, vertical asymptotes can occur at numbers at which the rational function is not defined.

4 EXAMPLE 4 Finding Vertical Asymptotes

Find the vertical asymptotes, if any, of the rational function

$$R(x) = \frac{x^2}{x - 4}$$

SOLUTION The domain of the rational function R is $\{x \mid x \neq 4\}$. To examine the behavior of the graph of R near 4, where R is not defined, we look at

$$\lim_{x \to 4} R(x) = \lim_{x \to 4} \frac{x^2}{x-4}$$

This will require that we examine the one-sided limits of R at 4.

$\lim_{x \to 4^-} R(x)$: Since $x \to 4^-$, we know $x < 4$, so $x - 4 < 0$. Since $x^2 \geq 0$, it follows that the expression $\dfrac{x^2}{x-4}$ is negative and becomes unbounded as $x \to 4^-$. That is,

$$\lim_{x \to 4^-} R(x) = \lim_{x \to 4^-} \frac{x^2}{x-4} = -\infty$$

$\lim_{x \to 4^+} R(x)$: Since $x \to 4^+$, we know $x > 4$, so $x - 4 > 0$. Since $x^2 \geq 0$, it follows that the expression $\dfrac{x^2}{x-4}$ is positive and becomes unbounded as $x \to 4^+$. That is,

$$\lim_{x \to 4^+} R(x) = \lim_{x \to 4^+} \frac{x^2}{x-4} = \infty$$

We conclude that the graph of R has a vertical asymptote at $x = 4$.

The graph of $R(x) = \dfrac{x^2}{x-4}$, based on the information obtained in Example 3, will exhibit the behavior shown in Figure 14 near $x = 4$. ▶

FIGURE 14

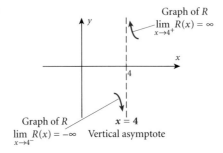

Graph of R
$\lim_{x \to 4^+} R(x) = \infty$

Graph of R
$\lim_{x \to 4^-} R(x) = -\infty$ $x = 4$ Vertical asymptote

NOW WORK PROBLEM 25.

Numbers at which a rational function is not defined are referred to as **points of discontinuity**. Example 4 demonstrates that a rational function can have a vertical asymptote at a point of discontinuity. Sometimes a rational function has a hole at a point of discontinuity.

5 EXAMPLE 5 Analyzing the Graph of a Rational Function at Points of Discontinuity

(a) Determine the numbers at which the rational function

$$R(x) = \frac{x-2}{x^2 - 6x + 8}$$

is continuous.
(b) Use limits to analyze the graph of R at any points of discontinuity.
(c) Graph R.

SOLUTION (a) Since $R(x) = \dfrac{x-2}{(x-2)(x-4)}$, the domain of R is $\{x | x \neq 2, x \neq 4\}$.

We conclude that R is discontinuous at both 2 and 4. (Condition 1 of the definition is violated.) Since R is a rational function, R is continuous at every number, except 2 and 4.

(b) To determine the behavior of the graph at the points of discontinuity, 2 and 4, we look at $\lim\limits_{x \to 2} R(x)$ and $\lim\limits_{x \to 4} R(x)$.

For $\lim\limits_{x \to 2} R(x)$, we have

$$\lim_{x \to 2} R(x) = \lim_{x \to 2} \frac{\cancel{x-2}}{\cancel{(x-2)}(x-4)} = \lim_{x \to 2} \frac{1}{x-4} = -\frac{1}{2}$$

As x gets closer to 2, the graph of R gets closer to $-\dfrac{1}{2}$. Since R is not defined at 2, the graph will have a hole at $\left(2, -\dfrac{1}{2}\right)$.

For $\lim\limits_{x \to 4} R(x)$, we have

$$\lim_{x \to 4} R(x) = \lim_{x \to 4} \frac{\cancel{x-2}}{\cancel{(x-2)}(x-4)} = \lim_{x \to 4} \frac{1}{x-4}$$

Since the limit of the denominator is 0, we use one-sided limits to investigate $\lim\limits_{x \to 4} \dfrac{1}{x-4}$.

If $x < 4$ and x is getting closer to 4, the value of $\dfrac{1}{x-4} < 0$ and is becoming unbounded; that is, $\lim\limits_{x \to 4^-} R(x) = -\infty$.

If $x > 4$ and x is getting closer to 4, the value of $\dfrac{1}{x-4} > 0$ and is becoming unbounded; that is, $\lim\limits_{x \to 4^+} R(x) = \infty$.

The graph of R will have a vertical asymptote at $x = 4$.

(c) It is easiest to graph R by observing that

$$\text{if } x \neq 2, \quad \text{then } R(x) = \frac{\cancel{x-2}}{\cancel{(x-2)}(x-4)} = \frac{1}{x-4}$$

So the graph of R is the graph of $y = \dfrac{1}{x}$ shifted to the right 4 units with a hole at $\left(2, -\dfrac{1}{2}\right)$. See Figure 15. ◼

Example 4 illustrates the following general result.

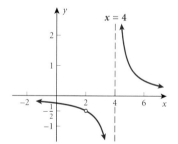

FIGURE 15

The graph of a rational function will have either a vertical asymptote or a hole at numbers at which it is not defined.

NOW WORK PROBLEM 29.

Next we list certain properties of the exponential function and the logarithmic function that involve limits at infinity and infinite limits.

Based on Figure 16, we conclude that:

$\lim\limits_{x \to -\infty} e^x = 0$: The line $y = 0$ (the x-axis) is a horizontal asymptote as $x \to -\infty$.

$\lim\limits_{x \to \infty} e^x = \infty$: The graph of $y = e^x$ becomes unbounded as $x \to \infty$.

See Figure 17. We conclude that:

$\lim\limits_{x \to 0^-} \ln x = -\infty$: The graph of $y = \ln x$ has a vertical asymptote as $x \to 0^+$.

$\lim\limits_{x \to \infty} \ln x = \infty$: The graph of $y = \ln x$ becomes unbounded as $x \to \infty$.

FIGURE 16

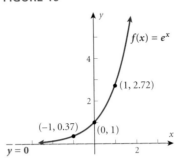

FIGURE 17

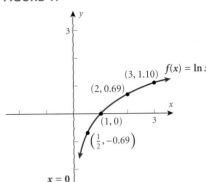

Application

EXAMPLE 6   Analyzing an Average Cost Function

A company estimates that the fixed costs for producing a new toy are $50,000 and the variable costs are $3 per toy.

(a) Express the cost C of producing x toys as a function $C = C(x)$.
(b) Find the domain of C.
(c) Find the average cost of producing x toys; that is, find $\overline{C}(x) = \dfrac{C(x)}{x}$.
(d) Find the domain of $\overline{C}$.
(e) Find $\lim\limits_{x \to 0^+} \overline{C}(x)$ and interpret the answer.
(f) Find $\lim\limits_{x \to \infty} \overline{C}(x)$ and interpret the answer.

SOLUTION   (a) The cost C of producing x toys is

$$C(x) = 50{,}000 + 3x$$

(b) The domain of C is $\{x \mid x \geq 0\}$.
(c) The average cost function $\overline{C}(x)$ is

$$\overline{C}(x) = \frac{C(x)}{x} = \frac{50{,}000 + 3x}{x} = \frac{50{,}000}{x} + 3$$

(d) The domain of $\overline{C}$ is $\{x \mid x > 0\}$.

(e) $\lim\limits_{x \to 0^+} \overline{C}(x) = \lim\limits_{x \to 0^+} \left(\dfrac{50{,}000}{x} + 3 \right) = \infty$

The average cost of producing close to zero toys will be unbounded. Notice that

$$C(1) = \$50{,}003 \quad \text{and} \quad C(2) = \$25{,}003$$

As you would expect, the average cost of producing a few toys is very high due to fixed costs.

FIGURE 18

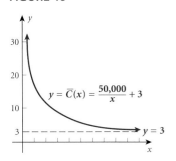

(f) $\lim_{x \to \infty} \overline{C}(x) = \lim_{x \to \infty} \left(\dfrac{50{,}000}{x} + 3 \right) = \lim_{x \to \infty} \dfrac{50{,}000}{x} + \lim_{x \to \infty} 3 = 0 + 3 = 3$

The line $y = 3$ is a horizontal asymptote to the graph of $\overline{C} = \overline{C}(x)$. This means the average cost of producing x toys will never go lower than \$3 per toy. This is expected since the cost of producing each toy is \$3. In other words, the more toys produced, the closer the average cost gets to the unit cost.

Figure 18 shows the graph of $\overline{C} = \overline{C}(x)$. As the graph illustrates, the more units that are produced, the closer the average cost will get to the unit cost, and the less important the fixed cost becomes.

 NOW WORK PROBLEM 39.

SUMMARY

Function Name	End Behavior	Asymptotes
Polynomial Function $P(x) = a_n x^n + a_{n-1} x^{n-1} + \ldots + a_1 x + a_0,\ a \ne 0$ Domain: all real numbers	Behaves like the graph of $y = a_n x^n$	None
Rational Function $R(x) = \dfrac{p(x)}{q(x)},\ p, q$ polynomials Domain: $\{x \mid q(x) \ne 0\}$	Either unbounded or has a horizontal asymptote	If $\lim\limits_{x \to \infty} R(x) = N$, then $y = N$ is a horizontal asymptote. At numbers where $q(x) = 0$ either a hole or a vertical asymptote occurs.
Exponential Function $f(x) = e^x$ Domain: all real numbers	Unbounded as $x \to \infty$; $\lim\limits_{x \to \infty} e^x = \infty$	The line $y = 0$ (the x-axis) is a horizontal asymptote as $x \to -\infty$; $\lim\limits_{x \to -\infty} e^x = 0$
Logarithmic Function $f(x) = \ln x$ Domain: Positive real numbers	Unbounded as $x \to \infty$; $\lim\limits_{x \to \infty} \ln x = \infty$	The line $x = 0$ (the y-axis) is a vertical asymptote as $x \to 0^+$; $\lim\limits_{x \to 0^+} \ln x = -\infty$

EXERCISE 3.4

In Problems 1–12, find each limit at infinity.

1. $\lim\limits_{x \to \infty} \dfrac{x^3 + x^2 + 2x - 1}{x^3 + x + 1}$
2. $\lim\limits_{x \to \infty} \dfrac{2x^2 - 5x + 2}{5x^2 + 7x - 1}$
3. $\lim\limits_{x \to \infty} \dfrac{2x + 4}{x - 1}$
4. $\lim\limits_{x \to \infty} \dfrac{x + 1}{x}$

5. $\lim\limits_{x \to \infty} \dfrac{3x^2 - 1}{x^2 + 4}$
6. $\lim\limits_{x \to -\infty} \dfrac{x^3 - 2x^2 + 1}{4x^3 + 5x + 4}$
7. $\lim\limits_{x \to -\infty} \dfrac{5x^3 - 1}{x^4 + 1}$
8. $\lim\limits_{x \to -\infty} \dfrac{x^2 + 1}{x^3 - 1}$

9. $\lim\limits_{x \to \infty} \dfrac{5x^3 + 3}{x^2 + 1}$
10. $\lim\limits_{x \to \infty} \dfrac{6x^2 + x}{x - 3}$
11. $\lim\limits_{x \to -\infty} \dfrac{4x^5}{x^2 + 1}$
12. $\lim\limits_{x \to -\infty} \dfrac{3x^6}{4x^3 - 1}$

In Problems 13–20, find each limit.

13. $\lim\limits_{x \to 2^+} \dfrac{1}{x - 2}$
14. $\lim\limits_{x \to -1^+} \dfrac{4}{x + 1}$
15. $\lim\limits_{x \to 1^-} \dfrac{x}{(x - 1)^2}$
16. $\lim\limits_{x \to -1^+} \dfrac{x^2}{(x + 1)^2}$

17. $\lim\limits_{x \to 1^+} \dfrac{x^2 + 1}{x^3 - 1}$
18. $\lim\limits_{x \to 3^-} \dfrac{6x^2 + x}{x - 3}$
19. $\lim\limits_{x \to 2^-} \dfrac{1 - x}{3x - 6}$
20. $\lim\limits_{x \to 5^+} \dfrac{2 - x}{5 - x}$

In Problems 21–26, locate all horizontal and vertical asymptotes, if any, of the function f.

21. $f(x) = 3 + \dfrac{1}{x^2}$

22. $f(x) = 2 - \dfrac{1}{x^2}$

23. $f(x) = \dfrac{2x^2}{(x-1)^2}$

24. $f(x) = \dfrac{3x-1}{x+1}$

25. $f(x) = \dfrac{x^2}{x^2-4}$

26. $f(x) = \dfrac{x}{x^2-1}$

27. Use the graph of f below for parts (a)–(p).

   (a) What is the domain of f?
   (b) What is the range of f?
   (c) What are the intercepts, if any, of f?
   (d) What is $f(-2)$?
   (e) What is x if $f(x) = 4$?
   (f) Where is f discontinuous?
   (g) List the vertical asymptotes, if any.
   (h) List the horizontal asymptotes, if any.
   (i) List any local maxima.
   (j) List any local minima.
   (k) Where is f increasing?
   (l) Where is f decreasing?
   (m) What is $\lim\limits_{x \to -\infty} f(x)$?
   (n) What is $\lim\limits_{x \to \infty} f(x)$?
   (o) What is $\lim\limits_{x \to 6^-} f(x)$?
   (p) What is $\lim\limits_{x \to 6^+} f(x)$?

28. Use the graph of f below for parts (a)–(p).

   (a) What is the domain of f?
   (b) What is the range of f?
   (c) What are the intercepts, if any, of f?
   (d) What is $f(6)$?
   (e) What is x if $f(x) = -2$?
   (f) Where is f discontinuous?
   (g) List the vertical asymptotes, if any.
   (h) List the horizontal asymptotes, if any.
   (i) List any local maxima.
   (j) List any local minima.
   (k) Where is f increasing?
   (l) Where is f decreasing?
   (m) What is $\lim\limits_{x \to -\infty} f(x)$?
   (n) What is $\lim\limits_{x \to \infty} f(x)$?
   (o) What is $\lim\limits_{x \to -6^-} f(x)$?
   (p) What is $\lim\limits_{x \to -6^+} f(x) = $?

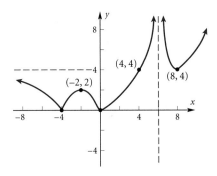

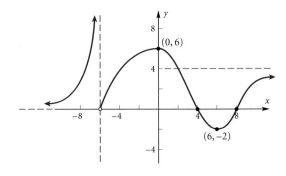

In Problems 29–32, R is discontinuous at c. Use limits to analyze the graph of R at c.

29. $R(x) = \dfrac{x-1}{x^2-1}$ $\quad c = -1$ and $c = 1$

30. $R(x) = \dfrac{3x+6}{x^2-4}$ $\quad c = -2$ and $c = 2$

31. $R(x) = \dfrac{x^2+x}{x^2-1}$ $\quad c = -1$ and $c = 1$

32. $R(x) = \dfrac{x^2+4x}{x^2-16}$ $\quad c = -4$ and $c = 4$

In Problems 33–38, determine where each rational function is undefined. Determine whether an asymptote or a hole appears at such numbers.

33. $R(x) = \dfrac{x^3 - x^2 + x - 1}{x^4 - x^3 + 8x - 8}$

34. $R(x) = \dfrac{x^3 + x^2 + 3x + 3}{x^4 + x^3 + 8x + 8}$

35. $R(x) = \dfrac{x^3 - 2x^2 + 4x - 8}{x^2 + x - 6}$

36. $R(x) = \dfrac{x^3 - x^2 + 3x - 3}{x^2 + 3x - 4}$

37. $R(x) = \dfrac{x^3 + 2x^2 + x}{x^4 + x^3 + x + 1}$

38. $R(x) = \dfrac{x^3 - 3x^2 + 4x - 12}{x^4 - 3x^3 + x - 3}$

39. A company has fixed costs of $79,000 to produce x calculators at a cost of $10 per calculator.
    (a) Express the cost C of producing x calculators as a function $C = C(x)$.
    (b) Find the domain of C.
    (c) Find the average cost of producing x calculators; that is, find $\overline{C}(x) = \dfrac{C(x)}{x}$.
    (d) Find the domain of $\overline{C}$.
    (e) Find $\lim\limits_{x \to 0^+} \overline{C}(x)$ and interpret the answer.
    (f) Find $\lim\limits_{x \to \infty} \overline{C}(x)$ and interpret the answer.

40. A company has fixed costs of $85,000 to produce x cell phones at a cost of $4 per cell phone.
    (a) Express the cost C of producing x cell phones as a function $C = C(x)$.
    (b) Find the domain of C.
    (c) Find the average cost of producing x cell phones; that is, find $\overline{C}(x) = \dfrac{C(x)}{x}$.
    (d) Find the domain of $\overline{C}$.
    (e) Find $\lim\limits_{x \to 0^+} \overline{C}(x)$ and interpret the answer.
    (f) Find $\lim\limits_{x \to \infty} \overline{C}(x)$ and interpret the answer.

41. **Pollution Control** The cost C, in thousands of dollars, for removal of a pollutant from a certain lake is
$$C(x) = \dfrac{5x}{100 - x} \qquad 0 \le x < 100$$
where x is the percent of pollutant removed.
    (a) Find $\lim\limits_{x \to 100^-} C(x)$.
    (b) Is it possible to remove 100% of the pollutant? Explain.

42. **Drug Concentration** The concentration C of a certain drug in a patient's bloodstream t hours after injection is given by
$$C(t) = \dfrac{0.3t}{t^2 + 2}$$
milligrams per cubic centimeter.
    (a) Find the horizontal asymptote of $C(t)$.
    (b) Interpret your answer.

43. Draw the graph of a function $y = f(x)$ that has the following characteristics:
Domain: all real numbers, except 2
$\lim\limits_{x \to -\infty} f(x) = 0 \quad \lim\limits_{x \to \infty} f(x) = \infty \quad \lim\limits_{x \to 2^-} f(x) = -\infty$
$\lim\limits_{x \to 2^+} f(x) = 5 \quad f(0) = 0$
local maximum of 5 at $x = -2$
local minimum of 3 at $x = 4$
increasing on $(-\infty, -2)$ and on $(4, \infty)$
decreasing on $(-2, 2)$ and on $(2, 4)$

44. Draw the graph of a function $y = f(x)$ that has the following characteristics:
Domain: all real numbers, except -3 and 4
$\lim\limits_{x \to -\infty} f(x) = 3 \quad \lim\limits_{x \to \infty} f(x) = 4$
$\lim\limits_{x \to -3^-} f(x) = -\infty \quad \lim\limits_{x \to -3^+} f(x) = 0$
$f(-8) = 0 \quad f(0) = 0 \quad f(10) = 0$
local maximum of 5 at $x = -2$
increasing on $(-3, -2)$ and on $(4, \infty)$
decreasing on $(-\infty, -3)$ and on $(-2, 4)$

Chapter 3 Review

OBJECTIVES

Section	You should be able to	Review Exercises
3.1	1 Find a limit using a table	1, 2
	2 Find a limit using a graph	3, 4, 47(j), (k)
3.2	1 Find the limit of a sum, a difference, and a product	5, 6 15, 16
	2 Find the limit of a polynomial	5, 6
	3 Find the limit of a power or a root	7–10, 13, 14
	4 Find the limit of a quotient	17–20, 23–26
	5 Find the limit of an average rate of change	48–52
3.3	1 Find the one-sided limits of a function	11, 12, 21, 22, 47(g), (h), (i)
	2 Determine whether a function is continuous	35–42, 47(l), (m), (n), (o), (p), (q)

3.4	1 Find limits at infinity	27, 28, 31, 32, 34, 47(t)
	2 Find infinite limits	29, 30, 33, 47(k)
	3 Find horizontal asymptotes	43–46, 47(v)
	4 Find vertical asymptotes	43–46, 47(v)
	5 Analyze the graph of a rational function at points of discontinuity	53–56

THINGS TO KNOW

Limit (p. 237)

$\lim\limits_{x \to c} f(x) = N$ As x gets closer to c, $x \neq c$, the value of f gets closer to N.

Limit Formulas (p. 243)

$\lim\limits_{x \to c} b = b$ The limit of a constant is the constant.

$\lim\limits_{x \to c} x = c$ The limit of x as x approaches c is c.

Limit Properties

$\lim\limits_{x \to c} [f(x) + g(x)] = \lim\limits_{x \to c} f(x) + \lim\limits_{x \to c} g(x)$ (p. 244)   The limit of a sum equals the sum of the limits.

$\lim\limits_{x \to c} [f(x) - g(x)] = \lim\limits_{x \to c} f(x) - \lim\limits_{x \to c} g(x)$ (p. 245)   The limit of a difference equals the difference of the limits.

$\lim\limits_{x \to c} [f(x) \cdot g(x)] = \left[\lim\limits_{x \to c} f(x)\right] \cdot \left[\lim\limits_{x \to c} g(x)\right]$ (p. 245)   The limit of a product equals the product of the limits.

$\lim\limits_{x \to c} \left[\dfrac{f(x)}{g(x)}\right] = \dfrac{\lim\limits_{x \to c} f(x)}{\lim\limits_{x \to c} g(x)}$ (p. 248)   The limit of a quotient equals the quotient of the limits, provided that the limit of the denominator is not zero.

provided that $\lim\limits_{x \to c} g(x) \neq 0$

Limit of a Polynomial Function (p. 247)

$\lim\limits_{x \to c} P(x) = P(c)$, where P is a polynomial

Limit of a Power (p. 247)

$\lim\limits_{x \to c} [f(x)]^n = \left[\lim\limits_{x \to c} f(x)\right]^n$

Limit of a Root (p. 247)

$\lim\limits_{x \to c} \sqrt[n]{f(x)} = \sqrt[n]{\lim\limits_{x \to c} f(x)}$

Continuous Function (p. 254)

$\lim\limits_{x \to c} f(x) = f(c)$

Horizontal Asymptote (p. 260)

If $\lim\limits_{x \to -\infty} f(x) = M$, then $y = M$ is a horizontal asymptote for the graph of f as $x \to -\infty$

If $\lim\limits_{x \to \infty} f(x) = N$, then $y = N$ is a horizontal asymptote for the graph of f as $x \to \infty$

Vertical Asymptote (p. 261)

If $\lim\limits_{x \to c^-} f(x) = -\infty$, then $x = c$ is a vertical asymptote for the graph of f as $x \to c^-$

If $\lim\limits_{x \to c^+} f(x) = -\infty$, then $x = c$ is a vertical asymptote for the graph of f as $x \to c^+$

If $\lim\limits_{x \to c^-} f(x) = \infty$, then $x = c$ is a vertical asymptote for the graph of f as $x \to c^-$

If $\lim\limits_{x \to c^+} f(x) = \infty$, then $x = c$ is a vertical asymptote for the graph of f as $x \to c^+$

Chapter 3 Review

TRUE–FALSE ITEMS

T F **1.** The limit of the sum of two functions equals the sum of their limits, provided that each limit exists.

T F **2.** The limit of a function f as x approaches c always equals $f(c)$.

T F **3.** $\lim_{x \to 4} \dfrac{x^2 - 16}{x - 4} = 8$

T F **4.** The function $f(x) = \dfrac{5x^2}{x^2 + 4}$ is continuous at $x = -2$.

T F **5.** The limit of a quotient of two functions equals the quotient of their limits, provided that each limit exists and the limit of the denominator is not zero.

T F **6.** The graph of a rational function might have both asymptotes and holes.

T F **7.** The graph of an exponential function has an asymptote.

FILL-IN-THE-BLANKS

1. The notation _____ may be described by saying, "For x approximately equal to c, but $x \neq c$, the value $f(x)$ is approximately equal to N."

2. If $\lim_{x \to c} f(x) = N$ and f is continuous at c, then $f(c)$ _____ N.

3. If there is no single number that the value of f approaches when x is close to c, then $\lim_{x \to c} f(x)$ does _____.

4. When $\lim_{x \to c} f(x) = f(c)$, we say that f is _____ at c.

5. $\lim_{x \to c} \dfrac{f(x)}{g(x)} = \dfrac{\lim_{x \to c} f(x)}{\lim_{x \to c} g(x)}$, provided that $\lim_{x \to c} f(x)$ and $\lim_{x \to c} g(x)$ each exist and $\lim_{x \to c} g(x)$ _____ 0.

6. If $\lim_{x \to c^-} f(x) = L$ and $\lim_{x \to c^+} f(x) = R$, then $\lim_{x \to c} f(x)$ exists provided that L _____ R.

7. If, for a function $y = f(x)$, $\lim_{x \to \infty} f(x) = 2$, then _____ is a _____ asymptote of the graph of f.

REVIEW EXERCISES Blue problem numbers indicate the author's suggestions for use in a practice test.

1. Use a table to find $\lim_{x \to 2} \dfrac{x^3 - 8}{x - 2}$

2. Use a table to find $\lim_{x \to 2} \left(\dfrac{1}{3}\right)^x$

3. Use a graph to find $\lim_{x \to 0} f(x)$, where
$$f(x) = \begin{cases} x^2 & \text{if } x < 0 \\ 2 & \text{if } x = 0 \\ e^x - 1 & \text{if } x > 0 \end{cases}$$

4. Use a graph to find $\lim_{x \to 2} g(x)$, where
$$g(x) = \begin{cases} x^2 + 1 & \text{if } x < 2 \\ 5 & \text{if } x = 2 \\ 3x - 2 & \text{if } x > 2 \end{cases}$$

In Problems 5–34, find the limit.

5. $\lim_{x \to 2} (3x^2 - 2x + 1)$

6. $\lim_{x \to 1} (-2x^3 + x + 4)$

7. $\lim_{x \to -2} (x^2 + 1)^2$

8. $\lim_{x \to -2} (x^3 + 1)^2$

9. $\lim_{x \to 3} \sqrt{x^2 + 7}$

10. $\lim_{x \to -2} \sqrt[3]{x + 10}$

11. $\lim_{x \to 1^-} \sqrt{1 - x^2}$

12. $\lim_{x \to 2^+} \sqrt{3x - 2}$

13. $\lim_{x \to 2} (5x + 6)^{3/2}$

14. $\lim_{x \to -3} (15 - 3x)^{-3/2}$

15. $\lim_{x \to -1} (x^2 + x + 2)(x^2 - 9)$

16. $\lim_{x \to 3} (3x + 4)(x^2 + 1)$

17. $\lim_{x \to 1} \dfrac{x - 1}{x^3 - 1}$

18. $\lim_{x \to -1} \dfrac{x^2 - 1}{x^2 + x}$

19. $\lim_{x \to -3} \dfrac{x^2 - 9}{x^2 - x - 12}$

20. $\lim_{x \to -3} \dfrac{x^2 + 2x - 3}{x^2 - 9}$

21. $\lim_{x \to -1^-} \dfrac{x^2 - 1}{x^3 - 1}$

22. $\lim_{x \to 2^+} \dfrac{x^2 - 4}{x^3 - 8}$

23. $\lim_{x \to 2} \dfrac{x^3 - 8}{x^3 - 2x^2 + 4x - 8}$

24. $\lim_{x \to 1} \dfrac{x^3 - 1}{x^3 - x^2 + 3x - 3}$

25. $\lim_{x \to 3} \dfrac{x^4 - 3x^3 + x - 3}{x^3 - 3x^2 + 2x - 6}$

26. $\lim\limits_{x \to -1} \dfrac{x^4 + x^3 + 2x + 2}{x^3 + x^2}$

27. $\lim\limits_{x \to \infty} \dfrac{5x^4 - 8x^3 + x}{3x^4 + x^2 + 5}$

28. $\lim\limits_{x \to \infty} \dfrac{8x^3 - x^2 - 5}{2x^3 - 10x + 1}$

29. $\lim\limits_{x \to 3^-} \dfrac{x^2}{x - 3}$

30. $\lim\limits_{x \to 2^+} \dfrac{5x}{x - 2}$

31. $\lim\limits_{x \to \infty} \dfrac{8x^4 - x^2 + 2}{-4x^3 + 1}$

32. $\lim\limits_{x \to \infty} \dfrac{8x^2 + 2}{4x^3 + x}$

33. $\lim\limits_{x \to -3^+} \dfrac{1 - 9x^2}{x^2 - 9}$

34. $\lim\limits_{x \to -\infty} \dfrac{1 - 9x^2}{1 - 4x^2}$

In Problems 35–42, determine whether f is continuous at c.

35. $f(x) = 3x^4 - x^2 + 2 \quad c = 5$

36. $f(x) = \dfrac{x^2 - 9}{x + 10} \quad c = 2$

37. $f(x) = \dfrac{x^2 - 4}{x + 2} \quad c = -2$

38. $f(x) = \dfrac{x^2 + 6x}{x^2 - 6x} \quad c = 0$

39. $f(x) = \begin{cases} \dfrac{x^2 - 4}{x + 2} & \text{if } x \neq -2 \\ 4 & \text{if } x = -2 \end{cases} \quad c = -2$

40. $f(x) = \begin{cases} \dfrac{x^2 + 6x}{x^2 - 6x} & \text{if } x \neq 0 \\ 1 & \text{if } x = 0 \end{cases} \quad c = 0$

41. $f(x) = \begin{cases} \dfrac{x^2 - 4}{x + 2} & \text{if } x \neq -2 \\ -4 & \text{if } x = -2 \end{cases} \quad c = -2$

42. $f(x) = \begin{cases} \dfrac{x^2 + 6x}{x^2 - 6x} & \text{if } x \neq 0 \\ -1 & \text{if } x = 0 \end{cases} \quad c = 0$

In Problems 43–46, locate all horizontal asymptotes and all vertical asymptotes of each function.

43. $f(x) = \dfrac{3x}{x^2 - 1}$

44. $f(x) = \dfrac{4x^2 - 2x + 1}{x^2 - 4}$

45. $f(x) = \dfrac{5x}{x + 2}$

46. $f(x) = \dfrac{x^3}{x - 3}$

47. In Problem 47, use the accompanying graph of $y = f(x)$.

(a) What is the domain of f?
(b) What is the range of f?
(c) Find the x-intercept(s), if any, of f.
(d) Find the y-intercept(s), if any, of f.
(e) Find $f(-6)$ and $f(-4)$.
(f) Find $f(-2)$ and $f(6)$.
(g) Find $\lim\limits_{x \to -4^-} f(x)$ and $\lim\limits_{x \to -4^+} f(x)$.
(h) Find $\lim\limits_{x \to -2^-} f(x)$ and $\lim\limits_{x \to -2^+} f(x)$.
(i) Find $\lim\limits_{x \to 5^-} f(x)$ and $\lim\limits_{x \to 5^+} f(x)$.
(j) Does $\lim\limits_{x \to 0} f(x)$ exist? If it does, what is it?
(k) Does $\lim\limits_{x \to 2} f(x)$ exist? If it does, what is it?
(l) Is f continuous at -2?
(m) Is f continuous at -4?
(n) Is f continuous at 0?
(o) Is f continuous at 2?
(p) Is f continuous at 4?
(q) Is f continuous at 5?
(r) Where is f increasing?
(s) Where is f decreasing?
(t) Find $\lim\limits_{x \to -\infty} f(x)$ and $\lim\limits_{x \to \infty} f(x)$.
(u) List any local maxima and local minima.
(v) List any horizontal or vertical asymptotes.

In Problems 48–52, find the limit as $x \to c$ of the average rate of change of $f(x)$ from c to x.

48. $c = 5; \ f(x) = 1 - x^2$

49. $c = -2; \ f(x) = 2x^2 - 3x$

50. $c = 2; \ f(x) = 4 - 3x + x^2$

51. $c = 3; f(x) = \dfrac{x}{x - 1}$

52. $c = 2; f(x) = \dfrac{x - 1}{x}$

In Problems 53 and 54, R is discontinuous at c. Use limits to analyze the graph of R at c.

53. $R(x) = \dfrac{x+4}{x^2 - 16}$ at $c = -4$ and $c = 4$

54. $R(x) = \dfrac{3x^2 + 6x}{x^2 - 4}$ at $c = -2$ and $c = 2$

In Problems 55 and 56, determine where each rational function is undefined. Determine whether an asymptote or a hole appears at such numbers.

55. $R(x) = \dfrac{x^3 - 2x^2 + 4x - 8}{x^2 - 11x + 18}$

56. $R(x) = \dfrac{x^3 + 3x^2 - 2x - 6}{x^2 + x - 6}$

57. Draw the graph of a function $y = f(x)$ that has the following characteristics:

Domain: all real numbers, except -2 and 4

$\lim\limits_{x \to -\infty} f(x) = \infty$

$\lim\limits_{x \to \infty} f(x) = 5$

$\lim\limits_{x \to -2^-} f(x) = \infty$

$\lim\limits_{x \to -2^+} f(x) = -\infty$

$\lim\limits_{x \to 4^-} f(x) = 0$

$\lim\limits_{x \to 4^+} f(x) = 0$

$f(0) = 1$

local maximum of 5 at $x = 2$

local minimum of 3 at $x = -4$

increasing on $(-4, -2)$, on $(-4, 2)$, and on $(4, \infty)$

decreasing on $(-\infty, -4)$ and on $(2, 4)$

58. A company has fixed costs of $158,000 to produce x scooters at a cost of $115 per scooter.

(a) Express the cost C of producing x scooters as a function $C = C(x)$.

(b) Find the domain of C.

(c) Find the average cost of producing x scooters; that is, find $\overline{C}(x) = \dfrac{C(x)}{x}$.

(d) Find the domain of $\overline{C}$.

(e) Find $\lim\limits_{x \to 0^+} \overline{C}(x)$ and interpret the answer.

(f) Find $\lim\limits_{x \to \infty} \overline{C}(x)$ and interpret the answer.

59. Advertising The sale of a new product over a period of time is expected to follow the relationship

$$S(x) = \dfrac{2000x^2}{3.5x^2 + 1000}$$

where x is the amount of money spent on advertising.

(a) Evaluate $\lim\limits_{x \to \infty} S(x)$.

(b) Interpret your answer.

Chapter 3 Project

TAX RATES AND CONTINUOUS FUNCTIONS

Even though it may seem that the notion of continuity is too abstract for any real application, in some cases we really need a function to be continuous. In other cases, discontinuities may be either unavoidable or even desirable. In this Chapter Project, we look at tax rates.

We consider two functions: the function that gives the tax rate and the function that gives the amount of tax paid.

On page 272 is the 2003 tax rate schedule for single taxpayers.

If Taxable Income Is		Then The Tax Is		
Over	But Not Over	This Amount	Plus This %	Of The Excess Over
$0	$7,000	$0.00	10%	$0.00
$7,000	$28,400	$700.00	15%	$7,000.00
$28,400	$68,800	$3,910.00	25%	$28,400.00
$68,800	$143,500	$14,010.00	28%	$68,800.00
$143,500	$311,950	$34,926.00	33%	$143,500.00
$311,950	—	$90,514.00	35%	$311,950.00

Source: Internal Revenue Service.

1. Construct a function $R = R(x)$ that gives the tax rate R as a function of taxable income x.

   Hint: Use a piecewise defined function

2. Graph $R = R(x)$.

3. Is R continuous? If not, where does it fail to be continuous?

4. Explain why R has to be discontinuous if 'tax brackets' are required.

5. Construct a function $A = A(x)$ that gives the amount A of tax due as a function of taxable income x.

6. Graph the function $A = A(x)$.

7. Is A continuous? If not, where does it fail to be continuous?

8. What if the third row of the tax table were changed slightly to:

If Taxable Income Is		Then The Tax Is		
Over	But Not Over	This Amount	Plus This %	Of The Excess Over
$28,400	$68,800	**$4,000.00**	25%	$28,400.00

   Show that $A = A(x)$ now has a discontinuity at $x = \$28,400$. Would a tax system using this method of taxation be fair? Why or why not? Why is continuity a desirable property for $A = A(x)$ to have?

9. Here is a partial 2003 tax rate schedule for married taxpayers filing jointly. Fill in the blanks with appropriate values to make the amount of tax paid a continuous function of taxable income.

If Taxable Income Is		Then The Tax Is		
Over	But Not Over	This Amount	Plus This %	Of The Excess Over
$0	$14,000	$0.00	10%	$0.00
$14,000	$56,800		15%	$14,000.00
$56,800	$114,650		25%	$56,800.00
$114,650	$174,700		28%	$114,650.00
$174,700	$311,950		33%	$174,700.00
$311,950	—		35%	$311,950.00

MATHEMATICAL QUESTIONS FROM PROFESSIONAL EXAMS*

1. **Actuary Exam—Part I** $\lim_{x \to 3} \dfrac{x^2 - x - 6}{x^2 - 9}$

   (a) 0   (b) $\dfrac{5}{6}$ (c) 1   (d) $\dfrac{5}{3}$ (e) Undefined

2. **Actuary Exam—Part I** For which of the following functions does $\lim_{x \to 0} f(x)$ exist?

   I. $f(x) = \begin{cases} -1 & \text{if } x < 0 \\ 0 & \text{if } x = 0 \\ 1 & \text{if } x > 0 \end{cases}$

   II. $f(x) = \begin{cases} |x| & \text{if } x \neq 0 \\ 1 & \text{if } x = 0 \end{cases}$

   III. $f(x) = \begin{cases} \dfrac{x^2 + x}{x} & \text{if } x \neq 0 \\ 1 & \text{if } x = 0 \end{cases}$

   (a) I   (b) II   (c) III   (d) I, III   (e) II, III

3. **Actuary Exam—Part I** What is the value of $\lim_{h \to 0} \dfrac{\sqrt{2 + h} - \sqrt{2}}{h}$?

   (a) Undefined   (b) 0   (c) $\dfrac{\sqrt{2}}{2}$ (d) $\dfrac{\sqrt{2}}{4}$ (e) $+\infty$

4. **Actuary Exam—Part I** $f(x) = \dfrac{x^3 + 1}{x + 1}$. For what values of x is f continuous at x?

   (a) $x \leq -1$ (b) $x \geq -1$ (c) $x > -1$
   (d) $x < -1$ (e) $x \neq -1$

*Copyright © 1998, 1999 by the Society of Actuaries. Reprinted with permission.

CHAPTER 4
The Derivative of a Function

OUTLINE

- 4.1 The Definition of a Derivative
- 4.2 The Derivative of a Power Function; Sum and Difference Formulas
- 4.3 Product and Quotient Formulas
- 4.4 The Power Rule
- 4.5 The Derivatives of the Exponential and Logarithmic Functions; the Chain Rule
- 4.6 Higher-Order Derivatives
- 4.7 Implicit Differentiation
- 4.8 The Derivative of $f(x) = x^{p/q}$
  - **Chapter Review**
  - **Chapter Project**
  - **Mathematical Questions from Professional Exams**

If you had to give a blood test to 3000 people to test for the presence of Hepatitis C, how would you do it? Would you test each person individually, which would require 3000 tests? That could be expensive, especially if each test were to cost $100. How can you use fewer tests? One way is to pool the samples. What sample size would you use? If you pooled the blood of 100 people and the test came out negative, then one test was used instead of 100. But if the test came out positive, then you would need to test each one separately, which now requires $1 + 100 = 101$ tests. Maybe the sample size to use is 200. Could it be 300? And the likelihood of a positive test must play a role as well. Sounds like a very hard problem. But with the discussion in this chapter and the Chapter Project to guide you, you can find the best sample size to use so cost is least.

Chapter 4 The Derivative of a Function

A LOOK BACK, A LOOK FORWARD

In Chapter 1, we discussed various properties that functions have, such as intercepts, even/odd, increasing/decreasing, local maxima and minima, and average rate of change. In Chapter 2 we discussed classes of functions and listed some properties that these classes possess. In Chapter 3 we began our study of the calculus by discussing limits of functions and continuity of functions. Now we are ready to define another property of functions: the *derivative of a function*.

The cofounders of calculus are generally recognized to be Gottfried Wilhelm von Leibniz (1646–1716) and Sir Isaac Newton (1642–1727). Newton approached calculus by solving a physics problem involving falling objects, while Leibniz approached calculus by solving a geometry problem. Surprisingly, the solution of these two problems led to the same mathematical concept: the derivative. We shall discuss the physics problem later in this chapter. We shall address the geometry problem, referred to as *The Tangent Problem*, now.

4.1 The Definition of a Derivative

PREPARING FOR THIS SECTION   *Before getting started, review the following:*

> Average Rate of Change (Chapter 1, Section 1.3, pp. 130–131)
> Secant Line (Chapter 1, Section 1.3, pp. 130–131)
> Factoring (Chapter 0, Section 0.3, pp. 29–34)
> Point–slope Form of a Line (Chapter 0, Section 0.9, p. 88)
> Difference Quotient (Chapter 1, Section 1.2, pp. 111–112)

OBJECTIVES
1. Find an equation of the tangent line to the graph of a function
2. Find the derivative of a function at a number c
3. Find the derivative of a function using the difference quotient
4. Find the instantaneous rate of change of a function
5. Find marginal cost and marginal revenue

The Tangent Problem

The geometry question that motivated the development of calculus was "What is the slope of the tangent line to the graph of a function $y = f(x)$ at a point P on its graph?" See Figure 1.

FIGURE 1

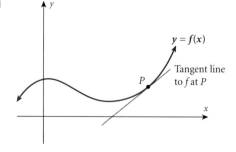

FIGURE 2

We first need to define what we mean by a *tangent* line. In high school geometry, the tangent line to a circle is defined as the line that intersects the graph in exactly one point. Look at Figure 2. Notice that the tangent line just touches the graph of the circle.

This definition, however, does not work in general. Look at Figure 3. The lines L_1 and L_2 only intersect the graph in one point P, but neither touches the graph at P. Additionally, the tangent line L_T shown in Figure 4 touches the graph of f at P, but also intersects the graph elsewhere. So how should we define the tangent line to the graph of f at a point P?

FIGURE 3 **FIGURE 4**

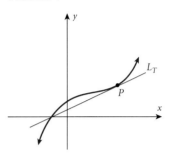

The tangent line L_T to the graph of a function $y = f(x)$ at a point P necessarily contains the point P. To find an equation for L_T using the point–slope form of the equation of a line, it remains to find the slope $m_{\tan}$ of the tangent line.

Suppose that the coordinates of the point P are $(c, f(c))$. Locate another point $Q = (x, f(x))$ on the graph of f. The line containing P and Q is a secant line. (Refer to Section 1.3.) The slope $m_{\sec}$ of the secant line is

$$m_{\sec} = \frac{f(x) - f(c)}{x - c}$$

FIGURE 5

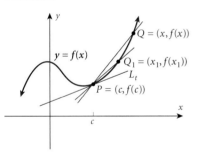

Now look at Figure 5.

As we move along the graph of f from Q toward P, we obtain a succession of secant lines. The closer we get to P, the closer the secant line is to the tangent line. The limiting position of these secant lines is the tangent line. Therefore, the limiting value of the slopes of these secant lines equals the slope of the tangent line. But, as we move from Q toward P, the values of x get closer to c. Therefore,

$$m_{\tan} = \lim_{x \to c} m_{\sec} = \lim_{x \to c} \frac{f(x) - f(c)}{x - c}$$

The **tangent line** to the graph of a function $y = f(x)$ at a point $P = (c, f(c))$ on its graph is defined as the line containing the point P whose slope is

$$m_{\tan} = \lim_{x \to c} \frac{f(x) - f(c)}{x - c} \tag{1}$$

provided that this limit exists.

If $m_{\tan}$ exists, an equation of the tangent line is

$$y - f(c) = m_{\tan}(x - c) \tag{2}$$

Chapter 4 The Derivative of a Function

EXAMPLE 1 Finding an Equation of the Tangent Line

Find an equation of the tangent line to the graph of $f(x) = \dfrac{x^2}{4}$ at the point $\left(1, \dfrac{1}{4}\right)$. Graph the function and the tangent line.

SOLUTION The tangent line contains the point $\left(1, \dfrac{1}{4}\right)$. The slope of the tangent line to the graph of $f(x) = \dfrac{x^2}{4}$ at $\left(1, \dfrac{1}{4}\right)$ is

$$m_{\tan} = \lim_{x \to 1} \frac{f(x) - f(1)}{x - 1} = \lim_{x \to 1} \frac{\dfrac{x^2}{4} - \dfrac{1}{4}}{x - 1} = \lim_{x \to 1} \frac{\dfrac{x^2 - 1}{4}}{x - 1} = \lim_{x \to 1} \frac{(x - 1)(x + 1)}{4(x - 1)}$$

$$= \lim_{x \to 1} \frac{x + 1}{4} = \frac{1}{2}$$

An equation of the tangent line is

$$y - \frac{1}{4} = \frac{1}{2}(x - 1) \qquad y - f(c) = m_{\tan}(x - c)$$

$$y = \frac{1}{2}x - \frac{1}{4}$$

Figure 6 shows the graph of $y = \dfrac{x^2}{4}$ and the tangent line at $\left(1, \dfrac{1}{4}\right)$.

FIGURE 6

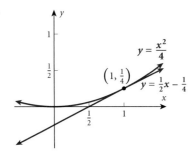

NOW WORK PROBLEM 3.

The limit in Formula (1) has an important generalization: it is called the *derivative of f at c*.

The Derivative of a Function at a Number c.

Let $y = f(x)$ denote a function f. If c is a number in the domain of f, the **derivative of f at c**, denoted by $f'(c)$, read "f prime of c," is defined as

$$f'(c) = \lim_{x \to c} \frac{f(x) - f(c)}{x - c} \qquad (3)$$

provided that this limit exists.

The steps for finding the derivative of a function are listed below:

Steps for Finding the Derivative of a Function at c

STEP 1 Find $f(c)$.
STEP 2 Subtract $f(c)$ from $f(x)$ to get $f(x) - f(c)$ and form the quotient

$$\frac{f(x) - f(c)}{x - c}$$

STEP 3 Find the limit (if it exists) of the quotient found in Step 2 as $x \to c$:

$$f'(c) = \lim_{x \to c} \frac{f(x) - f(c)}{x - c}$$

EXAMPLE 2 Finding the Derivative of a Function at a Number c

Find the derivative of $f(x) = 2x^2 - 5x$ at 2. That is, find $f'(2)$.

SOLUTION Step 1: $f(2) = 2(4) - 5(2) = -2$

Step 2: $\dfrac{f(x) - f(2)}{x - 2} = \dfrac{(2x^2 - 5x) - (-2)}{x - 2} = \dfrac{2x^2 - 5x + 2}{x - 2} = \dfrac{(2x - 1)(x - 2)}{x - 2}$

Step 3: The derivative of f at 2 is

$$f'(2) = \lim_{x \to 2} \frac{f(x) - f(2)}{x - 2} = \lim_{x \to 2} \frac{(2x - 1)(x - 2)}{x - 2} = 3$$

NOW WORK PROBLEM 13.

Example 2 provides a way of finding the derivative at 2 analytically. Graphing utilities have built-in procedures to approximate the derivative of a function at any number c. Consult your owner's manual for the appropriate keystrokes.

EXAMPLE 3 Finding the Derivative of a Function Using a Graphing Utility

Use a graphing utility to find the derivative of $f(x) = 2x^2 - 5x$ at 2. That is, find $f'(2)$.

SOLUTION Figure 7 shows the solution using a TI-83 Plus graphing calculator.

FIGURE 7

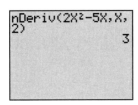

So $f'(2) = 3$.

NOW WORK PROBLEM 45.

EXAMPLE 4 Finding the Derivative of a Function at c

Find the derivative of $f(x) = x^2$ at c. That is, find $f'(c)$.

SOLUTION Since $f(c) = c^2$, we have

$$\frac{f(x) - f(c)}{x - c} = \frac{x^2 - c^2}{x - c} = \frac{(x + c)(x - c)}{x - c}$$

The derivative of f at c is

$$f'(c) = \lim_{x \to c} \frac{f(x) - f(c)}{x - c} = \lim_{x \to c} \frac{(x + c)(x - c)}{x - c} = 2c$$

As Example 4 illustrates, the derivative of $f(x) = x^2$ exists and equals $2c$ for any number c. In other words, the derivative is itself a function and, using x for the independent variable, we can write $f'(x) = 2x$. The function f' is called the **derivative function of f** or the **derivative of f**. We also say that f is **differentiable**. The instruction "differentiate f" means "find the derivative of f".

It is usually easier to find the derivative function by using another form. We derive this alternate form as follows:

Formula (3) for the derivative of f at c is

$$f'(c) = \lim_{x \to c} \frac{f(x) - f(c)}{x - c} \qquad \text{Formula (3).}$$

Let $h = x - c$. Then $x = c + h$ and

$$\frac{f(x) - f(c)}{x - c} = \frac{f(c + h) - f(c)}{h}$$

Since $h = x - c$, then, as $x \to c$, it follows that $h \to 0$. As a result,

$$f'(c) = \lim_{x \to c} \frac{f(x) - f(c)}{x - c} = \lim_{h \to 0} \frac{f(c + h) - f(c)}{h} \qquad (4)$$

Now replace c by x in (4). This gives us the following formula for finding the derivative of f at any number x.

Formula for the Derivative of a Function $y = f(x)$ at x

$$f'(x) = \lim_{h \to 0} \frac{f(x + h) - f(x)}{h} \qquad (5)$$

That is, the derivative of the function f is the limit as $h \to 0$ of its difference quotient.

EXAMPLE 5 Using the Difference Quotient to Find a Derivative

(a) Use Formula (5) to find the derivative of $f(x) = x^2 + 2x$.
(b) Find $f'(0), f'(-1), f'(3)$.

SOLUTION (a) First, we find the difference quotient of $f(x) = x^2 + 2x$.

$$\frac{f(x+h) - f(x)}{h} = \frac{[(x+h)^2 + 2(x+h)] - [x^2 + 2x]}{h}$$

$$= \frac{x^2 + 2xh + h^2 + 2x + 2h - x^2 - 2x}{h}$$

$$= \frac{2xh + h^2 + 2h}{h} \qquad \text{Simplify.}$$

$$= \frac{h(2x + h + 2)}{h} \qquad \text{Factor out } h.$$

$$= 2x + h + 2 \qquad \text{Cancel the } h\text{'s.}$$

The derivative of f is the limit of the difference quotient as $h \to 0$. That is,

$$f'(x) = \lim_{h \to 0} \frac{f(x+h) - f(x)}{h} = \lim_{h \to 0}(2x + h + 2) = 2x + 2$$

(b) Since

$$f'(x) = 2x + 2$$

we have

$$f'(0) = 2 \cdot 0 + 2 = 2$$
$$f'(-1) = 2(-1) + 2 = 0$$
$$f'(3) = 2(3) + 2 = 8$$

NOW WORK PROBLEM 59.

Instantaneous Rate of Change

In Chapter 1 we defined the average rate of change of a function f from c to x as

$$\frac{\Delta y}{\Delta x} = \frac{f(x) - f(c)}{x - c}$$

The limit as x approaches c of the average rate of change of f, based on Formula (3), is the derivative of f at c. As a result, we call the derivative of f at c the **instantaneous rate of change of f with respect to x at c.** That is,

$$\left(\begin{array}{c}\text{Instantaneous rate of} \\ \text{change of } f \text{ with respect to } x \text{ at } c\end{array}\right) = f'(c) = \lim_{x \to c} \frac{f(x) - f(c)}{x - c} \qquad (6)$$

4 EXAMPLE 6 Finding the Instantaneous Rate of Change

During a month-long advertising campaign, the total sales S of a magazine were given by the function

$$S(x) = 5x^2 + 100x + 10{,}000$$

where x represents the number of days of the campaign, $0 \leq x \leq 30$.

(a) What is the average rate of change of sales from $x = 10$ to $x = 20$ days?
(b) What is the instantaneous rate of change of sales when $x = 10$ days?

SOLUTION (a) Since $S(10) = 11{,}500$ and $S(20) = 14{,}000$, the average rate of change of sales from $x = 10$ to $x = 20$ is

$$\frac{\Delta S}{\Delta x} = \frac{S(20) - S(10)}{20 - 10} = \frac{14{,}000 - 11{,}500}{10} = 250 \text{ magazines per day}$$

(b) The instantaneous rate of change of sales when $x = 10$ is the derivative of S at 10.

$$S'(10) = \lim_{x \to 10} \frac{S(x) - S(10)}{x - 10} = \lim_{x \to 10} \frac{[(5x^2 + 100x + 10{,}000) - 11{,}500]}{x - 10}$$

$$= \lim_{x \to 10} \frac{5(x^2 + 20x - 300)}{x - 10}$$

$$= 5 \lim_{x \to 10} \frac{(x + 30)(x - 10)}{x - 10} = 5 \lim_{x \to 10} (x + 30) = 5 \cdot 40 = 200$$

The instantaneous rate of change of S at 10 is 200 magazines per day. ▶

We interpret the results of Example 6 as follows: The fact that the average rate of sales from $x = 10$ to $x = 20$ is $\dfrac{\Delta S}{\Delta x} = 250$ magazines per day indicates that on the 10th day of the campaign, we can expect to average 250 magazines per day of additional sales if we continue the campaign for 10 more days. The fact that $S'(10) = 200$ magazines per day indicates that on the 10th day of the campaign, one more day of advertising will result in additional sales of 200 magazines per day.

NOW WORK PROBLEM 59.

Find marginal cost and marginal revenue

5 Application to Economics: Marginal Analysis

Economics is one of the many fields in which calculus has been used to great advantage. Economists have a special name for the application of derivatives to problems in economics—it is **marginal analysis**. Whenever the term *marginal* appears in a discussion involving cost functions or revenue functions, it signals the presence of derivatives in the background.

> **Marginal Cost**
>
> Suppose $C = C(x)$ is the cost of producing x units. Then the derivative $C'(x)$ is called the **marginal cost.**

We interpret the marginal cost as follows. Since

$$C'(x) = \lim_{h \to 0} \frac{C(x + h) - C(x)}{h}$$

it follows, for small values of h, that

$$C'(x) \approx \frac{C(x + h) - C(x)}{h}$$

That is to say,

$$C'(x) \approx \frac{\text{cost of increasing production from } x \text{ to } x + h}{h}$$

In most practical situations x is very large. Because of this, many economists let $h = 1$, which is small compared to large x. Then, marginal cost may be interpreted as

$$C'(x) = C(x + 1) - C(x) = \text{cost of increasing production by one unit}$$

EXAMPLE 7 Finding Marginal Cost

Suppose that the cost in dollars for a weekly production of x tons of steel is given by the function:

$$C(x) = \frac{1}{10}x^2 + 5x + 1000$$

(a) Find the marginal cost.
(b) Find the cost and marginal cost when $x = 1000$ tons.
(c) Interpret $C'(1000)$.

SOLUTION **(a)** The marginal cost is the derivative $C'(x)$. We use the difference quotient of $C(x)$ to find $C'(x)$.

$$C'(x) = \lim_{h \to 0} \frac{C(x+h) - C(x)}{h} = \lim_{h \to 0} \frac{\left[\frac{1}{10}(x+h)^2 + 5(x+h) + 1000\right] - \left[\frac{1}{10}x^2 + 5x + 1000\right]}{h}$$

$$= \lim_{h \to 0} \frac{\frac{1}{10}(x^2 + 2xh + h^2) + 5x + 5h - \frac{1}{10}x^2 - 5x}{h}$$

$$= \lim_{h \to 0} \frac{\frac{1}{5}xh + \frac{1}{10}h^2 + 5h}{h} = \lim_{h \to 0} \left(\frac{1}{5}x + \frac{1}{10}h + 5\right) = \frac{1}{5}x + 5$$

(b) We evaluate $C(x)$ and $C'(x)$ at $x = 1000$. The cost when $x = 1000$ tons is

$$C(1000) = \frac{1}{10}(1000)^2 + 5 \cdot 1000 + 1000 = \$106{,}000$$

The marginal cost when $x = 1000$ tons is

$$C'(1000) = \frac{1}{5} \cdot 1000 + 5 = \$205 \text{ per ton}$$

(c) $C'(1000) = \$205$ per ton means that the cost of producing one additional ton of steel after 1000 tons have been produced is $205.

The average cost of producing one more ton of steel after the 1000th ton is

$$\frac{\Delta C}{\Delta x} = \frac{C(1001) - C(1000)}{1001 - 1000}$$

$$= \left(\frac{1}{10} \cdot 1001^2 + 5 \cdot 1001 + 1000\right) - \left(\frac{1}{10} \cdot 1000^2 + 5 \cdot 1000 + 1000\right)$$

$$= \$205.10/\text{ton}$$

Observe that the average cost differs from the marginal cost by only 0.1 dollar/ton, which is less than $\frac{1}{20}$th of 1%. Since the marginal cost is usually easier to compute than the average cost, the marginal cost is often used to approximate the cost of producing one additional unit.

The money received by our hypothetical steel producer when he sells his product is the revenue. Specifically, let $R = R(x)$ be the total revenue received from selling x tons. Then the derivative $R'(x)$ is called the **marginal revenue.** For this example, marginal revenue, like marginal cost, is measured in dollars per ton. An approximate value for $R'(x)$ is obtained by noting again that

$$R'(x) \approx \frac{R(x+h) - R(x)}{h}$$

When x is large, then $h = 1$ is small by comparison, so that

$$R'(x) = R(x+1) - R(x) = \text{revenue resulting from the sale of one additional unit}$$

This is the interpretation many economists give to marginal revenue.

EXAMPLE 8 Finding Marginal Revenue

Suppose that the weekly revenue R for the sale of x tons of steel is given by the formula

$$R = x^2 + 5x$$

(a) Find the marginal revenue.
(b) Find the revenue and marginal revenue when $x = 1000$ tons.
(c) Interpret $R'(1000)$.

SOLUTION **(a)** The marginal revenue is the derivative $R'(x)$. We use the difference quotient of $R(x)$ to find $R'(x)$.

$$R'(x) = \lim_{h \to 0} \frac{R(x+h) - R(x)}{h} = \lim_{h \to 0} \frac{[(x+h)^2 + 5(x+h)] - [x^2 + 5x]}{h}$$

$$= \lim_{h \to 0} \frac{x^2 + 2xh + h^2 + 5x + 5h - x^2 - 5x}{h}$$

$$= \lim_{h \to 0} \frac{2xh + h^2 + 5h}{h} = \lim_{h \to 0} (2x + h + 5) = 2x + 5$$

(b) The revenue when $x = 1000$ tons is

$$R(1000) = (1000)^2 + 5(1000) = \$1{,}005{,}000$$

The marginal revenue when $x = 1000$ tons is

$$R'(1000) = 2(1000) + 5 = \$2005 \text{ per ton}$$

(c) $R'(1000) = \$2005/\text{ton}$ means that the revenue obtained from selling one additional ton of steel after 1000 tons have been sold is \$2005.

The average revenue derived from selling one additional ton after 1000 tons have been sold is

$$\frac{\Delta R}{\Delta x} = \frac{R(1001) - R(1000)}{1001 - 1000} = 1{,}007{,}006 - 1{,}005{,}000 = \$2006/\text{ton}$$

Observe that the actual average revenue differs from the marginal revenue by only $1/ton, or 0.05%. Since the marginal revenue is usually easier to compute than the average revenue, the marginal revenue is often used to approximate the revenue from selling one additional unit.

NOW WORK PROBLEM 63.

SUMMARY The derivative of a function $y = f(x)$ at c is defined as

$$f'(c) = \lim_{x \to c} \frac{f(x) - f(c)}{x - c}$$

The derivative $f'(x)$ of a function $y = f(x)$ is

$$f'(x) = \lim_{h \to 0} \frac{f(x + h) - f(x)}{h}$$

In geometry, $f'(c)$ equals the slope of the tangent line to the graph of f at the point $(c, f(c))$.

In applications, if two variables are related by the function $y = f(x)$, then $f'(c)$ equals the instantaneous rate of change of f with respect to x at c.

In economics, the derivative of a cost function is the marginal cost and the derivative of a revenue function is the marginal revenue.

EXERCISE 4.1

In Problems 1–12, find the slope of the tangent line to the graph of f at the given point. What is an equation of the tangent line? Graph f and the tangent line.

1. $f(x) = 3x + 5$ at $(1, 8)$
2. $f(x) = -2x + 1$ at $(-1, 3)$
3. $f(x) = x^2 + 2$ at $(-1, 3)$
4. $f(x) = 3 - x^2$ at $(1, 2)$
5. $f(x) = 3x^2$ at $(2, 12)$
6. $f(x) = -4x^2$ at $(-2, -16)$
7. $f(x) = 2x^2 + x$ at $(1, 3)$
8. $f(x) = 3x^2 - x$ at $(0, 0)$
9. $f(x) = x^2 - 2x + 3$ at $(-1, 6)$
10. $f(x) = -2x^2 + x - 3$ at $(1, -4)$
11. $f(x) = x^3 + x^2$ at $(-1, 0)$
12. $f(x) = x^3 - x^2$ at $(1, 0)$

In Problems 13–24, find the derivative of each function at the given number.

13. $f(x) = -4x + 5$ at 3
14. $f(x) = -4 + 3x$ at 1
15. $f(x) = x^2 - 3$ at 0
16. $f(x) = 2x^2 + 1$ at -1
17. $f(x) = 2x^2 + 3x$ at 1
18. $f(x) = 3x^2 - 4x$ at 2
19. $f(x) = x^3 + 4x$ at 0
20. $f(x) = 2x^3 - x^2$ at 0
21. $f(x) = x^3 + x^2 - 2x$ at 1
22. $f(x) = x^3 - 2x^2 + x$ at 1
23. $f(x) = \dfrac{1}{x}$ at 1
24. $f(x) = \dfrac{1}{x^2}$ at 1

In Problems 25–36, find the derivative of f using the difference quotient.

25. $f(x) = 2x$
26. $f(x) = 3x$
27. $f(x) = 1 - 2x$
28. $f(x) = 5 - 3x$
29. $f(x) = x^2 + 2$
30. $f(x) = 2x^2 - 3$
31. $f(x) = 3x^2 - 2x + 1$
32. $f(x) = 2x^2 + x + 1$
33. $f(x) = x^3$
34. $f(x) = \dfrac{1}{x}$
35. $f(x) = mx + b$
36. $f(x) = ax^2 + bx + c$

In Problems 37–44, find

(a) The average rate of change as x changes from 1 to 3.
(b) The instantaneous rate of change at 1.

37. $f(x) = 3x + 4$ **38.** $f(x) = 2x - 6$ **39.** $f(x) = 3x^2 + 1$ **40.** $f(x) = 2x^2 + 1$
41. $f(x) = x^2 + 2x$ **42.** $f(x) = x^2 - 4x$ **43.** $f(x) = 2x^2 - x + 1$ **44.** $f(x) = 2x^2 + 3x - 2$

In Problems 45–54, find the derivative of each function at the given number using a graphing utility.

45. $f(x) = 3x^3 - 6x^2 + 2$ at -2 **46.** $f(x) = -5x^4 + 6x^2 - 10$ at 5

47. $f(x) = \dfrac{-x^3 + 1}{x^2 + 5x + 7}$ at 8 **48.** $f(x) = \dfrac{-5x^4 + 9x + 3}{x^3 + 5x^2 - 6}$ at -3

49. $f(x) = xe^x$ at 0 **50.** $f(x) = xe^x$ at 1 **51.** $f(x) = x^2 e^x$ at 1
52. $f(x) = x^2 e^x$ at 0 **53.** $f(x) = xe^{-x}$ at 1 **54.** $f(x) = x^2 e^{-x}$ at 2

---

55. Does the tangent line to the graph of $y = x^2$ at $(1, 1)$ pass through the point $(2, 5)$?

56. Does the tangent line to the graph of $y = x^3$ at $(1, 1)$ pass through the point $(2, 5)$?

57. A dive bomber is flying from right to left along the graph of $y = x^2$. When a rocket bomb is released, it follows a path that approximately follows the tangent line. Where should the pilot release the bomb if the target is at $(1, 0)$?

58. Answer the question in Problem 57 if the plane is flying from right to left along the graph of $y = x^3$.

59. Ticket Sales The cumulative ticket sales for the 12 days preceding a popular concert is given by

$$S = 4x^2 + 50x + 5000$$

where x, $1 \leq x \leq 12$, represents the number of days before the concert.

(a) What is the average rate of change in sales from day 1 to day 5?
(b) What is the average rate of change in sales from day 1 to day 10?
(c) What is the average rate of change in sales from day 5 to day 10?
(d) What is the instantaneous rate of change in sales on day 5?
(e) What is the instantaneous rate of change in sales on day 10?

60. Computer Sales The weekly revenue R, in dollars, due to selling x computers is

$$R(x) = -20x^2 + 1000x$$

(a) Find the average rate of change in revenue obtained from selling 5 additional computers after the 20th has been sold.
(b) Find the marginal revenue.
(c) Find the marginal revenue at $x = 20$.
(d) Interpret the answers found in (a) and (c).
(e) For what value of x is $R'(x) = 0$?

61. Supply and Demand Suppose $S(x) = 50x^2 - 50x$ is the supply function describing the number of crates of grapefruit a farmer is willing to supply to the market for x dollars per crate.

(a) How many crates is the farmer willing to supply for $10 per crate?
(b) How many crates is the farmer willing to supply for $13 per crate?
(c) Find the average rate of change in supply from $10 per crate to $13 per crate.
(d) Find the instantaneous rate of change in supply at $x = 10$.
(e) Interpret the answers found in (c) and (d).

62. Glucose Conversion In a metabolic experiment, the mass M of glucose decreases over time t according to the formula

$$M = 4.5 - 0.03t^2$$

(a) Find the average rate of change of the mass from $t = 0$ to $t = 2$.
(b) Find the instantaneous rate of change of mass at $t = 0$.
(c) Interpret the answers found in (a) and (b).

63. Cost and Revenue Functions For a certain production facility, the cost function is

$$C(x) = 2x + 5$$

and the revenue function is

$$R(x) = 8x - x^2$$

where x is the number of units (in thousands) produced and sold and R and C are measured in millions of dollars. Find:

(a) The marginal revenue.
(b) The marginal cost.
(c) The break-even point(s) [the number(s) x for which $R(x) = C(x)$].
(d) The number x for which marginal revenue equals marginal cost.
(e) Graph $C(x)$ and $R(x)$ on the same set of axes.

64. Cost and Revenue Functions For a certain production facility, the cost function is

$$C(x) = x + 5$$

and the revenue function is

$$R(x) = 12x - 2x^2$$

where x is the number of units (in thousands) produced and sold and R and C are measured in millions of dollars. Find:

(a) The marginal revenue.
(b) The marginal cost.
(c) The break-even point(s) [the number(s) x for which $R(x) = C(x)$].
(d) The number x for which marginal revenue equals marginal cost.
(e) Graph $C(x)$ and $R(x)$ on the same set of axes.

65. Demand Equation The price p per ton of cement when x tons of cement are demanded is given by the equation

$$p = -10x + 2000$$

dollars. Find:

(a) The revenue function $R = R(x)$.
   Hint: $R = xp$, where p is the unit price.
(b) The marginal revenue.
(c) The marginal revenue at $x = 100$ tons.
(d) The average rate of change in revenue from $x = 100$ to $x = 101$ tons.
(e) Interpret the answers found in (c) and (d).

66. Demand Equation The cost function and demand equation for a certain product are

$$C(x) = 50x + 40{,}000 \quad \text{and} \quad p = 100 - 0.01x$$

Find:

(a) The revenue function.
(b) The marginal revenue.
(c) The marginal cost.
(d) The break-even point(s).
(e) The number x for which marginal revenue equals marginal cost.

67. Demand Equation A certain item can be produced at a cost of $10 per unit. The demand equation for this item is

$$p = 90 - 0.02x$$

where p is the price in dollars and x is the number of units. Find:

(a) The revenue function.
(b) The marginal revenue.
(c) The marginal cost.
(d) The break-even point(s).
(e) The number x for which marginal revenue equals marginal cost.

68. Instantaneous Rate of Change A circle of radius r has area $A = \pi r^2$ and circumference $C = 2\pi r$. If the radius changes from r to $(r + h)$, find the:

(a) Change in area.
(b) Change in circumference.
(c) Average rate of change of area with respect to the radius.
(d) Average rate of change of the circumference with respect to the radius.
(e) Instantaneous rate of change of area with respect to the radius.
(f) Instantaneous rate of change of the circumference with respect to the radius.

69. Instantaneous Rate of Change The volume V of a right circular cylinder of height 3 feet and radius r feet is $V = V(r) = 3\pi r^2$. Find the instantaneous rate of change of the volume with respect to the radius r at $r = 3$.

70. Instantaneous Rate of Change The surface area S of a sphere of radius r feet is $S = S(r) = 4\pi r^2$. Find the instantaneous rate of change of the surface area with respect to the radius r at $r = 2$ feet.

4.2 The Derivative of a Power Function; Sum and Difference Formulas

OBJECTIVES
1. Find the derivative of a power function
2. Find the derivative of a constant times a function
3. Find the derivative of a polynomial function

In the previous section, we found the derivative $f'(x)$ of a function $y = f(x)$ by using the difference quotient:

$$f'(x) = \lim_{h \to 0} \frac{f(x+h) - f(x)}{h} \qquad (1)$$

We use this form for the derivative to derive formulas for finding derivatives.

We begin by considering the constant function $f(x) = b$, where b is a real number. Since the graph of the constant function f is a horizontal line (see Figure 8), the tangent line to f at any point is also a horizontal line. Since the derivative equals the slope of the tangent line to the graph of a function f at a point, then the derivative of f should be 0.

Analytically, the derivative is obtained by using Formula (1). The difference quotient of $f(x) = b$ is

$$\frac{f(x+h) - f(x)}{h} = \frac{b - b}{h} = \frac{0}{h} = 0$$

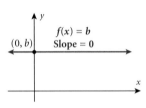

FIGURE 8

The derivative of $f(x) = b$ is

$$f'(x) = \lim_{h \to 0} \frac{f(x+h) - f(x)}{h} = \lim_{h \to 0} 0 = 0$$

> **Derivative of the Constant Function**
>
> For the constant function $f(x) = b$, the derivative is $f'(x) = 0$. In other words, the derivative of a constant is 0.

Besides the **prime notation** f', there are several other ways to denote the derivative of a function $y = f(x)$. The most common ones are

$$y' \quad \text{and} \quad \frac{dy}{dx}$$

The notation $\dfrac{dy}{dx}$, often referred to as the **Leibniz notation,** may also be written as

$$\frac{dy}{dx} = \frac{d}{dx} y = \frac{d}{dx} f(x)$$

where $\dfrac{d}{dx} f(x)$ is an instruction to compute the derivative of the function f with respect to its independent variable x. A change in the symbol used for the independent variable does not affect the meaning. If $s = f(t)$ is a function of t, then $\dfrac{ds}{dt}$ is an instruction to differentiate f with respect to t.

In terms of the Leibniz notation, if b is a constant, then

$$\frac{d}{dx} b = 0 \qquad (2)$$

EXAMPLE 1 Finding the Derivative of a Constant Function

(a) If $f(x) = 5$, then $f'(x) = 0$.
(b) If $y = -1.7$, then $y' = 0$.
(c) If $y = \frac{2}{3}$, then $\frac{dy}{dx} = 0$.
(d) If $s = f(t) = \sqrt{5}$, then $\frac{ds}{dt} = f'(t) = 0$.

In subsequent work with derivatives we shall use the prime notation or the Leibniz notation, or sometimes a mixture of the two, depending on which is more convenient.

NOW WORK PROBLEM 1.

Derivative of a Power Function

We now investigate the derivative of the power function $f(x) = x^n$, where n is a positive integer, to see if a pattern appears.

For $f(x) = x$, $n = 1$, we have

$$f'(x) = \lim_{h \to 0} \frac{f(x+h) - f(x)}{h} = \lim_{h \to 0} \frac{(x+h) - x}{h} = \lim_{h \to 0} \frac{h}{h} = \lim_{h \to 0} 1 = 1$$

For $f(x) = x^2$, $n = 2$, we have

$$f'(x) = \lim_{h \to 0} \frac{f(x+h) - f(x)}{h} = \lim_{h \to 0} \frac{(x+h)^2 - x^2}{h} = \lim_{h \to 0} \frac{x^2 + 2xh + h^2 - x^2}{h}$$

$$= \lim_{h \to 0} \frac{2xh + h^2}{h} = \lim_{h \to 0} \frac{h(2x + h)}{h}$$

$$= \lim_{h \to 0} (2x + h) = 2x$$

For $f(x) = x^3$, $n = 3$, we have

$$f'(x) = \lim_{h \to 0} \frac{f(x+h) - f(x)}{h} = \lim_{h \to 0} \frac{(x+h)^3 - x^3}{h} = \lim_{h \to 0} \frac{x^3 + 3x^2h + 3xh^2 + h^3 - x^3}{h}$$

$$= \lim_{h \to 0} \frac{3x^2h + 3xh^2 + h^3}{h} = \lim_{h \to 0} \frac{h(3x^2 + 3xh + h^2)}{h} = \lim_{h \to 0} (3x^2 + 3xh + h^2) = 3x^2$$

In the Leibniz notation, these results take the form

$$\frac{d}{dx} x = 1 \qquad \frac{d}{dx} x^2 = 2x \qquad \frac{d}{dx} x^3 = 3x^2$$

This pattern suggests the following formula:

Derivative of $f(x) = x^n$

For the power function $f(x) = x^n$, n a positive integer, the derivative is $f'(x) = nx^{n-1}$. That is,

$$\frac{d}{dx} x^n = nx^{n-1} \tag{3}$$

Formula (3) may be stated in words as follows:

The derivative with respect to x of x raised to the power n, where n is a positive integer, is n times x raised to the power $n - 1$.

Problems 73 and 74 outline proofs of Formula (3).

EXAMPLE 2 Finding the Derivative of a Power Function

(a) If $f(x) = x^6$, then $f'(x) = 6x^{6-1} = 6x^5$

(b) $\dfrac{d}{dt} t^5 = 5t^4$

(c) $\dfrac{d}{dx} x = 1$ $\dfrac{d}{dx} x^1 = 1 \cdot x^{1-1} = 1 \cdot x^0 = 1 \cdot 1 = 1$

 NOW WORK PROBLEM 3.

EXAMPLE 3 Finding the Derivative of a Power Function at a Number

Find $f'(4)$ if $f(x) = x^3$

SOLUTION We use Formula (3).

$f'(x) = 3x^2$ Use Formula (3).
$f'(4) = 3(4)^2 = 48$ Substitute 4 for x.

Formula (3) allows us to compute some derivatives with ease. However, do not forget that a derivative is, in actuality, the limit of a difference quotient.

The next formula is used often.

Derivative of a Constant Times a Function

The derivative of a constant times a function equals the constant times the derivative of the function. That is, if c is a constant and f is a differentiable function, then

$$\frac{d}{dx} [cf(x)] = c \frac{d}{dx} f(x) \tag{4}$$

The Derivative of a Power Function; Sum and Difference Formulas

Proof We prove Formula (4) as follows.

$$\frac{d}{dx}[cf(x)] = \lim_{h \to 0} \frac{cf(x+h) - cf(x)}{h} \quad \text{Use the difference quotient of } cf(x).$$

$$= \lim_{h \to 0} c \frac{f(x+h) - f(x)}{h} \quad \text{Factor out } c.$$

$$= \lim_{h \to 0} c \cdot \lim_{h \to 0} \frac{f(x+h) - f(x)}{h} \quad \text{The limit of a product is the product of the limits.}$$

$$= c \frac{d}{dx} f(x) \quad \text{The limit of a constant is the constant;}$$

$$\frac{d}{dx} f(x) = \lim_{h \to 0} \frac{f(x+h) - f(x)}{h} \quad \blacktriangleright$$

The usefulness and versatility of this formula are often overlooked, especially when the constant appears in the denominator. Note that

$$\frac{d}{dx}\left[\frac{f(x)}{c}\right] = \frac{d}{dx}\left[\frac{1}{c}f(x)\right] = \frac{1}{c}\frac{d}{dx}[f(x)]$$

Always be on the lookout for constant factors *before* differentiating.

2 EXAMPLE 4 Finding the Derivative of a Constant Times a Function

(a) If $f(x) = 10x^3$, then

$$f'(x) = \frac{d}{dx}(10x^3) = 10\frac{d}{dx}x^3 = 10 \cdot 3x^2 = 30x^2$$

(b) $\dfrac{d}{dx}\left(\dfrac{x^5}{10}\right) = \dfrac{1}{10}\dfrac{d}{dx}x^5 = \dfrac{1}{10} \cdot 5x^4 = \dfrac{1}{2}x^4$

(c) $\dfrac{d}{dt}(6t) = 6\dfrac{d}{dt}t = 6 \cdot 1 = 6$

(d) $\dfrac{d}{dx}\left(\dfrac{2\sqrt{3}}{3}x^3\right) = \dfrac{2\sqrt{3}}{3}\dfrac{d}{dx}x^3 = \dfrac{2\sqrt{3}}{3} \cdot 3x^2 = 2\sqrt{3}\,x^2$ ▶

NOW WORK PROBLEM 7.

Sum and Difference Formulas

Derivative of a Sum

The derivative of the sum of two differentiable functions equals the sum of their derivatives. That is,

$$\boxed{\frac{d}{dx}[f(x) + g(x)] = \frac{d}{dx}f(x) + \frac{d}{dx}g(x)} \tag{5}$$

A proof is given at the end of this section.

This formula states that functions that are sums can be differentiated "term by term."

EXAMPLE 5 Finding the Derivative of a Function

Find the derivative of: $f(x) = x^2 + 4x$

SOLUTION The function f is the sum of the two functions x^2 and $4x$. We can differentiate term by term.

$$\frac{d}{dx}f(x) = \underbrace{\frac{d}{dx}(x^2 + 4x)}_{\text{Formula (5)}} = \underbrace{\frac{d}{dx}x^2 + \frac{d}{dx}(4x) = 2x + 4\frac{d}{dx}x}_{\text{Formulas (3) and (4)}} = \underbrace{2x + 4}_{\frac{d}{dx}x = 1}$$

Derivative of a Difference

The derivative of the difference of two differentiable functions equals the difference of their derivatives. That is,

$$\boxed{\frac{d}{dx}[f(x) - g(x)] = \frac{d}{dx}f(x) - \frac{d}{dx}g(x)} \tag{6}$$

Formulas (5) and (6) extend to sums and differences of more than two functions. Since a polynomial function is a sum (or difference) of power functions, we can find the derivative of any polynomial function by using a combination of Formulas (2), (3), (4), (5), and (6).

3 EXAMPLE 6 Finding the Derivative of a Polynomial Function

Find the derivative of: $f(x) = 6x^4 - 3x^2 + 10x - 8$

SOLUTION

$$f'(x) = \frac{d}{dx}(6x^4 - 3x^2 + 10x - 8)$$

$$= \frac{d}{dx}(6x^4) - \frac{d}{dx}(3x^2) + \frac{d}{dx}(10x) - \frac{d}{dx}8 \quad \text{Use Formulas (5) and (6).}$$

$$= 6\frac{d}{dx}x^4 - 3\frac{d}{dx}x^2 + 10\frac{d}{dx}x - 0 \quad \text{Use Formulas (2) and (4).}$$

$$= 24x^3 - 6x + 10 \quad \text{Use Formula (3); Simplify.}$$

NOW WORK PROBLEM 21.

EXAMPLE 7 Finding the Derivative of a Polynomial Function

If $f(x) = -\dfrac{x^4}{2} - 2x + 3$, find

(a) $f'(x)$ (b) $f'(-1)$

SOLUTION (a) $f'(x) = -\dfrac{4x^3}{2} - 2 + 0 = -2x^3 - 2$

(b) $f'(-1) = -2(-1)^3 - 2 = 0$

NOW WORK PROBLEM 33.

EXAMPLE 8 Analyzing a Cost Function

The daily cost C, in dollars, of producing dishwashers is

$$C(x) = 1000 + 72x - 0.06x^2 \qquad 0 \le x \le 60$$

where x represents the number of dishwashers produced.

(a) Find the daily cost of producing 50 dishwashers.
(b) Find the marginal cost function.
(c) Find $C'(50)$ and interpret its meaning.
(d) Use the marginal cost to estimate the cost of producing 51 dishwashers.
(e) Find the actual cost of producing 51 dishwashers. Compare the actual cost of making 51 dishwashers to the estimated cost of producing 51 dishwashers found in part (d).
(f) Find the actual cost of producing the 51st dishwasher.
(g) The average cost function is defined as $\overline{C}(x) = \dfrac{C(x)}{x}$, $0 < x \le 60$. Find the average cost function for producing x dishwashers.
(h) Find the average cost of producing 51 dishwashers.

SOLUTION (a) The daily cost of producing 50 dishwashers is

$$C(50) = 1000 + 72(50) - 0.06(50)^2 = \$4450$$

(b) The marginal cost function is

$$C'(x) = \frac{d}{dx}(1000 + 72x - 0.06x^2) = 72 - 0.12x$$

(c) $C'(50) = 72 - 0.12(50) = \$66$.
The marginal cost of producing 50 dishwashers may be interpreted as the cost to produce the 51st dishwasher.

(d) From part (a) the cost to produce 50 dishwashers is \$4450. If the 51st costs \$66, then the cost to produce 51 dishwashers will be

$$\$4450 + \$66 = \$4516$$

(e) The actual cost to produce 51 dishwashers is

$$C(51) = \$1000 + 72(51) - 0.06(51)^2 = \$4515.94$$

There is a difference of \$0.06 between the actual cost and the cost obtained using the marginal cost.

(f) The actual cost of producing the 51st dishwasher is

$$C(51) - C(50) = \$4515.94 - \$4450 = \$65.94$$

(g) The average cost function is

$$\overline{C}(x) = \frac{C(x)}{x} = \frac{1000 + 72x - 0.06x^2}{x} = \frac{1000}{x} + 72 - 0.06x$$

(h) The average cost of producing 51 dishwashers is

$$\overline{C}(51) = \frac{1000}{51} + 72 - 0.06(51) = \$88.55$$

NOW WORK PROBLEM 65.

Proof of the Sum Formula We prove Formula (5) as follows. To compute

$$\frac{d}{dx}[f(x) + g(x)]$$

we need to find the limit of the difference quotient of $f(x) + g(x)$.

$$\frac{d}{dx}[f(x) + g(x)] = \lim_{h \to 0} \frac{[f(x+h) + g(x+h)] - [f(x) + g(x)]}{h}$$

$$= \lim_{h \to 0} \frac{[f(x+h) - f(x)] + [g(x+h) - g(x)]}{h}$$

$$= \lim_{h \to 0} \left[\frac{f(x+h) - f(x)}{h} + \frac{g(x+h) - g(x)}{h} \right]$$

$$= \lim_{h \to 0} \left[\frac{f(x+h) - f(x)}{h} \right] + \lim_{h \to 0} \left[\frac{g(x+h) - g(x)}{h} \right]$$

$$= \frac{d}{dx} f(x) + \frac{d}{dx} g(x)$$

■

Proof of the Difference Formula The proof uses Formulas (4) and (5).

$$\frac{d}{dx}[f(x) - g(x)] = \frac{d}{dx}[f(x) + (-1)g(x)]$$

$$= \frac{d}{dx} f(x) + \frac{d}{dx}[(-1)g(x)]$$

$$= \frac{d}{dx} f(x) + (-1) \frac{d}{dx} g(x)$$

$$= \frac{d}{dx} f(x) - \frac{d}{dx} g(x)$$

■

EXERCISE 4.2

In Problems 1–20, find the derivative of each function.

1. $f(x) = 4$
2. $f(x) = -2$
3. $f(x) = x^5$
4. $f(x) = x^4$
5. $f(x) = 6x^2$
6. $f(x) = -8x^3$
7. $f(t) = \dfrac{t^4}{4}$
8. $f(t) = \dfrac{t^3}{6}$
9. $f(x) = x^2 + x$
10. $f(x) = x^2 - x$
11. $f(x) = x^3 - x^2 + 1$
12. $f(x) = x^4 - x^3 + x$
13. $f(t) = 2t^2 - t + 4$
14. $f(t) = 3t^3 - t^2 + t$
15. $f(x) = \dfrac{1}{2}x^8 + 3x + \dfrac{2}{3}$
16. $f(x) = \dfrac{2}{3}x^6 - \dfrac{1}{2}x^4 + 2$
17. $f(x) = \dfrac{1}{3}(x^5 - 8)$
18. $f(x) = \dfrac{x^3 + 2}{5}$
19. $f(x) = ax^2 + bx + c$
    a, b, c are constants
20. $f(x) = ax^3 + bx^2 + cx + d$
    a, b, c, d are constants

In Problems 21–28, find the indicated derivative.

21. $\dfrac{d}{dx}(-6x^2 + x + 4)$
22. $\dfrac{d}{dx}(8x^3 - 6x^2 + 2x)$
23. $\dfrac{d}{dt}(-16t^2 + 80t)$
24. $\dfrac{d}{dt}(-16t^2 + 64t)$
25. $\dfrac{dA}{dr}$ if $A = \pi r^2$
26. $\dfrac{dC}{dr}$ if $C = 2\pi r$
27. $\dfrac{dV}{dr}$ if $V = \dfrac{4}{3}\pi r^3$
28. $\dfrac{dP}{dt}$ if $P = 0.2t$

In Problems 29–38, find the value of the derivative at the indicated number.

29. $f(x) = 4x^2$ at $x = -3$
30. $f(x) = -10x^3$ at $x = -2$
31. $f(x) = 2x^2 - x$ at $x = 4$
32. $f(x) = x^4 - 2x^2$ at $x = 2$
33. $f(t) = -\dfrac{1}{3}t^3 + 5t$ at $t = 3$
34. $f(t) = -\dfrac{1}{4}t^4 + \dfrac{1}{2}t^2 + 4$ at $t = 1$
35. $f(x) = \dfrac{1}{2}(x^6 - x^4)$ at $x = 1$
36. $f(x) = \dfrac{1}{3}(x^6 + x^3 + 1)$ at $x = -1$
37. $f(x) = ax^2 + bx + c$ at $x = -\dfrac{b}{2a}$
a, b, c are constants
38. $f(x) = ax^3 + bx^2 + cx + d$ at $x = 0$
a, b, c, d are constants

In Problems 39–48, find the value of $\dfrac{dy}{dx}$ at the indicated point.

39. $y = x^4$ at $(1, 1)$
40. $y = x^4$ at $(2, 16)$
41. $y = x^2 - 14$ at $(4, 2)$
42. $y = x^3 + 1$ at $(3, 28)$
43. $y = 3x^2 - x$ at $(-1, 4)$
44. $y = x^2 - 3x$ at $(-1, 4)$
45. $y = \dfrac{1}{2}x^2$ at $\left(1, \dfrac{1}{2}\right)$
46. $y = x^3 - x^2$ at $(1, 0)$
47. $y = 2 - 2x + x^3$ at $(2, 6)$
48. $y = 2x^2 - \dfrac{1}{2}x + 3$ at $(0, 3)$

In Problems 49–50, find the slope of the tangent line to the graph of the function at the indicated point. What is an equation of the tangent line?

49. $f(x) = x^3 + 3x - 1$ at $(0, -1)$
50. $f(x) = x^4 + 2x - 1$ at $(1, 2)$

In Problems 51–56, find those x, if any, at which $f'(x) = 0$.

51. $f(x) = 3x^2 - 12x + 4$
52. $f(x) = x^2 + 4x - 3$
53. $f(x) = x^3 - 3x + 2$
54. $f(x) = x^4 - 4x^3$
55. $f(x) = x^3 + x$
56. $f(x) = x^5 - 5x^4 + 1$

---

57. Find the point(s), if any, on the graph of the function $y = 9x^3$ at which the tangent line is parallel to the line $3x - y + 2 = 0$.

58. Find the points(s), if any, on the graph of the function $y = 4x^2$ at which the tangent line is parallel to the line $2x - y - 6 = 0$.

59. Two lines through the point $(1, -3)$ are tangent to the graph of the function $y = 2x^2 - 4x + 1$. Find the equations of these two lines.

60. Two lines through the point $(0, 2)$ are tangent to the graph of the function $y = 1 - x^2$. Find the equations of these two lines.

61. Marginal Cost The cost per day, $C(x)$, in dollars, of producing x pairs of eyeglasses is
$$C(x) = 0.2x^2 + 3x + 1000$$
(a) Find the average cost due to producing 10 additional pairs of eyeglasses after 100 have been produced.
(b) Find the marginal cost.
(c) Find the marginal cost at $x = 100$.
(d) Interpret $C'(100)$.

62. Toy Truck Sales At Dan's Toy Store, the revenue R, in dollars, derived from selling x electric trucks is
$$R(x) = -0.005x^2 + 20x$$
(a) What is the average rate of change in revenue obtained from selling 10 additional trucks after 1000 have been sold?
(b) What is the marginal revenue?
(c) What is the marginal revenue at $x = 1000$?
(d) Interpret $R'(1000)$.
(e) For what value of x is $R'(x) = 0$?

63. Medicine The French physician Poiseville discovered that the volume V of blood (in cubic centimeters) flowing through a clogged artery with radius R (in centimeters) can be modeled by
$$V(R) = kR^4$$
where k is a positive constant.
(a) Find the derivative $V'(R)$.
(b) Find the rate of change of volume for a radius of 0.3 cm.
(c) Find the rate of change of volume for a radius of 0.4 cm.
(d) If the radius of a clogged artery is increased from 0.3 cm to 0.4 cm, estimate the effect on the volume of blood flowing through the enlarged artery.

64. Respiration Rate A human being's respiration rate R (in breaths per minute) is given by
$$R = -10.35p + 0.59p^2$$
where p is the partial pressure of carbon dioxide in the lungs. Find the rate of change in respiration rate when $p = 50$.

65. Analyzing a Cost Function The daily cost C of producing microwave ovens is
$$C(x) = 2000 + 50x - 0.05x^2, \quad 0 \le x \le 50$$
where x represents the number of microwave ovens produced.

(a) Find the daily cost of producing 40 microwave ovens.
(b) Find the marginal cost function.
(c) Find $C'(40)$ and interpret its meaning.
(d) Use the marginal cost to estimate the cost of producing 41 microwave ovens.
(e) Find the actual cost of producing 41 microwave ovens. Compare the actual cost of making 41 microwave ovens to the estimated cost of producing 41 microwave ovens.
(f) Find the actual cost of producing the 41st microwave oven.
(g) The average cost function is defined as $\overline{C}(x) = \dfrac{C(x)}{x}$, $0 < x \le 50$. Find the average cost function for producing x microwave ovens.
(h) Find the average cost of producing 41 microwave ovens.
(i) Compare your answers from parts (c) and (f). Give explanations for the differences.

66. Analyzing a Cost Function The daily cost C of producing small televisions is
$$C(x) = 1500 + 25x - 0.05x^2, \quad 0 \le x \le 100$$
where x represents the number of televisions produced.

(a) Find the daily cost of producing 70 televisions.
(b) Find the marginal cost function.
(c) Find $C'(70)$ and interpret its meaning.
(d) Use the marginal cost to estimate the cost of producing 71 televisions.
(e) Find the actual cost of producing 71 televisions. Compare the actual cost of making 71 televisions to the estimated cost of producing 71 televisions.
(f) Find the actual cost of manufacturing the 71st television.
(g) The average cost function is defined as $\overline{C}(x) = \dfrac{C(x)}{x}$, $0 < x \le 100$. Find the average cost function for producing x television.
(h) Find the average cost of producing 71 televisions.
(i) Compare your answers from parts (c) and (f). Give explanations for the differences.

67. Price of Beans The price in dollars per cwt for beans from 1993 through 2002 can be modeled by the polynomial function $p(t) = 0.007t^3 - 0.63t^2 + 0.005t + 6.123$, where t is in years, and $t = 0$ corresponds to 1993.

(a) Find the marginal price of beans for the year 1995.
(b) Find the marginal price for beans for the year 2002.
(c) How do you interpret the two marginal prices? What is the trend?

68. Price of Beans The price in dollars per cwt for beans from 1993 through 2002 can also be modeled by the polynomial function, $p(t) = -0.002t^4 + 0.044t^3 - 0.335t^2 + 0.750t + 5.543$, where t is in years and $t = 0$ corresponds to 1993.

(a) Find the marginal price for beans for the year 1995.
(b) Find the marginal price for beans for the year 2002.
(c) How do you interpret the two marginal prices? What is the trend?
(d) Explain why there might be two different functions that can model the price of beans.

69. Instantaneous Rate of Change The volume V of a sphere of radius r feet is $V = V(r) = \frac{4}{3}\pi r^3$. Find the instantaneous rate of change of the volume with respect to the radius r at $r = 2$ feet.

70. Instantaneous Rate of Change The volume V of a cube of side x meters is $V = V(x) = x^3$. Find the instantaneous rate of change of the volume with respect to the side x at $x = 3$ meters.

71. Work Output The relationship between the amount $A(t)$ of work output and the elapsed time t, $t \ge 0$, was found through empirical means to be
$$A(t) = a_3 t^3 + a_2 t^2 + a_1 t + a_0$$
where a_0, a_1, a_2, a_3 are constants. Find the instantaneous rate of change of work output at time t.

72. Consumer Price Index The consumer price index (CPI) of an economy is described by the function
$$I(t) = -0.2t^2 + 3t + 200 \qquad 0 \le t \le 10$$
where $t = 0$ corresponds to the year 2000.

(a) What was the average rate of increase in the CPI over the period from 2000 to 2003?
(b) At what rate was the CPI of the economy changing in 2003? in 2006?

73. Use the binomial theorem to prove Formula (3), p. 288.

[Hint: $(x + h)^n - x^n = x^n + nx^{n-1}h + \dfrac{n(n-1)}{2}x^{n-2}h^2 + \cdots + h^n - x^n = nx^{n-1}h + h^2 \cdot$ (terms involving x and h). Now apply Formula (1), page 286.]

74. Use the following factoring rule to prove Formula (3), p. 288.
$$f(x) = x^n - c^n$$
$$= (x - c)(x^{n-1} + x^{n-2}c + x^{n-3}c^2 + \cdots + c^{n-1})$$

Now apply Formula (3), page 276, to find $f'(c)$.

4.3 Product and Quotient Formulas

OBJECTIVES
1. Find the derivative of a product
2. Find the derivative of a quotient
3. Find the derivative of $f(x) = x^n$, n is a negative integer

The Derivative of a Product

In the previous section we learned that the derivative of the sum or the difference of two functions is simply the sum or the difference of their derivatives. The natural inclination at this point may be to assume that differentiating a product or quotient of two functions is as simple. But this is not the case, as illustrated for the case of a product of two functions. Consider

$$F(x) = f(x) \cdot g(x) = (3x^2 - 3)(2x^3 - x) \tag{1}$$

where $f(x) = 3x^2 - 3$, and $g(x) = 2x^3 - x$. The derivative of $f(x)$ is $f'(x) = 6x$ and the derivative of $g(x)$ is $g'(x) = 6x^2 - 1$. The product of these derivatives is

$$f'(x) \cdot g'(x) = 6x(6x^2 - 1) = 36x^3 - 6x \tag{2}$$

To see if this is equal to the derivative of the product, we first multiply out the right side of equation (1) and then differentiate. Then

$$F(x) = (3x^2 - 3)(2x^3 - x) = 6x^5 - 9x^3 + 3x$$

so that

$$F'(x) = 30x^4 - 27x^2 + 3 \tag{3}$$

Since equations (2) and (3) are not equal, we conclude that the derivative of a product *is not* equal to the product of the derivatives.

The formula for finding the derivative of the product of two functions is given below:

Derivative of a Product

The derivative of the product of two differentiable functions equals the first function times the derivative of the second plus the second function times the derivative of the first. That is,

$$\boxed{\frac{d}{dx}[f(x)g(x)] = f(x)\frac{d}{dx}g(x) + g(x)\frac{d}{dx}f(x)} \tag{4}$$

The following version of Formula (4) may help you remember it.

$$\boxed{\frac{d}{dx}(\text{first} \cdot \text{second}) = \text{first} \cdot \frac{d}{dx}\text{second} + \text{second} \cdot \frac{d}{dx}\text{first}}$$

296 Chapter 4 The Derivative of a Function

EXAMPLE 1 Finding the Derivative of a Product

Find the derivative of: $F(x) = (x^2 + 2x - 5)(x^3 - 1)$

SOLUTION The function F is the product of the two functions $f(x) = x^2 + 2x - 5$ and $g(x) = x^3 - 1$ so that, by Formula (4), we have

$$F'(x) = (x^2 + 2x - 5)\left[\frac{d}{dx}(x^3 - 1)\right] + (x^3 - 1)\left[\frac{d}{dx}(x^2 + 2x - 5)\right] \quad \text{Use Formula (4).}$$

$$= (x^2 + 2x - 5)(3x^2) + (x^3 - 1)(2x + 2) \quad \text{Differentiate.}$$

$$= 3x^4 + 6x^3 - 15x^2 + 2x^4 + 2x^3 - 2x - 2 \quad \text{Simplify.}$$

$$= 5x^4 + 8x^3 - 15x^2 - 2x - 2 \quad \text{Simplify.}$$

Now that you know the formula for the derivative of a product, be careful not to use it unnecessarily. When one of the factors is a constant, you should use the formula for the derivative of a constant times a function. For example, it is easier to work

$$\frac{d}{dx}[5(x^2 + 1)] = 5\frac{d}{dx}(x^2 + 1) = (5)(2x) = 10x$$

than it is to work

$$\frac{d}{dx}[5(x^2 + 1)] = 5\left[\frac{d}{dx}(x^2 + 1)\right] + (x^2 + 1)\left(\frac{d}{dx}5\right)$$

$$= (5)(2x) + (x^2 + 1)(0) = 10x$$

NOW WORK PROBLEM 1.

The Derivative of a Quotient

As in the case of a product, the derivative of a quotient is *not* the quotient of the derivatives.

Derivative of a Quotient

The derivative of the quotient of two differentiable functions is equal to the denominator times the derivative of the numerator minus the numerator times the derivative of the denominator, all divided by the square of the denominator.

$$\frac{d}{dx}\left[\frac{f(x)}{g(x)}\right] = \frac{g(x)\frac{d}{dx}f(x) - f(x)\frac{d}{dx}g(x)}{[g(x)]^2} \quad \text{where } g(x) \neq 0 \quad (5)$$

You may want to memorize the following version of Formula (5):

Product and Quotient Formulas   297

$$\frac{d}{dx}\frac{\text{numerator}}{\text{denominator}} = \frac{(\text{denominator})\frac{d}{dx}(\text{numerator}) - (\text{numerator})\frac{d}{dx}(\text{denominator})}{(\text{denominator})^2}$$

EXAMPLE 2   Finding the Derivative of a Quotient

Find the derivative of: $F(x) = \dfrac{x^2 + 1}{x - 3}$

SOLUTION Here, the function F is the quotient of $f(x) = x^2 + 1$ and $g(x) = x - 3$. We use Formula (5) to get

$$\frac{d}{dx}\left(\frac{x^2 + 1}{x - 3}\right) = \frac{(x - 3)\frac{d}{dx}(x^2 + 1) - (x^2 + 1)\frac{d}{dx}(x - 3)}{(x - 3)^2} \qquad \text{Use Formula (5).}$$

$$= \frac{(x - 3)(2x) - (x^2 + 1)(1)}{(x - 3)^2} \qquad \text{Differentiate.}$$

$$= \frac{2x^2 - 6x - x^2 - 1}{(x - 3)^2} \qquad \text{Simplify.}$$

$$= \frac{x^2 - 6x - 1}{(x - 3)^2} \qquad \text{Simplify.}$$

NOW WORK PROBLEM 9.

We shall follow the practice of leaving our answers in factored form as shown in Example 2.

EXAMPLE 3   Finding the Derivative of a Quotient

Find the derivative of: $y = \dfrac{(1 - 3x)(2x + 1)}{3x - 2}$

SOLUTION We shall solve the problem in two ways.

Method 1 Use the formula for the derivative of a quotient right away.

$$y' = \frac{d}{dx}\frac{(1 - 3x)(2x + 1)}{3x - 2}$$

$$= \frac{(3x - 2)\frac{d}{dx}[(1 - 3x)(2x + 1)] - (1 - 3x)(2x + 1)\frac{d}{dx}(3x - 2)}{(3x - 2)^2} \qquad \text{Use Formula (5).}$$

$$= \frac{(3x - 2)\left[(1 - 3x)\frac{d}{dx}(2x + 1) + (2x + 1)\frac{d}{dx}(1 - 3x)\right] - (1 - 3x)(2x + 1)\cdot 3}{(3x - 2)^2} \qquad \text{Use Formula (4); Differentiate.}$$

$$= \frac{(3x-2)[(1-3x)(2)+(2x+1)(-3)]-(-6x^2-x+1)(3)}{(3x-2)^2} \quad \text{Differentiate, Simplify.}$$

$$= \frac{(3x-2)[2-6x-6x-3]-(-18x^2-3x+3)}{(3x-2)^2} \quad \text{Simplify.}$$

$$= \frac{(3x-2)(-12x-1)-(-18x^2-3x+3)}{(3x-2)^2} \quad \text{Simplify.}$$

$$= \frac{-36x^2+21x+2+18x^2+3x-3}{(3x-2)^2} \quad \text{Simplify.}$$

$$= \frac{-18x^2+24x-1}{(3x-2)^2} \quad \text{Simplify.}$$

Method 2 First, multiply the factors in the numerator and then apply the formula for the derivative of a quotient.

$$y = \frac{(1-3x)(2x+1)}{3x-2} = \frac{-6x^2-x+1}{3x-2}$$

Now use Formula (5):

$$y' = \frac{d}{dx}\frac{-6x^2-x+1}{3x-2}$$

$$= \frac{(3x-2)\frac{d}{dx}(-6x^2-x+1)-(-6x^2-x+1)\frac{d}{dx}(3x-2)}{(3x-2)^2} \quad \text{Formula (5).}$$

$$= \frac{(3x-2)(-12x-1)-(-6x^2-x+1)(3)}{(3x-2)^2} \quad \text{Differentiate.}$$

$$= \frac{-36x^2+21x+2+18x^2+3x-3}{(3x-2)^2} \quad \text{Simplify.}$$

$$= \frac{-18x^2+24x-1}{(3x-2)^2} \quad \text{Simplify.} \quad \blacktriangleright$$

As you can see from this example, looking at alternative methods may make the differentiation easier. Which method did you find easier?

The Derivative of $f(x) = x^n$, n a Negative Integer

Find the derivative of $f(x) = x^n$, n a negative integer   **3**   In the previous section, we learned that the derivative of a power function $f(x) = x^n$, $n \geq 1$ an integer, is $f'(x) = nx^{n-1}$.

The formula for the derivative of x raised to a negative integer exponent follows the same form.

The derivative of $f(x) = x^n$, where n is *any* integer, is n times x to the $n-1$ power. That is,

$$\frac{d}{dx}x^n = nx^{n-1} \quad \text{for any integer } n \qquad (6)$$

The proof is left as an exercise. See Problem 54.

EXAMPLE 4 **Using Formula (6)**

(a) $\dfrac{d}{dx} x^{-3} = -3x^{-4} = \dfrac{-3}{x^4}$

(b) $\dfrac{d}{dx} \dfrac{4}{x^2} = \dfrac{d}{dx}(4x^{-2}) = 4\dfrac{d}{dx} x^{-2} = 4(-2x^{-3}) = \dfrac{-8}{x^3}$

(c) $\dfrac{d}{dx}\left(x + \dfrac{2}{x}\right) = \dfrac{d}{dx}(x + 2x^{-1}) = \dfrac{d}{dx} x + 2\dfrac{d}{dx} x^{-1} = 1 + 2(-1)x^{-2} = 1 - \dfrac{2}{x^2}$

NOW WORK PROBLEM 17.

EXAMPLE 5 **Finding the Derivative of a Function**

Find the derivative of: $g(x) = \left(1 - \dfrac{1}{x^2}\right)(x + 1)$

SOLUTION Since $g(x)$ is the product of two simpler functions, we begin by applying the formula for the derivative of a product:

$g'(x) = \left(1 - \dfrac{1}{x^2}\right)\dfrac{d}{dx}(x + 1) + (x + 1)\dfrac{d}{dx}\left(1 - \dfrac{1}{x^2}\right)$ Derivative of a product.

$= \left(1 - \dfrac{1}{x^2}\right)(1) + (x + 1)\dfrac{d}{dx}(1 - x^{-2})$ Differentiate; $\dfrac{1}{x^2} = x^{-2}$.

$= 1 - \dfrac{1}{x^2} + (x + 1)(2x^{-3})$ Differentiate.

$= 1 - \dfrac{1}{x^2} + \dfrac{2(x + 1)}{x^3}$ Simplify.

$= 1 - \dfrac{1}{x^2} + \dfrac{2x}{x^3} + \dfrac{2}{x^3}$ Simplify.

$= 1 + \dfrac{1}{x^2} + \dfrac{2}{x^3}$ Simplify.

Alternatively, we could have solved Example 5 by multiplying the factors first. Then

$$g(x) = \left(1 - \dfrac{1}{x^2}\right)(x + 1) = x + 1 - \dfrac{1}{x} - \dfrac{1}{x^2}$$

so

$g'(x) = \dfrac{d}{dx}\left(x + 1 - \dfrac{1}{x} - \dfrac{1}{x^2}\right) = \dfrac{d}{dx} x + \dfrac{d}{dx} 1 - \dfrac{d}{dx}\dfrac{1}{x} - \dfrac{d}{dx}\dfrac{1}{x^2}$

$= 1 + 0 - \dfrac{d}{dx} x^{-1} - \dfrac{d}{dx} x^{-2} = 1 - (-1)x^{-2} - (-2)x^{-3} = 1 + \dfrac{1}{x^2} + \dfrac{2}{x^3}$

EXAMPLE 4 **Application**

The value $V(t)$, in dollars, of a car t years after its purchase is given by the equation

$$V(t) = \dfrac{8000}{t} + 5000 \qquad 1 \le t \le 5$$

Graph the function $V = V(t)$.
Then find:

(a) The average rate of change in value from $t = 1$ to $t = 4$.
(b) The instantaneous rate of change in value.
(c) The instantaneous rate of change in value after 1 year.
(d) The instantaneous rate of change in value after 3 years.
(e) Interpret the answers to (c) and (d).

SOLUTION The graph of $V = V(t)$ is given in Figure 9.

FIGURE 9

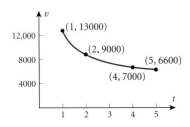

(a) The average rate of change in value from $t = 1$ to $t = 4$ is given by

$$\frac{V(4) - V(1)}{4 - 1} = \frac{7000 - 13{,}000}{3} = -2000$$

So the average rate of change in value from $t = 1$ to $t = 4$ is $-\$2000$ per year. That is, the value of the car is decreasing at the rate of $\$2000$ per year.

(b) The derivative $V'(t)$ of $V(t)$ equals the instantaneous rate of change in the value of the car.

$$V'(t) = \frac{d}{dt}\left(\frac{8000}{t} + 5000\right) = \frac{d}{dt}\frac{8000}{t} + \frac{d}{dt}(5000)$$

$$= \frac{d}{dt}8000t^{-1} + 0 = 8000(-1)t^{-2} = -\frac{8000}{t^2}$$

Notice that $V'(t) < 0$; we interpret this to mean that the value of the car is decreasing over time.

(c) After 1 year, $V'(1) = -\dfrac{8000}{1} = -\$8000/\text{year}$

(d) After 3 years, $V'(3) = -\dfrac{8000}{9} = -\$888.89/\text{year}$

(e) $V'(1) = -\$8000$ means that the value of the car after 1 year will decline by approximately $\$8000$ over the next year; $V'(3) = -\$888.89$ means that the value of the car after 3 years will decline by approximately $\$888.89$ over the next year. ▸

  NOW WORK PROBLEM 41.

SUMMARY Each of the derivative formulas given so far can be written without reference to the independent variable of the function. If f and g are differentiable functions, we have the following formulas:

Derivative of a constant times a function	$(cf)' = cf'$
Derivative of a sum	$(f + g)' = f' + g'$
Derivative of a difference	$(f - g)' = f' - g'$
Derivative of a product	$(f \cdot g)' = f \cdot g' + g \cdot f'$
Derivative of a quotient	$\left(\dfrac{f}{g}\right)' = \dfrac{g \cdot f' - f \cdot g'}{g^2}$

EXERCISE 4.3

In Problems 1–8, find the derivative of each function by using the formula for the derivative of a product.

1. $f(x) = (2x + 1)(4x - 3)$
2. $f(x) = (3x - 4)(2x + 5)$
3. $f(t) = (t^2 + 1)(t^2 - 4)$
4. $f(t) = (t^2 - 3)(t^2 + 4)$
5. $f(x) = (3x - 5)(2x^2 + 1)$
6. $f(x) = (3x^2 - 1)(4x + 1)$
7. $f(x) = (x^5 + 1)(3x^3 + 8)$
8. $f(x) = (x^6 - 2)(4x^2 + 1)$

In Problems 9–20, find the derivative of each function.

9. $f(x) = \dfrac{x}{x + 1}$
10. $f(x) = \dfrac{x + 4}{x^2}$
11. $f(x) = \dfrac{3x + 4}{2x - 1}$
12. $f(x) = \dfrac{3x - 5}{4x + 1}$
13. $f(x) = \dfrac{x^2}{x - 4}$
14. $f(x) = \dfrac{x}{x^2 - 4}$
15. $f(x) = \dfrac{2x + 1}{3x^2 + 4}$
16. $f(x) = \dfrac{2x^2 - 1}{5x + 2}$
17. $f(t) = \dfrac{-2}{t^2}$
18. $f(t) = \dfrac{4}{t^3}$
19. $f(x) = 1 + \dfrac{1}{x} + \dfrac{1}{x^2}$
20. $f(x) = 1 - \dfrac{1}{x} + \dfrac{1}{x^2}$

In Problems 21–24, find the slope of the tangent line to the graph of the function f at the indicated point. What is an equation of the tangent line?

21. $f(x) = (x^3 - 2x + 2)(x + 1)$ at $(1, 2)$
22. $f(x) = (2x^2 - 5x + 1)(x - 3)$ at $(1, 4)$
23. $f(x) = \dfrac{x^3}{x + 1}$ at $\left(1, \dfrac{1}{2}\right)$
24. $f(x) = \dfrac{x^2}{x - 1}$ at $\left(-1, -\dfrac{1}{2}\right)$

In Problems 25–28, find those x, if any, at which $f'(x) = 0$.

25. $f(x) = (x^2 - 2)(2x - 1)$
26. $f(x) = (3x^2 - 3)(2x^3 - x)$
27. $f(x) = \dfrac{x^2}{x + 1}$
28. $f(x) = \dfrac{x^2 + 1}{x}$

In Problems 29–40, find y'.

29. $y = x^2(3x - 2)$
30. $y = (x^2 + 2)(x - 1)$
31. $y = (x^2 + 4)(4x^2 + 3)$
32. $y = (2x + 3)(x^3 + x^2)$
33. $y = \dfrac{2x + 3}{3x + 5}$
34. $y = \dfrac{3x - 2}{4x - 3}$
35. $y = \dfrac{x^2}{x^2 - 4}$
36. $y = \dfrac{x^3}{x - 1}$
37. $y = \dfrac{(3x + 4)(2x - 3)}{2x + 1}$
38. $y = \dfrac{(2 - 3x)(1 - x)}{x + 2}$
39. $y = \dfrac{4x^3}{x^2 + 4}$
40. $y = \dfrac{3x^4}{x^3 + 1}$

---

41. Value of a Car The value V of a car after t years is

$$V(t) = \dfrac{10{,}000}{t} + 6000 \quad 1 \le t \le 6$$

Graph $V = V(t)$.
(a) What is the average rate of change in value from $t = 2$ to $t = 5$?
(b) What is the instantaneous rate of change in value?
(c) What is the instantaneous rate of change after 2 years?
(d) What is the instantaneous rate of change after 5 years?
(e) Interpret the answers found in (c) and (d).

42. Value of a Painting The value V of a painting t years after it is purchased is

$$V(t) = \dfrac{100t^2 + 50}{t} + 400 \quad 1 \le t \le 5$$

(a) What is the average rate of change in value from $t = 1$ to $t = 3$?
(b) What is the instantaneous rate of change in value?
(c) What is the instantaneous rate of change after 1 year?
(d) What is the instantaneous rate of change after 3 years?
(e) Interpret the answers found in (c) and (d).

43. Demand Equation The demand equation for a certain commodity is

$$p = 10 + \frac{40}{x} \quad 1 \leq x \leq 10$$

where p is the price in dollars when x units are demanded. Find:

(a) The revenue function.
(b) The marginal revenue.
(c) The marginal revenue for $x = 4$.
(d) The marginal revenue for $x = 6$.

44. Cost Function The cost of fuel in operating a luxury yacht is given by the equation

$$C(s) = \frac{-3s^2 + 1200}{s}$$

where s is the speed of the yacht. Find the rate at which the cost is changing when $s = 10$.

45. Price–Demand Function The price–demand function for calculators is given by

$$D(p) = \frac{100{,}000}{p^2 + 10p + 50} \quad 5 \leq p \leq 20$$

where D is the quantity demanded and p is the unit price in dollars.

(a) Find $D'(p)$, the rate of change of demand with respect to price.
(b) Find $D'(5)$, $D'(10)$, and $D'(15)$.
(c) Interpret the results found in part (b).

46. Height of a Balloon The height, in kilometers, that a balloon will rise in t hours is given by the formula

$$s = s(t) = \frac{t^2}{2 + t}$$

Find the rate at which the balloon is rising after
(a) 10 minutes
(b) 20 minutes.

47. Population Growth A population of 1000 bacteria is introduced into a culture and grows in number according to the formula

$$P(t) = 1000\left(1 + \frac{4t}{100 + t^2}\right)$$

where t is measured in hours. Find the rate at which the population is growing when

(a) $t = 1$ (b) $t = 2$ (c) $t = 3$ (d) $t = 4$

48. Drug Concentration The concentration of a certain drug in a patient's bloodstream t hours after injection is given by

$$C(t) = \frac{0.4t}{2t^2 + 1}$$

Find the rate at which the concentration of the drug is changing with respect to time. At what rate is the concentration changing

(a) 10 minutes after the injection?
(b) 30 minutes after the injection?
(c) 1 hour after the injection?
(d) 3 hours after the injection?

49. Intensity of Illumination The intensity of illumination I on a surface is inversely proportional to the square of the distance r from the surface to the source of light. If the intensity is 1000 units when the distance is 1 meter, find the rate of change of the intensity with respect to the distance when the distance is 10 meters.

50. Cost Function The cost C, in thousands of dollars, for removal of pollution from a certain lake is

$$C(x) = \frac{5x}{110 - x}$$

where x is the percent of pollutant removed. Find:

(a) $C'(x)$, the rate of change of cost with respect to the amount of pollutant removed.
(b) Compute $C'(10)$, $C'(20)$, $C'(70)$, $C'(90)$.
(c) Interpret the answers found in part (b).

51. Cost Function An airplane crosses the Atlantic Ocean (3000 miles) with an airspeed of 500 miles per hour. The cost C (in dollars) per person is

$$C(x) = 100 + \frac{x}{10} + \frac{36{,}000}{x}$$

where x is the ground speed (airspeed $\pm$ wind). Find:

(a) The marginal cost.
(b) The marginal cost at a ground speed of 500 mph.
(c) The marginal cost at a ground speed of 550 mph.
(d) The marginal cost at a ground speed of 450 mph.

52. Average Cost Function If C is the total cost function then $\overline{C}(x) = \dfrac{C(x)}{x}$ is defined as the **average cost function**, that is, the cost per unit produced. Suppose a company estimates that the total cost of producing x units of a certain product is given by

$$C(x) = 400 + 0.02x + 0.0001x^2$$

Then the average cost is given by

$$\overline{C}(x) = \frac{C(x)}{x} = \frac{400}{x} + 0.02 + 0.0001x$$

(a) Find the marginal average cost $\overline{C}'(x)$.
(b) Find the marginal average cost at $x = 200, 300,$ and 400.
(c) Interpret your results.

53. Satisfaction and Reward The relationship between satisfaction S and total reward r has been found to be

$$S(r) = \frac{ar}{g-r}$$

where $g \geq 0$ is the predetermined goal level and $a > 0$ is the perceived justice per unit of reward.

(a) Show that the instantaneous rate of change of satisfaction with respect to reward is inversely proportional to the square of the difference between the personal goal of the individual and the amount of reward received.

(b) Interpret the equation obtained in part (a).

54. Prove Formula (6).

Hint: If $n < 0$, then $-n > 0$. Now use the fact that

$$\frac{d}{dx} x^n = \frac{d}{dx} \frac{1}{x^{-n}}$$

and use the quotient formula.

4.4 The Power Rule

OBJECTIVES
1. Find derivatives using the Power Rule
2. Find derivatives using the Power Rule and other derivative formulas

When a function is of the form $y = [g(x)]^n$, n an integer, the formula used to find the derivative y' is called the *Power Rule*. Let's see if we can guess this formula by finding the derivative of $y = [g(x)]^n$ when $n = 2$, $n = 3$, and $n = 4$.

If $n = 2$,

$$\frac{d}{dx}[g(x)]^2 = \underset{\underset{\text{Derivative of a product}}{\uparrow}}{\frac{d}{dx}[g(x)g(x)]} = g(x)\,g'(x) + g(x)g'(x) = 2g(x)g'(x)$$

If $n = 3$,

$$\frac{d}{dx}[g(x)]^3 = \frac{d}{dx}\{[g(x)]^2 g(x)\} = [g(x)]^2 g'(x) + g(x)\left\{\frac{d}{dx}[g(x)]^2\right\}$$
$$= [g(x)]^2 g'(x) + g(x)[2g(x)g'(x)] = 3[g(x)]^2 g'(x)$$

If $n = 4$,

$$\frac{d}{dx}[g(x)]^4 = \frac{d}{dx}\{[g(x)]^3 g(x)\} = [g(x)]^3 g'(x) + g(x)\left\{\frac{d}{dx}[g(x)]^3\right\}$$
$$= [g(x)]^3 g'(x) + g(x)\{3[g(x)]^2 g'(x)\} = 4[g(x)]^3 g'(x)$$

Let's summarize what we've found:

$$\frac{d}{dx}[g(x)]^2 = 2g(x)g'(x)$$

$$\frac{d}{dx}[g(x)]^3 = 3[g(x)]^2 g'(x)$$

$$\frac{d}{dx}[g(x)]^4 = 4[g(x)]^3 g'(x)$$

These results suggest the following formula:

> **The Power Rule**
>
> If g is a differentiable function and n is any integer, then
>
> $$\frac{d}{dx}[g(x)]^n = n[g(x)]^{n-1} g'(x) \tag{1}$$

Note the similarity between the Power Rule and the formula for the derivative of a power function:

$$\frac{d}{dx} x^n = n x^{n-1}$$

The main difference between these formulas is the factor $g'(x)$. Be sure to remember to include $g'(x)$ when using Formula (1).

EXAMPLE 1 Using the Power Rule to Find a Derivative

Find the derivative of the function: $f(x) = (x^2 + 1)^3$

SOLUTION We could, of course, expand the right-hand side and proceed according to techniques discussed earlier. However, the usefulness of the Power Rule is that it enables us to find derivatives of functions like this without resorting to tedious (and sometimes impossible) computation.

The function $f(x) = (x^2 + 1)^3$ is the function $g(x) = x^2 + 1$ raised to the power 3. Using the Power Rule,

$$\frac{d}{dx} f(x) = \frac{d}{dx}(x^2+1)^3 = 3(x^2+1)^2 \frac{d}{dx}(x^2+1)$$
$$\uparrow$$
Use the Power Rule: $g(x) = x^2 + 1$

$$= 3(x^2+1)^2 (2x) = 6x(x^2+1)^2$$

NOW WORK PROBLEM 1.

EXAMPLE 2 Using the Power Rule

Find the derivative $f'(x)$.

(a) $f(x) = \dfrac{1}{(x^3+4)^5}$ (b) $f(x) = \dfrac{1}{(x^2+4)^3}$

SOLUTION (a) We write $f(x)$ as $f(x) = (x^3+4)^{-5}$. Then we use the Power Rule:

$$f'(x) = \frac{d}{dx}(x^3+4)^{-5} = -5(x^3+4)^{-6} \frac{d}{dx}(x^3+4)$$
$$\uparrow$$
Use the Power Rule.

$$= -5(x^3+4)^{-6}(3x^2) = \frac{-15x^2}{(x^3+4)^6}$$

(b) $f'(x) = \dfrac{d}{dx} \dfrac{1}{(x^2+4)^3} = \dfrac{d}{dx}(x^2+4)^{-3} = -3(x^2+4)^{-4} \dfrac{d}{dx}(x^2+4)$

$\qquad\qquad\qquad\qquad\qquad\qquad\qquad\uparrow$
$\qquad\qquad\qquad\qquad\qquad\qquad$ Use the Power Rule

$\qquad\qquad = -3(x^2+4)^{-4} \cdot 2x = \dfrac{-6x}{(x^2+4)^4}$

Often, we must use at least one other derivative formula along with the Power Rule to differentiate a function. Here are two examples.

EXAMPLE 3 Using the Power Rule with Other Derivative Formulas

Find the derivative of the function: $f(x) = x(x^2+1)^3$

SOLUTION The function f is the product of x and $(x^2+1)^3$. We begin by using the formula for the derivative of a product. That is,

$f'(x) = x \dfrac{d}{dx}(x^2+1)^3 + (x^2+1)^3 \dfrac{d}{dx} x \qquad$ Formula for the derivative of a product.

We continue by using the Power Rule:

$f'(x) = x\left[3(x^2+1)^2 \dfrac{d}{dx}(x^2+1)\right] + (x^2+1)^3 \cdot 1 \qquad$ Power Rule; $\dfrac{d}{dx} x = 1$.

$\qquad = x[3(x^2+1)^2(2x)] + (x^2+1)^3 \qquad$ Differentiate.

$\qquad = (x^2+1)^2(6x^2) + (x^2+1)^2(x^2+1) \qquad$ Simplify.

$\qquad = (x^2+1)^2[6x^2 + (x^2+1)] \qquad$ Factor.

$\qquad = (x^2+1)^2(7x^2+1) \qquad$ Simplify.

NOW WORK PROBLEM 7.

EXAMPLE 4 Using the Power Rule with Other Derivative Formulas

Find the derivative of the function: $f(x) = \left(\dfrac{3x+2}{4x^2-5}\right)^5$

SOLUTION Here, f is the quotient $\dfrac{3x+2}{4x^2-5}$ raised to the power 5. We begin by using the Power Rule and then use the formula for the derivative of a quotient:

$f'(x) = 5\left(\dfrac{3x+2}{4x^2-5}\right)^4 \left[\dfrac{d}{dx}\left(\dfrac{3x+2}{4x^2-5}\right)\right] \qquad$ Power Rule.

$\qquad = 5\left(\dfrac{3x+2}{4x^2-5}\right)^4 \left[\dfrac{(4x^2-5)\dfrac{d}{dx}(3x+2) - (3x+2)\dfrac{d}{dx}(4x^2-5)}{(4x^2-5)^2}\right] \qquad$ Formula for the derivative of a quotient.

$\qquad = 5\left(\dfrac{3x+2}{4x^2-5}\right)^4 \left[\dfrac{(4x^2-5)(3) - (3x+2)(8x)}{(4x^2-5)^2}\right] \qquad$ Differentiate.

$\qquad = \dfrac{5(3x+2)^4(-12x^2-16x-15)}{(4x^2-5)^6} \qquad$ Simplify.

NOW WORK PROBLEM 19.

Application

The revenue $R = R(x)$ derived from selling x units of a product at a price p per unit is

$$R = xp$$

where $p = d(x)$ is the demand equation, namely, the equation that gives the price p when the number x of units demanded is known. The marginal revenue is then the derivative of R with respect to x:

$$R'(x) = \frac{d}{dx}(xp) = p + x\frac{dp}{dx} \qquad (2)$$

It is sometimes easier to find the marginal revenue by using Formula (2) instead of differentiating the revenue function directly.

EXAMPLE 5 Finding the Marginal Revenue

Suppose the price p in dollars per ton when x tons of polished aluminum are demanded is given by the equation

$$p = \frac{2000}{x + 20} - 10, \quad 0 < x < 60$$

Find:

(a) The rate of change of price with respect to x.
(b) The revenue function.
(c) The marginal revenue.
(d) The marginal revenue at $x = 20$ and $x = 40$.
(e) Interpret the answers found in part (d).

SOLUTION **(a)** The rate of change of price with respect to x is the derivative $\frac{dp}{dx}$.

$$\frac{dp}{dx} = \frac{d}{dx}\left(\frac{2000}{x+20} - 10\right) = \frac{d}{dx}2000(x+20)^{-1} - \frac{d}{dx}10$$

$$= \underset{\uparrow}{-2000(x+20)^{-2}}\frac{d}{dx}(x+20) - 0 = \frac{-2000}{(x+20)^2}$$

Power Rule

(b) The revenue function is

$$R(x) = xp = x\left[\frac{2000}{x+20} - 10\right] = \frac{2000x}{x+20} - 10x$$

(c) Using Formula (2), the marginal revenue is

$$R'(x) = p + x\frac{dp}{dx} \qquad \text{Formula (2).}$$

$$= \left[\frac{2000}{x+20} - 10\right] + x\left(\frac{-2000}{(x+20)^2}\right) \qquad \text{Use the result from (a).}$$

$$= \frac{2000}{x+20} - 10 - \frac{2000x}{(x+20)^2} \qquad \text{Simplify.}$$

(d) Using the result from part (c), we find

$$R'(20) = \frac{2000}{40} - 10 - \frac{2000(20)}{(40)^2} = \$15.00/\text{ton}$$

$$R'(40) = \frac{2000}{60} - 10 - \frac{2000(40)}{(60)^2} = \$1.11/\text{ton}$$

(e) $R'(20) = \$15.00/\text{ton}$ means that the next ton of aluminum sold will generate $15.00 in revenue.
$R'(40) = \$1.11/\text{ton}$ means that the next ton of aluminum sold will generate $1.11 in revenue.

NOW WORK PROBLEM 31.

EXERCISE 4.4

In Problems 1–28, find the derivative of each function using the Power Rule.

1. $f(x) = (2x - 3)^4$
2. $f(x) = (5x + 4)^3$
3. $f(x) = (x^2 + 4)^3$
4. $f(x) = (x^2 - 1)^4$
5. $f(x) = (3x^2 + 4)^2$
6. $f(x) = (9x^2 + 1)^2$
7. $f(x) = x(x + 1)^3$
8. $f(x) = x(x - 4)^2$
9. $f(x) = 4x^2(2x + 1)^4$
10. $f(x) = 3x^2(x^2 + 1)^3$
11. $f(x) = [x(x - 1)]^3$
12. $f(x) = [x(x + 4)]^4$
13. $f(x) = (3x - 1)^{-2}$
14. $f(x) = (2x + 3)^{-3}$
15. $f(x) = \dfrac{4}{x^2 + 4}$
16. $f(x) = \dfrac{3}{x^2 - 9}$
17. $f(x) = \dfrac{-4}{(x^2 - 9)^3}$
18. $f(x) = \dfrac{-2}{(x^2 + 2)^4}$
19. $f(x) = \left(\dfrac{x}{x + 1}\right)^3$
20. $f(x) = \left(\dfrac{x^2}{x + 5}\right)^4$
21. $f(x) = \dfrac{(2x + 1)^4}{3x^2}$
22. $f(x) = \dfrac{(3x + 4)^3}{9x}$
23. $f(x) = \dfrac{(x^2 + 1)^3}{x}$
24. $f(x) = \dfrac{(3x^2 + 4)^2}{2x}$
25. $f(x) = \left(x + \dfrac{1}{x}\right)^3$
26. $f(x) = \left(x - \dfrac{1}{x}\right)^4$
27. $f(x) = \dfrac{3x^2}{(x^2 + 1)^2}$
28. $g(x) = \dfrac{2x^3}{(x^2 - 4)^2}$

---

29. Car Depreciation A certain car depreciates according to the formula

$$V(t) = \frac{29{,}000}{1 + 0.4t + 0.1t^2}$$

where V is the value of the car in dollars at time t in years. Find the rate at which the car is depreciating:

(a) 1 year after purchase. (b) 2 years after purchase.
(c) 3 years after purchase. (d) 4 years after purchase.

30. Demand Function The demand function for a certain calculator is given by

$$d(x) = \frac{100}{0.02x^2 + 1} \qquad 0 \le x \le 20$$

where x (in thousands of units) is the quantity demanded per week and $d(x)$ is the unit price in dollars.

(a) Find $d'(x)$.
(b) Find $d'(10)$, $d'(15)$, and $d'(20)$ and interpret your results.
(c) Find the revenue function.
(d) Find the marginal revenue.

31. Demand Equation The price p in dollars per pound when x pounds of a certain commodity are demanded is

$$p = \frac{10{,}000}{5x + 100} - 5 \qquad 0 < x < 90$$

Find:

(a) The rate of change of price with respect to x.
(b) The revenue function.
(c) The marginal revenue.
(d) The marginal revenue at $x = 10$ and at $x = 40$.
(e) Interpret the answers to (d).

32. Revenue Function The weekly revenue R in dollars resulting from the sale of x DVD players is

$$R(x) = \frac{100x^5}{(x^2 + 1)^2} \qquad 0 \le x \le 100$$

Find:

(a) The marginal revenue.
(b) The marginal revenue at $x = 40$.
(c) The marginal revenue at $x = 60$.
(d) Interpret the answers to (b) and (c).

33. Amino Acids A protein disintegrates into amino acids according to the formula

$$M = \frac{28}{t+2}$$

where M, the mass of the protein, is measured in grams and t is time measured in hours.

(a) Find the average rate of change in mass from $t = 0$ to $t = 2$ hours.
(b) Find $M'(0)$.
(c) Interpret the answers to (a) and (b).

4.5 The Derivatives of the Exponential and Logarithmic Functions; the Chain Rule

PREPARING FOR THIS SECTION *Before getting started, review the following:*

> Exponential Functions (Chapter 2, Section 2.3, pp. 186–197)
> Change-of-Base Formula (Chapter 2, Section 2.5, p. 220)
> Logarithmic Functions (Chapter 2, Section 2.4, pp. 202–210)

OBJECTIVES
1. Find the derivative of functions involving e^x
2. Find a derivative using the Chain Rule
3. Find the derivative of functions involving $\ln x$
4. Find the derivative of functions involving $\log_a x$ and a^x

Up to now, our discussion of finding derivatives has been focused on polynomial functions (derivative of a sum or difference of power functions), rational functions (derivative of a quotient of two polynomials), and these functions raised to an integer power (the Power Rule). In this section we present formulas for finding the derivative of the exponential and logarithmic functions.

The Derivative of $f(x) = e^x$

We begin the discussion of the derivative of $f(x) = e^x$ by considering the function

$$f(x) = a^x \qquad a > 0, \qquad a \neq 1$$

To find the derivative of $f(x) = a^x$, we use the formula for finding the derivative of f at x using the difference quotient, namely:

$$f'(x) = \lim_{h \to 0} \frac{f(x+h) - f(x)}{h}$$

For $f(x) = a^x$, we have

$$f'(x) = \frac{d}{dx} a^x = \lim_{h \to 0} \frac{a^{x+h} - a^x}{h} = \lim_{h \to 0} \left[a^x \left(\frac{a^h - 1}{h} \right) \right] = a^x \lim_{h \to 0} \frac{a^h - 1}{h}$$

$\uparrow$
Factor out a^x

Suppose we seek $f'(0)$. Assuming the limit on the right exists and equals some number, it follows (since $a^0 = 1$) that the derivative of $f(x) = a^x$ at 0 is

$$f'(0) = \lim_{h \to 0} \frac{a^h - 1}{h}$$

This limit equals the slope of the tangent line to the graph of $f(x) = a^x$ at the point $(0, 1)$. The value of this limit depends upon the choice of a. Observe in Figure 10 that the slope of the tangent line to the graph of $f(x) = 2^x$ at $(0, 1)$ is less than 1, and that the slope of the tangent line to the graph of $f(x) = 3^x$ at $(0, 1)$ is greater than 1.

FIGURE 10

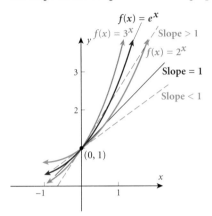

From this, we conclude there is a number a, $2 < a < 3$, for which the slope of the tangent line to the graph of $f(x) = a^x$ at $(0, 1)$ is exactly 1. The function $f(x) = a^x$ for which $f'(0) = 1$ is the function $f(x) = e^x$, whose base is the number e, that we introduced in Chapter 2. A further property of the number e is that

$$\lim_{h \to 0} \frac{e^h - 1}{h} = 1$$

Using this result, we find that

$$\frac{d}{dx} e^x = \lim_{h \to 0} \frac{e^{x+h} - e^x}{h} = \lim_{h \to 0} \frac{e^x(e^h - 1)}{h} = e^x \lim_{h \to 0} \frac{e^h - 1}{h} = e^x \cdot 1 = e^x$$

Derivative of $f(x) = e^x$

The derivative of the exponential function $f(x) = e^x$ is e^x. That is,

$$\boxed{\frac{d}{dx} e^x = e^x} \tag{1}$$

The simple nature of Formula (1) is one of the reasons the exponential function $f(x) = e^x$ appears so frequently in applications.

1 **EXAMPLE 1** **Finding the Derivative of Functions Involving e^x**

Find the derivative of each function:

(a) $f(x) = x^2 + e^x$ (b) $f(x) = xe^x$ (c) $f(x) = \dfrac{e^x}{x}$

SOLUTION (a) Use the formula for the derivative of a sum. Then

$$f'(x) = \frac{d}{dx}(x^2 + e^x) = \frac{d}{dx} x^2 + \frac{d}{dx} e^x = 2x + e^x$$

(b) Use the formula for the derivative of a product. Then

$$f'(x) = \frac{d}{dx}(xe^x) = x\frac{d}{dx}e^x + e^x\frac{d}{dx}x = xe^x + e^x \cdot 1 = e^x(x+1)$$

(c) Use the formula for the derivative of a quotient. Then

$$f'(x) = \frac{d}{dx}\frac{e^x}{x} = \frac{x\frac{d}{dx}e^x - e^x\frac{d}{dx}x}{x^2} = \frac{xe^x - e^x \cdot 1}{x^2} = \frac{(x-1)e^x}{x^2}$$

$\qquad\qquad\quad\uparrow\qquad\qquad\qquad\qquad\uparrow\qquad\qquad\qquad\uparrow$
$\quad$Derivative of a quotient$\qquad$Differentiate$\qquad\quad$Factor

  NOW WORK PROBLEM 3.

To find the derivative of other functions involving e^x and to find the derivative of the logarithmic function requires a formula called *the Chain Rule*.

The Chain Rule

The Power Rule is a special case of a more general, and more powerful formula, called the *Chain Rule*. This formula enables us to find the derivative of a *composite function*.

Consider the function $y = (2x + 3)^2$. If we write $y = f(u) = u^2$ and $u = g(x) = 2x + 3$, then, by a substitution process, we can obtain the original function, namely, $y = f(u) = f(g(x)) = (2x + 3)^2$. This process is called **composition** and the function $y = (2x + 3)^2$ is called the **composite function** of $y = f(u) = u^2$ and $u = g(x) = 2x + 3$.

EXAMPLE 2 **Finding a Composite Function**

Find the composite function of

$$y = f(u) = \sqrt{u} \quad \text{and} \quad u = g(x) = x^2 + 4$$

SOLUTION The composite function is

$$y = f(u) = \sqrt{u} = \sqrt{g(x)} = \sqrt{x^2 + 4}$$

The Chain Rule will require that we find the components of a composite function.

EXAMPLE 3 **Decomposing a Composite Function**

(a) If $y = (5x + 1)^3$, then $y = u^3$ and $u = 5x + 1$.
(b) If $y = (x^2 + 1)^{-2}$, then $y = u^{-2}$ and $u = x^2 + 1$.
(c) If $y = \dfrac{5}{(2x + 3)^3}$, then $y = \dfrac{5}{u^3}$ and $u = 2x + 3$.

In the above examples, the composite function was "broken up" into simpler functions. The Chain Rule provides a way to use these simpler functions to find the derivative of the composite function.

The Chain Rule

Suppose f and g are differentiable functions. If $y = f(u)$ and $u = g(x)$, then, after substitution, y is a function of x. The Chain Rule states that the derivative of y with respect to x is the derivative of y with respect to u times the derivative of u with respect to x. That is,

$$\frac{dy}{dx} = \frac{dy}{du} \cdot \frac{du}{dx} \qquad (2)$$

EXAMPLE 4 Finding a Derivative Using the Chain Rule

Use the Chain Rule to find the derivative of $y = (5x + 1)^3$

SOLUTION We break up y into simpler functions: If $y = (5x + 1)^3$, then $y = u^3$ and $u = 5x + 1$. To find $\frac{dy}{dx}$, we first find $\frac{dy}{du}$ and $\frac{du}{dx}$:

$$\frac{dy}{du} = \frac{d}{du} u^3 = 3u^2 \quad \text{and} \quad \frac{du}{dx} = \frac{d}{dx}(5x + 1) = 5$$

By the Chain Rule,

$$\frac{dy}{dx} = \frac{dy}{du} \cdot \frac{du}{dx} = 3u^2 \cdot 5 = 15u^2 \underset{\underset{u = 5x + 1}{\uparrow}}{=} 15(5x + 1)^2$$

Notice that when using the Chain Rule, we must substitute for u in the expression for $\frac{dy}{du}$ so that we obtain a function of x.

NOW WORK PROBLEM 9.

EXAMPLE 5 Finding a Derivative Using the Chain Rule

Find the derivative of $y = e^{x^2}$

SOLUTION We break up y into simpler functions. If $y = e^{x^2}$, then $y = e^u$ and $u = x^2$. Now use the Chain Rule to find $y' = \frac{dy}{dx}$. Since $\frac{dy}{du} = e^u$ and $\frac{du}{dx} = 2x$, we have

$$\frac{dy}{dx} = \frac{dy}{du} \cdot \frac{du}{dx} = e^u \cdot 2x \underset{\underset{u = x^2}{\uparrow}}{=} 2xe^{x^2}$$

The result of Example 5 can be generalized.

Derivative of $y = e^{g(x)}$

The derivative of a composite function $y = e^{g(x)}$, where g is a differentiable function, is

$$\frac{d}{dx} e^{g(x)} = e^{g(x)} \frac{d}{dx} g(x) = e^{g(x)} g'(x) \qquad (3)$$

The proof is left as an exercise. See Problem 76.

EXAMPLE 6 Finding the Derivative of Functions of the Form $e^{g(x)}$

Find the derivative of each function:

(a) $f(x) = 4e^{2x}$ (b) $f(x) = e^{x^2+1}$

SOLUTION (a) Use Formula (3) with $g(x) = 2x$. Then

$$f'(x) = \frac{d}{dx}(4e^{2x}) = 4\frac{d}{dx}e^{2x} \underset{\text{Formula (3)}}{=} 4 \cdot e^{2x}\frac{d}{dx}(2x) = 4e^{2x}(2) = 8e^{2x}$$

(b) Use Formula (3) with $g(x) = x^2 + 1$. Then

$$f'(x) = \frac{d}{dx}e^{x^2+1} \underset{\text{Formula (3)}}{=} e^{x^2+1}\frac{d}{dx}(x^2+1) = e^{x^2+1}(2x) = 2xe^{x^2+1}$$

  NOW WORK PROBLEM 23.

EXAMPLE 7 Finding the Derivative of Functions Involving e^x

Find the derivative of each function:

(a) $f(x) = xe^{x^2}$ (b) $f(x) = \dfrac{x}{e^x}$ (c) $f(x) = (e^x)^2$

SOLUTION (a) The function f is the product of two simpler functions, so we start with the formula for the derivative of a product.

$$f'(x) = \frac{d}{dx}(xe^{x^2}) \underset{\text{Derivative of a Product}}{=} x\frac{d}{dx}e^{x^2} + e^{x^2}\frac{d}{dx}x \underset{\text{Formula (3); }\frac{d}{dx}x=1}{=} x \cdot e^{x^2} \cdot \frac{d}{dx}x^2 + e^{x^2} \cdot 1 = xe^{x^2} \cdot 2x + e^{x^2} \underset{\text{Factor}}{=} e^{x^2}(2x^2+1)$$

(b) We could use the formula for the derivative of a quotient, but it is easier to rewrite f in the form $f(x) = xe^{-x}$ and use the formula for the derivative of a product.

$$f'(x) = \frac{d}{dx}(xe^{-x}) \underset{\text{Derivative of a Product}}{=} x\frac{d}{dx}e^{-x} + e^{-x}\frac{d}{dx}x \underset{\text{Formula (3)}}{=} x \cdot e^{-x}\frac{d}{dx}(-x) + e^{-x} \cdot 1 \underset{\frac{d}{dx}(-x)=-1}{=} xe^{-x}(-1) + e^{-x} \underset{\text{Factor}}{=} e^{-x}(1-x)$$

(c) Here the function f is e^x raised to the power 2. We first apply a Law of Exponents and write $f(x) = (e^x)^2 = e^{2x}$. Then we can use Formula (3).

$$f'(x) = \frac{d}{dx}e^{2x} = e^{2x}\frac{d}{dx}(2x) = e^{2x} \cdot 2 = 2e^{2x}$$

CAUTION: Notice the difference between e^{x^2} and $(e^x)^2$. In e^{x^2}, e is raised to the power x^2; in $(e^x)^2$, the parentheses tell us e^x is raised to the power 2.

  NOW WORK PROBLEM 29.

The Derivative of $f(x) = \ln x$

To find the derivative of $f(x) = \ln x$, we observe that if $y = \ln x$, then $e^y = x$. That is,

$$e^{\ln x} = x$$

If we differentiate both sides with respect to x, we obtain

$$\frac{d}{dx} e^{\ln x} = \frac{d}{dx} x$$

$$e^{\ln x} \frac{d}{dx} \ln x = 1 \qquad \text{Apply Formula (3) on the left.}$$

$$\frac{d}{dx} \ln x = \frac{1}{e^{\ln x}} \qquad \text{Solve for } \frac{d}{dx} \ln x.$$

$$\frac{d}{dx} \ln x = \frac{1}{x} \qquad e^{\ln x} = x.$$

We have proved the following formula:

Derivative of $f(x) = \ln x$

If $f(x) = \ln x$, then $f'(x) = \frac{1}{x}$. That is,

$$\boxed{\frac{d}{dx} \ln x = \frac{1}{x}} \qquad (4)$$

EXAMPLE 8   **Finding the Derivative of Functions Involving ln x**

Find the derivative of each function.

(a) $f(x) = x^2 + \ln x$ **(b)** $f(x) = x \ln x$

SOLUTION   **(a)** Use the formula for the derivative of a sum. Then

$$f'(x) = \frac{d}{dx}(x^2 + \ln x) = \frac{d}{dx} x^2 + \frac{d}{dx} \ln x = 2x + \frac{1}{x}$$

(b) Use the formula for the derivative of a product. Then

$$f'(x) = \frac{d}{dx}(x \ln x) = x \frac{d}{dx} \ln x + \ln x \frac{d}{dx} x$$

$$= x \left(\frac{1}{x} \right) + (\ln x)(1) = 1 + \ln x$$

NOW WORK PROBLEM 35.

To differentiate the natural logarithm of a function $g(x)$, namely, $\ln g(x)$, use the following formula.

Derivative of $\ln g(x)$

The formula for finding the derivative of the composite function $f(x) = \ln g(x)$, where g is a differentiable function, is

$$\boxed{\frac{d}{dx} \ln g(x) = \frac{\frac{d}{dx} g(x)}{g(x)} = \frac{g'(x)}{g(x)}} \qquad (5)$$

The proof uses the Chain Rule and is left as an exercise. See Problem 77.

EXAMPLE 9   Finding the Derivative of Functions Involving ln x

Find the derivative of each function.

(a) $f(x) = \ln(x^2 + 1)$ (b) $f(x) = (\ln x)^2$

SOLUTION   (a) The function $f(x) = \ln(x^2 + 1)$ is of the form $f(x) = \ln g(x)$. We use Formula (5) with $g(x) = x^2 + 1$. Then,

$$f'(x) = \frac{d}{dx} \ln(x^2+1) \underset{\text{Formula (5)}}{=} \frac{\frac{d}{dx}(x^2+1)}{x^2+1} = \frac{2x}{x^2+1}$$

(b) The function $f(x)$ is ln x raised to the power 2. We use the Power Rule. Then

$$f'(x) = \frac{d}{dx}(\ln x)^2 \underset{\text{Power Rule}}{=} 2\ln x\left(\frac{d}{dx}\ln x\right) = (2\ln x)\cdot\frac{1}{x} = \frac{2\ln x}{x}$$

   NOW WORK PROBLEM 45.

Find the derivative of functions involving $\log_a x$ and a^x

4   The Derivative of $f(x) = \log_a x$ and $f(x) = a^x$

To find the derivative of the logarithm function $f(x) = \log_a x$ for any base a, we use the Change-of-Base Formula. Then

$$f(x) = \log_a x = \frac{\log_e x}{\log_e a} = \frac{\ln x}{\ln a}$$

Since ln a is a constant, we have

$$f'(x) = \frac{d}{dx}\log_a x = \frac{d}{dx}\frac{\ln x}{\ln a} = \frac{1}{\ln a}\cdot\frac{d}{dx}\ln x = \frac{1}{\ln a}\cdot\frac{1}{x} = \frac{1}{x\ln a}$$

We have the formula

> **Derivative of $f(x) = \log_a x$**
>
> If $f(x) = \log_a x$, then $f'(x) = \dfrac{1}{x\ln a}$. That is,
>
> $$\frac{d}{dx}\log_a x = \frac{1}{x\ln a} \qquad (6)$$

EXAMPLE 10   Finding the Derivative of $\log_2 x$

Find the derivative of: $f(x) = \log_2 x$

SOLUTION   Using Formula (6), we have

$$f'(x) = \frac{d}{dx}\log_2 x \underset{\text{Formula (6)}}{=} \frac{1}{x\ln 2}$$

   NOW WORK PROBLEM 47.

To find the derivative of $f(x) = a^x$, where $a > 0$, $a \neq 1$, is any real constant, we use the definition of a logarithm and the Change-of-Base Formula. If $y = a^x$, we have

$$x = \log_a y \qquad \text{Definition of a logarithm.}$$

$$x = \frac{\ln y}{\ln a} \qquad \text{Apply the Change-of-Base Formula.}$$

$$x = \frac{\ln a^x}{\ln a} \qquad \text{Substitute } y = a^x.$$

Now, we differentiate both sides with respect to x:

$$\frac{d}{dx} x = \frac{d}{dx} \frac{\ln a^x}{\ln a}$$

$$1 = \frac{1}{\ln a} \cdot \frac{d}{dx} \ln a^x \qquad \text{ln } a \text{ is a constant.}$$

$$1 = \frac{1}{\ln a} \cdot \frac{\frac{d}{dx} a^x}{a^x} \qquad \text{Use Formula (5).}$$

$$1 = \frac{\frac{d}{dx} a^x}{a^x \ln a} \qquad \text{Simplify.}$$

$$\frac{d}{dx} a^x = a^x \ln a \qquad \text{Solve for } \frac{d}{dx} a^x.$$

We have derived the formula:

Derivative of $f(x) = a^x$

The derivative of $f(x) = a^x$, $a > 0$, $a \neq 1$, is $f'(x) = a^x \ln a$. That is,

$$\boxed{\frac{d}{dx} a^x = a^x \ln a} \qquad (7)$$

EXAMPLE 11 **Finding the Derivative of 2^x**

Find the derivative of: $f(x) = 2^x$

SOLUTION Using Formula (7), we have

$$f'(x) = \underset{\underset{\text{Formula (7)}}{\uparrow}}{\frac{d}{dx} 2^x} = 2^x \ln 2$$

NOW WORK PROBLEM 51.

EXAMPLE 12 **Maximizing Profit**

At a Notre Dame football weekend, the demand for game-day t-shirts is given by

$$p = 30 - 5 \ln\left(\frac{x}{100} + 1\right)$$

where p is the price of the shirt in dollars and x is the number of shirts demanded.

(a) At what price can 1000 t-shirts be sold?
(b) At what price can 5000 t-shirts be sold?
(c) Find the marginal demand for 1000 t-shirts and interpret the answer.
(d) Find the marginal demand for 5000 t-shirts and interpret the answer.
(e) Find the revenue function $R = R(x)$.
(f) Find the marginal revenue from selling 1000 t-shirts and interpret the answer.
(g) Find the marginal revenue from selling 5000 t-shirts and interpret the answer.
(h) If each t-shirt costs $4, find the profit function $P = P(x)$.
(i) What is the profit if 1000 t-shirts are sold?
(j) What is the profit if 5000 t-shirts are sold?
(k) Use the TABLE feature of a graphing utility to find the quantity x (to the nearest hundred) that maximizes profit.
(l) What price should be charged for a t-shirt to maximize profit?

SOLUTION (a) For $x = 1000$, the price p is

$$p = 30 - 5\ln\left(\frac{1000}{100} + 1\right) = \$18.01$$

(b) For $x = 5000$, the price p is

$$p = 30 - 5\ln\left(\frac{5000}{100} + 1\right) = \$10.34$$

(c) The marginal demand for x shirts is

$$p'(x) = \frac{dp}{dx} = \frac{d}{dx}\left[30 - 5\ln\left(\frac{x}{100} + 1\right)\right] = -5 \cdot \frac{\frac{1}{100}}{\frac{x}{100} + 1} = \frac{-5}{x + 100}$$

$$\text{Use Formula (5)} \qquad \text{Multiply by } \frac{100}{100}$$

For $x = 1000$,

$$p'(1000) = \frac{-5}{1000 + 100} = -\$0.0045$$

This means that another t-shirt will be demanded if the price is reduced by $0.0045.

(d) Use the result for $p'(x)$ found in part (c). Then for $x = 5000$, we have

$$p'(5000) = \frac{-5}{5000 + 100} = -\$0.00098$$

This means that another t-shirt will be demanded if the price is reduced by $0.00098.

(e) The revenue function $R = R(x)$ is

$$R = xp = x\left[30 - 5\ln\left(\frac{x}{100} + 1\right)\right]$$

(f) The marginal revenue is

$$R'(x) = \frac{d}{dx}[xp(x)] = xp'(x) + p(x) \qquad \text{Derivative of a product.}$$

$$= x \cdot \frac{-5}{x + 100} + 30 - 5\ln\left(\frac{x}{100} + 1\right) \qquad \text{Use the result of part (c).}$$

$$= \frac{-5x}{x + 100} + 30 - 5\ln\left(\frac{x}{100} + 1\right) \qquad \text{Simplify.}$$

If $x = 1000$,

$$R'(1000) = \frac{-5000}{1100} + 30 - 5\ln 11 = \$13.47$$

The revenue received for selling the 1001st t-shirt is $13.47.

(g) If $x = 5000$

$$R'(5000) = \frac{-25,000}{5100} + 30 - 5\ln 51 = \$5.44$$

The revenue received for selling the 5001st t-shirt is $5.44.

(h) The cost C for x t-shirts is $C = 4x$, so the profit function P is

$$P = P(x) = R(x) - C(x) = x\left[30 - 5\ln\left(\frac{x}{100} + 1\right)\right] - 4x$$

$$= 26x - 5x\ln\left(\frac{x}{100} + 1\right)$$

(i) If $x = 1000$, the profit is

$$P(1000) = 26(1000) - 5(1000)\ln\left(\frac{1000}{100} + 1\right) = \$14,010.52$$

(j) If $x = 5000$, the profit is

$$P(5000) = 26(5000) - 5(5000)\ln\left(\frac{5000}{100} + 1\right) = \$31,704.36$$

FIGURE 11

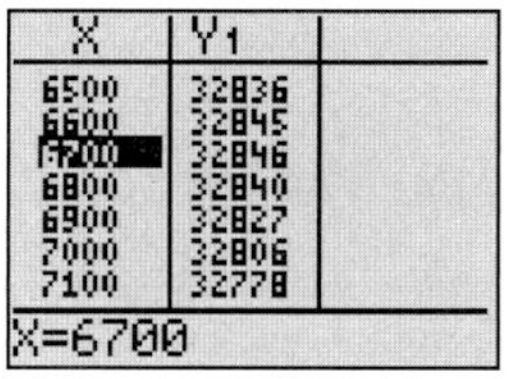

(k) See Figure 11. For $x = 6700$ t-shirts, the profit is largest. ($32,846).

(l) If $x = 6700$, the price p is

$$p(6700) = 30 - 5\ln\left(\frac{6700}{100} + 1\right) = \$8.90$$

SUMMARY

$$\frac{d}{dx}e^x = e^x \qquad \frac{d}{dx}e^{g(x)} = e^{g(x)}g'(x) \qquad \frac{d}{dx}a^x = a^x \ln a$$

$$\frac{d}{dx}\ln x = \frac{1}{x} \qquad \frac{d}{dx}\ln g(x) = \frac{g'(x)}{g(x)} \qquad \frac{d}{dx}\log_a x = \frac{1}{x \ln a}$$

EXERCISE 4.5

In Problems 1–8, find the derivative of each function.

1. $f(x) = x^3 - e^x$ **2.** $f(x) = 2e^x - x$ **3.** $f(x) = x^2 e^x$ **4.** $f(x) = x^3 e^x$

5. $f(x) = \dfrac{e^x}{x^2}$ **6.** $f(x) = \dfrac{5x}{e^x}$ **7.** $f(x) = \dfrac{4x^2}{e^x}$ **8.** $f(x) = \dfrac{3x^3}{e^x}$

In Problems 9–20, form the composite function $y = f(x)$. Then find $\dfrac{dy}{dx}$ using the Chain Rule.

9. $y = u^5$, $u = x^3 + 1$ **10.** $y = u^3$, $u = 2x + 5$ **11.** $y = \dfrac{u}{u+1}$, $u = x^2 + 1$

12. $y = \dfrac{u-1}{u}, \quad u = x^2 - 1$
13. $y = (u+1)^2, \quad u = \dfrac{1}{x}$
14. $y = (u^2 - 1)^3, \quad u = \dfrac{1}{x+2}$
15. $y = (u^3 - 1)^5, \quad u = x^{-2}$
16. $y = (u^2 + 4)^4, \quad u = x^{-2}$
17. $y = u^3, \quad u = e^x$
18. $y = 4u^2, \quad u = e^x$
19. $y = e^u, \quad u = x^3$
20. $y = e^u, \quad u = \dfrac{1}{x}$

21. Find the derivative y' of $y = (x^3 + 1)^2$ by:
   (a) Using the Chain Rule.
   (b) Using the Power Rule.
   (c) Expanding and then differentiating.

22. Follow the directions in Problem 21 for the function $y = (x^2 - 2)^3$.

In Problems 23–54, find the derivative of each function.

23. $f(x) = e^{5x}$
24. $f(x) = e^{-3x}$
25. $f(x) = 8e^{-x^2}$
26. $f(x) = -e^{3x^2}$
27. $f(x) = x^2 e^{x^2}$
28. $f(x) = x^3 e^{x^2}$
29. $f(x) = 5(e^x)^3$
30. $f(x) = 4(e^x)^4$
31. $f(x) = \dfrac{x^2}{e^x}$
32. $f(x) = \dfrac{8x}{e^{-x}}$
33. $f(x) = \dfrac{(e^x)^2}{x}$
34. $f(x) = \dfrac{e^{-2x}}{x^2}$
35. $f(x) = x^2 - 3 \ln x$
36. $f(x) = 5 \ln x - 2x$
37. $f(x) = x^2 \ln x$
38. $f(x) = x^3 \ln x$
39. $f(x) = 3 \ln (5x)$
40. $f(x) = -2 \ln (3x)$
41. $f(x) = x \ln (x^2 + 1)$
42. $f(x) = x^2 \ln (x^2 + 1)$
43. $f(x) = x + 8 \ln (3x)$
44. $f(x) = 3 \ln (2x) - 5x$
45. $f(x) = 8(\ln x)^3$
46. $f(x) = 2(\ln x)^4$
47. $f(x) = \log_3 x$
48. $f(x) = x + \log_4 x$
49. $f(x) = x^2 \log_2 x$
50. $f(x) = x^3 \log_3 x$
51. $f(x) = 3^x$
52. $f(x) = x + 4^x$
53. $f(x) = x^2 \cdot 2^x$
54. $f(x) = x^3 \cdot 3^x$

In Problems 55–62, find an equation of the tangent line to the graph of each function at the given point.

55. $f(x) = e^{3x}$ at $(0, 1)$
56. $f(x) = e^{4x}$ at $(0, 1)$
57. $f(x) = \ln x$ at $(1, 0)$
58. $f(x) = \ln (3x)$ at $(1, 0)$
59. $f(x) = e^{3x-2}$ at $\left(\dfrac{2}{3}, 1\right)$
60. $f(x) = e^{-x}$ at $\left(1, \dfrac{1}{e}\right)$
61. $f(x) = x \ln x$ at $(1, 0)$
62. $f(x) = \ln x^2$ at $(1, 0)$

63. Find the equation of the tangent line to $y = e^x$ that is parallel to the line $y = x$.

64. Find the equation of the tangent line to $y = e^{3x}$ that is parallel to the line $y = -\dfrac{1}{2}x$.

65. **Weber–Fechner Law** When a certain drug is administered, the reaction R to the dose x is given by the **Weber–Fechner law**:
$$R = 5.5 \ln x + 10$$
   (a) Find the reaction rate for a dose of 5 units.
   (b) Find the reaction rate for a dose of 10 units.
   (c) Interpret the results of parts (a) and (b).

66. **Marginal Cost** The cost (in dollars) of producing x units (measured in thousands) of a certain product is found to be
$$C(x) = 20 + \ln(x+1)$$
Find the marginal cost.

67. **Atmospheric Pressure** The atmospheric pressure at a height of x meters above sea level is $P(x) = 10^4 e^{-0.00012x}$ kilograms per square meter. What is the rate of change of the pressure with respect to the height at $x = 500$ meters? At $x = 700$ meters?

68. **Revenue** Revenue sales analysis of a new toy by Toys Inc. indicates that the relationship between the unit price p and the monthly sales x of its new toy is given by the equation
$$p = 10e^{-0.04x}$$
Find
   (a) The revenue function $R = R(x)$.
   (b) The marginal revenue R when $x = 200$.

69. Market Penetration The function

$$A(t) = 102 - 90e^{-0.21t}$$

expresses the relationship between A, the percentage of the market penetrated by DVD players, and t, the time in years, where $t = 0$ corresponds to the year 2000.

(a) Find the rate of change of A with respect to time.
(b) Evaluate $A'(5)$ and interpret your result.
(c) Evaluate $A'(10)$ and interpret your result.
(d) Evaluate $A'(30)$ and interpret your result.

70. Sales Because of lack of promotion, the yearly sales S of a product decline according to the equation

$$S(t) = 3000e^{-0.80t}$$

where t is the time. Find

(a) The rate of change of sales with respect to time.
(b) The rate of change of sales at $t = 0.5$.
(c) The rate of change of sales at $t = 2$.
(d) Interpret the results of (b) and (c). Explain the difference.

71. Advertising The function

$$S(x) = 100{,}000 + 400{,}000 \ln x$$

expresses the relation between sales (in dollars) of a product and the advertising for the product, where x is in thousands of dollars. Find

(a) The rate of change of S with respect to x.
(b) $S'(10)$.
(c) $S'(20)$.
(d) Interpret $S'(10)$ and $S'(20)$. Explain the difference.

72. Depreciation of a Car A car depreciates according to the function

$$V(t) = 35{,}000\, e^{-0.25t}$$

where t is measured in years and V represents the value of the car in dollars.

(a) What is the value of the car after 1 year?
(b) What is the value of the car after 5 years?
(c) Find $V'(t)$, the rate of depreciation.
(d) Interpret the result. What do you think the sign of $V'(t)$ represents?
(e) What is the depreciation rate after 1 year?
(f) What is the depreciation rate after 5 years?
(g) Interpret your answers to parts (e) and (f).

73. Maximizing Profit At the Super Bowl, the demand for game-day t-shirts is given by

$$p = 50 - 4 \ln\left(\frac{x}{100} + 1\right)$$

where p is the price of the shirt in dollars and x is the number of shirts demanded.

(a) At what price can 1000 t-shirts be sold?
(b) At what price can 5000 t-shirts be sold?
(c) Find the marginal demand for 1000 t-shirts and interpret the answer.
(d) Find the marginal demand for 5000 t-shirts and interpret the answer.
(e) Find the revenue function $R = R(x)$.
(f) Find the marginal revenue from selling 1000 t-shirts and interpret the answer.
(g) Find the marginal revenue from selling 5000 t-shirts and interpret the answer.
(h) If each t-shirt costs $4, find the profit function $P = P(x)$.
(i) What is the profit if 1000 t-shirts are sold?
(j) What is the profit if 5000 t-shirts are sold?
(k) Use the TABLE feature of a graphing utility to find the quantity x that maximizes profit.
(l) What price should be charged for a t-shirt to maximize profit?

74. Mean Earnings The mean earnings of workers 18 years old and over are given in the table below.

Year	1975	1980	1985	1990	1995	2000
Mean Earnings	8,552	12,665	17,181	21,793	26,792	32,604

Source: U.S. Bureau of the Census, Current Population Survey.

The data can be modeled by the function

$$E(t) = 8550 + 280t \ln t$$

where t is the number of years since 1974. Find

(a) The rate of change of E with respect to t.
(b) The rate of change at $t = 21$ (year 1995).
(c) The rate of change at $t = 26$ (year 2000).
(d) The rate of change at $t = 31$ (year 2005).
(e) Compare the answer to parts (b), (c), and (d). Explain the differences.

75. Price of Tomatoes The price of one pound of tomatoes from 1998 through 2003 are given in the table:

Year	1998	1999	2000	2001	2002	2003
Price	$0.473	$0.489	$0.490	$0.500	$0.509	$0.526

Source: The Bureau of Labor Statistics.

The price of tomatoes can be modeled by the function
$$p(t) = 0.470 + 0.026 \ln t$$

where t is the number of years since 1997. Find

(a) The rate of change of p with respect to t.
(b) The rate of change at $t = 5$ (year 2002).
(c) The rate of change at $t = 10$ (year 2007).
(d) Interpret the answers to parts (b) and (c). Explain the difference.

76 Prove Formula (3).
Hint: Use the Chain Rule with $y = e^u$, $u = g(x)$.

77. Prove Formula (5).
Hint: Use the Chain Rule with $y = \ln u$, $u = g(x)$.

4.6 Higher-Order Derivatives

OBJECTIVES
1. Find the first derivative and the second derivative of a function
2. Solve applied problems involving velocity and acceleration

The derivative of a function $y = f(x)$ is also a function. For example, if
$$f(x) = 6x^3 - 3x^2 + 2x - 5$$
(a polynomial function of degree 3), then
$$f'(x) = 18x^2 - 6x + 2$$
(a polynomial function of degree 2).

The derivative of the function $f'(x)$ is called the **second derivative of f** and is denoted by $f''(x)$. For the function f above,
$$f''(x) = \frac{d}{dx} f'(x) = \frac{d}{dx} (18x^2 - 6x + 2) = 36x - 6$$

By continuing in this fashion, we can find the third derivative $f'''(x)$, the fourth derivative $f^{(4)}(x)$, and so on, provided that these derivatives exist.*

NOW WORK PROBLEM 3.

The first, second, and third derivatives of the function
$$f(x) = 3x^3 - 2x^2 + 5x - 6$$

*The symbols $f'(x), f''(x)$, and so on for higher-order derivatives have several parallel notations. If $y = f(x)$, we may write

$$y' = f'(x) = \frac{dy}{dx} = \frac{d}{dx} f(x)$$

$$y'' = f''(x) = \frac{d^2y}{dx^2} = \frac{d^2}{dx^2} f(x)$$

$$y''' = f'''(x) = \frac{d^3y}{dx^3} = \frac{d^3}{dx^3} f(x)$$

$$\vdots$$

$$y^{(n)} = f^{(n)}(x) = \frac{d^ny}{dx^n} = \frac{d^n}{dx^n} f(x)$$

Higher-Order Derivatives

are

$$f'(x) = 9x^2 - 4x + 5$$
$$f''(x) = \frac{d}{dx}f'(x) = 18x - 4$$
$$f'''(x) = \frac{d}{dx}f''(x) = 18$$

For this function f, observe that $f^{(4)}(x) = 0$ and that all derivatives of order 5 or more also equal 0.

The result obtained in this example can be generalized:

For a polynomial function f of degree n, we have

$$f(x) = a_n x^n + a_{n-1} x^{n-1} + \cdots + a_1 x + a_0, \quad a_n \neq 0$$
$$f'(x) = na_n x^{n-1} + (n-1)a_{n-1} x^{n-2} + \cdots + a_1$$

The first derivative of a polynomial function of degree n is a polynomial function of degree $n - 1$. By continuing the differentiation process, it follows that the nth-order derivative of f

$$f^{(n)}(x) = n(n-1)(n-2) \cdot \ldots \cdot (3)(2)(1)a_n = n!a_n$$

is a polynomial of degree 0, a constant, so all derivatives of order greater than n will equal 0.

NOW WORK PROBLEM 35.

In some applications it is important to find both the first and second derivatives of a function and to solve for those numbers x that make these derivatives equal 0.

1   EXAMPLE 1   Finding the First and Second Derivatives of a Function

For $f(x) = 4x^3 - 12x^2 + 2$, find those numbers x, if any, at which the derivative $f'(x) = 0$. For what numbers x will $f''(x) = 0$?

SOLUTION $f'(x) = 12x^2 - 24x = 12x(x - 2) = 0$ when $x = 0$ or $x = 2$
$f''(x) = 24x - 24 = 24(x - 1) = 0$ when $x = 1$ ▶

NOW WORK PROBLEM 27.

Velocity and Acceleration

Solve applied problems involving velocity and acceleration 2

We mentioned at the beginning of this chapter that Sir Isaac Newton discovered calculus by solving a physics problem involving falling objects. We take up the problem of analyzing falling objects next.

We begin with the definition of *average velocity*:

Average Velocity

The **average velocity** is the ratio of the change in distance to the change in time. If s denotes distance and t denotes time, we have

$$\text{Average velocity} = \frac{\text{total distance}}{\text{elapsed time}} = \frac{\Delta s}{\Delta t}$$

EXAMPLE 2 Finding the Average Velocity

Mr. Doody and his family left on a car trip Saturday morning at 5 A.M. and arrived at their destination at 11 A.M. When they began the trip, the car's odometer read 26,700 kilometers, and when they arrived it read 27,000 kilometers. What was the average velocity for the trip?

SOLUTION
$$\text{Average velocity} = \frac{\Delta s}{\Delta t} = \frac{\text{total distance}}{\text{elapsed time}}$$

The total distance is $27{,}000 - 26{,}700 = 300$ kilometers and the elapsed time is $11 - 5 = 6$ hours. The average velocity is

$$\text{Average velocity} = \frac{300}{6} = 50 \text{ kilometers per hour}$$

Sometimes Mr. Doody will be traveling faster and sometimes slower, but the *average* velocity is 50 kilometers per hour. ▶

Average velocity provides information about velocity over an interval of time, but provides little information about the velocity at a particular instant of time. To get such information we require the *instantaneous velocity*.

Instantaneous Velocity

The rate of change of distance with respect to time is called **(instantaneous) velocity**. If $s = s(t)$ is a function that describes the position s of a particle at time t, the velocity of the particle at time t is

$$v = \frac{ds}{dt} = s'(t)$$

EXAMPLE 3 Finding the Instantaneous Velocity

In physics it is shown that the height s of a ball thrown straight up with an initial speed of 80 feet per second (ft/sec) from a rooftop 96 feet high is

$$s = s(t) = -16t^2 + 80t + 96$$

where t is the elapsed time that the ball is in the air. The ball misses the rooftop on its way down and eventually strikes the ground. See Figure 12.

(a) When does the ball strike the ground? That is, how long is the ball in the air?
(b) At what time t will the ball pass the rooftop on its way down?
(c) What is the average velocity of the ball from $t = 0$ to $t = 2$?
(d) What is the instantaneous velocity of the ball at time t?
(e) What is the instantaneous velocity of the ball at $t = 2$?
(f) When is the instantaneous velocity of the ball equal to zero?
(g) What is the instantaneous velocity of the ball as it passes the rooftop on the way down?
(h) What is the instantaneous velocity of the ball when it strikes the ground?

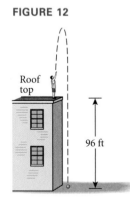

FIGURE 12

SOLUTION (a) The ball strikes the ground when $s = s(t) = 0$.

$$-16t^2 + 80t + 96 = 0 \qquad s(t) = 0$$
$$t^2 - 5t - 6 = 0 \qquad \text{Divide each side by } -16.$$
$$(t - 6)(t + 1) = 0 \qquad \text{Factor.}$$
$$t = 6 \quad \text{or} \quad t = -1 \qquad \text{Apply the Zero-Product Property; solve for } t.$$

We discard the solution $t = -1$. The ball strikes the ground after 6 seconds.

(b) The ball passes the rooftop when $s = s(t) = 96$.

$$-16t^2 + 80t + 96 = 96 \qquad s(t) = 96$$
$$t^2 - 5t = 0 \qquad \text{Simplify.}$$
$$t(t - 5) = 0 \qquad \text{Factor.}$$
$$t = 0 \quad \text{or} \quad t = 5 \qquad \text{Apply the Zero-Product Property; solve for } t.$$

We discard the solution $t = 0$. The ball passes the rooftop on the way down after 5 seconds.

(c) The average velocity of the ball from $t = 0$ to $t = 2$ is

$$\frac{\Delta s}{\Delta t} = \frac{s(2) - s(0)}{2 - 0} = \frac{192 - 96}{2} = 48 \text{ ft/sec}$$

(d) The instantaneous velocity of the ball at time t is the derivative $s'(t)$; that is,

$$s'(t) = \frac{d}{dt}(-16t^2 + 80t + 96) = -32t + 80 = -16(2t - 5) \text{ ft/sec}$$

(e) At $t = 2$ sec, the instantaneous velocity of the ball is

$$s'(2) = -16(4 - 5) = 16 \text{ ft/sec}$$

(f) The instantaneous velocity of the ball is zero when

$$s'(t) = 0$$
$$-16(2t - 5) = 0$$
$$t = \frac{5}{2} = 2.5 \text{ seconds}$$

(g) From part (b), the ball passes the rooftop on the way down when $t = 5$ seconds. The instantaneous velocity at $t = 5$ is

$$s'(5) = -16(10 - 5) = -80 \text{ ft/sec}$$

At $t = 5$ seconds, the ball is traveling -80 ft/sec. When the instantaneous velocity is negative, it means that the direction of the object is downward. The ball is traveling 80 ft/sec in the downward direction when $t = 5$ seconds.

(h) The ball strikes the ground when $t = 6$. The instantaneous velocity when $t = 6$ is

$$s'(6) = -16(12 - 5) = -112 \text{ ft/sec}$$

The velocity of the ball at $t = 6$ sec is -112 ft/sec. Again, the negative value implies that the ball is traveling downward. ▶

EXPLORATION: Determine the vertex of the quadratic function $s = s(t)$ given in Example 3. What do you conclude about instantaneous velocity when $s(t)$ is a maximum? ▶

Acceleration

The **acceleration** a of a particle is defined as the instantaneous rate of change of velocity with respect to time. That is,

$$a = \frac{dv}{dt} = \frac{d}{dt}v = \frac{d}{dt}\left(\frac{ds}{dt}\right) = \frac{d^2s}{dt^2} = s''(t)$$

In other words, acceleration is the second derivative of the function $s = s(t)$ with respect to time.

EXAMPLE 4   Analyzing the Motion of an Object

A ball is thrown vertically upward from ground level with an initial velocity of 19.6 meters per second. The distance s (in meters) of the ball above the ground is

$$s = -4.9t^2 + 19.6t$$

where t is the number of seconds elapsed from the moment the ball is thrown.

(a) What is the velocity of the ball at the end of 1 second?
(b) When will the ball reach its highest point?
(c) What is the maximum height the ball reaches?
(d) What is the acceleration of the ball at any time t?
(e) How long is the ball in the air?
(f) What is the velocity of the ball upon impact?
(g) What is the total distance traveled by the ball?

SOLUTION   (a) The velocity is

$$v = s'(t) = \frac{d}{dt}(-4.9t^2 + 19.6t) = -9.8t + 19.6$$

At $t = 1$, $v = s'(1) = 9.8$ meters per second.

(b) The ball will reach its highest point when it is stationary; that is, when $v = 0$.

$$v = -9.8t + 19.6 = 0 \quad \text{when } t = \frac{19.6}{9.8} = 2 \text{ seconds}$$

The ball reaches its maximum height 2 seconds after it is thrown.

(c) At $t = 2$, $s = s(2) = -4.9(4) + 19.6(2) = 19.6$ meters.

(d) $a = s''(t) = \dfrac{dv}{dt} = -9.8$ meters per second per second.

(e) We can answer this question in two ways. First, since the ball starts at ground level and it takes 2 seconds for the ball to reach its maximum height, it follows that it will take another 2 seconds to reach the ground, for a total time of 4 seconds in the air. The second way is to set $s = 0$ and solve for t:

$$-4.9t^2 + 19.6t = 0 \quad s(t) = 0$$
$$-4.9t(t - 4) = 0 \quad \text{Factor.}$$
$$t = 0 \quad \text{or} \quad t = 4 \quad \text{Apply the Zero-Product Property; solve for } t.$$

The ball is at ground level when $t = 0$ and when $t = 4$.

FIGURE 13

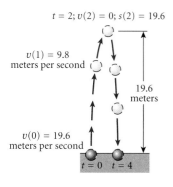

(f) Upon impact, $t = 4$. When $t = 4$,

$$v = s'(4) = (-9.8)(4) + 19.6 = -19.6 \text{ meters per second}$$

The minus sign here indicates that the direction of velocity is downward.

(g) The total distance traveled is

$$\text{Distance up} + \text{distance down} = 19.6 + 19.6 = 39.2 \text{ meters}$$

See Figure 13 for an illustration.

 NOW WORK PROBLEM 59.

In Example 4, the acceleration of the ball is constant, -9.8 meters/second/second. This is approximately true for all falling bodies provided air resistance is ignored. In fact, the constant is the same for all falling bodies, as Galileo (1564–1642) discovered in the sixteenth century. We can use calculus to see this. Galileo found by experimentation that all falling bodies obey the law that the distance they fall when they are dropped is proportional to the square of the time t it takes to fall that distance. Of importance is the fact that the constant of proportionality c is the same for all bodies. Thus, Galileo's law states that the distance s a body falls in time t is given by

$$s = -ct^2$$

The reason for the minus sign is that the body is falling and we have chosen our coordinate system so that the positive direction is up, along the vertical axis.

The velocity v of this freely falling body is

$$v = \frac{ds}{dt} = -2ct$$

and its acceleration a is

$$a = \frac{dv}{dt} = \frac{d^2s}{dt^2} = -2c$$

That is, the acceleration of a freely falling body is a constant. Usually, we denote this constant by $-g$ so that

$$a = -g$$

The number g is called the **acceleration of gravity.** For our planet, g may be approximated by 32 feet per second per second or 9.8 meters per second per second.* On the planet Jupiter, $g \approx 26$ meters per second per second, and on our moon, $g \approx 1.6$ meters per second per second.

EXAMPLE 5 **How High Can a Pitcher Throw a Ball?**

Safeco Field, home of the Seattle Mariners, has a 215-foot high retractable roof. Could a major league pitcher throw a ball up to the roof?

SOLUTION To answer this question, we make the following assumptions:

(i) We assume the pitcher can throw the ball with an initial velocity v_0 of around 95 miles per hour ≈ 140 feet per second.

---

* *The Earth is not perfectly round; it bulges slightly at the equator. But neither is it perfectly oval, and its mass is not distributed uniformly. As a result, the acceleration of any freely falling body varies slightly from these constants.*

(ii) The ball is thrown upward from an initial height of 6 feet (h_0) (average height of a pitcher).

The equation

$$s(t) = -16t^2 + v_0 t + h_0$$

is a formula describing the path of the ball, where $s(t)$ is the height (in feet) at time t, h_0 is the height of the object at time $t = 0$, and v_0 is the initial velocity of the ball ($t = 0$). Based on the assumptions, we have

$$s(t) = -16t^2 + 140t + 6 \qquad v_0 = 140; h_0 = 6$$

The velocity $v = v(t)$ of the ball at time t is

$$v(t) = s'(t) = -32t + 140$$

The ball is at its highest when $v(t) = 0$

$$-32t + 140 = 0. \qquad v(t) = 0$$

$$t = \frac{140}{32} = 4.375 \text{ seconds} \qquad \text{Solve for } t.$$

The height of the ball when $t = 4.375$ seconds is

$$s(4.375) = -16(4.375)^2 + 140(4.375) + 6 = 312.25 \text{ feet}.$$

So, a baseball can be thrown up to the roof of Safeco Field.

EXERCISE 4.6

In Problems 1–24, find the first derivative and the second derivative of each function.

1. $f(x) = 2x + 5$
2. $f(x) = 3x + 2$
3. $f(x) = 3x^2 + x - 2$
4. $f(x) = 5x^2 + 1$
5. $f(x) = -3x^4 + 2x^2$
6. $f(x) = -4x^3 + x^2 - 1$
7. $f(x) = \frac{1}{x}$
8. $f(x) = \frac{1}{x^2}$
9. $f(x) = x + \frac{1}{x}$
10. $f(x) = x - \frac{1}{x}$
11. $f(x) = \frac{x}{x+1}$
12. $f(x) = \frac{x+1}{x}$
13. $f(x) = e^x$
14. $f(x) = e^{-x}$
15. $f(x) = (x^2 + 4)^3$
16. $f(x) = (x^2 - 1)^4$
17. $f(x) = \ln x$
18. $f(x) = \ln (2x)$
19. $f(x) = xe^x$
20. $f(x) = x \ln x$
21. $f(x) = (e^x)^2$
22. $f(x) = (e^{-x})^2$
23. $f(x) = \frac{1}{\ln x}$
24. $f(x) = \frac{1}{e^{2x}}$

In Problems 25–32, for each function f, find:

(a) The domain of f.
(b) The derivative $f'(x)$.
(c) The domain of f'.
(d) Any numbers x for which $f'(x) = 0$.
(e) Any numbers x in the domain of f for which $f'(x)$ does not exist.
(f) The second derivative $f''(x)$.
(g) The domain of f''.

25. $f(x) = x^2 - 4$
26. $f(x) = x^2 + 2x$
27. $f(x) = x^3 - 9x^2 + 27x - 27$
28. $f(x) = x^3 - 6x^2 + 12x - 8$

29. $f(x) = 3x^4 - 12x^3 + 2$ **30.** $f(x) = x^4 - 4x + 2$ **31.** $f(x) = \dfrac{x}{x^2 - 4}$ **32.** $f(x) = \dfrac{x^2}{x^2 - 1}$

In Problems 33–38, find the indicated derivative.

33. $f^{(4)}(x)$ if $f(x) = x^3 - 3x^2 + 2x - 5$
34. $f^{(5)}(x)$ if $f(x) = 4x^3 + x^2 - 1$
35. $\dfrac{d^{20}}{dx^{20}}(8x^{19} - 2x^{14} + 2x^5)$
36. $\dfrac{d^{14}}{dx^{14}}(x^{13} - 2x^{10} + 5x^3 - 1)$
37. $\dfrac{d^8}{dx^8}\left(\dfrac{1}{8}x^8 - \dfrac{1}{7}x^7 + x^5 - x^3\right)$
38. $\dfrac{d^6}{dx^6}(x^6 + 5x^5 - 2x + 4)$

In Problems 39–42, find the velocity v and acceleration a of an object whose position s at time t is given.

39. $s = 16t^2 + 20t$ **40.** $s = 16t^2 + 10t + 1$ **41.** $s = 4.9t^2 + 4t + 4$ **42.** $s = 4.9t^2 + 5t$

In Problems 43–54, find a formula for the nth derivative of each function.

43. $f(x) = e^x$ **44.** $f(x) = e^{2x}$ **45.** $f(x) = \ln x$ **46.** $f(x) = \ln(2x)$
47. $f(x) = x \ln x$ **48.** $f(x) = (\ln x)^2$ **49.** $f(x) = (2x + 3)^n$ **50.** $f(x) = (4 - 3x)^n$
51. $f(x) = e^{ax}$ **52.** $f(x) = e^{-ax}$ **53.** $f(x) = \ln(ax)$ **54.** $f(x) = x^n \ln x$

---

55. If $y = e^{2x}$, find $y'' - 4y$

56. If $y = e^{-2x}$, find $y'' - 4y$

57. Find the second derivative of: $f(x) = x^2 g(x)$, where g' and g'' exist.

58. Find the second derivative of: $f(x) = \dfrac{g(x)}{x}$, where g' and g'' exist.

59. Falling Body A ball is thrown vertically upward with an initial velocity of 80 feet per second. The distance s (in feet) of the ball from the ground after t seconds is given by $s = s(t) = 6 + 80t - 16t^2$.
(a) What is the velocity of the ball after 2 seconds?
(b) When will the ball reach its highest point?
(c) What is the maximum height the ball reaches?
(d) What is the acceleration of the ball at any time t?
(e) How long is the ball in the air?
(f) What is the velocity of the ball upon impact?
(g) What is the total distance traveled by the ball?

60. Falling Body An object is propelled vertically upward with an initial velocity of 39.2 meters per second. The distance s (in meters) of the object from the ground after t seconds is $s = s(t) = -4.9t^2 + 39.2t$.

(a) What is the velocity of the object at any time t?
(b) When will the object reach its highest point?
(c) What is the maximum height?
(d) What is the acceleration of the object at any time t?
(e) How long is the object in the air?
(f) What is the velocity of the object upon impact?
(g) What is the total distance traveled by the object?

61. Ballistics A bullet is fired horizontally into a bale of paper. The distance s (in meters) the bullet travels in the bale of paper in t seconds is given by $s = 8 - (2 - t)^3$ for $0 \le t \le 2$. Find the velocity of the bullet after 1 second. Find the acceleration of the bullet at any time t.

62. Falling Rocks on Jupiter If a rock falls from a height of 20 meters on the planet Jupiter, then its height H after t seconds is approximately.

$$H(t) = 20 - 13t^2$$

(a) What is the average velocity of the rock from $t = 0$ to $t = 1$?
(b) What is the instantaneous velocity at time $t = 1$?
(c) What is the acceleration of the rock?
(d) When does the rock hit the ground?

63. Falling Body A rock is dropped from a height of 88.2 meters. In t seconds the rock falls $4.9t^2$ meters.

(a) How long does it take for the rock to hit the ground?
(b) What is the average velocity of the rock during the time it is falling?
(c) What is the average velocity of the rock for the first 3 seconds?
(d) What is the velocity of the rock when it hits the ground?

64. Falling Body A ball is thrown upward. The heights in feet of the ball is given by $s(t) = 100t - 16t^2$, where t is the time elapsed in seconds.

(a) What is the velocity of the ball when $t = 0$, $t = 1$, and $t = 4$ seconds?
(b) At what time does the ball strike the ground?
(c) At what time does the ball each its highest point?

4.7 Implicit Differentiation

PREPARING FOR THIS SECTION *Before getting started, review the following:*

>> Implicit Form of a Function (Chapter 1, Section 1.2, pp. 112–113)

OBJECTIVES 1 Find the derivative of a function defined implicitly

So far we have only discussed the derivative of a function that is given explicitly in the form $y = f(x)$. This expression of the relationship between x and y is said to be in *explicit form* because we have solved for the dependent variable y. For example, the equations

$$y = 7x - 2, \qquad s = -16t^2 + 10t + 100, \qquad v = 4h^2 - h$$

are all written in explicit form.

If the functional relationship between the independent variable x and the dependent variable y is not of this form, we say that x and y are related *implicitly*. For example, x and y are related implicitly in the expression

$$x^3 - y^4 - 3y + x = 6$$

In this equation, it is very difficult to find y as a function of x. How, then, do we go about finding the derivative $\dfrac{dy}{dx}$ in such a case?

The procedure for finding the derivative of y with respect to x when the functional relationship between x and y is given implicitly is called **implicit differentiation.** The procedure is to think of y as a function f of x, without actually expressing y in terms of x. If this requires differentiating terms like y^4, which we think of as $(f(x))^4$, then we use the Power Rule. The derivative of y^4 or $(f(x))^4$ is then

$$4(f(x))^3 f'(x) \qquad \text{or} \qquad 4y^3 \frac{dy}{dx}$$

Let's look at an example.

1 EXAMPLE 1 Differentiating Implicitly

Find $\dfrac{dy}{dx}$ if

$$3x + 4y - 5 = 0$$

SOLUTION We begin by assuming that there is a differentiable function $y = f(x)$ implied by the above relationship. Then,

$$3x + 4f(x) - 5 = 0$$

We differentiate both sides of the equality with respect to x. Then

$$\frac{d}{dx}[3x + 4f(x) - 5] = \frac{d}{dx} 0$$

$$\frac{d}{dx}(3x) + \frac{d}{dx}[4f(x)] - \frac{d}{dx} 5 = \frac{d}{dx} 0 \qquad \text{Derivative of a sum.}$$

$$3 + 4\frac{d}{dx}f(x) = 0 \qquad \text{Differentiate.}$$

Solving for $\frac{d}{dx} f(x)$, we find

$$\frac{d}{dx} f(x) = -\frac{3}{4}$$

Replacing $\frac{d}{dx} f(x)$ by $\frac{dy}{dx}$, we have

$$\frac{dy}{dx} = -\frac{3}{4}$$

For the function in Example 1, it is possible to solve for y as a function of x by algebraically solving for y.

$$3x + 4y - 5 = 0$$
$$4y = 5 - 3x$$
$$y = \frac{1}{4}(5 - 3x) = \frac{5}{4} - \frac{3}{4}x$$

Then

$$\frac{dy}{dx} = -\frac{3}{4}$$

which agrees with the result obtained using implicit differentiation. Often, though, it is very difficult, or even impossible, to actually solve for y in terms of x.

EXAMPLE 2 Differentiating Implicitly

Find $\frac{dy}{dx}$ if
$$3x^2 + 4y^2 = 2x$$

SOLUTION We again assume there is a differentiable function $y = f(x)$ implied by the above equation. We proceed to differentiate both sides of this equation with respect to x:

$$\frac{d}{dx}(3x^2 + 4y^2) = \frac{d}{dx}(2x)$$

$$\frac{d}{dx}(3x^2) + \frac{d}{dx}(4y^2) = 2$$

$$6x + 4\left[\frac{d}{dx} y^2\right] = 2$$

Using the Power Rule, $\frac{d}{dx} y^2 = 2y \frac{dy}{dx}$. Then

$$6x + 4\left(2y \frac{dy}{dx}\right) = 2$$

$$6x + 8y \frac{dy}{dx} = 2$$

This is a linear equation in $\frac{dy}{dx}$. Solving for $\frac{dy}{dx}$, we have

$$8y \frac{dy}{dx} = 2 - 6x$$

$$\frac{dy}{dx} = \frac{2 - 6x}{8y} = \frac{1 - 3x}{4y} \qquad \text{provided} \quad y \neq 0.$$

330   Chapter 4   The Derivative of a Function

> **Steps for Differentiating Implicitly**
>
> **STEP 1:** To find $\dfrac{dy}{dx}$ when x and y are related implicitly, assume that y is a differentiable function of x.
>
> **STEP 2:** Differentiate both sides of the equation with respect to x by employing the Power Rule or the Chain Rule or other differentiation formulas.
>
> **STEP 3:** Solve the resulting equation, which is linear in $\dfrac{dy}{dx}$, for $\dfrac{dy}{dx}$.

  NOW WORK PROBLEM 1.

EXAMPLE 3   **Differentiating Implicitly**

Find $\dfrac{dy}{dx}$ if

$$x^2 + y^2 = e^x + e^y$$

SOLUTION

$\dfrac{d}{dx}(x^2 + y^2) = \dfrac{d}{dx}(e^x + e^y)$ Differentiate both sides with respect to x.

$2x + \dfrac{d}{dx}y^2 = e^x + \dfrac{d}{dx}e^y$ Apply the sum formula; $\dfrac{d}{dx}x^2 = 2x$; $\dfrac{d}{dx}e^x = e^x$.

$2x + 2y\dfrac{dy}{dx} = e^x + e^y\dfrac{dy}{dx}$ Apply the Chain Rule on the right and the Power Rule on the left.

We proceed to solve for $\dfrac{dy}{dx}$. First bring the terms involving $\dfrac{dy}{dx}$ to the left side and bring any other terms to the right side.

$$2y\dfrac{dy}{dx} - e^y\dfrac{dy}{dx} = e^x - 2x$$

$$(2y - e^y)\dfrac{dy}{dx} = e^x - 2x \quad \text{Factor.}$$

$$\dfrac{dy}{dx} = \dfrac{e^x - 2x}{2y - e^y} \quad \text{Divide both sides by } 2y - e^y.$$

  NOW WORK PROBLEM 21.

EXAMPLE 4   **Finding the Equation of a Tangent Line**

Find the equation of the tangent line to the graph of $x^3 + xy + y^3 = 5$ at the point $(-1, 2)$.

SOLUTION   The slope of the tangent line is $\dfrac{dy}{dx}$, which can be found by differentiating implicitly. We differentiate both sides with respect to x, obtaining

$$\frac{d}{dx}(x^3 + xy + y^3) = \frac{d}{dx}5$$

$$\frac{d}{dx}x^3 + \frac{d}{dx}(xy) + \frac{d}{dx}y^3 = 0$$

$$3x^2 + \left(x\frac{dy}{dx} + y\right) + 3y^2\frac{dy}{dx} = 0$$

$$(3x^2 + y) + (x + 3y^2)\frac{dy}{dx} = 0$$

$$(x + 3y^2)\frac{dy}{dx} = -(3x^2 + y)$$

Solving for $\frac{dy}{dx}$, we find

$$\frac{dy}{dx} = \frac{-(3x^2 + y)}{x + 3y^2} \quad \text{provided} \quad x + 3y^2 \neq 0$$

The derivative $\frac{dy}{dx}$ equals the slope of the tangent line to the graph at any point (x, y) for which $x + 3y^2 \neq 0$. In particular, for $x = -1$ and $y = 2$, we find the slope of the tangent line to the graph at $(-1, 2)$ to be

$$\frac{dy}{dx} = \frac{-(3 + 2)}{-1 + 12} = -\frac{5}{11} \qquad \frac{dy}{dx} = \frac{-(3x^2 + y)}{x + 3y^2}; x = -1, y = 2$$

The equation of the tangent line at the point $(-1, 2)$ is

$$y - y_1 = m(x - x_1) \qquad \text{Point-slope form of a line.}$$

$$y - 2 = -\frac{5}{11}(x + 1) \qquad m = -\frac{5}{11}, x_1 = -1, y_1 = 2$$

$$y - 2 = -\frac{5}{11}x - \frac{5}{11}$$

$$y = -\frac{5}{11}x + \frac{17}{11}$$

▶

NOW WORK PROBLEM 35.

The prime notation y', y'', and so on, is usually used in finding higher-order derivatives for implicitly defined functions.

EXAMPLE 5 **Finding First and Second Derivatives Implicitly**

Using implicit differentiation, find y' and y'' in terms of x and y if

$$xy + y^2 - x^2 = 5$$

SOLUTION

$$\frac{d}{dx}(xy + y^2 - x^2) = \frac{d}{dx}5 \qquad \text{Differentiate both sides with respect to } x.$$

$$\frac{d}{dx}(xy) + \frac{d}{dx}y^2 - \frac{d}{dx}x^2 = 0$$

$$(xy' + y) + 2yy' - 2x = 0 \qquad (1)$$

$$(x + 2y)y' = 2x - y$$

$$y' = \frac{2x - y}{x + 2y} \qquad \text{provided} \quad x + 2y \neq 0 \qquad (2)$$

It is easier to find y'' by differentiating (1) than by using (2):

$$\frac{d}{dx}(xy' + y + 2yy' - 2x) = \frac{d}{dx}0$$

$$\frac{d}{dx}(xy') + \frac{d}{dx}y + \frac{d}{dx}(2yy') - \frac{d}{dx}(2x) = 0$$

$$xy'' + y' + y' + 2y'(y') + 2yy'' - 2 = 0$$

$$y''(x + 2y) = 2 - 2y' - 2(y')^2$$

$$y'' = \frac{2 - 2y' - 2(y')^2}{x + 2y}$$

provided $x + 2y \neq 0$. To express y'' in terms of x and y, use (2). Then

$$y'' = \frac{2 - 2\left(\frac{2x - y}{x + 2y}\right) - 2\left(\frac{2x - y}{x + 2y}\right)^2}{x + 2y} \qquad y' = \frac{2x - y}{x + 2y}$$

$$= \frac{2(x + 2y)^2 - 2(2x - y)(x + 2y) - 2(2x - y)^2}{(x + 2y)^3} \qquad \text{Multiply by } \frac{(x + 2y)^2}{(x + 2y)^2}.$$

$$= \frac{2x^2 + 8xy + 8y^2 - 4x^2 - 6xy + 4y^2 - 8x^2 + 8xy - 2y^2}{(x + 2y)^3} \qquad \text{Simplify.}$$

$$= \frac{-10x^2 + 10xy + 10y^2}{(x + 2y)^3} \qquad \text{Simplify.}$$

$$= \frac{-10(x^2 - xy - y^2)}{(x + 2y)^3} = \frac{50}{(x + 2y)^3}$$
$$\uparrow$$
$$x^2 - xy - y^2 = -5$$

NOW WORK PROBLEM 31.

Application

EXAMPLE 6 Finding Marginal Revenue

For a particular commodity, the demand equation is

$$3x^2 + 4p^2 = 1200 \qquad 0 < x < 20 \qquad 0 < p < 10\sqrt{3}$$

where x is the amount demanded and p is the price (in dollars). Find the marginal revenue when $x = 8$.

SOLUTION The revenue function is

$$R = xp$$

We could solve for p in the price demand function and then compute $\frac{dR}{dx}$. However, the technique introduced earlier is easier.

We differentiate $R = xp$ with respect to x remembering that p is a function of x. Then, by the rule for differentiating a product, we obtain the marginal revenue.

$$R'(x) = p + x\frac{dp}{dx} \qquad \frac{d}{dx}(xp) = x\frac{dp}{dx} + p\frac{d}{dx}x \qquad (1)$$

To find $\frac{dp}{dx}$, we differentiate the demand equation implicitly:

$$3x^2 + 4p^2 = 1200$$

$$\frac{d}{dx}(3x^2 + 4p^2) = \frac{d}{dx}1200$$

$$6x + 8p\frac{dp}{dx} = 0$$

Solving for $\frac{dp}{dx}$, we have

$$\frac{dp}{dx} = \frac{-6x}{8p} = -\frac{3x}{4p}$$

Now substitute $\frac{dp}{dx}$ into equation (1). Then $R'(x)$ is

$$R'(x) = p + x\left(-\frac{3x}{4p}\right) = \frac{4p^2 - 3x^2}{4p}$$

When $x = 8$,

$$4p^2 = 1200 - 3x^2 = 1200 - 3(64) = 1200 - 192 = 1008$$

So $p^2 = \frac{1008}{4} = 252$.

Then $p = \sqrt{252} = 15.87$, so that the marginal revenue at $x = 8$ is

$$R'(8) = \frac{1008 - 192}{4(15.87)} = \$12.85 \text{ per unit}$$

EXERCISE 4.7

In Problems 1–30, find $\frac{dy}{dx}$ by using implicit differentiation.

1. $x^2 + y^2 = 4$
2. $3x^2 - 2y^2 = 6$
3. $x^2y = 8$
4. $x^3y = 5$
5. $x^2 + y^2 - xy = 2$
6. $x^2y + xy^2 = x + 1$
7. $x^2 + 4xy + y^2 = y$
8. $x^2 + 2xy + y^2 = x$
9. $3x^2 + y^3 = 1$
10. $y^4 - 4x^2 = 5$
11. $4x^3 + 2y^3 = x^2$
12. $5x^2 + xy - y^2 = 0$
13. $\frac{1}{x^2} - \frac{1}{y^2} = 4$
14. $\frac{1}{x^2} + \frac{1}{y^2} = 6$
15. $\frac{1}{x} + \frac{1}{y} = 2$
16. $\frac{1}{x} - \frac{1}{y} = 4$
17. $x^2 + y^2 = ye^x$
18. $x^2 + y^2 = xe^y$
19. $\frac{x}{y} + \frac{y}{x} = 6e^x$
20. $x^2 + y^2 = 2ye^x$
21. $x^2 = y^2 \ln x$
22. $x^2 + y^2 = 2y^2 \ln x$
23. $(2x + 3y)^2 = x^2 + y^2$
24. $x^2 + y^2 = (3x - 4y)^2$
25. $(x^2 + y^2)^2 = (x - y)^3$
26. $(x^2 - y^2)^2 = (x + y)^3$
27. $(x^3 + y^3)^2 = x^2y^2$
28. $(x^3 - y^3)^2 = xy^2$
29. $y = e^{x^2 + y^2}$
30. $x = \ln(x^2 + y^2)$

In Problems 31–34, find y' and y'' in terms of x and y.

31. $x^2 + y^2 = 4$
32. $x^2 - y^2 = 1$
33. $xy + yx^2 = 2$
34. $4xy = x^2 + y^2$

In Problems 35–38, find the slope of the tangent line at the indicated point. Write an equation for this tangent line.

35. $x^2 + y^2 = 5$ at $(1, 2)$
36. $x^2 - y^2 = 8$ at $(3, 1)$
37. $e^{xy} = 1$ at $(0, 0)$
38. $\ln(x^2 + y^2) = 1$ at $(0, \sqrt{e})$

In Problems 39–42, find those points (x, y), if there are any, where the tangent line is horizontal $\left(\dfrac{dy}{dx} = 0\right)$.

39. $x^2 + y^2 = 4$ **40.** $xy + y^2 - x^2 = 4x$ **41.** $y^2 + 4x^2 = 16$ **42.** $y^2 = 4x^2 + 4$

---

43. Given the equation $x + xy + 2y^2 = 6$:
 (a) Find an expression for the slope of the tangent line at any point (x, y) on the graph.
 (b) Write an equation for the tangent line to the graph at the point $(2, 1)$.
 (c) Find the coordinates of all points (x, y) on the graph at which the slope of the tangent line equals the slope of the tangent line at $(2, 1)$.

44. The graph of the function $(x^2 + y^2)^2 = x^2 - y^2$ contains exactly four points at which the tangent line is horizontal. Find them.

45. Gas Pressure For ideal gases, **Boyle's law** states that pressure is inversely proportional to volume. A more realistic relationship between pressure P and volume V is given by **van der Waals equation**

$$P + \frac{a}{V^2} = \frac{C}{V - b}$$

where C is the constant of proportionality, a is a constant that depends on molecular attraction, and b is a constant that depends on the size of the molecules. Find the compressibility of the gas, which is measured by $\dfrac{dV}{dP}$.

46. Cost Function If the relationship between the cost C (in dollars) and the number x (in thousands) of units produced is

$$\frac{x^2}{9} - C^2 = 1 \quad x > 0, \quad C > 0$$

find the marginal cost by using implicit differentiation. For what number of units produced (approximately) does $C'(x) = 1$.

47. Master's Degrees in the U.S. In the 2000–01 academic year there were 430,164 Master's degrees granted in the United States. Assume the future relationship between the number N of Master's degrees granted and the years t since the 1999–2000 academic year is given by

$$e^{N(t)} = 430{,}163t + \frac{3t}{t^2 + 2}$$

 (a) Find $\dfrac{dN}{dt}$.
 (b) Evaluate $N'(t)$ for $t = 2$ and for $t = 4$.
 (c) Interpret the answers to (b) and explain the difference.

48. Farm Income In 2002 the income from farming in the United States was 32.4 billion dollars. Suppose the relationship between farm income I in the United States and time t, where $t = 1$ corresponds to the year 2002, is given by

$$I(t) = \sqrt{32.4\, t}$$

 (a) Find $I'(t)$.
 [Hint: Square both sides and differentiate implicitly].
 (b) Find $I'(3)$ and $I'(5)$.
 (c) Interpret the answers found in part (b). Explain the difference.

Source: Bureau of Economic Analysis, 2002.

4.8 The Derivative of $f(x) = x^{p/q}$

PREPARING FOR THIS SECTION *Before getting started, review the following:*

>> n^{th} Roots; Rational Exponents (Chapter 0, Section 0.6, pp. 64–69)

OBJECTIVES **1** Differentiate functions involving fractional exponents or radicals
 2 Use the Chain Rule and the Power Rule with fractional exponents or radicals

So far we have developed formulas for finding the derivative of polynomials, rational functions, exponential functions, and logarithmic functions. In addition, with the Chain Rule and the Power Rule we can differentiate each of these functions when they are raised to an integer power. In this section we develop a formula that will handle powers that are rational numbers.

If p and q, $q \geq 2$, are integers, then $x^{p/q} = \sqrt[q]{x^p}$. As a result, once we know how to find derivatives involving rational exponents, we will also know how to handle derivatives involving radicals.

We begin by restating the formula for finding the derivative of x^n, where n is any integer, namely,

$$\frac{d}{dx} x^n = nx^{n-1}$$

In this section we show this result is true even when n is a rational number. That is, if $n = \frac{p}{q}$, where p and $q \neq 0$ are integers, then

$$\frac{d}{dx} x^{p/q} = \frac{p}{q} x^{p/q-1}$$

Let's look at an example.

EXAMPLE 1 Differentiating $f(x) = x^{3/2}$

Find $\frac{d}{dx} x^{3/2}$.

SOLUTION If we let $y = x^{3/2}$, then by squaring both sides, we find
$$y^2 = x^3$$

Differentiate implicitly to obtain
$$2yy' = 3x^2$$

Solving for y', we find that
$$y' = \frac{3x^2}{2y}$$

But $y = x^{3/2}$ so
$$y' = \frac{3x^2}{2x^{3/2}} = \frac{3}{2} x^{1/2}$$

That is,
$$\frac{d}{dx} x^{3/2} = \frac{3}{2} x^{1/2}$$

Example 1 illustrated that
$$\frac{d}{dx} x^{3/2} = \frac{3}{2} x^{(3/2)-1} = \frac{3}{2} x^{1/2}$$

To show that this formula is valid for any rational exponent, we proceed as follows.
Start with the function
$$y = x^{p/q}$$

If we raise both sides to the power q, we obtain
$$y^q = (x^{p/q})^q = x^p$$

We now differentiate implicitly to find
$$\frac{d}{dx} y^q = \frac{d}{dx} x^p$$
$$qy^{q-1} y' = px^{p-1}$$
$$y' = \frac{px^{p-1}}{qy^{q-1}} = \underset{\underset{y \,=\, x^{p/q}}{\uparrow}}{\frac{px^{p-1}}{q(x^{p/q})^{(q-1)}}} = \frac{px^{p-1}}{qx^{p-(p/q)}}$$
$$= \frac{p}{q} x^{p-1-p+(p/q)}$$
$$= \frac{p}{q} x^{(p/q)-1}$$

336 Chapter 4 The Derivative of a Function

Derivative of $f(x) = x^{p/q}$

The derivative of $f(x) = x^{p/q}$, where p and $q \neq 0$ are integers, is $f'(x) = \dfrac{p}{q} x^{p/q-1}$. That is,

$$\frac{d}{dx} x^{p/q} = \frac{p}{q} x^{p/q-1} \tag{1}$$

EXAMPLE 2 Finding the Derivative of Functions Involving Fractional Exponents or Radicals

(a) $\dfrac{d}{dx} x^{3/2} = \dfrac{3}{2} x^{(3/2)-1} = \dfrac{3}{2} x^{1/2}$

(b) $\dfrac{d}{dx} \sqrt{x} = \dfrac{d}{dx} x^{1/2} = \dfrac{1}{2} x^{(1/2)-1} = \dfrac{1}{2} x^{-1/2} = \dfrac{1}{2x^{1/2}} = \dfrac{1}{2\sqrt{x}}$

(c) $\dfrac{d}{dx} x^{-2/3} = -\dfrac{2}{3} x^{(-2/3)-1} = -\dfrac{2}{3} x^{-5/3} = \dfrac{-2}{3x^{5/3}}$

NOW WORK PROBLEM 1.

EXAMPLE 3 Finding the Derivative of a Function Involving Fractional Exponents

Find the derivative of: $f(x) = \dfrac{x^{1/2} - 2}{x^{1/2}}$

SOLUTION We shall solve the problem in two ways.

Method 1 Use the formula for the derivative of a quotient:

$$f'(x) = \frac{d}{dx} \frac{x^{1/2} - 2}{x^{1/2}} = \frac{x^{1/2} \dfrac{d}{dx}(x^{1/2} - 2) - (x^{1/2} - 2) \dfrac{d}{dx} x^{1/2}}{(x^{1/2})^2}$$

$$= \frac{x^{1/2} \cdot \dfrac{1}{2} x^{-1/2} - (x^{1/2} - 2) \cdot \dfrac{1}{2} x^{-1/2}}{x} = \frac{\dfrac{1}{2} - \dfrac{1}{2} + x^{-1/2}}{x} = \frac{x^{-1/2}}{x} = \frac{1}{x^{3/2}}$$

Method 2 Simplify first. Then the problem becomes that of finding the derivative of:

$$f(x) = \frac{x^{1/2} - 2}{x^{1/2}} = \frac{x^{1/2}}{x^{1/2}} - \frac{2}{x^{1/2}} = 1 - \frac{2}{x^{1/2}} = 1 - 2x^{-1/2}$$

Then,

$$f'(x) = \frac{d}{dx}(1 - 2x^{-1/2}) = \frac{d}{dx} 1 - 2 \frac{d}{dx} x^{-1/2} = 0 - 2\left(-\frac{1}{2}\right) x^{-3/2} = x^{-3/2} = \frac{1}{x^{3/2}}$$

EXAMPLE 4 Finding the Derivative of a Function Involving Radicals

Find the derivative of: $f(x) = (2\sqrt{x} + 1)(\sqrt[3]{x} - 2)$

SOLUTION First we change each radical to its fractional exponent equivalent:

$$f(x) = (2x^{1/2} + 1)(x^{1/3} - 2)$$

Using the formula for the derivative of a product, we find

$$f'(x) = \frac{d}{dx}[(2x^{1/2} + 1)(x^{1/3} - 2)]$$

$$= (2x^{1/2} + 1)\frac{d}{dx}(x^{1/3} - 2) + (x^{1/3} - 2)\frac{d}{dx}(2x^{1/2} + 1) \quad \text{Derivative of a product.}$$

$$= (2x^{1/2} + 1)\left(\frac{1}{3}x^{-2/3}\right) + (x^{1/3} - 2)(x^{-1/2}) \quad \frac{d}{dx}(2x^{1/2} + 1) = 2 \cdot \frac{1}{2} \cdot x^{1/2-1} = x^{-1/2}$$

$$= \frac{2x^{1/2} + 1}{3x^{2/3}} + \frac{x^{1/3} - 2}{x^{1/2}} \quad \text{Simplify.}$$

$$= \frac{x^{1/2}(2x^{1/2} + 1) + 3x^{2/3}(x^{1/3} - 2)}{3x^{2/3}x^{1/2}} \quad \text{Add the quotients.}$$

$$= \frac{2x + x^{1/2} + 3x - 6x^{2/3}}{3x^{7/6}} \quad \text{Simplify.}$$

$$= \frac{5x + x^{1/2} - 6x^{2/3}}{3x^{7/6}} \quad \text{Simplify.}$$

NOW WORK PROBLEM 29.

The Power Rule can be extended as follows:

The Power Rule

If p and $q \neq 0$ are integers and g is a differentiable function, then

$$\frac{d}{dx}[g(x)]^{p/q} = \frac{p}{q}[g(x)]^{p/q-1} \cdot g'(x)$$

2 **EXAMPLE 5** **Using the Power Rule with Fractional Exponents and Radicals**

(a) $\dfrac{d}{dx}(x^2 + 4)^{5/2} = \dfrac{5}{2}(x^2 + 4)^{3/2}\dfrac{d}{dx}(x^2 + 4)$

$\qquad = \dfrac{5}{2}(x^2 + 4)^{3/2} \cdot 2x = 5x(x^2 + 4)^{3/2}$

(b) $\dfrac{d}{dx}\sqrt{x^2 + 4} = \dfrac{d}{dx}(x^2 + 4)^{1/2} = \dfrac{1}{2}(x^2 + 4)^{-1/2}\dfrac{d}{dx}(x^2 + 4)$

$\qquad = \dfrac{1}{2}(x^2 + 4)^{-1/2} \cdot 2x = \dfrac{x}{\sqrt{x^2 + 4}}$

(c) $\dfrac{d}{dx}\dfrac{1}{\sqrt{4x + 3}} = \dfrac{d}{dx}(4x + 3)^{-1/2} = -\dfrac{1}{2}(4x + 3)^{-3/2}\dfrac{d}{dx}(4x + 3)$

$\qquad = -\dfrac{1}{2}(4x + 3)^{-3/2} \cdot 4 = \dfrac{-2}{(4x + 3)^{3/2}}$

NOW WORK PROBLEM 7.

EXAMPLE 6 Using Implicit Differentiation

Find $\dfrac{dy}{dx}$ if $xy^2 - x + y^3\sqrt{x} + 5y = 10$

SOLUTION We use implicit differentiation to find $\dfrac{dy}{dx}$.

$$\frac{d}{dx}(xy^2 - x + y^3\sqrt{x} + 5y) = \frac{d}{dx}10$$

$$\frac{d}{dx}(xy^2) - \frac{d}{dx}x + \frac{d}{dx}(y^3\sqrt{x}) + \frac{d}{dx}(5y) = 0$$

Now use the formula for the derivative of a product to obtain

$$x\frac{d}{dx}y^2 + y^2\frac{d}{dx}x - 1 + y^3\frac{d}{dx}\sqrt{x} + \sqrt{x}\frac{d}{dx}y^3 + 5\frac{dy}{dx} = 0$$

$$x \cdot 2y\frac{dy}{dx} + y^2 - 1 + y^3\frac{1}{2\sqrt{x}} + \sqrt{x}\cdot 3y^2\frac{dy}{dx} + 5\frac{dy}{dx} = 0 \qquad \frac{d}{dx}\sqrt{x} = \frac{d}{dx}x^{1/2} = \frac{1}{2}x^{-1/2} = \frac{1}{2\sqrt{x}}$$

$$(2xy + 3\sqrt{x}\,y^2 + 5)\frac{dy}{dx} = 1 - y^2 - \frac{y^3}{2\sqrt{x}}$$

Then, if $2xy + 3\sqrt{x}\,y^2 + 5 \neq 0$,

$$\frac{dy}{dx} = \frac{1 - y^2 - \dfrac{y^3}{2\sqrt{x}}}{2xy + 3\sqrt{x}\,y^2 + 5} = \frac{2\sqrt{x}(1 - y^2) - y^3}{2\sqrt{x}(2xy + 3\sqrt{x}\,y^2 + 5)}$$

◼

NOW WORK PROBLEM 33.

EXERCISE 4.8

In Problems 1–32, find the derivative of each function.

1. $f(x) = x^{4/3}$
2. $f(x) = x^{5/2}$
3. $f(x) = x^{2/3}$
4. $f(x) = x^{3/4}$
5. $f(x) = \dfrac{1}{x^{1/2}}$
6. $f(x) = \dfrac{1}{x^{1/3}}$
7. $f(x) = (2x + 3)^{3/2}$
8. $f(x) = (3x + 4)^{4/3}$
9. $f(x) = (x^2 + 4)^{3/2}$
10. $f(x) = (x^3 + 1)^{4/3}$
11. $f(x) = \sqrt{2x + 3}$
12. $f(x) = \sqrt{4x - 5}$
13. $f(x) = \sqrt{9x^2 + 1}$
14. $f(x) = \sqrt{4x^2 + 1}$
15. $f(x) = 3x^{5/3} - 6x^{1/3}$
16. $f(x) = 4x^{5/4} - 8x^{1/4}$
17. $f(x) = x^{1/3}(x^2 - 4)$
18. $f(x) = x^{2/3}(x - 8)$
19. $f(x) = \dfrac{x}{\sqrt{x^2 - 4}}$
20. $f(x) = \dfrac{x^2}{\sqrt{x - 1}}$
21. $f(x) = \sqrt{e^x}$
22. $f(x) = e^{\sqrt{x}}$
23. $f(x) = \sqrt{\ln x}$
24. $f(x) = \ln\sqrt{x}$
25. $f(x) = e^{\sqrt[3]{x}}$
26. $f(x) = \sqrt[3]{e^x}$
27. $f(x) = \sqrt[3]{\ln x}$
28. $f(x) = \ln\sqrt[3]{x}$
29. $f(x) = \sqrt{x}\,e^x$
30. $f(x) = \sqrt{x}\,\ln x$
31. $f(x) = e^{2x}\sqrt{x^2 + 1}$
32. $f(x) = \ln\left[x\sqrt{x^2 + 1}\right]$

In Problems 33–40, use implicit differentiation to find y'.

33. $\sqrt{x} + \sqrt{y} = 4$
34. $\sqrt{x} - \sqrt{y} = 1$
35. $\sqrt{x^2 + y^2} = x$
36. $\sqrt{x^2 - y^2} = y$
37. $x^{1/3} + y^{1/3} = 1$
38. $x^{2/3} + y^{2/3} = 1$
39. $e^{\sqrt{x}} + e^{\sqrt{y}} = 4$
40. $\ln\sqrt{x^2 + y^2} = 4x$

In Problems 41–50, for each function f, find:

(a) The domain of f.
(b) The derivative $f'(x)$.
(c) The domain of f'.
(d) Any numbers x for which $f'(x) = 0$.
(e) Any numbers x in the domain of f for which $f'(x)$ does not exist.
(f) The second derivative $f''(x)$.
(g) The domain of f''.

41. $f(x) = \sqrt{x}$
42. $f(x) = \sqrt[3]{x}$
43. $f(x) = x^{2/3}$
44. $f(x) = x^{4/3}$
45. $f(x) = x^{2/3} + 2x^{1/3}$
46. $f(x) = x^{2/3} - 2x^{1/3}$
47. $f(x) = (x^2 - 1)^{2/3}$
48. $f(x) = (x^2 - 1)^{4/3}$
49. $f(x) = x\sqrt{1 - x^2}$
50. $f(x) = x^2\sqrt{4 - x}$

---

51. Enrollment Projection The Office of Admissions estimates that the total student enrollment in the University Division will be given by:

$$N(t) = -\frac{10{,}000}{\sqrt{1 + 0.1t}} + 11{,}000$$

where $N(t)$ denotes the number of students enrolled in the division t years from now.

(a) Find an expression for $N'(t)$.
(b) How fast will student enrollment be increasing 10 years from now?

52. Learning Curve The psychologist L. L. Thurstone suggested the following equation for the time T it takes to memorize a list of n words:

$$T = f(n) = Cn\sqrt{n - b}$$

where C and b are constants depending upon the person and the task.

(a) Compute $\dfrac{dT}{dn}$ and interpret the result.
(b) Suppose that for a certain person and a certain task, $C = 2$ and $b = 2$. Compute $f'(10)$ and $f'(30)$.
(c) Interpret the results found in part (b).

53. Production Function The production of commodities sometimes requires several resources, such as land, labor, machinery, and the like. If there are two inputs that require the amounts x and y, then the output z is given by a function of two variables: $z = f(x, y)$. Here z is called a **production function**. For example, if we use x to represent land and y to represent capital, and z to be the amount of a particular commodity produced, a possible production function is

$$z = x^{0.5} y^{0.4}$$

Set z equal to a fixed amount produced and show that $\dfrac{dy}{dx} = -\dfrac{5y}{4x}$. This shows that the rate of change of capital with respect to land is always negative when the amount produced is fixed.

54. Price Function It is estimated that t months from now, the average price (in dollars) of a personal computer will be given by

$$P(t) = \frac{300}{1 + \frac{1}{6}\sqrt{t}} + 100 \qquad 0 \le t \le 60$$

(a) Find an expression for $P'(t)$.
(b) Compute $P'(0)$, $P'(16)$, and $P'(49)$.
(c) Interpret the answers found in part (b).

57. Pollution The amount of pollution in a certain lake is found to be

$$A(t) = (t^{1/4} + 3)^3$$

where t is measured in years and $A(t)$ is measured in appropriate units.

(a) What is the instantaneous rate of change of the amount of pollution?
(b) At what rate is the amount of pollution changing after 16 years?

59. A large container is being filled with water. After t hours there are $8t - 4t^{1/2}$ liters of water in the container. At what rate is the water filling the container (in liters per hour) when $t = 4$?

60. A young child travels s feet down a slide in t seconds, where $s = t^{3/2}$.

(a) What is the child's velocity after 1 second?
(b) If the slide is 8 feet long, with what velocity does the child strike the ground?

Chapter 4 Review

OBJECTIVES

Section	You should be able to	Review Exercises
4.1	1 Find an equation of the tangent line to the graph of a function	85
	2 Find the derivative of a function at a number c	1–8
	3 Find the derivative of a function using the difference quotient	9–12
	4 Find the instantaneous rate of change of a function	89
	5 Find marginal cost and marginal revenue	95, 96
4.2	1 Find the derivative of a power function	13–16
	2 Find the derivative of a constant times a function	15, 16, 19, 20
	3 Find the derivative of a polynomial function	17–20
4.3	1 Find the derivative of a product	21–23
	2 Find the derivative of a quotient	24–26, 31, 32, 87
	3 Find the derivative of $f(x) = x^n$, n a negative integer	27–30
4.4	1 Find derivatives using the Power Rule	33, 34, 90
	2 Find derivatives using the Power Rule and other derivative formulas	35–42
4.5	1 Find the derivative of functions involving e^x	43, 44, 47, 49, 51, 71, 87, 92
	2 Find a derivative using the Chain Rule	45, 46, 48, 50, 52
	3 Find the derivative of functions involving $\ln x$	53–58
	4 Find the derivative of functions involving $\log_a x$ and a^x	59–62
4.6	1 Find the first derivative and the second derivative of a function	75–80
	2 Solve applied problems involving velocity and acceleration	91, 93, 94
4.7	1 Find the derivative of a function defined implicitly	81–84, 88
4.8	1 Differentiate functions involving fractional exponents or radicals	63–66
	2 Use the Chain Rule and the Power Rule with fractional exponents or radicals	67–74

THINGS TO KNOW

Slope of the Tangent Line to $y = f(x)$ at the Point $(c, f(c))$ (p. 275)

$$m_{\tan} = \lim_{x \to c} \frac{f(x) - f(c)}{x - c}, \quad \text{provided the limit exists}$$

Equation of the Tangent Line at a Point $(c, f(c))$ (p. 275)

$$y - f(c) = f'(c)(x - c) \quad \text{provided } f'(c) \text{ exists}$$

Derivative of a Function at a Number c (p. 276)

$$f'(c) = \lim_{x \to c} \frac{f(x) - f(c)}{x - c}, \quad \text{provided the limit exists}$$

Derivative of a Constant (pp. 286–287)

$$\frac{d}{dx} b = 0, \quad b \text{ a constant}$$

Derivative of a Constant Times a Function (p. 288)

$$\frac{d}{dx}[cf(x)] = c \frac{d}{dx} f(x), \quad c \text{ a constant}$$

Derivative of a Difference (p. 290)

$$\frac{d}{dx}[f(x) - g(x)] = \frac{d}{dx} f(x) - \frac{d}{dx} g(x)$$

Derivative of a Quotient (p. 296)

$$\frac{d}{dx}\left[\frac{f(x)}{g(x)}\right] = \frac{g(x) \frac{d}{dx} f(x) - f(x) \frac{d}{dx} g(x)}{[g(x)]^2}$$

Derivative of $f(x) = e^x$ (p. 309)

$$\frac{d}{dx} e^x = e^x$$

Derivative of $f(x) = \ln x$ (p. 313)

$$\frac{d}{dx} \ln x = \frac{1}{x}$$

Derivative of $f(x) = \ln g(x)$ (p. 313)

$$\frac{d}{dx} \ln g(x) = \frac{\frac{d}{dx} g(x)}{g(x)} = \frac{g'(x)}{g(x)}$$

Derivative of $f(x) = a^x$ (p. 315)

$$\frac{d}{dx} a^x = a^x \ln a$$

Derivative of f at x (p. 278)

$$f'(x) = \lim_{h \to 0} \frac{f(x+h) - f(x)}{h} \quad \text{provided the limit exists}$$

Derivative of $f(x) = x^n$ (pp. 288, 298, 336)

$$\frac{d}{dx} x^n = n x^{n-1}, \quad n \text{ any rational number}$$

Derivative of a Sum (p. 289)

$$\frac{d}{dx}[f(x) + g(x)] = \frac{d}{dx} f(x) + \frac{d}{dx} g(x)$$

Derivative of a Product (p. 295)

$$\frac{d}{dx}[f(x) \cdot g(x)] = f(x) \frac{d}{dx} g(x) + g(x) \frac{d}{dx} f(x)$$

Power Rule (p. 304)

$$\frac{d}{dx}[g(x)]^n = n[g(x)]^{n-1} g'(x), \quad n \text{ any rational number}$$

Chain Rule (pp. 310–311)

If $y = f(u)$ and $u = g(x)$, then

$$\frac{dy}{dx} = \frac{dy}{du} \cdot \frac{du}{dx}$$

Derivative of $f(x) = e^{g(x)}$ (p. 311)

$$\frac{d}{dx} e^{g(x)} = e^{g(x)} \frac{d}{dx} g(x) = e^{g(x)} g'(x)$$

Derivative of $f(x) = \log_a x$ (p. 314)

$$\frac{d}{dx} \log_a x = \frac{1}{x \ln a}$$

TRUE–FALSE ITEMS

T F **1.** The derivative of a function is the limit of a difference quotient.

T F **2.** The derivative of a product equals the product of the derivatives.

T F **3.** If $f(x) = \dfrac{1}{x}$, then $f'(x) = -\dfrac{1}{x^2}$.

T F **4.** The expression "rate of change of a function" means the derivative of the function.

T F **5.** Every function has a derivative at each number in its domain.

T F **6.** If $x^3 - y^3 = 1$, then $\dfrac{dy}{dx} = 3x^2 - 3y^2$.

T F **7.** The derivative of a function is the limit of an average rate of change.

FILL-IN-THE-BLANKS

1. The derivative of f at c equals the slope of the _____ line to f at c.

2. If $C = C(x)$ denotes the cost C of producing x items, then $C'(x)$ is called the _____ _____.

3. The derivative of $f(x) = (x^2 + 1)^{3/2}$ may be obtained using either the _____ _____ or the _____ _____.

4. The acceleration of an object equals the rate of change of _____ with respect to time.

5. The fifth-order derivative of a polynomial of degree 4 equals _____.

6. The derivative of $x^3 - y^4x + 3y = 5$ is obtained using _____ differentiation.

REVIEW EXERCISES Blue problem numbers indicate the author's suggestions for use in a practice test.

In Problems 1–8, find the derivative of each function at the given number.

1. $f(x) = 2x + 15$ at 2
2. $f(x) = 4x - 6$ at -5
3. $f(x) = x^2 - 5$ at 2
4. $f(x) = 4x^2 + 1$ at -1
5. $f(x) = x^2 - 2x$ at 1
6. $f(x) = 3x^2 + 7x$ at 0
7. $f(x) = e^{3x}$ at 0
8. $f(x) = \ln x$ at 4

In Problems 9–12, find the derivative of f using the difference quotient.

9. $f(x) = 4x + 3$
10. $f(x) = x - 7$
11. $f(x) = 2x^2 + 1$
12. $f(x) = 7 - 3x^2$

In Problems 13–74, find the derivative of each function.

13. $f(x) = x^5$
14. $f(x) = x^3$
15. $f(x) = \dfrac{x^4}{4}$
16. $f(x) = -6x^2$
17. $f(x) = 2x^2 - 3x$
18. $f(x) = 3x^3 + \dfrac{2}{3}x^2 - 5x + 7$
19. $f(x) = 7(x^2 - 4)$
20. $f(x) = \dfrac{5(x + 6)}{7}$
21. $f(x) = 5(x^2 - 3x)(x - 6)$
22. $f(x) = (2x^3 + x)(x^2 - 5)$
23. $f(x) = 12x(8x^3 + 2x^2 - 5x + 2)$
24. $f(x) = \dfrac{6x^4 - 9x^2}{3x^2}$
25. $f(x) = \dfrac{2x + 2}{5x - 3}$
26. $f(x) = \dfrac{7x}{x - 5}$
27. $f(x) = 2x^{-12}$
28. $f(x) = 2x^3 + 5x^{-2}$
29. $f(x) = 2 + \dfrac{3}{x} + \dfrac{4}{x^2}$
30. $f(x) = \dfrac{1}{x} - \dfrac{1}{x^3}$
31. $f(x) = \dfrac{3x - 2}{x + 5}$
32. $f(x) = \dfrac{2x + 3}{x + 2}$
33. $f(x) = (3x^2 - 2x)^5$
34. $f(x) = (x^3 - 1)^3$
35. $f(x) = 7x(x^2 + 2x + 1)^2$
36. $f(x) = x(2x + 5)^2$
37. $f(x) = \left(\dfrac{x + 1}{3x + 2}\right)^2$
38. $f(x) = \left(\dfrac{5x}{x + 1}\right)^3$
39. $f(x) = \dfrac{7}{(x^3 + 4)^2}$
40. $f(x) = \dfrac{3}{(x^2 - 3x)^2}$
41. $f(x) = \left(3x + \dfrac{4}{x}\right)^3$
42. $f(x) = \left(\dfrac{2x^2 + 1}{x}\right)^4$
43. $f(x) = 3e^x + x^2$
44. $f(x) = 1 - e^x$
45. $f(x) = e^{3x+1}$
46. $f(x) = 2e^{x^2}$
47. $f(x) = e^x(2x^2 + 7x)$
48. $f(x) = (x + 1)e^{2x}$
49. $f(x) = \dfrac{1 + x}{e^x}$
50. $f(x) = \dfrac{e^{3x}}{x}$
51. $f(x) = \left(\dfrac{e^x}{3x}\right)^2$

52. $f(x) = (2xe^x)^3$

53. $f(x) = \ln(4x)$

54. $f(x) = 3\ln(3x) - 15x$

55. $f(x) = x^2 \ln x$

56. $f(x) = e^x \ln x$

57. $f(x) = \ln(2x^3 + 1)$

58. $f(x) = 7[e^{3x} + \ln(x+2)]$

59. $f(x) = 2^x + x^2$

60. $f(x) = 10^x(x+5)$

61. $f(x) = x + \log x$

62. $f(x) = x^4 \log_2 x$

63. $f(x) = \sqrt{x}$

64. $f(x) = \sqrt[3]{x} + 2x$

65. $f(x) = 3x^{5/3} + 5$

66. $f(x) = 2x^{2/3} + x^2$

67. $f(x) = \sqrt{x^2 - 3x}$

68. $f(x) = \sqrt{2x^3 + x^2}$

69. $f(x) = \dfrac{x+1}{\sqrt{x+5}}$

70. $f(x) = \dfrac{3x^2}{\sqrt{x^2-1}}$

71. $f(x) = (1+x)\sqrt{e^x}$

72. $f(x) = \ln\sqrt{1+x}$

73. $f(x) = \sqrt{x}\ln x$

74. $f(x) = \sqrt{e^x} \cdot \ln x^{1/2}$

In Problems 75–80, find the first derivative and the second derivative of each function.

75. $f(x) = x^3 - 8$

76. $f(x) = (5x+3)^2$

77. $f(x) = e^{-3x}$

78. $f(x) = \ln(x^2)$

79. $f(x) = \dfrac{x}{2x+1}$

80. $f(x) = e^x \ln x$

In Problems 81–84, find $\dfrac{dy}{dx}$ by using implicit differentiation.

81. $xy + 3y^2 = 10x$

82. $y^3 + y = x^2$

83. $xe^y = 4x^2$

84. $\ln(x+y) = 8x$

In Problems 85–87, find the slope of the tangent line to the graph of f at the given point. What is an equation of the tangent line?

85. $f(x) = 2x^2 + 3x - 7$ at $(-1, -8)$

86. $f(x) = \dfrac{x^2+1}{x-1}$ at $(2, 5)$

87. $f(x) = x^2 + e^x$ at $(0, 1)$

---

88. Slope of a Tangent Line Find the slope of the tangent line at the indicated point:

$$4xy - y^2 = 3 \quad \text{at } (1, 3)$$

Write an equation of the tangent line.

89. For $f(x) = x^3 + 3x$, find:

(a) The average rate of change as x changes from 0 to 2.
(b) The instantaneous rate of change at $x = 2$.

90. For $f(x) = (4x - 5)^2$, find:

(a) The average rate of change as x changes from 2 to 5.
(b) The instantaneous rate of change at $x = 5$.

91. Falling Rocks A stone is dropped from a bridge that is 100 feet above the water. Its height h after t seconds is

$$h(t) = -16t^2 + 100$$

(a) How much time elapses before the stone hits the water?
(b) What is the average velocity of the stone during its fall?
(c) What is the instantaneous velocity of the stone as it hits the water?

92. Population Growth A small population is growing at the rate of

$$P(t) = 500\, e^{\frac{\ln 2}{3}t}$$

creatures per year.

(a) What is the average rate of growth of the population from year 0 to year 3?
(b) What is the instantaneous rate of growth of the population at year 3?

93. Throwing a Ball A major league pitcher throws a ball vertically upward with an initial velocity of 128 feet per second (87.3 miles per hour) from a height of 6 feet. The distance between the ball and the ground t seconds after the ball is thrown is given by

$$s(t) = -16t^2 + 128t + 6$$

(a) When does the ball reach its highest point?
(b) What is the maximum height the ball reaches?
(c) What is the total distance the ball travels?
(d) What is the velocity of the ball at any time t?

(e) At what time (if ever) is the velocity of the ball zero? Explain your answer.
(f) For how long is the ball in the air? (Round your answer to the nearest tenth of a second.)
(g) What is the velocity of the ball (rounded to the nearest tenth) when it hits the ground?
(h) What is the acceleration of the ball at any time t?
(i) What is the velocity of the ball when it has been in the air for 2 seconds? What is its velocity when it has been in the air for 6 seconds?
(j) Interpret the meaning of the different signs in the answers to part (i).

94. **Throwing Balls on the Moon** The same pitcher as in Problem 93 stands on the moon and again throws the ball vertically upward. This time the measurements are made in meters. The distance between the ball and the ground t seconds after the ball is thrown is given by

$$s(t) = -0.8t^2 + 39t + 2$$

(a) When does the ball reach its highest point?
(b) What is the maximum height the ball reaches?
(c) What is the total distance the ball travels?
(d) What is the velocity of the ball at any time t?
(e) At what time (if ever) is the velocity of the ball zero? Explain your answer.
(f) For how long is the ball in the air? (Round your answer to the nearest tenth of a second.)
(g) What is the velocity of the ball (rounded to the nearest tenth) when it hits the ground?
(h) What is the acceleration of the ball at any time t?
(i) What is the velocity of the ball when it has been in the air for 2 seconds? What is its velocity when it has been in the air for 6 seconds?
(j) Interpret the meaning of the different signs in the answers to part (i).

95. **Business Analysis** The cost function and the demand equation for a certain product are

$$C(x) = 15x + 550 \quad \text{and} \quad p = -0.50x + 75$$

Find:
(a) The revenue function.
(b) The marginal revenue.
(c) The marginal cost.
(d) The break even point(s).
   Hint: the numbers for which $R(x) = C(x)$.
(e) The number x for which marginal revenue equals marginal cost.

96. **Analyzing Revenue** The cost function and the demand equation for a certain product are

$$C(x) = 60x + 7200 \qquad p = -2x + 300$$

Find:
(a) The revenue function.
(b) The marginal revenue.
(c) The marginal cost.
(d) The break even point(s).
   Hint: the numbers for which $R(x) = C(x)$.
(e) The number x for which marginal revenue equals marginal cost.

Chapter 4 Project

TESTING BLOOD EFFICIENTLY

Suppose you are in charge of blood testing for a large organization (for example, drug testing for a corporation or disease testing for the armed forces). You have N people you need to test. One way to do this is to test each person individually. Then you must perform N tests, which could be costly.

You hit upon a clever idea: Take a sample of blood from each of x people, mix the samples, and test the mixture. This technique is called **pooling the sample**. If this "pooled" test comes out negative, then you're done with those x people. If the "pooled" test is positive, you would then test each of the x people individually. With this plan, to test x people, you would use either 1 test (potentially saving money in the process) or $x + 1$ tests (with a slight increase in cost from the earlier scheme). Is it worth changing to the new method? Let's try to find out by examining the cost of the new method.

1. Let p be the probability that the test of a single person is positive. Then $q = 1 - p$ is the probability that the test is negative. We assume that these probabilities are the same for all N people in the group to be tested. Let p_- be the probability that the test of the mixed sample of x people is negative and let p_+ be the probability that the test of the mixed sample is positive. Explain why $p_- = q^x$ and $p_+ = 1 - q^x$.

2. The number of tests we will have to do depends on the makeup of our group of x people. However, we can compute the long-term average of the number of tests we will have to do to get a result for each of the x people. Explain why the long-term average number of tests we will do to get a result for each of the x people is

$$1 \cdot p_- + (x+1) \cdot p_+ = x(1 - q^x) + 1$$

3. For simplicity assume that x evenly divides N. Then to find a result for all N people, we will have to do $\dfrac{N}{x}$ groupings. Explain why the long-term average number of tests we will do to get a result for all N people is

$$\dfrac{N}{x}[x(1-q^x) + 1] = N\left(1 - q^x + \dfrac{1}{x}\right)$$

4. Assume that each test costs K dollars to analyze. Explain why the cost function $C = C(x)$ for using the pooling method is

$$C(x) = KN\left(1 - q^x + \dfrac{1}{x}\right)$$

5. Suppose that you are testing for the presence of HIV in an adult population in North America. According to the Centers for Disease Control, 0.56% of this population carries HIV so the chance of a positive individual test is 0.0056. Then $q = 1 - 0.0056 = 0.9944$. The cost function is

$$C(x) = KN\left(1 - 0.9944^x + \dfrac{1}{x}\right)$$

Show that the marginal cost function is

$$C'(x) = KN\left(-(0.9944)^x \ln 0.9944 - \dfrac{1}{x^2}\right)$$
$$= KN\left(0.0056(0.9944)^x - \dfrac{1}{x^2}\right)$$

6. We will want to determine whether $C'(x)$ is positive or negative: if $C'(x)$ is negative, then increasing x by 1 person should cause a decrease in the cost, while if $C'(x)$ is positive, then increasing x by 1 person should cause an increase in the cost. Unfortunately, it is difficult to solve these inequalities using algebra. Since K and N will not effect whether $C'(x)$ is positive or negative, we need only consider the values of the expression $.0056(.9944)^x - \dfrac{1}{x^2}$. The following table gives values for this expression.

x	1	2	3	4	5	6	7	8	9	10
$0.0056(0.9944)^x - \dfrac{1}{x^2}$	−0.99442	−0.24445	−0.10559	−0.05701	−0.03454	−0.02235	−0.01501	−0.01026	−0.00701	−0.00469

x	11	12	13	14	15	16	17	18	19	20
$0.0056(0.9944)^x - \dfrac{1}{x^2}$	−0.00299	−0.00169	−0.00070	0.00009	0.00072	0.00123	0.00164	0.00199	0.00228	0.00252

Find the most likely pair of x values for which the cost might be a minimum, and compare the cost at each of these x values. Which value of x makes the cost smaller?

7. How will the value of N change your answer?

8. How many tests do you expect to run?

9. If each test costs five dollars to administer and you are testing 1000 people, how much money do you expect to save by using the pooling procedure instead of testing all 1000 people?

10. Approximately 8.3% of the adult population of sub-Saharan Africa carries HIV. Use an analysis similar to that above to find the pool size x which minimizes the cost of testing N people.

MATHEMATICAL QUESTIONS FROM PROFESSIONAL EXAMS*

1. **Actuary Exam—Part I** $\dfrac{d}{dx}(x^2 e^{x^2}) = ?$

   (a) $2xe^{x^2}$ (b) $\dfrac{x^3}{3}e^{x^2}$
   (c) $4x^2 e^{x^2}$ (d) $2xe^{x^2} + x^4 e^{x^2-1}$
   (e) $2x^3 e^{x^2} + 2xe^{x^2}$

2. **Actuary Exam—Part I** If $f(x) = |x + 3|$, then $f'(x)$ is continuous for what values of x?

   (a) $x < 3$ (b) $x > 3$ (c) $x \neq 3$
   (d) $x \neq -3$ (e) All x

3. **Actuary Exam—Part I** If $y = be^{c^2 + x^2}$, $\dfrac{dy}{dx} = ?$

   (a) $be^{c^2} 2x$ (b) $2xbe^{x^2}$ (c) $b(c^2 + x^2)e^{c^2+x^2-1}$
   (d) $2bxe^{c^2+x^2}$ (e) $be^{c^2+x^2}$

4. **Actuary Exam—Part I** $\dfrac{d}{dx}(e^{x^2}) = ?$

   (a) $2xe^{x^2}$ (b) $\dfrac{x^3 e^{x^2}}{3}$ (c) $4x^2 e^{x^2}$
   (d) $3xe^{x^2} + x^4 e^{x^2-1}$ (e) $2xe^{x^2} + 2x^3 e^{x^2}$

5. **Actuary Exam—Part I** If $f'(x) = x\, f(x)$ for all real x and $f(-2) = 3$, then $f''(-2) =$

   (a) -18 (b) -1 (c) 1
   (d) 12 (e) 15

6. **Actuary Exam—Part I** A particle moves along the x-axis so that at any time $t \geq 0$ the position of the particle is given by $f(t) = \dfrac{t^3}{3} + t^2 - 2t - 2$. At what time t will the acceleration and the velocity of the particle be the same?

   (a) $t = 0$ (b) $t = 1$ (c) $t = 2$
   (d) $t = 2\sqrt{3}$ (e) $t = 6$

*Copyright © 1998, 1999 by the Society of Actuaries. Reprinted with permission.

CHAPTER 5

Applications: Graphing Functions; Optimization

OUTLINE

- **5.1** Horizontal and Vertical Tangent Lines; Continuity and Differentiability
- **5.2** Increasing and Decreasing Functions; the First Derivative Test
- **5.3** Concavity; the Second Derivative Test
- **5.4** Optimization
- **5.5** Elasticity of Demand
- **5.6** Related Rates
- **5.7** The Differential; Linear Approximations
  - **Chapter Review**
  - **Chapter Project**
  - **Mathematical Questions from Professional Exams**

Most businesses stock items that are either required for the business or are to be sold. This is usually referred to as *inventory*. There is always the issue of how much to have in inventory. The choices range from keeping inventory levels high and not reordering very often to keeping inventory levels low and ordering frequently. There are costs associated with maintaining inventory and there are costs associated with ordering, such as shipping charges. The Chapter Project at the end of this chapter will walk you through an analysis of inventory control and ways of minimizing costs.

Chapter 5 Applications: Graphing Functions; Optimization

A LOOK BACK, A LOOK FORWARD

In Chapter 4 we derived various formulas for finding the derivative of a function. In particular, we now know how to find the derivative of a polynomial function, a rational function, an exponential function, and a logarithm function. We also learned to differentiate implicitly, find the derivative of functions raised to a rational power, and use the Chain Rule. Now we are ready for applications involving the derivative of a function.

We shall use the derivative as a tool for obtaining the graph of a function as well as for finding the maximum and minimum values of a function. Then we shall see how the derivative can be used to solve certain problems in which related variables vary over time. Finally, we shall use the derivative as an approximation tool.

5.1 Horizontal and Vertical Tangent Lines; Continuity and Differentiability

PREPARING FOR THIS SECTION *Before getting started, review the following:*

> Solving Equations (Chapter 0, Section 0.4, pp. 40–49)
> Library of Functions (Chapter 1, Section 1.4, pp. 137–143)
> Tangent Lines; Derivative of a function at a Number c (Chapter 4, Section 4.1, pp. 274–276)
> Lines (Chapter 0, Section 0.9, pp. 81–91)
> Continuous Functions (Chapter 3, Section 3.3, pp. 253–255)

OBJECTIVES
1. Find horizontal tangent lines
2. Find vertical tangent lines
3. Discuss the graph of a function f where the derivative of f does not exist

We have seen that the slope of the tangent line to the graph of a function f equals the derivative. In particular, if the tangent line is horizontal, then its slope is zero, so the derivative of f will also be zero. See Figure 1. The following result is a consequence of this fact.

FIGURE 1

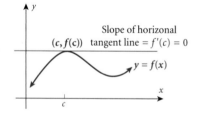

Criterion for a Horizontal Tangent Line

The tangent line to the graph of a differentiable function f at a point $(c, f(c))$ on the graph of f is horizontal if and only if $f'(c) = 0$.

To determine where the tangent line to the graph of $y = f(x)$ is horizontal, we need to solve the equation $f'(x) = 0$.

EXAMPLE 1 Finding Horizontal Tangent Lines

At what point(s) is the tangent line to the graph of

$$f(x) = x^3 + 3x^2 - 24x$$

horizontal?

SOLUTION We first find the derivative $f'(x)$:

$$f'(x) = 3x^2 + 6x - 24$$

Horizontal tangent lines occur where $f'(x) = 0$, so we solve the equation

$$\begin{aligned}
3x^2 + 6x - 24 &= 0 && f'(x) = 0 \\
3(x^2 + 2x - 8) &= 0 && \text{Factor.} \\
3(x + 4)(x - 2) &= 0 && \text{Factor.} \\
x + 4 = 0 \quad \text{or} \quad x - 2 &= 0 && \text{Zero-Product Property.} \\
x = -4 \quad \text{or} \quad x &= 2 && \text{Solve for } x.
\end{aligned}$$

Now evaluate the function f at each of these numbers.

$$f(-4) = (-4)^3 + 3(-4)^2 - 24(-4) = 80 \quad f(x) = x^3 + 3x^2 - 24x$$

and

$$f(2) = 2^3 + 3(2)^2 - 24(2) = -28$$

The tangent line to the graph of f is horizontal at the points $(-4, f(-4)) = (-4, 80)$ and $(2, f(2)) = (2, -28)$ on the graph of f.

Notice that we substitute $x = -4$ and $x = 2$ into the function $y = f(x)$ to find the y-coordinate of the point on the graph of f at which the tangent line is horizontal. Be careful not to substitute these values of x into the derivative $f'(x)$ because this will give the value of the derivative rather than the value of the function.

NOW WORK PROBLEM 3.

Vertical Tangent Lines

Recall that the slope of a vertical line is undefined. Look at Figure 2.

FIGURE 2

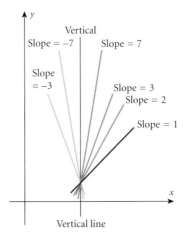

As a line moves toward being vertical, the absolute value of its slope becomes larger and larger, that is, it increases without bound. For vertical lines, the slope is unbounded. Since the slope of the tangent line to the graph of a function equals the derivative of the function, and since vertical lines have unbounded slope, we can formulate the following result about vertical tangent lines:

Conditions for a Vertical Tangent Line

The tangent line to the graph of a continuous function f at a point $(c, f(c))$ on the graph of f is vertical if $f'(x)$ is unbounded at $x = c$.

Two conditions must be present for a vertical tangent line at a point $(c, f(c))$ on the graph of a continuous function f:

1. There must be a point on the graph of f corresponding to $x = c$. That is, $x = c$ must be in the domain of the function f.
2. The derivative $f'(x)$ must be unbounded at $x = c$.

EXAMPLE 2 Finding Vertical Tangent Lines

At what point(s) is the tangent line to the graph of $f(x) = \sqrt[3]{x}$ vertical?

SOLUTION The cube root function f can be written as $f(x) = \sqrt[3]{x} = x^{1/3}$. Then, the derivative is

$$f'(x) = \frac{d}{dx} x^{1/3} = \frac{1}{3} x^{-2/3} = \frac{1}{3x^{2/3}} \qquad \frac{d}{dx} x^n = nx^{n-1}$$

The derivative $f'(x)$ is unbounded at $x = 0$. At $x = 0$, the value of f is $f(0) = 0^{1/3} = 0$. Since the conditions for a vertical tangent line have been met, there is a vertical tangent line to the graph of f at the point $(0, 0)$. See Figure 3 for the graph.

FIGURE 3

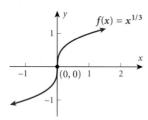

NOW WORK PROBLEM 7.

EXAMPLE 3 Finding the Horizontal and the Vertical Tangent Lines

At what point(s) is the tangent line to the graph of $f(x) = \dfrac{x^{2/3}}{x - 1}$ horizontal? At what point(s) is it vertical?

SOLUTION First, we determine that the domain of f is $\{x \mid x \neq 1\}$. Next we use the formula for the derivative of a quotient to find the derivative $f'(x)$:

$$f(x) = \frac{x^{2/3}}{x-1}$$

$$f'(x) = \frac{(x-1)\frac{d}{dx}x^{2/3} - x^{2/3} \cdot \frac{d}{dx}(x-1)}{(x-1)^2} \quad \text{Formula for the derivative of a quotient.}$$

$$= \frac{(x-1) \cdot \frac{2}{3}x^{-1/3} - x^{2/3} \cdot 1}{(x-1)^2} \quad \text{Differentiate.}$$

$$= \frac{\frac{2(x-1)}{3x^{1/3}} - x^{2/3}}{(x-1)^2} \quad \text{Simplify.}$$

$$= \frac{\frac{2(x-1) - 3x}{3x^{1/3}}}{(x-1)^2} \quad \text{Write the numerator as a single quotient.}$$

$$= \frac{-x-2}{3x^{1/3}(x-1)^2} \quad \text{Simplify.}$$

$$= \frac{-(x+2)}{3x^{1/3}(x-1)^2} \quad \text{Factor out } -1 \text{ in the numerator.}$$

We see that $f'(x) = 0$ if $x = -2$. We also see that $f'(x)$ is unbounded if $x = 0$ or if $x = 1$. However, since $x = 1$ is not in the domain of f, we disregard it.

Since

$$f(-2) = \frac{(-2)^{2/3}}{-2-1} = \frac{1.59}{-3} \approx -0.53 \qquad f(x) = \frac{x^{2/3}}{x-1}$$

and

$$f(0) = \frac{0^{2/3}}{0-1} = 0$$

we conclude that f has a horizontal tangent line at $(-2, -0.53)$ and a vertical tangent line at $(0, 0)$.

CHECK: Graph $y = \dfrac{x^{2/3}}{x-1}$. Can you see where the graph has a horizontal and a vertical tangent line?

Continuity and Differentiability

Discuss the graph of a function f where the derivative of f does not exist ③

In Chapter 3 we discussed an important property of functions, the property of being continuous, and in Chapter 4 we discussed another important property of functions, the derivative. As it turns out, there is an important result that relates these properties.

Continuity and Differentiability

Suppose c is a number in the domain of f. If f has a derivative at the number c, then f is continuous at c.

To put it another way, if a function f is not continuous at a number c, then it will have no derivative at c.

However, a function can be continuous at c, but not have a derivative at c. Let's look at an example.

EXAMPLE 4 A Function Continuous at 0, For Which $f'(0)$ Does Not Exist

Discuss the derivative of the absolute value function $f(x) = |x|$ at 0.

SOLUTION The absolute value function $f(x) = |x|$ is in our library of functions. Its graph is given in Figure 4. Notice that the absolute value function is continuous at 0.

To find the derivative of $f(x) = |x|$ at 0, we use the definition given in Chapter 4, page 276.

$$f'(0) = \lim_{x \to 0} \frac{f(x) - f(0)}{x - 0} = \lim_{x \to 0} \frac{|x| - |0|}{x - 0} = \lim_{x \to 0} \frac{|x|}{x}$$

We examine the one-sided limits. Remember if $x < 0$, then $|x| = -x$ and if $x > 0$, then $|x| = x$.

$$\lim_{x \to 0^-} \frac{|x|}{x} = \lim_{x \to 0^-} \frac{-x}{x} = -1 \quad \text{and} \quad \lim_{x \to 0^+} \frac{|x|}{x} = \lim_{x \to 0^+} \frac{x}{x} = 1$$

Since the one-sided limits are unequal, we conclude that $\lim_{x \to 0} \frac{|x|}{x}$ does not exist and so $f'(0)$ does not exist.

Notice in Figure 4 the sharp point at $(0, 0)$. To the left of $(0, 0)$ the slope of the line is -1 and to the right of $(0, 0)$ it is 1. At $(0, 0)$, there is no tangent line since the derivative $f'(0)$ does not exist. ▶

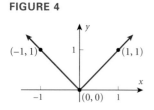

FIGURE 4

If f is continuous at c, the derivative at c may or may not exist. Figure 5 shows how the graph of a continuous function might look at points where the derivative does not exist.

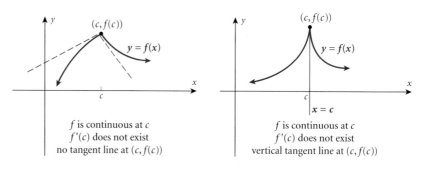

FIGURE 5

EXAMPLE 5 Discussing Continuity and Differentiability

For the function $f(x) = \begin{cases} 2x^2 & \text{if } x \leq 2 \\ x^3 & \text{if } x > 2 \end{cases}$

(a) Is f continuous at 2?
(b) Does $f'(2)$ exist? If it does, what is its value?
(c) If f is continuous at 2 but $f'(2)$ does not exist, is there a vertical tangent line or no tangent line at 2?
(d) Graph f.

SOLUTION **(a)** $f(2) = 2 \cdot 2^2 = 8$. The one-sided limits are

$$\lim_{x \to 2^-} f(x) = \lim_{x \to 2^-} (2x^2) = 8, \quad \lim_{x \to 2^+} f(x) = \lim_{x \to 2^+} x^3 = 8$$

Since $\lim_{x \to 2} f(x) = 8$ and $f(2) = 8$, the function f is continuous at 2.

(b) The derivative of f at 2 is $f'(2) = \lim_{x \to 2} \dfrac{f(x) - f(2)}{x - 2} = \lim_{x \to 2} \dfrac{f(x) - 8}{x - 2}$.

We look at the one-sided limits:

$$\lim_{x \to 2^-} \dfrac{f(x) - 8}{x - 2} = \lim_{x \to 2^-} \dfrac{2x^2 - 8}{x - 2} = \lim_{x \to 2^-} \left[2 \cdot \dfrac{x^2 - 4}{x - 2} \right] = 2 \lim_{x \to 2^-} \dfrac{(x-2)(x+2)}{x-2} = 2 \lim_{x \to 2^-} (x + 2) = 2 \cdot 4 = 8$$

$$\lim_{x \to 2^+} \dfrac{f(x) - 8}{x - 2} = \lim_{x \to 2^+} \dfrac{x^3 - 8}{x - 2} = \lim_{x \to 2^+} \dfrac{(x-2)(x^2 + 2x + 4)}{x - 2} = \lim_{x \to 2^+} (x^2 + 2x + 4) = 12$$

We conclude that $f'(2)$ does not exist.

(c) Since the one-sided limits in (b) are unequal, there is no tangent line at the point $(2, 8)$.

(d) See Figure 6 for the graph of f.

FIGURE 6

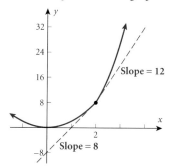

NOW WORK PROBLEM 31.

EXERCISE 5.1

In Problems 1–26, find any points at which the graph of f has either a horizontal or a vertical tangent line.

1. $f(x) = x^2 - 4x$
2. $f(x) = x^2 + 2x$
3. $f(x) = -x^2 + 8x$
4. $f(x) = -x^2 - 12x + 1$
5. $f(x) = -2x^2 + 8x + 1$
6. $f(x) = -3x^2 + 12x$
7. $f(x) = 3x^{2/3} + 1$
8. $f(x) = 3x^{1/3} - 1$
9. $f(x) = -x^3 + 3x + 1$
10. $f(x) = -2x^3 + 6x^2 + 1$
11. $f(x) = 4x^{3/4} - 2$
12. $f(x) = 8x^{1/4} - 1$
13. $f(x) = x^5 - 10x^4$
14. $f(x) = x^5 + 5x$
15. $f(x) = 3x^5 + 20x^3 - 1$
16. $f(x) = 3x^5 - 5x^3 - 1$
17. $f(x) = x^{2/3} + 2x^{1/3}$
18. $f(x) = x^{2/3} - 2x^{1/3}$
19. $f(x) = x^{2/3}(x - 10)$
20. $f(x) = x^{2/3}(x - 15)$
21. $f(x) = x^{2/3}(x^2 - 16)$
22. $f(x) = x^{2/3}(x^2 - 4)$
23. $f(x) = \dfrac{x^{2/3}}{x - 2}$
24. $f(x) = \dfrac{x^{2/3}}{x + 1}$
25. $f(x) = \dfrac{x^{1/3}}{x - 1}$
26. $f(x) = \dfrac{x^{1/3}}{x + 1}$

In Problems 27–36, answer the following questions about the function f at c.

(a) Is f continuous at c?
(b) Does $f'(c)$ exist? If it does, what is its value?
(c) If f is continuous at c but $f'(c)$ does not exist, is there a vertical tangent line or no tangent line at c?
(d) For Problems 31–36, graph f.

27. $f(x) = x^{2/3}$ at 0
28. $f(x) = x^{2/5}$ at 0
29. $f(x) = \dfrac{x}{x - 1}$ at 1
30. $f(x) = \dfrac{x^2}{x - 4}$ at 4
31. $f(x) = \begin{cases} 3x & \text{if } x < 0 \\ x^2 & \text{if } x \geq 0 \end{cases}$ at 0
32. $f(x) = \begin{cases} 6x & \text{if } x < 1 \\ x^2 + 5 & \text{if } x \geq 1 \end{cases}$ at 1
33. $f(x) = \begin{cases} 4x & \text{if } x \leq 2 \\ x^2 & \text{if } x > 2 \end{cases}$ at 2
34. $f(x) = \begin{cases} -2x^2 & \text{if } x < 0 \\ x + 1 & \text{if } x \geq 0 \end{cases}$ at 0
35. $f(x) = \begin{cases} x^2 & \text{if } x < 0 \\ x^3 & \text{if } x \geq 0 \end{cases}$ at 0
36. $f(x) = \begin{cases} x^2 & \text{if } x < 0 \\ x^4 & \text{if } x \geq 0 \end{cases}$ at 0

5.2 Increasing and Decreasing Functions; the First Derivative Test

PREPARING FOR THIS SECTION *Before getting started, review the following:*

> Increasing and Decreasing Functions (Chapter 1, Section 1.3, pp. 127–128)
> Solving Inequalities (Chapter 0, Section 0.5, pp. 51–61)
> Local Maxima and Local Minima (Chapter 1, Section 1.3, pp. 128–130)
> End Behavior; Asymptotes (Chapter 3, Section 3.4, pp. 260–265)

OBJECTIVES
1. Determine where a function is increasing and where it is decreasing
2. Use the first derivative test
3. Graph functions

Consider the graph of the function $y = f(x)$ given in Figure 7. The function is increasing on the intervals (a, b) and (c, d) while the function is decreasing on the intervals (b, c) and (d, e). Notice also that f has a local maximum at b and at d and a local minimum at c.

FIGURE 7

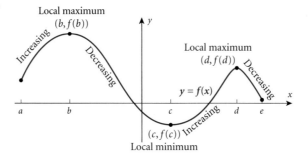

NOW WORK PROBLEMS 3 AND 7.

In this section we will give a test for differentiable functions that provides a straightforward way to determine the intervals on which the function is increasing or decreasing. This, in turn, will give us a way to locate the local maxima and the local minima.

Look at the graph of the function f given in Figure 8(a). On the interval (a, b), where f is increasing, we have drawn several tangent lines. Notice that each tangent line has a positive slope. This is characteristic of tangent lines of increasing functions. Since the derivative of f equals the slope of the tangent line, it follows that wherever the derivative is positive, the function f will be increasing.

Similarly, as Figure 8(b) illustrates, the tangent lines of a decreasing function have negative slope. Whenever the derivative of f is negative, the function f is decreasing. This leads us to the following test for an increasing or decreasing function.

FIGURE 8

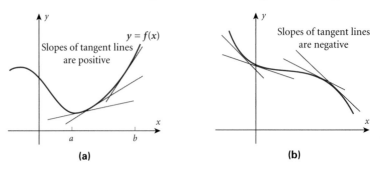

Increasing and Decreasing Functions; the First Derivative Test

Test for an Increasing or Decreasing Function

A differentiable function $y = f(x)$ is:

(a) Increasing on the interval (a, b) if $f'(x) > 0$ for all x in the interval (a, b).
(b) Decreasing on the interval (c, d) if $f'(x) < 0$ for all x in the interval (c, d).

To determine where the graph of a function is increasing or where it is decreasing, follow these steps:

Steps for Finding Where a Function Is Increasing and Where It Is Decreasing

STEP 1 Find the derivative $f'(x)$ of the function $y = f(x)$.
STEP 2 Set up a table that solves the two inequalities:

$$f'(x) > 0 \quad \text{and} \quad f'(x) < 0$$

EXAMPLE 1 Determining Where a Function Is Increasing and Where It Is Decreasing

Determine where the function

$$f(x) = x^3 - 6x^2 + 9x - 2$$

is increasing and where it is decreasing.

SOLUTION We follow the steps given above.

STEP 1 $f'(x) = 3x^2 - 12x + 9$
STEP 2 To set up the table, we first solve the equation $f'(x) = 0$.

$$f'(x) = 3x^2 - 12x + 9 = 3(x^2 - 4x + 3) = 3(x - 1)(x - 3)$$

The solutions of the equation $f'(x) = 3(x - 1)(x - 3) = 0$ are 1 and 3. These numbers separate the real number line into three parts: $-\infty < x < 1$, $1 < x < 3$, and $3 < x < \infty$. We construct Figure 9, using 0, 2, and 4 as test numbers to find the sign of $f'(x)$ on each interval.

FIGURE 9

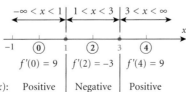

We conclude that the function f is increasing on the intervals $(-\infty, 1)$ and $(3, \infty)$ and is decreasing on the interval $(1, 3)$.

SEEING THE CONCEPT: Graph $y = x^3 - 6x^2 + 9x - 2$ and use TRACE to determine where f is increasing and where f is decreasing.

Local Maximum and Local Minimum

We have already observed that if a function f is increasing to the left of a point A on the graph of f and is decreasing to the right of A, then at A there is a local maximum since the graph of f is higher at A than at nearby points. Similarly, if f is decreasing to the left of a point B on the graph of f and is increasing to the right of B, then at B there is a local minimum. Since the first derivative of a function supplies information about where the function is increasing or decreasing, we call the test for locating local maxima and local minima the **First Derivative Test.**

> ### First Derivative Test
>
> Let f denote a differentiable function. Find the derivative of f and set up a table to determine where f is increasing and where it is decreasing.
>
> 1. If f is increasing to the left of a point A on the graph of f and is decreasing to the right of A, then at the point A there is a local maximum.
> 2. If f is decreasing to the left of a point B on the graph of f and is increasing to the right of B, then at the point B there is a local minimum.

EXAMPLE 2 Using the First Derivative Test

Use the First Derivative Test to locate the local maxima and local minima, if any, of

$$f(x) = x^3 - 6x^2 + 9x - 2$$

SOLUTION This is the same function discussed in Example 1. If we refer back to Figure 9, we see that f is increasing for $-\infty < x < 1$ and is decreasing for $1 < x < 3$. When $x = 1$, we have $y = f(1) = 2$. By the First Derivative Test, f has a local maximum at the point $(1, 2)$.

Similarly, f is decreasing for $1 < x < 3$ and is increasing for $3 < x < \infty$. When $x = 3$, we have $y = f(3) = -2$. By the First Derivative Test, f has a local minimum at the point $(3, -2)$.

SEEING THE CONCEPT: Graph $y = x^3 - 6x^2 + 9x - 2$ and use MAXIMUM/MINIMUM to find the local maximum and the local minimum.

Graphing Functions

Graph functions **3** We can use the First Derivative Test to graph functions. In graphing a function $y = f(x)$, we follow these steps:

> ### Steps for Graphing Functions
>
> **STEP 1** Find the domain of f.
> **STEP 2** Locate the intercepts of f (skip the x-intercepts if they are too hard to find).

STEP 3 Determine where the graph of f is increasing and where it is decreasing.
STEP 4 Find any local maxima or local minima of f by using the First Derivative Test.
STEP 5 Locate all points on the graph of f at which the tangent line is either horizontal or vertical.
STEP 6 Determine the end behavior and locate any asymptotes.

EXAMPLE 3 Graphing a Function

Graph the function
$$f(x) = x^3 - 12x$$

SOLUTION **STEP 1** The domain of f is all real numbers.
STEP 2 Let $x = 0$. Then $y = f(0) = 0$. The y-intercept is $(0, 0)$.
To find the x-intercepts, if any, let $y = 0$. Then

$$x^3 - 12x = 0 \quad y = f(x) = 0$$
$$x(x^2 - 12) = 0 \quad \text{Factor.}$$
$$x(x + 2\sqrt{3})(x - 2\sqrt{3}) = 0 \quad \text{Factor.}$$
$$x = 0 \text{ or } x + 2\sqrt{3} = 0 \text{ or } x - 2\sqrt{3} = 0 \quad \text{Apply the Zero-Product Property.}$$
$$x = 0 \text{ or } \qquad x = -2\sqrt{3} \text{ or } \quad x = 2\sqrt{3} \quad \text{Solve for } x.$$

The x-intercepts are $(0, 0)$, $(-2\sqrt{3}, 0)$, and $(2\sqrt{3}, 0)$.

STEP 3 To determine where the graph of f is increasing and where it is decreasing, we find $f'(x)$.

$$f(x) = x^3 - 12x$$
$$f'(x) = 3x^2 - 12 = 3(x^2 - 4) = 3(x + 2)(x - 2)$$
$$\qquad\qquad\qquad\quad\uparrow\qquad\qquad\quad\uparrow$$
$$\qquad\qquad\qquad\text{Factor}\qquad\quad\text{Factor}$$

The solutions of the equation $f'(x) = 3(x + 2)(x - 2) = 0$ are -2 and 2. These numbers separate the number line into three parts:

$$-\infty < x < -2 \qquad -2 < x < 2 \qquad 2 < x < \infty$$

We construct Figure 10, using -3, 0, and 3 as test numbers for $f'(x)$.

FIGURE 10

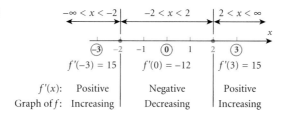

We conclude that the graph of f is increasing on the intervals $(-\infty, -2)$ and $(2, \infty)$. It is decreasing on the interval $(-2, 2)$.

STEP 4 From Figure 10, the graph of f is increasing for $-\infty < x < -2$ and is decreasing for $-2 < x < 2$. At $x = -2$, the graph changes from increasing to decreasing. Consequently, at the point $(-2, f(-2)) = (-2, 16)$ there is a local maximum. Similarly at $x = 2$, the graph changes from decreasing to increasing. At the point $(2, f(2)) = (2, -16)$ there is a local minimum.

STEP 5 The derivative of f is $f'(x) = 3x^2 - 12 = 3(x + 2)(x - 2)$. We see that $f'(x) = 0$ if $x = -2$ or if $x = 2$. The graph of f has a horizontal tangent line at the points $(-2, f(-2)) = (-2, 16)$ and $(2, f(2)) = (2, -16)$. There are no vertical tangent lines.

STEP 6 Since f is a polynomial function, its end behavior is like that of the power function $y = x^3$. Polynomial functions have no asymptotes.

To graph f we plot the intercepts, the local maximum, the local minimum, the points at which the tangent line is horizontal, and connect these points with a smooth curve. See Figure 11.

FIGURE 11

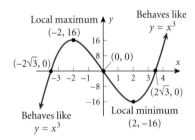

 CHECK: Use a graphing utility to graph $f(x) = x^3 - 12x$. Be sure to choose a viewing rectangle that will show a complete graph. [Use Figure 11 as a guide.] Use TRACE to confirm where f is increasing and where f is decreasing. Use MAXIMUM and MINIMUM to confirm the local maximum and the local minimum. Use ZERO (or ROOT) to locate the x-intercepts. Use a TABLE to confirm the end behavior.

 NOW WORK PROBLEM 9.

EXAMPLE 4 **Graphing a Function**

Graph the function
$$f(x) = x^4 - 4x^3$$

SOLUTION **STEP 1** f is a polynomial function so the domain of f is all real numbers.

STEP 2 Let $x = 0$. Then $y = f(0) = 0$. The y-intercept is $(0, 0)$.

To find the x-intercepts, if any, let $y = 0$. Then

$$x^4 - 4x^3 = 0 \quad y = f(x) = 0$$
$$x^3(x - 4) = 0 \quad \text{Factor.}$$
$$x^3 = 0 \quad \text{or} \quad x - 4 = 0 \quad \text{Apply the Zero-Product Property.}$$
$$x = 0 \quad \text{or} \quad x = 4 \quad \text{Solve for } x.$$

The x-intercepts are $(0, 0)$ and $(4, 0)$.

STEP 3 To determine where the graph of f is increasing and where it is decreasing, we find $f'(x)$:

$$f(x) = x^4 - 4x^3$$
$$f'(x) = 4x^3 - 12x^2 = 4x^2(x - 3)$$

Increasing and Decreasing Functions; the First Derivative Test

The solutions of the equation $f'(x) = 4x^2(x-3) = 0$ are $x = 0$ and $x = 3$. These numbers separate the number line into three parts:

$$-\infty < x < 0 \qquad 0 < x < 3 \qquad 3 < x < \infty$$

We construct Figure 12, using -1, 1 and 4 as test numbers for $f'(x)$.

FIGURE 12

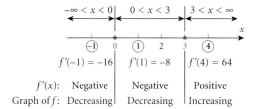

We conclude that the graph of f is decreasing on the intervals $(-\infty, 0)$ and $(0, 3)$. It is increasing on the interval $(3, \infty)$.

STEP 4 From Figure 12, the graph of f is decreasing for $-\infty < x < 3$ and is increasing for $3 < x < \infty$. At the point $(3, f(3)) = (3, -27)$, the graph changes from decreasing to increasing. Consequently, at the point $(3, -27)$ there is a local minimum.

STEP 5 The derivative of f is $f'(x) = 4x^3 - 12x^2 = 4x^2(x - 3)$. We see that $f'(x) = 0$ if $x = 0$ or if $x = 3$. The graph of f has a horizontal tangent line at the points $(0, f(0)) = (0, 0)$ and $(3, f(3)) = (3, -27)$. There are no vertical tangent lines.

STEP 6 Since f is a polynomial function, its end behavior is like that of the power function $y = x^4$. Polynomial functions have no asymptotes.

To graph f, we plot the intercepts, the local minimum, the point $(0, 0)$, and the points at which the tangent line is horizontal, and connect these points with a smooth curve. See Figure 13.

FIGURE 13

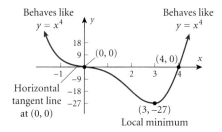

CHECK: Use a graphing utility to graph $f(x) = x^4 - 4x^3$. Be sure to choose a viewing rectangle that will show a complete graph. [Use Figure 13 as a guide.] Use TRACE to confirm where f is increasing and where f is decreasing. Use MINIMUM to confirm the local minimum. Use ZERO (or ROOT) to locate the x-intercepts. Use a TABLE to confirm the end behavior.

NOW WORK PROBLEM 19.

EXAMPLE 5 Graphing a Function

Graph the function
$$f(x) = 2x^{5/3} - 5x^{2/3}$$

SOLUTION **STEP 1** The domain of f is all real numbers.
STEP 2 Let $x = 0$. Then $y = f(0) = 0$. The y-intercept is $(0, 0)$.

To find the x-intercepts, if any, let $y = 0$. Then

$$2x^{5/3} - 5x^{2/3} = 0 \qquad y = f(x) = 0$$
$$x^{2/3}(2x - 5) = 0 \qquad \text{Factor out } x^{2/3}.$$
$$x^{2/3} = 0 \quad \text{or} \quad 2x - 5 = 0 \qquad \text{Apply the Zero-Product Property.}$$
$$x = 0 \quad \text{or} \quad x = \frac{5}{2} \qquad \text{Solve for } x.$$

The x-intercepts are $(0, 0)$ and $\left(\dfrac{5}{2}, 0\right)$.

STEP 3 To determine where the graph of f is increasing and where it is decreasing, we find $f'(x)$:

$$f(x) = 2x^{5/3} - 5x^{2/3}$$
$$f'(x) = 2 \cdot \frac{5}{3}x^{2/3} - 5 \cdot \frac{2}{3}x^{-1/3} = \frac{10x^{2/3}}{3} - \frac{10}{3x^{1/3}}$$
$$= \frac{10x - 10}{3x^{1/3}} = \frac{10(x - 1)}{3x^{1/3}} \qquad \text{Write with a common denominator.}$$

The factors that appear in the numerator and the denominator of $f'(x)$, $x - 1$ and $x^{1/3}$, equal 0 when $x = 1$ or $x = 0$. We use the numbers 0 and 1 to separate the number line into three parts:

$$-\infty < x < 0 \qquad 0 < x < 1 \qquad 1 < x < \infty$$

We construct Figure 14, using -1, $\dfrac{1}{2}$, and 2 as test numbers for $f'(x)$.

FIGURE 14

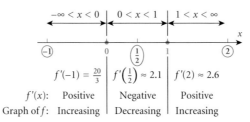

We conclude that the graph of f is increasing on the intervals $(-\infty, 0)$ and $(1, \infty)$; f is decreasing on the interval $(0, 1)$.

STEP 4 From Figure 14, we conclude that the graph of f is increasing for $-\infty < x < 0$ and is decreasing for $0 < x < 1$. At the point $(0, f(0)) = (0, 0)$, the graph changes from increasing to decreasing. Consequently, at the point $(0, 0)$ there is a local maximum.

Similarly, for $0 < x < 1$, the graph of f is decreasing and for $1 < x < \infty$, the graph of f is increasing. At the point $(1, f(1)) = (1, -3)$ there is a local minimum.

STEP 5 The derivative of f is $f'(x) = \dfrac{10(x - 1)}{3x^{1/3}}$. We see that $f'(x) = 0$ if $x = 1$. The graph of f has a horizontal tangent line at the point $(1, f(1)) = (1, -3)$. Also, $f'(x)$ is unbounded if $x = 0$. So there is a vertical tangent line at the point $(0, f(0)) = (0, 0)$.

STEP 6 For the end behavior of f, we look at the two limits at infinity:

$$\lim_{x \to -\infty} f(x) = \lim_{x \to -\infty} (2x^{5/3} - 5x^{2/3}) = \lim_{x \to -\infty} (2x^{5/3}) = -\infty$$
$$\lim_{x \to \infty} f(x) = \lim_{x \to \infty} (2x^{5/3} - 5x^{2/3}) = \lim_{x \to \infty} (2x^{5/3}) = \infty$$

The graph of f becomes unbounded in the negative direction as $x \to -\infty$ and unbounded in the positive direction as $x \to \infty$. There are no asymptotes.

To graph f, we plot the intercepts, the local maximum, the local minimum, and the points at which the tangent line is horizontal and vertical, and connect these points. See Figure 15. Notice how the vertical tangent line and local maximum at $(0, 0)$ are shown in the graph.

FIGURE 15

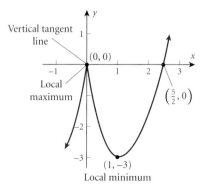

CHECK: Use a graphing utility to graph $f(x) = 2x^{5/3} - 5x^{2/3}$. Be sure to choose a viewing rectangle that will show a complete graph. [Use Figure 15 as a guide.] Use TRACE to confirm where f is increasing and where f is decreasing. Use MAXIMUM and MINIMUM to confirm the local maximum and the local minimum. Use ZERO (or ROOT) to locate the x-intercepts. Use a TABLE to confirm the end behavior.

NOW WORK PROBLEM 23.

EXAMPLE 6 Graphing a Function

Graph: $f(x) = \dfrac{x^2}{x^2 - 1}$

SOLUTION **STEP 1** The domain of f is $\{x \mid x \neq -1, x \neq 1\}$.
STEP 2 Let $x = 0$. Then $y = f(0) = 0$. The y-intercept is $(0, 0)$. Now let $y = 0$. Then $\dfrac{x^2}{(x^2 - 1)} = 0$, so $x = 0$. The x-intercept is also $(0, 0)$.
STEP 3 To find where the graph is increasing or decreasing, we find $f'(x)$:

$$f(x) = \frac{x^2}{x^2 - 1}$$

$$f'(x) = \frac{(x^2 - 1)\dfrac{d}{dx}x^2 - x^2\dfrac{d}{dx}(x^2 - 1)}{(x^2 - 1)^2} \qquad \text{Formula for the derivative of a quotient.}$$

$$= \frac{(x^2 - 1)(2x) - x^2(2x)}{(x^2 - 1)^2} \qquad \text{Differentiate.}$$

$$= \frac{-2x}{(x^2 - 1)^2} \qquad \text{Simplify.}$$

The numerator is zero for $x = 0$ and the denominator is zero for $x = -1$ and $x = 1$ so we use the numbers $-1, 0, 1$ to separate the number line into four parts:

$$-\infty < x < -1 \quad -1 < x < 0 \quad 0 < x < 1 \quad 1 < x < \infty$$

We construct Figure 16 using $-2, -\dfrac{1}{2}, \dfrac{1}{2}$, and 2 as test numbers for $f'(x)$.

FIGURE 16

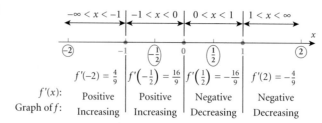

STEP 4 From Figure 16, we conclude the graph of f is increasing for $-\infty < x < -1$ and for $-1 < x < 0$. It is decreasing for $0 < x < 1$ and for $1 < x < \infty$. At the point $(0, f(0)) = (0, 0)$ there is a local maximum.

STEP 5 Since $f'(x) = \dfrac{-2x}{(x^2 - 1)}$, we conclude that $f'(0) = 0$ so the graph of f has a horizontal tangent line at $(0, 0)$. Since $x = -1$ and $x = 1$ are not in the domain of f, there are no vertical tangent lines at $x = -1$ or $x = 1$, even though $f'(x)$ is unbounded at $x = -1$ and at $x = 1$.

STEP 6 For the end behavior of f, we look at the limits at infinity. Since f is an even function $[f(-x) = f(x)]$, the graph is symmetric with respect to the y-axis. So we will only look at the limit as $x \to \infty$.

$$\lim_{x \to \infty} f(x) = \lim_{x \to \infty} \frac{x^2}{x^2 - 1} = \lim_{x \to \infty} \frac{x^2}{x^2} = 1$$

The line $y = 1$ is a horizontal asymptote to the graph of f as $x \to -\infty$ and as $x \to \infty$.

Since f is a rational function and f becomes unbounded for $x = -1$ and $x = 1$, the graph of f will have vertical asymptotes at $x = -1$ and at $x = 1$.

Further,

$$\lim_{x \to 1^-} f(x) = \lim_{x \to 1^-} \frac{x^2}{x^2 - 1} = \lim_{x \to 1^-} \frac{x^2}{x + 1} \cdot \lim_{x \to 1^-} \frac{1}{x - 1} = \frac{1}{2} \cdot \lim_{x \to 1^-} \frac{1}{x - 1} = -\infty$$

$$\lim_{x \to 1^+} f(x) = \lim_{x \to 1^+} \frac{x^2}{x^2 - 1} = \lim_{x \to 1^+} \frac{x^2}{x + 1} \cdot \lim_{x \to 1^+} \frac{1}{x - 1} = \frac{1}{2} \cdot \lim_{x \to 1^+} \frac{1}{x - 1} = \infty$$

Figure 17 shows the graph of f. The graph to the left of $(0, 0)$ is obtained using symmetry.

FIGURE 17

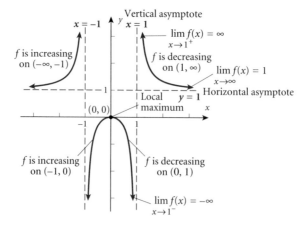

CHECK: Use a graphing utility to graph $f(x) = \dfrac{x^2}{x^2 - 1}$. Be sure to choose a viewing rectangle that will show a complete graph. [Use Figure 17 as a guide.] Use TRACE to

Increasing and Decreasing Functions; the First Derivative Test

confirm where f is increasing and where f is decreasing. Use MAXIMUM and MINIMUM to confirm the local maximum and the local minimum. Use ZERO (or ROOT) to locate the x-intercept. Use a TABLE to confirm the end behavior.

SUMMARY The box below summarizes what the first derivative tells us about the graph of a function.

Derivative of f	Graph of f
$f'(c) = 0$	Horizontal tangent line at the point $(c, f(c))$
$f'(c)$ is unbounded	Vertical tangent line at the point $(c, f(c))$
$f'(x) > 0$ for $a < x < b$	Increasing on the interval (a, b)
$f'(x) < 0$ for $a < x < b$	Decreasing on the interval (a, b)

EXERCISE 5.2

In Problems 1–8, use the graph of $y = f(x)$ given on the right.

1. What is the domain of f?
2. List the intercepts of f.
3. On what intervals, if any, is the graph of f increasing?
4. On what intervals, if any, is the graph of f decreasing?
5. For what values of x does $f'(x) = 0$?
6. For what values of x is $f'(x)$ not defined?
7. List the point(s) at which f has a local maximum.
8. List the point(s) at which f has a local minimum.

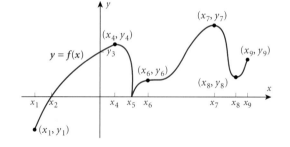

In Problems 9–34, follow the six steps on pages 356 and 357 to graph f.

9. $f(x) = -2x^2 + 4x - 2$
10. $f(x) = -3x^2 + 12x$
11. $f(x) = x^3 - 9x^2 + 27x - 27$
12. $f(x) = x^3 - 6x^2 + 12x - 8$
13. $f(x) = 2x^3 - 15x^2 + 36x$
14. $f(x) = 2x^3 + 6x^2 + 6x$
15. $f(x) = -x^3 + 3x - 1$
16. $f(x) = -2x^3 + 6x^2 + 1$
17. $f(x) = 3x^4 - 12x^3 + 2$
18. $f(x) = x^4 - 4x + 2$
19. $f(x) = x^5 - 5x + 1$
20. $f(x) = x^5 + 5x^4 + 1$
21. $f(x) = 3x^5 - 20x^3 + 1$
22. $f(x) = 3x^5 - 5x^3 - 1$
23. $f(x) = x^{2/3} + 2x^{1/3}$
24. $f(x) = x^{2/3} - 2x^{1/3}$
25. $f(x) = (x^2 - 1)^{2/3}$
26. $f(x) = (x^2 - 1)^{4/3}$
27. $f(x) = \dfrac{8}{x^2 - 16}$
28. $f(x) = \dfrac{2}{x^2 - 4}$
29. $f(x) = \dfrac{x}{x^2 - 9}$
30. $f(x) = \dfrac{x - 1}{x^2}$
31. $f(x) = \dfrac{x^2}{x^2 - 4}$
32. $f(x) = \dfrac{x^2}{x^2 - 1}$
33. $f(x) = x \ln x$
34. $f(x) = \dfrac{\ln x}{x}$

35. **Ticket Sales** The cumulative ticket sales for the 10 days preceding a popular concert is given by

$$S(x) = 4x^2 + 50x + 5000$$

where x represents the 10 days leading up to the concert and $1 \le x \le 10$. Show that S is an increasing function.

36. **Production Cost** The cost C per day, in dollars, of producing x pairs of eyeglasses is

$$C(x) = 0.2x^2 + 3x + 1000$$

Show that C is an increasing function.

37. Sales of Toy Trucks At Dan's Toy Store, the revenue R, in dollars, derived from selling x electric trucks is

$$R(x) = -0.005x^2 + 20x$$

(a) Determine where the graph of R is increasing and where it is decreasing.
(b) How many trucks need to be sold to maximize revenue?
(c) What is the maximum revenue?
(d) Graph the function R.

38. Sales of Calculators. The weekly revenue R, in dollars, from selling x calculators is

$$R(x) = -20x^2 + 1000x$$

(a) Determine where the graph of R is increasing and where it is decreasing.
(b) How many calculators need to be sold to maximize revenue?
(c) What is the maximum revenue?
(d) Graph the function R.

39. Wheat Acreage in the US The amount of wheat planted annually in the United States is given in the table.

Year	1990	1991	1992	1993	1994	1995	1996
Thousand Acres	77,041	69,881	72,219	72,168	70,349	69,031	75,105

Year	1997	1998	1999	2000
Thousand Acres	70,412	65,821	62,714	62,629

Source: United States Department of Agriculture.

If 1990 is taken as year 0, the number of thousand acres of wheat planted in the United States can be approximated by

$$A(t) = -119.2t^2 + 113.4t + 73{,}367$$

(a) On [0, 10], where is this function increasing?
(b) According to this model, will the acreage of wheat planted from 2004 to 2008 be increasing or decreasing?

40. SAT Scores in Math The data for the mathematics scores on the SAT can be approximated by

$$S(t) = -0.04t^3 + 0.43t^2 + 0.24t + 506,$$

where t is the number of years since 1994.

(a) On [0, 10], where is this function increasing?
(b) According to this model, will SAT scores in math be increasing or decreasing from 2004 to 2008?

41. Corn Production The von Liebig model states that the yield of a plant, $f(x)$, measured in bushels, will respond to the amount of nitrogen in a fertilizer in the following fashion:

$$f(x) = -0.057 - 0.417x + 0.852\sqrt{x}$$

where x is the amount of the nitrogen in the fertilizer.

(a) For what amounts of nitrogen will the yield be increasing?
(b) For what amounts x of nitrogen will the yield be decreasing?
Source: Amer. Agr. Econ. **74**:1019–1028.

42. Cost Function Suppose the cost C of producing x items is given by the function $C(x) = \sqrt{x}$.

(a) Find the marginal cost function.
(b) Show that the marginal cost is a decreasing function.
(c) Find the average cost function $\overline{C}(x) = \dfrac{C(x)}{x}$.
(d) Show that the average cost function is a decreasing function.

In Problems 43–46, use the discussion that follows.

Rolle's Theorem If a function $y = f(x)$ has the following three properties:

1. It is continuous on the closed interval $[a, b]$
2. It is differentiable on the open interval (a, b)
3. $f(a) = f(b)$

then there is at least one number c, $a < c < b$, at which $f'(c) = 0$. This result is called **Rolle's theorem.** See the figure.

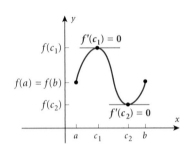

Verify Rolle's theorem by finding the number(s) c for each function on the interval indicated.

43. $f(x) = 2x^2 - 2x$ on $[0, 1]$

44. $f(x) = x^2 + 2x$ on $[-2, 0]$

45. $f(x) = x^4 - 1$ on $[-1, 1]$

46. $f(x) = x^4 - 2x^2 - 8$ on $[-2, 2]$

In Problems 47–50, use the discussion that follows.

Mean Value Theorem If $y = f(x)$ is a continuous function on the closed interval $[a, b]$ and is differentiable on the open interval (a, b), there is at least one number in the interval (a, b) at which the slope of the tangent line equals the slope of the line joining the points $(a, f(a))$ and $(b, f(b))$. That is, there is a number c, $a < c < b$, at which

$$f'(c) = \frac{f(b) - f(a)}{b - a}$$

This result is called the **Mean Value Theorem**. See the figure.

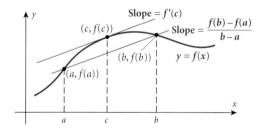

Verify the Mean Value Theorem by finding the number(s) c for each function on the interval indicated.

47. $f(x) = x^2$ on $[0, 3]$ **48.** $f(x) = x^3$ on $[0, 2]$ **49.** $f(x) = \dfrac{1}{x^2}$ on $[1, 2]$ **50.** $f(x) = x^{3/2}$ on $[0, 1]$

5.3 Concavity; the Second Derivative Test

OBJECTIVES
1. Determine the concavity of a graph
2. Find inflection points
3. Graph functions
4. Use the second derivative test
5. Solve applied problems

Consider the graphs of $y = x^2$, $x \geq 0$, and $y = \sqrt{x}$ as shown in Figure 18. Each graph starts at $(0, 0)$, is increasing, and passes through the point $(1, 1)$. However, the graph of $y = x^2$ increases rapidly while the graph of $y = \sqrt{x}$ increases slowly. We use the terms *concave up* or *concave down* to describe these characteristics of graphs.

FIGURE 18

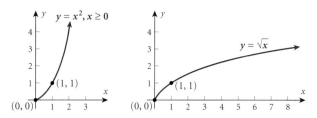

Now look at Figure 19.

FIGURE 19

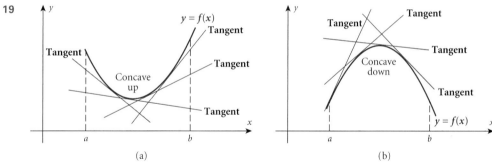

(a)   (b)

The graph of the function in Figure 19(a) is concave up on the interval (a, b), while the graph in Figure 19(b) is concave down on the interval (a, b). Notice that the tangent lines to the graph in Figure 19(a) lie below the graph, while the tangent lines to the graph in Figure 19(b) lie above the graph. This observation is the basis of the following definition.

> **Concave Up; Concave Down**
>
> Let f denote a function that is differentiable on the interval (a, b).
>
> 1. The graph of f is **concave up** on (a, b) if, throughout (a, b), the tangent lines to the graph of f lie below the graph.
> 2. The graph of f is **concave down** on (a, b), if throughout (a, b), the tangent lines to the graph of f lie above the graph.

NOW WORK PROBLEM 9.

Figure 19 also provides a test for determining whether a graph is concave up or concave down. Notice that in Figure 19(a), as one proceeds from left to right along the graph of f, the slopes of the tangent lines are increasing, starting off very negative and ending up very positive. Since the derivative $f'(x)$ equals the slope of a tangent line and these slopes are increasing, it follows that $f'(x)$ is an increasing function. So, its derivative $f''(x)$ must be positive on (a, b). Similarly, in Figure 19(b), the slopes of the tangent lines to the graph of f are decreasing. It follows that $f'(x)$ is a decreasing function and so $f''(x)$ must be negative on (a, b). As the preceding discussion shows, the second derivative provides information about the concavity of a function.

> **Test for Concavity**
>
> Let $y = f(x)$ be a function and let $f''(x)$ be its second derivative.
>
> 1. If $f''(x) > 0$ for all x in the interval (a, b), then the graph of f is concave up on (a, b).
> 2. If $f''(x) < 0$ for all x in the interval (a, b), then the graph of f is concave down on (a, b).

1 EXAMPLE 1   Determining the Concavity of a Graph

Determine where the graph of $f(x) = x^3 - 12x^2$ is concave up or concave down.

SOLUTION   We proceed to find $f''(x)$.

$$f(x) = x^3 - 12x^2$$
$$f'(x) = 3x^2 - 24x$$
$$f''(x) = 6x - 24$$
$$= 6(x - 4)$$

We need to solve the inequalities $f''(x) > 0$ and $f''(x) < 0$. We see that if $x < 4$, then $f''(x) < 0$. The graph of f is concave down on the interval $(-\infty, 4)$. If $x > 4$, then $f''(x) > 0$. The graph of f is concave up on the interval $(4, \infty)$.

EXAMPLE 2 Graphing Functions

Graph the function
$$f(x) = x^3 - 12x^2$$

SOLUTION We follow the 6 steps for graphing a function given on pp. 356–357.

STEP 1 The domain of f is all real numbers.

STEP 2 Let $x = 0$. Then $y = f(0) = 0$. The y-intercept is $(0, 0)$.
To find the x-intercepts, if any, let $y = 0$. Then

$$x^3 - 12x^2 = 0 \qquad y = f(x) = 0.$$
$$x^2(x - 12) = 0 \qquad \text{Factor.}$$
$$x^2 = 0 \quad \text{or} \quad x - 12 = 0 \qquad \text{Apply the Zero-Product Property.}$$
$$x = 0 \quad \text{or} \quad x = 12 \qquad \text{Solve for } x.$$

The x-intercepts are $(0, 0)$ and $(12, 0)$.

STEP 3 To determine where the graph of f is increasing and where it is decreasing, we find $f'(x)$:

$$f(x) = x^3 - 12x^2$$
$$f'(x) = 3x^2 - 24x \qquad \text{Differentiate.}$$
$$= 3x(x - 8) \qquad \text{Factor.}$$

The solutions of the equation $f'(x) = 3x(x - 8) = 0$ are 0 and 8. These numbers separate the number line into three parts:

$$-\infty < x < 0 \qquad 0 < x < 8 \qquad 8 < x < \infty$$

We construct Figure 20, using -1, 1, and 9 as test numbers for $f'(x)$. We conclude that the graph of f is increasing on the intervals $(-\infty, 0)$ and $(8, \infty)$; f is decreasing on the interval $(0, 8)$.

FIGURE 20

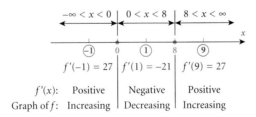

STEP 4 From Figure 20, the graph of f is increasing for $-\infty < x < 0$ and is decreasing for $0 < x < 8$. At $x = 0$, the graph changes from increasing to decreasing. Consequently, at the point $(0, f(0)) = (0, 0)$ there is a local maximum.

Similarly, at $x = 8$, the graph changes from decreasing to increasing. At the point $(8, f(8)) = (8, -256)$ there is a local minimum.

STEP 5 The derivative of f is $f'(x) = 3x(x - 8)$. We see that $f'(x) = 0$ if $x = 0$ or $x = 8$. The graph of f has a horizontal tangent line at the points $(0, 0)$ and $(8, -256)$. There are no vertical tangent lines.

STEP 6 Since f is a polynomial function, the graph of f behaves like that of $y = x^3$ for x unbounded. Polynomial functions have no asymptotes.

Finally, we use the result of Example 1, namely, that the graph of f is concave down on $(-\infty, 4)$ and is concave up on $(4, \infty)$.

To graph f, we plot the intercepts, the local maximum, the local minimum, the points at which the graph has a horizontal tangent line, and connect these points with a smooth curve, keeping the concavity of the graph in mind. See Figure 21.

FIGURE 21

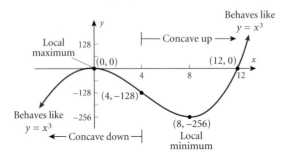

CHECK: Use a graphing utility to graph $f(x) = x^3 - 12x^2$. Be sure to choose a viewing rectangle that will show a complete graph. [Use Figure 22 as a guide.] Use TRACE to confirm where f is increasing and where f is decreasing. Use MAXIMUM and MINIMUM to confirm the local maximum and the local minimum. Use ZERO (or ROOT) to locate the x-intercepts. Use a TABLE to confirm the end behavior. Can you see where the graph is concave up and where it is concave down?

Look again at Figure 21. The point $(4, -128)$ on the graph of $f(x) = x^3 - 12x^2$ is the point at which the concavity of f changed from down to up. This point is called an *inflection point*.

> **Inflection Point**
>
> An **inflection point** of a function f is a point on the graph of f at which the concavity of f changes.

See Figure 22. The point $(b, f(b))$ is an inflection point of f.

FIGURE 22

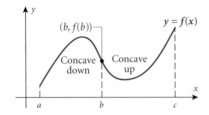

EXAMPLE 3 Finding Inflection Points

Find all the inflection points of

$$f(x) = x^2 - \frac{1}{x}$$

SOLUTION The domain of f is $\{x \mid x \neq 0\}$. To find the inflection points, we need to determine where the graph of f is concave up and where it is concave down. That is, we need to find $f''(x)$:

$$f(x) = x^2 - \frac{1}{x}$$

$$f'(x) = 2x + \frac{1}{x^2} \qquad \frac{d}{dx}\frac{1}{x} = \frac{d}{dx}x^{-1} = -1x^{-2} = -\frac{1}{x^2}$$

$$f''(x) = 2 - \frac{2}{x^3} \qquad \frac{d}{dx}\frac{1}{x^2} = \frac{d}{dx}x^{-2} = -2x^{-3} = \frac{-2}{x^3}$$

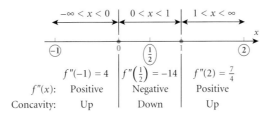

The factors in the numerator and the denominator of $f''(x)$, $x-1$, x^2+x+1, x^3, equal 0 if $x = 1$ or $x = 0$. The equation $x^2 + x + 1 = 0$ has no real solution. (Do you see why?) We use the numbers 0 and 1 to separate the number line into three parts:

$$-\infty < x < 0 \qquad 0 < x < 1 \qquad 1 < x < \infty$$

We construct Figure 23, using -1, $\frac{1}{2}$, and 2 as test numbers for $f''(x)$.

FIGURE 23

We conclude that the graph of f is concave up on the intervals $(-\infty, 0)$ and $(1, \infty)$. It is concave down on the interval $(0, 1)$. The only inflection point occurs at the point $(1, f(1)) = (1, 0)$, where the concavity changes from down to up. Note that even though the concavity changes from up to down at $x = 0$, there is no inflection point since 0 is not in the domain of f.

NOW WORK PROBLEM 13.

To the six steps we used earlier to graph a function, we now add Step 7:

> **Steps for Graphing Functions**
>
> **STEP 1** Find the domain of f.
> **STEP 2** Locate the intercepts of f. (Skip the x-intercepts if they are too hard to find.)
> **STEP 3** Determine where the graph of f is increasing and where it is decreasing.
> **STEP 4** Find any local maxima or local minima of f by using the First Derivative Test.
> **STEP 5** Locate all points on the graph of f at which the tangent line is either horizontal or vertical.
> **STEP 6** Determine the end behavior and locate any asymptotes.
> **STEP 7** Locate the inflection points, if any, of the graph by determining the concavity of the graph.

EXAMPLE 4 Graphing Functions

Graph the function
$$f(x) = 3x^5 - 5x^4$$

SOLUTION **STEP 1** The domain of f is all real numbers.

STEP 2 Let $x = 0$. Then $y = f(0) = 0$. The y-intercept is $(0, 0)$.
To find the x-intercepts, if any, let $y = 0$. Then

$$3x^5 - 5x^4 = 0 \qquad y = f(x) = 0.$$
$$x^4(3x - 5) = 0 \qquad \text{Factor.}$$
$$x^4 = 0 \quad \text{or} \quad 3x - 5 = 0 \qquad \text{Apply the Zero-Product Property.}$$
$$x = 0 \quad \text{or} \quad x = \frac{5}{3} \qquad \text{Solve for } x.$$

The x-intercepts are $(0, 0)$ and $\left(\dfrac{5}{3}, 0\right)$.

STEP 3 To determine where the graph of f is increasing and where it is decreasing, we find $f'(x)$:

$$f(x) = 3x^5 - 5x^4$$
$$f'(x) = 15x^4 - 20x^3 \qquad \text{Differentiate.}$$
$$= 5x^3(3x - 4) \qquad \text{Factor.}$$

The solutions of the equation $f'(x) = 5x^3(3x - 4) = 0$ are 0 and $\dfrac{4}{3}$.
These numbers separate the number line into three parts:

$$-\infty < x < 0 \qquad 0 < x < \frac{4}{3} \qquad \frac{4}{3} < x < \infty$$

We construct Figure 24, using -1, 1, and 2 as test numbers for $f'(x)$.

FIGURE 24

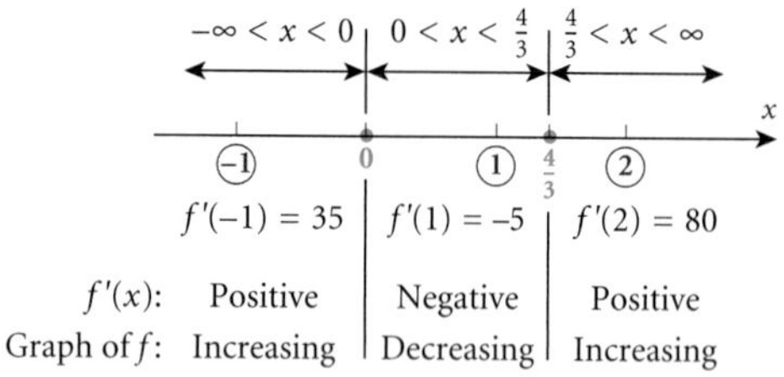

$-\infty < x < 0 \mid 0 < x < \frac{4}{3} \mid \frac{4}{3} < x < \infty$

$f'(-1) = 35 \quad f'(1) = -5 \quad f'(2) = 80$

| $f'(x)$: | Positive | Negative | Positive |
| Graph of f: | Increasing | Decreasing | Increasing |

We conclude that the graph of f is increasing on the intervals $(-\infty, 0)$ and $\left(\dfrac{4}{3}, \infty\right)$. It is decreasing on the interval $\left(0, \dfrac{4}{3}\right)$.

STEP 4 From Figure 24, the graph of f is increasing for $-\infty < x < 0$ and is decreasing for $0 < x < \dfrac{4}{3}$. At $x = 0$, the graph changes from increasing to decreasing. Consequently, at the point $(0, f(0)) = (0, 0)$ there is a local maximum.

Similarly, at $x = \dfrac{4}{3}$, the graph changes from decreasing to increasing. At the point $\left(\dfrac{4}{3}, f\left(\dfrac{4}{3}\right)\right) = \left(\dfrac{4}{3}, -\dfrac{256}{81}\right)$ there is a local minimum.

STEP 5 The derivative of f is $f'(x) = 5x^3(3x - 4)$. We see that $f'(x) = 0$ if $x = 0$ or $x = \dfrac{4}{3}$. The graph of f has a horizontal tangent line at the points $(0, 0)$ and $\left(\dfrac{4}{3}, -\dfrac{256}{81}\right)$. There are no vertical tangent lines.

STEP 6 The function f is a polynomial function. Its graph will behave like $y = 3x^5$ for x unbounded. Polynomial functions have no asymptotes.

STEP 7 To locate the inflection points, if any, of f, we find $f''(x)$:

$$f(x) = 3x^5 - 5x^4$$
$$f'(x) = 15x^4 - 20x^3 \quad \text{Differentiate.}$$
$$f''(x) = 60x^3 - 60x^2 \quad \text{Differentiate.}$$
$$= 60x^2(x - 1) \quad \text{Factor.}$$

We use the numbers 0 and 1 to separate the number line into three parts:

$$-\infty < x < 0 \qquad 0 < x < 1 \qquad 1 < x < \infty$$

We construct Figure 25. Using -1, $\dfrac{1}{2}$, and 2 as test numbers for $f''(x)$.

FIGURE 25

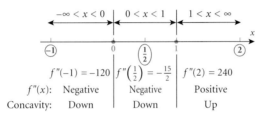

We conclude that the graph of f is concave down on $(-\infty, 1)$ and is concave up on $(1, \infty)$. Further, since the concavity changes at the point $(1, f(1)) = (1, -2)$, we conclude that $(1, -2)$ is an inflection point.

To graph f, we plot the intercepts, the local maximum, the local minimum, and the inflection point, and connect these points with a smooth curve. See Figure 26.

FIGURE 26

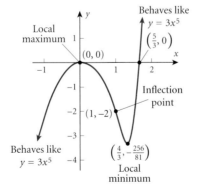

CHECK: Use a graphing utility to graph $f(x) = 3x^5 - 5x^4$. Be sure to choose a viewing rectangle that will show a complete graph. [Use Figure 26 as a guide.] Use TRACE to confirm where f is increasing and where f is decreasing. Use MAXIMUM and MINIMUM to confirm the local maximum and the local minimum. Use ZERO (or ROOT) to locate the x-intercepts. Use a TABLE to confirm the end behavior. Can you see where the graph is concave down and where it is concave up?

NOW WORK PROBLEM 29.

EXAMPLE 5 Graphing a Function

Graph the function: $f(x) = xe^{-x}$

SOLUTION **STEP 1** The domain of f is all real numbers.
STEP 2 Let $x = 0$. Then $y = f(0) = 0$, so $(0, 0)$ is the y-intercept.
Let $y = 0$. Then $xe^{-x} = 0$. Since $e^{-x} > 0$ for all x, the only solution is $x = 0$. The x-intercept is also $(0, 0)$.
STEP 3 To find where the graph of f is increasing and where it is decreasing, we find $f'(x)$:

$$f'(x) = \frac{d}{dx}(xe^{-x}) = x\frac{d}{dx}e^{-x} + e^{-x}\frac{d}{dx}x = x \cdot e^{-x}(-1) + e^{-x} \cdot 1$$
$$\uparrow \text{Derivative of a product} \qquad \uparrow \frac{d}{dx}e^{-x} = e^{-x}\frac{d}{dx}(-x) = e^{-x}(-1)$$
$$= e^{-x}(-x + 1) = (1 - x)e^{-x}$$

Since $e^{-x} > 0$ for all x, it follows that $f'(x) > 0$ if $-\infty < x < 1$, and $f'(x) < 0$ if $1 < x < \infty$. The graph of f is increasing on the interval $(-\infty, 1)$ and is decreasing on the interval $(1, \infty)$.
STEP 4 Using the First Derivative Test, we conclude that at the point $(1, e^{-1}) = (1, 0.368)$ there is a local maximum.
STEP 5 Since $f'(1) = 0$, the tangent line is horizontal at the point $(1, 0.368)$.
STEP 6 To find the end behavior, we look at the limits at infinity and use a table. See Table 1. We conclude that

$$\lim_{x \to -\infty} f(x) = \lim_{x \to -\infty}(xe^{-x}) = -\infty \quad \text{and} \quad \lim_{x \to \infty} f(x) = \lim_{x \to \infty}(xe^{-x}) = 0$$

This meas the graph of f becomes unbounded in the negative direction as $x \to -\infty$ and that $y = 0$ is a horizontal asymptotes as $x \to \infty$.
STEP 7 To locate any inflection points, we find $f''(x)$.

$$f''(x) = \frac{d}{dx}f'(x) = \frac{d}{dx}[(1-x)e^{-x}] = (1-x)\frac{d}{dx}e^{-x} + e^{-x}\frac{d}{dx}(1-x)$$
$$\uparrow \text{Derivative of a product}$$
$$= (1-x) \cdot e^{-x}(-1) + e^{-x}(-1) = e^{-x}[(1-x)(-1) + (-1)]$$
$$= (x-2)e^{-x} \qquad \uparrow \text{Factor out } e^{-x}$$

It follows that $f''(x) < 0$ if $x < 2$, and $f''(x) > 0$ if $x > 2$. The graph of f is concave down on the interval $(-\infty, 2)$ and is concave up on the interval $(2, \infty)$. The point $(2, 2e^{-2}) = (2, 0.27)$ is an inflection point.
See Figure 27 for the graph.

TABLE 1

x	$f(x) = xe^{-x}$
10	$4.5 \, E^{-4}$
100	$3.7 \, E^{-42}$
1000	0
$x \to \infty$	$f(x) \to 0$
-10	$-2.2 \, E^5$
-100	$-2.7 \, E^{45}$
-1000	$-1.4 \, E^{89}$
$x \to -\infty$	$f(x) \to -\infty$

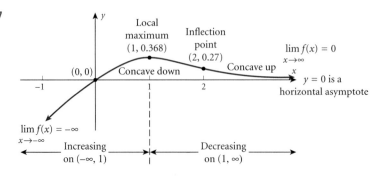

FIGURE 27

CHECK: Use a graphing utility to graph $f(x) = xe^{-x}$. Be sure to choose a viewing rectangle that will show a complete graph. [Use Figure 27 as a guide.] Use TRACE to confirm where f is increasing and where f is decreasing. Use MAXIMUM to confirm the local maximum. Use a TABLE to confirm the end behavior. Can you see where the graph is concave down and where it is concave up?

NOW WORK PROBLEM 45.

The Second Derivative Test

There is another test for finding the local maximum and local minimum that is sometimes easier to use than the First Derivative Test. This test is based on the geometric observation that the graph of a function has a local maximum at a point where the tangent line is horizontal and the graph is concave down. See Figure 28(a).

Similarly, the graph of a function has a local minimum at a point where the tangent line is horizontal and the graph is concave up. See Figure 28(b). This leads us to formulate the Second Derivative Test.

FIGURE 28

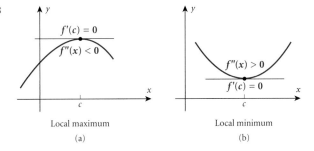

Second Derivative Test

Let $y = f(x)$ be a function that is differentiable on an open interval I and suppose that the second derivative $f''(x)$ exists on I. Also suppose c is a number in I for which $f'(c) = 0$.

1. If $f''(c) < 0$, then at the point $(c, f(c))$ there is a local maximum.
2. If $f''(c) > 0$, then at the point $(c, f(c))$ there is a local minimum.
3. If $f''(c) = 0$, the test is inconclusive and the First Derivative Test must be used.

Notice that the Second Derivative Test is used to locate those local maxima and local minima that occur where the tangent line is horizontal. Since local maxima and local minima can also occur at points on the graph at which the derivative does not exist, this test is usually only used for functions that are differentiable. Also note that the test provides no information if the second derivative is zero at c. Whenever either of these situations arise, you must use the First Derivative Test.

EXAMPLE 6 Using the Second Derivative Test

Use the Second Derivative Test to find the local maxima and local minima of
$$f(x) = x^3 - 12x^2$$

SOLUTION First we find those numbers at which $f'(x) = 0$:

$$f(x) = x^3 - 12x^2$$
$$f'(x) = 3x^2 - 24x$$
$$= 3x(x - 8)$$

$f'(x) = 0$ at $x = 0$ and $x = 8$.

Next we evaluate $f''(x)$ at these numbers:

$$f''(x) = 6x - 24 \qquad \frac{d}{dx}f'(x) = \frac{d}{dx}(3x^2 - 24x) = 6x - 24$$

At $x = 0$, $f''(0) = -24 < 0$. By the Second Derivative Test, f has a local maximum at $(0, 0)$.

At $x = 8$, $f''(8) = 24 > 0$. By the Second Derivative Test, f has a local minimum at $(8, -256)$.

[The graph of $f(x) = x^3 - 12x^2$ was given earlier in Figure 21.]

NOW WORK PROBLEM 47.

EXAMPLE 7 Finding a Graph

Sketch the graph of a differentiable function $y = f(x)$, $0 \leq x \leq 6$, which has the following properties:

1. The points $(0, 4)$, $(1, 3)$, $(3, 5)$, $(5, 7)$, and $(6, 6)$ are on the graph.
2. $f'(1) = 0$ and $f'(5) = 0$; $f'(x)$ is not 0 anywhere else.
3. $f''(x) > 0$ for $x < 3$ and $f''(x) < 0$ for $x > 3$.

SOLUTION First we plot the points $(0, 4)$, $(1, 3)$, $(3, 5)$, $(5, 7)$, and $(6, 6)$. Since we are told that $f'(1) = 0$ and $f'(5) = 0$, we know the tangent lines to the graph at $(1, 3)$ and at $(5, 7)$ are horizontal. See Figure 29(a).

Since $f''(x) > 0$ if $x < 3$, it follows that $f''(1) > 0$. By the Second Derivative Test, at the point $(1, 3)$ there is a local minimum. Similarly, since $f''(x) < 0$ if $x > 3$, it follows that $f''(5) < 0$, so that at the point $(5, 7)$ there is a local maximum. See Figure 29(b).

Finally, since the graph is concave up for $x < 3$ and concave down for $x > 3$, it follows that the point $(3, 5)$ is an inflection point. See Figure 29(c).

FIGURE 29

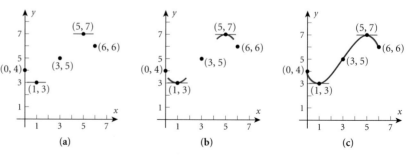

NOW WORK PROBLEM 55.

Applications

Interpreting an Inflection Point

EXAMPLE 8 **Interpreting an Inflection Point**

The sales $S(x)$, in thousands of dollars, of the Big Apple, a manufacturer of computers, is related to the amount x, in thousands of dollars, that the company spends on advertising by the function

$$S(x) = -0.02x^3 + 1.2x^2 + 1000 \qquad 0 \leq x \leq 60$$

SOLUTION Find the inflection point of the function S and discuss its significance.

The first and second derivatives of S are

$$S'(x) = -0.06x^2 + 2.4x = -0.06x(x - 40)$$
$$S''(x) = -0.12x + 2.4 = -0.12(x - 20)$$

To determine the concavity, we solve $S''(x) > 0$ and $S''(x) < 0$. Then,

$$S''(x) > 0 \quad \text{for} \quad x < 20 \qquad \text{and} \qquad S''(x) < 0 \quad \text{for} \quad x > 20$$

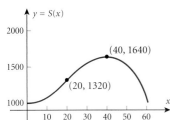

FIGURE 30

The point $(20, 1320)$ is an inflection point of the function S. Figure 30 illustrates the graph of S.

Notice that the sales of the company increase very slowly at the beginning, but as advertising money increases, sales increase rapidly. This rapid sales growth indicates that consumers are responding to the advertisement. However, there is a point on the graph at which the *rate of growth* of sales changes from positive to negative—where the rate of growth changes from increasing to decreasing—resulting in a slower rate of increased sales. This point, commonly known as the point of *diminishing returns*, is the point of inflection of S, the point $(20, 1320)$. ▶

Average Cost If the cost C of producing x units is given by the cost function $C = C(x)$, the **average cost** $\overline{C}(x)$ of producing x units is defined as

$$\overline{C}(x) = \frac{C(x)}{x}$$

EXAMPLE 9 **Analyzing Marginal Cost and Average Cost**

Consider the cost function $C(x) = 1000 + \dfrac{1}{10}x^2$, where x is the number of units manufactured.

(a) What is the average cost?
(b) What is the minimum average cost?
(c) Find the marginal cost.
(d) Where does average cost equal marginal cost?
(e) Graph the average cost function and the marginal cost function using the same coordinate system.

SOLUTION (a) The average cost $\overline{C}(x)$ is

$$\overline{C}(x) = \frac{C(x)}{x} = \frac{1000 + \frac{1}{10}x^2}{x} = \frac{1000}{x} + \frac{1}{10}x$$

(b) To find the minimum average cost, we use the Second Derivative Test. We begin by finding $\overline{C}'(x)$:

$$\overline{C}'(x) = \frac{-1000}{x^2} + \frac{1}{10} = \frac{-10{,}000 + x^2}{10x^2}$$

$$\overline{C}'(x) = 0 \quad \text{when } x^2 - 10{,}000 = 0 \quad \text{or} \quad x = \pm 100$$

We disregard $x = -100$, since the number of units manufactured must be positive. Next, we find the second derivative of $\overline{C}(x)$. Since

$$\overline{C}'(x) = \frac{-1000}{x^2} + \frac{1}{10}$$

we have

$$\overline{C}''(x) = \frac{d}{dx}\left(\frac{-1000}{x^2} + \frac{1}{10}\right) = \frac{d}{dx}\frac{-1000}{x^2} + 0 = -1000\frac{d}{dx}x^{-2} = -1000 \cdot (-2)x^{-3} = \frac{2000}{x^3}$$

Since $\overline{C}''(100) = \dfrac{2000}{1{,}000{,}000} > 0$, it follows by the Second Derivative Test that $\overline{C}(x)$ has a local minimum at $x = 100$. The minimum average cost is therefore

$$\overline{C}(100) = \frac{C(100)}{100} = \frac{2000}{100} = 20$$

(c) The cost function is $C(x) = 1000 + \dfrac{1}{10}x^2$. The marginal cost is

$$C'(x) = \frac{d}{dx}\left(1000 + \frac{1}{10}x^2\right) = \frac{1}{10}(2x) = \frac{1}{5}x$$

(d) Average cost equals marginal cost if

$$\overline{C}(x) = C'(x)$$

$$\frac{1000}{x} + \frac{1}{10}x = \frac{1}{5}x$$

$$\frac{1000}{x} = \frac{1}{10}x$$

$$x^2 = 10{,}000$$

$$x = 100$$

When 100 items are produced, average cost equals marginal cost.

(e) See Figure 31 for the graph.

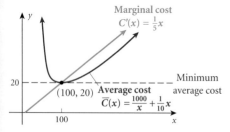

FIGURE 31

Notice that the minimum average cost occurs at the intersection of the average cost and the marginal cost. In fact, this is always the case.

> The minimum average cost occurs where the average cost and marginal cost are equal.

NOW WORK PROBLEM 61.

Concavity: the Second Derivative Test

LOGISTIC CURVES Many models in business and economics require the use of a *logistic curve*. Curves that describe a situation in which the rate of growth is slow at first, increases to a maximum rate, and then decreases are called **logistic curves,** or **saturation curves.** These curves are best characterized by their "S" shape. Figure 32 illustrates a typical general logistic curve.

FIGURE 32

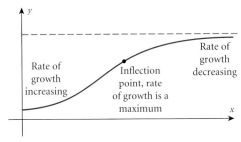

Logistic curve

EXAMPLE 10 Analyzing a Logistic Curve

The annual sales of a 42" plasma TV are expected to follow the logistic curve

$$f(x) = \frac{10{,}000}{1 + 100e^{-x}}$$

where x is measured in years and $x = 0$ corresponds to 2000, the year production begins. Analyze the graph of this function and determine the year in which a maximum sales rate is achieved.

SOLUTION We follow the 7 steps for graphing a function. (p. 369).

STEP 1 The domain of this function is $\{x | x \geq 0\}$
STEP 2 If $x = 0$, then

$$y = f(0) = \frac{10{,}000}{1 + 100} = 99.01$$

The y-intercept is $(0, 99.01)$. There is no x-intercept. The y-intercept represents the predicted number of plasma TV's sold when production begins.

STEP 3 The derivative of the function f is

$$f'(x) = \frac{d}{dx}[10{,}000(1 + 100e^{-x})^{-1}] = 10{,}000 \cdot (-1)(1 + 100e^{-x})^{-2} \cdot 100(-1)e^{-x} = \frac{1{,}000{,}000 e^{-x}}{(1 + 100e^{-x})^2}$$

Since $e^{-x} > 0$ for all x, it follows that $f'(x) > 0$ for $x \geq 0$. The function f is increasing, which means that sales are increasing each year.

STEP 4 There are no local maxima since the graph is increasing.
STEP 5 There are no vertical or horizontal tangent lines.
STEP 6 As $x \to \infty$, we have $e^{-x} \to 0$. As a result,

$$\lim_{x \to \infty} f(x) = \lim_{x \to \infty} \frac{10{,}000}{1 + 100e^{-x}} = 10{,}000$$

This means $y = 10{,}000$ is a horizontal asymptote as x becomes unbounded in the positive direction. This number represents the upper estimate for sales.

STEP 7 To locate any inflection points, we find $f''(x)$. You should verify that

$$f''(x) = 1{,}000{,}000 e^{-x} \left[\frac{100 e^{-x} - 1}{(1 + 100 e^{-x})^3} \right]$$

The sign of $f''(x)$ is controlled by the numerator since $1 + 100 e^{-x} > 0$ for all x. Now $100 e^{-x} - 1 = 0$ if $e^x = 100$, which happens if $x = \ln 100 = 4.6$. If $x < 4.6$, we have $100 e^{-x} - 1 > 0$, so $f''(x) > 0$, and the graph of f is concave up. If $x > 4.6$, we have $100 e^{-x} - 1 < 0$, so $f''(x) < 0$, and the graph of f is concave down. At $(4.6, 5000)$ there is an inflection point. At this point, the first derivative, f', achieves its maximum value. That is, at 4.6 years, around the middle of 2004, the rate of growth in sales is a maximum.

The graph is given in Figure 33.

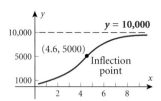

FIGURE 33

 CHECK: Graph $y = \dfrac{10{,}000}{1 + 100 e^{-x}}$ and compare the result with Figure 33.

EXERCISE 5.3

In Problems 1–12, use the graph of $y = f(x)$ given on the right.

1. What is the domain of f?
2. List the intercepts of f.
3. On what intervals, if any, is the graph of f increasing?
4. On what intervals, if any, is the graph of f decreasing?
5. For what values of x does $f'(x) = 0$?
6. For what values of x does $f'(x)$ not exist?
7. List the point(s) at which f has a local maximum.
8. List the point(s) at which f has a local minimum.
9. On what intervals, if any, is the graph of f concave up?
10. On what intervals, if any, is the graph of f concave down?
11. List any asymptotes.
12. List any vertical tangent lines.

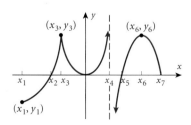

In Problems 13–28, determine the intervals on which the graph of f is concave up and concave down. List any inflection points.

13. $f(x) = x^3 - 6x^2 + 1$
14. $f(x) = x^3 + 3x^2 + 2$
15. $f(x) = x^4 - 2x^3 + 6x - 1$
16. $f(x) = x^4 + 2x^3 - 8x + 8$
17. $f(x) = 3x^5 - 5x^4 + 60x + 10$
18. $f(x) = 3x^5 + 5x^4 + 20x - 4$
19. $f(x) = 3x^5 - 10x^3 + 10x + 10$
20. $f(x) = 3x^5 - 10x^3 - 8x + 8$
21. $f(x) = x^5 - 10x^2 + 4$

22. $f(x) = x^5 + 10x^2 - 8x + 4$

23. $f(x) = 3x^{1/3} + 9x + 2$

24. $f(x) = 3x^{2/3} - 8x + 4$

25. $f(x) = x^{2/3}(x - 10)$

26. $f(x) = x^{2/3}(x - 15)$

27. $f(x) = x^{2/3}(x^2 - 16)$

28. $f(x) = x^{2/3}(x^2 - 4)$

In Problems 29–46, follow the seven steps on page 369 to graph f.

29. $f(x) = x^3 - 6x^2 + 1$

30. $f(x) = x^3 + 6x^2 + 2$

31. $f(x) = x^4 - 2x^2 + 1$

32. $f(x) = 2x^4 - 4x^2 + 2$

33. $f(x) = x^5 - 10x^4$

34. $f(x) = x^5 + 5x$

35. $f(x) = x^6 - 3x^5$

36. $f(x) = x^6 + 3x^5$

37. $f(x) = 3x^4 - 12x^3$

38. $f(x) = 3x^4 + 12x^3$

39. $f(x) = x^5 - 10x^2 + 4$

40. $f(x) = x^5 + 10x^2 + 2$

41. $f(x) = x^{2/3}(x - 10)$

42. $f(x) = x^{2/3}(x - 15)$

43. $f(x) = x^{2/3}(x^2 - 16)$

44. $f(x) = x^{2/3}(x - 4)$

45. $f(x) = xe^x$

46. $f(x) = x^2 e^x$

In Problems 47–54, determine where $f'(x) = 0$. Use the Second Derivative Test to determine the local maxima and local minima of each function.

47. $f(x) = x^3 - 3x + 2$

48. $f(x) = x^3 - 12x - 4$

49. $f(x) = 3x^4 + 4x^3 - 3$

50. $f(x) = 3x^4 - 6x^2 + 4$

51. $f(x) = x^5 - 5x^4 + 2$

52. $f(x) = 3x^5 - 20x^3$

53. $f(x) = x + \dfrac{1}{x}$

54. $f(x) = 2x + \dfrac{1}{x^2}$

---

55. Finding a Graph Sketch the graph of a differentiable function $y = f(x)$ that has the following properties:

1. $(0, 10), (6, 15)$, and $(10, 0)$ are on the graph.
2. $f'(6) = 0$ and $f'(10) = 0$; $f'(x)$ is not 0 anywhere else.
3. $f''(x) < 0$ for $x < 9$, $f''(9) = 0$, and $f''(x) > 0$ for $x > 9$.

56. Finding a Graph Sketch the graph of a differentiable function $y = f(x)$ that has the following properties:

1. $(-1, 3), (1, 5)$, and $(3, 7)$ are on the graph.
2. $f'(3) = 0$ and $f'(-1) = 0$; $f'(x)$ is not 0 anywhere else.
3. $f''(x) > 0$ for $x < 1$, $f''(1) = 0$, and $f''(x) > 0$ for $x > 1$.

57. Finding a Graph Sketch the graph of a differentiable function $y = f(x)$ that has the following properties:

1. $(1, 5), (2, 3)$ and $(3, 1)$ are on the graph.
2. $f'(1) = 0$ and $f'(3) = 0$; $f'(x)$ is not 0 anywhere else.
3. $f''(x) < 0$ for $x < 2$, $f''(2) = 0$, and $f''(x) > 0$ for $x > 2$.

58. Finding a Graph Sketch the graph of a differentiable function $y = f(x)$ that has the following properties:

1. Domain of $f(x)$ is $x \geq 0$.
2. $(0, 0)$ and $(6, 7)$ are on the graph.
3. $f'(x) > 0$ for $x > 0$.
4. $f''(x) < 0$ for $x < 6$, $f''(6) = 0$, and $f''(x) > 0$ for $x > 6$.

59. For the function $f(x) = ax^3 + bx^2$, determine a and b so that the point $(1, 6)$ is a point of inflection of $f(x)$.

60. Let $f(x) = ax^2 + bx + c$, where $a \neq 0$, b, and c are real numbers. Is it possible for $f(x)$ to have an inflection point? Explain your answer.

61. Average Cost The cost function for producing x items is

$$C(x) = 2x^2 + 50$$

(a) Find the average cost function.
(b) What is the minimum average cost?
(c) Find the marginal cost function.
(d) Graph the average cost function and the marginal cost function on the same set of axes. Label their point of intersection.
(e) Interpret the point of intersection.

62. Average Cost The cost function for producing x items is

$$C(x) = x^2 - 3x + 625$$

(a) Find the average cost function.
(b) What is the minimum average cost?
(c) Find the marginal cost function.
(d) Graph the average cost function and the marginal cost function on the same set of axes. Label their point of intersection.
(e) Interpret the point of intersection.

63. Average Cost The cost function for producing x items is

$$C(x) = 500 + 10x + \frac{x^2}{500}$$

(a) Find the average cost function.
(b) What is the minimum average cost?
(c) Find the marginal cost function.
(d) Graph the average cost function and the marginal cost function on the same set of axes. Label their point of intersection.
(e) Interpret the point of intersection.

64. Average Cost The cost function for producing x items is

$$C(x) = 800 + 0.04x + 0.0002x^2$$

(a) Find the average cost function.
(b) What is the minimum average cost?
(c) Find the marginal cost function.
(d) Graph the average cost function and the marginal cost function on the same set of axes. Label their point of intersection.
(e) Interpret the point of intersection.

65. Spread of Rumor In a city of 50,000 people, the number of people who have heard a certain rumor after t days obeys

$$N(t) = \frac{50,000}{1 + 49,999e^{-t}}$$

(a) Find the domain of N.
(b) Locate the y-intercept and the t-intercepts, if any.
(c) Determine where N is increasing and where it is decreasing.
(d) Determine where N is concave up and where it is concave down.
(e) Locate any inflection points.
(f) Graph N.
(g) At what time is the rumor spreading at the greatest rate?

66. Analyze the Function:

$$f(x) = \frac{2000}{1 + 4e^{-x}}$$

(a) Find the domain of f.
(b) Locate the y-intercept and the x-intercepts, if any.

(c) Determine where f is increasing and where it is decreasing.
(d) Determine where f is concave up and where it is concave down.
(e) Locate any inflection points.
(f) Graph f.

67. Bacteria Growth Rate A bacteria population grows from an initial population of 800 to a population $p(t)$ at time t (measured in days) according to the logistic curve

$$p(t) = \frac{800e^t}{1 + \frac{1}{10}(e^t - 1)}$$

(a) Determine the growth rate $p'(t)$.
(b) When is the growth rate a maximum?
(c) Determine $\lim_{t \to \infty} p(t)$, the equilibrium population.
(d) Graph the function.

68. Logistic Curve The sales of a new stereo system over a period of time are expected to follow the logistic curve

$$f(x) = \frac{5000}{1 + 5e^{-x}}$$

where x is measured in years.

(a) Determine the year in which the sales rate is a maximum.
(b) Graph the function.

69. Logistic Curve The sales of a new car model over a period of time are expected to follow the relationship

$$f(x) = \frac{20,000}{1 + 50e^{-x}}$$

where x is measured in months.

(a) Determine the month in which the sales rate is a maximum.
(b) Graph the function.

70. Spread of Disease In a town of 50,000 people, the number of people at time t who have influenza is

$$N(t) = \frac{10,000}{1 + 9999e^{-t}}$$

where t is measured in days. [Note that the flu is spread by the one person who has it at $t = 0$.]

(a) At what time t is the rate of spreading the greatest?
(b) Graph the function.

5.4 Optimization

PREPARING FOR THIS SECTION *Before getting started, review the following:*

>> Geometry Review (Chapter 0, Section 0.7, pp. 71–74)

OBJECTIVES 1 Find the absolute maximum and the absolute minimum of a function
              2 Solve applied problems

Absolute Maximum and Absolute Minimum

If you look again at the examples of the previous two sections, you will notice that the local maximum and local minimum, when they existed, always occurred at a number c at which $f'(c) = 0$ or at which $f'(c)$ did not exist.

Necessary Condition for a Local Maximum or Local Minimum

If a function f has a local maximum or a local minimum at c, then either $f'(c) = 0$ or $f'(c)$ does not exist.

Because of this result, we call the numbers c for which $f'(c) = 0$ or $f'(c)$ does not exist **critical numbers of f**. The corresponding points $(c, f(c))$ on the graph of f are called **critical points**.

The local maximum (and local minimum) of a function is a point on the graph that is higher (or lower) than nearby points on the graph. However, the value of the function at such points may not be the largest (or smallest) value of the function on its domain.

The largest value, if one exists, of a function on its domain is called the **absolute maximum** of the function. The smallest value, if it exists, of a function on its domain is called the **absolute minimum** of the function.

Let's look at some of the possibilities. See Figure 34.

FIGURE 34

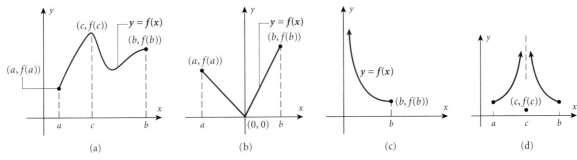

In Figure 34(a), the function f is continuous on the closed interval $[a, b]$. The absolute maximum of f is $f(c)$. Notice that c is a critical number since $f'(c) = 0$. The absolute minimum of f is $f(a)$.

In Figure 34(b), the function f is continuous on the closed interval $[a, b]$. The absolute maximum of f is $f(b)$. The absolute minimum of f is $f(0) = 0$. Notice that 0 is a critical number since $f'(0)$ does not exist.

In Figure 34(c), the function f is continuous on the interval $(0, b]$. The function f has no absolute maximum since there is no largest value of f on $(0, b]$. The absolute minimum is $f(b)$.

In Figure 34(d), the function f, whose domain is the closed interval $[a, b]$, is discontinuous at c. The function f has no absolute maximum since there is no largest value of f on $[a, b]$. The absolute minimum is $f(c)$.

The following result, which we state without proof, gives a condition under which a function will have an absolute maximum and an absolute minimum.

Condition for a Function to Have an Absolute Maximum and an Absolute Minimum

If a continuous function has as its domain a closed interval, the absolute maximum and absolute minimum exist.

Continuous functions defined on a closed interval will have an absolute maximum and an absolute minimum. Look again at Figures 34(a) and (b). We see that each function f is continuous on a closed interval $[a, b]$. Note that the absolute maximum and the absolute minimum occur either at a critical number or at an endpoint. This leads us to formulate the following test for finding the absolute maximum and the absolute minimum.

Test for Absolute Maximum and Absolute Minimum

If a continuous function $y = f(x)$ has a closed interval $[a, b]$ as its domain, we can find the absolute maximum (minimum) by choosing the largest (smallest) value from among the following:

1. Values of f at the critical numbers in the open interval (a, b)
2. $f(a)$
3. $f(b)$

If critical numbers of $y = f(x)$ are found that are not in the interval $[a, b]$, these critical numbers should be ignored since we are concerned only with the function on the interval $[a, b]$.

EXAMPLE 1   Finding the Absolute Maximum and the Absolute Minimum

Consider the function $f(x) = x^3 - 3x$. If the domain of f is $[0, 2]$, find the absolute maximum and the absolute minimum of f.

SOLUTION The function $f(x) = x^3 - 3x$, is a polynomial function so it is continuous on $[0, 2]$. To find the absolute maximum and absolute minimum, we first find $f'(x)$ so we can find any critical numbers of f,

$$f'(x) = 3x^2 - 3 = 3(x^2 - 1) = 3(x + 1)(x - 1)$$

The critical numbers of f are those values of x for which $f'(x)$ does not exist or for which $f'(x) = 0$. There are no values of x at which $f'(x)$ does not exist. The values of x for which $f'(x) = 0$ are found as follows;

$$f'(x) = 3(x + 1)(x - 1) = 0$$
$$x = -1 \quad \text{and} \quad x = 1$$

We ignore the critical number $x = -1$ since it is not in the domain, $0 \leq x \leq 2$. For the critical number $x = 1$, we have

$$f(1) = -2 \quad f(x) = x^3 - 3x$$

The values of f at the endpoints 0 and 2 of the interval $[0, 2]$ are

$$f(0) = 0 \quad \text{and} \quad f(2) = 2$$

We choose the largest and smallest of -2, 0, and 2. The absolute maximum of f on $[0, 2]$ is 2 and the absolute minimum is -2.

NOW WORK PROBLEM 1.

EXAMPLE 2 Finding the Absolute Maximum and the Absolute Minimum

Find the absolute maximum and absolute minimum of the function $f(x) = x^{2/3}$ on the interval $[-1, 8]$. Graph f.

SOLUTION The function $f(x) = x^{2/3}$ is continuous on $[-1, 8]$. First, we locate the critical numbers, if any, by finding $f'(x)$.

$$f'(x) = \frac{d}{dx} x^{2/3} = \frac{2}{3} x^{-1/3} = \frac{2}{3x^{1/3}}$$

The only critical numbers are those for which $f'(x)$ does not exist, so the only critical number is $x = 0$, for which $f(0) = 0$.

Next, we find the values of f at the endpoints:

$$f(-1) = 1 \quad \text{and} \quad f(8) = 4$$

We choose the largest and smallest of 0, 1, and 4. The absolute maximum of $f(x) = x^{2/3}$ on $[-1, 8]$ is 4 and the absolute minimum is 0.

The graph of f is given in Figure 35.

FIGURE 35

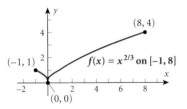

Solve applied problems 2

Applications

In general, each type of problem we will discuss requires some quantity to be minimized or maximized. We assume that the quantity we want to optimize can be represented by a function. Once this function is determined, the problem can be reduced to the question of determining at what number the function assumes its absolute maximum or absolute minimum.

Even though each applied problem has its unique features, it is possible to outline in a rough way a procedure for obtaining a solution. This five-step procedure is:

> **Steps for Solving Applied Problems**
>
> **STEP 1** Identify the quantity for which a maximum or a minimum value is to be found.
> **STEP 2** Assign symbols to represent other variables in the problem. If possible, use an illustration to assist you.
> **STEP 3** Determine the relationships among these variables.
> **STEP 4** Express the quantity to be optimized as a function of one of these variables. Be sure to state the domain.
> **STEP 5** Apply the test for absolute maximum and absolute minimum to this function.

The following examples illustrate this procedure.

EXAMPLE 3 Maximizing Volume

From each corner of a square piece of sheet metal 18 centimeters on a side, remove a small square of side x centimeters and turn up the edges to form an open box. What should be the dimensions of the box so as to maximize the volume?

SOLUTION **STEP 1** The quantity to be maximized is the volume. Denote it by V.
STEP 2 Denote the dimensions of the side of the small square by x, as shown in Figure 36. Although the area of the sheet metal is fixed, the sides of the square can be changed and so are variables. Let y denote the portion left after cutting the x's to make the square.
STEP 3 Then

$$y = 18 - 2x$$

STEP 4 The height of the box is x, while the area of the base of the box is y^2. The volume V is therefore

$$V = xy^2$$

To express V as a function of one variable, we substitute $y = 18 - 2x$ in the formula for V. This gives

$$V = V(x) = x(18 - 2x)^2 \qquad V = xy^2, y = 18 - 2x$$

This is the function to be maximized. Its domain is the set of real numbers. However, physically, the only numbers x that make sense are those between 0 and 9. We want to find the absolute maximum of

$$V(x) = x(18 - 2x)^2 \qquad 0 \le x \le 9$$

STEP 5 To find the number x that maximizes V, we differentiate and find the critical numbers, if any:

$$V'(x) = x \frac{d}{dx}(18 - 2x)^2 + (18 - 2x)^2 \frac{d}{dx} x \qquad \text{Derivative of a product.}$$
$$V'(x) = 2x(18 - 2x)(-2) + (18 - 2x)^2 \qquad \text{Differentiate.}$$
$$= (18 - 2x)[-4x + (18 - 2x)] \qquad \text{Factor.}$$
$$= (18 - 2x)(18 - 6x) \qquad \text{Factor.}$$

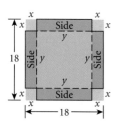

FIGURE 36

Now we set $V'(x) = 0$ and solve for x:

$$(18 - 2x)(18 - 6x) = 0 \qquad V'(x) = 0$$
$$18 - 2x = 0 \quad \text{or} \quad 18 - 6x = 0 \qquad \text{Apply the Zero-Product Property.}$$
$$x = 9 \quad \text{or} \quad x = 3 \qquad \text{Solve for } x.$$

The only critical number in the interval $(0, 9)$ is $x = 3$. We calculate the values of $V(x)$ at this critical number and at the endpoints of the interval $[0, 9]$:

$$V(0) = 0 \qquad V(3) = 3(18 - 6)^2 = 432 \qquad V(9) = 0$$

The maximum volume is 432 cubic centimeters and the dimensions of the box that yield the maximum volume are $x = 3$ centimeters deep by $y = 18 - 2(3) = 12$ centimeters on each side.

NOW WORK PROBLEM 25.

EXAMPLE 4 Maximizing Area

PLAYPEN PROBLEM* A manufacturer of playpens makes a model that can be opened at one corner and attached at right angles to a wall or the side of a house. If each side is 3 feet in length, the open configuration doubles the available area in which a child can play from 9 square feet to 18 square feet. See Figure 37(a).

Suppose hinges are placed at adjoining sides of the playpen to allow for a configuration like the one shown in Figure 37(b). If x is the distance between the two parallel sides of the playpen, then what value of x will maximize the area A? What is the maximum area?

FIGURE 37

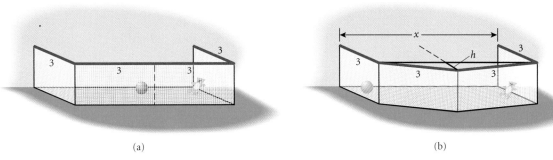

(a) (b)

SOLUTION The variable x has domain $0 \leq x \leq 6$. Do you see why? The area A consists of the area of a rectangle (with width 3 and length x) and the area of an isosceles triangles (with base x and two equal sides of length 3). The height h of the triangle is obtained using the Pythagorean theorem:

$$h^2 = 3^2 - \left(\frac{x}{2}\right)^2 \qquad\qquad h^2 + \left(\frac{x}{2}\right)^2 = 3^2$$

$$h = \sqrt{9 - \frac{x^2}{4}} = \sqrt{\frac{36 - x^2}{4}} = \frac{1}{2}\sqrt{36 - x^2}$$

The area A enclosed by the playpen is

$$A = \text{area of rectangle} + \text{area of triangle}$$
$$= 3x + \frac{1}{2} x h = 3x + \frac{1}{2} x \left(\frac{1}{2}\sqrt{36 - x^2}\right)$$
$$A(x) = 3x + \frac{1}{4} x \sqrt{36 - x^2}$$

*Adapted from Proceedings, Summer Conference for College Teachers on Applied Mathematics, *University of Missouri–Rolla*, 1971.

To find the absolute maximum of $A = A(x)$, $0 \leq x \leq b$, we first find the critical numbers by examining $A'(x)$.

$$A'(x) = \frac{d}{dx}[3x + \frac{1}{4}x\sqrt{36 - x^2}]$$

$$= \frac{d}{dx}(3x) + \frac{1}{4}\frac{d}{dx}[x(36 - x^2)^{1/2}] \quad \text{Derivative of a sum.}$$

$$= 3 + \frac{1}{4}\left[x\frac{d}{dx}(36 - x^2)^{1/2} + (36 - x^2)^{1/2}\frac{d}{dx}x\right] \quad \text{Derivative of a product.}$$

$$= 3 + \frac{1}{4}\left[x \cdot \frac{1}{2} \cdot (36 - x^2)^{-1/2}(-2x) + (36 - x^2)^{1/2} \cdot 1\right] \quad \text{Use the Power Rule.}$$

$$= 3 - \frac{1}{4} \cdot \frac{x^2}{\sqrt{36 - x^2}} + \frac{1}{4} \cdot \sqrt{36 - x^2} \quad \text{Simplify.}$$

$$= \frac{12\sqrt{36 - x^2} - x^2 + (36 - x^2)}{4\sqrt{36 - x^2}} \quad \text{Write as a single quotient.}$$

$$= \frac{12\sqrt{36 - x^2} + 36 - 2x^2}{4\sqrt{36 - x^2}} \quad \text{Simplify.}$$

Set $A'(x) = 0$. Then

$$\frac{12\sqrt{36 - x^2} + 36 - 2x^2}{4\sqrt{36 - x^2}} = 0$$

$$12\sqrt{36 - x^2} + 36 - 2x^2 = 0 \quad \text{Set the numerator equal to 0.}$$

$$12\sqrt{36 - x^2} = 2x^2 - 36 \quad \text{Isolate the square root.}$$

$$144(36 - x^2) = 4x^4 - 144x^2 + 1296 \quad \text{Square both sides.}$$

$$4x^4 = 3888 \quad \text{Simplify.}$$

$$x = \sqrt[4]{\frac{3888}{4}} \approx 5.58 \quad \text{Solve for } x.$$

$A'(x)$ does not exist at $x = -6$ and $x = 6$, but these values are not in the interval $(0, 6)$. The only critical number in $(0, 6)$ is 5.58.

Evaluate $A(x)$ at the endpoints $x = 0$ and $x = 6$, and at the critical number $x = 5.58$. The results are

$$A(0) = 0 \qquad A(6) = 18 \qquad A(5.58) = 19.82$$

A wall of length $2x = 2(5.58) = 11.16$ will maximize the area, and a configuration like the one in Figure 37(b), increases the play area by about 10% (from 18 square feet to 19.82 square feet). ▸

The next two examples illustrate how to solve optimization problems in which the function to be optimized has a domain that is not a closed interval.

EXAMPLE 5 **Minimizing Cost**

A can company wants to produce a cylindrical container with a capacity of 1000 cubic centimeters. The top and bottom of the container must be made of material that costs $0.05 per square centimeter, while the sides of the container can be made of material costing $0.03 per square centimeter. Find the dimensions that will minimize the total cost of the container.

SOLUTION Figure 38 shows a cylindrical container and the area of its top, bottom, and lateral surfaces.

FIGURE 38

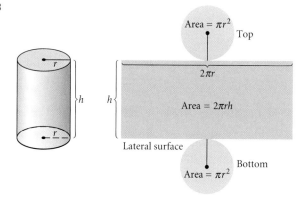

As indicated in the figure, if we let h stand for the height of the can and let r stand for the radius of the top and bottom, then the total area of the bottom and top is $2\pi r^2$ and the area of the lateral surface of the can is $2\pi rh$. The total cost C of manufacturing the can is

$$C = (\$0.05)(2\pi r^2) + (\$0.03)(2\pi rh) = 0.1\pi r^2 + 0.06\pi rh$$

This is the function we want to minimize.

The cost function is a function of two variables, h and r. But there is a relationship between h and r since the volume of the cylinder, $V = \pi r^2 h$, is fixed at 1000 cubic centimeters. That is,

$$\pi r^2 h = 1000$$

$$h = \frac{1000}{\pi r^2} \quad \text{Divide both sides by } \pi r^2.$$

Substituting this expression for h into the cost function C, we obtain

$$C = C(r) = 0.1\pi r^2 + 0.06\pi r\left(\frac{1000}{\pi r^2}\right) \quad C = 0.1\pi r^2 + 0.06\pi rh; h = \frac{1000}{\pi r^2}$$

$$= 0.1\pi r^2 + \frac{60}{r} \quad \text{Simplify.}$$

The domain of C is $\{r \mid r > 0\}$. The derivative of C with respect to r is

$$C'(r) = \frac{d}{dr}\left(0.1\pi r^2 + \frac{60}{r}\right) = 0.2\pi r - \frac{60}{r^2} \quad \frac{d}{dr}\frac{60}{r} = 60\frac{d}{dr}r^{-1} = -60r^{-2} = \frac{-60}{r^2}$$

$$= \frac{0.2\pi r^3 - 60}{r^2}$$

Set $C'(r) = 0$. Then

$$0.2\pi r^3 - 60 = 0 \quad \text{Set the numerator equal to 0.}$$

$$r^3 = \frac{300}{\pi} \quad \text{Solve for } r^3.$$

$$r = \sqrt[3]{\frac{300}{\pi}} \approx 4.57 \quad \text{Solve for } r.$$

Since $r > 0$, the only critical number is $r \approx 4.57$.

To use the Second Derivative Test, we need to evaluate C'' at this critical number.

$$C''(r) = \frac{d}{dr}\left(0.2\pi r - \frac{60}{r^2}\right) = 0.2\pi + \frac{120}{r^3}$$

and

$$C''\left(\sqrt[3]{\frac{300}{\pi}}\right) = 0.2\pi + \frac{120\pi}{300} > 0$$

By the Second Derivative Test, the cost is a local minimum for $r = \sqrt[3]{\frac{300}{\pi}} \approx 4.57$ centimeters.

Since the only physical constraint is that r be positive, this local minimum value is the absolute minimum. The corresponding height of this can is

$$h = \frac{1000}{\pi r^2} \approx 15.24 \text{ centimeters}$$

These are the dimensions that will minimize the cost of the material. ▶

If the cost of the material is the same for the top, bottom, and lateral surfaces of a cylindrical container, then the minimum cost occurs when the surface area is minimum. It can be shown that for any fixed volume, the minimum surface area is obtained when the height equals twice the radius. See Problem 39.

NOW WORK PROBLEM 29.

SEEING THE CONCEPT: Graph the function $y = 0.1\pi x^2 + \frac{60}{x}$. Use MINIMUM to verify the result of Example 5. ▶

MAXIMIZING TAX REVENUE In determining the tax rate on cars, telephones, etc., the government is always faced with the following problem: How large should the tax be so that the tax revenue will be as large as possible? Let's examine this situation. When the government places a tax on a product, the price of this product for the consumer may increase and the quantity demanded may decrease accordingly. A very large tax may cause the quantity demanded to diminish to zero with the result that no tax revenue is collected. On the other hand, if no tax is levied, there will be no tax revenue at all. The problem is to find the tax rate that optimizes tax revenue. (Tax revenue is the product of the tax rate times the actual quantity consumed.)

Let's assume that because of long-time experience in levying taxes, the government is able to determine that the relationship between the quantity q consumed of a certain product and the related tax rate t is

$$t = \sqrt{27 - 3q^2}$$

See Figure 39.

Notice that the relationship between tax rate and quantity consumed conforms to the restrictions discussed earlier. For example, when the tax rate $t = 0$, the quantity consumed is $q = 3$; when the tax is at a maximum ($t = 5.2\%$), the quantity consumed is zero.

The revenue R due to the tax rate t is the product of tax rate and the quantity consumed:

$$R = qt = q(27 - 3q^2)^{1/2}$$

where R is measured in millions of dollars. Since both q and t are assumed to be nonnegative, the domain of R is $0 \leq q \leq 3$.

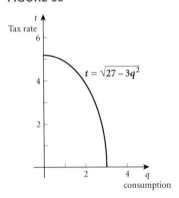

FIGURE 39

To find the absolute maximum, we differentiate R with respect to q using the formula for the derivative of a product and the Power Rule:

$$\begin{aligned}
R'(q) &= \frac{d}{dq}[q(27 - 3q^2)^{1/2}] \\
&= q\frac{d}{dq}(27 - 3q^2)^{1/2} + (27 - 3q^2)^{1/2}\frac{d}{dq}q \quad &\text{Derivative of a product.} \\
&= q \cdot \frac{1}{2}(27 - 3q^2)^{-1/2}(-6q) + (27 - 3q^2)^{1/2} \quad &\text{Apply the Power Rule.} \\
&= \frac{-3q^2}{(27 - 3q)^{1/2}} + (27 - 3q^2)^{1/2} \quad &\text{Simplify.} \\
&= \frac{-3q^2 + 27 - 3q^2}{(27 - 3q^2)^{1/2}} = \frac{27 - 6q^2}{(27 - 3q^2)^{1/2}}
\end{aligned}$$

The critical numbers obey

$$27 - 6q^2 = 0 \quad \text{and} \quad 27 - 3q^2 = 0$$
$$q^2 = 4.5 \quad \quad q^2 = 9$$
$$q = \pm\sqrt{4.5} \quad \quad q = \pm 3$$

The only critical number in the interval $(0, 3)$ is $q = \sqrt{4.5} \approx 2.12$.

To find the absolute maximum, we compare the values of R at the endpoints $q = 0$ and $q = 3$ with its value at $q = 2.12$:

$$R(0) = 0 \quad R(\sqrt{4.5}) = \sqrt{4.5}\sqrt{13.5} = (2.12)(3.67) = 7.79 \quad R(3) = 0$$

The revenue is maximized at $q = 2.12$. The tax rate corresponding to maximum revenue is

$$t = \sqrt{27 - 3q^2} = \sqrt{13.5} = 3.67$$

This means that, for a tax rate of 3.67%, a maximum revenue $R = 7.79$ million dollars is generated.

NOW WORK PROBLEM 37.

CONTRACTION OF WINDPIPE WHILE COUGHING Coughing is caused by an increase in pressure in the lungs and is accompanied by a decrease in the diameter of the windpipe. From physics, the amount V of air flowing through the windpipe is related to the radius r of the windpipe and pressure difference p at each end by the equation $V = kpr^4$, where k is a constant. The radius r will decrease with increased pressure p according to the formula $r_0 - r = cp$, where r_0 is the radius of the windpipe when there is no difference in pressure and c is a positive constant. We wish to find the radius r that allows the most air to flow through the windpipe.

We shall restrict r so that

$$0 < \frac{r_0}{2} \leq r \leq r_0$$

The amount V of air flowing through the windpipe is given by

$$V = kpr^4$$

where $cp = r_0 - r$ from which $p = \dfrac{r - r_0}{c}$.

We wish to maximize

$$V(r) = k\left(\frac{r_0 - r}{c}\right)r^4 = \frac{k}{c}r_0 r^4 - \frac{k}{c}r^5 \qquad \frac{r_0}{2} \leq r \leq r_0$$

The derivative is

$$V'(r) = 4\frac{k}{c}r_0 r^3 - 5\frac{k}{c}r^4 = \frac{k}{c}r^3(4r_0 - 5r)$$

The only critical number is $r = \frac{4r_0}{5}$. (We exclude $r = 0$ because $0 < \frac{r_0}{2} \leq r \leq r_0$.)

Using the test for an absolute maximum, we find

$$V\left(\frac{r_0}{2}\right) = \frac{k}{c}r_0 \cdot \frac{r_0^4}{16} - \frac{k}{c} \cdot \frac{r_0^5}{32} = \frac{kr_0^5}{32c} \qquad V(r_0) = 0 \qquad V\left(\frac{4r_0}{5}\right) = \frac{k}{c}r_0 \cdot \frac{256 r_0^4}{625} - \frac{k}{c} \cdot \frac{1024}{3125 r_0^5} = \frac{256\, kr_0^5}{3125c}$$

Since $\frac{256 kr_0^5}{3125c} > \frac{kr_0^5}{32c}$, the maximum air flow is obtained when the radius of the windpipe is $\frac{4}{5}r_0$; that is, the windpipe contracts by a factor of $\frac{1}{5}$.

MARGINAL ANALYSIS The revenue derived from selling x units is $R(x) = x\, d(x)$, where $p = d(x)$ is the price function. If $C(x)$ is the cost of producing x units, the **profit function P,** assuming whatever is produced can be sold, is

$$P(x) = R(x) - C(x)$$

What quantity x will maximize profit?

To maximize $P(x)$, we find the critical numbers of P:

$$\frac{d}{dx}P(x) = \frac{d}{dx}[R(x) - C(x)] = \frac{d}{dx}R(x) - \frac{d}{dx}C(x) = R'(x) - C'(x) = 0$$

$$R'(x) = C'(x)$$

We apply the Second Derivative Test to the function P:

$$P''(x) = R''(x) - C''(x)$$

The profit P has a local maximum at a number x if $P''(x) < 0$. This will occur at a number x for which the marginal revenue function equals the marginal cost and $R''(x) < C''(x)$. The numbers x are restricted to a closed interval in which the endpoints should be tested separately.

Maximizing Profit

The equality

$$\boxed{R'(x) = C'(x)}$$

is the basis for the classical economic criterion for maximum profit—that marginal revenue and marginal cost be equal.

EXERCISE 5.4

In Problems 1–24, find the absolute maximum and absolute minimum of each function f on the indicated interval.

1. $f(x) = x^2 + 2x$ on $[-3, 3]$
2. $f(x) = x^2 - 8x$ on $[-1, 10]$
3. $f(x) = 1 - 6x - x^2$ on $[0, 4]$
4. $f(x) = 4 - 2x - x^2$ on $[-2, 2]$
5. $f(x) = x^3 - 3x^2$ on $[1, 4]$
6. $f(x) = x^3 - 6x$ on $[-2, 2]$
7. $f(x) = x^4 - 2x^2 + 1$ on $[0, 1]$
8. $f(x) = 3x^4 - 4x^3$ on $[-2, 0]$
9. $f(x) = x^{2/3}$ on $[-1, 1]$
10. $f(x) = x^{1/3}$ on $[-1, 1]$
11. $f(x) = 2\sqrt{x}$ on $[1, 4]$
12. $f(x) = 4 - \sqrt{x}$ on $[0, 4]$
13. $f(x) = x\sqrt{1 - x^2}$ on $[-1, 1]$
14. $f(x) = x^2\sqrt{2 - x}$ on $[0, 2]$
15. $f(x) = \dfrac{x^2}{x - 1}$ on $\left[-1, \dfrac{1}{2}\right]$
16. $f(x) = \dfrac{x}{x^2 - 1}$ on $\left[-\dfrac{1}{2}, \dfrac{1}{2}\right]$
17. $f(x) = (x + 2)^2(x - 1)^{2/3}$ on $[-4, 5]$
18. $f(x) = (x - 1)^2(x + 1)^3$ on $[-2, 7]$
19. $f(x) = \dfrac{(x - 4)^{1/3}}{x - 1}$ on $[2, 12]$
20. $f(x) = \dfrac{(x + 3)^{2/3}}{x + 1}$ on $[-4, -2]$
21. $f(x) = xe^x$ on $[-10, 10]$
22. $f(x) = xe^{-x}$ on $[-10, 10]$
23. $f(x) = \dfrac{\ln x}{x}$ on $[1, 3]$
24. $f(x) = x \ln x$ on $[1, 5]$

---

25. **Best Dimensions for a Box** An open box with a square base is to be made from a square piece of cardboard 12 centimeters on a side by cutting out a square from each corner and turning up the sides. Find the dimensions of the box that yield the maximum volume.

26. **Best Dimensions for a Box** An open box with a square base is to be made from a square piece of cardboard 24 centimeters on a side by cutting out a square from each corner and turning up the sides. Find the dimensions of the box that yield the maximum volume.

27. **Best Dimensions for a Box** A box, open at the top with a square base, is to have a volume of 8000 cubic centimeters. What should the dimensions of the box be if the amount of material used is to be a minimum?

28. **Best Dimensions for a Closed Box** If the box in Problem 27 is to be closed on top, what should the dimensions of the box be if the amount of material used is to be a minimum?

29. **Best Dimensions for a Can** A cylindrical container is to be produced that will have a capacity of 4000 cubic centimeters. The top and bottom of the container are to be made of material that costs $0.50 per square centimeter, while the side of the container is to be made of material costing $0.40 per square centimeter. Find the dimensions that will minimize the total cost of the container.

30. **Best Dimensions for a Can** A cylindrical container is to be produced that will have a capacity of 10 cubic meters. The top and bottom of the container are to be made of a material that costs $2 per square meter, while the side of the container is to be made of material costing $1.50 per square meter. Find the dimensions that will minimize the total cost of the container.

31. **Placing Telephone Boxes** A telephone company is asked to provide telephone service to a customer whose house is located 2 kilometers away from the road along which the telephone lines run. The nearest telephone box is located 5 kilometers down this road. If the cost to connect the telephone line is $50 per kilometer along the road and $60 per kilometer away from the road, where along the road from the box should the company connect the telephone line so as to minimize construction cost?

Hint: Let x denote the distance from the box to the connection so that $5 - x$ is the distance from this point to the point on the road closest to the house.

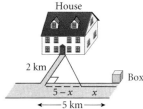

32. **Minimizing Travel Time** A small island is 3 kilometers from the nearest point P on the straight shoreline of a large lake. If a woman on the island can row her boat 2.5 kilometers per hour and can walk 4 kilometers per hour, where should she land her boat in order to arrive in the shortest time at a town 12 kilometers down the shore from P?

Hint: Let x be the distance from P to the landing point.

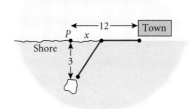

33. **Most Economical Speed** A truck has a top speed of 75 miles per hour and, when traveling at the rate of x miles per hour, consumes gasoline at the rate of $\frac{1}{200}\left[\left(\frac{1600}{x}\right) + x\right]$ gallon per mile. If the length of the trip is 200 miles and the price of gasoline is $1.60 per gallon, the cost is

$$C(x) = 1.60\left(\frac{1600}{x} + x\right)$$

where $C(x)$ is measured in dollars. What is the most economical speed for the truck? Use the interval $[10, 75]$.

34. If the driver of the truck in Problem 33 is paid $8 per hour, what is the most economical speed for the truck?

35. **Page Layout** A printer plans on having 50 square inches of printed matter per page and is required to allow for margins of 1 inch on each side and 2 inches on the top and bottom. What are the most economical dimensions for each page if the cost per page depends on the area of the page?

36. **Dimensions for a Window** A window is to be made in the shape of a rectangle surmounted by a semicircle with diameter equal to the width of the rectangle. See the figure. If the perimeter of the window is 22 feet, what dimensions will let in the most light?

37. **Tax Revenue** On a particular product, government economists determine that the relationship between tax rate t and the quantity q consumed is

$$t + 3q^2 = 18$$

Graph this relationship and explain how it could be justified. Find the optimal tax rate and the revenue generated by this tax rate.

38. **Most Economical Speed** A truck has a top speed of 75 miles per hour and, when traveling at the rate of x miles per hour, consumes gasoline at the rate of $\frac{1}{200}\left[\left(\frac{1600}{x}\right) + x\right]$

gallon per mile. This truck is to be taken on a 200 mile trip by a driver who is to be paid at the rate of b dollars per hour plus a commission of c dollars. Since the time required for this trip at x miles per hour is $\frac{200}{x}$, the total cost, if gasoline costs a dollars per gallon, is

$$C(x) = \left(\frac{1600}{x} + x\right)a + \frac{200}{x}b + c$$

Find the most economical possible speed under each of the following sets of conditions:

(a) $b = 0$, $c = 0$ (b) $a = 1.50$, $b = 8.00$, $c = 500$
(c) $a = 1.60$, $b = 10.00$, $c = 0$

39. Prove that a cylindrical container of fixed volume V requires the least material (minimum surface area) when its height is twice its radius.

40. **Transatlantic Crossing** An airplane crosses the Atlantic Ocean (3000 miles) with an airspeed of 500 mph.

(a) Find the time saved with a 25 mile per hour tail wind.
(b) Find the time lost with a 50 mile per hour head wind.
(c) If the cost per passenger is

$$C(x) = 100 + \frac{x}{10} + \frac{36{,}000}{x}$$

where x is the ground speed and $C(x)$ is the cost in dollars, what is the cost per passenger when there is no wind?

(d) What is the cost with a tail wind of 25 miles per hour?
(e) What is the cost with a head wind of 50 miles per hour?
(f) What ground speed minimizes the cost?
(g) What is the minimum cost per passenger?

41. **Drug Concentration** The concentration C of a certain drug in the bloodstream t hours after injection into muscle tissue is given by

$$C(t) = \frac{2t}{16 + t^3}$$

When is the concentration greatest?

42. **Profit Function** The cost C in dollars of producing n machines is $C(n) = 10n + 0.05n^2 + 150{,}000$. If a machine is priced at m, the estimated number that would be sold is $n = 2000 - \frac{m}{5}$. At what price would the profit be maximum?

5.5 Elasticity of Demand

OBJECTIVE 1 Determine the elasticity of demand

In this section we will study how economists describe the effect that changes in price have on demand and revenue. Recall that a demand equation expresses the market price p which will generate a demand for exactly x units. Suppose the price p and the quantity x demanded for a certain product are related by the following demand equation:

$$p = 200 - 0.02x \tag{1}$$

The equation states that in order to sell x units, the price must be set at $200 - 0.02x$ dollars. For example, to sell 5000 units, the price must be set at $200 - 0.02(5000) = \$100$ per unit.

In problems involving revenue, sales, and profit, it is customary to use the demand equation to express price as a function of demand. Since we are now interested in the effects that changes in price have on demand, it is more practical to express demand as a function of price. Equation (1) may be solved for x in terms of p to yield

$$x = \frac{1}{0.02}(200 - p) \quad \text{or} \quad x = f(p) = 50(200 - p)$$

This equation expresses quantity x as a function of the price. Since x and p must be nonnegative, we restrict p so that $0 \leq p \leq 200$.

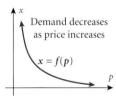

FIGURE 40

Usually, increasing the price of a commodity lowers the quantity demanded, while decreasing the price results in higher demand. Therefore, the typical demand function $x = f(p)$ is decreasing. See Figure 40. If $x = f(p)$ is a differentiable function, then $f'(p) < 0$ and $f'(p)$ equals the rate of change in quantity demanded with respect to price.

Elasticity measures the ratio of the relative change of the quantity demanded to the relative change of price. For the demand function $x = f(p)$ and the price p we have:

Elasticity

$$\text{The relative change in quantity demanded} = \frac{\Delta x}{x} = \frac{f(p + \Delta p) - f(p)}{f(p)} \tag{2}$$

$$\text{The relative change in price} = \frac{\Delta p}{p}$$

$$\text{Elasticity} = \frac{\text{Relative change in quantity demanded}}{\text{Relative change in price}} = \frac{\frac{\Delta x}{x}}{\frac{\Delta p}{p}} \tag{3}$$

Economists use elasticity to study the effect of price change on quantity demanded. Since elasticity depends on p and Δp, if we let $\Delta p \to 0$, we obtain an expression for the elasticity of demand at price p, denoted by $E(p)$:

$$E(p) = \lim_{\Delta p \to 0} \frac{\frac{\Delta x}{x}}{\frac{\Delta p}{p}}$$

$$= \lim_{\Delta p \to 0} \frac{\frac{f(p + \Delta p) - f(p)}{f(p)}}{\frac{\Delta p}{p}} \qquad \begin{array}{l} x = f(p); \\ \Delta x = f(p + \Delta p) - f(p) \end{array}$$

$$= \lim_{\Delta p \to 0} \left[\frac{f(p + \Delta p) - f(p)}{f(p)} \cdot \frac{p}{\Delta p} \right]$$

$$= \lim_{\Delta p \to 0} \left[\frac{p}{f(p)} \cdot \frac{f(p + \Delta p) - f(p)}{\Delta p} \right] \qquad \text{Rearrange terms.}$$

$$= \lim_{\Delta p \to 0} \frac{p}{f(p)} \cdot \lim_{\Delta p \to 0} \frac{f(p + \Delta p) - f(p)}{\Delta p} \qquad \begin{array}{l} \text{Limit of a product equals} \\ \text{the product of the limits.} \end{array}$$

$$= \frac{p}{f(p)} \lim_{\Delta p \to 0} \frac{f(p + \Delta p) - f(p)}{\Delta p} \qquad \lim_{\Delta p \to 0} \frac{p}{f(p)} = \frac{p}{f(p)}$$

$$= \frac{p}{f(p)} \cdot f'(p)$$

> **Elasticity of Demand**
>
> The **elasticity of demand** $E(p)$ **at a price** p for the demand function $x = f(p)$ is
>
> $$E(p) = \frac{pf'(p)}{f(p)}$$

Since $f'(p)$ is always negative for a typical demand function, the quantity $\dfrac{pf'(p)}{f(p)}$ will be negative for all values of p. For convenience, economists prefer to work with positive numbers. Therefore the **price elasticity of demand** is taken to be $|E(p)|$. For a given price p, if $|E(p)| > 1$, the demand is said to be **elastic**. If $|E(p)| < 1$, the demand is said to be **inelastic**. If $|E(p)| = 1$, that is, if $E(p) = -1$, the demand is said to be **unit elastic**.

EXAMPLE 1 Determine the Elasticity of Demand

Suppose $x = f(p) = 5000 - 30p^2$ is the demand function for a certain commodity, where p is the price per pound and x is the quantity demanded in pounds.

(a) What quantity can be sold at $10 per pound?
(b) Determine the elasticity of demand function $E(p)$.
(c) Interpret the elasticity of demand at $p = \$5$.
(d) Interpret the elasticity of demand at $p = \$10$.
(e) At what price is the elasticity of demand equal to -1? That is, at what price is demand unit elastic? Interpret unit elasticity.

SOLUTION (a) At $p = \$10$, $f(10) = 5000 - 30(10)^2 = 2000$. Therefore, 2000 pounds of the commodity can be sold at a price of \$10.

(b) $E(p) = \dfrac{p f'(p)}{f(p)} = \dfrac{p(-60p)}{5000 - 30p^2} = \dfrac{-60p^2}{5000 - 30p^2}$ $f'(p) = \dfrac{d}{dp}(5000 - 30p^2) = -60p$

(c) The elasticity of demand at the price $p = \$5$ is $E(5)$.

$$E(5) = \dfrac{-60(5)^2}{5000 - 30(5)^2} = -\dfrac{1500}{4250} = -0.353$$

$$|E(5)| = 0.353$$

When the price is set at \$5 per pound, a small increase in price will result in a relative decrease in quantity demanded of about 0.353 times the relative increase in price. For instance, if the price is increased from \$5 by 10%, then the quantity demanded will decrease by $(0.353)(10\%) = 0.0353 = 3.53\%$.

(d) When $p = \$10$, we have

$$E(10) = \dfrac{-60(10)^2}{5000 - 30(10)^2} = -3$$

or $|E(10)| = 3$

When the price is set at \$10, a small increase in price will result in a relative decrease in quantity demanded of 3 times the relative increase of price. For instance, a 10% price increase will result in a decrease in quantity demanded of approximately $3(10\%) = 30\%$.

(e) The demand is unit elastic if $E(p) = -1$.

$\dfrac{-60p^2}{5000 - 30p^2} = -1$ $E(p) = \dfrac{-60p^2}{5000 - 30p^2}$ from part (b).

$-60p^2 = -5000 + 30p^2$ Clear fractions.

$-90p^2 = -5000$ Simplify.

$p^2 = 55.5556$ Divide by -90.

$p = 7.45$ Solve for p.

If the price is set at \$7.45, a small increase in price will result in the same decrease in quantity demanded. For example, at this price, a 2% increase in price results in a 2% decrease in quantity demanded. ▶

NOW WORK PROBLEM 1.

Revenue and Elasticity of Demand

The concept of elasticity has an interesting relationship to the total revenue $R(p)$:

(a) If the demand is elastic, then an increase in the price per unit will result in a decrease in total revenue.
(b) If the demand is inelastic, then an increase in the price per unit will result in an increase in total revenue.
(c) If the demand is unit elastic, then an increase in price will not change the total revenue.

Chapter 5 Applications: Graphing Functions; Optimization

We establish these relationships as follows:
Since total revenue is given by $R(p) = xp = f(p) \cdot p$, we calculate the marginal revenue to be

$$R'(p) = f(p) + pf'(p) \qquad \frac{d}{dp}[f(p) \cdot p] = f(p) \cdot \frac{d}{dp}p + p\frac{d}{dp}f(p) = f(p) + pf'(p)$$

$$= f(p)\left[1 + \frac{pf'(p)}{f(p)}\right] \qquad \text{Factor out } f(p); pf'(p) = f(p) \cdot \frac{pf'(p)}{f(p)}$$

$$= f(p)[1 + E(p)] \qquad E(p) = \frac{pf'(p)}{f(p)}$$

We know that $f(p) > 0$.

If the demand is elastic, then $|E(p)| > 1$. Since $E(p) < 0$, this means that $E(p) < -1$, so that $1 + E(p) < 0$. As a result, $R'(p) < 0$. In other words, $R(p)$ is decreasing. This means an increase in price will result in a decrease in total revenue when the demand is elastic.

If the demand is inelastic, then $|E(p)| < 1$, so that $E(p) > -1$ or $1 + E(p) > 0$. In this case, $R'(p) > 0$ so $R(p)$ is increasing. This implies that an increase in price will result in an increase in total revenue when the demand is inelastic.

Finally, if demand is unit elastic, then $|E(p)| = 1$. Then $E(p) = -1$ and $1 + E(p) = 0$. Then $R'(p) = 0$. When the first derivative of a function equals 0, the function is neither increasing nor decreasing. So this implies that when demand has unit elasticity, a small increase (or decrease) in price results in no change in total revenue.

NOW WORK PROBLEM 27.

EXERCISE 5.5

1. Given the demand equation $p + \frac{1}{100}x = 40$

   (a) Express the demand x as a function of p.
   (b) Find the elasticity of demand $E(p)$.
   (c) What is the elasticity of demand when $p = \$5$? If the price is increased by 10%, what is the approximate change in demand?
   (d) What is the elasticity of demand when $p = \$15$? If the price is increased by 10%, what is the approximate change in demand?
   (e) What is the elasticity of demand when $p = \$20$? If the price is increased by 10%, what is the approximate change in demand?

2. Repeat Problem 1 for the demand equation

   $$p + \frac{1}{200}x = 80$$

3. Given the demand equation $p + \frac{1}{200}x = 50$

   (a) Express the demand x as a function of p.
   (b) Find the elasticity of demand $E(p)$.
   (c) What is the elasticity of demand when $p = \$10$? If the price is increased by 5%, what is the approximate change in demand?
   (d) What is the elasticity of demand when $p = \$25$? If the price is increased by 5%, what is the approximate change in demand?
   (e) What is the elasticity of demand when $p = \$35$? If the price is increased by 5%, what is the approximate change in demand?

4. Repeat Problem 3 for the demand function

   $$p + \frac{1}{200}x = 100$$

Elasticity of Demand

In Problems 5–16, a demand function is given. Find E(p) and determine if demand is elastic, inelastic, or unit elastic at the indicated price.

5. $x = f(p) = 600 - 3p$ at $p = 50$

6. $x = f(p) = 700 - 4p$ at $p = 40$

7. $x = f(p) = \dfrac{600}{p+4}$ at $p = 10$

8. $x = f(p) = \dfrac{500}{p+6}$ at $p = 10$

9. $x = f(p) = 10{,}000 - 10p^2$ at $p = 10$

10. $x = f(p) = 2250 - p^2$ at $p = 15$

11. $x = f(p) = \sqrt{100 - p}$ at $p = 10$

12. $x = f(p) = \sqrt{2500 - 2p^2}$ at $p = 25$

13. $x = f(p) = 40(4 - p)^3$ at $p = 2$

14. $x = f(p) = 20(8 - p)^3$ at $p = 4$

15. $x = f(p) = 20 - 3\sqrt{p}$ at $p = 4$

16. $x = f(p) = 30 - 4\sqrt{p}$ at $p = 20$

In Problems 17–20, use implicit differentiation to find the elasticity of demand at the indicated values of x and p.

17. $x^{1/2} + 2px + p^2 = 148$; $x = 16$, $p = 4$

18. $x^{3/2} + 2px + p^3 = 1088$; $x = 4$, $p = 10$

19. $2x^2 + 3px + 10p^2 = 600$; $x = 10$, $p = 5$

20. $3x^3 + x^2p^2 + 10p^3 = 3480$; $x = 10$; $p = 2$

The discussion about elasticity assumed that the demand x is a function of price p. However, in economics it is common to use x as the independent variable. Thus, if $p = F(x)$ is the demand equation, then it can be shown the elasticity of demand is given by

$$E(x) = \dfrac{F(x)}{xF'(x)}$$

In Problems 21–26, use this formula to find the elasticity of demand at the given value of x.

21. $p = F(x) = 10 - \dfrac{1}{20}x$ at $x = 5$

22. $p = F(x) = 40 - \dfrac{1}{10}x$ at $x = 4$

23. $p = F(x) = 10 - 2x^2$ at $x = 2$

24. $p = F(x) = 20 - 4x^2$ at $x = 4$

25. $p = F(x) = 50 - 2\sqrt{x}$ at $x = 100$

26. $p = F(x) = 20 - 4\sqrt{x}$ at $x = 400$

---

27. **Revenue and Elasticity** A movie theater has a capacity of 1000 people. The number of people attending the show at a price of $p per ticket is $x = f(p) = \dfrac{6000}{p} - 500$. Currently the price is $4 per ticket.

(a) Determine whether the demand is elastic or inelastic at $4.
(b) If the price is increased, will revenue increase or decrease or remain the same?

28. **Revenue and Elasticity** The demand function for a rechargeable hand-held vacuum cleaner is given by

$$x = \dfrac{1}{5}(300 - p^2)$$

where x (measured in units of 100) is the quantity demanded per week and p is the unit price in dollars. The manufacturer would like to increase revenue.

(a) Is demand elastic, inelastic or unit elastic at $p = \$15$?
(b) Should the price of the unit be raised or lowered to increase revenue?

29. **Revenue and Elasticity** The demand function for digital watches is

$$x = \sqrt{300 - 6p}$$

where x is measured in hundreds of units.

(a) Is demand elastic or inelastic at $p = \$10$?
(b) If the price is lowered slightly, will revenue increase or decrease or remain the same.

30. **Revenue and Elasticity** A company wishes to increase its revenue by lowering the price of its product. The demand function for this product is

$$x = \dfrac{10{,}000}{p^2}$$

(a) Compute $E(p)$.
(b) Will the company succeed in raising its revenue?

31. Revenue and Elasticity When a wholesaler sold a certain product at $15 per unit, sales were 2000 units each week. However, after a price rise of $3, the average number of units sold decreased to 1800 per week. Assume that the demand function is linear.

(a) Determine the demand function.
(b) Find the elasticity of demand at the new price of $18.
(c) Approximate the change in demand if the new price is increased by 5%.
(d) Will the price increase cause the revenue to increase or decrease or remain the same?

5.6 Related Rates

PREPARING FOR THIS SECTION *Before getting started, review the following:*

> Geometry Review (Chapter 0, Section 0.7, pp. 71–74)
> Chain Rule (Chapter 4, Section 4.5, pp. 310–311)
> Implicit Differentiation (Chapter 4, Section 4.7, pp. 328–333)

OBJECTIVE 1 Solve related rate problems

In all of the natural sciences and many of the social and behavioral sciences, quantities that are related, but vary with time, are encountered. For example, the pressure of an ideal gas of fixed volume is proportional to temperature, yet each of these quantities may change over a period of time. Problems involving rates of related variables are referred to as **related rate problems.** In such problems we normally want to find the rate at which one of the variables is changing at a certain time, when the rates at which the other variables are changing are known.

The usual procedure in such problems is to write an equation that relates all the time-dependent variables involved. Such a relationship is often obtained by investigating the geometric and/or physical conditions imposed by the problem. When this relationship is differentiated with respect to the time t, a new equation that involves the variables and their rates of change with respect to time is obtained.

For example, suppose x and y are two differentiable functions of time t; that is, $x = x(t), y = y(t)$. And suppose they obey the equation

$$x^3 - y^3 + 2y - x - 199 = 0$$

Differentiate both sides of this equation with respect to the time t. Since x and y are functions of time t, we use the Chain Rule to obtain

$$3x^2 \frac{dx}{dt} - 3y^2 \frac{dy}{dt} + 2\frac{dy}{dt} - \frac{dx}{dt} = 0$$

This equation is valid for all times t under consideration and involves the derivatives of x and y with respect to t, as well as the variables themselves. Because the derivatives are related by this equation, we call them **related rates.** We can solve for one of these rates once the value of the other rate and the values of the variables are known. For example, if in the above equation at a specific time t, we know that $x = 5$, $y = 3$, and $\frac{dx}{dt} = 2$, then by direct substitution we can find that $\frac{dy}{dt} = \frac{148}{25}$. The following examples illustrate how related rates can be used to solve certain types of practical problems.

NOW WORK PROBLEM 1.

1 EXAMPLE 1 Solving a Related Rate Problem

A child throws a stone into a still millpond causing a circular ripple to spread. If the radius of the circle increases at the constant rate of 0.5 feet per second, how fast is the area of the ripple increasing when the radius of the ripple is 30 feet? See Figure 41.

FIGURE 41

SOLUTION   The variables involved are:

$t = $ time (in seconds) elapsed from the time the stone hits the water
$r = $ radius of the ripple (in feet) after t seconds
$A = $ area of the ripple (in square feet) after t seconds

The rates involved are:

$$\frac{dr}{dt} = \text{the rate at which the radius is increasing with time}$$

$$\frac{dA}{dt} = \text{the rate at which the area is increasing with time}$$

We wish to find $\frac{dA}{dt}$ when $r = 30$; that is, we seek the rate at which the area of the ripple is increasing at the instant when $r = 30$. The relationship between A and r is given by the formula for the area of a circle:

$$A = \pi r^2 \tag{1}$$

Since A and r are functions of t, we differentiate both sides of (1) with respect to t, using the Chain Rule, to obtain

$$\frac{dA}{dt} = 2\pi r \frac{dr}{dt} \quad \frac{dA}{dt} = \frac{dA}{dr} \cdot \frac{dr}{dt}; \frac{dA}{dr} = \frac{d}{dr}(\pi r^2) = 2\pi r \tag{2}$$

Since the radius increases at the rate of 0.5 feet per second, we know that

$$\frac{dr}{dt} = 0.5 \tag{3}$$

By substituting (3) into (2), we get

$$\frac{dA}{dt} = 2\pi r(0.5) = \pi r$$

When $r = 30$, the area of the ripple is increasing at the rate

$$\frac{dA}{dt} = \pi(30) = 30\pi \approx 94.25 \text{ square feet per second} \quad \blacktriangleright$$

Example 1 illustrates some general guidelines that will prove helpful for solving related rate problems:

Chapter 5 Applications: Graphing Functions; Optimization

> **Steps for Solving Related Rate Problems**
>
> **STEP 1** If possible, draw a picture illustrating the problem.
> **STEP 2** Identify the variables and assign symbols to them.
> **STEP 3** Identify and interpret rates of change as derivatives.
> **STEP 4** Write down what is known. Express all relationships among the variables by equations.
> **STEP 5** Obtain additional relationships among the variables and their derivatives by differentiating.
> **STEP 6** Substitute numerical values for the variables and the derivatives. Solve for the unknown rate.

Note: It is important to remember that the substitution of numerical values must occur after the differentiation process (Step 5). Also rates must be represented in appropriate units. Finally, remember a positive rate means the quantity is increasing; a negative rate means it is decreasing.

EXAMPLE 2 Solving a Related Rate Problem

A balloon in the form of a sphere is being inflated at the rate of 20 cubic feet per minute. Find the rate at which the surface area of the sphere is increasing at the instant when the radius of the sphere is 6 feet.

SOLUTION **STEP 1** See Figure 42.

FIGURE 42

STEP 2 The variables of the problem are:

t = time (in minutes) measured from the moment inflation of the balloon begins
r = length (in feet) of the radius of the balloon at time t
V = volume (in cubic feet) of the balloon at time t
S = surface area (in square feet) of the balloon at time t

STEP 3 The rates of change are:

$$\frac{dr}{dt} = \text{the rate of change of radius with respect to time}$$

$$\frac{dV}{dt} = \text{the rate of change of volume with respect to time}$$

$$\frac{dS}{dt} = \text{the rate of change of surface area with respect to time}$$

STEP 4 We are given that $\dfrac{dV}{dt} = 20$ cubic feet per minute, and we seek $\dfrac{dS}{dt}$ when $r = 6$ feet. At any time t, the volume V of the balloon (a sphere) is $V = \dfrac{4}{3}\pi r^3$ and the surface area S of the balloon is $S = 4\pi r^2$.

STEP 5 Differentiate each of these equations with respect to the time t. Using the Chain Rule, we have

$$\dfrac{dV}{dt} = 4\pi r^2 \dfrac{dr}{dt} \qquad \dfrac{dV}{dt} = \dfrac{dV}{dr}\cdot\dfrac{dr}{dt}; \ \dfrac{dV}{dr} = \dfrac{d}{dr}\left(\dfrac{4}{3}\pi r^3\right) = 4\pi r^2$$

$$\dfrac{dS}{dt} = 8\pi r \dfrac{dr}{dt} \qquad \dfrac{dS}{dt} = \dfrac{dS}{dr}\cdot\dfrac{dr}{dt}; \ \dfrac{dS}{dr} = \dfrac{d}{dr}(4\pi r^2) = 8\pi r$$

In the equation for $\dfrac{dV}{dt}$, we solve for $\dfrac{dr}{dt}$ and substitute this quantity into the equation for $\dfrac{dS}{dt}$. Then

$$\dfrac{dV}{dt} = 4\pi r^2 \dfrac{dr}{dt}$$

$$\dfrac{dr}{dt} = \dfrac{1}{4\pi r^2}\dfrac{dV}{dt} \quad \text{Solve for } \dfrac{dr}{dt}.$$

Then

$$\dfrac{dS}{dt} = 8\pi r \dfrac{dr}{dt} = 8\pi r \cdot \dfrac{1}{4\pi r^2}\dfrac{dV}{dt} = \dfrac{2}{r}\dfrac{dV}{dt}$$

STEP 6 At $r = 6$ and $\dfrac{dV}{dt} = 20$, we have

$$\dfrac{dS}{dt} = \left(\dfrac{2}{6}\right)(20) = 6.67 \text{ square feet per minute}$$

The surface area is increasing at the rate of 6.67 square feet per minute when the radius is 6 feet.

NOW WORK PROBLEM 13.

EXAMPLE 3 Solving a Related Rate Problem

A rectangular swimming pool 50 feet long and 25 feet wide is 15 feet deep at one end and 5 feet deep at the other. If water is pumped into the pool at the rate of 300 cubic feet per minute, at what rate is the water level rising when it is 7.5 feet deep at the deep end?

SOLUTION **STEP 1** A cross-sectional view of the pool is illustrated in Figure 43.

FIGURE 43

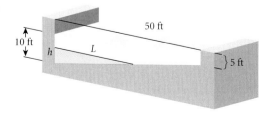

STEP 2 The variables involved are:

t = the time (in minutes) measured from the moment water begins to flow into the pool
h = the water level (in feet) measured at the deep end
L = the distance (in feet) from the deep end toward the short end measured at water level
V = the volume (in cubic feet) of water in the pool

STEP 3 The rates of change are:

$\dfrac{dV}{dt}$ = the rate of increase in volume at a given instant

$\dfrac{dh}{dt}$ = the rate of increase in height at a given instant

STEP 4 The volume V is related to L and h by the formula

$$V = \text{(Cross-sectional triangular area)(width)} = \left(\frac{1}{2}Lh\right)(25) \text{ cubic feet} \quad (4)$$

Using similar triangles, we see from Figure 43 that L and h are related by the equation

$$\frac{L}{h} = \frac{50}{10}$$

$$L = 5h$$

By replacing L by $5h$ in equation (4), we have

$$V = \frac{1}{2}(5h)(h)(25) = \frac{125}{2}h^2 \text{ cubic feet} \quad (5)$$

Here, V and h are each functions of time t.

STEP 5 By differentiating (5) with respect to t, we obtain

$$\frac{dV}{dt} = 125h\frac{dh}{dt} \text{ cubic feet per minute} \qquad \frac{dV}{dt} = \frac{dV}{dh} \cdot \frac{dh}{dt}; \frac{dV}{dh} = \frac{d}{dh}\frac{125}{2}h^2 = 125h \quad (6)$$

STEP 6 We seek the rate at which the water level is rising, $\dfrac{dh}{dt}$, when $h = 7.5$ and the rate of water pumped into the pool is $\dfrac{dV}{dt} = 300$ cubic feet per minute.

Using equation (6), we find

$$300 = 125(7.5)\frac{dh}{dt}$$

$$\frac{dh}{dt} = \frac{300}{125(7.5)} = 0.32 \text{ feet per minute}$$

The water level is rising at the rate of 0.32 feet per minute when the height is 7.5 feet. ▶

NOW WORK PROBLEM 15.

EXAMPLE 4 Solving a Related Rate Problem

Suppose that for a company manufacturing digital watches, the cost, revenue, and profit functions are given by

Cost function $\quad C(x) = 10{,}000 + 3x$

Revenue function $\quad R(x) = 5x - \dfrac{x^2}{2000}$

Profit function $\quad P(x) = R(x) - C(x)$

where x is the daily production of digital watches. Production is increasing at the rate of 50 watches per day when production is 1000 watches.

(a) Find the rate of increase in cost when production is 1000 watches.
(b) Find the rate of increase in revenue when production is 1000 watches.
(c) Find the rate of increase in profit when production is 1000 watches.

SOLUTION **STEP 1** No illustration here.

STEP 2 The variables involved are

$t =$ the time in days
$x(t) =$ the production x as a function of time
$C(t) =$ the cost as a function of time
$R(t) =$ the revenue as a function of time
$P(t) =$ the profit as a function of time

STEP 3 The rates involved are

$\dfrac{dx}{dt} = 50$, the rate at which production is increasing when $x = 1000$

$\dfrac{dC}{dt} =$ the rate at which cost is increasing

$\dfrac{dR}{dt} =$ the rate at which revenue is increasing

$\dfrac{dP}{dt} =$ the rate at which profit is increasing

(a) STEP 4 $C(x) = 10{,}000 + 3x$

STEP 5 $\dfrac{dC}{dt} = \dfrac{d}{dt}(10{,}000 + 3x) = \dfrac{d}{dt}(10{,}000) + \dfrac{d}{dt}(3x) = 3\dfrac{dx}{dt}$

STEP 6 Since

$\dfrac{dx}{dt} = 50 \quad \text{when} \quad x = 1000$

then

$\dfrac{dC}{dt} = 3(50) = \$150 \text{ per day}$

The cost is increasing at the rate of $150 per day when production is 1000 watches.

(b) STEP 4 $R(x) = 5x - \dfrac{x^2}{2000}$

STEP 5 $\dfrac{dR}{dt} = \dfrac{d}{dt}\left(5x - \dfrac{x^2}{2000}\right) = \dfrac{d}{dt}(5x) - \dfrac{d}{dt}\left(\dfrac{x^2}{2000}\right) = 5\dfrac{dx}{dt} - \dfrac{x}{1000}\dfrac{dx}{dt}$

STEP 6 Since

$$\frac{dx}{dt} = 50 \quad \text{when} \quad x = 1000$$

then

$$\frac{dR}{dt} = 5(50) - \frac{1000}{1000}(50) = \$200 \text{ per day}$$

Revenue is increasing at the rate of $200 per day when production is 1000 watches.

(c) STEP 4 $P = R - C$

STEP 5 $\dfrac{dP}{dt} = \dfrac{dR}{dt} - \dfrac{dC}{dt} = \$200 - \$150 = \$50$ per day

Profit is increasing at the rate of $50 per day when production is 1000 watches. ▶

NOW WORK PROBLEM 19.

EXERCISE 5.6

In Problems 1–4, assume x and y are differentiable functions of t. Find $\dfrac{dx}{dt}$ when $x = 2$, $y = 3$, and $\dfrac{dy}{dt} = 2$.

1. $x^2 + y^2 = 13$ 2. $x^2 - y^2 = -5$ 3. $x^3 y^2 = 72$ 4. $x^2 y^3 = 108$

5. Suppose h is a differentiable function of t and suppose that when $h = 3$, then $\dfrac{dh}{dt} = \dfrac{1}{12}$. Find $\dfrac{dV}{dt}$ if $V = 80h^2$.

6. Suppose x is a differentiable function of t and suppose that when $x = 15$, then $\dfrac{dx}{dt} = 3$. Find $\dfrac{dy}{dt}$ if $y^2 = 625 - x^2$.

7. Suppose h is a differentiable function of t and suppose that $\dfrac{dh}{dt} = \dfrac{5}{16}\pi$ when $h = 8$. Find $\dfrac{dV}{dt}$ if $V = \dfrac{1}{12}\pi h^3$.

8. Suppose x and y are differentiable functions of t and suppose that when $t = 20$, then $\dfrac{dx}{dt} = 5$, $\dfrac{dy}{dt} = 4$, $x = 150$, and $y = 80$. Find $\dfrac{ds}{dt}$ if $s^2 = x^2 + y^2$.

9. **Changing Volume** If each edge of a cube is increasing at the constant rate of 3 centimeters per second, how fast is the volume increasing when x, the length of an edge, is 10 centimeters long?

10. **Changing Volume** If the radius of a sphere is increasing at 1 centimeter per second, find the rate of change of its volume when the radius is 6 centimeters.

11. Consider a right triangle with hypotenuse of (fixed) length 45 centimeters and variable legs of lengths x and y, respectively. If the leg of length x increases at the rate of 2 centimeters per minute, how fast is y changing when x is 4 centimeters long?

12. **Increasing Surface Area** Air is pumped into a balloon with a spherical shape at the rate of 80 cubic centimeters per second. How fast is the surface area of the balloon increasing when the radius is 10 centimeters?

13. **Decreasing Surface Area** A spherical balloon filled with gas has a leak that permits the gas to escape at a rate of 1.5 cubic meters per minute. How fast is the surface area of the balloon shrinking when the radius is 4 meters?

14. **Heating a Plate** When a metal plate is heated, it expands. If the shape of the metal is circular and if its radius, as a result of expansion, increases at the rate of 0.02 centimeter per second, at what rate is the area of the surface increasing when the radius is 3 centimeters?

15. **Filling a Pool** A public swimming pool has a rectangular shape with the following dimensions: length 30 meters, width 15 meters, depth 3 meters at the adult side and 1 meter at the children's side. If water is pumped into the pool at the rate of 15 cubic meters per minute, how fast is the water level rising when it is 2 meters deep at the adult side?

16. **Cost, Revenue, Profit Functions** Suppose that for a company manufacturing computers, the cost, revenue, and profit functions are given by

$$C(x) = 85{,}000 + 300x$$
$$R(x) = 400x - \frac{x^2}{20}$$
$$P(x) = R(x) - C(x)$$

where x is the weekly production of computers. If production is increasing at the rate of 400 computers per week

when production output is 5000 computers, find the rate of increase in:
(a) Cost (b) Revenue (c) Profit

17. **Pollution** Assume that oil spilled from a ruptured tanker forms a circular oil slick whose radius increases at a constant rate of 0.42 feet per minute $\left(\dfrac{dr}{dt} = 0.42\right)$. Estimate the rate $\dfrac{dA}{dt}$ (in square feet per minute) at which the area of the spill is increasing when the radius of the spill is 120 feet ($r = 120$).

18. **Demand, Revenue Functions** The marketing department of a computer manufacturer estimates that the demand q (in thousands of units per year) for a laptop is related to price by the demand equation $q = 200 - 0.9p$. Because of efficiency and technological advances, the prices are falling at a rate of \$30 per year $\left(\dfrac{dp}{dt} = -30\right)$. The current price of a laptop is \$650. At what rate $\dfrac{dR}{dt}$ are revenues changing?

19. **Cost, Revenue, Profit** The cost C and revenue R of a company are given by

$$C(x) = 5x + 5000 \qquad R(x) = 15x - \dfrac{x^2}{10{,}000}$$

where x is the daily production. Production is increasing at the rate of 100 units per day at a production level of 1000 units. Find:
(a) The rate of change in daily cost when production is 1000 units.
(b) The rate of change in daily revenue when production is 1000 units.
(c) Whether revenue is increasing or decreasing when production is 1000 units.
(d) The profit function.
(e) The rate of change in profit when production is 1000 units.

20. Rework Problem 19 if production is decreasing at the rate of 40 units per day at a production level of 2000 units.

21. **Demand, Revenue Functions** The marketing department of a manufacturing company estimates that the demand q (in thousands of units per year) for a plasma television is related to price by the demand equation $q = 10{,}000 - 0.9p$. Because of efficiency and technological advances, the prices are falling at a rate of \$100 per year $\left(\dfrac{dp}{dt} = -100\right)$. The current price of a television is \$7000. At what rate $\dfrac{dR}{dt}$ is revenue changing?

5.7 The Differential; Linear Approximations

PREPARING FOR THIS SECTION *Before getting started, review the following:*

>> Definition of a Derivative (Chapter 4, Section 4.1, p. 278)

OBJECTIVES
1 Find differentials
2 Find linear approximations
3 Solve applied problems involving linear approximations

The Differential

In studying the derivative of a function $y = f(x)$, we use the notation $\dfrac{dy}{dx}$ to represent the derivative. The symbols dy and dx, called *differentials*, which appear in this notation may also be given their own meanings. To pursue this, recall that for a differentiable function f, the derivative is defined as

$$\dfrac{dy}{dx} = f'(x) = \lim_{\Delta x \to 0} \dfrac{\Delta y}{\Delta x} = \lim_{\Delta x \to 0} \dfrac{f(x + \Delta x) - f(x)}{\Delta x}$$

That is, the derivative f' is the limit of the ratio of the change in y to the change in x as Δx approaches 0, but $\Delta x \neq 0$. In other words, for Δx sufficiently close to 0, we can make $\dfrac{\Delta y}{\Delta x}$ as close as we please to $f'(x)$. We express this fact by writing

$$\dfrac{\Delta y}{\Delta x} \approx f'(x) \qquad \text{when} \qquad \Delta x \approx 0 \; (\Delta x \neq 0) \qquad (1)$$

Another way of writing (1) is to write

$$\Delta y \approx f'(x)\Delta x \quad \text{when} \quad \Delta x \approx 0 \, (\Delta x \neq 0)$$

The quantity $f'(x)\,\Delta x$ is given a special name, the *differential of y*.

> **Differential**
>
> Let f denote a differentiable function and let Δx denote a change in x.
>
> (a) The **differential of y**, denoted by dy, is defined as $\quad dy = f'(x)\,\Delta x$.
> (b) The **differential of x**, denoted by dx, is defined as $\quad dx = \Delta x \neq 0$.

Using the notation of differentials, we can write

$$dy = f'(x)\,dx \tag{2}$$

Since $dx \neq 0$, (2) can be written as

$$\frac{dy}{dx} = f'(x) \tag{3}$$

The expression in (3) should look very familiar. Interestingly enough, we have given an independent meaning to the symbols dy and dx in such a way that, when dy is divided by dx, their quotient will be equal to the derivative. That is, the differential of y divided by the differential of x is equal to the derivative $f'(x)$. For this reason, *we may formally regard the derivative as a quotient of differentials.*

Note that the differential dy is a function of both x and dx. For example, the differential dy of the function $y = x^3$ is

$$dy = 3x^2\,dx$$

so that

if $x = 1$ and $dx = 0.2$, then $dy = 3(1)^2(0.2) = 0.6$
if $x = 0.5$ and $dx = 0.1$, then $dy = 3(0.5)^2(0.1) = 0.075$
if $x = 2$ and $dx = -0.5$, then $dy = 3(2)^2(-0.5) = -6$

1 EXAMPLE 1 Finding Differentials

(a) If $y = x^2 + 3x - 5$ then $dy = (2x + 3)\,dx$. $\quad \frac{d}{dx}(x^2 + 3x - 5) = 2x + 3$

(b) If $y = \sqrt{x^2 + 4}$ then $dy = \dfrac{x}{\sqrt{x^2 + 4}}\,dx$. $\quad \dfrac{d}{dx}\sqrt{x^2+4} = \dfrac{d}{dx}(x^2+4)^{1/2} = \dfrac{1}{2}(x^2+4)^{-1/2} \cdot 2x = \dfrac{x}{\sqrt{x^2+4}}$

▶

NOW WORK PROBLEM 1.

Differential Formulas

All the formulas derived earlier for finding derivatives carry over to differentials. The list below gives formulas for differentials next to the corresponding derivative formulas.

	Derivative	Differential
1.	$\dfrac{d}{dx} c = 0$	1'. $dc = 0$ if c is constant
2.	$\dfrac{d}{dx}(kx) = k$	2'. $d(kx) = k\, dx$ if k is constant
3.	$\dfrac{d}{dx}(u + v) = \dfrac{du}{dx} + \dfrac{dv}{dx}$	3'. $d(u + v) = du + dv$
4.	$\dfrac{d}{dx}(uv) = u\dfrac{dv}{dx} + v\dfrac{du}{dx}$	4'. $d(uv) = u\, dv + v\, du$
5.	$\dfrac{d}{dx}\left(\dfrac{u}{v}\right) = \dfrac{v\dfrac{du}{dx} - u\dfrac{dv}{dx}}{v^2}$	5'. $d\left(\dfrac{u}{v}\right) = \dfrac{v\, du - u\, dv}{v^2}$
6.	$\dfrac{d}{dx} x^r = rx^{r-1}$	6'. $d(x^r) = rx^{r-1}\, dx$ r is a rational number
7.	$\dfrac{d}{dx} e^x = e^x$	7'. $d(e^x) = e^x\, dx$
8.	$\dfrac{d}{dx} \ln x = \dfrac{1}{x}$	8'. $d(\ln x) = \dfrac{1}{x}\, dx$

From now on, to find the differential of a function $y = f(x)$, either find the derivative $\dfrac{dy}{dx}$ and then multiply by dx or use formulas (1') through (8').

For example, if $y = x^3 + 2x + 1$, then the derivative is

$$\frac{dy}{dx} = 3x^2 + 2 \qquad \text{so that} \qquad dy = (3x^2 + 2)\, dx$$

Using the differential formulas,

$$dy = d(x^3 + 2x + 1) = d(x^3) + d(2x) + d(1)$$
$$= 3x^2\, dx + 2\, dx + 0 = (3x^2 + 2)\, dx$$

Caution: The use of dy on the left side of an equation requires dx on the right side. That is, $dy = 3x^2 + 2$ is incorrect. The symbol d is an instruction to take the differential!

EXAMPLE 2   Finding Differentials

(a) $d(x^2 - 3x) = (2x - 3)dx$ $\dfrac{d}{dx}(x^2 - 3x) = 2x - 3$

(b) $d(3y^4 - 2y + 4) = (12y^3 - 2)\, dy$ $\dfrac{d}{dy}(3y^4 - 2y + 4) = 12y^3 - 2$

(c) $d(\sqrt{z^2 + 1}) = \dfrac{z}{\sqrt{z^2 + 1}}\, dz$ $\dfrac{d}{dz}\sqrt{z^2 + 1} = \dfrac{d}{dz}(z^2 + 1)^{1/2} = \dfrac{1}{2}(z^2 + 1)^{-1/2} \cdot 2z = \dfrac{z}{\sqrt{z^2 + 1}}$

(d) $d(xe^x) = xe^x dx + e^x dx = (x + 1)e^x dx$ $\dfrac{d}{dx}(xe^x) = x\dfrac{d}{dx}e^x + e^x\dfrac{d}{dx}x = xe^x + e^x = (x + 1)e^x$

▶

The differential can be used to find the derivative of a function that is defined implicitly.

EXAMPLE 3 Using Differentials to Find Derivatives

Find $\dfrac{dy}{dx}$ and $\dfrac{dx}{dy}$ if $x^2 + y^2 = 2xy^2$.

SOLUTION We take the differential of each side:

$$d(x^2 + y^2) = d(2xy^2)$$

$2x\,dx + 2y\,dy = 2(y^2\,dx + 2xy\,dy)$ Differential of a sum on the left; differential of a product on the right.

$x\,dx + y\,dy = y^2\,dx + 2xy\,dy$ Cancel the 2's.

$(y - 2xy)dy = (y^2 - x)\,dx$ Rearrange terms.

$\dfrac{dy}{dx} = \dfrac{y^2 - x}{y - 2xy}$ provided $y - 2xy \neq 0$ Solve for $\dfrac{dy}{dx}$.

$\dfrac{dx}{dy} = \dfrac{y - 2xy}{y^2 - x}$ provided $y^2 - x \neq 0$ Solve for $\dfrac{dx}{dy}$.

NOW WORK PROBLEM 5.

Geometric Interpretation

We use Figure 44 to arrive at a geometric interpretation of the differentials dx and dy and their relationship to Δx and Δy. From the definition, the differential dx and the change Δx are equal. Therefore, we concentrate on the relationship between dy and Δy.

In Figure 44(a), $P = (x, y)$ is a point on the graph of $y = f(x)$ and $Q = (x + \Delta x, y + \Delta y)$ is a nearby point that is also on the graph of f. The slope of the tangent line to the graph of f at P is $f'(x)$. From the figure, it follows that

$$f'(x) = \frac{dy}{\Delta x} = \frac{dy}{dx} \quad \text{or} \quad dy = f'(x)\,dx$$

Figure 44(a) illustrates the case for which $dy < \Delta y$ and $\Delta x > 0$. The case for which $dy > \Delta y$ and $\Delta x > 0$ is illustrated in Figure 44(b). The remaining cases, in which $\Delta x = dx < 0$, have similar graphical representations.

FIGURE 44

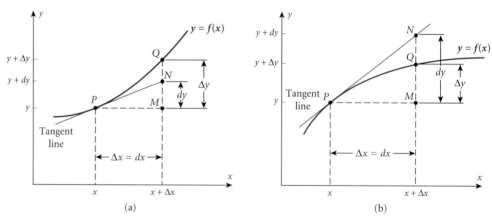

Linear Approximations

We now examine the relationship between Δy and dy. In Figure 44(a), the increment Δy is represented by the length of the line segment $|MQ|$. That is, $\Delta y - dy$ is the length of the line segment $|NQ|$. The size of $|NQ|$ equals the amount by which the graph departs from its tangent line. In fact, for $dx = \Delta x$ sufficiently small, the graph does not depart very much from its tangent line. As a result, the function whose graph is this tangent line is referred to as the *linear approximation to f near P*.

Approximating Δy

For $dx = \Delta x$ sufficiently small, the differential dy is a good approximation to Δy. That is,

$$\boxed{\Delta y \approx dy \quad \text{if} \quad \Delta x \approx 0} \tag{4}$$

We can use (4) to obtain the linear approximation to a function f near a point $P = (x_0, y_0)$ on f. Since

$$dy = f'(x_0)\, dx = f'(x_0)\, \Delta x = f'(x_0)(x - x_0)$$
$$ \underset{dx = \Delta x}{\uparrow} \quad \underset{\Delta x = x - x_0}{\uparrow}$$

we find from (4) that

$$\Delta y \approx dy$$
$$f(x) - f(x_0) \approx f'(x_0)(x - x_0)$$
$$f(x) \approx f(x_0) + f'(x_0)(x - x_0) \quad \text{Add } f(x_0) \text{ to both sides.}$$

Linear Approximation to f Near x_0

For $dx = \Delta x$ sufficiently small, that is, for x close to x_0,

$$\boxed{f(x) \approx f(x_0) + f'(x_0)(x - x_0)} \tag{5}$$

The line $y = f(x_0) + f'(x_0)(x - x_0)$ is called the **linear approximation to f near x_0**.

EXAMPLE 4 Finding a Linear Approximation

Find the linear approximation to $f(x) = x^2 + 2x$ near $x = 1$. Graph f and the linear approximation.

SOLUTION First, $f(1) = 3$. Next, $f'(x) = 2x + 2$ so that $f'(1) = 2(1) + 2 = 4$. By (5), the linear approximation to f near $x = 1$ is

$$f(x) \approx f(1) + f'(1)(x - 1) = 3 + 4(x - 1) = 4x - 1$$

Figure 45 illustrates the graph of f and the linear approximation $y = 4x - 1$ to f near 1.

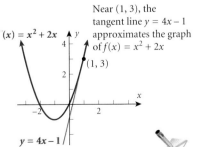

FIGURE 45 Near $(1, 3)$, the tangent line $y = 4x - 1$ approximates the graph of $f(x) = x^2 + 2x$

NOW WORK PROBLEM 15.

The next two examples use (4).

EXAMPLE 5 Using Linear Approximations

A bearing with a spherical shape has a radius of 3 centimeters when it is new. Find the approximate volume of the metal lost after it wears down to a radius of 2.971 centimeters.

SOLUTION The exact volume of metal lost equals the change, ΔV, in the volume V of the sphere, where $V = \frac{4}{3}\pi r^3$. The change in the radius r is $\Delta r = 2.971 - 3 = -0.029$ centimeter. Since the change Δr is small, we can use the differential dV of volume to approximate the change ΔV in volume. Therefore,

$$\Delta V \approx dV = 4\pi r^2 dr = (4\pi)(9)(-0.029) \approx -3.28$$

$\uparrow$ $\quad\uparrow$ $\qquad\uparrow$
Formula (4) $\quad dV = V'(r)dr \quad r = 3$
$\qquad\qquad\qquad\qquad dr = \Delta r = -0.029$

The approximate loss in volume is 3.28 cubic centimeters.

NOW WORK PROBLEM 23.

The use of dy to approximate Δy when dx is small may also be helpful in approximating errors.

> If Q is the quantity to be measured and if ΔQ is the change in Q, we define
>
> $$\text{Relative error in } Q = \frac{|\Delta Q|}{Q} \qquad \text{Percentage error in } Q = \frac{|\Delta Q|}{Q}(100\%)$$

For example, if $Q = 50$ units and the change ΔQ in Q is measured to be 5 units, then

$$\text{Relative error in } Q = \frac{5}{50} = 0.10$$

$$\text{Percentage error in } Q = 10\%$$

EXAMPLE 6 Using Linear Approximations

Suppose a company manufactures spherical ball bearings with radius 3 centimeters, and the percentage error in the radius must be no more than 1%. What is the approximate percentage error for the surface area of the ball bearing?

SOLUTION If S is the surface area of a sphere of radius r, then $S = 4\pi r^2$ and $\frac{\Delta r}{r} = 0.01$. The relative error $\frac{\Delta S}{S}$ we seek may be approximated by the use of differentials. That is,

$$\frac{\Delta S}{S} \approx \frac{dS}{S} = \frac{8\pi r\, dr}{4\pi r^2} = 2\frac{dr}{r} = 2\frac{\Delta r}{r} = 2(0.01) = 0.02$$

Formula (4) $dS = S'(r)dr$ Simplify $dr = \Delta r$ $\frac{\Delta r}{r} = 0.01$

The percentage error in the surface area is 2%.

In Example 6 the percentage error of 1% in the radius of the sphere means the radius will lie somewhere between 2.97 and 3.03 centimeters. But the percentage error of 2% in the surface area means the surface area lies within a factor of $\pm(0.02)$ of $S = 4\pi r^2 = 36\pi$; that is, it lies between $(0.98)(36\pi) = 35.28\pi = 110.84$ and $(1.02)(36\pi) = 36.72\pi = 115.36$ square centimeters. A rather small error in the radius results in a more significant range of possibilities for the surface area!

NOW WORK PROBLEM 29.

EXERCISE 5.7

In Problems 1–4, find the differential dy.

1. $y = x^3 - 2x + 1$ **2.** $y = 4(x^2 + 1)^{3/2}$ **3.** $y = \dfrac{x - 1}{x^2 + 2x - 8}$ **4.** $y = \sqrt{x^2 - 1}$

In Problems 5–10, find $\dfrac{dy}{dx}$ and $\dfrac{dx}{dy}$ using differentials.

5. $xy = 6$ **6.** $3x^2y + 2x - 10 = 0$ **7.** $x^2 + y^2 = 16$

8. $4xy^2 + yx^2 + 6 = 0$ **9.** $x^3 + y^3 = 3x^2y$ **10.** $2x^2 + y^3 = xy^2$

In Problems 11–14, find the indicated differential.

11. $d(\sqrt{x - 2})$ **12.** $d\left(\dfrac{1-x}{1+x}\right)$ **13.** $d(x^3 - x - 4)$ **14.** $d(x^2 + 5)^{2/3}$

In Problems 15–20, find the linear approximation to f near x_0. Graph f and the linear approximation.

15. $f(x) = x^2 - 2x + 1$; $x_0 = 2$ **16.** $f(x) = x^3 - 1$; $x_0 = 0$ **17.** $f(x) = \sqrt{x}$; $x_0 = 4$

18. $f(x) = x^{2/3}$; $x_0 = 1$ **19.** $f(x) = e^x$; $x_0 = 0$ **20.** $f(x) = \ln x$; $x_0 = 1$

---

21. Use equation (5) to find the approximate change in:
(a) $y = f(x) = x^2$ as x changes from 3 to 3.001
(b) $y = f(x) = \dfrac{1}{x + 2}$ as x changes from 2 to 1.98

22. Use equation (5) to find the approximate change in:
(a) $y = x^3$ as x changes from 3 to 3.01
(b) $y = \dfrac{1}{x - 1}$ as x changes from 2 to 1.98

23. A circular plate is heated and expands. If the radius of the plate increases from $r = 10$ centimeters to $r = 10.1$ centimeters, find the approximate increase in area of the top surface.

24. In a wooden block 3 centimeters thick, an existing circular hole with a radius of 2 centimeters is enlarged to a hole with a radius of 2.2 centimeters. Approximately what volume of wood is removed?

25. Find the approximate change in volume of a spherical balloon of radius 3 meters as the balloon swells to a radius of 3.1 meters.

26. A bee flies around the circumference of a circle traced on a ball with a radius of 7 centimeters at a constant distance of 2 centimeters from the ball. An ant travels along the circumference of the same circle on the ball. Approximately how many more centimeters does the bee travel in one trip around than does the ant?

27. If the percentage error in measuring the edge of a cube is 2%, what is the percentage error in computing its volume?

28. The radius of a spherical ball is computed by measuring the volume of the sphere (by finding how much water it displaces). The volume is found to be 40 cubic centimeters, with a percentage error of 1%. Compute the corresponding percentage error in the radius (due to the error in measuring the volume).

29. A manufacturer produces paper cups in the shape of a right circular cone with radius equal to one-fourth its height. Specifications call for the cups to have a diameter of 4 centimeters. After production, it is discovered that the diameters measure 3.9 centimeters. Assuming that the radius is still one-fourth of the height, what is the approximate loss in the capacity of the cup?

    Hint: The volume V of a right circular cone of height h and radius r is $V = \frac{1}{3}\pi r^2 h$.

30. The oil pan of a car is shaped in the form of a hemisphere with a radius of 8 centimeters. The depth h of the oil is found to be 3 centimeters, with a percentage error of 10%. Approximate the percentage error in the volume. [Hint: The volume V for a spherical segment is $V = \frac{1}{3}\pi h^2(3r - h)$, where r is the radius.]

31. To find the height of a building, the length of the shadow of a 3 meter pole placed 9 meters from the building is measured. This measurement is found to be 1 meter, with a percentage error of 1%. What is the estimated height of the building? What is the percentage error in the estimate? See the figure.

32. The period of the pendulum of a grandfather clock is

$$T = 2\pi\sqrt{\frac{l}{g}},$$

where l is the length (in meters) of the pendulum, T is the period (in seconds), and g is the acceleration due to gravity (9.8 meters per second per second). Suppose the length of the pendulum, a thin wire, increases by 1% due to an increase in temperature. What is the corresponding percentage error in the period? How much time will the clock lose each day?

33. Refer to Problem 32. If the pendulum of a grandfather clock is normally 1 meter long and the length is increased by 10 centimeters, how many minutes will the clock lose each day?

34. What is the approximate volume enclosed by a hollow sphere if its inner radius is 2 meters and its outer radius is 2.1 meters?

Chapter 5 Review

OBJECTIVES

Section	You should be able to	Review Exercises
5.1	1 Find horizontal tangent lines	1–4, 15(e)–22(e)
	2 Find vertical tangent lines	1–4, 15(e)–22(e)
	3 Discuss the graph of a function f where the derivative of f does not exist	5–8
5.2	1 Determine where a function is increasing and where it is decreasing	9(a)–14(a), 15(c)–22(c)
	2 Use the first derivative test	9(b)–14(b), 15(d)–22(d)
	3 Graph functions	15–22
5.3	1 Determine the concavity of a graph	15(g)–22(g)
	2 Find inflection points	15(g)–22(g), 66
	3 Graph functions	15–22
	4 Use the second derivative test	23–28
	5 Solve applied problems	53, 54

5.4	1	Find the absolute maximum and the absolute minimum of a function	29–34
	2	Solve applied problems	55–58
5.5	1	Determine the elasticity of demand	35–40
5.6	1	Solve related rate problems	47–52, 64
5.7	1	Find differentials	41–44
	2	Find linear approximations	45, 46
	3	Solve applied problems involving linear approximations	59–63

THINGS TO KNOW

Tangent Line
  Horizontal (p. 348) — f is continuous; $(c, f(c))$ is a point on the graph of c.
    $f'(c) = 0$.
  Vertical (p. 350) — $f'(x)$ is unbounded at $x = c$.

Theorem on Continuity and Differentiability (p. 351) — If c is in the domain of f and if $f'(c)$ exists, then f is continuous at c.

Increasing Function on (a, b) (p. 355) — f is differentiable and $f'(x) > 0$ for all x in (a, b).

Decreasing Function on (c, d) (p. 355) — f is differentiable and $f'(x) < 0$ for all x in (c, d).

First Derivative Test (p. 356) — f is differentiable.
  If f is increasing to the left of a point A and decreasing on the right of A, then at A there is a local maximum.
  If f is decreasing to the left of a point B and decreasing on the right of B, then at B there is a local minimum.

Test for Concavity (p. 366) — If $f''(x) > 0$ for all x in (a, b), then f is concave up on (a, b).
  If $f''(x) < 0$ for all x in (c, d), then f is concave down on (c, d).

Inflection Point (p. 368) — A point on the graph at which the concavity changes.

Second Derivative Test (p. 374) — f is differentiable on an open interval I.
  $f''(x)$ exists on I.
  c is a number in I and $f'(c) = 0$.
  If $f''(c) < 0$, then at $(c, f(c))$ there is a local maximum.
  If $f''(c) > 0$, then at $(c, f(c))$ there is a local minimum.
  If $f''(c) = 0$ the test is inconclusive.

Average Cost Function (p. 375) — $\overline{C}(x) = \dfrac{C(x)}{x}$, where C is the cost function.

Test for Absolute Maximum and Absolute Minimum (p. 382) — f is continuous on a closed interval $[a, b]$.
  List the values of f at each critical number in the open interval (a, b).
  List $f(a)$ and $f(b)$.
  The largest of these is the absolute maximum; the smallest of these is the absolute minimum.

Elasticity of Demand (p. 394) — $E(p) = \dfrac{pf'(p)}{f(p)}$ where p is the price and $x = f(p)$ is the demand function.

Differential of $y = f(x)$ (p. 406) — f is differentiable; $\Delta x =$ change in x.
  $dx = \Delta x;\ dy = f'(x)\Delta x = f'(x)dx$.

Linear Approximation to f near x_0 (p. 408) — For x close to x_0, $f(x) \approx f(x_0) + f'(x_0)(x - x_0)$.

TRUE–FALSE ITEMS

T F 1. If the derivative of a function f does not exist at c, then f has a vertical tangent line at c.

T F 2. A differentiable function f is increasing on (a, b) if $f'(x) > 0$ throughout (a, b).

T F 3. The absolute maximum of a function equals the value of the function at a critical number.

T F 4. If $f''(x) > 0$ for all x in (a, b), then the graph of f is concave down on (a, b).

T F 5. If x and y are two differentiable functions of t, then after differentiating $x^2 - y^3 + 4y - x = 100$, we obtain

$$2x\left(\frac{dx}{dt}\right) - 3y^2\left(\frac{dy}{dt}\right) + 4\left(\frac{dy}{dt}\right) - \left(\frac{dx}{dt}\right) = 0$$

T F 6. If $y = x^2 + 2x$, then $dy = 2x + 2$.

FILL IN THE BLANKS

1. The function $f(x)$ is _____ on (a, b) if $f'(x) < 0$ for all x in (a, b).

2. At a point $(c, f(c))$ there is a local minimum if the graph of the function is _____ to the left of the point and _____ to the right of the point.

3. A differentiable function is _____ _____ on (a, b) if the tangent lines to the graph at every point lie below its graph.

4. At a point $(c, f(c))$ on the graph of a differentiable function f for which $f'(c) = 0$ there is a _____ tangent line.

5. At an inflection point, the graph exhibits a change in _____.

6. If $y = f(x)$ is a differentiable function, then the differential of y is _____.

7. The function $y = f(x_0) + f'(x_0)(x - x_0)$ is called the _____ _____ to f at x_0.

REVIEW EXERCISES Blue problem numbers indicate the author's suggestions for a practice test.

In Problems 1–4, locate any points at which the graph of f has a horizontal or a vertical tangent line.

1. $f(x) = x^3 - x^2 + x + 15$
2. $f(x) = x^5 - 15x^3 + 5$
3. $f(x) = \dfrac{x^{1/3}}{x + 4}$
4. $f(x) = x^{4/5}(x^2 - 14)$

In Problems 5–8, answer the following questions about the function f at c.

(a) Is f continuous at c?
(b) Does $f'(c)$ exist? If it does, what is its value?
(c) If f is continuous at c, but $f'(c)$ does not exist, is there a vertical tangent line or no tangent line at c?

5. $f(x) = 3x^{1/5}$ at $c = 0$

6. $f(x) = 5x^{4/5} - 2x$ at $c = 0$

7. $f(x) = \begin{cases} 3x + 1 & x < 3 \\ x^2 + 1 & x \geq 3 \end{cases}$ at $c = 3$

8. $f(x) = \begin{cases} 4x^2 & x \leq 1 \\ 5x^2 - 1 & x > 1 \end{cases}$ at $c = 1$

In Problems 9–14,
(a) *Determine the intervals on which the graph of f is increasing and the intervals on which it is decreasing.*
(b) *Use the First Derivative Test to identify any local maxima and minima (if they exist).*

9. $f(x) = \dfrac{1}{5}x^5 - x^3 - 4x$

10. $f(x) = 2x^4 - 9x^2$

11. $f(x) = \dfrac{x^2}{x^2 - 8}$

12. $f(x) = \dfrac{6x + 1}{x^2 + 5}$

13. $f(x) = 1 + 3e^{-x}$

14. $f(x) = xe^x$

In Problems 15–22, graph f by following the 7 steps listed.
  (a) STEP 1 Find the domain of f.
  (b) STEP 2 Locate the intercepts of f.
  (c) STEP 3 Determine where the graph is increasing and where it is decreasing.
  (d) STEP 4 Find any local maxima or minima using the First Derivative Test.
  (e) STEP 5 Locate all points on the graph of f at which the tangent line is either horizontal or vertical.
  (f) STEP 6 Determine the end behavior and locate any asymptotes.
  (g) STEP 7 Locate the inflection points, if any, of the graph by determining the concavity of the graph.

15. $f(x) = x^3 - 3x^2 + 3x - 1$
16. $f(x) = 2x^3 - x^2 + 2$
17. $f(x) = x^5 - 5x$
18. $f(x) = x^5 + 5x^4$
19. $f(x) = x^{4/3} + 4x^{1/3}$
20. $f(x) = x^{4/3} - 4x^{1/3}$
21. $f(x) = \dfrac{2x}{x^2 + 1}$
22. $f(x) = \dfrac{4x}{x^2 + 4}$

In Problems 23–28, use the Second Derivative Test to determine the local maxima and local minima of each function.

23. $f(x) = 4x^3 - 3x$
24. $f(x) = 5x^4 + 7x^3 + 20$
25. $f(x) = x^4 - 2x^2$
26. $f(x) = x^4 + 2x^2$
27. $f(x) = xe^x$
28. $f(x) = xe^{-x}$

In Problems 29–34, find the absolute maximum and absolute minimum of each function on the given interval.

29. $f(x) = x^3 - 3x^2 + 3x - 1$ on $[0, 3]$
30. $f(x) = 2x^3 - x^2 + 2$ on $[0, 1]$
31. $f(x) = x^4 - 4x^3 + 4x^2$ on $[1, 3]$
32. $f(x) = x^4 - 2x^2$ on $[-1, 1]$
33. $f(x) = x^{4/3} - 4x^{1/3}$ on $[-1, 8]$
34. $f(x) = x^{4/3} + 4x^{1/3}$ on $[-1, 1]$

In Problems 35–38, a demand function is given. Find $E(p)$ and determine if demand is elastic or inelastic or unit elastic at the indicated price.

35. $x = 1000 - 2p^2$ at $p = 20$
36. $x = \dfrac{1000}{p + 10}$ at $p = 12$
37. $x = \sqrt{500 - p^2}$ at $p = 10$
38. $x = \sqrt{2200 - 2p^2}$ at $p = 30$

39. **Pricing DVD Recorders** The demand function for a DVD player/recorder is $x = 40 - 2\sqrt{p}$ where p is the price of a unit.
  (a) Is the demand function elastic, inelastic, or unit elastic at $p = \$300$?
  (b) If the price is raised to $310 will revenue increase, decrease, or remain the same?

40. **Cost of Cellular Service** The demand equation for a certain cellular phone package is given by $x = 60{,}000 - 1200p$, where p is the price of the package in dollars.
  (a) Is demand elastic, inelastic, or unit elastic at $p = \$35$?
  (b) If the price is raised to $37, will revenue increase, decrease, or remain the same?

In Problems 41–44, find the differential dy.

41. $y = 3x^4 - 2x^3 + x$
42. $y = 3(x^2 - 1)^5$
43. $y = \dfrac{3 - 2x}{1 + x}$
44. $y = \sqrt[3]{2 + x^4}$

In Problems 45 and 46, find the linear approximation of f near x_0. Graph f and the linear approximation.

45. $f(x) = x^2 - 9$ $x_0 = 3$
46. $f(x) = \dfrac{1}{x}$ $x_0 = 1$

47. Suppose that x and y are both differentiable functions of t and $x^2 + y^2 = 8$. Find $\dfrac{dx}{dt}$ when $x = 2$, $y = 2$, and $\dfrac{dy}{dt} = 3$.

48. Suppose that x and y are both differentiable functions of t and $x^2 - y^2 = 5$. Find $\dfrac{dy}{dt}$ when $x = 4$, $y = 3$, and $\dfrac{dx}{dt} = 2$.

49. Suppose that x and y are both differentiable functions of t and $xy + 6x + y^3 = -2$. Find $\dfrac{dy}{dt}$ when $x = 2$, $y = -3$ and $\dfrac{dx}{dt} = 3$.

50. Suppose that x and y are both differentiable functions of t and $y^2 - 6x^2 = 3$. Find $\dfrac{dx}{dt}$ when $x = 2$, $y = -3$, and $\dfrac{dy}{dt} = 3$.

51. A balloon in the form of a sphere is being inflated at the rate of 10 cubic meters per minute. Find the rate at which the surface area of the sphere is increasing at the instant when the radius of the sphere is 3 meters.

    Hint: The volume of a sphere is $V = \dfrac{4}{3}\pi r^3$; the surface area is $S = 4\pi r^2$.

52. A child throws a stone into a still millpond, causing a circular ripple to spread. If the radius of the circle increases at the constant rate of 0.5 meter per second, how fast is the area of the ripple increasing when the radius of the ripple is 20 meters?

53. **Average Cost** The cost function for producing x items is
$$C(x) = 5x^2 + 1125$$
    (a) Find the average cost function.
    (b) What is the minimum average cost?
    (c) Find the marginal cost function.
    (d) Graph the average cost function and the marginal cost function on the same set of axes. Label their point of intersection.
    (e) Interpret the point of intersection.

54. **Average Cost** The cost function for producing x items is
$$C(x) = 1000 + x + x^2$$
    (a) Find the average cost function.
    (b) What is the minimum average cost?
    (c) Find the marginal cost function.
    (d) Graph the average cost function and the marginal cost function on the same set of axes. Label their point of intersection.
    (e) Interpret the point of intersection.

55. **Maximizing Profit** The price function of a certain mobile home producer is
$$p(x) = 62{,}402.50 - 0.5x^2$$
where p is the price (in dollars) and x is the number of units sold. The cost of production for x units is
$$C(x) = 48{,}002.50 + 1500$$
How many units need to be sold to maximize profit?

56. **Maximizing Profit** A company's history shows that profit increases, as a result of advertising, according to
$$P(x) = 150 + 120x - 3x^2$$
where x is the number of dollars, in thousands, spent on advertising. How much should be spent on advertising to maximize profit?

57. **Best Dimensions of a Can** A beer can is cylindrical and holds 500 cubic centimeters of beer. If the cost of the material used to make the sides, top, and bottom is the same, what dimensions should the can have to minimize cost?

58. **Setting Refrigerator Prices** A distributor of refrigerators has average monthly sales of 1500 refrigerators, each selling for $300. From past experience, the distributor knows that a special month-long promotion will enable them to sell 200 additional refrigerators for each $15 decrease in price. What should be charged for each refrigerator during the month of promotion in order to maximize revenue?

59. **Size of a Burn** A burn on a person's skin is in the shape of a circle, so that if r is the radius of the burn and A is the area of the burn, then $A = \pi r^2$. Use the differential to approximate the decrease in the area of the burn when the radius decreases from 10 to 8 millimeters.

60. **Related Rate** A spherical ball is being inflated. Find the approximate change in volume if the radius increases from 3 to 3.1 cm.

61. **Demand Function** The demand for peanuts (in hundreds of pounds) at a price of x dollars is
$$D(x) = -4x^3 - 3x^2 + 2000$$
Approximate the change in demand as the price changes from

    (a) $1.50 to $2.00.   (b) $2.50 to $3.50.

62. **Growth of a Tumor** A tumor is approximately spherical in shape. If the radius of the tumor changes from 11 to 13 millimeters, find the approximate change in volume.

63. Drug Concentration The concentration of a certain drug in the bloodstream x hours after being administered is

$$c(x) = \frac{3x}{4 + 2x^2}$$

Approximate the change in concentration as x changes from

(a) 1.2 to 1.3. (b) 2 to 2.25.

64. Wound Healing A wound on a person's skin is in the shape of a circle and is healing at the rate of 30 square millimeters per day. How fast is the radius r of the wound decreasing when $r = 10$ millimeters?

65. If a function f is differentiable for all x except 3, and if f has a local maximum at $(-1, 4)$ and a local minimum at $(3, -2)$, which of the following statements must be true?

(a) The graph of f has a point of inflection somewhere between $x = -1$ and $x = 3$.
(b) $f'(-1) = 0$
(c) The graph of f has a horizontal asymptote.
(d) The graph of f has a horizontal tangent line at $(3, -2)$.
(e) The graph of f intersects both axes.

66. Let $f(x)$ be a function with derivative

$$f'(x) = \frac{1}{1 + x^2}$$

Show that the graph of $f(x)$ has an inflection point at $x = 0$. Note that $f'(x) > 0$ for all x.

Chapter 5 Project

INVENTORY CONTROL

Suppose that you are employed by a local department store, and you are placed in charge of ordering vacuum cleaners. Based on past experience, you know that the store will sell 500 vacuum cleaners per year. You must decide how many times a year to order vacuum cleaners and how many to get with each order. You could order all 500 at the beginning of the year, but there will be a cost (the holding cost) for storing the unsold vacuum cleaners. You could order 5 at a time and place 100 orders over the course of the year, but there are costs for paperwork and shipping for each order you place. Perhaps there is an amount to order somewhere between 5 and 500 that minimizes the total cost: the holding cost plus the reorder cost. This number is called the lot size.

To simplify the mathematics, we make two assumptions:

1. Demand for vacuum cleaners remains constant through the year.
2. Stock is immediately replenished exactly when the inventory level of vacuum cleaners reaches zero.

With these assumptions, the graph of the number of vacuum cleaners in stock at any time will look like the graph shown. Here x is the lot size, the amount ordered each time inventory reaches zero. The graph signifies that the inventory is decreasing at a constant rate (lines with negative slope). The average number of items in stock is $\frac{x}{2}$.

Let D be the annual demand for the item. [For the vacuum cleaners, $D = 500$.] Our goal is to find a function $C = C(x)$,

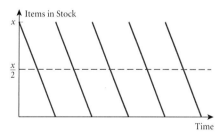

where C represents the holding costs plus the reorder costs and x is the lot size. We seek the value of x that minimizes C.

(a) Let H be the holding cost per year for each item you have in stock. Explain why your average holding costs for the year will be $\frac{Hx}{2}$.

(b) Let R be the cost for each reorder. Explain why you will make $\frac{D}{x}$ reorders during the year, and why your reorder cost will be $R \cdot \frac{D}{x} = \frac{DR}{x}$.

(c) What is the cost function $C = C(x)$?

(d) Show that $C(x)$ is minimized when $x = \sqrt{\frac{2DR}{H}}$. This value of x is called the "Wilson-Harris lot size."

(e) Apply the Wilson-Harris lot size formula to the vacuum cleaner problem. Assume that annual demand is 500 vacuum cleaners, that the holding cost is $10 per vacuum cleaner, and that the reorder cost $40. Find

the lot size that will minimize the total cost. How many orders will you place over the course of a year?
(f) Suppose that in part (e) the holding cost is reduced to $3 per vacuum cleaner. What lot size will now minimize the total cost? How many orders will you place?
(g) In many cases, the shipping charges (which are part of the reorder cost) vary with how many items are ordered.

Suppose that the reorder cost is now $R + Sx$, where R is the cost of reordering exclusive of shipping costs, and S is the shipping charge per item. The cost function is now

$$C(x) = \frac{Hx}{2} + \frac{D(R + SX)}{x}.$$

Find the lot size that will now minimize the total cost.

MATHEMATICAL QUESTIONS FROM PROFESSIONAL EXAMS*†

1. **CPA Exam** The mathematical notation for the total cost for a business is $2X^3 + 4X^2 + 3X + 5$, where X equals production volume. Which of the following is the mathematical notation for the marginal cost function for this business?

   (a) $2(X^3 + 2X^2 + 1.5X + 2.5)$ (b) $6X^2 + 8X + 3$
   (c) $2X^3 + 4X^2 + 3X$ (d) $3X + 5$

2. **CPA Exam** The mathematical notation for the total cost function for a business is $4X^3 + 6X^2 + 2X + 10$ where X equals production volume. Which of the following is the mathematical notation for the average cost function for that business?

   (a) $2(2X^2 + 3X + 2)$
   (b) $2X^3 + 3X^2 + X + 5$
   (c) $0.4X^3 + 0.6X^2 + 0.2X + 1$
   (d) $4X^2 + 6X + 2 + \dfrac{10}{X}$

3. **CPA Exam** The mathematical notation for the average cost function for a business is $6X^3 + 4X^2 + 2X + 8 + 2/X$, where X equals production volume. What would be the mathematical notation for the total cost function for the business?

   (a) The average cost function multiplied by X.
   (b) The average cost function divided by X.
   (c) The average cost function divided by $X/2$.
   (d) The first derivative of the average cost function.

4. **CPA—Review** To find a minimum cost point given a total cost equation, the initial steps are to find the first derivative, set this derivative equal to zero, and solve the equation. Using the solution(s) so derived, what additional steps must be taken, and what result indicates a minimum?

   (a) Substitute the solution(s) in the first derivative equation and a positive solution indicates a minimum.
   (b) Substitute the solution(s) in the first derivative equation and a negative solution indicates a minimum.
   (c) Substitute the solution(s) in the second derivative equation and a positive solution indicates a minimum.
   (d) Substitute the solution(s) in the second derivative equation and a negative solution indicates a minimum.

5. **Actuary Exam—Part I** Figure A could represent the graph of which of the following?

   (a) $y = x^3 e^{x^2}$ (b) $y = xe^x$
   (c) $y = xe^{-x}$ (d) $y = x^2 e^x$
   (e) $y = x^2 e^{x^2}$

   FIGURE A

   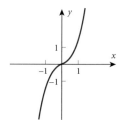

6. **Actuary Exam—Part I** According to classical economic theory, the business cycle peaks when employment reaches a maximum, relative to adjacent time periods. Employment can be approximated as a function of time, t, by a differentiable function $E(t)$. The graph of $E'(t)$ is pictured below.

   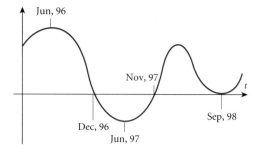

   Which of the following points represents a peak in the business cycle?

   (a) Jun, 96 (b) Dec, 96 (c) Jun, 97
   (d) Nov, 97 (e) Sep, 98

*Copyright © 1998, 1999 by the American Institute of Certified Public Accountants, Inc. Reprinted with permission.
†Copyright © 1998, 1999 by the Society of Actuaries. Reprinted with permission.

7. An economist defines an "index of economic health," D, as follows:

$$D = E^2(100 - I)$$

where

E is the percent of the working-age population that is employed and I is the rate of inflation (expressed as a percent).

On June 30, 1996, employment is at 95% and is increasing at a rate of 2% per year, and the rate of inflation is at 6% and is increasing at a rate of 3% per year. Calculate the rate of change of D on June 30, 1996.

(a) $-9,503$ per year (b) $-9,500$ per year
(c) 0 per year (d) $8,645$ per year
(e) $17,860$ per year

8. **Actuary Exam—Part I** What values of x produce a relative minimum and relative maximum, respectively, for $f(x) = 2x^3 + 3x^2 - 12x - 5$?

(a) $-5, 0$ (b) $-2, 1$
(c) $1, -3$ (d) $1, -2$
(e) $2, -1$

9. **Actuary Exam—Part I** On which of the following intervals is the function $f(x) = \frac{1}{6}x^3 - 2x$ decreasing and concave upward?

(a) $(-2, 2)$ (b) $(-2, 0)$
(c) $(0, 2)$ (d) $(2, 3)$
(e) $(-2, 4)$

10. **Actuary Exam—Part I** What is the maximum value of $f(x) = \dfrac{x}{1 + x + x^2}$ on the closed interval $[-2, 2]$?

(a) $\dfrac{1}{4}$ (b) $\dfrac{2}{7}$ (c) $\dfrac{1}{3}$ (d) $\dfrac{1}{2}$ (e) 1

11. **Actuary Exam—Part I** If $f(x) = \dfrac{xe^{3x}}{1 + x}$ for $x \neq -1$, then $f'(1) =$

(a) $\dfrac{3e^3}{4}$ (b) $\dfrac{5e^3}{4}$ (c) $\dfrac{7e^3}{4}$ (d) $\dfrac{9e^3}{4}$
(e) $4e^3$

12. **Actuary Exam—Part I** Let $f(x) = 5 + 6x + 12x^2 - 2x^3 - x^4$, and let $g(x) = f'(x)$ for $-\infty < x < \infty$. At what value of x is $g(x)$ increasing most rapidly?

(a) -2 (b) $-\dfrac{1}{2}$ (c) $\dfrac{1}{2}$ (d) 1 (e) 2

13. **Actuary Exam—Part I** A water tank in the shape of a right circular cone has a height of 10 feet. The top rim of the tank is a circle with a radius of 4 feet. If water is being pumped into the tank at the rate of 2 cubic feet per minute, what is the rate of change of the water depth, in feet per minute, when the depth is 5 feet? (The volume V of a right circular cone is $V = \dfrac{1}{3}\pi r^2 h$, where r is its radius and h is its height.)

(a) $\dfrac{1}{2\pi}$ (b) $\dfrac{1}{\pi}$ (c) $\dfrac{3}{2\pi}$ (d) $\dfrac{2}{\pi}$ (e) $\dfrac{5}{2\pi}$

14. **Actuary Exam—Part I** What is the y-coordinate of the point on the curve $y = 2x^2 - 3x$ at which the slope of the tangent line is the same as that of the secant line between $x = 1$ and $x = 2$?

(a) -1 (b) 0 (c) 1 (d) 3 (e) 9

15. **Actuary Exam—Part I** Two vehicles start at time $t = 0$ from the same point A and travel along a straight line so that at each moment t the distance between the second vehicle and A equals the square of the distance between the first vehicle and A. After $t = 3$ hours, the first vehicle is moving at 3 kilometers per hour and is 9 kilometers away from A. What is the speed, in kilometers per hour, of the second vehicle at $t = 3$ hours?

(a) 9 (b) 18 (c) 27 (d) 54 (e) 81

16. **Actuary Exam—Part I** A cube of ice melts without changing shape at the uniform rate of 4 cm³/min. What is the rate of change of surface area of the cube, in cm²/min, when the volume of the cube is 125 cm³?

(a) -4 (b) $\dfrac{-16}{5}$ (c) $\dfrac{-16}{6}$ (d) $\dfrac{60}{19}$
(e) $\dfrac{16}{5}$

CHAPTER 6

The Integral of a Function and Applications

Different families have different incomes. We all know that! But how is that income distributed? If we rank families in order of income they earn, will the lowest 20% have earnings equal to 20% of the total income earned by all families? Will the first 80% of these families have incomes that equal 80% of the total income earned by families? The answer is no. If we rank families by income earned, then the lowest 20% will have incomes that are less than 20% of the total income earned by all families.

We can determine what percent of total income of all families is earned by the lowest 20%. For example, in 2001, the percent of income earned by the lowest 20% was 3.5% of the total. The percent of income earned by the first 80% was 49.8%. (*Source:* U.S. Census Bureau.) An important question to ask is, What is the trend over time in these percents? And how can we measure the trend? The Chapter Project at the end of this chapter provides a methodology for measuring the distribution of income.

OUTLINE

6.1 Antiderivatives; the Indefinite Integral; Marginal Analysis

6.2 Integration Using Substitution

6.3 Integration by Parts

6.4 The Definite Integral; Learning Curves; Total Sales over Time

6.5 Finding Areas; Consumer's Surplus, Producer's Surplus; Maximizing Profit over Time

6.6 Approximating Definite Integrals

6.7 Differential Equations

- **Chapter Review**
- **Chapter Project**
- **Mathematical Questions from Professional Exams**

A LOOK BACK, A LOOK FORWARD

In Chapter 1 we discussed various properties that functions have, like increasing or decreasing, the average rate of change, the difference quotient, and so on. Chapter 2 identified classes of functions and properties they possessed. With Chapter 3 we began our study of the calculus and found limits of functions. Then in Chapter 4 we introduced the derivative of a functions, developing formulas for finding derivatives. We used these formulas in Chapter 5 to do applications.

Now in Chapter 6 we introduce the second important idea of the calculus, *the integral*. There are two types of integral: the *indefinite integral* and the *definite integral*. The indefinite integral involves the inverse process of differentiation, usually referred to as *finding the antiderivative* of a function. The definite integral plays a major role in applications to geometry (finding areas), to business and economics (marginal analysis, consumer's surplus, maximization of profit), and to solving differential equations.

6.1 Antiderivatives; the Indefinite Integral; Marginal Analysis

PREPARING FOR THIS SECTION *Before getting started, review the following:*

> Derivative of a Power Function (Chapter 4, Section 4.2, p. 288)
> Derivative of $f(x) = x^{p/q}$ (Chapter 4, Section 4.8, p. 336)
> Derivative of $f(x) = e^x$ (Chapter 4, Section 4.5, p. 309)
> Derivative of $f(x) = \ln x$ (Chapter 4, Section 4.5, p. 313)

OBJECTIVES
1. Find an antiderivative
2. Use integration formulas
3. Evaluate indefinite integrals
4. Find cost and revenue functions

Antiderivatives

Find an antiderivative **1** We have already learned that to each differentiable function f there corresponds a derivative function f'. It is also possible to ask the following question: If a function f is given, can we find a function F whose derivative is f? That is, is it possible to find a function F so that $F'(x) = f(x)$? If such a function F can be found, it is called *an antiderivative of f*.

> **Antiderivative**
>
> A function F is called an **antiderivative** of the function f if
>
> $$F'(x) = f(x)$$

EXAMPLE 1 Finding an Antiderivative

An antiderivative of $f(x) = 2x$ is x^2, since

$$\frac{d}{dx} x^2 = 2x$$

Another antiderivative of $2x$ is $x^2 + 3$ since

$$\frac{d}{dx}(x^2 + 3) = 2x$$

This example leads us to suspect that the function $f(x) = 2x$ has an unlimited number of antiderivatives. Indeed, any of the functions x^2, $x^2 + \frac{1}{2}$, $x^2 + 2$, $x^2 + \sqrt{5}$, $x^2 - \pi$, $x^2 - 1$, $x^2 + K$, where K is a constant, has the property that its derivative is $2x$. So all functions of the form $F(x) = x^2 + K$, where K is a constant, are antiderivatives of $f(x) = 2x$. Are there others? The answer is no! As the following result tells us, *all* the antiderivatives of $f(x) = 2x$ are of the form $F(x) = x^2 + K$, where K is a constant.

Antiderivatives of f

If F is an antiderivative of f, then all the antiderivatives of f are of the form

$$\boxed{F(x) + K}$$

where K is a constant.

EXAMPLE 2 **Finding All the Antiderivatives of a Function**

All the antiderivatives of $f(x) = x^5$ are of the form

$$\frac{x^6}{6} + K$$

where K is a constant.

We check the answer by finding its derivative. Since

$$\frac{d}{dx}\left(\frac{x^6}{6} + K\right) = x^5$$

the answer checks.

EXAMPLE 3 **Finding All the Antiderivatives of a Function**

Find all the antiderivatives of: $f(x) = x^{1/2}$

SOLUTION The derivative of the function $\frac{2}{3}x^{3/2}$ is

$$\frac{d}{dx}\left(\frac{2}{3}x^{3/2}\right) = \frac{2}{3}\frac{d}{dx}x^{3/2} = \frac{2}{3}\left(\frac{3}{2}x^{1/2}\right) = x^{1/2}$$

So, $\frac{2}{3}x^{3/2}$ is an antiderivative of $f(x) = x^{1/2}$. All the antiderivatives of $f(x) = x^{1/2}$ are of the form

$$\frac{2}{3}x^{3/2} + K$$

where K is a constant.

In Example 3, you may ask how we knew to choose the function $\frac{2}{3}x^{3/2}$. First, we know that, for n a rational number,

$$\frac{d}{dx}x^n = nx^{n-1}$$

That is, differentiation of a power of x reduces the exponent by 1. Antidifferentiation is the inverse process, so it should increase the exponent by 1. This is how we obtained the $x^{3/2}$ part of $\frac{2}{3}x^{3/2}$. Second, the $\frac{2}{3}$ factor is needed so that, when we differentiate, we get $x^{1/2}$ and not $\frac{3}{2}x^{1/2}$.

Because

$$\frac{d}{dx}x^{n+1} = (n+1)x^n$$

for n a rational number, if $n \neq -1$, it follows that:

Antiderivatives of $f(x) = x^n$

All the antiderivatives of $f(x) = x^n$ are of the form

$$F(x) = \frac{x^{n+1}}{n+1} + K$$

where n is any rational number except -1, and K is a constant.

Notice that $n = -1$ is excluded from the formula. That is, we cannot use this formula to find the antiderative of $f(x) = x^{-1} = \frac{1}{x}$. The antiderivative of this function requires special attention and is considered a little later.

NOW WORK PROBLEM 1.

Indefinite Integrals

We use a special symbol to represent all the antiderivatives of a function — the **integral sign,** $\int$.

Indefinite Integral

Let F be an antiderivative of the function f. The **indefinite integral of f,** denoted by $\int f(x)\,dx$, is defined as

$$\int f(x)\,dx = F(x) + K$$

where K is a constant.

In other words, the indefinite integral of f equals all the antiderivatives of f.

In the expression $\int f(x)\,dx$, the integral sign $\int$ indicates that the operation of antidifferentiation is to be performed on the function f, and the dx reinforces the fact that the operation is to be performed with respect to the variable x. The function f is called the **integrand,** and the process of antidifferentiation is called **integration.**

Basic Integration Formulas

Based on the relationship between the process of differentiation and that of integration, or antidifferentiation, we can construct a list of formulas. These formulas may be verified by differentiating the right-hand side.

Basic Integration Formulas

If c is a real number and K is a constant,

$$\int c\,dx = cx + K \tag{1}$$

$$\int x^n\,dx = \frac{x^{n+1}}{n+1} + K \qquad n \neq -1 \tag{2}$$

$$\int [f(x) + g(x)]\,dx = \int f(x)\,dx + \int g(x)\,dx \tag{3}$$

$$\int [f(x) - g(x)]\,dx = \int f(x)\,dx - \int g(x)\,dx \tag{4}$$

$$\int cf(x)\,dx = c\int f(x)\,dx \tag{5}$$

As a special case of Formula (1), let $c = 1$. Then we find that

$$\int dx = \int 1 \cdot dx = 1 \cdot x + K = x + K$$

where K is a constant.

Formulas (3) and (4) state that the integral of a sum or a difference equals the sum or difference of the integrals.

Formula (5) states that a constant factor can be moved across an integral sign. Be careful! A variable factor cannot be moved across an integral sign.

2 EXAMPLE 4 Using Integration Formulas

(a) $\int \underset{\text{Formula (1)}}{5}\,dx = 5x + K$

(b) $\int \underset{\text{Formula (2)}}{x^5}\,dx = \frac{x^{5+1}}{5+1} + K = \frac{x^6}{6} + K$

Antiderivatives; the Indefinite Integral; Marginal Analysis

(c) $\displaystyle\int 3x^4\,dx \underset{\text{Formula (5)}}{=} 3\int x^4\,dx \underset{\text{Formula (2)}}{=} 3\cdot\frac{x^5}{5} + K = \frac{3}{5}x^5 + K$

(d) $\displaystyle\int (x^2 + x^3)\,dx \underset{\text{Formula (3)}}{=} \int x^2\,dx + \int x^3\,dx \underset{\text{Formula (2)}}{=} \frac{x^3}{3} + K_1 + \frac{x^4}{4} + K_2$

$\displaystyle = \frac{x^3}{3} + \frac{x^4}{4} + K$

where $K = K_1 + K_2$.

NOW WORK PROBLEMS 9, 13, AND 19.

Formulas (3) and (4) can be combined and used for sums and differences of three or more functions.

EXAMPLE 5 Using Integration Formulas

$\displaystyle\int\left(7x^5 + \frac{1}{2}x^2 - x\right)dx = \int 7x^5\,dx + \int \frac{1}{2}x^2\,dx - \int x\,dx$ Formulas (3) and (4).

$\displaystyle = 7\int x^5\,dx + \frac{1}{2}\int x^2\,dx - \int x\,dx$ Formula (5).

$\displaystyle = \left(7\cdot\frac{x^6}{6} + K_1\right) + \left(\frac{1}{2}\cdot\frac{x^3}{3} + K_2\right) - \left(\frac{x^2}{2} + K_3\right)$ Formula (2).

$\displaystyle = \frac{7}{6}x^6 + \frac{1}{6}x^3 - \frac{1}{2}x^2 + K$ Simplify.

where $K = K_1 + K_2 - K_3$.

As Example 5 illustrates, we can now integrate any polynomial function.

NOW WORK PROBLEM 23.

Sometimes, it is necessary to use algebra to put the integrand in a form that matches one of the Basic Integration Formulas.

EXAMPLE 6 Using Integration Formulas

(a) $\displaystyle\int\frac{1}{\sqrt{x}}\,dx \underset{\text{Algebra}}{=} \int x^{-1/2}\,dx \underset{\text{Formula(2)}}{=} \frac{x^{(-1/2)+1}}{-\frac{1}{2}+1} + K = \frac{x^{1/2}}{\frac{1}{2}} + K = 2x^{1/2} + K$

(b) $\displaystyle\int 4\sqrt[3]{x^5}\,dx \underset{\text{Formula (5)}}{=} 4\int\sqrt[3]{x^5}\,dx \underset{\text{Algebra}}{=} 4\int x^{5/3}\,dx \underset{\text{Formula (2)}}{=} 4\cdot\frac{x^{5/3+1}}{\frac{5}{3}+1} + K = \frac{3}{2}x^{8/3} + K$

(c) $\displaystyle\int\frac{15\,dx}{x^5} \underset{\text{Formula (5)}}{=} 15\int\frac{1}{x^5}\,dx \underset{\text{Algebra}}{=} 15\int x^{-5}\,dx \underset{\text{Formula (2)}}{=} 15\cdot\frac{x^{-4}}{-4} + K = \frac{-15}{4x^4} + K$

(d) $\displaystyle\int \frac{x^{3/2} - 2x}{\sqrt{x}}\,dx = \int \underset{\text{Algebra}}{\frac{x^{3/2} - 2x}{x^{1/2}}}\,dx = \int \underset{\text{Algebra}}{\left(\frac{x^{3/2}}{x^{1/2}} - \frac{2x}{x^{1/2}}\right)}\,dx$

$\displaystyle = \int \underset{\text{Algebra}}{(x - 2x^{1/2})}\,dx = \underset{\text{Formulas (4) and (5)}}{\int x\,dx - 2\int x^{1/2}\,dx}$

$\displaystyle = \underset{\text{Formula (2)}}{\frac{x^2}{2} - \frac{2x^{3/2}}{\frac{3}{2}} + K} = \frac{x^2}{2} - \frac{4}{3}x^{3/2} + K$

  NOW WORK PROBLEM 29.

Indefinite Integrals Involving Exponential and Logarithmic Functions

The next three integration formulas involve the exponential and logarithmic functions. Each one is a direct result of formulas developed in Chapter 4.

$$\int e^x\,dx = e^x + K \qquad (6)$$

$$\int e^{ax}\,dx = \frac{1}{a}e^{ax} + K \qquad a \neq 0 \qquad (7)$$

$$\int \frac{1}{x}\,dx = \ln|x| + K \qquad (8)$$

where K is a constant.

Proof Formulas (6) and (7) follow from the facts that

$$\frac{d}{dx}e^x = e^x \quad \text{and} \quad \frac{d}{dx}\left[\frac{1}{a}e^{ax}\right] = \frac{1}{a}\cdot ae^{ax} = e^{ax}$$

Formula (8) requires more attention. To prove it, we need to show that

$$\frac{d}{dx}\ln|x| = \frac{1}{x}$$

Since $x \neq 0$ (do you see why?), we consider two cases: $x > 0$ and $x < 0$.

Case 1 $x > 0$:

$$\frac{d}{dx}\ln|x| = \underset{\substack{\uparrow \\ |x| = x \\ \text{since } x > 0}}{\frac{d}{dx}\ln x} = \frac{1}{x}$$

Case 2 $x < 0$:

$$\frac{d}{dx}\ln|x| = \underset{\substack{\uparrow \\ |x| = -x \\ \text{since } x < 0}}{\frac{d}{dx}\ln(-x)} = \underset{\substack{\uparrow \\ \frac{d}{dx}\ln g(x) = \frac{\frac{d}{dx}g(x)}{g(x)}}}{\frac{1}{-x}\frac{d}{dx}(-x)} = \frac{1}{-x}(-1) = \frac{1}{x}$$

In each case,

$$\frac{d}{dx}\ln|x| = \frac{1}{x}$$

so that

$$\int \frac{1}{x}\,dx = \ln|x| + K$$

Formula (8) takes care of finding $\int \frac{1}{x}\,dx = \int x^{-1}\,dx$. Now we can find $\int x^n\,dx$ for any rational number n.

We have still not discussed the indefinite integral of $\ln x$. We postpone a discussion of $\int \ln x\,dx$ until Section 6.3, where we introduce *integration by parts*.

EXAMPLE 7 Evaluating Indefinite Integrals

Evaluate each of the following indefinite integrals

(a) $\displaystyle\int \frac{e^x + e^{-x}}{2}\,dx$ (b) $\displaystyle\int \frac{3x^7 - 4x}{x^2}\,dx$

SOLUTION (a) $\displaystyle\int \frac{e^x + e^{-x}}{2}\,dx = \frac{1}{2}\int (e^x + e^{-x})\,dx$ $\qquad \int cf(x)\,dx = c\int f(x)\,dx$

$\qquad\qquad = \dfrac{1}{2}\left[\int e^x\,dx + \int e^{-x}\,dx\right]$ $\quad \int [f(x) + g(x)]\,dx = \int f(x)\,dx + \int g(x)\,dx$

$\qquad\qquad = \dfrac{1}{2}[e^x + (-e^{-x})] + K$ $\qquad$ Formula (7).

$\qquad\qquad = \dfrac{e^x - e^{-x}}{2} + K$ $\qquad$ Simplify.

(b) $\displaystyle\int \frac{3x^7 - 4x}{x^2}\,dx = \int \left(\frac{3x^7}{x^2} - \frac{4x}{x^2}\right)dx$ $\qquad$ Algebra.

$\qquad\qquad = \displaystyle\int \left(3x^5 - \frac{4}{x}\right)dx$ $\qquad$ Algebra.

$\qquad\qquad = \displaystyle\int 3x^5\,dx - \int \frac{4}{x}\,dx$ $\quad \int[f(x) - g(x)]\,dx = \int f(x)\,dx - \int g(x)\,dx$

$\qquad\qquad = 3\displaystyle\int x^5\,dx - 4\int \frac{1}{x}\,dx$ $\quad \int cf(x)\,dx = c\int f(x)\,dx$

$\qquad\qquad = 3\cdot\dfrac{x^6}{6} - 4\ln|x| + K$ $\qquad$ Formulas (2) and (8).

$\qquad\qquad = \dfrac{x^6}{2} - 4\ln|x| + K$ $\qquad$ Simplify.

NOW WORK PROBLEM 35.

Application to Marginal Analysis

Find cost and revenue functions 4 As we discussed in Chapter 4, marginal revenue and marginal cost are defined as the derivative of the revenue function and cost function, respectively. As a result, if the marginal revenue (or marginal cost) is a known function, the revenue function R (or cost function C) may be found by using the process of integration. That is,

> **Finding Revenue and Cost from Marginal Revenue and Marginal Cost**
>
> $$\int R'(x)\, dx = R(x) + K \quad \text{and} \quad \int C'(x)\, dx = C(x) + K$$
>
> where K is a constant.

The presence of a constant K in the integral of a function plays an important role in applications and will equal the particular value that the situation demands.

EXAMPLE 8 Finding a Cost Function

By experimenting with various production techniques, a manufacturer finds that the marginal cost of production is given by the function

$$C'(x) = 2x + 6$$

where x is the number of units produced and C' is the marginal cost in dollars. The fixed cost of production is known to be $9. Find the cost of production.

SOLUTION The antiderivative of the marginal cost C' of production is the cost C of production. That is,

$$C(x) = \int C'(x)\, dx = \int (2x + 6)\, dx = x^2 + 6x + K$$

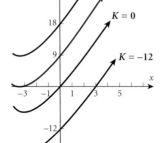

FIGURE 1

where K is a constant. We can find the value of the constant K by noting that of all the cost functions with derivative $2x + 6$, only one has a fixed cost of production of $9, namely, the one whose cost is 9 when $x = 0$. We use this requirement to find the constant K.

$$C(x) = x^2 + 6x + K$$

$C(0) = 0^2 + 6 \cdot 0 + K = 9$ $C(0) = 9$

$K = 9$ Solve for K.

As a result, the cost function C is

$$C(x) = x^2 + 6x + 9$$

Figure 1 illustrates various cost functions whose marginal cost is $2x + 6$. The one having a fixed cost of $9 is shown in color.

 NOW WORK PROBLEM 43.

EXAMPLE 9 Finding a Profit Function

Suppose the manufacturer in Example 8 receives a price of $60 per unit. This means the marginal revenue is $60. That is,

$$R'(x) = 60$$

(a) Find the revenue function R.
(b) Find the profit function P.
(c) Find the sales volume that yields maximum profit.
(d) What is the profit at this sales volume?

SOLUTION (a) The revenue function R is the antiderivative of the marginal revenue function R'. That is,

$$R(x) = \int R'(x)\, dx = \int 60\, dx = 60x + K$$

Now, of all these revenue functions, there is only one for which revenue equals zero for $x = 0$ units sold. To find it, we need to find K. We find K as follows:

$$R(0) = 60(0) + K = 0 \quad R(0) = 0$$
$$K = 0 \quad \text{Solve for } K.$$

This means the revenue function is

$$R(x) = 60x$$

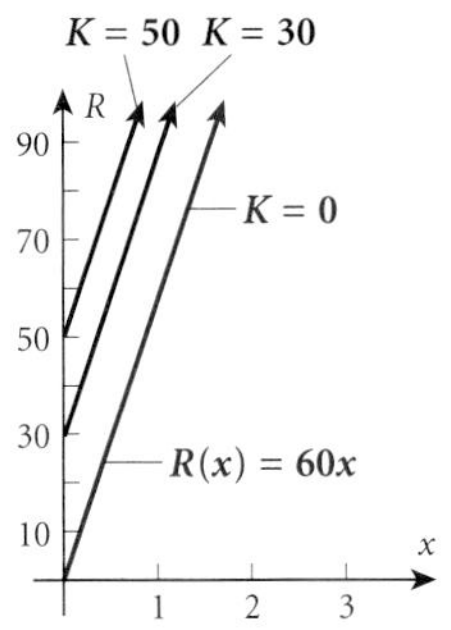

See Figure 2.

(b) The profit function P is the difference between revenue and cost. Since $R(x) = 60x$ and $C(x) = x^2 + 6x + 9$ (from Example 8), we have

$$P(x) = R(x) - C(x) = 60x - (x^2 + 6x + 9) = -x^2 + 54x - 9$$

(c) The maximum profit is obtained when marginal revenue equals marginal cost:

$$R'(x) = C'(x)$$
$$60 = 2x + 6$$
$$2x = 54$$
$$x = 27$$

When sales total 27 units, a maximum profit is obtained.

(d) The profit for sales of 27 units is

$$P(27) = -(27)^2 + 54(27) - 9 = \$720$$

FIGURE 2

NOW WORK PROBLEM 47.

EXAMPLE 10 Predicting the Size of a Mosquito Population

The size of a mosquito population is changing at the rate of $432t^2 - 5t^4$ per month where t is the time in months. If the current population is 40, what will the population size be 5 months from now?

SOLUTION If $P(t)$ is the population of mosquitoes at time t, then $P'(t)$ represents the rate of change of population with respect to time. That is,

$$P'(t) = 432t^2 - 5t^4$$

It follows that $P(t)$ is the antiderivative of $432t^2 - 5t^4$. That is,

$$P(t) = \int P'(t)\, dt = \int (432t^2 - 5t^4)\, dt = 144t^3 - t^5 + K$$

The constant K is determined from the condition that at $t = 0$ the population is 40. That is, $P(0) = 40$.

$$40 = 144(0^3) - 0^5 + K \quad P(0) = 40.$$
$$K = 40 \quad \text{Solve for } K.$$

As a result,

$$P(t) = 144t^3 - t^5 + 40$$

The population 5 months from now will be

$$P(5) = 144(5^3) - 5^5 + 40 = 14{,}915$$

mosquitoes.

EXERCISE 6.1

In Problems 1-8, find all the antiderivatives of f.

1. $f(x) = x^3$ **2.** $f(x) = 6x^2$ **3.** $f(x) = 2x + 3$ **4.** $f(x) = 10 - 3x^2$

5. $f(x) = \dfrac{4}{x}$ **6.** $f(x) = e^x$ **7.** $f(x) = \sqrt[3]{x}$ **8.** $f(x) = e^{3x}$

In Problems 9–38, evaluate each indefinite integral.

9. $\int 3\, dx$ **10.** $\int -4\, dx$ **11.** $\int x\, dx$ **12.** $\int x^2\, dx$ **13.** $\int x^{1/3}\, dx$

14. $\int x^{4/3}\, dx$ **15.** $\int x^{-2}\, dx$ **16.** $\int x^{-3}\, dx$ **17.** $\int x^{-1/2}\, dx$ **18.** $\int x^{-2/3}\, dx$

19. $\int (2x^3 + 5x)\, dx$ **20.** $\int (3x^2 - 4x)\, dx$ **21.** $\int (x^2 + 2e^x)\, dx$ **22.** $\int (3x + 5e^x)\, dx$

23. $\int (x^3 - 2x^2 + x - 1)\, dx$ **24.** $\int (2x^4 + x^2 - 5)\, dx$ **25.** $\int \left(\dfrac{x-1}{x}\right) dx$ **26.** $\int \left(\dfrac{x+1}{x}\right) dx$

27. $\int \left(2e^x - \dfrac{3}{x}\right) dx$ **28.** $\int \left(\dfrac{8}{x} - e^{-x}\right) dx$ **29.** $\int \left(\dfrac{3\sqrt{x}+1}{\sqrt{x}}\right) dx$ **30.** $\int \left(\dfrac{2\sqrt{x}-4}{\sqrt{x}}\right) dx$

31. $\int \dfrac{x^2-4}{x+2}\, dx$ **32.** $\int \dfrac{x^2-1}{x-1}\, dx$ **33.** $\int x(x-1)\, dx$ **34.** $\int x(x+2)\, dx$

35. $\int \dfrac{3x^5+2}{x}\, dx$ **36.** $\int \dfrac{x^6+x^2+1}{x^3}\, dx$ **37.** $\int \dfrac{4e^x+e^{2x}}{e^x}\, dx$ **38.** $\int \dfrac{3e^x+xe^{2x}}{xe^x}\, dx$

In Problems 39–42, find the revenue function R. Assume that revenue is zero when zero units are sold.

39. $R'(x) = 600$ **40.** $R'(x) = 350$ **41.** $R'(x) = 20x + 5$ **42.** $R'(x) = 50x - x^2$

In Problems 43–46, find the cost function C. Determine where the cost is a minimum.

43. $C'(x) = 14x - 2800$
Fixed cost = $4300

44. $C'(x) = 6x - 2400$
Fixed cost = $800

45. $C'(x) = 20x - 8000$
Fixed cost = $500

46. $C'(x) = 15x - 3000$
Fixed cost = $1000

---

47. Profit Function The marginal cost of production is found to be

$$C'(x) = 1000 - 20x + x^2$$

where x is the number of units produced. The fixed cost of production is $9000. The manufacturer sets the price per unit at $3400.

(a) Find the cost function.
(b) Find the revenue function.
(c) Find the profit function.
(d) Find the sales volume that yields maximum profit.
(e) What is the profit at this sales volume?
(f) Graph the revenue, cost, and profit functions.

48. Profit A company determines that the marginal cost of producing x units of a particular commodity during 1 day of operation is $C'(x) = 6x - 141$, where the production cost is in dollars. The selling price of the commodity is fixed at $9 per unit and the fixed cost is $1800 per day.

(a) Find the cost function.
(b) Find the revenue function.
(c) Find the profit function.
(d) What is the maximum profit that can be obtained in 1 day of operation?
(e) Graph the revenue, cost, and profit functions.

49. Prison Population In 1998 the total number of inmates in United States prisons was 592,462. Research indicates that this number will change at a rate of $7000t + 20,000$ where t represents the number of years elapsed since the year 1998. Predict the total number of inmates in United States prisons in the year 2008.

Source: Office of Justice Programs, United States Department of Justice.

50. Undergraduate Tuition Average annual undergraduate tuition paid by students at institutions of higher learning in the United States changes at the rate of $-0.14t + 225$ where t is the number of years elapsed since 1996. What will tuition be in 2007 if the cost of tuition in the year 2001 was $5442?

Source: National Council of Education Statistics, United States Department of Education.

51. Population Growth It is estimated that the population of a certain town changes at the rate of $2 + t^{4/5}$ people per month. If the current population is 20,000, what will the population be in 10 months?

52. Resource Depletion The water currently used from a lake is estimated to amount to 150 million gallons a month. Water usage (in millions of gallons) is expected to increase at the rate of $3 + 0.01x$ after x months. What will the water usage be a year from now?

53. Population Growth There are currently 20,000 citizens of voting age in a small town. Demographics indicate that the voting population will change at the rate of $2.2t - 0.8t^2$ (in thousands of voting citizens), where t denotes time in years. How many citizens of voting age will there be 3 years from now?

54. Air Pollution An environmental study of a certain town suggests that t years from now the level of carbon monoxide in the air will be changing at the rate of $0.2t + 0.2$ parts per million per year. If the current level of carbon monoxide in the air is 3.8 parts per million, what will the level be 5 years from now?

55. Chemical Reaction The end product of a chemical reaction is produced at the rate of $\dfrac{\sqrt{t} - 1}{t}$ milligrams per minute. If the reaction started at time $t = 1$, determine the amount produced during the first 4 minutes.

56. Free Fall An object dropped from an airplane 3200 feet above the ground falls at the rate of

$$v(t) = 32t \text{ feet per second}$$

where t is given in seconds. If the object has fallen 576 feet after 6 seconds, how long after it is dropped will the object hit the ground?

57. Water Depletion A water reservoir is being filled at the rate of 15,000 gallons per hour. Due to increased consumption, the water in the reservoir is decreasing at the rate of $\dfrac{5}{2}t$ gallons per hour at time t. When will the reservoir be empty if the initial water volume was 100,000 gallons?

58. Verify the following statements:

(a) $\displaystyle\int (x \cdot \sqrt{x})\, dx \neq \int x\, dx \cdot \int \sqrt{x}\, dx$

(b) $\displaystyle\int x(x^2 + 1)\, dx \neq x \int (x^2 + 1)\, dx$

(c) $\displaystyle\int \frac{x^2 - 1}{x - 1}\, dx \neq \frac{\displaystyle\int (x^2 - 1)\, dx}{\displaystyle\int (x - 1)\, dx}$

6.2 Integration Using Substitution

PREPARING FOR THIS SECTION *Before getting started, review the following:*

> The Power Rule (Chapter 4, Section 4.4, p. 304)
> Differentials (Chapter 5, Section 5.7, pp. 405–407)

OBJECTIVE 1 Integrate using substitution

Indefinite integrals that cannot be evaluated by using Formulas (1)–(8) on pages 424 and 426 of Section 6.1 may sometimes be evaluated by the *substitution method*. This method involves the introduction of a function that changes the integrand into a form to which the formulas of Section 6.1 apply.

The basic idea behind integration by substitution is the Power Rule. To see how integration by substitution works, consider the following example.

EXAMPLE 1 **Evaluating an Indefinite Integral**

Evaluate: $\int (x^2 + 5)^3 \, 2x \, dx$

SOLUTION By the Power Rule, we know that

$$\frac{d}{dx}[g(x)]^n = n[g(x)]^{n-1} g'(x)$$

With $g(x) = x^2 + 5$ and $n = 4$, we have

$$\frac{d}{dx}(x^2 + 5)^4 = 4(x^2 + 5)^3 \, 2x$$

Then

$$\int 4(x^2 + 5)^3 \, 2x \, dx = (x^2 + 5)^4 + K$$

where K is a constant. We apply Formula (5), on page 424 and move the 4 outside the integral.

$$4 \int (x^2 + 5)^3 \, 2x \, dx = (x^2 + 5)^4 + K$$

Now divide both sides by 4. Then

$$\int (x^2 + 5)^3 \, 2x \, dx = \frac{1}{4}(x^2 + 5)^4 + K_1$$

where $K_1 = \dfrac{K}{4}$.

We may simplify the procedure used in Example 1 by *changing the variables*. Introduce the variable u, defined as

$$u = x^2 + 5$$

The differential of $u = u(x)$ is $du = u'(x) \, dx$. Then

$$du = 2x \, dx \qquad \frac{d}{dx}(x^2 + 5) = 2x$$

Now we express the integrand $(x^2 + 5)^3\, 2x$ in terms of u and the differential dx in terms of u and du.

$$\int \underbrace{(x^2 + 5)^3}_{\substack{u = x^2 + 5}} \underbrace{2x\, dx}_{du = 2x\, dx} = \int u^3\, du = \underbrace{\frac{1}{4}u^4 + K}_{\text{Integrate}} = \underbrace{\frac{1}{4}(x^2 + 5)^4 + K}_{\substack{\text{Express in terms of } x; \\ u = x^2 + 5}}$$

EXAMPLE 2 Integration Using Substitution

Evaluate: $\displaystyle\int \frac{dx}{2x + 1}$

SOLUTION We try the substitution $u = 2x + 1$ to see if it simplifies the integral. Then

$$du = 2\, dx \qquad \text{so} \qquad dx = \frac{du}{2}$$

$$\int \underbrace{\frac{dx}{2x + 1}}_{\substack{u = 2x + 1 \\ dx = \frac{du}{2}}} = \int \underbrace{\frac{\frac{du}{2}}{u}}_{\text{Simplify}} = \int \underbrace{\frac{du}{2u}}_{\text{Move } \frac{1}{2} \text{ outside}} = \frac{1}{2}\int \underbrace{\frac{du}{u}}_{\text{Integrate}} = \frac{1}{2}\ln|u| + K = \underbrace{\frac{1}{2}\ln|2x + 1| + K}_{u = 2x + 1}$$

NOW WORK PROBLEM 1.

EXAMPLE 3 Integration Using Substitution

Evaluate: $\displaystyle\int x\sqrt{x^2 + 1}\, dx$

SOLUTION We try the substitution $u = x^2 + 1$ to see if it simplifies the integral. Then

$$du = 2x\, dx \qquad \text{so} \qquad x\, dx = \frac{du}{2}$$

Now

$$\int x\sqrt{x^2 + 1}\, dx = \int \sqrt{x^2 + 1}\, x\, dx$$

$$= \int \sqrt{u}\, \frac{du}{2} \qquad u = x^2 + 1,\ x\, dx = \frac{du}{2}$$

$$= \frac{1}{2}\int \sqrt{u}\, du \qquad \text{Move } \frac{1}{2} \text{ outside the integral.}$$

$$= \frac{1}{2}\int u^{1/2}\, du \qquad \text{Change to a rational exponent.}$$

$$= \frac{1}{2}\cdot\frac{u^{3/2}}{\frac{3}{2}} + K \qquad \text{Integrate.}$$

$$= \frac{(x^2 + 1)^{3/2}}{3} + K \qquad u = x^2 + 1$$

Notice that the substitution $u = x^2 + 1$ in Example 3 worked because the x in the integrand along with dx gives du, except for the constant 2. If we try this same substitution to evaluate $\int \sqrt{x^2 + 1}\, dx$, we obtain

$$\int \sqrt{x^2 + 1}\, dx = \int \sqrt{u}\, \frac{du}{2\sqrt{u-1}} = \int \frac{\sqrt{u}}{2\sqrt{u-1}}\, du$$

In this case, the substitution results in an integrand that is *more complicated* than the original one.

EXAMPLE 4 Integration Using Substitution

Evaluate: $\int x^2 e^{x^3}\, dx$

SOLUTION Let $u = x^3$. Then $du = 3x^2\, dx$.

$$\int x^2 e^{x^3}\, dx = \underset{\substack{\uparrow \\ u = x^3 \\ \frac{du}{3} = x^2\, dx}}{\int e^{x^3} \cdot x^2\, dx} = \underset{\uparrow}{\int e^u \frac{du}{3}} = \underset{\substack{\uparrow \\ \text{Move } \frac{1}{3} \text{ outside} \\ \text{the integral}}}{\frac{1}{3} \int e^u\, du} = \underset{\substack{\uparrow \\ \text{Integrate}}}{\frac{1}{3} e^u + K} = \underset{\substack{\uparrow \\ u = x^3}}{\frac{1}{3} e^{x^3} + K}$$

EXAMPLE 5 Integration Using Substitution

Evaluate: $\int \frac{dx}{x \ln x}$

SOLUTION Let $u = \ln x$. Then $du = \frac{1}{x} dx = \frac{dx}{x}$. Then

$$\int \frac{dx}{x \ln x} = \underset{\substack{\uparrow \\ \text{Substitute}}}{\int \frac{1}{\ln x} \frac{dx}{x}} = \underset{\substack{\uparrow \\ \text{Integrate}}}{\int \frac{1}{u}\, du} = \ln |u| + K = \underset{\substack{\uparrow \\ u = \ln x}}{\ln |\ln x| + K}$$

EXAMPLE 6 Integration Using Substitution

Evaluate: $\int \frac{dx}{2\sqrt{x}(1 + \sqrt{x})^3}$

SOLUTION We try the substitution

$$u = 1 + \sqrt{x} = 1 + x^{1/2}$$

Then

$$du = \frac{1}{2} x^{-1/2}\, dx = \frac{1}{2\sqrt{x}}\, dx = \frac{dx}{2\sqrt{x}}$$

Now we substitute.

$$\int \frac{dx}{2\sqrt{x}(1 + \sqrt{x})^3} = \underset{\substack{\uparrow \\ \text{Rearrange}}}{\int \frac{1}{(1+\sqrt{x})^3} \frac{dx}{2\sqrt{x}}} = \underset{\substack{\uparrow \\ \text{Substitute}}}{\int \frac{1}{u^3}\, du} = \int u^{-3}\, du = \underset{\substack{\uparrow \\ \text{Integrate}}}{\frac{u^{-2}}{-2} + K}$$

$$= \underset{\substack{\uparrow \\ \text{Simplify}}}{-\frac{1}{2u^2} + K} = \underset{\substack{\uparrow \\ u = 1 + \sqrt{x}}}{-\frac{1}{2(1+\sqrt{x})^2} + K}$$

NOW WORK PROBLEM 27.

Sometimes more than one substitution will work. In the next example we use two different substitutions to evaluate the same integral.

EXAMPLE 7 Integrating Using Substitution

Evaluate: $\displaystyle\int x\sqrt{4+x}\,dx$

SOLUTION   **Substitution I:** Let $u = 4 + x$. Then $du = dx$ and $x = u - 4$, so that

$$\int x\sqrt{4+x}\,dx = \int (u-4)\sqrt{u}\,du \qquad x = u - 4;\ u = 4 + x;\ dx = du$$

$$= \int (u^{3/2} - 4u^{1/2})\,du \qquad \sqrt{u} = u^{1/2};\ \text{multiply out.}$$

$$= \frac{u^{5/2}}{\frac{5}{2}} - \frac{4u^{3/2}}{\frac{3}{2}} + K \qquad \text{Integrate.}$$

$$= \frac{2(4+x)^{5/2}}{5} - \frac{8(4+x)^{3/2}}{3} + K \qquad u = 4 + x$$

Substitution II: Let $u^2 = 4 + x$. Then $x = u^2 - 4$, and $dx = 2u\,dx$. Now substitute into the integral.

$$\int x\sqrt{4+x}\,dx = \int (u^2 - 4)u\,2u\,du \qquad x = u^2 - 4;\ 4 + x = u^2;\ dx = 2u\,du$$

$$= 2\int (u^2 - 4)u^2\,du \qquad \text{Move 2 outside the integral.}$$

$$= 2\int (u^4 - 4u^2)\,du \qquad \text{Multiply out.}$$

$$= 2\int u^4\,du - 2\int 4u^2\,du \qquad \text{The integral of a difference is the difference of the integrals.}$$

$$= 2\int u^4\,du - 8\int u^2\,du \qquad \text{Move 4 outside the second integral.}$$

$$= 2\cdot\frac{u^5}{5} - 8\cdot\frac{u^3}{3} + K \qquad \text{Integrate.}$$

$$= \frac{2(4+x)^{5/2}}{5} - \frac{8(4+x)^{3/2}}{3} + K \qquad u^2 = 4 + x.$$

  NOW WORK PROBLEM 17.

The idea behind the substitution method is to obtain an integral $\displaystyle\int h(u)\,du$ that is simpler than the original integral $\displaystyle\int f(x)\,dx$. When a substitution does not simplify the integral, other substitutions should be tried. If these do not work, other integration methods must be applied. Since integration, unlike differentiation, has no prescribed method, a lot of practice is required.

A summary of the method of integration by substitution is given in the flowchart in Figure 3.

FIGURE 3

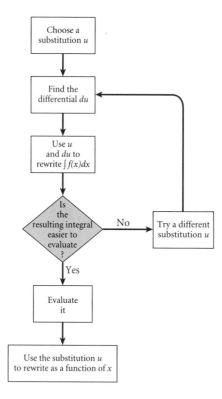

EXERCISE 6.2

In Problems 1–34, evaluate each indefinite integral. Use the substitution method.

1. $\int (2x+1)^5 \, dx$
2. $\int (3x-5)^4 \, dx$
3. $\int e^{2x-3} \, dx$
4. $\int e^{3x+4} \, dx$

5. $\int (-2x+3)^{-2} \, dx$
6. $\int (5-2x)^{-3} \, dx$
7. $\int x(x^2+4)^2 \, dx$
8. $\int x(x^2-2)^3 \, dx$

9. $\int e^{x^3+1} x^2 \, dx$
10. $\int e^{2x^2+1} x \, dx$
11. $\int (e^{x^2}+e^{-x^2}) x \, dx$
12. $\int (e^{x^2}-e^{-x^2}) x \, dx$

13. $\int x^2(x^3+2)^6 \, dx$
14. $\int x^2(x^3-1)^4 \, dx$
15. $\int \dfrac{x}{\sqrt[3]{1+x^2}} \, dx$
16. $\int \dfrac{x}{\sqrt[5]{1-x^2}} \, dx$

17. $\int x\sqrt{x+3} \, dx$
18. $\int x\sqrt{x-3} \, dx$
19. $\int \dfrac{e^x}{e^x+1} \, dx$
20. $\int \dfrac{e^{-x}}{e^{-x}+4} \, dx$

21. $\int \dfrac{e^{\sqrt{x}} \, dx}{\sqrt{x}}$
22. $\int \dfrac{e^{\sqrt[3]{x}} \, dx}{x^{2/3}}$
23. $\int \dfrac{(x^{1/3}-1)^6 \, dx}{x^{2/3}}$
24. $\int \dfrac{(x^{1/3}+2)^3}{x^{2/3}} \, dx$

25. $\int \dfrac{(x+1) \, dx}{(x^2+2x+3)^2}$
26. $\int \dfrac{(x+4) \, dx}{(x^2+8x+2)^3}$
27. $\int \dfrac{dx}{\sqrt{x}(1+\sqrt{x})^4}$
28. $\int \dfrac{(3-2\sqrt{x})^2}{\sqrt{x}} \, dx$

29. $\int \dfrac{dx}{2x+3}$
30. $\int \dfrac{dx}{3x-5}$
31. $\int \dfrac{x \, dx}{4x^2+1}$
32. $\int \dfrac{x \, dx}{5x^2-2}$

33. $\int \dfrac{x+1}{x^2+2x+2} \, dx$
34. $\int \dfrac{2x-1}{x^2-x+4} \, dx$

35. **Depreciation of a Car** The depreciation rate of a new car obeys the equation
$$V'(t) = -6000\, e^{-0.5t}$$
where V is the value of the car and t, in years, is the time from the date of purchase. If the car cost \$27,000 new, what is its value after 2 years? What is the value after 4 years?

36. **Depreciation of a Car** The depreciation rate of a new car obeys the equation
$$V'(t) = \frac{-16{,}000}{(2t+1)^2}, \quad 0 \le t \le 4$$
where V is the value of the car and t, in years, is the time from the date of purchase. If the car cost \$15,000 new, what is its value after 2 years? What is the value after 4 years?

37. **Department of Agriculture Budget** The annual budget outlay of the United States Department of Agriculture is growing at a rate of $B'(t) = 1.715 e^{0.025t}$ where B is in billions of dollars per year and t is the number of years elapsed since 2001. The budget for 2001 was 68.6 billion dollars.
    (a) Find an equation for $B(t)$.
    (b) If the current rate of growth continues, predict when the budget will exceed \$100 billion.

    Source: United States Department of Agriculture.

38. **Revenue** The marginal revenue (in thousands of dollars) from the sale of x tractors is
$$R'(x) = \frac{2x(x^2+10)^2}{1000}$$
Find the revenue function R if the revenue from the sale of 4 tractors is \$198,000.

39. **Work Force Growth** The number N of employees at the Ajex Steel Company is growing at a rate given by the equation
$$N'(t) = 20 e^{0.01t}$$
where N is the number of people employed at time t, in years. The number of employees currently is 400.
    (a) Find an equation for $N(t)$.
    (b) How long will it take the work force to reach 800 employees?

40. **Pollution** An oil tanker is leaking oil and producing a circular oil slick that is growing at the rate of
$$R'(t) = \frac{30}{\sqrt{t+4}} \quad t \ge 0$$
where R is the radius in meters of the slick after t minutes. Find the radius of the slick after 21 minutes if the radius is 0 when $t = 0$.

41. Derive the formula
$$\int (ax+b)^n\, dx = \frac{(ax+b)^{n+1}}{a(n+1)} + K \quad a \ne 0, n \ne -1$$
[**Hint:** Let $u = ax + b$.]

6.3 Integration by Parts

PREPARING FOR THIS SECTION *Before getting started, review the following:*

>> Formula for the Derivative of a Product (Chapter 4, Section 4.3, p. 295)

OBJECTIVE   1   Integrate by parts

Integrate by parts   1

Next, we discuss a method for evaluating indefinite integrals such as
$$\int x e^x\, dx \quad \text{and} \quad \int \ln x\, dx$$
for which the substitution technique does not work.

This method, called *integration by parts*, is based on the formula for the derivative of a product. Recall that if u and v are differentiable functions of x, then
$$\frac{d}{dx}(uv) = u\frac{dv}{dx} + v\frac{du}{dx}$$
By integrating both sides, we obtain
$$\int \frac{d}{dx}(uv)\, dx = \int \left(u\frac{dv}{dx} + v\frac{du}{dx} \right) dx$$

The integral on the left is just uv. On the right, use the fact that the integral of a sum is the sum of the integrals. Then,

$$uv = \int u\frac{dv}{dx}\,dx + \int v\frac{du}{dx}\,dx$$

Rearranging terms, we obtain

$$\int u\frac{dv}{dx}\,dx = uv - \int v\frac{du}{dx}\,dx$$

In abbreviated form, this formula may be written in the following way.

Integration by Parts Formula

$$\int u\,dv = uv - \int v\,du \qquad (1)$$

To use the integration by parts formula, we choose u and dv so $\int u\,dv$ is equal to the integral we seek. We find the differential of u to obtain du and integrate dv to obtain v. Then we substitute into Formula (1). If we can integrate $\int v\,du$, the problem is solved.

The goal of this procedure, then, is to choose u and dv so that the term $\int v\,du$ is easier to integrate than the original problem, $\int u\,dv$. As the examples will illustrate, this usually happens when u is simplified by differentiation.

EXAMPLE 1 **Using the Integration by Parts Formula**

Evaluate: $\int xe^x\,dx$

SOLUTION To use the integration by parts formula, we choose u and dv so that

$$\int u\,dv = \int xe^x\,dx$$

and $\int v\,du$ is easier to evaluate than $\int u\,dv$. In this example, we choose

$$u = x \quad \text{and} \quad dv = e^x\,dx \qquad (2)$$

With this choice, $\int u\,dv = \int xe^x\,dx$, as required. Now use equation (2) to find du and $v = \int dv$.

$$du = dx \quad \text{and} \quad v = \int dv = \int e^x\,dx = e^x \qquad (3)$$
$$\uparrow$$
$$dv = e^x\,dx$$

Notice in evaluating $\int dv$ that we only require a particular antiderivative of dv; we will add the constant of integration later.

Now substitute the results of (2) and (3) into (1). The result is

$$\int \overbrace{x}^{u} \overbrace{e^x \, dx}^{dv} = \overbrace{x}^{u} \overbrace{e^x}^{v} - \int \overbrace{e^x}^{v} \overbrace{dx}^{du} \quad \text{Substitute into } \int u \, dv = uv - \int v \, du$$

$$= xe^x - e^x + K \quad \text{Integrate } \int e^x \, dx; \text{ add the constant of integration.}$$

NOW WORK PROBLEM 1.

Let's look once more at Example 1. Suppose we had chosen u and dv differently as

$$u = e^x \quad \text{and} \quad dv = x \, dx$$

This choice would have resulted in $\int u \, dv = \int xe^x \, dx$, as required, but now

$$du = e^x \, dx \quad \text{and} \quad v = \int dv = \int x \, dx = \frac{x^2}{2}$$

so equation (1) would have yielded

$$\int \overbrace{xe^x \, dx}^{u \, dv} = \overbrace{e^x \frac{x^2}{2}}^{uv} - \int \overbrace{\frac{x^2}{2} e^x \, dx}^{v \, du}$$

As you can see, instead of obtaining an integral that is easier to evaluate, we obtain one that is more complicated than the original. This means that an unwise choice of u and dv has been made.

Unfortunately, there are no general directions for choosing u and dv. Some hints you can use are given next.

> **Hints for Using Integration by Parts**
>
> 1. Choose u and dv so that $\int u \, dv$ is the integral you seek. In choosing dv, dx is always a part of dv.
> 2. You must be able to find du and $v = \int dv$.
> 3. u and dv are chosen so that $\int v \, du$ is easier to evaluate than the original integral $\int u \, dv$. This often happens when u is simplified by differentiation.

In making an initial choice for u and dv, a certain amount of trial and error is used. If a selection appears to hold little promise, abandon it and try some other choice. If no choices work, it may be that some other technique of integration, such as substitution, should be tried.

Let's look at some more examples.

EXAMPLE 2 **Using Integration by Parts**

Evaluate: $\int x \ln x \, dx$

SOLUTION Choose

$$u = \ln x \quad \text{and} \quad dv = x \, dx \quad (4)$$

Then

$$\int u \, dv = \int (\ln x)(x \, dx) = \int x \ln x \, dx$$

as required.

Use (4) to find du and $\int v\, du$.

$$u = \ln x \qquad du = \frac{1}{x}\, dx \qquad \frac{d}{dx}\ln x = \frac{1}{x}$$

$$dv = x\, dx \qquad v = \frac{x^2}{2} \qquad v = \int dv = \int x\, dx = \frac{x^2}{2}$$

Integrate by parts.

$$\begin{aligned}\int x \ln x\, dx &= \ln x \cdot \frac{x^2}{2} - \int \frac{x^2}{2}\cdot\frac{1}{x}\, dx \qquad &&\int u\, dv = uv - \int v\, du\\ &= \frac{1}{2}x^2 \ln x - \frac{1}{2}\int x\, dx &&\text{Simplify. Move } \frac{1}{2} \text{ outside the integral.}\\ &= \frac{1}{2}x^2 \ln x - \frac{1}{2}\cdot\frac{x^2}{2} + K &&\text{Integrate.}\\ &= \frac{1}{2}x^2 \ln x - \frac{1}{4}x^2 + K\end{aligned}$$

  NOW WORK PROBLEM 3 AND 7.

Sometimes it is necessary to integrate by parts more than once to solve a particular problem, as illustrated by the next example.

EXAMPLE 3   Using Integration by Parts

Evaluate: $\int x^2 e^x\, dx$

SOLUTION Choose

$$u = x^2 \qquad \text{and} \qquad dv = e^x\, dx \qquad \int u\, dv = \int x^2 e^x\, dx$$

Then

$$du = 2x\, dx \qquad \text{and} \qquad v = e^x \qquad v = \int dv = \int e^x\, dx = e^x$$

Integrate by parts.

$$\begin{aligned}\int x^2 e^x\, dx &= x^2 e^x - \int e^x \cdot 2x\, dx \qquad &&\int u\, dv = uv - \int v\, du\\ &= x^2 e^x - 2\int xe^x\, dx &&\text{Move 2 outside the integral.}\end{aligned}$$

Although we must still evaluate $\int xe^x\, dx$, we can see that the original integral has been replaced by a simpler one. In fact, in Example 1 we found (using integration by parts) that

$$\int xe^x\, dx = xe^x - e^x + K \qquad \text{Example 1.}$$

Using this result, we have

$$\begin{aligned}\int x^2 e^x\, dx &= x^2 e^x - 2(xe^x - e^x) + K\\ &= x^2 e^x - 2xe^x + 2e^x + K\end{aligned}$$

EXAMPLE 4 Formula for $\int \ln x \, dx$

Find a formula for $\int \ln x \, dx$.

SOLUTION To find $\int \ln x \, dx$, we use the integration by parts formula. Choose $u = \ln x$ and $dv = dx$. Then $du = \dfrac{1}{x} dx$ and $v = x$, so

$$\int \ln x \, dx = x \ln x - \int x \cdot \frac{1}{x} dx \qquad \int u \, dv = uv - \int v \, du$$

$$= x \ln x - \int dx \qquad \text{Simplify.}$$

$$= x \ln x - x + K \qquad \text{Integrate.}$$

The integration by parts formula is useful for the evaluation of indefinite integrals that have integrands composed of e^x times a polynomial function of x or $\ln x$ times a polynomial function of x. It can also be used for other types of indefinite integrals that will not be discussed in this book.

EXERCISE 6.3

1. Evaluate $\int x e^{4x} dx$. Choose $u = x$ and $dv = e^{4x} dx$.

2. Evaluate $\int x e^{-x} dx$. Choose $u = x$ and $dv = e^{-x} dx$.

In Problems 3–16, evaluate each indefinite integral. Use integration by parts.

3. $\int x e^{2x} dx$
4. $\int x e^{-3x} dx$
5. $\int x^2 e^{-x} dx$
6. $\int x^2 e^{2x} dx$
7. $\int \sqrt{x} \ln x \, dx$
8. $\int x (\ln x)^2 dx$
9. $\int (\ln x)^2 dx$
10. $\int \dfrac{\ln x}{x^2} dx$
11. $\int x^2 \ln 3x \, dx$
12. $\int x^2 \ln 5x \, dx$
13. $\int x^2 (\ln x)^2 dx$
14. $\int x^3 (\ln x)^2 dx$
15. $\int \dfrac{\ln x}{x^3} dx$
16. $\int \sqrt{x} (\ln \sqrt{x})^2 dx$

17. **Population Growth** The growth rate of a colony of leaf-eater ants follows the equation

$$P'(t) = 90\sqrt{t} - 100 \, te^{-t}$$

where P is the population at time t measured in days. If the population of the colony at time $t = 0$ is 5000, what is the population after 4 days? What is it after one week?

18. **Population Growth** The growth rate of a colony of slugs follows the equation

$$P'(t) = 180t + 300te^{-2t}$$

where P is the population at time t measured in days. If the population of the colony at time $t = 0$ is 5000, what is the population after 4 days? What is it after one week?

19. **Depreciation of a Car** The depreciation rate of a new car obeys the equation

$$V'(t) = -8000te^{-0.8t}$$

where V is the value of the car and t, in years, is the time from the date of purchase. If the car cost $20,000 new, what is its value after 2 years? What is the value after 4 years?

6.4 The Definite Integral; Learning Curves; Total Sales over Time

OBJECTIVES
1. Evaluate a definite integral
2. Use properties of a definite integral
3. Solve applied problems involving definite integrals

We begin with an example illustrating the general idea of a *definite integral*.

EXAMPLE 1 Finding the Change in a Cost Function

The marginal cost of a certain firm is given by the equation

$$C'(x) = 4 - 0.2x \qquad 0 \leq x \leq 10$$

where C' is in thousands of dollars and the quantity x produced is in hundreds of units per day. If the number of units produced in a given day changes from two hundred to five hundred units, what is the change in cost?

SOLUTION If C is the cost function, the change in cost from $x = 2$ to $x = 5$ is

$$C(5) - C(2)$$

This is the number we seek. The cost C is an antiderivative of $C'(x) = 4 - 0.2x$ so

$$C(x) = \int C'(x)\, dx = \int (4 - 0.2x)\, dx = 4x - 0.1x^2 + K$$

We use this to compute $C(5) - C(2)$:

$$C(5) - C(2) = [4(5) - (0.1)(25) + K] - [4(2) - (0.1)(4) + K] = 9.9 \qquad (1)$$

The change in cost is 9.9 thousand dollars.

In this example, the change in C was computed by using an antiderivative of C', which is symbolized by $\int C'(x)\, dx$. To indicate that the change is from $x = 2$ to $x = 5$, we add to this notation as follows:

$$\text{Change in } C \text{ from 2 to 5} = \int_2^5 C'(x)\, dx$$

This form is called a *definite integral*.

Definite Integral

Let f denote a function that is continuous on a closed interval $[a, b]$. The **definite integral** from a to b of f equals the change from a to b in an antiderivative of f. If F is an antiderivative of f, the definite integral of f from a to b is

$$\int_a^b f(x)\, dx = F(b) - F(a) \qquad (2)$$

In $\int_a^b f(x)\, dx$, the numbers a and b are called the **lower limit of integration** and the **upper limit of integration**, respectively.

EXAMPLE 2 Evaluating a Definite Integral

Evaluate: $\int_{2}^{3} x^2 \, dx$

SOLUTION First we find an antiderivative of $f(x) = x^2$.
One such antiderivative is $F(x) = \dfrac{x^3}{3}$. Then, based on (2), we have

$$\int_{2}^{3} x^2 \, dx = F(3) - F(2) = \frac{3^3}{3} - \frac{2^3}{3} = \frac{27}{3} - \frac{8}{3} = \frac{19}{3} \qquad F(x) = \frac{x^3}{3}$$

 NOW WORK PROBLEM 1.

In computing $\int_{a}^{b} f(x) \, dx$, we find that the choice of an antiderivative of f does not matter. Look back at equation (1) in the solution to Example 1. The constant K drops out. Now look at Example 2. If we had used $F(x) = \dfrac{x^3}{3} + K$ as the antiderivative of x^2, we would have found that

$$\int_{2}^{3} x^2 \, dx = F(3) - F(2) = \left(\frac{27}{3} + K\right) - \left(\frac{8}{3} + K\right) = \frac{19}{3}$$

Again, the constant K drops out. This will always be the case.

Any antiderivative of f can be used to evaluate $\int_{a}^{b} f(x) \, dx$.

For convenience, we introduce new notation.

If F is an antiderivative of f, then

$$\int_{a}^{b} f(x) \, dx = F(x) \Big|_{a}^{b} = F(b) - F(a)$$

In terms of this new notation, to calculate $F(x)\big|_{a}^{b}$, first replace x by the upper limit b to obtain $F(b)$, and from this subtract $F(a)$, obtained by letting $x = a$.

EXAMPLE 3 Evaluating Definite Integrals

(a) $\displaystyle\int_{-1}^{5} 6x \, dx = (3x^2)\Big|_{-1}^{5} = 3(5)^2 - 3(-1)^2 = 75 - 3 = 72$

(b) $\displaystyle\int_{1}^{2} x^3 \, dx = \frac{x^4}{4}\Big|_{1}^{2} = \frac{2^4}{4} - \frac{1^4}{4} = \frac{16}{4} - \frac{1}{4} = \frac{15}{4}$

It is important to distinguish between the indefinite integral and the definite integral. The indefinite integral, a symbol for all the antiderivatives of a function, is a function. On the other hand, the definite integral is a number.

EXAMPLE 4 Evaluating a Definite Integral

Evaluate: $\int_{1}^{4} \sqrt{x}\, dx$

SOLUTION An antiderivative of $\sqrt{x} = x^{1/2}$ is

$$\frac{x^{3/2}}{\frac{3}{2}} = \frac{2}{3} x^{3/2}$$

Then

$$\int_{1}^{4} \sqrt{x}\, dx = \left(\frac{2}{3} x^{3/2} \right) \Big|_{1}^{4} = \frac{2}{3}(4)^{3/2} - \frac{2}{3}(1)^{3/2} = \frac{16}{3} - \frac{2}{3} = \frac{14}{3}$$

NOW WORK PROBLEM 5.

Properties of the Definite Integral

We list some properties of the definite integral below.

Properties of the Definite Integral

If f is a continuous function that has an antiderivative on the interval $[a, b]$, then

$$\int_{a}^{b} f(x)\, dx = -\int_{b}^{a} f(x)\, dx \tag{3}$$

$$\int_{a}^{a} f(x)\, dx = 0 \tag{4}$$

In Equation (3), notice that the limits of integration have been interchanged. In Equation (4), notice that the limits of integration are the same.

2 EXAMPLE 5 Using Properties of the Definite Integral

(a) $\int_{4}^{1} \sqrt{x}\, dx = -\int_{1}^{4} \sqrt{x}\, dx = -\frac{14}{3}$ (b) $\int_{1}^{1} x\, dx = 0$

↑ Property (3)   ↑ Example 4   ↑ Property (4)

Properties (3) and (4) are an immediate consequence of the definition of a definite integral. Specifically, if F is an antiderivative of f, then

$$\int_{a}^{b} f(x)\, dx = F(b) - F(a) = -[F(a) - F(b)] = -\int_{b}^{a} f(x)\, dx$$

and

$$\int_{a}^{a} f(x)\, dx = F(a) - F(a) = 0$$

NOW WORK PROBLEM 35.

Properties of the Definite Integral

If f is a continuous function that has an antiderivative on the interval $[a, b]$, and if c is between a and b, then

$$\int_a^b f(x)\, dx = \int_a^c f(x)\, dx + \int_c^b f(x)\, dx \tag{5}$$

If f is a continuous function that has an antiderivative on the interval $[a, b]$ and if c is a real number, then

$$\int_a^b cf(x)\, dx = c\int_a^b f(x)\, dx \tag{6}$$

EXAMPLE 6 Using Properties of the Definite Integral

Suppose it is known that

$$\int_1^3 f(x)\, dx = 5 \quad \text{and} \quad \int_3^6 f(x) = 7$$

Find:

(a) $\int_1^6 f(x)\, dx$ (b) $\int_1^3 16 f(x)\, dx$

SOLUTION (a) $\int_1^6 f(x)\, dx \underset{\text{Property (5)}}{=} \int_1^3 f(x)\, dx + \int_3^6 f(x)\, dx = 5 + 7 = 12$

(b) $\int_1^3 16 f(x)\, dx \underset{\text{Property (6)}}{=} 16 \int_1^3 f(x)\, dx = 16 \cdot 5 = 80$

▶

Properties of the Definite Integral

If f and g are continuous functions that have antiderivatives on the interval $[a, b]$, then

$$\int_a^b [f(x) \pm g(x)]\, dx = \int_a^b f(x)\, dx \pm \int_a^b g(x)\, dx \tag{7}$$

EXAMPLE 7 — Using Properties of the Definite Integral

Suppose it is known that

$$\int_1^4 f(x)\,dx = 5 \qquad \int_1^4 g(x)\,dx = -3$$

Find:

(a) $\int_1^4 [f(x) + g(x)]\,dx$ (b) $\int_1^4 [3f(x) + 4g(x)]\,dx$

SOLUTION (a) $\int_1^4 [f(x) + g(x)]\,dx \underset{\substack{\uparrow \\ \text{Property (7)}}}{=} \int_1^4 f(x)\,dx + \int_1^4 g(x)\,dx = 5 + (-3) = 2$

(b) $\int_1^4 [3f(x) + 4g(x)]\,dx \underset{\substack{\uparrow \\ \text{Property (3)}}}{=} \int_1^4 3f(x)\,dx + \int_1^4 4g(x)\,dx \underset{\substack{\uparrow \\ \text{Property (6)}}}{=} 3\int_1^4 f(x)\,dx + 4\int_1^4 g(x)\,dx$

$= 3 \cdot 5 + 4 \cdot (-3) = 3$ ▸

NOW WORK PROBLEM 43.

Applications

EXAMPLE 8 — Finding the Cost Due To An Increase in Production

The marginal cost function for producing x units is $3x^2 - 200x + 1500$ dollars. Find the increase in cost if production is increased from 90 to 100 units.

SOLUTION If C equals the cost of producing x units, then

$$C'(x) = 3x^2 - 200x + 1500$$

The increase in cost due to a production increase from 90 to 100 units is

$$C(100) - C(90) = \int_{90}^{100} C'(x)\,dx$$

$$= \int_{90}^{100} (3x^2 - 200x + 1500)\,dx$$

$$= \left(x^3 - 200\,\frac{x^2}{2} + 1500x \right) \Big|_{90}^{100}$$

$$= [1{,}000{,}000 - 100(10{,}000) + 1500(100)] - [(90)^3 - 100(8100) + 1500(90)]$$

$$= 96{,}000 \text{ dollars} \qquad ▸$$

We discuss below two additional applications involving definite integrals. The applications are independent of each other.

LEARNING CURVES Quite often, the managerial planning and control component of a production industry is faced with the problem of predicting labor time requirements and cost per unit of product. The tool used to achieve such predictions is the *learning curve*. The basic assumption made here is that, in certain production industries such as the assembling of televisions and cars, the worker learns from experience. As a result, the more often a worker repeats an operation, the more efficiently the job is performed and direct labor input per unit of product declines. If the *rate* of improvement is regular enough, the learning curve can be used to predict future reductions in labor requirements.

One function that might be used to model such a situation is

$$f(x) = cx^k$$

where $f(x)$ is the number of hours of direct labor required to produce the xth unit, $-1 \leq k < 0$, and $c > 0$. The choice of x^k, with $-1 \leq k < 0$, guarantees that, as the number of units x produced increases, the direct labor input decreases. See Figure 4.

The function $f(x) = cx^k$ describes a rate of learning per unit produced. This rate is measured in terms of labor-hours per unit. As Figure 4 illustrates, the number of direct labor-hours declines as more items are produced.

Once a learning curve has been determined for a gross production process, it can be used as a predictor to determine the number of production hours for future work.

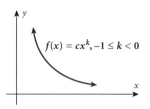

FIGURE 4

Learning Curves

For a learning curve $f(x) = cx^k$, the total number N of labor-hours required to produce units numbered a through b is

$$N = \int_a^b f(x)\, dx = \int_a^b cx^k\, dx$$

EXAMPLE 9 Applying Learning Curves

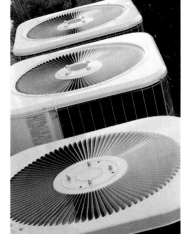

The Ace Air Conditioning Company manufactures air conditioners on an assembly line. From experience, it was determined that the first 100 air conditioners required 1272 labor-hours. For each 100 subsequent air conditioners (1 unit), fewer labor-hours were required according to the learning curve

$$f(x) = 1272 x^{-0.25}$$

where $f(x)$ is the rate of labor-hours required to assemble the xth unit (each unit being 100 air conditioners). This curve was determined after 30 units (3000 air conditioners) had been manufactured.

The company is in the process of bidding for a large contract involving 5000 additional air conditioners, or 50 additional units. The company can estimate the labor-hours required to assemble these units by evaluating

$$N = \int_{30}^{80} 1272 x^{-0.25}\, dx = \left.\frac{1272 x^{0.75}}{0.75}\right|_{30}^{80} = 1696(80^{0.75} - 30^{0.75})$$

$$= 1696(26.75 - 12.82) = 23{,}627$$

The company can bid estimating the total labor-hours needed as 23,627.

NOW WORK PROBLEM 49.

TOTAL SALES OVER TIME

Total Sales over Time

When the rate of sales of a product is a known function, say, $f(t)$, where t is the time, the total sales of this product over a time period T are

$$\text{Total sales over time } T = \int_0^T f(t)\, dt$$

For example, suppose the rate of sales per day of a new product is given by

$$f(t) = 100 - 90e^{-t}$$

where t is the number of days the product is on the market. The total sales during the first 4 days are

$$\int_0^4 f(t)\, dt = \int_0^4 (100 - 90e^{-t})\, dt = (100t + 90e^{-t})\Big|_0^4 = 400 + 90e^{-4} - 90 = 311.6 \text{ units}$$

EXAMPLE 10 Finding Total Sales

A company has current sales of $1,000,000 per month, and profit to the company averages 10% of sales. The company's past experience with a certain advertising strategy is that sales will increase by 2% per month over the length of the advertising campaign (12 months). The company now needs to decide whether to embark on a similar campaign that will have a total cost of $130,000. The decision will be yes, provided the increase in sales due to the campaign results in profits that exceed $13,000. (This is a 10% return on the advertising investment of $130,000.)

SOLUTION Without advertising, the company has sales of $12,000,000 over the 12 months. The monthly rate of sales during the advertising campaign obeys a growth curve of the form

$$\$1{,}000{,}000\, e^{0.02t}$$

where t is measured in months. The total sales after 12 months (the length of the campaign) are

$$\text{Total sales} = \int_0^{12} 1{,}000{,}000\, e^{0.02t}\, dt = \frac{1{,}000{,}000\, e^{0.02t}}{0.02}\Big|_0^{12}$$

$$= 50{,}000{,}000(e^{0.24} - 1) = \$13{,}562{,}458$$

The result is an increase in sales of $13{,}562{,}458 - 12{,}000{,}000 = 1{,}562{,}458$. The profit to the company is 10% of sales, so that the profit due to the increase in sales is

$$0.10(1{,}562{,}458) = \$156{,}246$$

This $156,246 profit was achieved through the expenditure of $130,000 in advertising. The advertising yielded a true profit of

$$\$156{,}246 - \$130{,}000 = \$26{,}246$$

Since this represents more than a 10% return on the cost of the advertising, the company should proceed with the advertising campaign.

NOW WORK PROBLEM 53.

EXERCISE 6.4

In Problems 1–34, evaluate each definite integral.

1. $\int_{1}^{2} (3x - 1)\, dx$
2. $\int_{1}^{2} (2x + 1)\, dx$
3. $\int_{0}^{1} (3x^2 + e^x)\, dx$
4. $\int_{-2}^{0} (e^x + x^2)\, dx$

5. $\int_{0}^{1} \sqrt{u}\, du$
6. $\int_{1}^{4} \sqrt{u}\, du$
7. $\int_{0}^{1} (t^2 - t^{3/2})\, dt$
8. $\int_{1}^{4} (\sqrt{x} - 4x)\, dx$

9. $\int_{-2}^{3} (x - 1)(x + 3)\, dx$
10. $\int_{0}^{1} (z^2 + 1)^2\, dz$
11. $\int_{1}^{2} \frac{x^2 - 1}{x^4}\, dx$
12. $\int_{1}^{3} \frac{2 - x^2}{x^4}\, dx$

13. $\int_{1}^{8} \left(\sqrt[3]{t^2} + \frac{1}{t}\right) dt$
14. $\int_{1}^{4} \left(\sqrt{u} + \frac{1}{u}\right) du$
15. $\int_{1}^{4} \frac{x + 1}{\sqrt{x}}\, dx$
16. $\int_{1}^{9} \frac{\sqrt{x} + 1}{x^2}\, dx$

17. $\int_{3}^{3} (5x^4 + 1)^{3/2}\, dx$
18. $\int_{-1}^{1} (x + 1)^3\, dx$
19. $\int_{-1}^{1} (x + 1)^2\, dx$
20. $\int_{-1}^{-1} \sqrt[3]{x^2 + 4}\, dx$

21. $\int_{1}^{e} \left(x - \frac{1}{x}\right) dx$
22. $\int_{1}^{e} \left(x + \frac{1}{x}\right) dx$
23. $\int_{0}^{1} e^{-x}\, dx$
24. $\int_{0}^{1} x^2 e^{x^3}\, dx$

25. $\int_{1}^{3} \frac{dx}{x + 1}$
26. $\int_{-2}^{2} e^{-7x/2}\, dx$
27. $\int_{0}^{1} \frac{\sqrt{x}}{x^{3/2} + 1}\, dx$
28. $\int_{2}^{3} \frac{dx}{x \ln x}$

29. $\int_{1}^{3} xe^{2x}\, dx$
30. $\int_{0}^{4} (1 + xe^{-x})\, dx$
31. $\int_{1}^{2} xe^{-3x}\, dx$
32. $\int_{1}^{3} x^2 \ln x\, dx$

33. $\int_{1}^{5} \ln x\, dx$
34. $\int_{1}^{2} x \ln x\, dx$

In Problems 35–38, evaluate each expression by applying properties of the definite integral.

35. $\int_{2}^{2} e^{x^2}\, dx$
36. $\int_{1}^{1} e^{-x^2}\, dx$
37. $\int_{0}^{1} e^{-x^2}\, dx + \int_{1}^{0} e^{-x^2}\, dx$
38. $\int_{1}^{2} e^{x^2}\, dx + \int_{2}^{1} e^{x^2}\, dx$

In Problems 39–46, evaluate each definite integral if it is known that

$$\int_{1}^{3} f(x)\, dx = 4 \qquad \int_{1}^{3} g(x)\, dx = -2 \qquad \int_{3}^{6} f(x)\, dx = 8 \qquad \int_{3}^{6} g(x)\, dx = 3$$

39. $\int_{1}^{3} [f(x) + g(x)]\, dx$
40. $\int_{1}^{3} [f(x) - g(x)]\, dx$
41. $\int_{3}^{6} 8f(x)\, dx$
42. $\int_{3}^{6} -2f(x)\, dx$

43. $\int_{3}^{6} [3f(x) + 4g(x)]\, dx$
44. $\int_{3}^{6} [2f(x) - 3g(x)]\, dx$
45. $\int_{1}^{6} f(x)\, dx$
46. $\int_{1}^{6} g(x)\, dx$

47. **Finding a Cost Function** The marginal cost function for producing x units is $6x^2 - 100x + 1000$ dollars. Find the increase in cost if production is increased from 100 to 110 units.

48. **Finding a Revenue Function** The marginal revenue function for selling x units is $10 - 4x$. Find the increase in revenue if selling is increased from 10 to 12 units.

49. **Learning Curve** (a) Rework Example 9 for the learning curve $f(x) = 1272\, x^{-0.35}$.
    (b) Rework Example 9 for the learning curve $f(x) = 1272\, x^{-0.15}$.
    (c) Based on the answers to (a) and (b), explain the role of k in the learning curve $f(x) = cx^k$.

50. **Learning Curve** (a) Rework Example 9 for the learning curve $f(x) = 1500\, x^{-0.25}$. [This means it was determined that

the first 100 air conditioners (1 unit) required 1500 labor-hours.]
(b) Rework Example 9 for the learning curve
$f(x) = 1000 x^{-0.25}$.
(c) Based on the answers to (a) and (b), explain the role of c in the learning curve $f(x) = cx^k$.

51. **US Budget Deficit** The monthly budget deficit of the United States government between October 2002 and March 2003 can be modeled by $D(x) = -8.93x + 70$, where $D(x)$ is in billions of dollars and x is the number of months elapsed since September 2002. If this rate continues, project the total budget deficit for the 2002–2003 fiscal year (October 2002–September 2003).

Source: Bureau of Public Debt, United States Department of Treasury.

52. **Interest Expense** The Fiscal Year Interest Expense measures the total interest that accumulates on debt outstanding for a given fiscal year. From 1997–2002 the increase on this amount was modeled by $f(x) = -2.64x^2 + 15x + 342$, where x is the number of years since 1997 and $f(x)$ is measured in billions of dollars. If this rate continues, find the total Fiscal Year Interest Expense from 2003 to 2010.

53. **Total Sales** The rate of sales of a new product is given by
$$f(x) = 1200 - 950e^{-x}$$
where x is the number of months the product is on the market. Find the total sales during the first year.

54. **Total Sales** In Example 10, what decision should the company make if sales due to advertising increase by only 1.5% per month?

55. **Learning Curve** After producing 35 units, a company determines that its production facility is following a learning curve of the form $f(x) = 1000x^{-0.5}$, where $f(x)$ is the rate of labor-hours required to assemble the xth unit. How many total labor-hours should the company estimate are required to produce an additional 25 units?

56. **Learning Curve** Danny's Auto Shop has found that, after tuning up 50 cars, a learning curve of the form $f(x) = 1000x^{-1}$ is being followed. How many total labor-hours should the shop estimate are required to tune up an additional 50 cars?

57. **Even Functions** It can be shown that if f is an even function, then
$$\int_{-a}^{a} f(x)\, dx = 2\int_{0}^{a} f(x)\, dx \qquad a > 0$$
Verify the above formula by evaluating the following definite integrals:
(a) $\int_{-1}^{1} x^2\, dx$ (b) $\int_{-1}^{1} (x^4 + x^2)\, dx$

58. **Odd Functions** It can be shown that if f is an odd function, then
$$\int_{-a}^{a} f(x)\, dx = 0 \qquad a > 0$$
Verify the above formula by evaluating the following definite integrals:
(a) $\int_{-1}^{1} x\, dx$ (b) $\int_{-1}^{1} x^3\, dx$

6.5 Finding Areas; Consumer's Surplus, Producer's Surplus; Maximizing Profit over Time

PREPARING FOR THIS SECTION *Before getting started, review the following:*

> Geometry Formulas (Chapter 0, Section 0.7, pp. 73–74)

> Interval Notation (Chapter 0, Section 0.5, pp. 51–53)

OBJECTIVES
1 Find the area under a graph
2 Find the area enclosed by two graphs
3 Solve applied problems involving definite integrals

The development of the integral, like that of the derivative, was originally motivated to a large extent by attempts to solve a basic problem in geometry—namely, the *area problem*. The question is: Given a nonnegative function f, whose domain is the closed interval $[a, b]$, what is the area enclosed by the graph of f, the x-axis, and the vertical lines $x = a$ and $x = b$? Figure 5 illustrates the area to be found. We will refer to this area as the **area under the graph of f from a to b.**

FIGURE 5

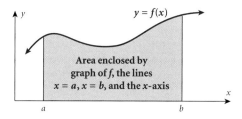

In plane geometry we learn how to find the area of certain geometric figures, such as squares, rectangles, and circles. For example, the area of a square with a side of length 3 feet is 9 square feet.

We also know that the area of a rectangle with length a units and width b units is ab square units.

All area problems have certain features in common. For example, whenever an area is computed, it is expressed as a number of square units; this number is never negative. One property of area is that it is nonnegative.

Consider the trapezoid shown in Figure 6. This trapezoid has been decomposed into two nonoverlapping geometric figures, a triangle (with area A_1) and a rectangle (with area A_2). The area of the trapezoid is the sum $A_1 + A_2$ of the two component areas. So, as long as two regions do not overlap (except perhaps for a common boundary), the total area can be found by adding the component areas. We sometimes call this the **additive property of area.**

FIGURE 6

Properties of Area

Two properties of area are:

   I. Nonnegative Property Area ≥ 0
   II. Additive Property If A and B are two nonoverlapping regions with areas that are known, then

$$\text{Total area of } A \text{ and } B = \text{area of } A + \text{area of } B$$

The next result gives a technique for evaluating areas such as the shaded region shown in Figure 5.

Area under a Graph

Suppose $y = f(x)$ is a continuous function defined on a closed interval I and $f(x) \geq 0$ for all points x in I. Then, for $a < b$ in I,

$$\text{Area under the graph of } f \text{ from } a \text{ to } b = \int_a^b f(x)\, dx \tag{1}$$

FIGURE 7

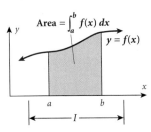

Figure 7 illustrates the above statement. A proof is given at the end of this section.

We are now able to find the area under the graph of $y = f(x)$, provided three conditions are met:

1. f is continuous on $[a, b]$.
2. f is nonnegative on $[a, b]$, that is, $f(x) \geq 0$ for $a \leq x \leq b$.
3. An antiderivative for f can be found.

1 EXAMPLE 1 Finding the Area under a Graph

Find the area under the graph $f(x) = x^2$, from $x = 0$ to $x = 1$.

SOLUTION First we graph f and show the area to be found. See Figure 8. Using Formula (1), the area we seek is given by the definite integral

FIGURE 8

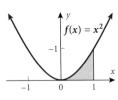

$$\int_0^1 x^2 \, dx = \frac{x^3}{3} \Big|_0^1 = \frac{1}{3}$$

The area illustrated in Figure 8 is $\frac{1}{3}$ square unit.

 NOW WORK PROBLEM 1.

Suppose a function f is continuous on the interval I, $a \leq x \leq b$ and has an antiderivative on I. Suppose $f(x) \leq 0$ for $a \leq x \leq c$ and $f(x) \geq 0$ for $c \leq x \leq b$. How do we compute the area enclosed by $y = f(x)$, the x-axis, $x = a$, and $x = b$? See Figure 9.

FIGURE 9

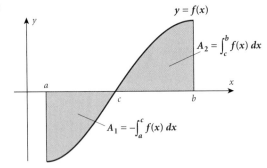

Notice in Figure 9 that the area A in question is composed of two nonoverlapping areas, A_1 and A_2, so that by the additive property of area,

$$A = A_1 + A_2$$

Also, we know that on the interval $[c, b]$, the function is nonnegative, so that

$$A_2 = \int_c^b f(x) \, dx$$

To find the area A_1, we note that, since $f(x) \leq 0$ on $a \leq x \leq c$, then $-f(x) \geq 0$, and, by symmetry, the area A_1 equals

$$A_1 = \int_a^c [-f(x)]\, dx = -\int_a^c f(x)\, dx$$

The total area A we seek is therefore

$$A = A_1 + A_2 = -\int_a^c f(x)\, dx + \int_c^b f(x)\, dx$$

The next example illustrates this procedure for calculating area.

EXAMPLE 2 Finding an Area

Find the area A enclosed by $f(x) = x^3$, the x-axis, $x = -1$, and $x = \dfrac{1}{2}$.

SOLUTION The desired area is indicated by the shaded region in Figure 10. Notice that it is composed of two nonoverlapping regions: A_1, in which $f(x) \leq 0$ over the interval $[-1, 0]$; and A_2, in which $f(x) \geq 0$ over the interval $[0, \tfrac{1}{2}]$.

FIGURE 10

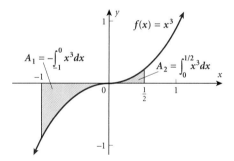

We use the additive property of area. Since $f(x) \leq 0$ for $-1 \leq x \leq 0$,

$$A_1 = -\int_{-1}^{0} x^3\, dx = -\dfrac{x^4}{4}\bigg|_{-1}^{0} = -\left[\dfrac{0^4}{4} - \dfrac{(-1)^4}{4}\right] = \dfrac{1}{4}$$

For the area A_2, we have

$$A_2 = \int_0^{1/2} x^3\, dx = \dfrac{x^4}{4}\bigg|_0^{1/2} = \dfrac{\left(\tfrac{1}{2}\right)^4}{4} - \dfrac{0^4}{4} = \dfrac{1}{64}$$

The total area A is

$$A = A_1 + A_2 = \dfrac{1}{4} + \dfrac{1}{64} = \dfrac{17}{64} \text{ square units}$$

Example 2 illustrates the necessity of graphing the function before any attempt is made to compute the area. In subsequent examples we shall always graph the function before doing anything else.

EXAMPLE 3 Finding an Area

Find the area A enclosed by $f(x) = x^2 - 4$ and the x-axis from $x = 0$ to $x = 4$.

FIGURE 11

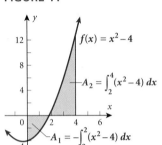

SOLUTION We begin by graphing f. See Figure 11. On the interval $[0, 4]$, the graph crosses the x-axis at $x = 2$ since $f(2) = 0$. Also, $f(x) \leq 0$ from $x = 0$ to $x = 2$, and $f(x) \geq 0$ from $x = 2$ to $x = 4$. The areas A_1 and A_2, as shown in Figure 11, are

$$A_1 = -\int_0^2 (x^2 - 4)\,dx = -\left(\frac{x^3}{3} - 4x\right)\Big|_0^2 = -\left(\frac{8}{3} - 8\right) = \frac{16}{3}$$

$$A_2 = \int_2^4 (x^2 - 4)\,dx = \left(\frac{x^3}{3} - 4x\right)\Big|_2^4 = \left(\frac{64}{3} - 16\right) - \left(\frac{8}{3} - 8\right)$$

$$= \frac{56}{3} - 8 = \frac{32}{3}$$

The area A is

$$A = A_1 + A_2 = \frac{16}{3} + \frac{32}{3} = \frac{48}{3} = 16 \text{ square units}$$

 NOW WORK PROBLEM 5.

Area Enclosed by Two Graphs

The next example illustrates how to find the area enclosed by the graphs of two functions.

EXAMPLE 4 Finding an Area Enclosed by Two Graphs

Find the area A enclosed by the graphs of the functions

$$f(x) = 2x^2 \quad \text{and} \quad g(x) = 2x + 4$$

FIGURE 12

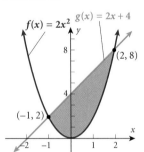

SOLUTION First, we graph each of the functions, as shown in Figure 12.

The area to be calculated (the shaded portion of Figure 12) lies under the graph of the line $g(x) = 2x + 4$ and above the graph of $f(x) = 2x^2$. To find this area, we first need to find the numbers x at which the graphs intersect, that is, all numbers x for which $f(x) = g(x)$. The solutions of this equation are obtained as follows:

$$2x^2 = 2x + 4 \qquad f(x) = g(x)$$
$$x^2 - x - 2 = 0 \qquad \text{Simplify.}$$
$$(x + 1)(x - 2) = 0 \qquad \text{Factor.}$$
$$x + 1 = 0 \quad \text{or} \quad x - 2 = 0 \qquad \text{Apply the Zero-Product Property.}$$
$$x = -1 \quad \text{or} \quad x = 2 \qquad \text{Solve for } x.$$

The points of intersection of the two graphs are $(-1, 2)$ and $(2, 8)$, as shown in Figure 12.

From Figure 12 we can see that if we subtract the area under $f(x) = 2x^2$, between $x = -1$ and $x = 2$, from the area under $g(x) = 2x + 4$, between $x = -1$ and $x = 2$, we will have the area A we seek. That is,

$$A = \int_{-1}^{2} g(x)\, dx - \int_{-1}^{2} f(x)\, dx \qquad A = \text{Area under } g - \text{Area under } f$$

$$= \int_{-1}^{2} (2x + 4)\, dx - \int_{-1}^{2} 2x^2\, dx \qquad g(x) = 2x + 4;\ f(x) = 2x^2$$

$$= \left(x^2 + 4x\right)\Big|_{-1}^{2} - 2 \cdot \left.\frac{x^3}{3}\right|_{-1}^{2} \qquad \text{Integrate.}$$

$$= [(4 + 8) - (1 - 4)] - 2 \cdot \left(\frac{8}{3} - \frac{-1}{3}\right)$$

$$= 9$$

NOW WORK PROBLEM 11.

The technique used in Example 4 can be used whenever we are asked to determine the area enclosed by the graphs of two continuous nonnegative functions f and g from $x = a$ to $x = b$.

Suppose, as depicted in Figure 13, $f(x) \geq g(x) \geq 0$ for x in $[a, b]$, and we wish to determine the area enclosed by the graphs of f and g and the lines $x = a$, and $x = b$. If we denote this area by A, the area under f by A_2, and the area under g by A_1, then

FIGURE 13

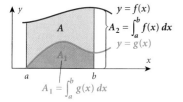

$$A = A_2 - A_1$$

$$= \int_a^b f(x)\, dx - \int_a^b g(x)\, dx \qquad A_2 = \int_a^b f(x)\, dx;\ A_1 = \int_a^b g(x)\, dx$$

$$= \int_a^b [f(x) - g(x)]\, dx$$

The next example illustrates this formula.

EXAMPLE 5 Finding the Area Enclosed By Two Graphs

FIGURE 14

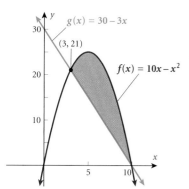

Find the area A enclosed by the graphs of the functions

$$f(x) = 10x - x^2 \quad \text{and} \quad g(x) = 30 - 3x$$

SOLUTION First, we graph the two functions. See Figure 14. The points of intersection of the two graphs were obtained by finding all numbers x for which $f(x) = g(x)$.

$$10x - x^2 = 30 - 3x \qquad f(x) = g(x)$$

$$x^2 - 13x + 30 = 0 \qquad \text{Place in standard form.}$$

$$(x - 3)(x - 10) = 0 \qquad \text{Factor.}$$

$$x - 3 = 0 \quad \text{or} \quad x - 10 = 0 \qquad \text{Apply the Zero-Product Property.}$$

$$x = 3 \quad \text{or} \quad x = 10 \qquad \text{Solve for } x.$$

The points where the two graphs intersect are $(3, 21)$ and $(10, 0)$. We also see that for $3 \leq x \leq 10$,
$$f(x) \geq g(x) \geq 0$$
The area A we seek, indicated by the shaded portion in Figure 14, is

$$\begin{aligned}
A &= \int_3^{10} [(10x - x^2) - (30 - 3x)]\, dx & A &= \int_3^{10} [f(x) - g(x)]\, dx \\
&= \int_3^{10} [-x^2 + 13x - 30]\, dx & &\text{Simplify the integrand.} \\
&= \left(-\frac{x^3}{3} + 13\frac{x^2}{2} - 30x \right)\bigg|_3^{10} & &\text{Integrate.} \\
&= \left[-\frac{10^3}{3} + 13\frac{10^2}{2} - 30(10) \right] - \left[-\frac{3^3}{3} + 13\frac{3^2}{2} - 30(3) \right] & &\text{Evaluate at the upper and lower limits.} \\
&= -\frac{1000}{3} + \frac{1300}{2} - 300 + 9 - \frac{117}{2} + 90 \\
&= \frac{343}{6}
\end{aligned}$$

▶

Remember, when computing area using the formula $\int_a^b [f(x) - g(x)]\, dx$, it must be true that $f(x) \geq g(x)$ on $[a, b]$. If this condition is not met, we break the area up into pieces on which the inequality does hold, compute each one separately, and use the Additive Property of area.

EXAMPLE 6 **Finding the Area Enclosed by Two Graphs**

Find the area A enclosed by the graphs of the functions
$$f(x) = x^3 \quad \text{and} \quad g(x) = -x^2 + 2x$$

SOLUTION First, we graph the two functions. See Figure 15.

FIGURE 15

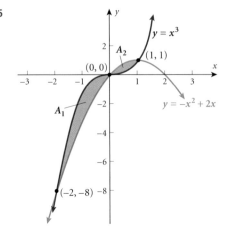

The points of intersection of the two graphs are found by solving the equation $f(x) = g(x)$.

$$\begin{aligned} x^3 &= -x^2 + 2x & f(x) &= g(x) \\ x^3 + x^2 - 2x &= 0 & &\text{Place in standard form.} \\ x(x^2 + x - 2) &= 0 & &\text{Factor.} \\ x(x+2)(x-1) &= 0 & &\text{Factor.} \\ x = 0 \quad \text{or} \quad x + 2 &= 0 \quad \text{or} \quad x - 1 = 0 & &\text{Apply the Zero-Product Property.} \\ x = 0 \quad \text{or} \quad x &= -2 \quad \text{or} \quad x = 1 & &\text{Solve for } x. \end{aligned}$$

The points of intersection are $(0, 0)$, $(-2, -8)$, and $(1, 1)$. Notice that $f(x) \geq g(x)$ on the interval $[-2, 0]$, while $g(x) \geq f(x)$ on the interval $[0, 1]$. To find the area A enclosed by the graphs of f and g, we compute the areas A_1 and A_2. Then,

$$\begin{aligned} A &= A_1 + A_2 \\ &= \int_{-2}^{0} [x^3 - (-x^2 + 2x)] dx + \int_{0}^{1} [(-x^2 + 2x) - x^3] dx \quad A_1 = \int_{-2}^{0} [f(x) - g(x)] dx; \ A_2 = \int_{0}^{1} [g(x) - f(x)] dx \\ &= \int_{-2}^{0} (x^3 + x^2 - 2x) dx + \int_{0}^{1} (-x^3 - x^2 + 2x) dx & &\text{Simplify each integrand.} \\ &= \left(\frac{x^4}{4} + \frac{x^3}{3} - x^2 \right) \Big|_{-2}^{0} + \left(\frac{-x^4}{4} - \frac{x^3}{3} + x^2 \right) \Big|_{0}^{1} & &\text{Integrate.} \\ &= 0 - \left(4 - \frac{8}{3} - 4 \right) + \left(-\frac{1}{4} - \frac{1}{3} + 1 \right) - 0 & &\text{Evaluate each expression at the upper limit and the lower limit.} \\ &= \frac{8}{3} + \frac{5}{12} = \frac{37}{12} & &\text{Simplify.} \end{aligned}$$

 NOW WORK PROBLEM 25.

Applications

Solve applied problems involving definite integrals 3 We discuss two applications in business that involve area. They are independent of each other.

Consumer's Surplus; Producer's Surplus Suppose the price p a consumer is willing to pay for a quantity x of a particular commodity is governed by the demand curve

$$p = D(x)$$

In general, the demand function D is a decreasing function, indicating that as the price of the commodity increases, the quantity the consumer is willing to buy declines.

Suppose the price p that a producer is willing to charge for a quantity x of a particular commodity is governed by the supply curve

$$p = S(x)$$

In general, the supply function S is an increasing function since, as the price p of a commodity increases, the more the producer is willing to supply the commodity.

The point of intersection of the demand curve and the supply curve is called the **equilibrium point** E. If the coordinates of E are (x^*, p^*), then p^*, the **market price,** is the price a consumer is willing to pay for the commodity and p^* is also the price at which a producer is willing to sell; x^*, the **demand level,** is the quantity of the commodity purchased by the consumer and sold by the producer. See Figure 16.

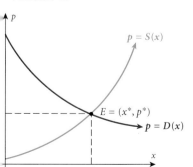

FIGURE 16

FIGURE 17

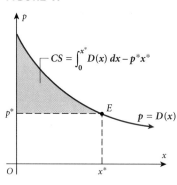

The revenue to the producer at a market price p^* and a demand level x^* is p^*x^* (the price per unit times the number of units). This revenue can be interpreted geometrically as the area of the rectangle Op^*Ex^* in Figure 16.

In a free market economy, there are times when some consumers would be willing to pay more for a commodity than the market price p^* that they actually do pay. The benefit of this to consumers—that is, the difference between what consumers *actually* paid and what they were *willing* to pay—is called the **consumer's surplus CS**. To obtain a formula for consumer's surplus CS, we use Figure 17 as a guide.

The quantity $\int_0^{x^*} D(x)\,dx$ is the area under the demand curve $D(x)$ from $x = 0$ to $x = x^*$ and represents the total revenue that would have been generated by the willingness of some consumers to pay more. By subtracting p^*x^* (the revenue actually achieved), the result is a surplus CS to the consumer. The formula for consumer's surplus is

$$CS = \int_0^{x^*} D(x)\,dx - p^*x^* \qquad (2)$$

FIGURE 18

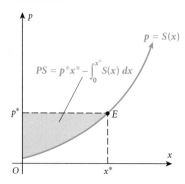

In a free market economy, there are also times when some producers would be willing to sell at a price below the market price p^* that the consumer actually pays. The benefit of this to the producer—that is, the difference between the revenue producers *actually* receive and what they would have been willing to receive—is called the **producer's surplus PS**. To obtain a formula for PS, we use Figure 18 as a guide.

The quantity $\int_0^{x^*} S(x)\,dx$ is the area under the supply curve $S(x)$ from $x = 0$ to $x = x^*$ and represents the total revenue that would have been generated by some producer's willingness to sell at a lower price. If we subtract this amount from p^*x^* (the revenue actually achieved), the result is a surplus to the producer, PS. The formula for producer's surplus is

$$PS = p^*x^* - \int_0^{x^*} S(x)\,dx \qquad (3)$$

Example 7 illustrates a situation in which both the supply and demand curves are linear.

EXAMPLE 7 **Finding Consumer's Surplus and Producer's Surplus**

Find consumer's surplus CS and producer's surplus PS for the demand curve

$$D(x) = 18 - 3x$$

and the supply curve

$$S(x) = 3x + 6$$

where

$$p^* = D(x^*) = S(x^*)$$

SOLUTION The equilibrium point is $E = (x^*, p^*)$, where $p^* = D(x^*) = S(x^*)$ is the market price and x^* is the demand level. To find x^*, we solve the equation

$$D(x^*) = S(x^*)$$
$$18 - 3x^* = 3x^* + 6$$
$$6x^* = 12$$
$$x^* = 2$$

To find p^*, we evaluate $D(x^*)$ or $S(x^*)$:

$$p^* = D(x^*) = D(2) = 18 - 6 = 12$$

To find CS and PS, we use Formulas (2) and (3) with

$$D(x) = 18 - 3x, \quad S(x) = 3x + 6, \quad x^* = 2, \quad p^* = 12$$

Then

$$CS = \int_0^{x^*} D(x)\,dx - x^*p^* = \int_0^2 (18 - 3x)\,dx - (2)(12)$$
$$= \left(18x - 3 \cdot \frac{x^2}{2}\right)\bigg|_0^2 - 24$$
$$= 36 - 6 - 24 = 6$$

$$PS = x^*p^* - \int_0^{x^*} S(x)\,dx = (2)(12) - \int_0^2 (3x + 6)\,dx$$
$$= 24 - \left(3 \cdot \frac{x^2}{2} + 6x\right)\bigg|_0^2$$
$$= 24 - (6 + 12) = 6$$

In this example, the consumer's surplus and the producer's surplus each equal $6. In general, PS and CS are unequal.

NOW WORK PROBLEM 35.

Maximizing Profit Over Time The model introduced here concerns business operations of a special character. In oil drilling, mining, and other depletion operations, the initial revenue rate is generally higher than the revenue rate after a period of time has passed. That is, the revenue rate, as a function of time, is a decreasing function, because depletion is occurring.

The cost rate of such operations generally increases with time because of inflation and other reasons. That is, the cost rate, as a function of time, is an increasing function. The problem that management faces is to determine the time t_{max} that maximizes the profit function $P = P(t)$.

To construct a model, we denote the cost function by $C = C(t)$ and the revenue function by $R = R(t)$, where t denotes time. This representation of cost and revenue deviates from the usual economic definitions of cost per unit times number of units, and price per unit times number of units. The derivatives C' and R', taken with respect to time, represent cost and revenue as time rates. Furthermore, we make the assumption that the revenue rate, say, dollars per week, is greater than the cost rate, also in dollars per week, at the beginning of the business operation under consideration. Also, as time goes on, we assume that the cost rate increases to the revenue rate and thereafter exceeds it. The optimum time at which the business operation should terminate is that point in time where the rates are equal. That is, the optimum time t_{max} obeys

$$C'(t_{max}) = R'(t_{max})$$

The profit rate P' is the difference between the revenue rate and the cost rate. That is,
$$P'(t) = R'(t) - C'(t)$$
Integrate each side with respect to t from 0 to t. Then
$$\int_0^t P'(t)\,dt = \int_0^t [R'(t) - C'(t)]\,dt$$
$$P(t) - P(0) = \int_0^t [R'(t) - C'(t)]\,dt$$

The maximum profit $P = P(t_{max})$ is obtained when $t = t_{max}$. Geometrically, the maximum profit $P(t_{max})$ is the area enclosed by the graphs of C' and R' from $t = 0$ to $t = t_{max}$. See Figure 19.

Notice that in Figure 19 the revenue rate function obeys the assumptions made in constructing the model: it is decreasing and it is high initially. Also, the cost rate function is increasing and is concave down, indicating that the cost rate eventually levels off.

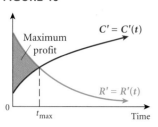

FIGURE 19

EXAMPLE 8 Maximizing Profit over Time

The G-B Oil Company's revenue rate (in millions of dollars per year) at time t years is
$$R'(t) = 9 - t^{1/3}$$
and the corresponding cost rate function (also in millions of dollars) is
$$C'(t) = 1 + 3t^{1/3}$$
Determine how long the oil company should continue to operate and what the total profit will be at the end of the operation.

SOLUTION   The time t_{max} of optimal termination is found when
$$R'(t) = C'(t)$$
$$9 - t^{1/3} = 1 + 3t^{1/3}$$
$$8 = 4t^{1/3}$$
$$2 = t^{1/3}$$
$$t_{max} = 8 \text{ years}$$

See Figure 20.

At $t_{max} = 8$, both the revenue and the cost rates are 7 million dollars per year. The profit $P(t_{max})$ is

$$P(t_{max}) = \int_0^8 [R'(t) - C'(t)]\,dt$$
$$= \int_0^8 [(9 - t^{1/3}) - (1 + 3t^{1/3})]\,dt$$
$$= \int_0^8 [8 - 4t^{1/3}]\,dt$$
$$= (8t - 3t^{4/3})\Big|_0^8 = 16 \text{ million dollars}$$

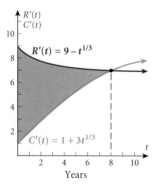

FIGURE 20

In Example 8 we were forced to overlook the *fixed* cost for the cost function at time $t = 0$. This is because if $C = C(x)$ contains a constant (the fixed cost), then it becomes zero when we take the derivative C'. In the final analysis of the problem, total profit should be reduced by the amount corresponding to the fixed cost.

Justification of Area As a Definite Integral

Look at Figure 21. Choose a number c in I so that $c < a$. Suppose x in I is an arbitrary number for which $x > c$. Let $A(x)$ denote the area enclosed by $y = f(x)$ and the x-axis from c to x. We want to show that $A'(x) = f(x)$ for all x in I, $x > c$.

Now, choose $h > 0$ so that $x + h$ is in I. Then $A(x + h)$ is the area under the graph of $y = f(x)$ from c to $x + h$. See Figure 22. The difference $A(x + h) - A(x)$ is just the area under the graph of $y = f(x)$ from x to $x + h$. See Figure 23.

Next, we construct a rectangle with base h and area $A(x + h) - A(x)$. The height of the rectangle is then

$$\frac{A(x + h) - A(x)}{h}$$

Now, we superimpose this rectangle on Figure 23 to obtain Figure 24.

FIGURE 21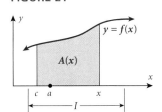

FIGURE 22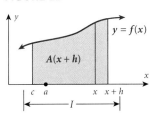

FIGURE 23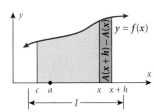

FIGURE 24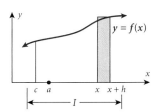

Since $y = f(x)$ is assumed to be a continuous function, and since both the rectangle and the shaded area have the same base and the same area, the upper edge of the rectangle must cross the graph of $y = f(x)$.

As we let $h \to 0^+$, the height of the rectangle tends to $f(x)$; that is,

$$\frac{A(x + h) - A(x)}{h} \to f(x) \quad \text{as} \quad h \to 0^+$$

A similar argument applies if we choose $h < 0$ and let $h \to 0^-$. As a result,

$$\lim_{h \to 0} \frac{A(x + h) - A(x)}{h} = f(x)$$

The limit on the left is the derivative of A. That is,

$$A'(x) = f(x)$$

Since the choice of x is arbitrary (except for the condition that $x > c$), it follows that

$$A'(x) = f(x) \qquad \text{for all } x \text{ in } I, x > c$$

In other words, we have shown that the area A is an antiderivative of f on I. As a result,

$$\int_a^b f(x)\, dx = A(x) \Big|_a^b = A(b) - A(a)$$

But the area we want to find is the area under the graph of $y = f(x)$ from a to b. Since a and b are in I, this is the quantity $A(b) - A(a)$. See Figure 25. That is, we have

FIGURE 25

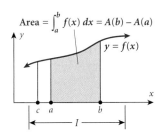

Area $= \int_a^b f(x)\, dx = A(b) - A(a)$

$$\boxed{\text{Area under the graph of } y = f(x) \text{ from } a \text{ to } b \text{ is } \int_a^b f(x)\, dx}$$

EXERCISE 6.5

In Problems 1–10, find the area described. Be sure to sketch the graph first.

1. Under the graph of $f(x) = 3x + 2$ from 2 to 6
2. Under the graph of $f(x) = 3 - x$ from 0 to 3
3. Under the graph of $f(x) = x^2$ from 0 to 2
4. Under the graph of $f(x) = x^2$, from -2 to 1
5. Under the graph of $f(x) = x^2 - 1$, from -2 to 1
6. Under the graph of $f(x) = x^2 - 9$, from 0 to 6
7. Under the graph of $f(x) = \sqrt[3]{x}$, from -1 to 8
8. Under the graph of $f(x) = \sqrt[3]{x}$, from -8 to 1
9. Under the graph of $f(x) = e^x$, from 0 to 1
10. Under the graph of $f(x) = x^3$, from 0 to 1

In Problems 11–28, find the area enclosed by the graphs of the given functions. Draw a graph first.

11. $f(x) = x$, $g(x) = 2x$, $x = 0$, $x = 1$
12. $f(x) = x$, $g(x) = 3x$, $x = 0$, $x = 3$
13. $f(x) = x^2$, $g(x) = x$
14. $f(x) = x^2$, $g(x) = 4x$
15. $f(x) = x^2 + 1$, $g(x) = x + 1$
16. $f(x) = x^2 + 1$, $g(x) = 4x + 1$
17. $f(x) = \sqrt{x}$, $g(x) = x^3$
18. $f(x) = x^2$, $g(x) = x^3$
19. $f(x) = x^2$, $g(x) = x^4$
20. $f(x) = \sqrt{x}$, $g(x) = x^2$
21. $f(x) = x^2 - 4x$, $g(x) = -x^2$
22. $f(x) = x^2 - 8x$, $g(x) = -x^2$
23. $f(x) = 4 - x^2$, $g(x) = x + 2$
24. $f(x) = 2 + x - x^2$, $g(x) = -x - 1$
25. $f(x) = x^3$, $g(x) = 4x$
26. $f(x) = x^3$, $g(x) = 16x$
27. $y = x^2$, $y = x$, $y = -x$
28. $y = x^2 - 1$, $y = x - 1$, $y = -x - 1$

In Problems 29–34, an integral is given.
(a) What area does the integral represent?
(b) Provide a graph that illustrates this area.
(c) Find the area.

29. $\int_0^4 (3x + 1)\, dx$
30. $\int_1^3 (-2x + 7)\, dx$
31. $\int_2^5 (x^2 - 1)\, dx$
32. $\int_0^4 (16 - x^2)\, dx$
33. $\int_0^2 e^x\, dx$
34. $\int_e^{2e} \ln x\, dx$

35. **Consumer's Surplus; Producer's Surplus** Find the consumer's surplus and the producer's surplus for the demand curve
$$D(x) = -5x + 20$$
and the supply curve
$$S(x) = 4x + 8$$
Sketch the graphs and show the equilibrium point.

36. **Consumer's Surplus; Producer's Surplus** Follow the same directions as in Problem 35 if
$$D(x) = -0.4x + 15 \quad \text{and} \quad S(x) = 0.8x + 0.5$$

37. **Maximizing Profit Over Time** The revenue and the cost rate of Gold Star mining operation are, respectively,
$$R'(t) = 19 - t^{1/2} \quad \text{and} \quad C'(t) = 3 + 3t^{1/2}$$
where t is measured in years and R and C are measured in millions of dollars. Determine how long the operation should continue to maximize profit. Find the profit that can be generated during this period. Ignore any fixed costs.

38. **Consumer's Surplus** Find the consumer's surplus for the demand curve
$$D(x) = 50 - 0.025x^2$$
if it is known that the demand level x^* is 20 units.

39. **Mean Value Theorem for Integrals** If $y = f(x)$ is continuous on a closed interval $a \le x \le b$, then there is a number c, $a < c < b$, so that
$$\int_a^b f(x)\, dx = f(c)(b - a)$$

The interpretation of this result is that there is a rectangle with base $b - a$ and height $f(c)$, whose area is numerically equal to the area $\int_a^b f(x)\, dx$. See the illustration.

Verify this result by finding c for the functions below. Graph each function.

(a) $f(x) = x^2$, $a = 0$, $b = 1$
(b) $f(x) = \dfrac{1}{x^2}$, $a = 1$, $b = 4$

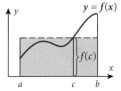

40. Show that the shaded area in the figure is $\frac{2}{3}$ of the area of the parallelogram $ABCD$. (This illustrates a result due to Archimedes concerning sectors of parabolas.)

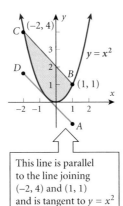

This line is parallel to the line joining $(-2, 4)$ and $(1, 1)$ and is tangent to $y = x^2$

41. If $y = f(x)$ is continuous on the interval I and if it has an antiderivative on I, then for some a in I,

$$\frac{d}{dx}\int_a^x f(t)\,dt = f(x) \qquad \text{for } x > a \text{ in } I$$

This result gives us a technique for finding the derivative of a definite integral in which the lower limit is fixed and the upper limit is variable. Use this result to find:

(a) $\dfrac{d}{dx}\displaystyle\int_1^x t^2\,dt$

(b) $\dfrac{d}{dx}\displaystyle\int_2^x \sqrt{t^2 - 2}\,dt$

(c) $\dfrac{d}{dx}\displaystyle\int_5^x \sqrt{t^t + 2t}\,dt$

6.6 Approximating Definite Integrals

PREPARING FOR THIS SECTION *Before getting started, review the following:*

>> Interval Notation (Chapter 0, Section 0.5, pp. 51–53)

OBJECTIVES 1 Approximate definite integrals using rectangles
2 Approximate definite integrals using a graphing utility

Up to now the evaluation of a definite integral

$$\int_a^b f(x)\,dx$$

has required that we find an antiderivative F of f so that

$$\int_a^b f(x)\,dx = F(x)\Big|_a^b = F(b) - F(a) \qquad \text{where} \qquad F'(x) = f(x)$$

But what if we can't find an antiderivative? In fact, sometimes it is impossible to find an antiderivative. In such situations, it is necessary to *approximate* the definite integral. One way is to use rectangles.

Approximate definite integrals using rectangles   1   We have already discussed the fact that when f is a continuous nonnegative function defined on the closed interval $[a, b]$, then the definite integral $\displaystyle\int_a^b f(x)\,dx$ equals the area under the graph of f from a to b. We will use this idea to obtain an approximation to $\displaystyle\int_a^b f(x)\,dx$.

Consider the graph of the function f in Figure 26(a) on page 464. The area under the graph of f from a to b is $\displaystyle\int_a^b f(x)\,dx$. Pick a number u in the interval $[a, b]$ and form the rectangle whose height is $f(u)$ and whose base is $b - a$. See Figure 26(b). The area of this rectangle provides a rough approximation to $\displaystyle\int_a^b f(x)\,dx$. That is,

$$\int_a^b f(x)\,dx \approx f(u)(b - a)$$

FIGURE 26

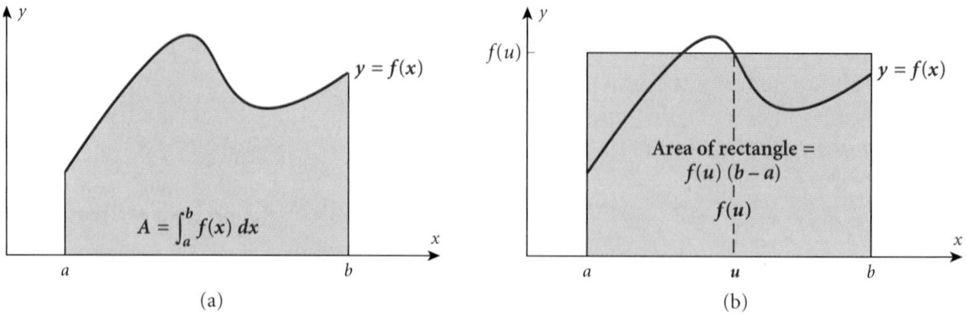

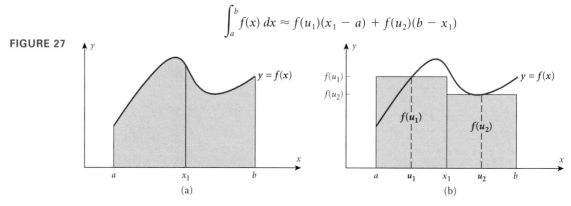

A better approximation to $\int_a^b f(x)\,dx$ can be obtained by dividing the interval $[a, b]$ into two subintervals of the same length. See Figure 27(a) where x_1 is the midpoint of $[a, b]$. Now pick a number u_1 between a and x_1, and a number u_2 between x_1 and b, and form two rectangles: One whose height is $f(u_1)$ and whose base is $x_1 - a$ and the other whose height is $f(u_2)$ and whose base is $b - x_1$. See Figure 27(b). The sum of the areas of these two rectangles provides an approximation to $\int_a^b f(x)\,dx$. That is,

$$\int_a^b f(x)\,dx \approx f(u_1)(x_1 - a) + f(u_2)(b - x_1)$$

FIGURE 27

To get an even better approximation, divide the interval $[a, b]$ into three subintervals, each of the same length Δx. See Figure 28(a) where x_1 and x_2 are chosen so that

$$\Delta x = x_1 - a = x_2 - x_1 = b - x_2$$

Now pick numbers u_1, u_2, u_3 in each subinterval and form three rectangles of heights $f(u_1), f(u_2), f(u_3)$, each with the same base Δx. See Figure 28(b). The sum of the areas of these rectangles approximates $\int_a^b f(x)\,dx$. That is,

FIGURE 28

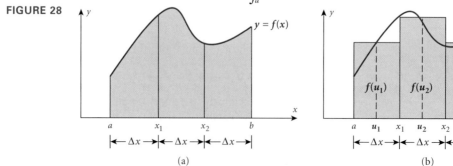

Approximating Definite Integrals

$$\int_a^b f(x)\,dx \approx f(u_1)\,\Delta x + f(u_2)\,\Delta x + f(u_3)\,\Delta x$$

The more subintervals we divide $[a, b]$ into, the better the approximation. Suppose we divide $[a, b]$ into n subintervals

$$[a, x_1], [x_1, x_2], \ldots, [x_{k-1}, x_k], \ldots, [x_{n-1}, b]$$

each of length $\Delta x = \dfrac{b - a}{n}$. See Figure 29(a). Now pick numbers $u_1, u_2, \ldots, u_n$ in each subinterval and form n rectangles of base Δx and heights $f(u_1), f(u_2), \ldots, f(u_n)$. See Figure 29(b). The sum of the areas of these rectangles approximates $\int_a^b f(x)\,dx$. That is,

$$\int_a^b f(x)\,dx \approx f(u_1)\,\Delta x + f(u_2)\,\Delta x + \cdots + f(u_n)\,\Delta x$$

FIGURE 29

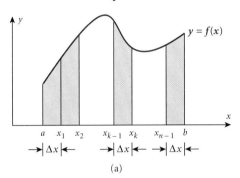
(a)

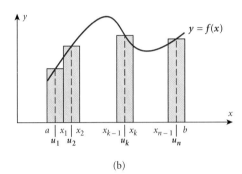
(b)

Let's summarize this process.

Steps for Approximating a Definite Integral

To approximate $\int_a^b f(x)\,dx$ follow these steps:

STEP 1 Divide the interval $[a, b]$ into n subintervals of equal length $\Delta x = \dfrac{b - a}{n}$. The larger n is, the better your approximation will be.

STEP 2 Pick a number u in each subinterval and evaluate $f(u)$.

STEP 3 The sum

$$f(u_1)\Delta x + f(u_2)\Delta x + \cdots + f(u_n)\Delta x$$

approximates

$$\int_a^b f(x)\,dx$$

EXAMPLE 1 Approximating a Definite Integral

Approximate $\int_0^4 (3 + 2x)\,dx$ by dividing the interval $[0, 4]$ into four subintervals of equal length. Pick u_i as the left endpoint of each subinterval. Compare the approximation to the exact value.

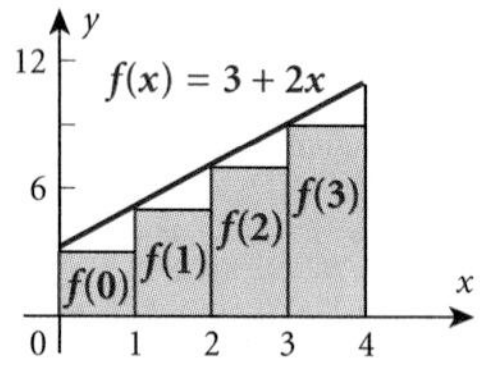

SOLUTION Figure 30 illustrates the graph of $f(x) = 3 + 2x$ on $[0, 4]$.

STEP 1 Divide $[0, 4]$ into four subintervals of equal length:
$$[0, 1], [1, 2], [2, 3], [3, 4]$$

STEP 2 Pick the left endpoint of each of these subintervals and evaluate f there:
$$f(0) = 3, \ f(1) = 5, \ f(2) = 7, \ f(3) = 9.$$

FIGURE 30

STEP 3 $\int_0^4 (3 + 2x)\,dx \approx f(0) \cdot 1 + f(1) \cdot 1 + f(2) \cdot 1 + f(3) \cdot 1$
$$= 3 + 5 + 7 + 9 = 24$$

This is an approximation of the integal. The exact value is
$$\int_0^4 (3 + 2x)\,dx = (3x + x^2)\Big|_0^4 = 12 + 16 = 28$$

▶

NOW WORK PROBLEM 1.

EXAMPLE 2 Approximating a Definite Integral

Approximate: $\int_0^8 x^2\,dx$

(a) By dividing the interval $[0, 8]$ into four subintervals of equal length and picking u_i as the left end point of each subinterval.
(b) By dividing the interval $[0, 8]$ into eight subintervals of equal length and picking u_i as the right endpoint of each subinterval.
(c) Find the exact value.

SOLUTION (a) We divide $[0, 8]$ into four subintervals of equal length:
$$[0, 2], [2, 4], [4, 6], [6, 8]$$

Now we evaluate f at the left end point of each subinterval:
$$f(0) = 0, \ f(2) = 4, \ f(4) = 16, \ f(6) = 36$$

Since each subinterval is of length $\Delta x = 2$, we find
$$\int_0^8 x^2\,dx \approx f(0) \cdot 2 + f(2) \cdot 2 + f(4) \cdot 2 + f(6) \cdot 2$$
$$= 0 + 8 + 32 + 72 = 112$$

(b) We divide $[0, 8]$ into eight subintervals of equal length:
$$[0, 1], [1, 2], [2, 3], [3, 4], [4, 5], [5, 6], [6, 7], [7, 8]$$

Now we evaluate f at the right end point of each of these subintervals.
$$f(1) = 1, \ f(2) = 4, \ f(3) = 9, \ f(4) = 16,$$
$$f(5) = 25, \ f(6) = 36, \ f(7) = 49, \ f(8) = 64$$

Since each subinterval is of length $\Delta x = 1$, we find

$$\int_0^8 x^2\,dx \approx f(1) + f(2) + f(3) + f(4) + f(5) + f(6) + f(7) + f(8) = 204$$

(c) The exact value of $\int_0^8 x^2\,dx$, is

$$\int_0^8 x^2\,dx = \frac{1}{3}x^3\Big|_0^8 = \frac{512}{3} = 170.7$$

   NOW WORK PROBLEM 7.

Approximate definite integrals using a graphing utility

   We can use a graphing utility to approximate a definite integral.

EXAMPLE 3   Using a Graphing Utility to Approximate a Definite Integral

FIGURE 31

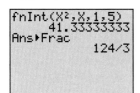

Use a graphing utility to approximate the area under the graph of $f(x) = x^2$ from 1 to 5. That is, evaluate the integral

$$\int_1^5 x^2\,dx$$

SOLUTION   Figure 31 shows the result using a TI–83 Plus calculator. Consult your owner's manual for the proper keystrokes.

   NOW WORK PROBLEM 19.

Riemann Sums

In the procedure for approximating a definite integral, suppose we subdivide each interval but don't require that each length be the same.

We begin with a continuous function f defined on a closed interval $[a, b]$. We partition, or divide, this interval into n subintervals, not necessarily of the same length. The point $a = x_0$ is the initial point, the first point of the subdivision is x_1, the second is $x_2, \ldots$, and the nth point is $b = x_n$. See Figure 32.

FIGURE 32

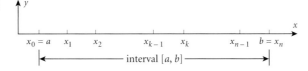

The original interval $[a, b]$ now consists of n subintervals; the length of each one is

First, $\Delta x_1 = x_1 - x_0$,   Second, $\Delta x_2 = x_2 - x_1$,   Third, $\ldots$, $\Delta x_3 = x_3 - x_2, \ldots$, kth, $\ldots$, $\Delta x_k = x_k - x_{k-1}, \ldots$, nth $\Delta x_n = x_n - x_{n-1}$

We use the symbol Δ to denote the largest such length, which depends on how the partition itself has been chosen. We call Δ the **norm of the partition**.

Next, we concentrate on the function f. Pick a number in each subinterval (you may select a number in the interval or either endpoint, if you wish) and evaluate f at this number. To fix our ideas, let u_k denote the number chosen from the kth subinterval. The corresponding value of the function is $f(u_k)$. This represents the height of the function at u_k.

Multiply $f(u_1)$ times $\Delta x_1 = x_1 - x_0$, $f(u_2)$ times $\Delta x_2 = x_2 - x_1, \ldots, f(u_n)$ times $\Delta x_n = x_n - x_{n-1}$, and add these products. The result is the sum

$$f(u_1)\,\Delta x_1 + f(u_2)\,\Delta x_2 + \cdots + f(u_n)\,\Delta x_n$$

This is called a **Riemann sum** for the function f on $[a, b]$.

Now take the limit of this sum as the norm $\Delta \to 0$. If this limit exists, it is the definite integral of $f(x)$ from a to b. That is,

$$\int_a^b f(x)\,dx = \lim_{\Delta \to 0} [f(u_1)\,\Delta x_1 + f(u_2)\,\Delta x_2 + \cdots + f(u_n)\,\Delta x_n] \quad (1)$$

where $\Delta = \max(\Delta x_1, \Delta x_2, \ldots, \Delta x_n)$.

The above formula, in a more formal course in calculus, is taken as the definition of a definite integral. Then it can be proven as a theorem, called the **Fundamental Theorem of Calculus**, that

$$\int_a^b f(x)\,dx = F(b) - F(a)$$

where F is an antiderivative of f.

EXERCISE 6.6

In Problems 1 and 2, refer to the illustration. The interval $[1, 3]$ is partitioned into two subintervals $[1, 2]$ and $[2, 3]$.

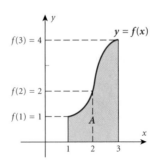

 1. Approximate the area A, choosing u as the left endpoint of each subinterval.

2. Approximate the area A, choosing u as the right endpoint of each subinterval.

Approximating Definite Integrals   469

In Problems 3 and 4, refer to the illustration. The interval [0, 8] is partitioned into four subintervals [0, 2], [2, 4], [4, 6], and [6, 8].

3. Approximate the area A, choosing u as the left endpoint of each subinterval.

4. Approximate the area A, choosing u as the right endpoint of each subinterval.

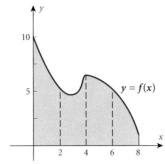

$f(0) = 10, f(2) = 6, f(4) = 7, f(6) = 5, f(8) = 1$

5. The function $f(x) = 3x$ is defined on the interval $[0, 6]$.
   (a) Graph f.
   In (b)–(e), approximate the area A under f from 0 to 6 as follows:
   (b) By partitioning $[0, 6]$ into three subintervals of equal length and choosing u as the left endpoint of each subinterval.
   (c) By partitioning $[0, 6]$ into three subintervals of equal length and choosing u as the right endpoint of each subinterval.
   (d) By partitioning $[0, 6]$ into six subintervals of equal length and choosing u as the left endpoint of each subinterval.
   (e) By partitioning $[0, 6]$ into six subintervals of equal length and choosing u as the right endpoint of each subinterval.
   (f) What is the actual area A?

6. Repeat Problem 5 for $f(x) = 4x$.

7. The function $f(x) = -3x + 9$ is defined on the interval $[0, 3]$.
   (a) Graph f.
   In (b)–(e), approximate the area A under f from 0 to 3 as follows:
   (b) By partitioning $[0, 3]$ into three subintervals of equal length and choosing u as the left endpoint of each subinterval.
   (c) By partitioning $[0, 3]$ into three subintervals of equal length and choosing u as the right endpoint of each subinterval.
   (d) By partitioning $[0, 3]$ into six subintervals of equal length and choosing u as the left endpoint of each subinterval.
   (e) By partitioning $[0, 3]$ into six subintervals of equal length and choosing u as the right endpoint of each subinterval.
   (f) What is the actual area A?

8. Repeat Problem 7 for $f(x) = -2x + 8$.

In Problems 9–16, a function f is defined over an interval $[a, b]$.
(a) Graph f, indicating the area A under f from a to b.
(b) Approximate the area A by partitioning $[a, b]$ into four subintervals of equal length and choosing u as the left endpoint of each subinterval.
(c) Approximate the area A by partitioning $[a, b]$ into eight subintervals of equal length and choosing u as the left endpoint of each subinterval.
(d) Express the area A as an integral.
(e) Evaluate the integral.

9. $f(x) = x^2 + 2$, $[0, 4]$ 10. $f(x) = x^2 - 4$, $[2, 6]$

11. $f(x) = x^3$, $[0, 4]$ 12. $f(x) = x^3$, $[1, 5]$

13. $f(x) = \dfrac{1}{x}$, $[1, 5]$ 14. $f(x) = \sqrt{x}$, $[0, 4]$

15. $f(x) = e^x$, $[-1, 3]$ 16. $f(x) = e^{-x}$, $[3, 7]$

In Problems 17–20, use a graphing utility to approximate each integral. Round your answer to two decimal places.

17. $\displaystyle\int_0^1 e^{x^2}\, dx$ 18. $\displaystyle\int_{-2}^4 e^{x^2}\, dx$

19. $\displaystyle\int_1^5 \dfrac{e^x}{x}\, dx$ 20. $\displaystyle\int_1^5 \dfrac{\ln x}{x}\, dx$

21. Consider the function $f(x) = \sqrt{1 - x^2}$ whose domain is the closed interval $[-1, 1]$.

   (a) Graph f.
   (b) Approximate the area under the graph of f from -1 to 1 by dividing $[-1, 1]$ into five subintervals, each of equal length. Choose u as the left end point.
   (c) Approximate the area under the graph of f from -1 to 1 by dividing $[-1, 1]$ into ten subintervals each of equal length. Choose u as the left end point.
   (d) Express the area as an integral.
   (e) Evaluate the integral using a graphing utility.
   (f) What is the actual area?
   [**Hint:** The graph of f is a semi-circle.]

6.7 Differential Equations

PREPARING FOR THIS SECTION *Before getting started, review the following:*

> Properties of Logarithms (Chapter 2, Section 2.5, pp. 214–221)
> Logarithmic Equations (Chapter 2, Section 2.4, pp. 207–208)
> Exponential Equations (Chapter 2, Section 2.3, pp. 195–196)
> Differentials (Chapter 5, Section 5.7, pp. 405–407)

OBJECTIVES
1. Find a particular solution of a differential equation
2. Solve applied problems involving population growth
3. Solve applied problems involving radioactive decay
4. Solve applied problems involving price-demand equations

In studies of physical, chemical, biological, and other natural phenomena, scientists attempt, on the basis of long observation, to deduce mathematical laws that will describe and predict nature's behavior. Such laws often involve the derivatives of some unknown function F, and it is required to find this unknown function F.

For example, for a given function f it may be required to find all functions $y = F(x)$ so that

$$\frac{dy}{dx} = f(x) \tag{1}$$

This equation is an example of what is called a **differential equation.** A function $y = F(x)$ for which $\frac{dy}{dx} = f(x)$ is a **solution** of the differential equation. The **general solution** of $\frac{dy}{dx} = f(x)$ consists of all the antiderivatives of f.

EXAMPLE 1 Finding the General Solution of a Differential Equation

The general solution of the differential equation

$$\frac{dy}{dx} = 5x^2 + 2 \tag{2}$$

is

$$y = \frac{5x^3}{3} + 2x + K$$

where K is a constant.

A **particular solution** of the differential equation $\frac{dy}{dx} = f(x)$ occurs when K is assigned a particular value. When a particular solution is required, we use a **boundary condition.**

1 EXAMPLE 2 Finding a Particular Solution of a Differential Equation

In the differential equation (2) we require the general solution to obey the boundary condition that $y = 5$ when $x = 3$. Find the particular solution.

SOLUTION In the general solution,

$$y = \frac{5x^3}{3} + 2x + K$$

let $x = 3$ and $y = 5$. Then

$$5 = \frac{5(27)}{3} + (2)(3) + K \qquad x = 3; y = 1$$

$$5 = 45 + 6 + K \qquad \text{Simplify.}$$

$$K = -46 \qquad \text{Solve for } K.$$

The particular solution of the differential equation (2) with the boundary condition that $y = 5$ when $x = 3$ is

$$y = \frac{5x^3}{3} + 2x - 46$$

EXAMPLE 3 Finding a Particular Solution of a Differential Equation

Solve the differential equation below with the boundary condition that $y = -1$ when $x = 3$.

$$\frac{dy}{dx} = x^2 + 2x + 1$$

SOLUTION The general solution of the differential equation is

$$y = \frac{x^3}{3} + x^2 + x + K$$

where K is a constant. To determine the number K, we use the boundary condition. Then

$$-1 = \frac{3^3}{3} + 3^2 + 3 + K \qquad x = 3; y = -1$$

$$-1 = 9 + 9 + 3 + K \qquad \text{Simplify.}$$

$$K = -22 \qquad \text{Solve for } K.$$

The particular solution of the differential equation with the boundary condition that $y = -1$ when $x = 3$ is

$$y = \frac{x^3}{3} + x^2 + x - 22$$

NOW WORK PROBLEM 1.

Applications

The statement below describes many situations in business, biology, and the natural sciences.

> The amount A of a substance varies with time t in such a way that the time rate of change of A is proportional to A itself.

The mathematical formulation of this statement is the differential equation

$$\frac{dA}{dt} = kA \qquad (3)$$

where $k \neq 0$ is a real number.

If $k > 0$, then equation (3) asserts that the time rate of change of A is positive, so that the amount A of the substance is increasing over time. If $k < 0$, then equation (3) asserts that the time rate of change of A is negative, so that the amount A of the substance is decreasing over time.

We seek a solution $A = A(t)$ to the differential equation (3). We begin by rewriting the equation in terms of differentials as

$$dA = kA\, dt$$

$$\frac{dA}{A} = k\, dt \qquad \text{Divide both sides by } A.$$

$$\int \frac{dA}{A} = \int k\, dt \qquad \text{Take the integral of each side.}$$

$$\ln A = kt + K \qquad \text{Integrate.}$$

To determine the constant K, we use the boundary condition that when $t = 0$, the initial amount present is $A_0 = A(0)$. When $t = 0$, we have

$$\ln A_0 = k(0) + K$$

$$K = \ln A_0$$

Replace K by $\ln A_0$ and proceed to solve for A.

$$\ln A = kt + \ln A_0 \qquad K = \ln A_0$$

$$\ln A - \ln A_0 = kt \qquad \text{Subtract } \ln A_0 \text{ from each side.}$$

$$\ln \frac{A}{A_0} = kt \qquad \log_a \frac{M}{N} = \log_a M - \log_a N$$

$$\frac{A}{A_0} = e^{kt} \qquad \text{Change to an exponential expression.}$$

The solution of the differential equation (3) is therefore

$$A = A_0 e^{kt} \qquad (4)$$

When a function $A = A(t)$ varies according to the law given by the differential equation (3), or its solution, equation (4), it is said to follow the **exponential law**, or the **law of uninhibited growth or decay**, or the **law of continuously compounded interest**.

As previously noted, the sign of k determines whether A is increasing ($k > 0$) or decreasing ($k < 0$). Figure 33 illustrates the graphs of $A = A(t) = A_0 e^{kt}$ for $k > 0$ and for $k < 0$.

FIGURE 33

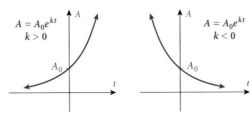

Applications

We present here three applications: to biology, radioactive decay, and business. They are independent of each other and may be covered in any order or omitted without loss of continuity.

Bacterial Growth Our first application is to bacterial growth.

EXAMPLE 4 Growth of Bacteria

Assume that the population of a colony of bacteria *increases at a rate proportional to the number present.** If the number of bacteria doubles in 5 hours, how long will it take for the bacteria to triple?

SOLUTION Let $N = N(t)$ be the number of bacteria present at time t. Then the assumption that this colony of bacteria increases at a rate proportional to the number present can be written as

$$\frac{dN}{dt} = kN \tag{5}$$

where k is a positive constant of proportionality. The solution of the differential equation (5) is

$$N(t) = N_0 e^{kt}$$

where N_0 is the initial number of bacteria in this colony. Since the number of bacteria doubles in 5 hours, we have

$$N(5) = 2N_0$$

But $N(5) = N_0 e^{k(5)}$, so that

$$N_0 e^{5k} = 2N_0$$
$$e^{5k} = 2 \quad\quad \text{Cancel the } N_0\text{'s.}$$
$$5k = \ln 2 \quad\quad \text{Write as a logarithm.}$$
$$k = \left(\frac{1}{5}\right)\ln 2 = 0.1386 \quad\quad \text{Solve for } k.$$

The function that gives the number N of bacteria present at time t is

$$N(t) = N_0 e^{0.1386t}$$

The time t that is required for this colony to triple must satisfy the equation

$$N(t) = 3N_0$$
$$N_0 e^{0.1386t} = 3N_0$$
$$e^{0.1386t} = 3 \quad\quad \text{Cancel the } N_0\text{'s.}$$
$$0.1386t = \ln 3 \quad\quad \text{Write as a logarithm.}$$
$$t = \frac{\ln 3}{0.1386} = 7.925 \text{ hours} \quad\quad \text{Solve for } t.$$

The bacteria will triple in size is just under 8 hours.

* This is a model of uninhibited growth. However, after enough time has passed, growth will not continue at a rate proportional to the number present. Other factors, such as lack of living space, dwindling food supply, and so on, will start to affect the rate of growth. The model presented accurately reflects the way growth occurs in the early stages.

NOW WORK PROBLEM 11.

Radioactive Decay Our second application is to radioactive decay, and, in particular, its use in carbon dating. For a radioactive substance, *the rate of decay is proportional to the amount present at a given time t.* That is, if $A = A(t)$ represents the amount of a radioactive substance at time t, we have

$$\frac{dA}{dt} = kA$$

where the constant k is negative and depends on the radioactive substance. The **half-life** of a radioactive substance is the time required for half of the substance to decay.

In carbon dating, we use the fact that all living organisms contain two kinds of carbon, carbon-12 (a stable carbon) and carbon-14 (a radioactive carbon). As a result, when an organism dies, the amount of carbon-12 present remains unchanged, while the amount of carbon-14 begins to decrease. This change in the amount of carbon-14 present relative to the amount of carbon-12 present makes it possible to calculate the time at which the organism lived.

EXAMPLE 5 Using Carbon Dating in Archaeology

In the skull of an animal found in an archaeological dig, it was determined that about 20% of the original amount of carbon-14 was still present. If the half-life of carbon-14 is 5600 years, find the approximate age of the animal.

SOLUTION Let $A = A(t)$ be the amount of carbon-14 present in the skull at time t. Then A satisfies the differential equation $\dfrac{dA}{dt} = kA$, whose solution is

$$A(t) = A_0 e^{kt}$$

where A_0 is the amount of carbon-14 present at time $t = 0$. To determine the constant k, we use the fact that when $t = 5600$, half of the original amount A_0 will remain. That is,

$$A(t) = A_0 e^{kt}$$

$$\frac{1}{2}A_0 = A_0 e^{5600k} \qquad A = \frac{1}{2}A_0 \text{ when } t = 5600.$$

$$\frac{1}{2} = e^{5600k} \qquad \text{Cancel the } A_0\text{'s.}$$

$$5600k = \ln\frac{1}{2} \qquad \text{Write as a logarithm.}$$

$$k = -0.000124 \qquad \text{Solve for } k.$$

The relationship between the amount A of carbon-14 and time t is therefore

$$A(t) = A_0 e^{-(0.000124)t}$$

If the amount A of carbon-14 is 20% of the original amount A_0, we have

$$0.2 A_0 = A_0 e^{-(0.000124)t}$$

$$0.2 = e^{-(0.000124)t} \qquad \text{Cancel the } A_0\text{'s.}$$

$$-(0.000124)t = \ln 0.2 \qquad \text{Change to a logarithm.}$$

$$t = \frac{1.6094}{0.000124} = 12{,}979 \text{ years} \qquad \text{Solve for } t.$$

The animal found in the dig lived approximately 13,000 years ago.

NOW WORK PROBLEM 15.

Price-Demand Equations Our third application is to business: Suppose the marginal price of a product, when the demand is x units, is proportional to its price, p. This situation can be represented by the differential equation

$$\frac{dp}{dx} = kp$$

Since demand usually decreases as price increases, the constant k will be negative.

EXAMPLE 6 Pricing a Product

A manufacturer finds that the marginal price of one of its products is proportional to the price. Past experience has shown that at a price of $125 per unit, zero units are sold; but at a price of $100 per unit, 100 units are sold.

(a) Find the price-demand equation.
(b) Find the approximate price that should be charged if the manufacturer wants to sell 50 units.

SOLUTION **(a)** If $p = p(x)$ represents the price when x units are demanded, then

$$\frac{dp}{dx} = kp, \quad k < 0$$

Rewriting the equation in terms of differentials we get

$$\frac{dp}{p} = k\, dx$$

We integrate each side to obtain

$$\ln p = kx + K$$

where K is a constant.

To determine the constant K we use the boundary condition that when $x = 0$, $p = 125$. Then

$$\ln 125 = k(0) + K$$
$$k = \ln 125$$

Substitute into $\ln p = \ln x + K$ and proceed to solve for p.

$$\ln p = kx + \ln 125$$
$$\ln p - \ln 125 = kx \quad \text{Subtract } \ln 125 \text{ from each side.}$$
$$\ln \frac{p}{125} = kx \quad \ln\left(\frac{M}{N}\right) = \ln M - \ln N$$
$$\frac{p}{125} = e^{kx} \quad \text{Change to exponential form.}$$
$$p = 125e^{kx} \quad \text{Solve for } p. \quad (6)$$

We use the second boundary condition that when 100 units are demanded the price is $100 to find k.

$$100 = 125e^{100k} \quad p = 125e^{kx}; p = \$100; x = 100$$
$$\frac{100}{125} = e^{100k} \quad \text{Divide by 125.}$$
$$\ln\left(\frac{100}{125}\right) = 100k \quad \text{Write as a logarithm.}$$
$$k = \frac{1}{100}\ln\left(\frac{100}{125}\right) \approx -0.00223 \quad \text{Solve for } k.$$

From (6), the demand equation is

$$p(x) = 125e^{-0.00223x}$$

(b) To sell 50 units, the manufacturer should set the price at

$$p = 125e^{(-0.00223)50} = \$111.80$$

EXERCISE 6.7

In Problems 1–10, solve each differential equation using the indicated boundary condition.

1. $\dfrac{dy}{dx} = x^2 - 1$,
   $y = 0$ when $x = 0$

2. $\dfrac{dy}{dx} = x^2 + 4$,
   $y = 1$ when $x = 0$

3. $\dfrac{dy}{dx} = x^2 - x$,
   $y = 3$ when $x = 3$

4. $\dfrac{dy}{dx} = x^2 + x$,
   $y = 5$ when $x = 3$

5. $\dfrac{dy}{dx} = x^3 - x + 2$,
   $y = 1$ when $x = -2$

6. $\dfrac{dy}{dx} = x^3 + x - 5$,
   $y = 1$ when $x = -2$

7. $\dfrac{dy}{dx} = e^x$,
   $y = 4$ when $x = 0$

8. $\dfrac{dy}{dx} = \dfrac{1}{x}$,
   $y = 0$ when $x = 1$

9. $\dfrac{dy}{dx} = \dfrac{x^2 + x + 1}{x}$,
   $y = 0$ when $x = 1$

10. $\dfrac{dy}{dx} = x + e^x$,
    $y = 4$ when $x = 0$

11. **Bacterial Growth** The rate of growth of bacteria is proportional to the amount present. If initially there are 100 bacteria and 5 minutes later there are 150 bacteria, how many bacteria will be present after 1 hour? How many are present after 90 minutes? How long will it take for the number of bacteria to reach 1,000,000?

12. Answer the questions given in Problem 11 if after 8 minutes the number of bacteria present grows from 100 to 150.

13. **Radioactive Decay** The half-life of radium is 1690 years. If 8 grams of radium are present now, how many grams will be present in 100 years?

14. **Radioactive Decay** If 25% of a radioactive substance disappears in 10 years, what is the half-life of the substance?

15. **Age of a Tree** A piece of charcoal is found to contain 30% of the carbon-14 it originally had. When did the tree from which the charcoal came die? Use 5600 years as the half-life of carbon-14.

16. **Age of a Fossil** A fossilized leaf contains 70% of a normal amount of carbon-14. How old is the fossil?

17. **Population Growth** The population growth of a colony of mosquitoes obeys the uninhibited growth equation. If there are 1500 mosquitoes initially, and there are 2500 mosquitoes after 24 hours, what is the size of the mosquito population after 3 days?

18. **Population Growth** The population of a suburb doubled in size in an 18-month period. If this growth continues and the current population is 8000, what will the population be in 4 years?

19. **Bacterial Growth** The number of bacteria in a culture is growing at a rate of $3000e^{2t/5}$. At $t = 0$, the number of bacteria present was 7500. Find the number present at $t = 5$.

20. **Bacterial Growth** At any time t, the rate of increase in the area of a culture of bacteria is twice the area of the culture. If the initial area of the culture is 10, then what is the area at time t?

21. **Bacterial Growth** The rate of change in the number of bacteria in a culture is proportional to the number present. In a certain laboratory experiment, a culture had 10,000 bacteria initially, 20,000 bacteria at time t_1 minutes, and 100,000 bacteria at $t_1 + 10$ minutes.

    (a) In terms of t only, find the number of bacteria in the culture at any time t minutes ($t \geq 0$).
    (b) How many bacteria were there after 20 minutes?
    (c) At what time were 20,000 bacteria observed? That is, find the value of t_1.

22. **Chemistry** Salt (NaCl) decomposes in water into sodium (Na^+) and chloride (Cl^-) ions at a rate proportional to its mass. If the initial amount of salt is 25 kilograms, and after 10 hours, 15 kilograms are left:

    (a) How much salt would be left after 1 day?
    (b) After how many hours would there be less than $\frac{1}{2}$ kilogram of salt left?

23. **Age of a Fossil** Radioactive beryllium is sometimes used to date fossils found in deep-sea sediment. The decay of radioactive beryllium satisfies the equation $\dfrac{dA}{dt} = -\alpha A$,

where $\alpha = 1.5 \times 10^{-7}$, and t is measured in years. What is the half-life of radioactive beryllium?

24. **Pressure** Atmospheric pressure P is a function of the altitude a above sea level and is given by the equation $\dfrac{dP}{da} = \beta P$, where β is a constant. The pressure is measured in millibars (mb). At sea level ($a = 0$), $P(0)$ is 1013.25 mb, which means that the atmosphere at sea level will support a column of mercury 1013.25 millimeters high at a standard temperature of 15 °C. At an altitude of $a = 1500$ meters, the pressure is 845.6 mb.

(a) What is the pressure at $a = 4000$ meters?
(b) What is the pressure at 10 kilometers?
(c) In California, the highest and lowest points are Mount Whitney (4418 meters) and Death Valley (86 meters below sea level). What is the difference in their atmospheric pressures?
(d) What is the atmospheric pressure at Mount Everest (elevation 8848 meters)?
(e) At what elevation is the atmospheric pressure equal to 1 mb?

25. **Pricing a Product** A manufacturer finds that the marginal price of one of its products is proportional to the price. The manufacturer also knows that at a price of $300 per unit, no units are sold. At a price of $150, a total of 200 units are sold.

(a) Find the price-demand equation.
(b) Find the price p that should be charged if the manufacturer wants to sell 300 units.
(c) What price should be charged if the manufacturer wants to sell 350 units?

26. **Pricing a Product** A manufacturer finds that the marginal price of one of its products is proportional to the price. The manufacturer also knows that at a price of $500 per unit, no units are sold. At a price of $300, a total of 150 units are sold.

(a) Find the price-demand equation.
(b) Find the price p that should be charged if the manufacturer wants to sell 200 units.
(c) What price should be charged if the manufacturer wants to sell 250 units?

Chapter 6 Review

OBJECTIVES

Section	You should be able to	Review Exercises
6.1	1 Find an antiderivative	1–6
	2 Use integration formulas	7–14, 17, 21, 22
	3 Evaluate indefinite integrals	7–30
	4 Find cost and revenue functions	31–36
6.2	1 Integrate using substitution	15, 16, 18–20, 23–26
6.3	1 Integration by parts	27–30
6.4	1 Evaluate a definite integral	37–48
	2 Use properties of a definite integral	49–52
	3 Solve applied problems involving definite integrals	63, 64, 66, 81–84
6.5	1 Find the area under a graph	53–56
	2 Find the area enclosed by two graphs	57–62
	3 Solve applied problems involving definite integrals	65, 85, 86
6.6	1 Approximate definite integrals using rectangles	67, 68
	2 Approximate definite integrals using a graphing utility	69, 70
6.7	1 Find a particular solution of a differential equation	71–76
	2 Solve applied problems involving population growth	77, 78
	3 Solve applied problems involving radioactive decay	79, 80
	4 Solve applied problems involving price-demand equations	87

Chapter 6 The Integral of a Function and Applications

THINGS TO KNOW

Indefinite Integral (p. 423)

$$\int f(x)\,dx = F(x) + K, \quad F'(x) = f(x) \text{ and } K \text{ is a constant}$$

Integration Formulas (pp. 424 and 426)
(c is a real number; K is a constant)

$$\int c\,dx = cx + K$$

$$\int x^n\,dx = \frac{x^{n+1}}{n+1} + K \quad n \neq -1 \text{ a rational number}$$

$$\int [f(x) \pm g(x)]\,dx = \int f(x)\,dx \pm \int g(x)\,dx$$

$$\int cf(x)\,dx = c\int f(x)\,dx$$

$$\int e^x\,dx = e^x + K$$

$$\int e^{ax}\,dx = \frac{1}{a}e^{ax} + K \quad a \neq 0$$

$$\int \frac{1}{x}\,dx = \ln|x| + K$$

Integration by Parts (p. 438)

$$\int u\,dv = uv - \int v\,du$$

Definite Integral (p. 442)

$$\int_a^b f(x)\,dx = F(b) - F(a), \quad F'(x) = f(x)$$

Properties of the Definite Integral (pp. 444–445)
(f and g are continuous functions on the interval $[a, b]$)

$$\int_a^b f(x)\,dx = -\int_b^a f(x)\,dx$$

$$\int_a^a f(x)\,dx = 0$$

$$\int_a^b f(x)\,dx = \int_a^c f(x)\,dx + \int_c^b f(x)\,dx \quad c \text{ is between } a \text{ and } b$$

$$\int_a^b cf(x)\,dx = c\int_a^b f(x)\,dx \quad c \text{ is a real number}$$

$$\int_a^b [f(x) \pm g(x)]\,dx = \int_c^b f(x)\,dx \pm \int_a^b g(x)\,dx$$

Area A under the graph of $y = f(x)$ from a to b. (p. 451)

$$A = \int_a^b f(x)\,dx \quad f(x) \geq 0 \text{ on } [a, b].$$

Approximating a Definite Integral (p. 465)

$$\int_a^b f(x)\,dx \approx f(u_1)\Delta x + f(u_2)\Delta x + \cdots + f(u_n)\Delta x, \quad \Delta x = \frac{b-a}{n}$$

TRUE–FALSE ITEMS

T F **1.** The integral of the sum of two functions equals the sum of their integrals.

T F **2.** $\int \left(\frac{1}{2}x^3 + x^{3/2} - 1\right) dx$
$= \frac{x^4}{6} + \frac{2x^{7/2}}{7} - x + K$

T F **3.** $\int \frac{x^2 + 4}{x}\,dx = \dfrac{\frac{x^3}{3} + 4x}{\frac{x^2}{2}} + K$

T F **4.** $\int \frac{\sqrt{x^2 + 1}}{x}\,dx = \frac{1}{x}\int \sqrt{x^2 + 1}\,dx$

T F **5.** $\int \ln x\,dx = \frac{1}{x} + K$

T F 6. Any antiderivative of $f(x)$ can be used to evaluate $\int_a^b f(x)\,dx$.

T F 7. The definite integral $\int_a^b f(x)\,dx$, if it exists, is a number.

T F 8. $\int_a^b f(x)\,dx + \int_b^a f(x)\,dx = 0$

T F 9. $\int_0^1 x^2\,dx = 3$

T F 10. The area under the graph of $f(x) = x^4$ from 0 to 2 equals $\int_0^2 x^4\,dx$.

FILL-IN-THE-BLANKS

1. A function F is called an antiderivative of the function f if _____.

2. The symbol _____ represents all the antiderivatives of a function f.

3. The formula $\int u\,dv = uv - \int v\,du$ is referred to as the _____ _____ formula.

4. In $\int_a^b f(x)\,dx$, the numbers a and b are called the _____ and _____ _____ of _____, respectively.

5. $\int_a^a f(x)\,dx = $ _____.

6. If f is continuous on $[a, b]$ and $F' = f$, then $\int_a^b f(x)\,dx = $ _____.

7. The area under the graph of $f(x) = \sqrt{x^2 + 1}$ from 0 to 2 may be symbolized by the integral _____.

REVIEW EXERCISES Blue problem numbers indicate the author's suggestions for use in a practice test.

In Problems 1–6, find all the antiderivatives of the given function.

1. $f(x) = 6x^5$
2. $f(x) = 3x^2$
3. $f(x) = x^3 + x$
4. $f(x) = x + 7$
5. $f(x) = \dfrac{1}{\sqrt{x}}$
6. $f(x) = \dfrac{1}{2\sqrt{x}}$

In Problems 7–30, evaluate each indefinite integral.

7. $\int 7\,dx$
8. $\int \dfrac{1}{2}\,dx$
9. $\int (5x^3 + 2)\,dx$
10. $\int (x^2 + 3x)\,dx$
11. $\int (x^4 - 3x^2 + 6)\,dx$
12. $\int 12(x^3 + 6x^2 - 2x - 1)\,dx$
13. $\int \dfrac{3}{x}\,dx$
14. $\int \left(x^2 + \dfrac{4}{x}\right)\,dx$
15. $\int \dfrac{2x}{x^2 - 1}\,dx$
16. $\int \dfrac{3x^2 + 5}{x^3 + 5x}\,dx$
17. $\int e^{3x}\,dx$
18. $\int 5xe^{2x^2}\,dx$
19. $\int (x^3 + 3x)^5(x^2 + 1)\,dx$
20. $\int \dfrac{x}{\sqrt{1 - x^2}}\,dx$
21. $\int 2x(x - 3)\,dx$
22. $\int (x + 1)(x - 1)\,dx$
23. $\int e^{3x^2 + x}(6x + 1)\,dx$
24. $\int \dfrac{1}{x^4}e^{x-3}\,dx$
25. $\int x\sqrt{x - 5}\,dx$
26. $\int x(x + 2)^{2/3}\,dx$
27. $\int xe^{4x}\,dx$

28. $\displaystyle\int 3x^2 e^{x+4}\, dx$

29. $\displaystyle\int x^{-2} \ln 2x\, dx$

30. $\displaystyle\int (x^5 + 7) \ln x\, dx$

In Problems 31 and 32, find the revenue function R. Assume that the revenue is zero if there are no sales.

31. $R'(x) = 5x + 2$

32. $R'(x) = 12x - 5\sqrt{x}$

In Problems 33 and 34, find the cost function C, and determine where the cost is minimum.

33. $C'(x) = 5x + 120{,}000$; fixed costs are $7500

34. $C'(x) = 1.3x - 2600$; fixed costs are $2735

35. Revenue analysis (a) If sales of televisions have a marginal revenue given by $R'(x) = 500 - 0.01x$, where x is the number of televisions sold, find the revenue obtained by selling x televisions. Assume that the revenue is 0 if there are no sales.
  (b) Find the number of televisions that need to be sold to maximize revenue.
  (c) What is the maximum revenue that can be obtained from the sales of these televisions?
  (d) If currently 35,000 televisions are sold, what is the total increase in revenue if sales increase to 40,000 units?

36. Cost analysis If the daily marginal cost of production of t-shirts has been measured to be $C'(x) = 0.06x - 6$ and fixed costs are known to be $1000 per day,
  (a) Find the cost function.
  (b) Find the daily production that minimizes cost. What is the minimum daily cost of production?
  (c) If the t-shirts are sold for $10 each, find the revenue function.
  (d) Find the daily break-even point.
  (e) Find the daily sales volume that maximizes profit. What is the daily maximum profit?

In Problems 37–48, evaluate each definite integral.

37. $\displaystyle\int_{-2}^{1} (x^2 + 3x - 1)\, dx$

38. $\displaystyle\int_{1}^{2} (x^3 - 1)\, dx$

39. $\displaystyle\int_{4}^{9} 8\sqrt{x}\, dx$

40. $\displaystyle\int_{1}^{8} \sqrt[3]{x^2}\, dx$

41. $\displaystyle\int_{0}^{1} (e^x - e^{-x})\, dx$

42. $\displaystyle\int_{0}^{2} \frac{1}{x+2}\, dx$

43. $\displaystyle\int_{0}^{4} \frac{dx}{(3x+2)^2}$

44. $\displaystyle\int_{0}^{\sqrt{15}} \frac{x}{\sqrt{x^2+1}}\, dx$

45. $\displaystyle\int_{-2}^{2} e^{3x}\, dx$

46. $\displaystyle\int_{2}^{3} \frac{e^{1/x}}{x^2}\, dx$

47. $\displaystyle\int_{0}^{1} (x+2)e^{-x}\, dx$

48. $\displaystyle\int_{1}^{4} x \ln 3x\, dx$

In Problems 49–52, use the properties of the definite integral to evaluate each integral if it is known that

$$\int_{0}^{5} f(x)\, dx = 3 \qquad \int_{0}^{5} g(x)\, dx = 8 \qquad \int_{5}^{9} f(x)\, dx = -2 \qquad \int_{5}^{9} g(x)\, dx = 10$$

49. $\displaystyle\int_{0}^{9} f(x)\, dx$

50. $\displaystyle\int_{0}^{5} [2f(x) + g(x)]\, dx$

51. $\displaystyle\int_{9}^{5} g(x)\, dx$

52. $\displaystyle\int_{5}^{9} -3f(x)\, dx$

In Problems 53–62, find the area described. Be sure to sketch the graph first.

53. Under the graph of $f(x) = x^2 + 4$ from -1 to 2.

54. Under the graph of $f(x) = x^2 + 3x + 2$ from 0 to 3.

55. Under the graph of $f(x) = e^x + x$ from 0 to 1.

56. Under the graph of $f(x) = x^3 + 1$ from -1 to 2.

57. Enclosed by the graph of $f(x) = x^2 - x - 2$ and the x-axis from 0 to 3.
58. Enclosed by the graph of $f(x) = x^3 - 8$ and the x-axis from 0 to 3.
59. Enclosed by the graphs of $f(x) = x^2 - 4$ and $x + y = 2$.
60. Enclosed by the graphs of $f(x) = \sqrt{x}$ and $g(x) = \dfrac{1}{2}x$.
61. Enclosed by the graphs of $f(x) = x^3$ and $g(x) = 4x$.
62. Enclosed by the graphs of $f(x) = x^3$ and $g(x) = \sqrt[3]{x}$.
63. **Profit from Jeans** A clothing manufacturer produces x pairs of jeans per month. The company's marginal profit from the sale of the jeans is given by $P'(x) = 9 - 0.004x$. Currently the company is manufacturing 2000 pairs of jeans per month. Find the increased monthly profit if production increases to 2500 pairs.
64. **Total Maintenance Cost** Maintenance costs for a piece of machinery usually increase with the age of the machine. From experience, the production manager at a factory found that the rate of change of expense to maintain a particular type of machine is given by $f'(t) = 12t^2 + 2500$, where t is the age of the machine in years, and $f(t)$ is the total cost of maintenance.
    (a) Find the total cost of maintenance for the years 0 through 5.
    (b) Find the total cost of maintenance for years 5 through 10.
    (c) Find the total cost of maintenance for the first 10 years of operation.
    (d) If company policy dictates replacing the machine when total maintenance costs equal $100,000, what is the useful life of the machine?
65. **Useful Life of Vending Machines** It has been found that owning and operating a vending machine has an increasing cost rate and a decreasing revenue rate. At time t a particular machine generates a marginal revenue $R'(t) = -10t$, and has a marginal maintenance cost $C'(t) = 2t - 12$.
    (a) Determine how long the owner should keep the machine to maximize profit.
    (b) What is the total profit that the machine will generate in this time?
66. **Total Sales** The rate of sales of a certain product obeys
$$f(t) = 1340 - 850e^{-t}$$
where t is the number of years the product is on the market. Find the total sales during the first 5 years.
67. **Approximating Area** The function $f(x) = -2x + 10$ is defined on $[0, 4]$.
    (a) Graph f.
    Approximate the area A under f from 0 to 4 as follows:
    (b) By partitioning $[0, 4]$ into 4 subintervals of equal length and choosing u as the left endpoint.
    (c) By partitioning $[0, 4]$ into 4 subintervals of equal length and choosing u as the right endpoint.
    (d) By partitioning $[0, 4]$ into 8 subintervals of equal length and choosing u as the left endpoint.
    (e) By partitioning $[0, 4]$ into 8 subintervals of equal length and choosing u as the right endpoint.
    (f) Express the area A as an integral.
    (g) Evaluate the integral.
68. **Approximating Area** The function $f(x) = x + 3$ is defined on $[-1, 2]$.
    (a) Graph f.
    Approximate the area A under f from -1 to 2 as follows:
    (b) By partitioning $[-1, 2]$ into 3 subintervals of equal length and choosing u as the left endpoint.
    (c) By partitioning $[-1, 2]$ into 3 subintervals of equal length and choosing u as the right endpoint.
    (d) By partitioning $[-1, 2]$ into 6 subintervals of equal length and choosing u as the left endpoint.
    (e) By partitioning $[-1, 2]$ into 6 subintervals of equal length and choosing u as the right endpoint.
    (f) Express the area A as an integral.
    (g) Evaluate the integral.
69. Approximate $\displaystyle\int_{1}^{10} x \ln x \, dx$ with a graphing utility.
70. Approximate $\displaystyle\int_{-1}^{1} \sqrt{1 - x^2} \, dx$ with a graphing utility.

In Problems 71–76, solve each differential equation using the indicated boundary condition.

71. $\dfrac{dy}{dx} = x^2 + 5x - 10$; $y = 1$ when $x = 0$

72. $\dfrac{dy}{dx} = x^4 - 2x^2 + 1$; $y = 10$ when $x = 0$

73. $\dfrac{dy}{dx} = e^{2x} - x$; $y = 3$ when $x = 0$

74. $\dfrac{dy}{dx} = \dfrac{2}{2x + 5}$; $y = 0$ when $x = -2$

75. $\dfrac{dy}{dx} = 10y$; $y = 1$ when $x = 0$

76. $\dfrac{dy}{dx} = 15\sqrt{5x}$; $y = 100$ when $x = 5$

77. **Bacterial Growth** Bacteria grown in a certain culture increase at a rate proportional to the amount present. If there are 2000 bacteria present initially and the amount triples in 2 hours, how many bacteria will there be in $4\frac{1}{2}$ hours?

78. **Bugs in the House** A house becomes infected with fleas, whose population growth obeys the law of uninhibited growth. If initially a nest of 100 fleas hatched and after 2 days there are 500, how many fleas will be present in one week (7 days)?

79. **Prehistoric Civilization** An ancient burial ground is discovered in a countryside field. Carbon dating is done on the bones in the graves. It is found that 40% of the carbon-14 is remaining. How old is the burial ground? (Use 5600 years as the half-life of carbon-14.)

80. **Age of an Animal** The skeleton of an animal is found to contain 35% of the original amount of carbon-14. What is the approximate age of this animal? (Use 5600 years as the half-life of carbon-14).

81. **Expanding Economy** It is known that the economy of a certain developing nation is growing at a rate of $E'(t) = 0.02t^2 + t$, where t is time measured in years. This year the nation's economy totals 5 million dollars.
    (a) Find the function that measures the economy at time t.
    (b) Assuming the economy continues to expand at the current rate, what will it equal in 5 years?

82. **Total Cost** The marginal cost function for producing x units of a product is
$$C'(x) = 200 - \frac{1}{5}x$$
Find the increase in total cost if production increases from 100 to 150 units.

83. **Learning Curve** Margo Manufacturing introduced a new line of tennis ball servers equipped with electronic timing and delivery devices. It took the production crew 700 labor-hours to produce the first 50 servers, but production time is decreasing according to the learning curve
$$f(x) = 700 \, x^{-0.1}$$
where $f(x)$ is the labor-hours needed to produce the xth production run of 50 servers and was determined after 8 production runs were completed. At that time a large order for 500 servers was received. How much time should Margo allow for labor to produce the additional 10 production runs of the tennis ball servers?

84. **Total Revenue** The marginal revenue function for selling x units in a week is
$$R'(x) = 50 - x$$
Find the increase in total revenue attained if sales increase from 30 to 40 units per week.

85. **Consumer's Surplus; Producer's Surplus** The demand curve is $D(x) = 12 - \dfrac{x}{50}$ and the supply curve is $S(x) = \dfrac{x}{20} + 5$.
    (a) Find the market price and the demand level of market equilibrium.
    (b) Find the consumer's surplus and the producer's surplus.
    (c) Sketch the graphs.

86. **Consumer's Surplus; Producer's Surplus** The demand curve is $D(x) = 20 \, e^{-0.01x}$ and the supply curve is $S(x) = 0.005x^2$.
    (a) Find the market price and the demand level of market equilibrium (rounded to the nearest dollar).
    (b) Find the consumer's surplus and the producer's surplus.
    (c) Sketch the graphs.

87. **Pricing a Product** A manufacturer finds that the marginal price of one of its products is proportional to the price. The manufacturer also knows that at a price of $800 per unit, no units are sold. At a price of $600, a total of 80 units are sold.
    (a) Find the price-demand equation.
    (b) Find the price p that should be charged if the manufacturer wants to sell 200 units.
    (c) What price should be charged if the manufacturer wants to sell 250 units?

Chapter 6 Project

INEQUALITY OF INCOME DISTRIBUTION

The fact that income is distributed unequally in our economy is pretty obvious. Yet how can this inequality be measured? Certainly it would be useful to track some measure of inequality over time to see whether the inequality was getting better or worse. One way of measuring inequality of income distribution is to construct a function known as a **Lorenz** curve. To find this curve, we first rank every household in order of increasing income. The point $\left(\dfrac{x}{100}, \dfrac{x}{100}\right)$ is on the Lorenz curve if the bottom x % of the households receives y% of the income. For example, since the bottom 0% of the households receives 0% of the income and the bottom 100%

of the households receive 100% of the income, the points (0, 0) and (1, 1) must lie on every Lorenz curve. If income distribution were perfectly equal, then the bottom x % of the households would receive x% of the income, and the point (x, x) would be on the the Lorenz curve for any x. That is, if income distribution were perfectly equal, the Lorenz curve would be the line $y = x$.

But the income distribution in the American economy is not equal. According to the U.S. Census Bureau, household income in the United States in 2001 was distributed according to the following table:

2001 Percent of Income Earned by Bottom . . .				
20%	40%	60%	80%	95%
3.5%	12.2%	26.8%	49.8%	77.6%

Source: U.S. Census Bureau.

The Lorenz curve for 2001 contains the points (0, 0), (0.2, 0.035), (0.4, 0.122), (0.6, 0.268), (0.8, 0.498), (0.95, 0.776), and (1,1). We do not know other points on the Lorenz curve, but we can play "connect the dots" to get a realistic Lorenz curve:

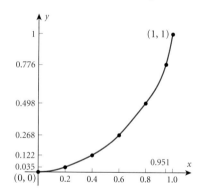

In fact, we can approximate this Lorenz curve by the polynomial function

$L(x) = 0.297x^2 + 7.053x^3 - 27.988x^4 + 36.115x^5 + 6.87x^6 - 45.13x^7 + 23.783x^8$.

Once we have the Lorenz curve, we can produce a number, called the **Gini coefficient,** which measures the inequality in the distribution of income. The Gini coefficient is defined to be the area between the line $y = x$ and the Lorenz curve $y = L(x)$ divided by the area under the line $y = x$. Using the figure, the Gini coefficient is

$G = \dfrac{\text{Area of the green region}}{\text{Area of the green region} + \text{Area of the blue region}}$

1. Express the area of the green region plus the area of the blue region as an integral.
2. Express the area of the blue region as an integral.

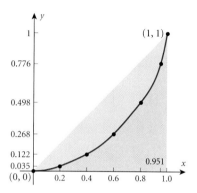

3. Show that the Gini coefficient is $G = 1 - 2\int_0^1 L(x)\,dx$.
4. Use the formula found in Problem 3 to find the Gini coefficient for the U.S. economy in 2001.
5. In 1997, income in the United States was distributed according to the following table.

1977 Percent of Income Earned by Bottom . . .				
20%	40%	60%	80%	95%
3.6%	12.5%	27.5%	50.7%	78.3%

In 1993, income in the United States was distributed according to the following table.

1993 Percent of Income Earned by Bottom . . .				
20%	40%	60%	80%	95%
3.6%	12.6%	27.7%	51.2%	79.0%

Just by looking at the distributions in the tables for 1993, 1997, and 2001, does it look like incomes are becoming less equally distributed or more equally distributed over this time span?

6. The Lorenz curve for 1997 may be approximated by

$L(x) = 0.423x^2 + 6.083x^3 - 25.006x^4 + 32.938x^5 + 5.920x^6 - 41.316x^7 + 21.958x^8$.

Find the Gini coefficient for the U.S. economy in 1997.

484 Chapter 6 The Integral of a Function and Applications

7. The Lorenz curve for 1993 may be approximated by
$$L(x) = .442x^2 + 5.8x^3 - 23.71x^4 + 31.036x^5 + 5.71x^6 - 38.842x^7 + 20.564x^8.$$
Find the Gini coefficients for the U.S. economy in 1993.

8. Compare the Gini coefficients you found in Problems 4, 6, and 7. Are incomes becoming less equally distributed or more equally distributed? Explain.

MATHEMATICAL QUESTIONS FROM PROFESSIONAL EXAMS*

1. **Actuary Exam—Part I** $\int_1^e \frac{1}{x} \ln x \, dx = ?$

   (a) $\dfrac{1}{e}$ (b) $\dfrac{1}{2}$ (c) 1

   (d) e (e) e^2

2. **Actuary Exam—Part I** $\int_0^1 x \ln x \, dx = ?$

   (a) $-\infty$ (b) -2 (c) -1

   (d) $-\dfrac{1}{4}$ (e) $-\dfrac{2}{9}$

3. **Actuary Exam—Part I** If $\int_1^b f(x) \, dx = b^2 e^b - e$ for all $b > 0$, then for all $x > 0$, $f(x) = ?$

   (a) $x^2 e^x$ (b) $\dfrac{x^3}{3} e^x$ (c) $x^2 e^x + 2xe^x$

   (d) $2xe^x$ (e) $x^2 e^x - e^{x-1}$

4. **Actuary Exam—Part I** If the area of the region bounded by $y = f(x)$, the x-axis, and the lines $x = a$ and $x = b$ is given by $\int_a^b f(x) \, dx$, which of the following must be true?

   (a) $a < b$ and $f(x) > 0$ (b) $a < b$ and $f(x) < 0$
   (c) $a > b$ and $f(x) > 0$ (d) $a > b$ and $f(x) < 0$
   (e) None of the above

5. **Actuary Exam—Part I** The rate of change of the population of a town in Pennsylvania at any time t is proportional to the population at time t. Four years ago, the population was 25,000. Now, the population is 36,000. Calculate what the population will be six years from now.

   (a) 43,200   (b) 52,500   (c) 62,208
   (d) 77,760   (e) 89,580

6. **Actuary Exam—Part I** For $x > 0$, $\int \dfrac{\ln (x^2)}{x} \, dx = ?$

   (a) $\ln x^2 - \ln x + C$ (b) $\dfrac{1}{2} (\ln x)^2 + C$

   (c) $(\ln x)^2 + C$ (d) $\ln (\ln x) + C$

   (e) $\dfrac{1}{2} \ln (\ln x) + C$

*Copyright © 1998, 1999 by the Society of Actuaries. Reprinted with Permission.

CHAPTER 7

Other Applications and Extensions of the Integral

OUTLINE

7.1 Improper Integrals

7.2 Average Value of a Function

7.3 Continuous Probability Functions

- **Chapter Review**
- **Chapter Project**
- **Mathematical Questions from Professional Exams**

It's Thanksgiving break and you are headed for the airport to fly home for turkey dinner. When you get to the airport, the check-in lines are backed up. As you wait in line, you wonder whether the number of people arriving equals the number that have been waited on. Or are the arrivals coming faster than they can be processed, making the lines even longer? If the airline knew how many people to expect in a given period of time, then additional employees could be used to handle high demand and fewer employees would be used when demand was low. How could the airline find out? The Chapter Project gives some insight into arrivals and how to predict waiting time.

Chapter 7 Other Applications and Extensions of the Integral

A LOOK BACK, A LOOK FORWARD

In Chapter 6 we discussed the integral of a function, defining the indefinite integral and the definite integral. We gave applications in geometry (finding the area under the graph of a function), in business (marginal analysis, sales over time, learning curves, and maximizing profit over time), and in economics (consumer's surplus and producer's surplus). We also used integration to solve differential equations.

In this chapter we give additional applications of the integral, beginning with a discussion of improper integrals, an extension of the definite integral.

7.1 Improper Integrals

PREPARING FOR THIS SECTION *Before getting started, review the following:*

> The Definite Integral (Chapter 6, Section 6.4, pp. 442–446)
> Infinite Limits; Limits at Infinity (Chapter 3, Section 3.4, pp. 258–260)
> Continuous Functions (Chapter 3, Section 3.3, pp. 251–255)

OBJECTIVE   1 Evaluate improper integrals

Recall that in defining $\int_a^b f(x)\,dx$ two basic assumptions are made:

1. The limits of integration a and b are both finite.
2. The function is continuous on $[a, b]$.

In many situations, one or both of these assumptions are not met. For example, one of the limits of integration might be infinity; or the function $y = f(x)$ might be discontinuous at some number in $[a, b]$. If either of the conditions (1) and (2) are not satisfied, then $\int_a^b f(x)\,dx$ is called an **improper integral.**

One Limit of Integration Is Infinite

We begin with an example.

EXAMPLE 1   Finding an Area

Find the area under the graph of $f(x) = \dfrac{1}{x^2}$ to the right of $x = 1$.

SOLUTION   First we graph $f(x) = \dfrac{1}{x^2}$. See Figure 1. The area to the right of $x = 1$ is shaded. To find this area, we pick a number b to the right of $x = 1$. The area under the graph of $f(x) = \dfrac{1}{x^2}$ from $x = 1$ to $x = b$ is

$$\int_1^b \frac{1}{x^2}\,dx = \left(-\frac{1}{x}\right)\Big|_1^b = -\frac{1}{b} + 1$$

This area depends on the choice of b. Now the area we seek is obtained by letting $b \to \infty$. Since

$$\lim_{b \to \infty}\left(-\frac{1}{b} + 1\right) = 1$$

FIGURE 1

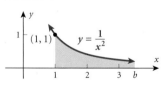

we conclude that the area under the graph of $f(x) = x^2$ to the right of $x = 1$ is 1.

The area we found in Example 1 can be represented symbolically by the *improper integral*

$$\int_1^\infty \frac{1}{x^2}\, dx$$

and it can be evaluated by finding

$$\lim_{b\to\infty} \int_1^b \frac{1}{x^2}\, dx$$

This leads us to formulate the following definition.

Improper Integral

Suppose a function f is continuous on the interval $[a, \infty)$. The **improper integral**, $\int_a^\infty f(x)\, dx$, is defined as

$$\int_a^\infty f(x)\, dx = \lim_{b\to\infty} \int_a^b f(x)\, dx$$

provided this limit exists and is a real number. If this limit does not exist or if it is infinite, the improper integral has no value.

Suppose a function f is continuous on the interval $(-\infty, b]$. The **improper integral**, $\int_{-\infty}^b f(x)\, dx$, is defined as

$$\int_{-\infty}^b f(x)\, dx = \lim_{a\to-\infty} \int_a^b f(x)\, dx$$

provided that this limit exists and is a real number. If this limit does not exist or if it is infinite, the improper integral has no value.

EXAMPLE 2 Evaluating Improper Integrals

Find the value, if there is one, of

(a) $\displaystyle\int_{-\infty}^0 e^x\, dx$ **(b)** $\displaystyle\int_1^\infty \frac{1}{x}\, dx$

SOLUTION **(a)** $\displaystyle\int_{-\infty}^0 e^x\, dx = \lim_{a\to-\infty} \int_a^0 e^x\, dx = \lim_{a\to-\infty} e^x \Big|_a^0$

$= \lim_{a\to-\infty} [1 - e^a] = 1 - \lim_{a\to-\infty} e^a = 1$

$\uparrow$

$\lim_{a\to-\infty} e^a = 0$

(b) $\displaystyle\int_1^\infty \frac{1}{x}\, dx = \lim_{b\to\infty} \int_1^b \frac{1}{x}\, dx = \lim_{b\to\infty} \ln x \Big|_1^b = \lim_{b\to\infty} [\ln b - \ln 1] = \lim_{b\to\infty} \ln b = \infty$

Since the limit is infinite, $\int_1^\infty \frac{1}{x} dx$ has no value.

NOW WORK PROBLEM 7.

The Integrand Is Discontinuous

The improper integrals studied in the above examples had infinity either as a lower limit of integration or as an upper limit of integration. A second type of improper integral occurs when the integrand f in $\int_a^b f(x)\, dx$ is discontinuous at either a or b, where a and b are both real numbers. Again, we illustrate the technique for evaluating this type of improper integral by an example.

EXAMPLE 3 Evaluating an Improper Integral

Evaluate: $\int_0^4 \frac{1}{\sqrt{x}}\, dx$

SOLUTION The function $f(x) = \frac{1}{\sqrt{x}}$ is not continuous at $x = 0$, but is continuous on the interval $(0, \infty)$. To evaluate the improper integral $\int_0^4 \frac{1}{\sqrt{x}}\, dx$ we proceed as follows:

$$\int_0^4 \frac{1}{\sqrt{x}}\, dx = \lim_{t \to 0^+} \int_t^4 x^{-1/2}\, dx = \lim_{t \to 0^+} \frac{x^{1/2}}{\frac{1}{2}}\bigg|_t^4$$

$$= \lim_{t \to 0^+} (2\sqrt{4} - 2\sqrt{t}) = 4 - \lim_{t \to 0^+} 2\sqrt{t} = 4$$

This leads to the following definition.

Improper Integral

If f is continuous on $[a, b)$, but is discontinuous at b, we define the **improper integral**, $\int_a^b f(x)\, dx$, to be

$$\boxed{\int_a^b f(x)\, dx = \lim_{t \to b^-} \int_a^t f(x)\, dx}$$

provided that this limit exists and is a real number. If this limit does not exist or if it is infinite, then $\int_a^b f(x)\, dx$ has no value.

If f is continuous on $(a, b]$, but is discontinuous at a, we define the **improper integral**, $\int_a^b f(x)\, dx$, to be

$$\boxed{\int_a^b f(x)\, dx = \lim_{t \to a^+} \int_t^b f(x)\, dx}$$

provided that this limit exists as a real number. If this limit does not exist or if it is infinite, then $\int_a^b f(x)\,dx$ has no value.

Observe that when the upper limit is the point of discontinuity, the limit is left-handed and when the lower limit is the point of discontinuity, the limit is right-handed.

NOW WORK PROBLEM 13.

EXAMPLE 4 Finding an Area

Find the area A between the graph of $f(x) = \dfrac{1}{\sqrt[3]{x}}$ and the x-axis from $x = -1$ to $x = 0$.

SOLUTION We begin by graphing $f(x) = \dfrac{1}{\sqrt[3]{x}}$. See Figure 2. Since the area lies below the x-axis, the area is $A = -\int_{-1}^{0} \dfrac{1}{\sqrt[3]{x}}\,dx$.

FIGURE 2

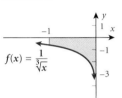

We observe that the integrand $f(x) = \dfrac{1}{\sqrt[3]{x}}$ is discontinuous at $x = 0$ (the upper limit of integration). So,

$$A = -\int_{-1}^{0} \dfrac{1}{\sqrt[3]{x}}\,dx = -\lim_{t \to 0^-} \int_{-1}^{t} \dfrac{1}{\sqrt[3]{x}}\,dx = -\lim_{t \to 0^-} \int_{-1}^{t} x^{-1/3}\,dx$$

$$= -\lim_{t \to 0^-} \dfrac{x^{2/3}}{\tfrac{2}{3}} \bigg|_{-1}^{t} = -\lim_{t \to 0^-} \dfrac{3}{2}(t^{2/3} - 1) = \dfrac{3}{2}$$

The area is $\dfrac{3}{2}$ square units. ▶

There are other types of improper integrals besides those discussed here. See Problems 20 and 21.

EXERCISE 7.1

In Problems 1–6, determine which of the integrals are improper. Explain why they are, or are not, improper.

1. $\displaystyle\int_0^{\infty} x^2\,dx$
2. $\displaystyle\int_2^3 \dfrac{dx}{x-1}$
3. $\displaystyle\int_0^1 \dfrac{1}{x}\,dx$
4. $\displaystyle\int_{-1}^1 \dfrac{x}{x^2+1}\,dx$
5. $\displaystyle\int_1^2 \dfrac{dx}{x-1}$
6. $\displaystyle\int_0^1 \dfrac{x}{x^2-1}\,dx$

In Problems 7–14, find the value, if any, of each improper integral.

7. $\displaystyle\int_1^{\infty} e^{-4x}\,dx$
8. $\displaystyle\int_{-\infty}^{-1} \dfrac{1}{x^3}\,dx$
9. $\displaystyle\int_0^{\infty} \sqrt{x}\,dx$
10. $\displaystyle\int_0^{\infty} xe^{-x}\,dx$
11. $\displaystyle\int_{-1}^{0} \dfrac{1}{\sqrt[5]{x}}\,dx$
12. $\displaystyle\int_2^4 \dfrac{x\,dx}{\sqrt{x^2-4}}$
13. $\displaystyle\int_0^1 \dfrac{1}{x}\,dx$
14. $\displaystyle\int_0^1 \dfrac{\ln x}{x}\,dx$

15. Find the area, if it exists, under the graph of $f(x) = \dfrac{1}{\sqrt{x}}$ from $x = 0$ to $x = 1$.

16. Find the area, if it exists, under the graph of $f(x) = \sqrt{x}$ to the right of $x = 0$.

17. **Capital Value of Rental Property** The capital value of a rental property, assuming it will last indefinitely, is given by the integral $\int_0^\infty Re^{-kt}\,dt$, where R is the annual rent and k is the current annual rate of interest on investments. Find the capital value of a typical apartment in the South Center Township area of Indianapolis, where annual rent is $5124 and the current rate of interest on investments is 5%.

    Source: CB Richard Ellis-Indianapolis.

18. **Waiting Time** The probability of waiting at least x minutes for Amtrak trains to arrive in Washington, D.C. between the hours of 7:30 A.M. and 1:30 A.M. is given by the integral $\int_x^\infty \frac{7}{3} e^{-7/3\,t}\,dt$, where x is measured in hours. Determine the probability that one will have to wait at least one hour for an Amtrak train to arrive.

    Source: Amtrak.

19. **Reaction Rate of a Drug** The rate of reaction to a given dose of a drug at time t hours after administration is given by $r(t) = te^{-t^2}$ (measured in appropriate units).

    (a) Why is it reasonable to define the **total reaction** as the area under the curve $y = r(t)$ from $t = 0$ to $t = \infty$?

    (b) Evaluate the total reaction to the given dose of the drug.

20. If $y = f(x)$ is continuous, the improper integral $\int_{-\infty}^\infty f(x)\,dx$ is defined as
    $$\int_{-\infty}^\infty f(x)\,dx = \int_{-\infty}^0 f(x)\,dx + \int_0^\infty f(x)\,dx$$
    provided that each of the improper integrals on the right has a value. Use this definition to find the value, if it exists, of

    (a) $\int_{-\infty}^\infty e^x\,dx$ (b) $\int_{-\infty}^\infty \frac{x\,dx}{(x^2+1)^2}$

21. If $y = f(x)$ is continuous on $[a, b]$, except at a point c, $a < c < b$, the integral $\int_a^b f(x)\,dx$ is improper and is defined by
    $$\int_a^b f(x)\,dx = \int_a^c f(x)\,dx + \int_c^b f(x)\,dx$$
    provided that each of the improper integrals on the right has a value. Use this definition to evaluate

    (a) $\int_{-1}^1 \frac{1}{x^2}\,dx$ (b) $\int_0^4 \frac{x\,dx}{\sqrt[3]{x^2-4}}$

7.2 Average Value of a Function

PREPARING FOR THIS SECTION *Before getting started, review the following:*

>> Approximating Definite Integrals
(Chapter 6, Section 6.6, pp. 463–468)

OBJECTIVES 1 Find the average value of a function
2 Use average value in applications

At the U.S. Weather Bureau, a continuous reading of the temperature over a 24-hour period is taken daily. To obtain the average daily temperature, 12 readings may be taken at 2-hour intervals beginning at midnight (0) and ending at 10 p.m. (22): $f(0), f(2), f(4), \ldots, f(20), f(22)$. The average temperature is then calculated as

$$\frac{f(0) + f(2) + f(4) + \cdots + f(20) + f(22)}{12}$$

This number represents a good approximation to the true average as long as there were no drastic temperature changes over any of the 2-hour intervals.

To improve the approximation, readings could be taken every hour. The average in this case would be

$$\frac{f(0) + f(1) + \cdots + f(22) + f(23)}{24}$$

An even better approximation would be obtained if readings were recorded every half hour.

In general, if $y = f(x)$ is a continuous function defined on the closed interval $[a, b]$, we can obtain the *average of f on* $[a, b]$ as follows:
Partition the interval $[a, b]$ into n subintervals

$$[a, x_1], \quad [x_1, x_2], \ldots, [x_{k-1}, x_k], \ldots, [x_{n-1}, b]$$

each of equal length $\Delta x = \dfrac{b-a}{n}$. This is the norm Δ of the partition. Pick a number in each subinterval and let these numbers be $u_1, u_2, \ldots, u_n$. An approximation of the average value of f over the interval $[a, b]$ is then the sum

$$\frac{f(u_1) + f(u_2) + \cdots + f(u_n)}{n} \tag{1}$$

If we multiply and divide the expression in (1) by $b - a$, we get

$$\frac{f(u_1) + f(u_2) + \cdots + f(u_n)}{n} = \frac{1}{b-a}\left[f(u_1)\frac{b-a}{n} + f(u_2)\frac{b-a}{n} + \cdots + f(u_n)\frac{b-a}{n}\right]$$

$$= \frac{1}{b-a}[f(u_1)\Delta x + f(u_2)\Delta x + \cdots + f(u_n)\Delta x]$$

This sum gives an approximation to the average value. As the norm $\Delta \to 0$, it provides a better and better approximation to the average value of f on $[a, b]$. Since this sum is a Riemann sum, its limit is a definite integral. This suggests the following definition:

Average Value of a Function over an Interval

The **average value** AV of a continuous function f over the interval $[a, b]$ is

$$AV = \frac{1}{b-a}\int_a^b f(x)\, dx \tag{2}$$

1 EXAMPLE 1 Finding the Average Value of a Function

The average value of $f(x) = x^3$ over the interval $[0, 2]$ is

$$AV = \frac{1}{2-0}\int_0^2 x^3\, dx = \frac{1}{2}\left.\frac{x^4}{4}\right|_0^2 = \frac{1}{2}(4) = 2$$

NOW WORK PROBLEM 1.

Geometric Interpretation

The average value AV of a function f, as defined in (2), has an interesting geometric interpretation. If we rearrange the formula for AV, we obtain

$$(AV)(b-a) = \int_a^b f(x)\, dx \tag{3}$$

If $f(x) \geq 0$ on $[a, b]$, the right side of (3) represents the area under the graph of $y = f(x)$, from $x = a$ to $x = b$. The left side of the equation can be interpreted as the

FIGURE 3

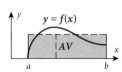

area of a rectangle of height AV and base $b - a$. As a result, equation (3) asserts that the average value of the function is the height of a rectangle with base $b - a$ and area equal to the area under the graph of f. See Figure 3.

Application

EXAMPLE 2 Using the Average Value of a Function

Suppose the current world population is $6 \cdot 10^9$ and the population in t years is assumed to grow exponentially at a 2% growth rate according to the law

$$P(t) = (6 \cdot 10^9)e^{0.02t}$$

What will be the average world population during the next 20 years?

SOLUTION The average value of the population during the next 20 years is

$$AV = \frac{1}{20 - 0} \int_0^{20} P(t)\, dt$$

$$= \frac{1}{20} \int_0^{20} (6 \cdot 10^9) e^{0.02t}\, dt$$

$$= \frac{6 \cdot 10^9}{20} \int_0^{20} e^{0.02t}\, dt$$

$$= 3 \cdot 10^8 \left. \frac{e^{0.02t}}{0.02} \right|_0^{20}$$

$$= 3 \cdot 10^8 \frac{(e^{0.4} - 1)}{0.02}$$

$$\approx 7.38 \cdot 10^9$$

▶

NOW WORK PROBLEM 11.

EXERCISE 7.2

In Problems 1–10, find the average value of the function f over the given interval.

1. $f(x) = x^2$, over $[0, 1]$
2. $f(x) = 2x^2$, over $[-4, 2]$
3. $f(x) = 1 - x^2$, over $[-1, 1]$
4. $f(x) = 16 - x^2$, over $[-4, 4]$
5. $f(x) = 3x$, over $[1, 5]$
6. $f(x) = 4x$, over $[-5, 5]$
7. $f(x) = -5x^4 + 4x - 10$, over $[-2, 2]$
8. $f(x) = 10x^4 - 2x + 7$, over $[-1, 2]$
9. $f(x) = e^x$, over $[0, 1]$
10. $f(x) = e^{-x}$, over $[0, 1]$

11. **Population Prediction** Rework Example 2 if the population function is given by $P(t) = (6 \cdot 10^9)e^{0.03t}$. (This is a 3% growth rate.)

12. **Population Prediction** Rework Example 2 if the growth rate is 1%.

13. **Average Temperature of a Rod** A rod 3 meters long is heated to $25x$ degrees Celsius, where x is the distance (in meters) from one end of the rod. Calculate the average temperature of the rod.

14. **Average Rainfall** The rainfall per day, measured in centimeters, x days after the beginning of the year is $0.00002(6511 + 366x - x^2)$. Estimate the average daily rainfall for the first 180 days of the year.

15. **Average Speed** A car starting from rest accelerates at the rate of 3 meters per second. Find its average speed over the first 8 seconds.

16. **Average Area** What is the average area of all circles with radii between 1 and 3 meters?

17. **Average Annual Revenue** The annual revenue for the Exxon-Mobil Corporation between 1997 and 2002 is given by $R(x) = -4.43x^3 + 46.17x^2 - 132.5x + 290$, where x is the number of years since 1996 and $R(x)$ is in billions of dollars. Find the average annual revenue during that time.

    Source: Exxon-Mobil Corporation Financial Report 2002.

18. **Deaths from Automobile Accidents** Find the average number of annual deaths from automobile accidents in the United States from the period 1990–1999 if the annual number of deaths is given by $y = 38929 + 1443 \ln x$, where x is the number of years since 1989.

    Source: National Highway Traffic Safety Administration's National Center for Statistical Analysis.

19. **Average Rainfall in Baton Rouge** For the first 90 days of the year the average daily rainfall in Baton Rouge, Louisiana is given by $r(x) = -0.000414x + 0.206748$, with $r(x)$ in inches and x the number of days since the beginning of the year. Find the average rainfall per day for the first 90 days of the year.

    Source: National Weather Service.

7.3 Continuous Probability Functions

PREPARING FOR THIS SECTION *Before getting started, review the following:*

>> Area Under a Graph (Chapter 6, Section 6.5, pp. 450–457)

OBJECTIVES
1. Verify probability density functions
2. Find probabilities using a probability density function
3. Use the uniform density function
4. Use the exponential density function
5. Find expected value

TABLE 1

	Number of Heads	Probability
HHH	3	$\frac{1}{8}$
HHT	2	$\frac{1}{8}$
HTH	2	$\frac{1}{8}$
THH	2	$\frac{1}{8}$
HTT	1	$\frac{1}{8}$
THT	1	$\frac{1}{8}$
TTH	1	$\frac{1}{8}$
TTT	0	$\frac{1}{8}$

Random Variables

Intuitively, a *random variable* is a quantity that is measured in connection with a random experiment. For example, if the random experiment involves weighing individuals, then the weights of the individuals would be random variables. As another example, if the random experiment is to determine the time between arrivals of customers at a gas station, then the time between arrivals of customers at the gas station would be a random variable.

Let's consider some examples that demonstrate how to obtain random variables from random experiments.

When we perform a simple experiment, we are often interested not in a particular outcome, but rather in some number associated with that outcome. For example, in tossing a coin three times, we may be interested in the number of heads obtained, regardless of the particular sequence in which the heads appear. Similarly, the gambler throwing a pair of dice is generally more interested in the sum of the faces than in the particular number on each face.

Table 1 summarizes the results of the simple experiment of flipping a fair coin three times. The first column in the table gives a sample space of this experiment. The second column shows the number of heads for each outcome, and the third column shows the probability associated with each outcome.

Suppose that in this experiment we are interested only in the total number of heads. This information is given in Table 2.

The role of the random variable is to transform the original sample space {*HHH*, *HHT*, *HTH*, *HTT*, *THH*, *THT*, *TTH*, *TTT*} into a new sample space that consists of the

TABLE 2

Number of Heads Obtained in Three Flips of a Coin	Probability
0	$\frac{1}{8}$
1	$\frac{3}{8}$
2	$\frac{3}{8}$
3	$\frac{1}{8}$

number of heads that occur: {0, 1, 2, 3}. If X denotes the random variable, then X may take on any of the values 0, 1, 2, 3. From Table 2, the probability that the random variable X assumes the value 2 is

$$\text{Probability } (X = 2) = \frac{3}{8}$$

Also,

$$\text{Probability } (X = 5) = 0$$

We see that a random variable indicates the rule of correspondence between any member of a sample space and a number assigned to it. Because of this, a random variable is a function.

> **Random Variable**
>
> A **random variable** is a function that assigns a numerical value to each outcome of a sample space S.

We shall use the capital letter X to represent a random variable. In the coin-flipping example, the random variable X is

$$X(HHH) = 3 \quad X(HHT) = 2 \quad X(HTH) = 2 \quad X(THH) = 2$$
$$X(HTT) = 1 \quad X(THT) = 1 \quad X(TTH) = 1 \quad X(TTT) = 0$$

The random variable X indicates a relationship between the first two columns of Table 1 and pairs each outcome of the experiment with the real numbers 0, 1, 2, or 3.

Random variables fall into two classes: those related to *discrete sample spaces* and those associated with *continuous sample spaces*.

> **Discrete Random Variable**
>
> A sample space is discrete if it contains a finite number of outcomes or as many outcomes as there are counting numbers. A random variable is said to be **discrete** if it is defined over a discrete sample space.

> **Continuous Random Variable**
>
> Whenever a random variable has values that consist of an entire interval of real numbers, it is called a **continuous random variable.** In such cases, we also say the sample space is **continuous.**

Any practical problem that measures variables such as height, weight, time, and age will utilize a continuous random variable. As a result, the sample space associated with such experiments is continuous.

Probability Functions

The function that has as its domain the value of a random variable and has as its range the corresponding probability, is called a **probability function.**

For example, look again at Table 2. The left column contains values of the random variable and the right column contains the corresponding probability. The probability function f for this experiment has domain $\{0, 1, 2, 3\}$ and range $\left\{\dfrac{1}{8}, \dfrac{3}{8}\right\}$. Moreover,

$$f(0) = Pr(X = 0) = \dfrac{1}{8} \qquad f(1) = Pr(X = 1) = \dfrac{3}{8}$$
$$f(2) = Pr(X = 2) = \dfrac{3}{8} \qquad f(3) = Pr(X = 3) = \dfrac{1}{8}$$

EXAMPLE 1 Finding the Value of a Probability Function

In the experiment of one toss of two fair dice, compute the value of the probability function f at $x = 7$ where x is the sum of the dots shown on the faces of the dice. Construct a table that shows every value of f.

SOLUTION The domain of the probability function f for this experiment is $\{2, 3, 4, 5, 6, 7, 8, 9, 10, 11, 12\}$. The value of f at 7 equals the probability that the sum of the faces equals 7. That is,

$$f(7) = Pr(X = 7) = Pr\{(1, 6), (2, 5), (3, 4), (4, 3), (5, 2), (6, 1)\} = \dfrac{6}{36} = \dfrac{1}{6}$$

We compute the values $f(2), f(3), \ldots, f(12)$ of f in a similar way. These values are given in the following table:

Values of x	2	3	4	5	6	7	8	9	10	11	12
Probability Function of x, $f(x)$	$\dfrac{1}{36}$	$\dfrac{2}{36}$	$\dfrac{3}{36}$	$\dfrac{4}{36}$	$\dfrac{5}{36}$	$\dfrac{6}{36}$	$\dfrac{5}{36}$	$\dfrac{4}{36}$	$\dfrac{3}{36}$	$\dfrac{2}{36}$	$\dfrac{1}{36}$

We distinguish between two types of probability function. A **discrete probability function** is one for which the random variable is discrete. A **continuous probability function** is one for which the random variable is continuous. We shall discuss only continuous probability functions here.

Probability Density Functions

Continuous probability functions have as their domain the values that a continuous random variable can assume. For example, if a random experiment involves weighing individuals, the weight of each individual is a continuous random variable. Likewise, in the random experiment of determining the time between arrival of customers at a gas station, the time between arrivals is a continuous random variable.

Suppose the distribution of a population by age is given by data grouped in 10-year intervals. See Figure 4.

In this illustration there are 30 million people in the age group between 0 and 10, 40 million between 10 and 20, and so on. We also know that the total population is 200 million.

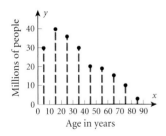

FIGURE 4

FIGURE 5

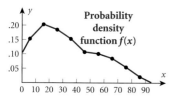

FIGURE 6

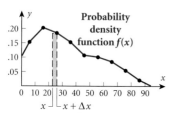

Since there are 40 million people in the age group 10–20, the probability that a person chosen at random is in this age group is $\frac{40}{200} = .20$. Figure 5 illustrates the distribution of probabilities for each age group. The function constructed by connecting the probability values by a smooth curve is an example of a *probability density function*.

When probabilities are associated with intervals, it is reasonable to assume that their values depend not only on the lengths of the intervals but also on their locations. For instance, there is no reason why the probability that a person is in the age group 10–20 should equal the probability that a person is in the age group 60–70, even though the two intervals have the same length. If we assume that there exists a function f with the values $f(x)$, then the probability that a person will be in a certain age group on a small interval from x to $x + \Delta x$ is approximately $f(x) \Delta x$. This approximation is given by the area of the shaded rectangle shown in Figure 6, where the height of the rectangle is $f(x)$ and the base is Δx.

In a similar manner, we obtain the probabilities of other age groups by computing the areas corresponding to different subintervals. The desired probability for the whole interval is approximately the sum of the areas

$$f(x_1)\, \Delta x + f(x_2)\, \Delta x + \cdots + f(x_n)\, \Delta x$$

where x_1 is in the first subinterval, x_2 is in the second, and so on. This leads us to the following definition.

> **Probability Density Function**
>
> A function f is a **probability density function** on the interval $[a, b]$ if two conditions are met:
>
> $$f(x) \geq 0 \quad \text{on } [a, b] \tag{1}$$
>
> and
>
> $$\int_a^b f(x)\, dx = 1 \tag{2}$$
>
> where the interval $[a, b]$, possibly the interval $(-\infty, \infty)$, contains all values that the random variable X can assume.

With condition (1) in mind, the rationale behind condition (2) becomes apparent. Since the interval $[a, b]$ contains all the values the random variable can assume, the probability that a random variable lies between a and b must equal 1.

Using the fact that the area under the graph of f is given by a definite integral, the probability that a person is between the ages c and d, denoted by $Pr(c \leq X \leq d)$, is given by

$$Pr(c \leq X \leq d) = \int_c^d f(x)\, dx$$

For example, for the density function illustrated in Figure 5, the probability that a person is between 22 and 24 years of age is

$$\int_{22}^{24} f(x)\, dx$$

The discussion presented here leads to the following result.

The probability that the outcome of an experiment results in a value of a random variable X between c and d is given by

$$Pr(c \leq X \leq d) = \int_c^d f(x)\, dx \qquad (3)$$

where $f(x)$ is the probability density function of the random variable.

EXAMPLE 1 Verifying a Probability Density Function

Show that the function $f(x) = \dfrac{3}{56}(5x - x^2)$ is a probability density function over the interval $[0, 4]$.

SOLUTION If f is indeed a probability density function, it has to satisfy conditions (1) and (2).
Condition (1) is satisfied since

$$f(x) = \frac{3}{56}(5x - x^2) = \frac{3}{56}x(5 - x) \geq 0$$

for all x in the interval $[0, 4]$.
To verify condition (2), we evaluate

$$\int_0^4 \frac{3}{56}(5x - x^2)\, dx = \frac{3}{56}\int_0^4 (5x - x^2)\, dx = \frac{3}{56}\left(\frac{5x^2}{2} - \frac{x^3}{3}\right)\bigg|_0^4 = \frac{3}{56}\left[\frac{(5)16}{2} - \frac{64}{3}\right] = \frac{3}{56}\left(\frac{56}{3}\right) = 1$$

Condition (2) is also satisfied.

As a result, $f(x) = \dfrac{3}{56}(5x - x^2)$, for x in $0 \leq x \leq 4$, is a probability density function.

NOW WORK PROBLEM 1.

EXAMPLE 2 Find Probabilities Using a Probability Density Function

Compute the probability that the random variable X with probability density function $f(x) = \dfrac{3}{56}(5x - x^2)$ assumes a value between 1 and 2.

SOLUTION To compute $Pr(1 \leq X \leq 2)$, we use equation (3):

$$Pr(1 \leq X \leq 2) = \int_1^2 \frac{3}{56}(5x - x^2)\, dx = \frac{3}{56}\left(\frac{5x^2}{2} - \frac{x^3}{3}\right)\bigg|_1^2$$

$$= \frac{3}{56}\left[\left(\frac{(5)(4)}{2} - \frac{8}{3}\right) - \left(\frac{5}{2} - \frac{1}{3}\right)\right] = \frac{3}{56} \cdot \frac{31}{6} = \frac{31}{112}$$

NOW WORK PROBLEM 25.

How do we obtain a probability density function f? For individual random experiments, it is possible to construct them as we indicated in the example on age probabilities. However, the construction of a probability density function is usually difficult and depends on the nature of the experiment. Fortunately, several relatively simple probability density functions are available that can be used to fit most random experiments. In every example we discuss, the probability density function is given.

Uniform Density Function

The **uniform density function,** or the **uniform distribution,** the simplest of probability density functions, is one in which the random variable assumes all its values with equal probability. The probability density function for a uniformly distributed random variable is

$$f(x) = \begin{cases} \dfrac{1}{b-a} & \text{if } a \leq x \leq b \\ 0 & \text{if } x < a \text{ or } x > b \end{cases}$$

The graph of f is given in Figure 7.

FIGURE 7

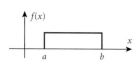

Notice that the function has the value 0 outside the interval $a \leq x \leq b$.

To verify that the uniform density function is a probability density function, we need to show that conditions (1) and (2) are satisfied. Since $b > a$, $f(x) = \dfrac{1}{b-a} > 0$ for $a \leq x \leq b$. Next,

$$\int_a^b f(x)\, dx = \int_a^b \frac{1}{b-a}\, dx = \frac{1}{b-a} \int_c^b dx = \frac{1}{b-a} \cdot x \Big|_a^b = \frac{1}{b-a} \cdot (b-a) = 1$$

The uniform density function satisfies conditions (1) and (2), and so is a probability density function.

3 EXAMPLE 3   Using the Uniform Density Function

Trains leave a terminal every 40 minutes. What is the probability that a passenger arriving at a random time to catch a train will have to wait more than 10 minutes? Use a uniform density function.

SOLUTION Let T (time) be a random variable and assume it is uniformly distributed for $0 \leq T \leq 40$. The probability that the passenger must wait more than 10 minutes is

$$\Pr(T \geq 10) = \int_{10}^{40} \frac{1}{40}\, dt = \frac{1}{40} \int_{10}^{40} dt = \frac{1}{40} \cdot t \Big|_{10}^{40} = \frac{1}{40}(40 - 10) = \frac{3}{4}$$

The probability is .75 that the passenger must wait more than 10 minutes. ▶

Exponential Density Function

Exponential Density Function

Let X be a continuous random variable. Then X is said to be **exponentially distributed** if X has the probability density function

$$f(x) = \begin{cases} \lambda e^{-\lambda x} & \text{if } x \geq 0 \\ 0 & \text{if } x < 0 \end{cases}$$

where λ* is a positive constant. The function f is called the **exponential density function.**

---

*The Greek letter lambda.

FIGURE 8

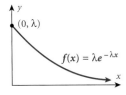

The graph of the exponential density function is given in Figure 8. The exponential density function is nonnegative for all x. Also,

$$\int_0^\infty f(x)\,dx = \int_0^\infty \lambda e^{-\lambda x}\,dx \underset{\text{Integral is improper}}{=} \lim_{b\to\infty}\int_0^b \lambda e^{-\lambda x}\,dx \underset{\substack{\text{Move }\lambda\text{ outside}\\ \text{the integral.}}}{=} \lim_{b\to\infty}\left[\lambda\int_0^b e^{-\lambda x}\,dx\right] \underset{\text{Integrate.}}{=} \lim_{b\to\infty}\left[\lambda\cdot\frac{1}{-\lambda}\cdot e^{-\lambda x}\Big|_0^b\right]$$

$$= \lim_{b\to\infty}[(-1)(e^{-\lambda b} - 1)] = 1$$

$$\lim_{b\to\infty} e^{-\lambda b} = 0,\ \lambda > 0$$

The exponential density function satisfies conditions (1) and (2) and so is a probability density function.

In general, any situation that deals with *waiting time* between successive events will lead to an exponential density function. In the exponential density function, the constant λ is the average number of arrivals per unit time and $\dfrac{1}{\lambda}$ is the average waiting time.

EXAMPLE 4 Using the Exponential Density Function

Airplanes arriving at an airport follow a pattern similar to the exponential density function with an average of $\lambda = 15$ arrivals per hour. Determine the probability of an arrival within 6 minutes (0.1 hour). Use an exponential density function.

SOLUTION If X is the random variable associated with this random experiment, the probability that $X \leq 0.1$ (6 minutes) is

$$Pr(X \leq 0.1) = \int_0^{0.1} 15 e^{-15x}\,dx = 15 \cdot \frac{1}{-15} e^{-15x}\Big|_0^{0.1} = -e^{-15x}\Big|_0^{0.1} = -e^{-1.5} + 1 = .777$$

The probability of an arrival within 6 minutes is .777.

 NOW WORK PROBLEM 27.

EXAMPLE 5 Waiting Time to Find a Defective Product

From past data it is known that a certain machine normally produces 1 defective product every hour (60 minutes). To detect defective products, an inspector walks up to the machine and tests a continuous stream of products until she detects a defective one. Use an exponential density function to find the probability that

(a) the inspector waits less than 30 minutes before finding a defective product.
(b) the inspector must wait more than an hour and 15 minutes (75 minutes) before finding a defective product.

SOLUTION The average defective rate is 1 every 60 minutes, so we let $\lambda = \dfrac{1}{60}$ in the exponential density function

$$f(x) = \lambda e^{-\lambda x}$$

Then $f(x) = \dfrac{1}{60} e^{-x/60}$, $x \geq 0$. Suppose X is the random variable associated with this random experiment.

(a) The inspector waits less than 30 minutes before finding a defective product with probability

$$\Pr(X \leq 30) = \int_0^{30} \frac{1}{60} e^{-x/60} \, dx = -e^{-x/60} \Big|_0^{30} = -e^{-1/2} + 1 = .393$$

The probability the inspector waits less than 30 minutes is .393.

(b) First we compute the probability the inspector waits less than 75 minutes before finding a defective product.

$$\Pr(X \leq 75) = \int_0^{75} \frac{1}{60} e^{-x/60} \, dx = -e^{-x/60} \Big|_0^{75} = -e^{-5/4} + 1 = .713$$

The probability the inspector waits more than 75 minutes is $1 - .713 = .287$. ▶

An alternative solution to part (b) of Example 5 is to evaluate

$$\Pr(X \geq 75) = \int_{75}^{\infty} \frac{1}{60} e^{-x/60} \, dx.$$

Try it and compare to the solution given above.

Expected Value

Expected Value for a Continuous Random Variable

If X is a continuous random variable with the probability density function $f(x)$, $a \leq x \leq b$, the **expected value of X** is

$$E(X) = \int_a^b x f(x) \, dx$$

The expected value of the random variable X is the average (mean) value of the random variable. For example, if X is a random variable measuring heights, then $E(X)$ is the average (mean) height of the population.

EXAMPLE 6 Finding the Expected Value

A passenger arrives at a train terminal where trains arrive every 40 minutes. Determine the expected waiting time using a uniform density function.

SOLUTION Let the random variable T measure waiting time with uniform density function $f(t) = \dfrac{1}{40}$, where $0 \leq T \leq 40$. The expected value $E(T)$ is then

$$E(T) = \int_0^{40} t \cdot \frac{1}{40} \, dt = \frac{1}{40} \int_0^{40} t \, dt = \frac{1}{40} \cdot \frac{t^2}{2} \Big|_0^{40} = \frac{1}{40} \cdot \frac{1600}{2} = 20 \text{ minutes}$$ ▶

NOW WORK PROBLEM 17.

EXAMPLE 7 The Expected Value of the Uniform Density Function

Show that the expected value of the random variable with a uniform density function
$$f(x) = \frac{1}{(b-a)}, \quad a \leq x \leq b, \quad \text{is } \frac{a+b}{2}.$$

SOLUTION Let X be a random variable with probability density function $f(x) = \frac{1}{b-a}$, $a \leq x \leq b$. Then

$$E(X) = \int_a^b x \frac{1}{b-a} dx = \frac{1}{b-a} \int_a^b x \, dx = \frac{1}{b-a} \cdot \frac{x^2}{2} \Big|_a^b$$

$$= \frac{1}{b-a} \cdot \left(\frac{b^2}{2} - \frac{a^2}{2} \right) = \frac{1}{2} \cdot \frac{b^2 - a^2}{b-a} = \frac{1}{2} \cdot \frac{(b-a)(b+a)}{b-a} = \frac{a+b}{2}$$

EXERCISE 7.3

In Problems 1–8, verify that each function is a probability density function over the indicated interval.

1. $f(x) = \frac{1}{2}$ over $[0, 2]$
2. $f(x) = \frac{1}{5}$ over $[0, 5]$
3. $f(x) = 2x$ over $[0, 1]$
4. $f(x) = \frac{1}{8}x$ over $[0, 4]$
5. $f(x) = \frac{3}{250}(10x - x^2)$ over $[0, 5]$
6. $f(x) = \frac{6}{27}(3x - x^2)$ over $[0, 3]$
7. $f(x) = \frac{1}{x}$ over $[1, e]$
8. $f(x) = \frac{4}{3(x+1)^2}$ over $[0, 3]$

If $f(x) \geq 0$ is not a probability density function, we can find a constant k such that $kf(x)$ satisfies the condition $\int_a^b kf(x) \, dx = 1$. For the functions in Problems 9–16, determine the constant k that will make each one a probability density function over the interval indicated.

9. $f(x) = 1$ over $[0, 3]$
10. $f(x) = 1$ over $[0, 4]$
11. $f(x) = x$ over $[0, 2]$
12. $f(x) = x$ over $\left[0, \frac{1}{2}\right]$
13. $f(x) = 10x - x^2$ over $[0, 5]$
14. $f(x) = 10x - x^2$ over $[0, 8]$
15. $f(x) = \frac{1}{x}$ over $[1, 2]$
16. $f(x) = \frac{1}{(x+1)^3}$ over $[3, 7]$

In Problems 17–24, compute the expected value for each probability density function.

17. $f(x) = \frac{1}{2}$ over $[0, 2]$
18. $f(x) = \frac{1}{5}$ over $[0, 5]$
19. $f(x) = 2x$ over $[0, 1]$
20. $f(x) = \frac{1}{8}x$ over $[0, 4]$
21. $f(x) = \frac{3}{250}(10x - x^2)$ over $[0, 5]$
22. $f(x) = \frac{6}{27}(3x - x^2)$ over $[0, 3]$
23. $f(x) = \frac{1}{x}$ over $[1, e]$
24. $f(x) = \frac{4}{3(x+1)^2}$ over $[0, 3]$

25. A number x is selected at random from the interval $[0, 5]$. The probability density function for x is

$$f(x) = \frac{1}{5} \quad \text{for } 0 \leq x \leq 5$$

Find the probability that a number is selected in the subinterval $[1, 3]$.

26. A number x is selected at random from the interval $[0, 10]$. The probability density function for x is

$$f(x) = \frac{1}{10} \quad \text{for } 0 \leq x \leq 10$$

Find the probability that a number is selected in the subinterval $[6, 9]$.

27. **Time Between Telephone Calls** The time between incoming telephone calls at a hotel switchboard has an exponential density function with $\lambda = 0.5$ minute. What is the probability that there is an interval of at least 6 minutes between incoming calls?

28. **Waiting Time on a Phone Call** The length of the wait to speak to a customer representative at an airline is a random variable X. The average time a person is on hold is 6 minutes and follows an exponential distribution.

   (a) Write the exponential density function which describes a caller's wait time.
   (b) What is the probability that a caller is on hold between 5 and 10 minutes?
   (c) What is the probability that a caller waits less than 1 minute for a customer representative?

29. **Time to Make a Choice** Let T be the random variable that a subject in a psychological testing program will make a certain choice after t seconds. If the probability density function is

$$f(t) = 0.4e^{-0.4t}$$

what is the probability that the subject will make the choice in less than 5 seconds?

30. **Waiting for a Bus** Buses on a certain route run every 50 minutes. What is the probability that a person arriving at a random stop along the route will have to wait at least 30 minutes? Assume that the random variable T is the time the person will have to wait and assume that T is uniformly distributed.

31. **Life of a Light Bulb** The length of time X a light bulb lasts is a random variable with an exponential probability distribution. Philips' 120 watt indoor spot lights have an average life of 2000 hours.

   (a) What is the probability that Philips' 120 watt indoor spot light lasts between 1800 and 2200 hours?
   (b) What is the probability that Philips' 120 watt indoor spot light lasts more than 2500 hours?

   Source: Philips Lighting Company, Somerset, NJ.

32. **Life of a Light Bulb** The length of time X a light bulb lasts is a random variable with an exponential probability distribution. GE 100 watt Soft White light bulbs have an average life of 750 hours.

   (a) What is the probability that a GE 100 watt Soft White light bulb lasts fewer than 500 hours?
   (b) What is the probability that a GE 100 watt Soft White light bulb lasts between 750 and 1000 hours?
   (c) What is the probability that a GE 100 watt Soft White light bulb lasts more than 1200 hours?
   (d) Suppose you worked as manager of quality control at the GE production plant, and found that the 100 watt Soft White light bulbs consistently lasted more than 1200 hours. What might you conclude? Write a paragraph outlining a proposal you will make at next week's manager's meeting concerning the bulbs.

   Source: General Electric Company, Cleveland, Ohio.

33. **Life of a Light Bulb** The length of time X a light bulb lasts is a random variable with an exponential probability distribution. GE 65 watt indoor spot lights have an average life of 2000 hours.

   (a) What is the probability that a GE 65 watt indoor spot light lasts fewer than 1500 hours?
   (b) What is the probability that a GE 65 watt indoor spot light lasts between 1750 and 2000 hours?
   (c) What is the probability that a GE 65 watt indoor spot light lasts more than 1900 hours?
   (d) Suppose you were the vice president of building maintenance, ordered a shipment of 1000 GE 65 watt bulbs, and found that after using half the bulbs, they consistently burned out before burning 1500 hours. What might you conclude? Explain.

   Source: General Electric Company, Cleveland, Ohio.

34. **Waiting for Lunch** At a local Wendy's the lunch-time wait for service at the drive-in counter averages 4.5 minutes (270 seconds). If the wait is an exponentially distributed random variable X, what is the probability that a customer waits more than 5 minutes for an order?

35. **Express Lunch Special** A Pizza Hut manager guarantees that the lunch-time express pizza will be served within 10 minutes or it is free. He has found that the time to prepare, bake, and serve the pizza is an exponentially distributed

random variable with $\lambda = \frac{1}{8}$ minutes. What is the probability that a customer ordering the lunch-time express pizza will not have to pay for the pizza? That is, what is the probability that the customer waits longer than 10 minutes to be served after ordering a pizza?

36. **Pizza Delivery** At one time, Domino's Pizza guaranteed that their pizzas would be delivered within 30 minutes after ordering or the pizza was free. (The offer has been discontinued because it was found that delivery persons were involved in too many car accidents while rushing to deliver.) Time management experts had learned that the time it took to prepare, bake, and deliver the pizza is a random variable X and followed an exponential distribution with $\lambda = \frac{1}{24}$.

    (a) What is the probability that a customer who calls and orders a pizza will get it free?
    (b) Suppose that it is found that between 5 and 7 p.m. on Friday evenings it takes an extra 2 minutes to deliver the pizza. What is the probability that a customer calling during this time will not have to pay for the pizza?

37. **Supermarket Lines** A local supermarket has received complaints that the checkout lines are too slow. Management hires an industrial engineer who determines that it takes an average of 10 minutes for a customer who enters the line to complete checking out.

    (a) Write the exponential density function which describes a customer's wait time.
    (b) What is the probability that a customer entering the line takes between 7 and 12 minutes to complete the checkout process.
    (c) What is the probability that a customer entering the line takes more than 15 minutes to checkout.
    (d) If a mother entering the line has to pick up her child in 20 minutes, and it takes her 5 minutes to get to the bus stop, what is the probability she can get there on time?

38. **Flight Arrivals** On May 2, 2003, New Orleans International Airport had 14 flights arriving between 5 P.M. and 6 P.M. Determine the probability that a flight arrived within 3 minutes (.05 hour). Use an exponential density function.

    Source: New Orleans International Airport.

39. **Waiting Time** Pairings at the 2003 HP Classic Golf Championship were scheduled to tee off every nine minutes. Determine the expected waiting time between pairings for spectators gathered on the 18th green.

    Source: Professional Golfers Association.

40. **Learning to Play a Game** A manufacturer of educational games for children finds through extensive psychological research that the average time it takes for a child in a certain age group to learn the rules of the game is predicted by a **beta probability density function.**

$$f(x) = \begin{cases} \frac{1}{4500}(30x - x^2) & \text{if } 0 \le x \le 30 \\ 0 & \text{if } x < 0 \text{ or } x > 30 \end{cases}$$

where x is the time in minutes.

(a) Show that f is a probability density function.
(b) What is the probability a child will learn how to play the game within 10 minutes?
(c) What is the probability a child will learn the game after 20 minutes?
(d) What is the probability the game is learned in at least 10 minutes, but no more than 20 minutes?

41. **Cost Estimates** The probability density function that gives the probability that an electrical contractor's cost estimate is off by x percent is

$$f(x) = \frac{3}{56}(5x - x^2)$$

for x in the interval $[0, 4]$. On average, by what percent can the contractor be expected to be off?

42. **Radioactive Decay** Plutonium 239 decays continuously at a rate of .002845% per year (based on a half-life of 24,360 years). If X is the time that a randomly chosen plutonium atom will decay, a probability density function for X is the following:

$$f(x) = \begin{cases} .00002845 e^{-.00002845x} & \text{if } x \ge 0 \\ 0 & \text{if } x < 0 \end{cases}$$

Use this probability density function to compute the probability that a plutonium atom will decay between 100 and 1000 years from now.

Source: Microsoft Encarta 97 Encyclopedia.

43. **Gestation Time** Suppose that X is the length of gestation in healthy humans. Then X is approximately normally distributed with a mean of 280 days and a standard deviation of 10 days. A probability density function for X is given by

$$f(x) = \frac{1}{10\sqrt{2\pi}} e^{-(x-280)^2/200}$$

(a) Use this probability density function to determine the probability that a healthy pregnant woman will have a pregnancy that lasts more than one week beyond the mean for the length of gestation in healthy humans.
(b) Determine the probability that a healthy pregnant woman will have a pregnancy such that the length of the

pregnancy is within one week of the mean for the length of gestation in healthy humans.

Source: School of Public Health and Health Sciences, University of Massachusetts.

44. Birth Weights Suppose that the random variable X is the weight of a baby at birth. The random variable X has a normal probability density function with mean approximately 114 ounces and standard deviation approximately 18 ounces.

A probability density function for X is

$$f(x) = \frac{1}{18\sqrt{2\pi}} e^{-(x-114)^2/648}$$

Use this probability density function to determine the probability that a baby's weight at birth will be between 100 ounces and 125 ounces.

Source: Bao-Feng Feng, Department of Mathematics, University of Kansas.

Problems 45–46, require the following discussion:
Triangular probability distributions are often used to model business situations. The graph of a possible triangular probability function, $f(x)$, is shown below. Such a distribution is used when we know only two pieces of information: the interval of possible values of the random variable X, that is, $a \leq X \leq b$, and the most likely value for the random variable X, that is, the value $X = c$ with the largest probability of occurring.

The probability density at $X = c$ can be calculated geometrically using the fact that

$$\int_a^b f(x)\, dx = 1$$

That is, the area of the triangle must be 1. Since $f(c)$ is the height of the triangle and $b - a$ is its base, we have

$$\frac{1}{2}(b - a) \cdot f(c) = 1$$

so

$$f(c) = \frac{2}{b - a}$$

Using this information, we can write the probability density function as a piecewise-defined function consisting of two line segments:

$$f(x) = \begin{cases} m_1 x + b_1 & \text{if } a \leq x \leq c \\ m_2 x + b_2 & \text{if } c < x \leq b \end{cases}$$

Use this model in Problems 45 and 46.

45. Price of a New Car A business analyst predicts that a new car will cost between \$10,000 and \$20,000, with the most likely price being \$17,000.

(a) Using the triangular model above, find the probability density function $f(x)$.

(b) What is the probability that the car will cost less than \$15,000?
(c) Graph the function $f(x)$.
(d) Estimate the expected price of the car from the graph.
(e) Evaluate the expected price of the car.
(f) Why is expected price different from \$17,000?

46. Stock Market Analysis A stock market analyst predicts that a stock will be worth between $15 and $19, with the most likely value being $16.

(a) Using the triangular model above, find the probability density function $f(x)$.

(b) What is the probability that the stock will be worth more than $18?
(c) Graph the function $f(x)$.
(d) Estimate the expected value of the stock from the graph.
(e) Evaluate the expected value of the stock.
(f) Why is the expected value of the stock different from $16?

Most graphing utilities have a random number function (usually RAND or RND) generating numbers between 0 and 1. Every time you use a random number function, a different number is selected. Check your user's manual to see how to use this feature of your graphing utility.

Sometimes probabilities are found by experimentation, that is, by performing an experiment. In Problems 47 and 48 use a random number function to perform an experiment.

47. Use a random number function to select a value for the random variable X. Repeat this experiment 50 times. [*Note:* Most calculators repeat the action of the last entry if you simply press the ENTER, or EXE, key again.] Count the number of times the random variable X is between 0.6 and 0.9.

(a) Calculate the ratio

$$R = \frac{\text{Number of times the random variable } X \text{ is between 0.6 and 0.9}}{50}$$

(b) Calculate the actual probability $Pr(0.6 \le X < 0.9)$ using a uniform density function.

48. Use a random number function to select a value for the random variable X. Repeat this experiment 50 times. [*Note:* Most calculators repeat the action of the last entry if you simply press the ENTER, or EXE, key again.] Count the number of times the random variable X is between 0.1 and 0.3.

(a) Calculate the ratio

$$R = \frac{\text{Number of times the random variable } X \text{ is between 0.1 and 0.3}}{50}$$

(b) Calculate the actual probability $Pr(0.1 \le X < 0.3)$ using a uniform density function.

49. The **variance** σ^2 associated with the probability density function f on $[a, b]$ is defined as

$$\sigma^2 = \int_a^b x^2 f(x)\, dx - [E(x)]^2$$

Verify that

$$\sigma^2 = \frac{(b-a)^2}{12}$$

for the uniform density function.

Chapter 7 Review

OBJECTIVES

Section	You should be able to	Review Exercises
7.1	1 Evaluate improper integrals	1–8
7.2	1 Find the average value of a function	9–14
	2 Use average value in applications	19–22
7.3	1 Verify probability density functions	15(a)–18(a), 25(a)–27(a)
	2 Find probabilities using a probability density function	25(b)–27(b), 28(a)(b), 30, 34
	3 Use the uniform density function	23, 29, 30, 34
	4 Use the exponential density function	24, 31–33, 35, 36
	5 Find expected value	15(b)–18(b), 23(c), 24(c), 25(d)–27(d), 28(c)

Chapter 7 Other Applications and Extensions of the Integral

THINGS TO KNOW

Improper Integrals (pp. 487, 488)

f is continuous on $[a, \infty)$ 	 $\int_a^\infty f(x)\,dx = \lim_{b \to \infty} \int_a^b f(x)\,dx$, provided the limit exists

f is continuous on $(-\infty, b]$ 	 $\int_{-\infty}^b f(x)\,dx = \lim_{a \to -\infty} \int_a^b f(x)\,dx$, provided the limit exists

f is continuous on $[a, b)$, and is not continuous at b 	 $\int_a^b f(x)\,dx = \lim_{t \to b^-} \int_a^t f(x)\,dx$, provided the limit exists

f is continuous on $(a, b]$, and is not continuous at a 	 $\int_a^b f(x)\,dx = \lim_{t \to a^+} \int_t^b f(x)\,dx$, provided the limit exists

Average Value (p. 491) 	 f is continuous on $[a, b]$ 	 $AV = \dfrac{1}{b-a} \int_a^b f(x)\,dx$

Probability Density Function (pp. 496–497) 	 $f(x) \geq 0$ and $\int_a^b f(x)\,dx = 1$ 	 $Pr(c \leq X \leq d) = \int_c^d f(x)\,dx$

Expected Value for a Continuous Random Variable X (p. 500) 	 $E(x) = \int_a^b x f(x)\,dx$

TRUE-FALSE ITEMS

T F **1.** The average value of a function f over the interval $[a, b]$ equals $\dfrac{f(b) - f(a)}{b - a}$.

T F **2.** The integral $\int_0^1 \dfrac{1}{x - 1}\,dx$ is improper.

T F **3.** A function f is a probability density function if $\int_a^b f(x)\,dx = 1$.

FILL-IN-THE-BLANKS

1. For a continuous function f defined on the interval $[a, b]$, the number $\dfrac{1}{b-a} \int_a^b f(x)\,dx$ is called the _____ of f.

2. A function that assigns a numerical value to each outcome of a sample space is called a _____.

3. The function constructed by connecting probability values by a smooth curve is called a _____ _____ function.

4. If f is continuous on $[2, \infty)$, then $\int_2^\infty f(x)\,dx =$ _____, provided this limit exists.

5. If f is discontinuous at 2, but continuous elsewhere, then $\int_0^2 f(x)\,dx =$ _____, provided this limit exists.

REVIEW EXERCISES Blue problem numbers indicate the author's suggestions for use in a practice test.

In Problems 1–6, find the value, if any, of each improper integral.

1. $\int_0^\infty 20 e^{-20x}\,dx$

2. $\int_0^\infty \dfrac{1}{3} e^{-1/3x}\,dx$

3. $\int_0^8 \dfrac{1}{\sqrt[3]{x}}\,dx$

4. $\displaystyle\int_1^{10} \frac{4}{\sqrt{x-1}}\, dx$

5. $\displaystyle\int_0^1 \frac{x+1}{x}\, dx$

6. $\displaystyle\int_0^1 \frac{x^4+1}{x^2}\, dx$

7. Find the area, if it exists, between the graph of $f(x) = -e^{-x}$ and the x-axis to the right of $x = 0$.

8. Find the area, if it exists, under the graph of $f(x) = \dfrac{1}{x^2}$ to the right of $x = 1$.

In Problems 9–14, find the average value of each function f over the given interval.

9. $f(x) = x^3$ over $[-1, 3]$

10. $f(x) = \dfrac{1}{x}$ over $[1, 2]$

11. $f(x) = x^2 + x$ over $[2, 6]$

12. $f(x) = e^{4x}$ over $[0, 1]$

13. $f(x) = 3x^2$ over $[-2, 2]$

14. $f(x) = \dfrac{1}{2}x$ over $[2, 10]$

In Problems 15–18, (a) verify that each function is a probability density function; (b) compute the expected value of each one.

15. $f(x) = \dfrac{8}{9}x$ over $\left[0, \dfrac{3}{2}\right]$

16. $f(x) = \dfrac{1}{4}x^3$ over $[0, 2]$

17. $f(x) = 12x^3(1 - x^2)$ over $[0, 1]$

18. $f(x) = \dfrac{15}{7}(2x^2 - x^4)$ over $[0, 1]$

---

19. Average Sales The rate of sales of a certain product obeys

$$f(t) = 1340 - 850e^{-t}$$

where t is the number of years the product is on the market. Find the average yearly sales for the first 10 years.

20. Profits from CD Sales The weekly profit from the sales of a new CD is described by the function,

$$P(t) = 1000\, t\, e^{-t}$$

in dollars, where t is the number of weeks the CD is on the market. Find the average profit for the first 10 weeks of sales.

21. Average Price of Sandals The demand curve for pairs of sandals is given by

$$D(x) = 50\, e^{-0.01x}$$

Find the average price over the demand interval from 100 to 150 pairs of sandals.

22. Average Price A producer's supply curve for a product is

$$S(x) = 0.005x^2$$

Find the average price if the producer supplies between 50 and 80 units.

23. A random variable X obeys the uniform probability distribution

$$f(x) = \dfrac{1}{12} \text{ for } -2 \le x \le 10$$

(a) What is the probability that $X \le 1$?
(b) What is the probability that $X \ge 5$?
(c) What is the expected value of X?

24. A random variable X has the exponential probability distribution

$$f(x) = \dfrac{1}{20}e^{-1/20x} \quad x \ge 0$$

(a) Find $Pr(10 < X < 30)$
(b) Find $Pr(X > 30)$
(c) What is the expected value of X?

25. Life Expectancy A man who is currently 20 years old wants to purchase life insurance. The insurance company is interested in determining at what age X (in years) he is likely to die. The likelihood of his dying at age X is given by

$$f(x) = \dfrac{3}{688,000}(-x^2 + 200x - 5000)$$

over $[20, 100]$.
(a) Show that f is a probability density function.
(b) Find the probability that the man is likely to die at or before the age of 40.
(c) What is the probability that he will die at or before the age of 60?
(d) What is the man's expected age of death?

26. An experiment follows the function $f(x) = 6(x - x^2)$ for outcomes between 0 and 1.

(a) Show that f is a probability density function.
(b) Determine the probability that an outcome lies between $\frac{1}{3}$ and $\frac{1}{2}$.
(c) Determine the probability that an outcome lies between 0 and $\frac{3}{4}$.
(d) Find the expected value of the outcome.

27. The outcome X of an experiment lies between 0 and 2, and follows the function
$$f(x) = \frac{1}{2}x$$
(a) Show that f is a probability density function.
(b) What is the probability that $X < 1$?
(c) What is the probability that X is between 1 and 1.5?
(d) What is the probability that X is greater than 1.5?
(e) What is the expected value of X?

28. X is a random variable with a probability density function
$$f(x) = 3x^2 \quad -1 \le x \le 0$$
(a) What is the probability that $X > -0.1$?
(b) What is the probability that $-0.5 < X < -0.75$?
(c) What is the expected value of X?

29. **Waiting for Clock Chimes** The Gastown Steam Clock in Vancouver, Canada, chimes every 15 minutes. A tourist walking around the city randomly arrives at the corner where the clock is located.

(a) What is the probability she will wait fewer than 3 minutes to hear the clock?
(b) What is the probability the tourist waits longer than 10 minutes?
(c) What is the expected wait time of a person arriving at the corner to hear the clock?

Use a uniform density function.

30. **Uniform Probability** Use a uniform density function. A number is randomly chosen from the interval $[0, 1]$.

(a) What is the probability that the first decimal digit will be a 1?
(b) What is the probability that the first decimal digit will be greater than 4?

31. **Fire Alarms** A fire department in a medium-sized city receives an average of 2.5 calls each minute. Use an exponential distribution to find the probability that the switchboard is idle (no calls are received) for more than one minute.

32. **Waiting for Lunch** At a fast-food counter it takes an average of 3 minutes to get served. The service time X for a customer has an exponential probability density function.

(a) What fraction of the customers is served within 2 minutes?
(b) What is the probability that a customer has to wait at least 3 minutes to be served?

33. **Life of a Light Bulb** The length of time X a light bulb lasts is a random variable with an exponential probability distribution. Philips' *Dura Max* 3-way long life light bulbs have an average life of 1750 hours.

(a) What is the probability that a Philips' *Dura Max* bulb lasts between 1500 and 2000 hours?
(b) What is the probability that Philips' *Dura Max* bulb lasts more than 2000 hours?

Source: Philips Lighting Company, Somerset, NJ.

34. **Quality Control** A toy manufacturing machine produces a toy every 2 minutes. An inspector arrives at a random time and must wait X minutes for a toy.

(a) Find the probability density function for X.
(b) Find the probability that the inspector has to wait at least 1 minute for a toy to be produced.
(c) Find the probability that the inspector has to wait no more than 1 minute for a toy.

Use a uniform density function.

35. **Collecting Tolls** Cars approach a remote toll station at a rate of 40 cars per hour. The time between arrivals is a random variable X, which obeys the exponential probability density function
$$f(t) = \frac{2}{3}e^{-2t/3} \quad t \ge 0$$
What is the probability that the attendant waits more than 1 minute for the first car?

36. **Length of a Phone Call** The length t of a telephone call to technical service at a computer software company follows an exponential distribution with $\lambda = 0.10$.

(a) What is the probability that a call lasts between 5 and 10 minutes?
(b) What is the probability that a call lasts more than 30 minutes?
(c) What is the expected length of a phone call?

Chapter 7 Project

A MATHEMATICAL MODEL FOR ARRIVALS

In this Chapter Project we try to see if the waiting times between arrivals really do have an exponential distribution. The following data set tabulates the dates from 1997 to 2001 on which a member of a large church in Charlotte, North Carolina died. Day 1 is January 1, 1997, while day 1461 is December 31, 2001.

{6, 47, 109, 113, 131, 152, 158, 178, 201, 206, 215, 250, 251, 252, 274, 276, 306, 311, 338, 382, 414, 429, 432, 440, 468, 478, 479, 498, 510, 530, 542, 554, 561, 597, 601, 602, 603, 605, 606, 629, 650, 653, 660, 666, 667, 669, 686, 732, 745, 760, 764, 782, 790, 803, 820, 828, 863, 870, 878, 881, 893, 896, 913, 960, 974, 998, 1004, 1005, 1050, 1068, 1088, 1094, 1095, 1146, 1151, 1171, 1185, 1273, 1274, 1282, 1288, 1334, 1348, 1364, 1388, 1412, 1423, 1437, 1441}

The first member of the church died on January 6, 1997, the second on February 16, 1997, and so on. Although it may seem a bit odd, this is a list of "arrivals." Notice that there were 89 deaths in 1461 days, which gives a rate of $\lambda = \dfrac{89}{1461} = 0.061$ arrivals per day. One can calculate that the waiting times between arrivals are

{6, 41, 62, 4, 18, 21, 6, 20, 23, 5, 9, 35, 1, 1, 22, 2, 30, 5, 27, 44, 32, 15, 3, 8, 28, 10, 1, 19, 12, 20, 12, 12, 7, 36, 4, 1, 1, 2, 1, 23, 21, 3, 7, 6, 1, 2, 17, 46, 13, 15, 4, 18, 8, 13, 17, 8, 35, 7, 8, 3, 12, 3, 17, 47, 14, 24, 6, 1, 45, 18, 20, 6, 1, 51, 5, 20, 14, 88, 1, 8, 6, 46, 14, 16, 24, 24, 11, 14, 4}

Notice how uneven these data are—in some cases a long time passes between deaths, while in other cases several deaths are bunched closely together.

1. Group the times t between deaths into 18 intervals: $0 \leq t < 5, 5 \leq t < 10, \ldots, 85 \leq t < 90$, and tally the number of deaths in each interval.

2. Build a histogram for the data tabulated in Problem 1. To build a histogram, we construct a set of adjoining rectangles, each having as base the size of the interval, 5, and as height the number of deaths in that particular interval.

3. Comment on the shape of the histogram built in Problem 2.

4. Refer to Problem 1. Find the relative frequency of the number of deaths by dividing the number of deaths in each time interval by 89, the total number of deaths that occurred in the four-year period.

5. Construct a relative frequency histogram. That is, a histogram, for which the height of each rectangle is the relative frequency of the interval.

6. Using $\lambda = 0.061$, and assuming that the waiting times are exponentially distributed, write the probability density function that will describe the time between deaths. Show that the distribution is a probability density function.

7. Using the probability density function from Problem 6, calculate the probability of the waiting time between deaths for each interval: $0 \leq t < 5$ days, $5 \leq t < 10$ days, ..., $85 \leq t < 90$ days.

8. Compare the results from Problems 4 and 6. Does it seem that the waiting times from the data are exponentially distributed with $\lambda = 0.061$?

9. Graph $f(t) = 0.061e^{-0.061t}$ on the same set of axes as the relative frequency histogram. Comment on the graphs.

MATHEMATICAL QUESTIONS FROM PROFESSIONAL EXAMS*

1. **Actuary Exam—Part I** $\displaystyle\int_0^1 x \ln x\, dx =$

   (a) $-\infty$ (b) -2 (c) -1 (d) $-\dfrac{1}{4}$ (e) $-\dfrac{2}{9}$

2. **Actuary Exam—Part I** $\displaystyle\int_{-1}^2 \dfrac{x}{\sqrt{x+1}}\, dx =$

   (a) $\dfrac{1}{2}$ (b) $\dfrac{2}{3}$ (c) 1 (d) $\dfrac{3}{2}$ (e) 0

*Copyright © 1998, 1999 by the Society of Actuaries. Reprinted with permission.

3. **Actuary Exam—Part I** $\int_0^\infty \dfrac{x+1}{(x^2+2x+2)^2}\,dx =$

   (a) 0   (b) $\dfrac{1}{4}$ (c) $\dfrac{1}{2}$ (d) 1   (e) ∞

4. **Actuary Exam—Part II** Two men patronize the same barber shop. If they both arrive independently between 3 P.M. and 4 P.M. on the same day and both stay for 15 minutes, what is the probability that they are in the barber shop for part or all of the same time?

   (a) $\dfrac{3}{8}$ (b) $\dfrac{7}{16}$ (c) $\dfrac{1}{2}$ (d) $\dfrac{9}{16}$ (e) $\dfrac{5}{8}$

5. **Actuary Exam—Part II** Each day, X arrives at a point A between 8:00 and 9:00 A.M., his times of arrival being uniformly distributed. Y arrives independently at A between 8:30 and 9:00 A.M., his times of arrival also being uniformly distributed. What is the probability that Y arrives before X?

   (a) $\dfrac{1}{8}$ (b) $\dfrac{1}{6}$ (c) $\dfrac{2}{9}$ (d) $\dfrac{1}{4}$ (e) $\dfrac{1}{2}$

CHAPTER 8

Calculus of Functions of Two or More Variables

OUTLINE

- 8.1 Rectangular Coordinates in Space
- 8.2 Functions and Their Graphs
- 8.3 Partial Derivatives
- 8.4 Local Maxima and Local Minima
- 8.5 Lagrange Multipliers
- 8.6 The Double Integral
  - **Chapter Review**
  - **Chapter Project**

The inventory control project given in Chapter 5 dealt with controlling the costs associated with the ordering and storage of vacuum cleaners. But stores would typically order and stock a variety of items, not just vacuum cleaners. How do we minimize costs when more than one item is involved? The Chapter Project at the end of this chapter will guide you to a solution when two items are involved.

8.1 Rectangular Coordinates in Space

PREPARING FOR THIS SECTION *Before getting started, review the following:*

> Rectangular Coordinates (in the plane) (Chapter 0, Section 0.8, pp. 76–77)
> Distance Formula (Chapter 0, Section 0.8, p. 78)
> Completing the Square (Chapter 0, Section 0.4, p. 45)
> Pythagorean Theorem (Chapter 0, Section 0.7, p. 72)

OBJECTIVES
1. Use the distance formula
2. Find the standard equation of a sphere

FIGURE 1

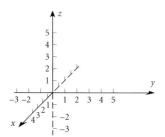

FIGURE 2

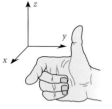

FIGURE 3

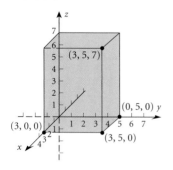

In Chapter 0 we established a correspondence between points on a line and real numbers. Then we showed that each point in a plane can be associated with an ordered pair of real numbers. Here we show that each point in (three-dimensional) space can be associated with an **ordered triple** of real numbers.

First we select a fixed point called the **origin.** Through the origin, we draw three mutually perpendicular lines. These are called the **coordinate axes,** and are labeled the **x-axis, y-axis,** and **z-axis.** On each of the three lines we choose one direction as positive and select an appropriate scale on each axis.

As indicated in Figure 1, we position the positive z-axis so that the system is **right-handed.** This conforms to the so-called **right-hand rule,** which asserts that if the index finger of the right hand points in the direction of the positive x-axis and the middle finger points in the direction of the positive y-axis, then the thumb will point in the direction of the positive z-axis.* See Figure 2.

Just as we did in one and two dimensions, we assign coordinates to each point P in space. Specifically, we identify each point P with an ordered triple of real numbers (x, y, z), and we refer to it as "the point (x, y, z)." So, "the point $(3, 5, 7)$" is the point for which $x = 3, y = 5, z = 7$, and, starting at the origin, we reach P by moving 3 units along the positive x-axis, then 5 units in the direction of the positive y-axis, and, finally, 7 units in the direction of the positive z-axis. Figure 3 illustrates the location of the point $(3, 5, 7)$ as well as the points $(3, 0, 0), (3, 5, 0)$, and $(0, 5, 0)$. Observe that any point on the x-axis will have the form $(x, 0, 0)$. Similarly, $(0, y, 0)$ and $(0, 0, z)$ represent points on the y-axis and the z-axis, respectively.

In addition, all points of the form $(x, y, 0)$ constitute a plane called the **xy-plane.** This plane is perpendicular to the z-axis. Similarly, the points $(0, y, z)$ form the **yz-plane,** which is perpendicular to the x-axis; and the points $(x, 0, z)$ form the **xz-plane,**

---

*Although there are left-handed systems and left-handed rules, we shall adopt the usual convention and only use a right-handed system.

FIGURE 4

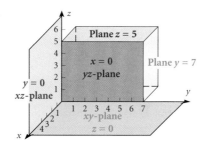

which is perpendicular to the y-axis (see Figure 4). Figure 4 also illustrates that points of the form (x, y, z), where $z = 5$, lie in a plane parallel to the xy-plane. Similarly, points (x, y, z), where $y = 7$, lie in a plane parallel to the xz-plane.

NOW WORK PROBLEMS 1 AND 13.

FIGURE 5

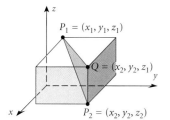

Distance in Space

To derive a formula for the distance $d = d(P_1, P_2)$ between two points $P_1 = (x_1, y_1, z_1)$ and $P_2 = (x_2, y_2, z_2)$ in space, we apply the Pythagorean theorem twice. As Figure 5 illustrates, we utilize the point $Q = (x_2, y_2, z_1)$. The first application of the Pythagorean theorem involves observing that the triangle QP_1P_2 is a right triangle in which the side P_1P_2 is the hypotenuse. As a result,

$$[d(P_1, P_2)]^2 = [d(P_1, Q)]^2 + [d(Q, P_2)]^2 \tag{1}$$

The points P_1 and Q lie in a plane parallel to the xy-plane (the plane $z = z_1$). So, we can use the formula for distance in two dimensions and obtain

$$[d(P_1, Q)]^2 = (x_2 - x_1)^2 + (y_2 - y_1)^2 \tag{2}$$

The points P_2 and Q lie along a line parallel to the z-axis so that

$$d(Q, P_2) = |z_2 - z_1| \quad \text{and} \quad [d(Q, P_2)]^2 = (z_2 - z_1)^2 \tag{3}$$

Using the results of equations (2) and (3) in equation (1), we arrive at a formula for the distance between two points in space.

Distance Formula

The distance $d = d(P_1, P_2)$ between two points $P_1 = (x_1, y_1, z_1)$ and $P_2 = (x_2, y_2, z_2)$ in space is given by the formula

$$\boxed{d = \sqrt{(x_2 - x_1)^2 + (y_2 - y_1)^2 + (z_2 - z_1)^2}} \tag{4}$$

EXAMPLE 1 **Using the Distance Formula**

If $P_1 = (-1, 4, 2)$ and $P_2 = (6, -2, 3)$, find the distance d between P_1 and P_2.

SOLUTION We use Formula (4), with $x_1 = -1, y_1 = 4, z_1 = 2$ and $x_2 = 6, y_2 = -2, z_2 = 3$. Then

$$d = \sqrt{[6 - (-1)]^2 + [-2 - 4]^2 + [3 - 2]^2}$$
$$= \sqrt{49 + 36 + 1}$$
$$= \sqrt{86}$$

NOW WORK PROBLEM 19.

The Sphere

FIGURE 6

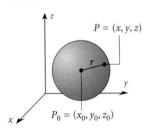

In space the collection of all points that are the same distance from some fixed point is called a **sphere**. The constant distance is called the **radius** and the fixed point is the **center** of the sphere. See Figure 6.

Any point $P = (x, y, z)$ on a sphere of radius r and center at the point $P_0 = (x_0, y_0, z_0)$ obeys

$$d(P, P_0) = r$$

By the Distance Formula (4), an equation of this sphere is

$$\sqrt{(x - x_0)^2 + (y - y_0)^2 + (z - z_0)^2} = r$$

Squaring both sides gives

$$(x - x_0)^2 + (y - y_0)^2 + (z - z_0)^2 = r^2$$

> **Standard Equation of a Sphere**
>
> The **standard equation of a sphere** with center at the point (x_0, y_0, z_0) and radius r is
>
> $$(x - x_0)^2 + (y - y_0)^2 + (z - z_0)^2 = r^2$$

2   EXAMPLE 2   Finding the Standard Equation of a Sphere

The standard equation of a sphere with radius 2 and center at $(-1, 2, 0)$, is

$$(x - x_0)^2 + (y - y_0)^2 + (z - z_0)^2 = r^2$$
$$(x + 1)^2 + (y - 2)^2 + z^2 = 4 \quad x_0 = -1, y_0 = 2, z_0 = 0, r = 2$$

NOW WORK PROBLEM 25.

When an equation of a sphere is given, it is sometimes necessary to complete the squares involving the variables in order to find the center and radius.

EXAMPLE 3   Finding the Center and Radius of a Sphere

Complete the squares in the expression

$$x^2 + y^2 + z^2 + 2x + 4y - 2z = 10$$

and show that it is the equation of a sphere. Find its center and radius.

SOLUTION Rewrite the given expression as

$$(x^2 + 2x) + (y^2 + 4y) + (z^2 - 2z) = 10$$

and complete the squares. The result is

$$(x^2 + 2x + 1) + (y^2 + 4y + 4) + (z^2 - 2z + 1) = 10 + 1 + 4 + 1$$
$$(x + 1)^2 + (y + 2)^2 + (z - 1)^2 = 16 = 4^2$$

This is the standard equation of a sphere with radius 4 and center at $(-1, -2, 1)$. ▶

NOW WORK PROBLEM 31.

EXERCISE 8.1

In Problems 1–6, plot each point.

1. $(1, 1, 1)$ **2.** $(0, 0, 1)$ **3.** $(0, 2, 5)$ **4.** $(-1, 5, 0)$ **5.** $(-3, 1, 0)$ **6.** $(4, -1, -3)$

In Problems 7–12, opposite vertices of a rectangular box whose edges are parallel to the coordinate axes are given. List the coordinates of the other six vertices of the box.

7. $(0, 0, 0); (2, 1, 3)$ **8.** $(0, 0, 0); (4, 2, 2)$ **9.** $(1, 2, 3); (3, 4, 5)$

10. $(5, 6, 1); (3, 8, 2)$ **11.** $(-1, 0, 2); (4, 2, 5)$ **12.** $(-2, -3, 0); (-6, 7, 1)$

In Problems 13–18, describe in words the set of all points (x, y, z) that satisfy the given conditions.

13. $y = 3$ **14.** $z = -3$ **15.** $x = 0$

16. $x = 1$ and $y = 0$ **17.** $z = 5$ **18.** $x = y$ and $z = 0$

In Problems 19–24, find the distance between each pair of points.

19. $(1, 3, 0)$ and $(4, 1, 2)$ **20.** $(3, 2, 1)$ and $(1, 2, 3)$ **21.** $(-1, 2, -3)$ and $(4, -2, 1)$

22. $(-2, 1, 3)$ and $(4, 0, -3)$ **23.** $(4, -2, -2)$ and $(3, 2, 1)$ **24.** $(2, -3, -3)$ and $(4, 1, -1)$

In Problems 25–28, find the standard equation of a sphere with radius r and center P_0.

25. $r = 1; P_0 = (3, 1, 1)$ **26.** $r = 2; P_0 = (1, 2, 2)$

27. $r = 3; P_0 = (-1, 1, 2)$ **28.** $r = 1; P_0 = (-3, 1, -1)$

In Problems 29–34, find the radius and center of each sphere.

29. $x^2 + y^2 + z^2 + 2x - 2y = 2$ **30.** $x^2 + y^2 + z^2 + 2x - 2z = -1$

31. $x^2 + y^2 + z^2 + 4x + 4y + 2z = 0$ **32.** $x^2 + y^2 + z^2 + 4x = 0$

33. $2x^2 + 2y^2 + 2z^2 - 8x + 4z = -2$ **34.** $3x^2 + 3y^2 + 3z^2 + 6x - 6y = 3$

In Problems 35–36, write the standard equation of the sphere described.

35. The endpoints of a diameter are $(-2, 0, 4)$ and $(2, 6, 8)$.

36. The endpoints of a diameter are $(1, 3, 6)$ and $(-3, 1, 4)$.

8.2 Functions and Their Graphs

PREPARING FOR THIS SECTION *Before getting started, review the following:*

> Definition of a Function (Chapter 1, Section 1.2, pp. 109–111)

> Completing the Square (Chapter 0, Section 0.4, p. 45)

OBJECTIVES
1. Evaluate a function of two variables
2. Find the domain of a function of two variables

So far we have considered only functions of one independent variable, usually expressed explicitly by an equation $y = f(x)$. Often, models require functions of more than one variable.

For example, the cost of producing a certain item may depend on variables such as labor and material. In economic theory, supply and demand of a commodity often depend not only on the commodity's own price, but also on the prices of related commodities and other factors (such as income level, time of year, etc.).

Definition of a Function of Two Variables

A **function f of two variables** x and y is a relation $z = f(x, y)$ that assigns a unique real number z to each ordered pair (x, y) of real numbers in a subset D of the xy-plane. The set D is called the **domain** of the function f.

In the equation $z = f(x, y)$ we refer to z as the **dependent variable** and to x and y as the **independent variables**. The **range** of f consists of all real numbers $f(x, y)$ where (x, y) is in D. Figure 7 illustrates one way of depicting $z = f(x, y)$.

FIGURE 7

D is the domain of f

EXAMPLE 1 Evaluating a Function of Two Variables

If $z = f(x, y) = x\sqrt{y} + xy^2$, find:

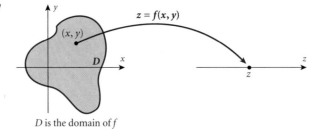

(a) $f(0, 0)$ (b) $f(2, 4)$ (c) $f(1, 9)$ (d) $f(x + \Delta x, y)$ (e) $f(x, y + \Delta y)$

SOLUTION (a) $f(0, 0) = 0\sqrt{0} + 0 \cdot 0^2 = 0$
(b) $f(2, 4) = 2\sqrt{4} + 2 \cdot 4^2 = 36$
(c) $f(1, 9) = 1\sqrt{9} + 1 \cdot 9^2 = 84$
(d) $f(x + \Delta x, y) = (x + \Delta x)\sqrt{y} + (x + \Delta x) \cdot y^2$
(e) $f(x, y + \Delta y) = x\sqrt{y + \Delta y} + x(y + \Delta y)^2$

NOW WORK PROBLEM 1.

As with functions of one variable, a function of two variables is usually given by an equation and, unless otherwise stated, the domain is taken to be the largest set of points in the plane for which this equation is defined in the real number system.

EXAMPLE 2   Finding the Domain of a Function of Two Variables

Find the domain of each function. Graph each domain.

(a) $z = f(x, y) = \sqrt{x^2 + y^2 - 1}$
(b) $z = g(x, y) = \ln(y - x^2)$

FIGURE 8

SOLUTION (a) The function f equals the square root of the expression $x^2 + y^2 - 1$. Since only square roots of nonnegative numbers are allowed in the real number system, we must have

$$x^2 + y^2 - 1 \geq 0$$
$$x^2 + y^2 \geq 1$$

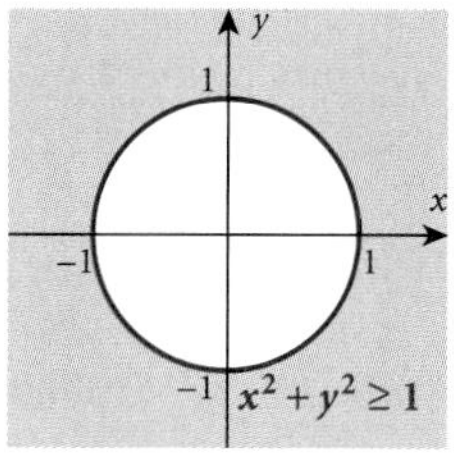

This inequality describes the domain of f, namely the set of points (x, y) that are either on the circle $x^2 + y^2 = 1$ or outside of it. See Figure 8. Notice that we use a solid rule to show that the points on the circle are part of the domain.

(b) The function g is the natural logarithm of the expression $y - x^2$. Since only logarithms of positive numbers are allowed, we must have

FIGURE 9

$$y - x^2 > 0$$

from which

$$y > x^2$$

This inequality describes the domain of g, namely, the set of points (x, y) "inside" the parabola $y = x^2$. See Figure 9. Notice that we use a dashed rule to show that the points on the parabola $y = x^2$ are not part of the domain.

FIGURE 10

NOW WORK PROBLEM 17.

Graphing Functions of Two Variables

The graph of a function $z = f(x, y)$ of two variables, called a **surface**, consists of all points (x, y, z) for which $z = f(x, y)$ and (x, y) is in the domain of f. See Figure 10 for an illustration of a surface for which $f(x, y) \geq 0$.

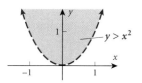

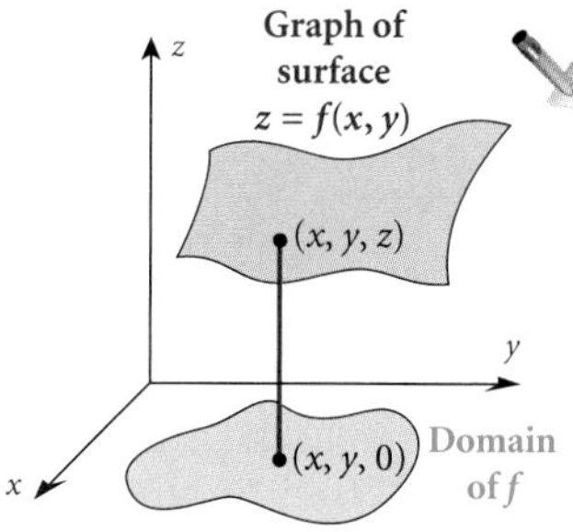

EXAMPLE 3 Describing the Graph of a Function of Two Variables

Describe the graph of the function

$$z = f(x, y) = \sqrt{4 - x^2 - y^2}$$

SOLUTION First, notice that the domain of f consists of all points (x, y) for which $x^2 + y^2 \leq 4$. Since $z = f(x, y)$ is defined as a square root, it follows that $z \geq 0$, so the graph of f will lie above the xy-plane. By squaring both sides of $z = \sqrt{4-x^2-y^2}$, we find that

$$z^2 = 4 - x^2 - y^2, \quad z \geq 0$$
$$x^2 + y^2 + z^2 = 4, \quad z \geq 0$$

This is the equation of a hemisphere, with center at $(0, 0, 0)$ and radius 2. Figure 11 illustrates the graph.

FIGURE 11

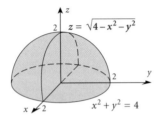

Obtaining the shape, or even a rough sketch, of the graph of most functions of two variables is a difficult task and is not taken up in this book. Graphing utilities may be used to generate *computer graphics* of surfaces in space.

For example, a portion of a surface $z = f(x, y)$ can be viewed in perspective and delineated by a rectangular mesh over a rectangular portion of the xy-plane. By changing the computer program, the point of view and other characteristics of the computer graphic may be altered. See Figure 12 for computer-generated graphs of several surfaces.

FIGURE 12

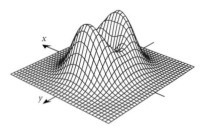
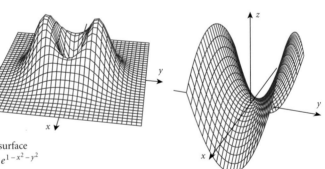

(a) Two views of the surface
$z = f(x, y) = (x^2 + 2y^2) e^{1-x^2-y^2}$

(b) The surface
$z = f(x, y) = 4y^2 - x^2$

(c) The surface
$z = f(x, y) = e^{x^2+y^2}$

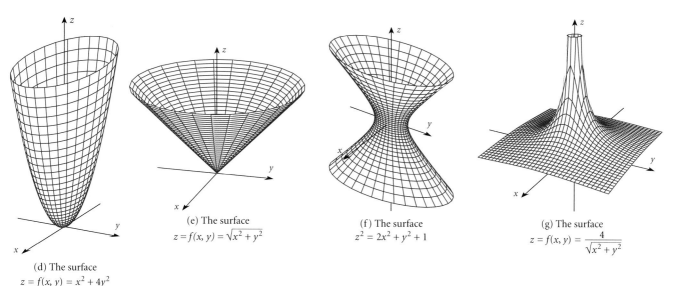

(d) The surface
$z = f(x, y) = x^2 + 4y^2$

(e) The surface
$z = f(x, y) = \sqrt{x^2 + y^2}$

(f) The surface
$z^2 = 2x^2 + y^2 + 1$

(g) The surface
$z = f(x, y) = \dfrac{4}{\sqrt{x^2 + y^2}}$

Functions of Three Variables

The function f defined by $w = f(x, y, z)$ is a function of the three independent variables x, y, and z. For each ordered triple (x, y, z) in the domain, the function f assigns a value to w, the dependent variable. In this case, the domain is a collection of points in space. Figure 13 illustrates a way of depicting the function $w = f(x, y, z)$.

FIGURE 13

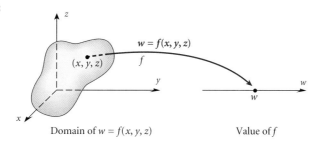

Domain of $w = f(x, y, z)$ Value of f

As with functions of two variables, a function of three variables is usually given by an equation, and, unless otherwise stated, the domain is taken to be the largest set of points in space for which this equation is defined in the real number system.

EXAMPLE 4   Finding the Domain of a Function of Three Variables

Find the domain of the function $w = f(x, y, z) = \sqrt{9 - x^2 - y^2 - z^2}$.

SOLUTION   Since square roots of negative numbers are not defined, the domain of this function consists of all points for which $x^2 + y^2 + z^2 - 9 \leq 0$. Therefore, the domain consists of all points inside and on the sphere with center at $(0, 0, 0)$ and radius 3. See Figure 14 for the graph of the domain.

FIGURE 14

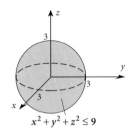

$x^2 + y^2 + z^2 \le 9$

The graph of a function $w = f(x, y, z)$ of three variables consists of all points (x, y, z, w) for which (x, y, z) is in the domain of f and $w = f(x, y, z)$. Because this requires locating points in a four-dimensional space, it is impossible for us to draw such a graph.

 NOW WORK PROBLEM 27.

EXERCISE 8.2

In Problems 1–10, evaluate $f(2, 1)$.

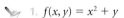 1. $f(x, y) = x^2 + y$
2. $f(x, y) = x - y^2$
3. $f(x, y) = \sqrt{xy}$
4. $f(x, y) = x\sqrt{y}$
5. $f(x, y) = \dfrac{1}{2x + y}$

6. $f(x, y) = \dfrac{x}{x - 3y}$
7. $f(x, y) = \dfrac{x^2 - y}{x - y}$
8. $f(x, y) = \dfrac{x + y^2}{x^2 - y^2}$
9. $f(x, y) = \sqrt{4 - x^2 y^2}$
10. $f(x, y) = \sqrt{9 - x^2 y^2}$

11. Let $f(x, y) = 3x + 2y + xy$. Find:
    (a) $f(1, 0)$ (b) $f(0, 1)$ (c) $f(2, 1)$
    (d) $f(x + \Delta x, y)$ (e) $f(x, y + \Delta y)$

12. Let $f(x, y) = x^2 y + x + 1$. Find:
    (a) $f(0, 0)$ (b) $f(0, 1)$ (c) $f(2, 1)$
    (d) $f(x + \Delta x, y)$ (e) $f(x, y + \Delta y)$

13. Let $f(x, y) = \sqrt{xy} + x$. Find:
    (a) $f(0, 0)$ (b) $f(0, 1)$ (c) $f(a^2, t^2); a > 0, t > 0$
    (d) $f(x + \Delta x, y)$ (e) $f(x, y + \Delta y)$

14. Let $f(x, y) = e^{x+y}$. Find:
    (a) $f(0, 0)$ (b) $f(1, -1)$ (c) $f(x + \Delta x, y)$
    (d) $f(x, y + \Delta y)$

15. Let $f(x, y, z) = x^2 y + y^2 z$. Find:
    (a) $f(1, 2, 3)$ (b) $f(0, 1, 2)$ (c) $f(-1, -2, -3)$

16. Let $f(x, y, z) = 3x^2 + y^2 - 2z^2$. Find:
    (a) $f(1, 2, 3)$ (b) $f(0, 1, 2)$ (c) $f(-1, -2, -3)$

In Problems 17–30, find the domain of each function and give a description of the domain. For Problems 17–26, graph the domain.

17. $z = f(x, y) = \sqrt{x}\sqrt{y}$
18. $z = f(x, y) = \sqrt{xy}$
19. $z = f(x, y) = \sqrt{9 - x^2 - y^2}$
20. $z = f(x, y) = \sqrt{x^2 + y^2 - 16}$
21. $z = f(x, y) = \dfrac{\ln x}{\ln y}$
22. $z = f(x, y) = \ln \dfrac{x}{y}$
23. $z = f(x, y) = \dfrac{3}{x^2 + y^2 - 4}$
24. $z = f(x, y) = \dfrac{4}{9 - x^2 - y^2}$
25. $z = f(x, y) = \ln(x^2 + y^2)$
26. $z = f(x, y) = \ln(4x - y^2)$
27. $w = f(x, y, z) = \sqrt{x^2 + y^2 + z^2 - 16}$
28. $w = f(x, y, z) = \sqrt{9 - (x^2 + y^2 + z^2)}$
29. $w = f(x, y, z) = \dfrac{4}{x^2 + y^2 + z^2}$
30. $w = f(x, y, z) = \ln(x^2 + y^2 + z^2)$

31. For the function $z = f(x, y) = 3x + 4y$, find:

(a) $f(x + \Delta x, y)$
(b) $f(x + \Delta x, y) - f(x, y)$
(c) $\dfrac{f(x + \Delta x, y) - f(x, y)}{\Delta x}, \quad \Delta x \neq 0$
(d) $\lim\limits_{\Delta x \to 0} \dfrac{f(x + \Delta x, y) - f(x, y)}{\Delta x}, \quad \Delta x \neq 0$

32. For the function $z = f(x, y) = 4x + 5y$, find:

(a) $f(x, y + \Delta y)$
(b) $f(x, y + \Delta y) - f(x, y)$
(c) $\dfrac{f(x, y + \Delta y) - f(x, y)}{\Delta y}, \quad \Delta y \neq 0$
(d) $\lim\limits_{\Delta y \to 0} \dfrac{f(x, y + \Delta y) - f(x, y)}{\Delta y}, \quad \Delta y \neq 0$

33. Cost of Construction The cost of the bottom and top of a cylindrical tank is \$300 per square meter and the cost of the sides is \$500 per square meter. Write the total cost of constructing such a tank as a function of the radius r and height h (both is meters).

34. Cost of Construction The cost per square centimeter of the material to be used for an open rectangular box is \$4 for the bottom and \$2 for the other sides. Write the total cost of constructing such a box as a function of its bottom and side dimensions.

35. Baseball A pitcher's earned run average is given by

$$A(N, I) = 9\left(\dfrac{N}{I}\right)$$

where N is the total number of earned runs given up in I innings of pitching. Find

(a) $A(3, 4)$ (b) $A(6, 3)$ (c) $A(2, 9)$ (d) $A(3, 18)$

36. Intelligence Quotient In psychology, intelligence quotient (IQ) is measured by

$$IQ = f(M, C) = 100\,\dfrac{M}{C}$$

where M is a person's mental age and C is the person's chronological or actual age, $0 < C \leq 16$.

(a) Find the IQ of a 12-year-old child whose mental age is 10.
(b) Find the IQ of a 10-year-old child whose mental age is 12.
(c) If a 10-year-old girl has an IQ of 120, what is her mental age?

37. Cell Phone Usage One version of the Cingular Family Plan for cellular phone usage has a monthly cost of $B(x, y) = 79.99 + 0.4x + 0.02y$, where x is the number of minutes in excess of 500 during weekdays between 7:00 A.M. and 9:00 P.M. and y is the number of minutes in excess of 1500 at all other times. Find the total monthly bill for using 650 minutes weekdays between 7:00 A.M. and 9:00 P.M. and 1600 minutes at all other times.

Source: Cingular Wireless Phone Company.

38. Field Goal Percentages in the NBA In the National Basketball Association (NBA), the Adjusted Field Goal Percentage is determined by $f(x, y, s) = \dfrac{x + 1.5y}{s}$, where x is the number of two-point field goals made, y is the number of three-point field goals made, and s is the sum of two-point and three-point field goal attempts. Find the Adjusted Field Goal Percentages for the following NBA players for the 2002–2003 season.

(a) Stromile Swift of the Memphis Grizzlies, who made 235 out of 489 two-point attempts and 0 of 2 three-point attempts.
(b) Kobe Bryant of the Los Angeles Lakers, who made 744 out of 1600 two point field goal attempts, and 124 out of 324 three-point field goal attempts.

Source: National Basketball Association.

39. Heat Index At the 2000 Olympic Games in Sydney, Australia, the following formula was used to calculate the heat index (apparent temperature):

$$H = -42.379 + 2.04901523\,t + 10.14333127\,r \\ - 0.22475541\,tr - 0.00683783\,t^2 - 0.05481717\,r^2 \\ + 0.00122874\,t^2 r + 0.00085282\,t r^2 - 0.00000199\,t^2 r^2$$

where H = the heat index (in °F)
t = the air temperature (in °F)
r = the % relative humidity (e.g., $r = 75$ when the relative humidity is 75%)

To alert athletes and others, the Australian Bureau of Meteorology published the following effects for various apparent temperatures, H:

Heat Index H	Physical Effects
90°F – 104°F	Heat cramps or heat exhaustion possible
105°F – 130°F	Heat cramps or heat exhaustion likely, heat-stroke possible
above 130°F	Heatstroke highly likely

(a) What is the heat index when the air temperature is 95°F and the relative humidity is 50%?

(b) If the air temperature is 97°F, what is the smallest relative humidity (to the nearest whole percent) that will result in a heat index of 105°F?

(c) If the air temperature is 102°F, what is the smallest relative humidity (to the nearest whole percent) that will result in a heat index of 130°F?

Source: Commonwealth Bureau of Meteorology, Melbourne, Australia.

40. **Cobb-Douglas Model** The production function for a toy manufacturer is given by the equation

$$Q(L, M) = 400L^{0.3}M^{0.7}$$

where Q is the output in units, L is the labor in hours, and M is the number of machine hours. Find

(a) $Q(19, 21)$ (b) $Q(21, 20)$

8.3 Partial Derivatives

PREPARING FOR THIS SECTION *Before getting started, review the following:*

>> The Definition of the Derivative of a Function (Chapter 4, Section 4.1, p. 278)

OBJECTIVES
1. Find partial derivatives
2. Find the slope of the tangent line to a surface
3. Interpret partial derivatives
4. Find higher-order partial derivatives

For a function $y = f(x)$ of one independent variable, we have defined the derivative f'. For a function $z = f(x, y)$ of two independent variables, we introduce the idea of a *partial derivative*. A function $z = f(x, y)$ of two variables x and y will have two partial derivatives: f_x, the partial derivative of f with respect to x; and f_y, the partial derivative of f with respect to y.

> The partial derivative of f with respect to x is found by differentiating f with respect to x while treating y as if it were a constant.
> The partial derivative of f with respect to y is found by differentiating f with respect to y while treating x as if it were a constant.

For example, if $z = f(x, y) = 2xy + 3xy^2$, then we can find f_x by differentiating $z = 2xy + 3xy^2$ with respect to x, while treating y as if it were a constant. The result is

$$f_x = 2y + 3y^2 \qquad \frac{dz}{dx} = \frac{d}{dx}(2xy + 3xy^2) = 2y + 3y^2 \text{ if } y \text{ is a constant}$$

Similarly, by treating x as if it were constant and differentiating with respect to y, we obtain

$$f_y = 2x + 6xy \qquad \frac{dz}{dy} = \frac{d}{dy}(2xy + 3xy^2) = 2x + 6xy \text{ if } x \text{ is a constant}$$

We define these partial derivatives as limits of certain difference quotients.

Partial Derivatives of $z = f(x, y)$

Let $z = f(x, y)$ be a function of two variables. Then the **partial derivative of f with respect to x and the partial derivative of f with respect to y** are functions f_x and f_y defined as follows:

$$f_x(x, y) = \lim_{\Delta x \to 0} \frac{f(x + \Delta x, y) - f(x, y)}{\Delta x} \qquad (1)$$

$$f_y(x, y) = \lim_{\Delta y \to 0} \frac{f(x, y + \Delta y) - f(x, y)}{\Delta y} \qquad (2)$$

provided these limits exist.

Observe the similarity between the above definitions and the definition of a derivative given in Chapter 4. Observe also that in $f_x(x, y)$, an increment Δx is given to x, while y is fixed; in $f_y(x, y)$, an increment Δy is given to y, while x is fixed.

Finding Partial Derivatives of $z = f(x, y)$

To find $f_x(x, y)$: Differentiate f with respect to x while treating y as a constant.

To find $f_y(x, y)$: Differentiate f with respect to y while treating x as a constant.

EXAMPLE 1 Finding Partial Derivatives

Find f_x and f_y for

$$z = f(x, y) = 4x^3 + 2x^2y + y^2$$

SOLUTION To find f_x, we treat y as a constant and differentiate $z = 4x^3 + 2x^2y + y^2$ with respect to x. The result is

$$f_x(x, y) = 12x^2 + 4xy$$

To find f_y, we treat x as a constant and differentiate $z = 4x^3 + 2x^2y + y^2$ with respect to y. The result is

$$f_y(x, y) = 2x^2 + 2y$$

EXAMPLE 2 Finding Partial Derivatives

(a) Find f_x and f_y for $z = f(x, y) = x^2 \ln y + ye^x$.
(b) Evaluate $f_x(2, 1)$ and $f_y(2, 1)$.

SOLUTION (a) $f_x(x, y) = 2x \ln y + ye^x$ $f_y(x, y) = x^2 \cdot \dfrac{1}{y} + e^x = \dfrac{x^2}{y} + e^x$

(b) $f_x(2, 1) = 2 \cdot 2 \cdot \ln 1 + 1 \cdot e^2 = e^2$ $f_y(2, 1) = \dfrac{2^2}{1} + e^2 = 4 + e^2$ ▶

NOW WORK PROBLEM 1.

Another Notation

There is another notation used for the partial derivatives f_x and f_y of a function $z = f(x, y)$, which we introduce here:

$$f_x(x, y) = \frac{\partial f}{\partial x} = \frac{\partial z}{\partial x} \qquad f_y(x, y) = \frac{\partial f}{\partial y} = \frac{\partial z}{\partial y}$$

The symbols $\dfrac{\partial}{\partial x}$ and $\dfrac{\partial}{\partial y}$ read "the partial with respect to x" and "the partial with respect to y" respectively, denote operations performed on a function to obtain the partial derivatives with respect to x in the case of $\dfrac{\partial}{\partial x}$ and with respect to y in the case of $\dfrac{\partial}{\partial y}$. Using these notations sometimes makes it a little easier to find partial derivatives.

EXAMPLE 3 Finding Partial Derivatives

(a) $\dfrac{\partial}{\partial x}(3e^x y^2 - xy) = \dfrac{\partial}{\partial x}(3e^x y^2) - \underbrace{\dfrac{\partial}{\partial x}(xy)}_{\text{Treat } y \text{ as a constant}} = 3y^2 \dfrac{\partial}{\partial x} e^x - y \dfrac{\partial}{\partial x} x = 3y^2 e^x - y$

(b) $\dfrac{\partial}{\partial y}(3e^x y^2 - xy) = \dfrac{\partial}{\partial y}(3e^x y^2) - \underbrace{\dfrac{\partial}{\partial y}(xy)}_{\text{Treat } x \text{ as a constant}} = 3e^x \dfrac{\partial}{\partial y} y^2 - x \dfrac{\partial}{\partial y} y = 6e^x y - x$ ▶

Geometric Interpretation

For a geometric interpretation of the partial derivatives of $z = f(x, y)$, we look at the graph of the surface $z = f(x, y)$. See Figure 15. In computing f_x, we hold y fixed, say, at $y = y_0$,

FIGURE 15

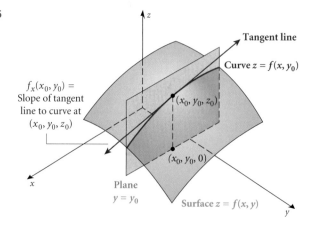

and then differentiate with respect to x. But holding y fixed at y_0 is equivalent to intersecting the surface $z = f(x, y)$ with the plane $y = y_0$, the result being the curve $z = f(x, y_0)$. The partial derivative f_x is the slope of the tangent line to this curve.

Geometric Interpretation of $f_x(x_0, y_0)$

The slope of the tangent line to the curve of intersection of the surface $z = f(x, y)$ and the plane $y = y_0$ at the point (x_0, y_0, z_0) on the surface equals $f_x(x_0, y_0)$.

Geometric Interpretation of $f_y(x_0, y_0)$

The slope of the tangent line to the curve of intersection of the surface $z = f(x, y)$ and the plane $x = x_0$ at the point (x_0, y_0, z_0) on the surface equals $f_y(x_0, y_0)$.

See Figure 16.

FIGURE 16

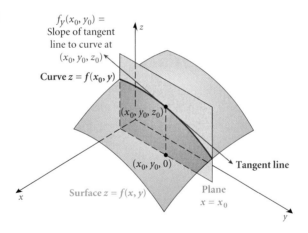

EXAMPLE 4 Finding the Slope of the Tangent Line

Find the slope of the tangent line to the curve of intersection of the surface $z = f(x, y) = 16 - x^2 - y^2$:

(a) With the plane $y = 2$ at the point $(1, 2, 11)$.
(b) With the plane $x = 1$ at the point $(1, 2, 11)$.

SOLUTION **(a)** The slope of the tangent line to the curve of intersection of the surface $z = 16 - x^2 - y^2$ and the plane $y = 2$ at any point is $f_x(x, y) = -2x$. At the point $(1, 2, 11)$, the slope is $f_x(1, 2) = -2(1) = -2$. See Figure 17(a) on page 526.

(b) The slope of the tangent line to the curve of intersection of the surface $z = 16 - x^2 - y^2$ and the plane $x = 1$ at any point is $f_y(x, y) = -2y$. At the point $(1, 2, 11)$, the slope is $f_y(1, 2) = -2(2) = -4$. See Figure 17(b).

FIGURE 17

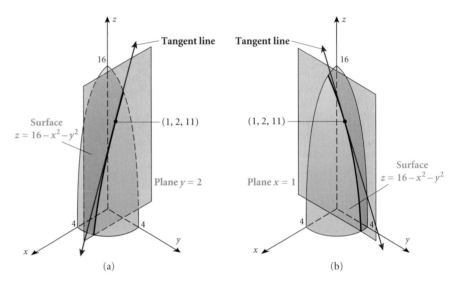

NOW WORK PROBLEM 31.

Applications

Marginal Analysis We have already noted the similarity of the definition of the partial derivatives of $z = f(x, y)$ and the definition of the derivative of a function of one variable. Let's look at how partial derivatives are interpreted in a business environment.

EXAMPLE 5 Interpreting Partial Derivatives

In many production processes, the total cost of manufacturing consists of a fixed cost and two variable costs: the cost of raw materials and the cost of labor. If the total cost is given by

$$C(x, y) = 180 + 18x + 40y$$

where x is the cost (in dollars) of raw materials and y is the cost (in dollars) of labor, find C_x and C_y and give an interpretation to your answer.

SOLUTION
$$C_x(x, y) = \frac{\partial C}{\partial x} = \frac{\partial}{\partial x}(180 + 18x + 40y) = 18$$

We interpret $C_x(x, y) = 18$ to mean that when the cost of labor y is held fixed, an increase of \$1 in the cost of raw materials causes an increase of \$18 in the total cost of the product. The partial derivative C_x measures the incremental cost due to an increase of \$1 in the cost of raw material, while labor costs are held fixed.

$$C_y(x, y) = \frac{\partial C}{\partial y} = \frac{\partial}{\partial y}(180 + 18x + 40y) = 40$$

The partial derivative $C_y(x, y) = 40$ means that when the cost of raw materials x is held fixed, an increase of \$1 in the cost of labor y causes a \$40 increase in the total cost of the product.

The partial derivative C_x is called the **marginal cost of raw material** and C_y is the **marginal cost of labor**.

NOW WORK PROBLEM 43.

Production Function The production of most goods requires the use of more than one component. If the quantity z of a good is produced by using the components x and

y, respectively, then the **production function**

$$z = f(x, y)$$

gives the amount of **output** z when the amounts x and y, the **inputs,** are used simultaneously.

> **Marginal Productivity**
>
> If a production function is given by $z = f(x, y)$, then the partial derivative $\dfrac{\partial z}{\partial x}$ of z with respect to x (with y held constant) is the **marginal productivity** of x or the **marginal product** of x. The partial derivative $\dfrac{\partial z}{\partial y}$ of z with respect to y (with x held constant) is the **marginal productivity** of y or the **marginal product** of y.

Notice that the marginal productivity of either input is the rate of increase of the total product as that input is increased, assuming that the amount of the other input remains constant.

Marginal productivity is usually positive—that is, as the amount of one input increases (with the amount of the other input held constant), the output also increases. However, as the input of one component increases, the output usually increases at a decreasing rate until the point is reached at which there is no further increase in output. Then, a decrease in total output occurs. This characteristic behavior of production functions is known as the **law of eventually diminishing marginal productivity.**

EXAMPLE 6 **Finding Marginal Productivity**

Let $z = 2x^{1/2}y^{1/2}$ be a production function. Find the marginal productivity of x and the marginal productivity of y.

SOLUTION The marginal productivity of x is

$$\frac{\partial z}{\partial x} = \frac{\partial}{\partial x}(2x^{1/2}y^{1/2}) = 2y^{1/2}\frac{\partial}{\partial x}x^{1/2} = 2y^{1/2} \cdot \frac{1}{2}x^{-1/2} = \frac{y^{1/2}}{x^{1/2}}$$

The marginal productivity of y is

$$\frac{\partial z}{\partial y} = \frac{\partial}{\partial y}(2x^{1/2}y^{1/2}) = 2x^{1/2}\frac{\partial}{\partial x}y^{1/2} = 2x^{1/2} \cdot \frac{1}{2}y^{-1/2} = \frac{x^{1/2}}{y^{1/2}}$$

Observe that $\dfrac{\partial z}{\partial x}$ is always positive, but decreases as x increases since y is held fixed. Similarly, $\dfrac{\partial z}{\partial y}$ is always positive, but decreases as y increases since x is held fixed.

Higher-Order Partial Derivatives

For a function $z = f(x, y)$ of two variables for which the limits (1) and (2) exist, there are two **first-order partial derivatives:** f_x and f_y. If it is possible to differentiate each of these partially with respect to x or partially with respect to y, there will result four **second-order partial derivatives,** namely,

$$f_{xx}(x, y) = \frac{\partial}{\partial x}f_x(x, y) = \frac{\partial}{\partial x}\frac{\partial z}{\partial x} = \frac{\partial^2 z}{\partial x^2} \qquad f_{xy}(x, y) = \frac{\partial}{\partial y}f_x(x, y) = \frac{\partial}{\partial y}\frac{\partial z}{\partial x} = \frac{\partial^2 z}{\partial y\,\partial x}$$

$$f_{yx}(x, y) = \frac{\partial}{\partial x}f_y(x, y) = \frac{\partial}{\partial x}\frac{\partial z}{\partial y} = \frac{\partial^2 z}{\partial x\,\partial y} \qquad f_{yy}(x, y) = \frac{\partial}{\partial y}f_y(x, y) = \frac{\partial}{\partial y}\frac{\partial z}{\partial y} = \frac{\partial^2 z}{\partial y^2}$$

The two second-order partial derivatives

$$\frac{\partial^2 z}{\partial x\, \partial y} = f_{yx}(x, y) \quad \text{and} \quad \frac{\partial^2 z}{\partial y\, \partial x} = f_{xy}(x, y)$$

are called **mixed partials.**

Observe the difference in the mixed partials. The notation f_{yx} means that first we should differentiate f partially with respect to y and then differentiate the result partially with respect to x—in that order! On the other hand, f_{xy} means we should differentiate with respect to x and then with respect to y.

For most functions the two mixed partials are equal.

Although there are functions for which the mixed partials are unequal, they are rare and will not be encountered in this book.

EXAMPLE 7 Finding Higher-Order Partial Derivatives

Find all second-order partial derivatives of: $\quad z = f(x, y) = x \ln y + y e^x$

SOLUTION First, we find the first-order partial derivatives $f_x(x, y)$ and $f_y(x, y)$.

$$f_x(x, y) = \frac{\partial}{\partial x}(x \ln y + y e^x) = \ln y + y e^x \qquad f_y(x, y) = \frac{\partial}{\partial y}(x \ln y + y e^x) = \frac{x}{y} + e^x$$

Then,

$$f_{xx} = \frac{\partial}{\partial x} f_x = \frac{\partial}{\partial x}(\ln y + y e^x) = y e^x \qquad f_{xy} = \frac{\partial}{\partial y} f_x = \frac{\partial}{\partial y}(\ln y + y e^x) = \frac{1}{y} + e^x$$

$$f_{yx} = \frac{\partial}{\partial x} f_y = \frac{\partial}{\partial x}\left(\frac{x}{y} + e^x\right) = \frac{1}{y} + e^x \qquad f_{yy} = \frac{\partial}{\partial y} f_y = \frac{\partial}{\partial y}\left(\frac{x}{y} + e^x\right) = \frac{-x}{y^2}$$

NOW WORK PROBLEM 7.

Functions of Three Variables

The idea of partial differentiation may be extended to a function of three variables. If $w = f(x, y, z)$ is a function of three variables, there will be three first-order partial derivatives: the partial derivative with respect to x is f_x; the partial derivative with respect to y is f_y; and the partial derivative with respect to z is f_z. Each of these is calculated by differentiating with respect to the indicated variable, while treating the other two as constants.

EXAMPLE 8 Finding Partial Derivatives

Find f_x, f_y, f_z, if $f(x, y, z) = 10x^2 y^3 z^4$.

SOLUTION
$$f_x(x, y, z) = \frac{\partial}{\partial x}(10x^2 y^3 z^4) = 10 y^3 z^4 \frac{\partial}{\partial x} x^2 = 10 y^3 z^4 \cdot 2x = 20 x y^3 z^4$$

$$f_y(x, y, z) = \frac{\partial}{\partial y}(10x^2 y^3 z^4) = 10 x^2 z^4 \frac{\partial}{\partial y} y^3 = 10 x^2 z^4 \cdot 3 y^2 = 30 x^2 y^2 z^4$$

$$f_z(x, y, z) = \frac{\partial}{\partial z}(10x^2 y^3 z^4) = 10 x^2 y^3 \frac{\partial}{\partial z} z^4 = 10 x^2 y^3 \cdot 4 z^3 = 40 x^2 y^3 z^3$$

NOW WORK PROBLEM 23.

EXERCISE 8.3

In Problems 1–6, find f_x, f_y, $f_x(2, -1)$, and $f_y(-2, 3)$.

1. $f(x, y) = 3x - 2y + 3y^3$
2. $f(x, y) = 2x^3 - 3y + x^2$
3. $f(x, y) = (x - y)^2$
4. $f(x, y) = (x - y)^3$
5. $f(x, y) = \sqrt{x^2 + y^2}$
6. $f(x, y) = \sqrt{x^2 - y^2}$

In Problems 7–16, find f_x, f_y, f_{xx}, f_{yy}, f_{yx}, and f_{xy}.

7. $f(x, y) = y^3 - 2xy + y^2 - 12x^2$
8. $f(x, y) = x^3 - xy + 10y^2x$
9. $f(x, y) = xe^y + ye^x + x$
10. $f(x, y) = xe^x + xe^y + y$
11. $f(x, y) = \dfrac{x}{y}$
12. $f(x, y) = \dfrac{y}{x}$
13. $f(x, y) = \ln(x^2 + y^2)$
14. $f(x, y) = \ln(x^2 - y^2)$
15. $f(x, y) = \dfrac{10 - x + 2y}{xy}$
16. $f(x, y) = \dfrac{5 + 3x - 2y}{xy}$

In Problems 17–22, verify that $f_{xy} = f_{yx}$.

17. $f(x, y) = x^3 + y^2$
18. $f(x, y) = x^2 - y^3$
19. $f(x, y) = 3x^4y^2 + 7x^2y$
20. $f(x, y) = 5x^3y - 8xy^2$
21. $f(x, y) = \dfrac{y}{x^2}$
22. $f(x, y) = \dfrac{x}{y^2}$

In Problems 23–30, find f_x, f_y, f_z.

23. $f(x, y, z) = x^2y - 3xyz + z^3$
24. $f(x, y, z) = 3xy + 4yz + 8z^2$
25. $f(x, y, z) = xe^y + ye^z$
26. $f(x, y, z) = x \ln y + y \ln z$
27. $f(x, y, z) = x \ln(yz) + y \ln(xz)$
28. $f(x, y, z) = e^{(3x + 4y + 5z)}$
29. $f(x, y, z) = \ln(x^2 + y^2 + z^2)$
30. $f(x, y, z) = e^{(x^2 + y^2 + z^2)}$

In Problems 31–38, find the slope of the tangent line to the curve of intersection of the surface $z = f(x, y)$ with the given plane at the indicated point.

31. $z = f(x, y) = 5x^2 + 3y^2$; plane: $y = 3$; point: $(2, 3, 47)$
32. $z = f(x, y) = 2x^2 - 4y^2$; plane: $x = 2$; point: $(2, 3, -28)$
33. $z = f(x, y) = \sqrt{16 - x^2 - y^2}$; plane: $x = 1$; point: $(1, 2, \sqrt{11})$
34. $z = f(x, y) = \sqrt{x^2 - y^2}$; plane: $y = 0$; point: $(4, 0, 4)$
35. $z = f(x, y) = e^x \ln y$; plane: $x = 0$; point: $(0, 1, 0)$
36. $z = f(x, y) = e^{2x + 3y}$; plane: $y = 0$; point: $(0, 0, 1)$
37. $z = f(x, y) = 2 \ln \sqrt{x^2 + y^2}$; plane: $x = 1$; point: $(1, 1, 2 \ln 2)$
38. $z = f(x, y) = e^{x^2 + y^2}$; plane: $y = 0$; point: $(1, 0, e)$

39. If $z = x^2 + 4y^2$, show that $x\dfrac{\partial z}{\partial x} + y\dfrac{\partial z}{\partial y} = 2z$

40. If $z = xy^2$, show that $x\dfrac{\partial z}{\partial x} + y\dfrac{\partial z}{\partial y} = 3z$

41. If $z = \ln \sqrt{x^2 + y^2}$, show that $\dfrac{\partial^2 z}{\partial x^2} + \dfrac{\partial^2 z}{\partial y^2} = 0$

42. If $z = e^{x \ln y}$, find $\dfrac{\partial^2 z}{\partial x^2}$, $\dfrac{\partial^2 z}{\partial y^2}$, and $\dfrac{\partial^2 z}{\partial x \, \partial y}$

43. **Demand for Butter** In a large town, the demand for butter (measured in pounds) is given by the formula

$$z = 1000 - 20x - 50y$$

where x (in dollars) is the average price per pound of butter and y (in dollars) is the average price per pound of margarine.

(a) Find the two first-order partial derivatives of z.
(b) Interpret each partial derivative.

44. **Marginal Productivity** The production function of a certain commodity is given by

$$P = 8I - I^2 + 3Ik + 50k - k^2$$

where I and k are the labor and capital inputs, respectively.

(a) Find the marginal productivities of I and k at $I = 2$ and $k = 5$.
(b) Interpret each marginal productivity.

45. Baseball A pitcher's earned run average is given by

$$A = 9\left(\frac{N}{I}\right)$$

where N is the total number of earned runs given up in I inning of pitching,

(a) Find both first-order partial derivatives.
(b) Evaluate the two-first order partial derivatives for the earned run average of Kevin Millwood of the Philadelphia Phillies who in the 2002 season gave up 78 earned runs in 217 innings.
(c) Interpret each partial derivative found in part (b).

46. Field Goal Percentage in the NBA The Adjusted Field Goal Percentage is determined by

$$f(x, y, z) = \frac{x + 1.5y}{s},$$

where x is the number of two-point field goals made, y is the number of three-point field goals made, and s is the total number of field goals attempted.

(a) Evaluate the three first-order partial derivatives for the Adjusted Field Goal Percentage function for Tracy McGrady, who made 656 of 1365 two point attempts and 173 out of 448 three-point attempts.
(b) Interpret each one of the three partial derivatives.

47. Heat Index At the 2000 Olympic Games in Sydney, Australia, the following formula was used to calculate the heat index (apparent temperature):

$$H = -42.379 + 2.04901523t + 10.14333127r \\ - 0.22475541tr - 0.00683783t^2 \\ - 0.05481717r^2 + 0.00122874t^2r \\ + 0.00085282tr^2 - 0.00000199t^2r^2$$

where H = the heat index (in °F)
  t = the air temperature (in °F)
  r = the % relative humidity (e.g., $r = 75$ when the relative humidity is 75%)

(a) Find $\frac{\partial H}{\partial t}$.
(b) Give an interpretation to $\frac{\partial H}{\partial t}$.
(c) Find $\frac{\partial H}{\partial r}$.
(d) Give an interpretation to $\frac{\partial H}{\partial r}$.

Source: Commonwealth Bureau of Meteorology, Melbourne, Australia.

48. The Cobb-Douglas Model In 1928 the mathematician Charles W. Cobb and the economist Paul H. Douglas empirically derived a production model for the manufacturing sector of the United States. The table on the right shows the data used by Cobb and Douglas to construct their original model. In the table, P represents the output, K the capital input, and L the labor input of the manufacturing sector of the United States for the period 1899–1922. The data are scaled so that for the year 1899, $P = K = L = 100$. Using the model $P = aK^bL^{1-b}$ and multiple regression techniques, Cobb and Douglas determined the following relationship between P, K, and L:

$$P = 1.014651K^{.254134}L^{.745866} \approx 1.01K^{.25}L^{.75}$$

(a) For the Cobb-Douglas production model, find the partial derivatives: $\frac{\partial P}{\partial K}$ and $\frac{\partial P}{\partial L}$.
(b) Evaluate each partial derivative for $K = 226$ and $L = 152$. (These are the values for K and L in 1912.)
(c) Interpret these partial derivatives in terms of the Cobb-Douglas production model.

Year	P	K	L
1899	100	100	100
1900	101	107	105
1901	112	114	110
1902	122	122	118
1903	124	131	123
1904	122	138	116
1905	143	149	125
1906	152	163	133
1907	151	176	138
1908	126	185	121
1909	155	198	140
1910	159	208	144
1911	153	216	145
1912	177	226	152
1913	184	236	154
1914	169	244	149
1915	189	266	154
1916	225	298	182
1917	227	335	196
1918	223	366	200
1919	218	387	193
1920	231	407	193
1921	179	417	147
1922	240	431	161

Source: Cobb and Douglas, 1928.

49. If you are told that a function $z = f(x, y)$ has the two partial derivatives $f_x(x, y) = 3x - y$ and $f_y(x, y) = x - 3y$, should you believe it? Explain.

8.4 Local Maxima and Local Minima

PREPARING FOR THIS SECTION *Before getting started, review the following:*

> First Derivative Test (Chapter 5, Section 5.2, pp. 355–356)

> Second Derivative Test (Chapter 5, Section 5.3, p. 373)

OBJECTIVES
1. Find critical points
2. Determine the character of a critical point
3. Solve applied problems

We saw in Chapter 5 that an important application of the derivative is to find the local maxima and the local minima of a function of one variable. In this section we learn that the partial derivatives of a function of two variables are used in a similar way to find the local maxima and the local minima of a function of two variables.

> **Local Maximum; Local Minimum**
>
> Let $z = f(x, y)$ denote a function of two variables. We say that f has a **local maximum** at a point (x_0, y_0) if $f(x_0, y_0) \geq f(x, y)$ for all points (x, y) close to the point (x_0, y_0). Similarly, f has a **local minimum** at (x_1, y_1) if $f(x_1, y_1) \leq f(x, y)$ for all points (x, y) close to (x_1, y_1).

Figure 18 illustrates the graph of a function $z = f(x, y)$ that has several local maxima and minima.

FIGURE 18

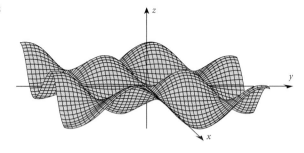

FIGURE 19

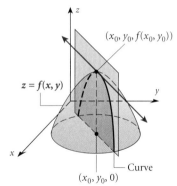

Suppose a function $z = f(x, y)$ has a local maximum at the point (x_0, y_0). Then the curve that results from the intersection of the surface $z = f(x, y)$ and any plane through the point $(x_0, y_0, 0)$ that is perpendicular to the xy-plane will have a local maximum at (x_0, y_0). See Figure 19.

In particular, the curve that results from intersecting $z = f(x, y)$ with the plane $x = x_0$ has this property. This means that if f_y exists at (x_0, y_0), then $f_y(x_0, y_0) = 0$. By a similar argument, we also have $f_x(x_0, y_0) = 0$. This leads us to formulate the following necessary condition for a local maximum and a local minimum:

A Necessary Condition for Local Maxima and Local Minima

Let $z = f(x, y)$ denote a function of two variables and let (x_0, y_0) be a point in its domain. Suppose f_x and f_y each exist at (x_0, y_0). If f has a local maximum at (x_0, y_0) or a local minimum at (x_0, y_0), then

$$f_x(x_0, y_0) = 0 \quad \text{and} \quad f_y(x_0, y_0) = 0 \tag{1}$$

From this theorem we see the importance of those points at which the partial derivatives exist and are zero simultaneously. Such points are called **critical points.**

We say that f has a **critical point** at (x_0, y_0) if

$$f_x(x_0, y_0) = 0 \quad \text{and} \quad f_y(x_0, y_0) = 0$$

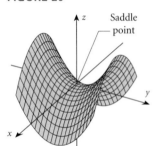

FIGURE 20

Saddle point

It can happen that f has a critical point at (x_0, y_0) but that the point is neither a local maximum nor a local minimum. Such points are called **saddle points.** Figure 20 illustrates a saddle point—and suggests justification for the name. Note that this saddle point is a maximum when viewed relative to x and a minimum when viewed relative to y.

EXAMPLE 1   Finding Critical Points

Find all the critical points of

$$z = f(x, y) = x^2 + y^4 - 2y^2 + 6$$

SOLUTION We compute the partial derivatives f_x and f_y, set each equal to zero, and solve the resulting system of equations.

$$f_x(x, y) = \frac{\partial}{\partial x}(x^2 + y^4 - 2y^2 + 6) = 2x = 0$$

$$f_y(x, y) = \frac{\partial}{\partial y}(x^2 + y^4 - 2y^2 + 6) = 4y^3 - 4y = 0$$

From the first equation, we obtain $x = 0$; from the second, we obtain

$$4y^3 - 4y = 0$$
$$4y(y^2 - 1) = 0 \quad \text{Factor.}$$
$$4y(y + 1)(y - 1) = 0 \quad \text{Factor.}$$
$$y = 0 \text{ or } y = -1 \text{ or } y = 1 \quad \text{Solve for } y.$$

We find that f has three critical points: $(0, 0)$, $(0, -1)$, and $(0, 1)$.

 NOW WORK PROBLEM 1.

Tests for Local Maxima, Local Minima, Saddle Points

We still don't know, though, the character of these critical points. What we need is a test, or tests, to tell us whether these critical points are in fact local maxima, local minima, or saddle points. For functions that possess both first- and second-order partial derivatives (this is true of all the ones we will encounter in this book), the following tests may be used.

Test for a Local Maximum

Suppose (x_0, y_0) is a critical point of $z = f(x, y)$. If

> 1. $f_{xx}(x_0, y_0) < 0$
> 2. $D = f_{xx}(x_0, y_0) \cdot f_{yy}(x_0, y_0) - [f_{xy}(x_0, y_0)]^2 > 0$

then the function has a local maximum at (x_0, y_0).

Test for a Local Minimum

Suppose (x_0, y_0) is a critical point of $z = f(x, y)$. If

> 1. $f_{xx}(x_0, y_0) > 0$
> 2. $D = f_{xx}(x_0, y_0) \cdot f_{yy}(x_0, y_0) - [f_{xy}(x_0, y_0)]^2 > 0$

then the function f has a local minimum at (x_0, y_0).

Test for Saddle Point

Suppose (x_0, y_0) is a critical point of $z = f(x, y)$. If

> $D = f_{xx}(x_0, y_0) \cdot f_{yy}(x_0, y_0) - [f_{xy}(x_0, y_0)]^2 < 0$

then the point (x_0, y_0) is a saddle point of f.

Two comments about using these tests:

1. If $D > 0$, then $f_{xx}(x_0, y_0)$ and $f_{yy}(x_0, y_0)$ are each of the same sign.
2. If $D = 0$, no information results.

EXAMPLE 2 Determining the Character of a Critical Point

Return to the function discussed in Example 1, $f(x, y) = x^2 + y^4 - 2y^2 + 6$, and determine the character of each critical point.

SOLUTION We need to compute the second-order partial derivatives. From Example 1, we have

$$f_x(x, y) = 2x \quad \text{and} \quad f_y(x, y) = 4y^3 - 4y$$

Then,

$$f_{xx}(x, y) = \frac{\partial}{\partial x} f_x(x, y) = \frac{\partial}{\partial x}(2x) = 2$$

$$f_{yy}(x, y) = \frac{\partial}{\partial y} f_y(x, y) = \frac{\partial}{\partial y}(4y^3 - 4y) = 12y^2 - 4$$

$$f_{xy}(x, y) = \frac{\partial}{\partial y} f_x(x, y) = \frac{\partial}{\partial y}(2x) = 0$$

Next, we evaluate these partial derivatives at each critical point to find the value of D.

$(0, 0)$: $f_{xx}(0, 0) = 2$ $f_{yy}(0, 0) = -4$ $f_{xy}(0, 0) = 0$

$$D = f_{xx}(0, 0) \cdot f_{yy}(0, 0) - [f_{xy}(0, 0)]^2 = 2 \cdot (-4) - 0^2 = -8 < 0$$

Since $D < 0$, the critical point $(0, 0)$ is a saddle point of f. That is, the point $(0, 0, 6)$ on the graph is a saddle point.

$(0, -1)$: $f_{xx}(0, -1) = 2$ $f_{yy}(0, -1) = 8$ $f_{xy}(0, -1) = 0$

$$D = f_{xx}(0, -1) \cdot f_{yy}(0, -1) - [f_{xy}(0, -1)]^2 = 2 \cdot 8 - 0^2 = 16 > 0$$

Since $f_{xx}(0, -1) = 2 > 0$, and $D > 0$, f has a local minimum at the critical point $(0, -1)$. That is, at the point $(0, -1, 5)$ on the graph, there is a local minimum.

$(0, 1)$: $f_{xx}(0, 1) = 2$ $f_{yy}(0, 1) = 8$ $f_{xy}(0, 1) = 0$

$$D = f_{xx}(0, 1) \cdot f_{yy}(0, 1) - [f_{xy}(0, 1)]^2 = 2 \cdot 8 - 0^2 = 16 > 0$$

Since $f_{xx}(0, 1) = 2 > 0$ and $D > 0$, f has a local minimum at the critical point $(0, 1)$. That is, at the point $(0, 1, 5)$ on the graph of f, there is a local minimum. ▶

NOW WORK PROBLEM 7.

EXAMPLE 3 **Determining the Character of a Critical Point**

For the function
$$z = f(x, y) = x^2 + xy + y^2 - 6x + 6$$
find all critical points. Determine the character of each one.

SOLUTION First, we compute the first-order partial derivatives of $z = f(x, y)$.

$$f_x(x, y) = \frac{\partial}{\partial x}(x^2 + xy + y^2 - 6x + 6) = 2x + y - 6$$

$$f_y(x, y) = \frac{\partial}{\partial y}(x^2 + xy + y^2 - 6x + 6) = x + 2y$$

The critical points, if there are any, obey the system of equations

$$f_x(x, y) = 2x + y - 6 = 0 \quad \text{and} \quad f_y(x, y) = x + 2y = 0$$

Now, we solve these equations simultaneously, using substitution. From the equation $2x + y - 6 = 0$ we find $y = 6 - 2x$. Substituting into the equation $x + 2y = 0$, we find

$$x + 2(6 - 2x) = 0$$
$$x + 12 - 4x = 0$$
$$-3x = -12$$
$$x = 4$$

Then $y = 6 - 2x = 6 - 2(4) = -2$. The only critical point of f is $(4, -2)$.

$$f_{xx}(x, y) = \frac{\partial}{\partial x} f_x(x, y) = \frac{\partial}{\partial x}(2x + y - 6) = 2$$

$$f_{xy}(x, y) = \frac{\partial}{\partial y} f_x(x, y) = \frac{\partial}{\partial y}(2x + y - 6) = 1$$

$$f_{yy}(x, y) = \frac{\partial}{\partial y} f_y(x, y) = \frac{\partial}{\partial y}(x + 2y) = 2$$

At the critical point $(4, -2)$, we have

$$f_{xx}(4, -2) = 2 \quad f_{xy}(4, -2) = 1 \quad f_{yy}(4, -2) = 2$$

and,

$$D = f_{xx}(4, -2) \cdot f_{yy}(4, -2) - [f_{xy}(4, -2)]^2 = 2 \cdot 2 - 1^2 = 3$$

Since $f_{xx}(4, -2) = 2 > 0$ and $D > 0$, f has a local minimum at the critical point $(4, -2)$. That is, at the point $(4, -2, -6)$ on the graph of f, there is a local minimum. ▶

Application

EXAMPLE 4 Maximizing Profit

The demand functions for two products are

$$p = 12 - 2x \quad \text{and} \quad q = 20 - y$$

where p and q are the respective prices (in thousands of dollars) for each product, and x and y are the respective amounts (in thousands of units) of each sold. Suppose the joint cost function is

$$C(x, y) = x^2 + 2xy + 2y^2$$

Find the revenue function and the profit function. Determine the prices and amounts that will maximize profit. What is the maximum profit?

SOLUTION The revenue function R is the sum of the revenues due to each product. That is,

$$R = R(x, y) = xp + yq = x(12 - 2x) + y(20 - y)$$

The profit function P is

$$\begin{aligned} P = P(x, y) &= R(x, y) - C(x, y) \\ &= x(12 - 2x) + y(20 - y) - (x^2 + 2xy + 2y^2) \\ &= 12x - 2x^2 + 20y - y^2 - x^2 - 2xy - 2y^2 \\ &= -3x^2 - 3y^2 - 2xy + 12x + 20y \end{aligned}$$

The first-order partial derivatives of P are

$$P_x(x, y) = \frac{\partial}{\partial x}(-3x^2 - 3y^2 - 2xy + 12x + 20y) = -6x - 2y + 12$$

$$P_y(x, y) = \frac{\partial}{\partial y}(-3x^2 - 3y^2 - 2xy + 12x + 20y) = -6y - 2x + 20$$

The critical points obey

$$-6x - 2y + 12 = 0 \quad \text{and} \quad -6y - 2x + 20 = 0$$

We solve the left equation $-6x - 2y + 12 = 0$ for y.

$$-2y = 6x - 12$$
$$y = -3x + 6$$

Substitute into the right equation and solve for x.

$$-6(-3x + 6) - 2x + 20 = 0 \quad\quad -6y - 2x + 20 = 0$$
$$18x - 36 - 2x + 20 = 0$$
$$16x = 16$$
$$x = 1$$

Then $y = -3x + 6 = -3(1) + 6 = 3$ so $(1, 3)$ is the critical point.
The second-order partial derivatives of P are

$$P_{xx}(x, y) = \frac{\partial}{\partial x}(-6x - 2y + 12) = -6$$

$$P_{xy}(x, y) = \frac{\partial}{\partial y}(-6x - 2y + 12) = -2$$

$$P_{yy}(x, y) = \frac{\partial}{\partial y}(-6y - 2x + 20) = -6$$

At the critical point $(1, 3)$, we see that

$$P_{xx}(1, 3) = -6 < 0$$
$$D = P_{xx}(1, 3) \cdot P_{yy}(1, 3) - [P_{xy}(1, 3)]^2 = (-6)(-6) - (-2)^2 = 32 > 0$$

We conclude that P has a local maximum at $(1, 3)$. For these quantities sold, namely, $x = 1000$ units and $y = 3000$ units, the corresponding prices p and q are $p = \$10,000$ and $q = \$17,000$. The maximum profit is $P(1, 3) = \$36,000$.

NOW WORK PROBLEM 25.

EXERCISE 8.4

In Problems 1–6, find all the critical points of each function.

1. $f(x, y) = x^4 - 2x^2 + y^2 + 15$
2. $f(x, y) = x^2 - y^2 + 6x - 2y + 14$
3. $f(x, y) = 4xy - x^4 - y^4 + 12$
4. $f(x, y) = x^3 + 6xy + 3y^2 + 8$
5. $f(x, y) = x^4 + y^4$
6. $f(x, y) = xy + \dfrac{2}{x} + \dfrac{4}{y}$

In Problems 7–24, find all the critical points of each function f and determine at each one whether there is a local maximum, a local minimum, or a saddle point of f.

7. $f(x, y) = 3x^2 - 2xy + y^2$
8. $f(x, y) = x^2 - 2xy + 3y^2$
9. $f(x, y) = x^2 + y^2 - 3x + 12$
10. $f(x, y) = x^2 + y^2 - 6y + 10$
11. $f(x, y) = x^2 - y^2 + 4x + 8y$
12. $f(x, y) = x^2 - y^2 - 2x + 4y$

13. $f(x, y) = x^2 + 4y^2 - 4x + 8y - 1$
14. $f(x, y) = x^2 + y^2 - 4x + 2y - 4$
15. $f(x, y) = x^2 + y^2 + xy - 6x + 6$
16. $f(x, y) = x^2 + y^2 + xy - 8y$
17. $f(x, y) = 2 + x^2 - y^2 + xy$
18. $f(x, y) = x^2 - y^2 + 2xy$
19. $f(x, y) = x^3 - 6xy + y^3$
20. $f(x, y) = x^3 - 3xy - y^3$
21. $f(x, y) = x^3 + x^2y + y^2$
22. $f(x, y) = 3y^3 - x^2y + x$
23. $f(x, y) = \dfrac{y}{x + y}$
24. $f(x, y) = \dfrac{x}{x + y}$

---

25. **Economics** The demand functions for two products are $p = 12 - x$ and $q = 8 - y$, where p and q are the respective prices (in thousands of dollars), and x and y are the respective amounts (in thousands of units) of each product sold. If the joint cost function is $C(x, y) = x^2 + 2xy + 3y^2$, determine the quantities x, y and prices p, q that maximize profit. What is the maximum profit?

26. **Economics** The labor cost of a firm is given by the function
$$Q(x, y) = x^2 + y^3 - 6xy + 3x + 6y - 5$$
where x is the number of days required by a skilled worker and y is the number of days required by a semiskilled worker. Find the values of x and y for which the labor cost is a minimum.

27. **Maximizing Profit** A steel manufacturer produces two grades of steel, x tons of grade A and y tons of grade B. The cost C and revenue R are given in dollars by the formulas
$$C = \dfrac{1}{20}x^2 + 700x + y^2 - 150y - \dfrac{1}{2}xy$$
$$R = 2700x - \dfrac{3}{20}x^2 + 1000y - y^2 + \dfrac{1}{2}xy + 10{,}000$$
If $P = \text{Profit} = R - C$, find the production (in tons) of grades A and B that maximizes the manufacturer's profit.

28. A certain mountain is in the shape of the surface
$$z = 2xy - 2x^2 - y^2 - 8x + 6y + 4$$
(The unit of distance is 1000 feet.) If sea level is the xy-plane, how high is the mountain?

29. **Reaction to Drugs** Two drugs are used simultaneously as a treatment for a certain disease. The reaction R (measured in appropriate units) to x units of the first drug and y units of the second drug is
$$R(x, y) = x^2y^2(a - x)(b - y) \qquad 0 \le x \le a, \ 0 \le y \le b$$
(a) For a fixed amount x of the first drug, what amount y of the second drug produces the maximum reaction?
(b) For a fixed amount y of the second drug, what amount x of the first drug produces the maximum reaction?
(c) If x and y are both variable, what amount of each maximizes the reaction?

30. **Reaction to Drugs** The reaction R to x units of a drug t hours after the drug has been administered is given by
$$R(x, t) = x^2(a - x)t^2e^{-t} \qquad 0 \le x \le a$$
For what amount x is the reaction as large as possible? When does the maximum reaction occur?

31. **Reaction to Drugs** The reaction y to an injection of x units of a certain drug, t hours after the injection, is given by
$$y = x^2(a - x)t \qquad 0 \le x \le a$$
Find the values of x and t, if any, that will maximize y.

32. **Metal Detector** A metal detector is used to locate an underground pipe. After several readings of the detector are taken, it is determined that the reading D at an arbitrary point (x, y), $x \ge 0$, $y \ge 0$, is given by
$$D = y(x - x^2) - x^2 \qquad \text{volts}$$
Determine the point (x, y) where the reading is largest.

33. **Parcel Post Regulations** United Parcel Service regulations state that individual packages can be up to 108 inches in length and 130 inches in combined length and girth (perimeter of a cross section) before additional charges apply.
(a) Find the dimensions of maximum volume of a rectangular box which can be sent without additional charges.
(b) What are the dimensions of maximum volume if the package is cylindrical?

Source: United Parcel Service.

Hint: The volume of a cylinder is $V = \pi r^2 h$, where r is the radius and h is the height.

34. **Parcel Post Regulations** The U.S. Post Office regulations state that the combined (sum) length and girth of a parcel post package being sent first-class in the United States may not exceed 84 inches. If this combined length and girth exceeds 84 inches, extra postage will be charged according to weight. Find:
(a) The length, width, and height of a rectangular box of maximum volume that can be mailed first class, subject to the 84 inch restriction.
(b) The dimensions of a circular tube of maximum volume that can be mailed first class, subject to the 84 inch restriction.

Source: United States Postal Service.

35. **Waste Management** A car manufacturer uses x tons of steel at the rate of y tons per week. It is found that the waste

W due to storage and interplant distribution amounts to

$$W = \frac{1}{100}\left[\frac{1}{20}x^2 + 25y^2 - x(y+4)\right] \text{ tons}$$

Determine the value of x and y for which waste is minimum.

36. **Expansion of Gas** The volume of a fixed amount of gas varies directly with the temperature and inversely with the pressure. That is, $V = k\left(\dfrac{T}{P}\right)$, where $k > 0$ is a constant and V, T, and P are the volume, temperature, and pressure, respectively.

(a) Calculate $\dfrac{\partial V}{\partial T}$ and $\dfrac{\partial V}{\partial P}$.

(b) Prove that

$$P \cdot \frac{\partial V}{\partial P} + T \cdot \frac{\partial V}{\partial T} = 0$$

8.5 Lagrange Multipliers

OBJECTIVE 1 Use the method of Lagrange multipliers

In the previous section we introduced a method to find the local maximum and the local minimum of a function of two variables without any constraints or conditions on the function or the variables. However, in many practical problems we are faced with maximizing or minimizing a function subject to conditions or constraints on the variables involved.

For example, a manufacturer may want to produce a box with a fixed volume so that the least amount of material is used.

Let's look again at an example we solved earlier.

EXAMPLE 1 Maximizing Area

A farmer wants to enclose a rectangular plot that borders on a straight river with a fence. He will not fence in the side along the river. If the farmer has 4000 meters of fencing, what is the largest area that can be enclosed?

SOLUTION Refer to Figure 21. If A is the area to be enclosed, then the problem is to find the maximum value of $A = xy$ subject to the condition that $2x + y = 4000$, that is, subject to the 4000 meters of fence that are available. To express the problem in terms of a single variable, we solve for y in the equation $2x + y = 4000$. Then $y = 4000 - 2x$ and the area A can be expressed in terms of x alone as

$$A = xy = x(4000 - 2x) = 4000x - 2x^2$$

This equation for A is easy to differentiate. We proceed to find the critical numbers of A:

$$A'(x) = \frac{d}{dx}(4000x - 2x^2) = 4000 - 4x = 0$$

$$x = 1000$$

Since $A''(x) = -4 < 0$, this critical number yields a maximum value for the area A, namely,

$$A = 4000(1000) - 2(1000)^2 = 2{,}000{,}000 \text{ square meters}$$

FIGURE 21

In Example 1, the problem required that we maximize $A = A(x, y) = xy$ subject to the side condition or *constraint* involving x and y, namely, that $2x + y = 4000$. We were able to the solve this problem using earlier techniques for two reasons:

1. In the equation $2x + y = 4000$ it was easy to solve for y in terms of x.
2. After substituting into $A = xy$, the area A became a function of the single variable x, which was easy to differentiate.

Suppose we want to maximize or minimize a function $z = f(x, y)$ subject to a constraint $g(x, y) = 0$ in which

1. It is *not* easy to solve the equation $g(x, y) = 0$ for x or for y, or
2. After substitution, the resulting function z of a single variable is not easy to differentiate.

In such cases, we can instead use the *method of Lagrange multipliers.* We describe the method below:

Consider a function $z = f(x, y)$ of two variables x and y, subject to a single constraint $g(x, y) = 0$. We introduce a new variable λ,* called the **Lagrange multiplier,** and construct the function

$$F(x, y, \lambda) = f(x, y) + \lambda g(x, y)$$

This new function F is a function of three variables x, y, and λ. The following result establishes the connection between the function F and the local maxima and the local minima of $z = f(x, y)$.

Method of Lagrange Multipliers

Suppose that, subject to the constraint $g(x, y) = 0$, the function $z = f(x, y)$ has a local maximum or a local minimum at the point (x_0, y_0). Form the function

$$F(x, y, \lambda) = f(x, y) + \lambda g(x, y)$$

Then there is a value of λ such that (x_0, y_0, λ) is a solution of the system of equations

$$\frac{\partial F}{\partial x} = \frac{\partial f}{\partial x} + \lambda \frac{\partial g}{\partial x} = 0$$

$$\frac{\partial F}{\partial y} = \frac{\partial f}{\partial y} + \lambda \frac{\partial g}{\partial y} = 0 \qquad (1)$$

$$\frac{\partial F}{\partial \lambda} = g(x, y) = 0$$

provided all the partial derivatives exist.

In other words, the above result tells us that if we find all the solutions of the system of equations (1), then among the solutions we find the points at which $z = f(x, y)$ may have a local maximum or a local minimum subject to the condition $g(x, y) = 0$.

The Greek letter lambda.

EXAMPLE 2 Using the Method of Lagrange Multipliers

Find the maximum value of

$$z = f(x, y) = xy$$

subject to the constraint

$$g(x, y) = x + y - 16 = 0$$

SOLUTION First, we construct the function F:

$$F(x, y, \lambda) = f(x, y) + \lambda g(x, y) = xy + \lambda(x + y - 16)$$

The system of equations (1) is

$$\frac{\partial F}{\partial x} = 0 \qquad \frac{\partial F}{\partial y} = 0 \qquad \frac{\partial F}{\partial \lambda} = 0$$

$$y + \lambda = 0 \qquad x + \lambda = 0 \qquad x + y - 16 = 0$$

Using the solutions of the first two equations, namely, $y = -\lambda$, $x = -\lambda$, in the third equation, we get

$$-\lambda - \lambda - 16 = 0 \quad \text{or} \quad \lambda = -8$$

Since $x = -\lambda$ and $y = -\lambda$, the only solution of the system is

$$x = 8 \qquad y = 8 \qquad \lambda = -8$$

We find that $z = f(x, y) = xy$ has a local maximum at $(8, 8)$; the maximum value is $z = f(8, 8) = 64$.

In Example 2, we used the Method of Lagrange Multipliers to find the maximum value of z subject to a constraint. The steps we followed are outlined below.

Steps for Using the Method of Langrange Multipliers

STEP 1 Write the function to be maximized (or minimized) and the constraint in the form:

Find the maximum (or minimum) value of

$$z = f(x, y)$$

subject to the constraint

$$g(x, y) = 0$$

STEP 2 Construct the function F:

$$F(x, y, \lambda) = f(x, y) + \lambda g(x, y)$$

STEP 3 Set up the system of equations

$$\frac{\partial F}{\partial x} = 0$$

$$\frac{\partial F}{\partial y} = 0$$

$$\frac{\partial F}{\partial \lambda} = g(x, y) = 0$$

STEP 4 Solve the system of equations for x, y, and λ.

STEP 5 Evaluate $z = f(x, y)$ at each solution (x_0, y_0, λ) found in Step 4. Choose the maximum (or minimum) value of z.

NOW WORK PROBLEM 1.

EXAMPLE 3 **Using the Method of Lagrange Multipliers**

Find the minimum value of

$$z = f(x, y) = xy$$

subject to the constraint

$$g(x, y) = x^2 + y^2 - 4 = 0$$

SOLUTION **STEP 1:** The problem is already in the desired form.
STEP 2: We construct the function F:

$$F(x, y, \lambda) = f(x, y) + \lambda g(x, y) = xy + \lambda(x^2 + y^2 - 4)$$

STEP 3: The system of equations (1) is

$$\frac{\partial F}{\partial x} = y + \lambda \cdot 2x = 0 \quad (1)$$

$$\frac{\partial F}{\partial y} = x + \lambda \cdot 2y = 0 \quad (2) \tag{2}$$

$$\frac{\partial F}{\partial \lambda} = x^2 + y^2 - 4 = 0 \quad (3)$$

STEP 4: We proceed to solve the system. From the first equation, we find that

$$y = -2x\lambda \tag{3}$$

Substituting into the second equation, we obtain

$$x + 2y\lambda = 0 \quad (2)$$
$$x + 2(-2x\lambda)\lambda = 0 \quad y = -2x\lambda$$
$$x - 4x\lambda^2 = 0 \quad \text{Simplify.}$$
$$x(1 - 4\lambda^2) = 0 \quad \text{Factor out } x.$$
$$x = 0 \quad \text{or} \quad 1 - 4\lambda^2 = 0 \quad \text{Apply the Zero-Product Property.}$$

If $x = 0$, then from (3) we find $y = 0$. But $x = 0$ and $y = 0$ do not satisfy the constraint $g(x, y) = x^2 + y^2 - 4 = 0$. We discard $x = 0$. Then $1 - 4\lambda^2 = 0$.

$$1 - 4\lambda^2 = 0$$

$$\lambda^2 = \frac{1}{4}$$

$$\lambda = \pm\frac{1}{2}$$

Substituting these values for λ into equation (3), $y = 2x\lambda$, we find

$$y = x \quad \text{or} \quad y = -x$$

Since x and y are subject to the constraint $g(x, y) = x^2 + y^2 - 4 = 0$, we must have

$$x^2 + y^2 - 4 = x^2 + x^2 - 4 = 0$$
$$2x^2 = 4$$
$$x = \pm\sqrt{2}$$

Since $y = x$ or $y = -x$, the solutions of the system are

$$(\sqrt{2}, \sqrt{2}), \quad (\sqrt{2}, -\sqrt{2}), \quad (-\sqrt{2}, \sqrt{2}), \quad (-\sqrt{2}, -\sqrt{2})$$

STEP 5: We find the value of $z = f(x, y) = xy$ at each of these points.

$$f(\sqrt{2}, \sqrt{2}) = \sqrt{2}\sqrt{2} = 2$$
$$f(\sqrt{2}, -\sqrt{2}) = \sqrt{2}(-\sqrt{2}) = -2$$
$$f(-\sqrt{2}, \sqrt{2}) = -\sqrt{2}\sqrt{2} = -2$$
$$f(-\sqrt{2}, -\sqrt{2}) = (-\sqrt{2})(-\sqrt{2}) = 2$$

We see that $z = f(x, y)$ attains its minimum value at the two points $(-\sqrt{2}, \sqrt{2})$ and $(\sqrt{2}, -\sqrt{2})$. The minimum value is -2.

NOW WORK PROBLEM 3.

Application

EXAMPLE 4 **Maximizing Profit**

A manufacturer produces two types of engines, x units of type I and y units of type II. The joint profit function is given by

$$P(x, y) = x^2 + 3xy - 6y$$

To maximize profit, how many engines of each type should be produced if there must be a total of 42 engines produced?

SOLUTION **STEP 1:** The condition of a total of 42 engines constitutes the constraint of the problem. The constraint is

$$g(x, y) = x + y - 42 = 0$$

The problem is to

$$\text{Maximize} \quad P(x, y) = x^2 + 3xy - 6y$$
$$\text{subject to the constraint} \quad g(x, y) = x + y - 42 = 0$$

STEP 2: The function F is
$$F(x, y, \lambda) = P(x, y) + \lambda g(x, y) = x^2 + 3xy - 6y + \lambda(x + y - 42)$$
STEP 3: The system of equations (1) is
$$\frac{\partial F}{\partial x} = 2x + 3y + \lambda = 0 \quad (1)$$
$$\frac{\partial F}{\partial y} = 3x - 6 + \lambda = 0 \quad (2)$$
$$\frac{\partial F}{\partial \lambda} = x + y - 42 = 0 \quad (3)$$

STEP 4: From the middle equation $3x - 6 + \lambda = 0$, we have $3x = 6 - \lambda$ so $x = \frac{1}{3}(6 - \lambda)$. We subsitute into the first equation $2x + 3y + \lambda = 0$ and proceed to solve for y.

$$2x + 3y + \lambda = 0 \quad (1)$$
$$2 \cdot \frac{1}{3}(6 - \lambda) + 3y + \lambda = 0 \qquad x = \frac{1}{3}(6 - \lambda)$$
$$4 - \frac{2}{3}\lambda + 3y + \lambda = 0 \qquad \text{Simplify.}$$
$$3y = -\frac{1}{3}\lambda - 4$$
$$y = -\frac{1}{9}\lambda - \frac{4}{3}$$

Now use the third equation.
$$x + y - 42 = 0 \quad (3)$$
$$\frac{1}{3}(6 - \lambda) + \left(-\frac{1}{9}\lambda - \frac{4}{3}\right) = 42 \qquad x = \frac{1}{3}(6 - \lambda); y = -\frac{1}{9}\lambda - \frac{4}{3}$$
$$2 - \frac{1}{3}\lambda - \frac{1}{9}\lambda - \frac{4}{3} = 42 \qquad \text{Simplify.}$$
$$-\frac{4}{9}\lambda = 42 - \frac{2}{3}$$
$$-\frac{4}{9}\lambda = \frac{124}{3}$$
$$\lambda = -93$$

Then $x = \frac{1}{3}(6 - \lambda) = \frac{1}{3}(99) = 33$ and $y = -\frac{1}{9}\lambda - \frac{4}{3} = \frac{93}{9} - \frac{4}{3} = 9$.

The solution of the system is
$$x = 33 \qquad y = 9 \qquad \lambda = -93$$

STEP 5: The maximum profit is achieved for a production of $x = 33$ type I engines and $y = 9$ type II engines. ▶

NOW WORK PROBLEM 17.

Function of Three Variables

One of the advantages of the method of Lagrange multipliers is that it extends easily to functions of three variables.

Method of Lagrange Multipliers

Suppose that, subject to the constraint $g(x, y, z) = 0$, the function $w = f(x, y, z)$ has a local maximum or a local minimum at the point (x_0, y_0, z_0). Form the function

$$F(x, y, z) = f(x, y, z) + \lambda g(x, y, z)$$

Then there is a value of λ so that (x_0, y_0, z_0, λ) is a solution of the system of equations:

$$F_x(x, y, z, \lambda) = f_x(x, y, z) + \lambda g_x(x, y, z) = 0$$
$$F_y(x, y, z, \lambda) = f_y(x, y, z) + \lambda g_y(x, y, z) = 0$$
$$F_z(x, y, z, \lambda) = f_z(x, y, z) + \lambda g_z(x, y, z) = 0$$
$$F_\lambda(x, y, z, \lambda) = g(x, y, z) = 0$$

provided each of the partial derivatives exist.

EXAMPLE 5 Minimizing Cost

The material for a rectangular container costs $3 per square foot for the bottom and $2 per square foot for the sides and top. Find the dimensions of the container so that its volume is 12 cubic feet and the cost is minimum.

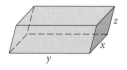

FIGURE 22

SOLUTION **STEP 1:** Refer to Figure 22. Let x and y (in feet) equal the length and width of the container and z (in feet) equal its height. The cost of the bottom is then $3xy$; the cost of the top is $2xy$; the cost of the sides is $2(2xz) + $2(2yz)$. The volume is constrained to be $xyz = 12$ cubic feet. The problem is:

Minimize

$$C(x, y, z) = \underbrace{3xy}_{\text{Bottom}} + \underbrace{2xy}_{\text{Top}} + \underbrace{4yz + 4xz}_{\text{Sides}} = 5xy + 4yz + 4xz$$

subject to the constraint

$$g(x, y, z) = xyz - 12 = 0$$

STEP 2: Form the function

$$F(x, y, z, \lambda) = C(x, y, z) + \lambda g(x, y, z)$$
$$= 5xy + 4yz + 4xz + \lambda(xyz - 12)$$

STEP 3: The system of equations to be solved is

$$F_x(x, y, z, \lambda) = 5y + 4z + \lambda yz = 0 \quad (1)$$
$$F_y(x, y, z, \lambda) = 5x + 4z + \lambda xz = 0 \quad (2)$$
$$F_z(x, y, z, \lambda) = 4y + 4x + \lambda xy = 0 \quad (3)$$
$$F_\lambda(x, y, z, \lambda) = xyz - 12 = 0 \quad (4)$$

STEP 4: Since $x > 0, y > 0, z > 0$, we can solve for λ in the first three equations to get

$$\lambda = \frac{-(5y + 4z)}{yz} \qquad \lambda = \frac{-(5x + 4z)}{xz} \qquad \lambda = \frac{-4(y + x)}{xy}$$

From the first two of these, we find that

$$\frac{5y + 4z}{yz} = \frac{5x + 4z}{xz}$$

$$x(5y + 4z) = y(5x + 4z) \qquad \text{Multiply both sides by } xyz \text{ and simplify.}$$

$$5xy + 4xz = 5xy + 4yz$$

$$4xz = 4yz \qquad \text{Cancel the } z\text{'s.}$$

$$x = y$$

Now use the second two

$$\frac{5x + 4z}{xz} = \frac{4y + 4x}{xy}$$

$$5xy + 4yz = 4yz + 4xz \qquad \text{Multiply both sides by } xyz \text{ and simplify.}$$

$$5xy = 4xz$$

$$5y = 4z \qquad \text{Cancel the } x\text{'s.}$$

$$y = \frac{4}{5}z \qquad \text{Solve for } y.$$

Using these results in equation (4), we get

$$xyz - 12 = 0 \quad (4)$$

$$\frac{4}{5}z \cdot \frac{4}{5}z \cdot z = 12 \qquad x = y = \frac{4}{5}z$$

$$z^3 = \frac{75}{4}$$

$$z = 2.657$$

STEP 5: The only solution is $x = y = \frac{4}{5}z = 2.125, z = 2.657$ feet. These are the dimensions of the container that minimize the cost of the container.

NOW WORK PROBLEM 23.

EXERCISE 8.5

In Problems 1–12, use the Method of Lagrange Multipliers.

1. Find the maximum value of $z = f(x, y) = 3x + 4y$ subject to the constraint $g(x, y) = x^2 + y^2 - 9 = 0$.

2. Find the maximum value of $z = f(x, y) = 3xy$ subject to the constraint $g(x, y) = x^2 + y^2 - 4 = 0$.

3. Find the minimum value of $z = f(x, y) = x^2 + y^2$ subject to the constraint $g(x, y) = x + y - 1 = 0$.

4. Find the minimum value of $z = f(x, y) = 3x + 4y$ subject to the constraint $g(x, y) = x^2 + y^2 - 9 = 0$.

5. Find the maximum value of $z = f(x, y) = 12xy - 3y^2 - x^2$ subject to the constraint $g(x, y) = x + y - 16 = 0$.

6. Find the maximum value of $z = f(x, y) = xy$ subject to the constraint $g(x, y) = x + y - 8 = 0$.

7. Find the minimum value of $z = f(x, y) = 5x^2 + 6y^2 - xy$ subject to the constraint $g(x, y) = x + 2y - 24 = 0$.

8. Find the minimum value of $z = f(x, y) = x^2 + y^2$ subject to the constraint $g(x, y) = 2x + 3y - 4 = 0$.

9. Find the maximum value of $w = f(x, y, z) = xyz$ subject to the constraint $g(x, y, z) = x + 2y + 2z - 120 = 0$.

10. Find the maximum value of $w = f(x, y, z) = x + y + z$ subject to the constraint $g(x, y, z) = x^2 + y^2 + z^2 - 12 = 0$.

11. Find the minimum value of
$$w = f(x, y, z) = x^2 + y^2 + z^2 - x - 3y - 5z$$
subject to the constraint $g(x, y, z) = x + y + 2z - 20 = 0$.

12. Find the minimum value of $w = f(x, y, z) = 4x + 4y + 2z$ subject to the constraint $g(x, y, z) = x^2 + y^2 + z^2 - 9 = 0$.

13. Find two numbers x and y so that their product is a maximum while their sum is 100.

14. Find two numbers x and y so that the sum of their squares is a minimum while their sum is 100.

15. Find three numbers x, y, and z so that their sum is a maximum while the sum of their squares is 25.

16. Find three numbers x, y, and z so that their sum is a minimum while the sum of their squares is 25.

17. **Joint Cost Function** Let x and y be two types of items produced by a factory, and let
$$C = 18x^2 + 9y^2$$
be the joint cost of production of x and y. If $x + y = 54$, find x and y that minimize cost.

18. **Production Function** The production function of ABC Manufacturing is
$$P(x, y) = x^2 + 3xy - 6x$$
where x and y represent two different types of input. Find the amounts of x and y that maximize production if $x + y = 40$.

19. **Checked Luggage Requirements** The linear measurements (length + width + height) for baggage checked onto Delta Airlines flights must not exceed 62 inches. Find the dimensions of the rectangular box of greatest volume that meets this requirement.

Source: Delta Airlines.

20. **Fencing in an Area** The A Vinyl Fence Company prices its Cape Cod Concave fence, which is three feet tall, at $14.11/ft. If a home builder has $3000 available to spend on fencing a rectangular region, determine the largest area which can be enclosed.

Source: A Vinyl Fence Co. Inc, San Jose, CA.

21. **The Cobb-Douglas Model** (Refer to Problem 48 of Section 8.3 on page 530.) Apply the Cobb-Douglas production model $P = 1.01K^{.25} L^{.75}$ as follows: Suppose that each unit of capital (K) has a value of $175 and each unit of labor (L) has a value of $125.

   (a) If there is a total of $125,000 to invest in the economy, use the method of Lagrange multipliers to find the number of units of capital and the number of units of labor that will maximize the total production in the manufacturing sector of the economy.

   (b) What is the maximum number of units of production that the manufacturing sector of the economy could generate under the given conditions?

22. **The Cobb-Douglas Model** Apply the Cobb-Douglas production model $P = 1.01K^{.25}L^{.75}$ as follows: Suppose that each unit of capital (K) has a value of $125 and each unit of labor (L) has a value of $175.

   (a) If there is a total of $125,000 to invest in the economy, use the method of Lagrange multipliers to find the number of units of capital and the number of units of labor that will maximize the total production in the manufacturing sector of the economy.

   (b) What is the maximum number of units of production that the manufacturing sector of the economy could generate under the given conditions?

   (c) Is this maximum value greater than, smaller than, or equal to the maximum value found in Problem 21? Explain.

23. **Minimizing Materials** A container producer wants to build a closed rectangular box with a volume of 175 cubic feet. Determine what the dimensions of the container should be so as to use the least amount of material in construction.

24. **Cost of a Box** A rectangular box, open at the top, is to be made from material costing $2 per square foot. If the volume is to be 12 cubic feet, what dimensions will minimize the cost?

25. **Cost of a Box** A rectangular box is to have a bottom made from material costing $2 per square foot while the top and sides are made from material costing $1 per square foot. If the volume of the box is to be 18 cubic feet, what dimensions will minimize the cost?

8.6 The Double Integral

PREPARING FOR THIS SECTION *Before getting started, review the following:*

>> The Definite Integral (Chapter 6, Section 6.4, pp. 441–445)

OBJECTIVES
1. Evaluate partial integrals
2. Evaluate iterated integrals
3. Evaluate double integrals
4. Find the volume of a solid

The definite integral of a function of a single variable can be extended to functions of two variables. Integrals of a function of two variables are called **double integrals.** Recall that the definite integral of a function of one variable is defined over an interval. Double integrals, on the other hand, involve integration over a region of the plane.

An example of a double integral is

$$\iint_R 2x^2 y \, dx \, dy$$

where the integrand is the function $f(x, y) = 2x^2 y$ and R is some region of the x-y plane.

For example, R might be the rectangular region $1 \leq x \leq 2$, $0 \leq y \leq 4$. See Figure 23.

The evaluation of a double integral of a function f of two variables over a rectangular region is equivalent to the evaluation of a pair of definite integrals in which one of the integrations is performed *partially*. Partial integration is merely the reverse of partial differentiation. The symbol $\int_a^b f(x, y) \, dx$ is an instruction to hold y fixed and integrate with respect to x. The result will be a function of y alone. Similarly, $\int_c^d f(x, y) \, dy$ is an instruction to hold x fixed and integrate with respect to y. The result here is a function of x alone.

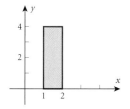

FIGURE 23

1 EXAMPLE 1 Evaluating Partial Integrals

Evaluate: **(a)** $\int_1^2 2x^2 y \, dx$ **(b)** $\int_0^4 2x^2 y \, dy$

SOLUTION **(a)** The dx tells us to integrate with respect to x, holding y as a constant. Then

$$\int_1^2 2x^2 y \, dx = 2y \int_1^2 x^2 \, dx = 2y \cdot \frac{x^3}{3} \Big|_1^2 = 2y \cdot \left(\frac{8}{3} - \frac{1}{3}\right) = 2y \cdot \frac{7}{3} = \frac{14y}{3}$$

        ↑   ↑
Treat $2y$ as a constant Integrate

(b) The dy tells us to integrate with respect to y, holding x as a constant. Then

$$\int_0^4 2x^2 y \, dy = 2x^2 \cdot \int_0^4 y \, dy = 2x^2 \cdot \frac{y^2}{2}\Big|_0^4 = 2x^2(8 - 0) = 16x^2$$

        ↑   ↑
Treat $2x^2$ as a constant Integrate

EXAMPLE 2 Evaluating Partial Integrals

Evaluate: **(a)** $\int_1^2 (6x^2y + 3y^2)\,dy$ **(b)** $\int_1^2 (6x^2y + 3y^2)\,dx$

SOLUTION **(a)** The dy tells us to integrate with respect to y, holding x as a constant. Then

$$\int_1^2 (6x^2y + 3y^2)\,dy = \underbrace{\int_1^2 6x^2y\,dy + \int_1^2 3y^2\,dy}_{\text{Integral of a sum is sum of the integrals}} = \underbrace{6x^2 \int_1^2 y\,dy + 3\int_1^2 y^2\,dy}_{\text{Treat } x \text{ as a constant}} = \underbrace{6x^2 \cdot \frac{y^2}{2}\Big|_1^2 + 3 \cdot \frac{y^3}{3}\Big|_1^2}_{\text{Integrate}}$$

$$= 6x^2\left(2 - \frac{1}{2}\right) + 3\left(\frac{8}{3} - \frac{1}{3}\right) = 9x^2 + 7$$

(b) The dx tells us to integrate with respect to x, holding y as a constant. Then

$$\int_1^2 (6x^2y + 3y^2)\,dx = \underbrace{\int_1^2 6x^2y\,dx + \int_1^2 3y^2\,dx}_{\text{Integral of a sum is sum of the integrals}} = \underbrace{6y\int_1^2 x^2\,dx + 3y^2\int_1^2 dx}_{\text{Treat } y \text{ as a constant}} = \underbrace{6y \cdot \frac{x^3}{3}\Big|_1^2 + 3y^2 \cdot x\Big|_1^2}_{\text{Integrate}}$$

$$= 6y\left(\frac{8}{3} - \frac{1}{3}\right) + 3y^2(2-1) = 14y + 3y^2$$

 NOW WORK PROBLEM 3.

Integrals of the form

$$\int_a^b \left[\int_c^d f(x,y)\,dy\right] dx \quad \text{and} \quad \int_c^d \left[\int_a^b f(x,y)\,dx\right] dy$$

are called **iterated integrals**. In the iterated integral on the left, the function f is integrated partially with respect to y from c to d, resulting in a function of x that is then integrated from a to b. In the iterated integral on the right, the function f is integrated partially with respect to x from a to b, resulting in a function of y that is then integrated from c to d.

EXAMPLE 3 Evaluating Iterated Integrals

Evaluate: **(a)** $\int_0^4 \left[\int_1^2 2x^2y\,dx\right] dy$ **(b)** $\int_1^2 \left[\int_0^4 2x^2y\,dy\right] dx$

SOLUTION **(a)** $\int_0^4 \left[\int_1^2 2x^2y\,dx\right] dy = \int_0^4 \frac{14y}{3}\,dy = \frac{7y^2}{3}\Big|_0^4 = \frac{112}{3}$

From Example 1, Part (a)

(b) $\int_1^2 \left[\int_0^4 2x^2y\,dy\right] dx = \int_1^2 16x^2\,dx = \frac{16x^3}{3}\Big|_1^2 = \frac{112}{3}$

From Example 1, Part (b)

EXAMPLE 4 Evaluating Iterated Integrals

Evaluate:

(a) $\displaystyle\int_1^2 \left[\int_0^1 (6x^2y + 8y^3)\,dy\right] dx$ (b) $\displaystyle\int_0^1 \left[\int_1^2 (6x^2y + 8y^3)\,dx\right] dy$

SOLUTION (a) We evaluate the partial integral inside the brackets first. Then

$$\int_0^1 (6x^2y + 8y^3)\,dy = \int_0^1 6x^2y\,dy + \int_0^1 8y^3\,dy = 6x^2 \int_0^1 y\,dy + 8\int_0^1 y^3\,dy$$

$$= 6x^2 \cdot \frac{y^2}{2}\Big|_0^1 + 8 \cdot \frac{y^4}{4}\Big|_0^1 = 6x^2\left(\frac{1}{2}\right) + 8\left(\frac{1}{4}\right) = 3x^2 + 2$$

Then

$$\int_1^2 \left[\int_0^1 (6x^2y + 8y^3)\,dy\right] dx = \int_1^2 (3x^2 + 2)\,dx = (x^3 + 2x)\Big|_1^2 = 12 - 3 = 9$$

(b) We evaluate the partial integral inside the brackets first. Then,

$$\int_1^2 (6x^2y + 8y^3)\,dx = \int_1^2 6x^2y\,dx + \int_1^2 8y^3\,dx = 6y\int_1^2 x^2\,dx + 8y^3\int_1^2 dx$$

$$= 6y \cdot \frac{x^3}{3}\Big|_1^2 + 8y^3 \cdot x\Big|_1^2 = 6y\left(\frac{8}{3} - \frac{1}{3}\right) + 8y^3(2-1)$$

$$= 14y + 8y^3$$

Then

$$\int_0^1 \left[\int_1^2 (6x^2y + 8y^3)\,dx\right] dy = \int_0^1 (14y + 8y^3)\,dy = (7y^2 + 2y^4)\Big|_0^1 = 9 - 0 = 9$$

NOW WORK PROBLEM 19.

Notice in Example 4 that the integrand of the iterated integral in part (a) is the same as the one in part (b). Also notice that the limits of integration for x, $1 \le x \le 2$, and the limits of integration for y, $0 \le y \le 1$, are also the same. The difference between part (a) and part (b) is the order of the integration. Yet the answer obtained is the same.

Now look at Example 3. The same circumstances there led to equal answers as well. Examples 3 and 4 are special cases of the following result.

If $f(x, y)$ is a function that is continuous over a rectangular region R, $a \le x \le b$, $c \le y \le d$, then

$$\int_a^b \left[\int_c^d f(x, y)\,dy\right] dx = \int_c^d \left[\int_a^b f(x, y)\,dx\right] dy$$

Furthermore, the double integral of f over R has the value

$$\iint_R f(x, y)\,dx\,dy = \int_a^b \left[\int_c^d f(x, y)\,dy\right] dx = \int_c^d \left[\int_a^b f(xy)\,dx\right] dy$$

EXAMPLE 5 Evaluating Double Integrals

Evaluate $\iint_R 2xy\, dx\, dy$ if R is the rectangular region $1 \le x \le 2, 0 \le y \le 1$.

SOLUTION We choose to evaluate the double integral as follows:

$$\iint_R 2xy\, dx\, dy = \int_0^1 \left[\int_1^2 2xy\, dx\right] dy = \int_0^1 2y \left[\int_1^2 x\, dx\right] dy = \int_0^1 2y \cdot \left[\frac{x^2}{2}\right]_1^2 dy = \int_0^1 2y \cdot \frac{3}{2}\, dy = \int_0^1 3y\, dy = \left(3\frac{y^2}{2}\right)\Big|_0^1 = \frac{3}{2}$$

$\underbrace{}_{\substack{1 \le x \le 2 \\ 0 \le y \le 1}}$

NOW WORK EXAMPLE 5 using the iterated integral $\int_1^2 \left[\int_0^1 2xy\, dy\right] dx$.

NOW WORK PROBLEM 27.

Finding Volume by Using Double Integrals

Find the volume of a solid ❹ One application of the definite integral $\int_a^b f(x)\, dx$ is to find the area under a curve. In a similar manner, the double integral is used to find the volume of a solid bounded above by the surface $z = f(x, y)$, below by the xy-plane, and on the sides by the vertical walls defined by the rectangular region R. See Figure 24.

FIGURE 24

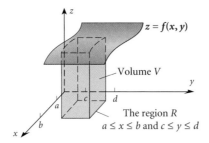

The region R
$a \le x \le b$ and $c \le y \le d$

Volume

If $f(x, y) \ge 0$ over a rectangular region R: $a \le x \le b, c \le y \le d$, then the volume V of the solid under the graph of f and over the region R is

$$V = \iint_R f(x, y)\, dx\, dy$$

The Double Integral 551

EXAMPLE 6 **Finding the Volume of a Solid**

Find the volume V of the solid under $f(x, y) = x^2 + y^2$ and over the rectangular region $0 \leq x \leq 2, 0 \leq y \leq 1$.

SOLUTION Figure 25 illustrates the volume we seek. The volume V is

$$V = \iint_R (x^2 + y^2)\, dx\, dy = \int_0^2 \left[\int_0^1 (x^2 + y^2)\, dy \right] dx$$

$$= \int_0^2 \left(x^2 y + \frac{y^3}{3} \right) \bigg|_0^1 dx$$

$$= \int_0^2 \left(x^2 + \frac{1}{3} \right) dx$$

$$= \left(\frac{x^3}{3} + \frac{x}{3} \right) \bigg|_0^2$$

$$= \frac{8}{3} + \frac{2}{3} = \frac{10}{3} \text{ cubic units}$$

FIGURE 25

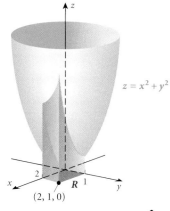

NOW WORK PROBLEM 31.

EXERCISE 8.6

In Problems 1–16, evaluate each partial integral.

1. $\int_0^2 (xy^3 + x^2)\, dx$

2. $\int_1^3 (xy^3 - x^2)\, dx$

3. $\int_2^4 (3x^2y + 2x)\, dy$

4. $\int_0^1 (6xy^2 - 2xy + 3)\, dy$

5. $\int_2^3 (x + 3y)\, dx$

6. $\int_1^3 (6xy + 12x^2y^3)\, dy$

7. $\int_2^4 (4x - 6y + 7)\, dy$

8. $\int_1^3 (4x - 6y + 7)\, dx$

9. $\int_0^1 \frac{x^2}{\sqrt{1 + y^2}}\, dx$

10. $\int_0^2 \frac{y^3}{\sqrt{1 + x^2}}\, dy$

11. $\int_0^2 e^{x+y}\, dx$

12. $\int_0^2 e^{x+6y}\, dx$

13. $\int_0^4 e^{x-4y}\, dx$

14. $\int_2^4 e^{x+y}\, dy$

15. $\int_0^2 \frac{x}{\sqrt{y + 6}}\, dx$

16. $\int_1^4 \frac{y}{\sqrt{x^2 + 9}}\, dy$

In Problems 17–26, evaluate each iterated integral.

17. $\int_0^2 \left[\int_0^4 y\, dx \right] dy$

18. $\int_1^2 \left[\int_3^4 x\, dy \right] dx$

19. $\int_1^2 \left[\int_1^3 (x^2 + y)\, dx \right] dy$

20. $\int_0^1 \left[\int_2^3 (x + y)\, dy \right] dx$

21. $\int_0^1 \left[\int_1^2 (x^2 + y)\, dx \right] dy$

22. $\int_0^3 \left[\int_1^2 (x - y^2)\, dy \right] dx$

23. $\int_1^2 \left[\int_3^4 (4x + 2y + 5)\, dx \right] dy$

24. $\int_1^2 \left[\int_3^4 (6x + 4y + 7)\, dy \right] dx$

25. $\int_2^4 \left[\int_0^1 (6xy^2 - 2xy + 3)\, dy \right] dx$

26. $\int_0^2 \left[\int_1^3 (6xy + 12x^2y^3)\, dy \right] dx$

In Problems 27–30, evaluate each double integral over the indicated rectangular region R.

27. $\iint_R (y + 3x^2)\, dx\, dy \quad R: 0 \leq x \leq 2, \quad 1 \leq y \leq 3$

28. $\iint_R (x + 3y^2)\, dx\, dy \quad R: 0 \leq x \leq 3, \quad 0 \leq y \leq 4$

29. $\iint_R (x + y)\, dy\, dx \quad R: 0 \leq x \leq 2, \quad 1 \leq y \leq 4$

30. $\iint_R (x^2 - 2xy)\, dy\, dx \quad R: 0 \leq x \leq 2, \quad 1 \leq y \leq 4$

In Problems 31 and 32, find the volume under the surface of $f(x, y)$ and above the indicated rectangle.

31. $f(x, y) = 2x + 3y + 4 \quad 1 \leq x \leq 2, \quad 3 \leq y \leq 4$

32. $f(x, y) = x + y - 1 \quad 0 \leq x \leq 1, \quad 0 \leq y \leq 1$

Chapter 8 Review

OBJECTIVES

Section	You should be able to	Review Exercises
8.1	1 Use the distance formula	1–8
	2 Find the standard equation of a sphere	9–14
8.2	1 Evaluate a function of two variables	15–18
	2 Find the domain of a function of two variables	19–24
8.3	1 Find partial derivatives	25–32, 33, 34
	2 Find the slope of the tangent line to a surface	35–38
	3 Interpret partial derivatives	63–70
	4 Find higher-order partial derivatives	25–32
8.4	1 Find critical points	39(a)–44(a)
	2 Determine the character of a critical point	39(b)–44(b)
	3 Solve applied problems	71–74
8.5	1 Use the method of Lagrange multipliers	45–48, 73, 74
8.6	1 Evaluate partial integrals	49–52
	2 Evaluate iterated integrals	53–56
	3 Evaluate double integrals	57–60
	4 Find the volume of a solid	61, 62

THINGS TO KNOW

Distance Formula (p. 513)
The distance d from $P_1 = (x_1, y_1, z_1)$ to $P_2 = (x_2, y_2, z_2)$ is
$$d = \sqrt{(x_2 - x_1)^2 + (y_2 - y_1)^2 + (z_2 - z_1)^2}$$

Sphere (p. 514)
The equation of a sphere of radius r with center at (x_0, y_0, z_0) is
$$(x - x_0)^2 + (y - y_0)^2 + (z - z_0)^2 = r^2$$

Partial Derivatives (p. 523)

$$\frac{\partial f}{\partial x} = f_x(x, y) = \lim_{\Delta x \to 0} \frac{f(x + \Delta x, y) - f(x, y)}{\Delta x}$$

$$\frac{\partial f}{\partial y} = f_y(x, y) = \lim_{\Delta y \to 0} \frac{f(x, y + \Delta y) - f(x, y)}{\Delta y}$$

Tests for Local Maxima, Local Minima, and Saddle Points (p. 533)

(x_0, y_0) is a critical point if $f_x(x_0, y_0) = f_y(x_0, y_0) = 0$. Let

$$D = f_{xx}(x_0, y_0) \cdot f_{yy}(x_0, y_0) - [f_{xy}(x_0, y_0)]^2$$

If $D > 0$ and $f_{xx}(x_0, y_0) > 0$, f has a local minimum at (x_0, y_0).
If $D > 0$ and $f_{xx}(x_0, y_0) < 0$, f has a local maximum at (x_0, y_0).
If $D < 0$, f has a saddle point at (x_0, y_0).

Method of Lagrange Multipliers (pp. 539 and 544)

To find a local maximum or a local minimum for $z = f(x, y)$, subject to $g(x, y) = 0$, solve the system of equations

$$\frac{\partial f}{\partial x} + \lambda \frac{\partial g}{\partial x} = 0 \quad \frac{\partial f}{\partial y} + \lambda \frac{\partial g}{\partial y} = 0 \quad g(x, y) = 0$$

Double Integrals (pp. 549 and 550)

$R: a \le x \le b, c \le y \le d$

$$\iint_R f(x, y) \, dx \, dy = \int_a^b \left[\int_c^d f(x, y) \, dy \right] dx$$

$$= \int_c^d \left[\int_a^b f(x, y) \, dx \right] dy$$

If $f(x, y) \ge 0$ on R, Volume $= V = \iint_R f(x, y) \, dx \, dy$

TRUE–FALSE ITEMS

T F **1.** The domain of a function of two variables is a set of points in the xy-plane.

T F **2.** The partial derivative $f_x(x, y)$ of $z = f(x, y)$ is

$$f_x(x, y) = \lim_{\Delta x \to 0} \frac{f(x + \Delta x, y + \Delta y) - f(x, y)}{\Delta x}$$

provided the limit exists.

T F **3.** For most functions in this book, $f_{xy} \ne f_{yx}$.

T F **4.** If (x_0, y_0) is a critical point of $z = f(x, y)$ and if $f_{xx}(x_0, y_0) > 0$ and

$$D = f_{xx}(x_0, y_0) \cdot f_{yy}(x_0, y_0) - [f_{xy}(x_0, y_0)]^2 < 0$$

then f has a local minimum at (x_0, y_0).

FILL-IN-THE-BLANKS

1. The graph of a function of two variables is called a _____.

2. If $f(x, y) = x^2 y - \sqrt{xy}$, then $f(1, 2) =$ _____.

3. The partial derivative $f_y(x_0, y_0)$ equals the slope of the tangent line to the curve of intersection of the surface $z = f(x, y)$ and the plane _____ at the point (x_0, y_0, z_0) on the surface.

4. A critical point that is neither a local maximum nor a local minimum is a _____ _____.

REVIEW EXERCISES Blue problem numbers indicate the author's suggestions for a practice test.

In Problems 1–6, find the distance between each pair of points.

1. $(2, 4, 0)$ and $(1, 6, -2)$
2. $(7, 2, 1)$ and $(1, 6, -2)$
3. $(6, 2, 1)$ and $(4, 6, 8)$
4. $(5, 8, 3)$ and $(7, 6, 2)$
5. $(0, 3, -1)$ and $(-3, 7, -1)$
6. $(6, 2, 3)$ and $(6, -10, -2)$

7. $(2, 2, 2)$ is the center of a sphere and $(3, 4, 0)$ is a point on its surface. Find the radius of the sphere.

8. The endpoints of the diameter of a sphere are $(3, 0, 2)$ and $(9, 0, -6)$. Find the radius of the sphere.

9. Find the standard equation of a sphere which has its center at $(-6, 3, 1)$ and has a radius of 2.

10. Find the standard equation of a sphere which has its center at $(0, 2, -1)$ and has a radius of 3.

11. What is the center and radius of the sphere described by $(x - 1)^2 + (y + 3)^2 + (z + 8)^2 = 25$?

12. What is the center and radius of the sphere described by $(x + 5)^2 + y^2 + z^2 = 16$?

In Problems 13 and 14, (a) find the standard equation of the sphere;
(b) list its center and radius.

13. $x^2 + y^2 + z^2 - 2x + 8y - 6z = 10$

14. $x^2 + y^2 + z^2 - 6y + 2z = 6$

In Problems 15–18, evaluate each function (a) at the point $(1, -3)$; and (b) at the point $(4, -2)$.

15. $f(x, y) = 2x^2 + 6xy - y^3$

16. $f(x, y) = 3x^2 y - x^2 + y^2$

17. $f(x, y) = \dfrac{x + 2y}{x - 3y}$

18. $f(x, y) = \dfrac{2x + y}{x^2 - y}$

In Problems 19–24, find the domain of each function. Give a description of its domain.

19. $z = f(x, y) = x^2 + 3y + 5$

20. $z = f(x, y) = 2xy - 5x + 10$

21. $z = f(x, y) = \ln(y - x^2 - 4)$

22. $z = f(x, y) = \ln(x - y^2)$

23. $z = f(x, y) = \sqrt{x^2 + y^2 + 4x - 5}$

24. $z = f(x, y) = \dfrac{25}{x^2 + y^2}$

In Problems 25–32, find $f_x(x, y)$, $f_y(x, y)$, $f_{xx}(x, y)$, $f_{xy}(x, y)$, $f_{yx}(x, y)$, and $f_{yy}(x, y)$ for each function.

25. $z = f(x, y) = x^2 y + 4x$

26. $z = f(x, y) = x^2 + y^2 + 2xy$

27. $z = f(x, y) = y^2 e^x + x \ln y$

28. $z = f(x, y) = \ln(x^2 + 3y)$

29. $z = f(x, y) = \sqrt{x^2 + y^2}$

30. $z = f(x, y) = \sqrt{x - 2y^2}$

31. $z = f(x, y) = e^x \ln(5x + 2y)$

32. $z = f(x, y) = (x + y^2)e^{3x}$

In Problems 33 and 34, find $f_x(x, y, z)$, $f_y(x, y, z)$, and $f_z(x, y, z)$.

33. $f(x, y, z) = 3x e^y + xy e^z - 12x^2 y$

34. $f(x, y, z) = \ln |2xy + z|$

In Problems 35–38, find the slope of the tangent line to the curve of intersection of the surface $z = f(x, y)$ with the given plane at the indicated point.

35. $z = f(x, y) = 3xy^2$; plane: $y = 2$; point: $(1, 2, 12)$

36. $z = f(x, y) = 2x^2 y + y \ln x$; plane: $y = 1$; point: $(1, 1, 2)$

37. $z = f(x, y) = xe^{xy}$; plane: $x = 1$; point: $(1, 0, 1)$

38. $z = f(x, y) = x \ln(xy)$; plane: $x = 1$; point: $(1, 1, 0)$

In Problems 39–44, (a) find all the critical points of each function;
(b) determine the character of each critical point found in part (a).

39. $z = f(x, y) = xy - 6x - x^2 - y^2$

40. $z = f(x, y) = x^2 + 2x + y^2 + 4y + 10$

41. $z = f(x, y) = 2x - x^2 + 4y - y^2 + 10$

42. $z = f(x, y) = xy$

43. $z = f(x, y) = x^2 - 9y + y^2$

44. $z = f(x, y) = xy + 2y - 3x - 2$

In Problems 45–48, use the Method of Lagrange Multipliers.

45. Find the maximum value of $f(x, y) = 5x^2 + 3y^2 + xy$ subject to the constraint $g(x, y) = 2x - y - 20 = 0$.

46. Find the maximum value of $f(x, y) = x\sqrt{y}$ subject to the constraint $g(x, y) = 2x + y - 3000 = 0$.

47. Find the minimum value of $f(x, y) = x^2 + y^2$ subject to the constraint $g(x, y) = 2x + y - 4 = 0$.

48. Find the minimum value of $f(x, y) = xy^2$ subject to the constraint $g(x, y) = x^2 + y^2 - 1 = 0$.

In Problems 49–52, evaluate each partial integral.

49. $\int_0^2 (4x^2y - 12y)\, dx$

50. $\int_0^2 (4x^2y - 12y)\, dy$

51. $\int_{-1}^3 (6x^2y + 2y)\, dy$

52. $\int_{-1}^3 (6x^2y + 2y)\, dx$

In Problems 53–56, evaluate each iterated integral.

53. $\int_1^2 \left[\int_0^3 (6x^2 + 2x)\, dy\right] dx$

54. $\int_0^3 \left[\int_1^2 (6x^2 + 2x)\, dx\right] dy$

55. $\int_0^2 \left[\int_1^8 (x^2 + 2xy - y^2)\, dx\right] dy$

56. $\int_1^8 \left[\int_0^2 (x^2 + 2xy - y^2)\, dy\right] dx$

In Problems 57–60, evaluate each double integral over the indicated rectangular region R.

57. $\iint_R (2x + 4y)\, dy\, dx$
R: $-1 \leq x \leq 1, 1 \leq y \leq 3$

58. $\iint_R (3x + 2)\, dy\, dx$
R: $0 \leq x \leq 2, 1 \leq y \leq 3$

59. $\iint_R (2xy)\, dy\, dx$
R: $0 \leq x \leq 3, 1 \leq y \leq 2$

60. $\iint_R (x + y)^3\, dy\, dx$
R: $0 \leq x \leq 4, 1 \leq y \leq 3$

In Problems 61 and 62, find the volume under the surface of $z = f(x, y)$ and above the indicated rectangle.

61. $f(x, y) = 2x + 2y + 1$; $1 \leq x \leq 8$, $0 \leq y \leq 6$

62. $f(x, y) = x^2 + y^2 - 4$; $0 \leq x \leq 2$, $0 \leq y \leq 2$

63. **Production Function** The Cobb-Douglas production function for a certain factory is

$$z = f(K, L) = 80\, K^{\frac{1}{4}} L^{\frac{3}{4}}$$

(a) Find: $\dfrac{\partial z}{\partial K}$ and $\dfrac{\partial z}{\partial L}$.

(b) Evaluate $\dfrac{\partial z}{\partial K}$ and $\dfrac{\partial z}{\partial L}$ when $K = \$800{,}000$ and $L = 20{,}000$ worker hours.

(c) To best improve productivity, should the factory increase the use of capital or labor? Explain your answer.

64. **Production Function** The productivity of a manufacturer of car parts approximately follows the Cobb-Douglas production function

$$z = f(K, L) = 50\, K^{\frac{2}{5}} L^{\frac{3}{5}}$$

(a) Find: $\dfrac{\partial z}{\partial K}$ and $\dfrac{\partial z}{\partial L}$.

(b) Evaluate $\dfrac{\partial z}{\partial K}$ and $\dfrac{\partial z}{\partial L}$ when $K = \$128{,}000$ and $L = 4000$ worker hours.

(c) To best improve productivity, should the manufacturer increase the use of capital or labor? Explain your answer.

65. **Marginal Cost of Vacuum Cleaners** The Vacitup Company manufactures two types of vacuum cleaners, the standard Vacu-Clean and deluxe Vacu-Clean Plus. The company's weekly cost function for manufacturing x standard and y deluxe vacuum cleaners is

$$C(x, y) = 1050 + 40x + 45y$$

Find $C_x(x, y)$ and $C_y(x, y)$ and interpret your answers.

66. **Marginal Cost of Lawn Mowers** A company manufactures two models, the standard gasoline powered lawn mower and a combination mower-mulcher. The monthly cost of producing x mowers and y mower-mulchers is given by the function

$$C(x, y) = 15{,}000 + 120x + 150y$$

Find $C_x(x, y)$ and $C_y(x, y)$ and interpret your answers.

67. **Analyzing Revenue** The demand functions for the vacuum cleaners manufactured in Problem 65 are given by

$$p = 350 - 6x + y$$
$$q = 400 + 2x - 8y$$

where p and q are the prices of the standard and deluxe vacuum cleaners respectively.

(a) Find the revenue function for the Vacitup Company.
(b) Find $R_x(x, y)$ and $R_y(x, y)$ and interpret your answers.

68. **Analyzing Revenue** The demand functions for the products manufactured in Problem 66, are given by

$$p = 1000 - 7x + y$$
$$q = 2500 + 2x - 50y$$

where p and q are the prices of the mower and the combination mower-mulcher respectively.

(a) Find the company's revenue function.
(b) Find $R_x(x, y)$ and $R_y(x, y)$ and interpret your answers.

69. **Analyzing Profit** Refer to Problems 65 and 67. If currently the Vacitup Company is manufacturing 50 standard and 30 deluxe vacuum cleaners each week,

(a) Find the profit function for the Vacitup Company.
(b) Evaluate $P_x(50, 30)$ and $P_y(50, 30)$ and interpret your answers.

70. **Analyzing Profit** Refer to Problems 66 and 68. If currently the company is manufacturing 100 lawn mowers and 40 combination mower-mulchers each month,

(a) Find the profit function.
(b) Evaluate $P_x(100, 40)$ and $P_y(100, 40)$ and interpret your answers.

71. **Maximum Profit** A supermarket sells two brands of refrigerated orange juice. The demand functions for the two products are

$$p = 9 - x \quad \text{and} \quad q = 21 - 2y$$

where p and q are the respective prices (in thousands of dollars), and x and y are the respective amounts (in thousands of units) of each brand. The joint cost to the supermarket is

$$C(x, y) = x + y + 225$$

(a) Determine the quantities x, y and the prices p, q that maximize profit.
(b) What is the maximum profit?

72. **Maximum Profit** A company produces two products at a total cost

$$C(x, y) = x^2 + 200x + y^2 + 100y - xy$$

where x and y represent the units of each product. The revenue function is

$$R(x, y) = 2000x - 2x^2 + 100y - y^2 + xy$$

(a) Find the number of units of each product that will maximize profit.
(b) What is the maximum profit?

73. **Maximizing Productivity** A manufacturer introduces a new product with a Cobb-Douglas production function of

$$P(x, y) = 10K^{0.3} L^{0.7}$$

where K represents the units of capital and L the units of labor needed to produce P units of the product. A total of $51,000 has been budgeted for production, and each unit of labor costs the manufacturer $100 and each unit of capital costs $50.

(a) How should the $51,000 be allocated between labor and capital to maximize production?
(b) What is the maximum number of units that can be produced?

74. **Minimizing Material** A rectangular cardboard box with an open top is to have a volume of 96 cubic feet. Find the dimensions of the box so that the amount of cardboard used is minimized.

Chapter 8 Project

MORE INVENTORY CONTROL

Due to your excellent work with the vacuum cleaner inventory in Chapter 5, you've been promoted to inventory specialist in the medium-sized household appliances department. You now need to order not only vacuum cleaners but also microwave ovens. Based on past experience, you know that the store will sell 500 vacuum cleaners and 800 microwave ovens per year, and you again must decide how many times a year to order each product, and how many to get with each order. There are two types of costs to consider. First, items must be stored when they arrive at the store, and this costs money. These costs are called the *holding costs associated with each product*. Second, each time you place an order for a product you incur costs for the paperwork, handling, and shipping. These costs we will call *reorder costs*. You will want to find the number of each product to order at a time (called the *lot size*) that minimizes the total of the holding costs and the reorder costs.

We assume that demand for each product is constant through the year, and that your stock is immediately replenished exactly when you run out of each product. If x is the lot size for vacuum cleaners and y is the lot size for microwave ovens, then the average number of vacuum cleaners you have in stock is $\frac{x}{2}$ and the average number of microwave ovens is $\frac{y}{2}$.

1. Assume that the annual holding costs are $30 for each vacuum cleaner and $15 for each microwave oven, and that the reorder costs are $40 for vacuum cleaners and $60 for microwave ovens. What is the total cost function?

2. Find the lot sizes x and y, give a minimum total cost, and confirm that this is indeed a local minimum.

3. To make your job slightly more difficult, you have discovered that your department only has access to 1000 cubic feet of storage space. You know that each vacuum cleaner uses 20 cubic feet of storage space and each microwave oven uses 10 cubic feet of storage space. How much space does your solution from Problem 2 require?

4. Since the solution in Problem 2 cannot be used, you will need to introduce a constraint function $g(x, y)$ to solve the problem. Such a function will equal 0 when all of the storage space is in use. Find $g(x, y)$.

5. Produce a system of equations that uses the Method of Lagrange Multipliers to solve the minimization problem.

6. Solve the system to find lot sizes x and y, which minimize the total cost function subject to the storage space constraint.

7. The solution will probably have noninteger solutions. Find integers x and y, which minimize the total cost function subject to the storage space constraint. How many orders will you make in the coming year for each item? How much of your storage space will you use on the average?

APPENDIX

Graphing Utilities

OUTLINE

A.1 The Viewing Rectangle
A.2 Using a Graphing Utility to Graph Equations
A.3 Square Screens
A.4 Using a Graphing Utility to Locate Intercepts and Check for Symmetry
A.5 Using a Graphing Utility to Solve Equations

A.1 The Viewing Rectangle

All graphing utilities, that is, all graphing calculators and all computer software graphing packages, graph equations by plotting points on a screen. The screen itself actually consists of small rectangles, called **pixels.** The more pixels the screen has, the better the resolution. Most graphing calculators have 48 pixels per square inch; most computer screens have 32 to 108 pixels per square inch. When a point to be plotted lies inside a pixel, the pixel is turned on (lights up). The graph of an equation is a collection of pixels. Figure 1 shows how the graph of $y = 2x$ looks on a TI-83 Plus graphing calculator.

The screen of a graphing utility will display the coordinate axes of a rectangular coordinate system. However, you must set the scale on each axis. You must also include the smallest and largest values of x and y that you want included in the graph. This is called **setting the viewing rectangle** or **viewing window.** Figure 2 illustrates a typical viewing window.

FIGURE 1 $y = 2x$

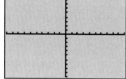

FIGURE 2

To select the viewing window, we must give values to the following expressions:

X min:	the smallest value of x
X max:	the largest value of x
X scl:	the number of units per tick mark on the x-axis
Y min:	the smallest value of y
Y max:	the largest value of y
Y scl:	the number of units per tick mark on the y-axis

Figure 3 illustrates these settings and their relation to the Cartesian coordinate system.

FIGURE 3

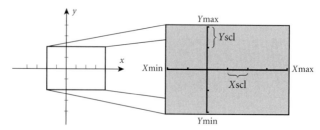

558

If the scale used on each axis is known, we can determine the minimum and maximum values of x and y shown on the screen by counting the tick marks. Look again at Figure 2. For a scale of 1 on each axis, the minimum and maximum values of x are -10 and 10, respectively; the minimum and maximum values of y are also -10 and 10. If the scale is 2 on each axis, then the minimum and maximum values of x are -20 and 20, respectively; and the minimum and maximum values of y are -20 and 20, respectively.

Conversely, if we know the minimum and maximum values of x and y, we can determine the scales being used by counting the tick marks displayed. We shall follow the practice of showing the minimum and maximum values of x and y in our illustrations so that you will know how the viewing window was set. See Figure 4.

FIGURE 4

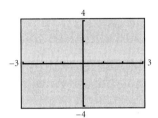

means

X min $= -3$ Y min $= -4$
X max $= 3$ Y max $= 4$
X scl $= 1$ Y scl $= 2$

FIGURE 5

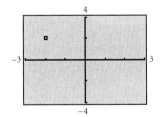

EXAMPLE 1   Finding the Coordinates of a Point Shown on a Graphing Utility Screen

Find the coordinates of the point shown in Figure 5. Assume that the coordinates are integers.

SOLUTION   First we note that the viewing window used in Figure 5 is

X min $= -3$ Y min $= -4$
X max $= 3$ Y max $= 4$
X scl $= 1$ Y scl $= 2$

The point shown is 2 tick units to the left on the horizontal axis (scale $= 1$) and 1 tick up on the vertical axis (scale $= 2$). The coordinates of the point shown are $(-2, 2)$.

EXERCISE A.1

In Problems 1–4, determine the coordinates of the points shown. Tell in which quadrant each point lies. Assume that the coordinates are integers.

1.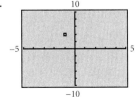

2.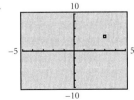

3.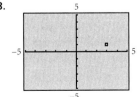

4.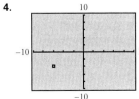

In Problems 5–10, determine the viewing window used.

5.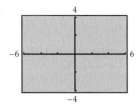

6.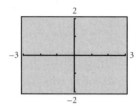

7.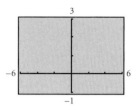

8.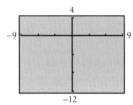

9.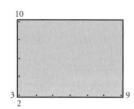

10.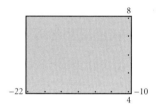

In Problems 11–16, select a setting so that each of the given points will lie within the viewing rectangle.

11. $(-10, 5)$, $(3, -2)$, $(4, -1)$

12. $(5, 0)$, $(6, 8)$, $(-2, -3)$

13. $(40, 20)$, $(-20, -80)$, $(10, 40)$

14. $(-80, 60)$, $(20, -30)$, $(-20, -40)$

15. $(0, 0)$, $(100, 5)$, $(5, 150)$

16. $(0, -1)$, $(100, 50)$, $(-10, 30)$

A.2 Using a Graphing Utility to Graph Equations

The graph of an equation in two variables can usually be obtained by plotting points in a rectangular coordinate system and connecting them. Graphing utilities perform these same steps when graphing an equation. For example, the T1-83 Plus determines 95 evenly spaced input values,* uses the equation to determine the output values, plots these points on the screen, and finally (if in the connected mode) draws a line between consecutive points.

To graph an equation in two variables x and y using a graphing utility requires that the equation be written in the form $y = \{\text{expression in } x\}$. If the original equation is not in this form, replace it by equivalent equations until the form $y = \{\text{expression in } x\}$ is obtained. In general, there are four ways to obtain equivalent equations.

---

*These input values depend on the values of X min and X max. For example, if X min $= -10$ and X max $= 10$, then the first input value will be -10 and the next input value will be $-10 + \dfrac{10 - (-10)}{94} = -9.7872$, and so on.

Procedures that Result in Equivalent Equations

1. Interchange the two sides of the equation:

    Replace $\quad 3x + 5 = y \quad$ by $\quad y = 3x + 5$

2. Simplify the sides of the equation by combining like terms, eliminating parentheses, and so on:

    Replace $\quad (2y + 2) + 6 = 2x + 5(x + 1)$
    by $\quad\quad\quad 2y + 8 = 7x + 5$

3. Add or subtract the same expression on both sides of the equation:

    Replace $\quad y + 3x - 5 = 4$
    by $\quad\quad y + 3x - 5 + 5 = 4 + 5$

4. Multiply or divide both sides of the equation by the same nonzero expression:

    Replace $\quad 3y = 6 - 2x$
    by $\quad \frac{1}{3} \cdot 3y = \frac{1}{3}(6 - 2x)$

EXAMPLE 1 **Expressing an Equation in the Form $y = \{$expression in $x\}$**

Solve for y: $2y + 3x - 5 = 4$

SOLUTION We replace the original equation by a succession of equivalent equations.

$$\begin{aligned}
2y + 3x - 5 &= 4 \\
2y + 3x - 5 + 5 &= 4 + 5 &&\text{Add 5 to both sides.} \\
2y + 3x &= 9 &&\text{Simplify.} \\
2y + 3x - 3x &= 9 - 3x &&\text{Subtract } 3x \text{ from both sides.} \\
2y &= 9 - 3x &&\text{Simplify.} \\
\frac{2y}{2} &= \frac{9 - 3x}{2} &&\text{Divide both sides by 2.} \\
y &= \frac{9 - 3x}{2} &&\text{Simplify.}
\end{aligned}$$

Now we are ready to graph equations using a graphing utility. Most graphing utilities require the following steps:

Steps for Graphing an Equation Using a Graphing Utility

STEP 1: Solve the equation for y in terms of x.

STEP 2: Get into the graphing mode of your graphing utility. The screen will usually display $y = $, prompting you to enter the expression involving x that you found in Step 1. (Consult your manual for the correct way to enter the expression; for example, $y = x^2$ might be entered as $x \wedge 2$ or as $x{*}x$ or as $x \, x^y \, 2$.)

STEP 3: Select the viewing window. Without prior knowledge about the behavior of the graph of the equation, it is common to select the **standard viewing window*** initially. The viewing window is then adjusted based on the graph that appears. In this text the standard viewing window will be

$$X\min = -10 \quad Y\min = -10$$
$$X\max = 10 \quad Y\max = 10$$
$$X\operatorname{scl} = 1 \quad Y\operatorname{scl} = 1$$

STEP 4: Graph.

STEP 5: Adjust the viewing window until a complete graph is obtained.

EXAMPLE 2 Graphing an Equation on a Graphing Utility

Graph the equation: $6x^2 + 3y = 36$

SOLUTION **STEP 1:** We solve for y in terms of x.

$$6x^2 + 3y = 36$$
$$3y = -6x^2 + 36 \quad \text{Subtract } 6x^2 \text{ from both sides of the equation.}$$
$$y = -2x^2 + 12 \quad \text{Divide both sides of the equation by 3 and simplify.}$$

STEP 2: From the graphing mode, enter the expression $-2x^2 + 12$ after the prompt $y = $.

STEP 3: Set the viewing window to the standard viewing window.

STEP 4: Graph. The screen should look like Figure 6.

STEP 5: The graph of $y = -2x^2 + 12$ is not complete. The value of $Y\max$ must be increased so that the top portion of the graph is visible. After increasing the value of $Y\max$ to 12, we obtain the graph in Figure 7. The graph is now complete.

FIGURE 6

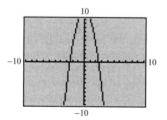

FIGURE 7

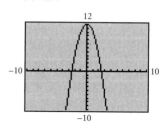

*Some graphing utilities have a ZOOM-STANDARD *feature that automatically sets the viewing window to the standard viewing window and graphs the equation.*

Look again at Figure 7. Although a complete graph is shown, the graph might be improved by adjusting the values of Xmin and Xmax. Figure 8 shows the graph of $y = -2x^2 + 12$ using Xmin $= -4$ and Xmax $= 4$. Do you think this is a better choice for the viewing window?

FIGURE 8

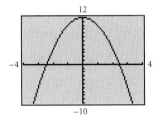

EXAMPLE 3 Creating a Table and Graphing an Equation

Create a table and graph the equation: $y = x^3$

SOLUTION Most graphing utilities have the capability of creating a table of values for an equation. (Check your manual to see if your graphing utility has this capability.) Table 1 illustrates a table of values for $y = x^3$ on a TI-83 Plus. See Figure 9 for the graph.

TABLE 1

FIGURE 9

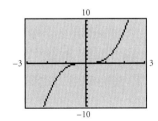

EXERCISE A.2

In Problems 1–16, graph each equation using the following viewing windows:

(a) X min $= -5$
    X max $= 5$
    X scl $= 1$
    Y min $= -4$
    Y max $= 4$
    Y scl $= 1$

(b) X min $= -10$
    X max $= 10$
    X scl $= 1$
    Y min $= -8$
    Y max $= 8$
    Y scl $= 1$

(c) X max $= -10$
    X max $= 10$
    X scl $= 2$
    Y min $= -8$
    Y max $= 8$
    Y scl $= 2$

(d) X max $= -5$
    X max $= 5$
    X scl $= 1$
    Y min $= -20$
    Y max $= 20$
    Y scl $= 5$

1. $y = x + 2$
2. $y = x - 2$
3. $y = -x + 2$
4. $y = -x - 2$
5. $y = 2x + 2$
6. $y = 2x - 2$
7. $y = -2x + 2$
8. $y = -2x - 2$
9. $y = x^2 + 2$
10. $y = x^2 - 2$
11. $y = -x^2 + 2$
12. $y = -x^2 - 2$
13. $3x + 2y = 6$
14. $3x - 2y = 6$
15. $-3x + 2y = 6$
16. $-3x - 2y = 6$

17.–32. For each of the above equations, create a table, $-3 \leq x \leq 3$, and list points on the graph.

A.3 Square Screens

FIGURE 10

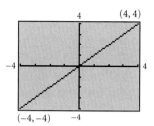

Most graphing utilities have a rectangular screen. Because of this, using the same settings for both x and y will result in a distorted view. For example, Figure 10 shows the graph of the line $y = x$ connecting the points $(-4, -4)$ and $(4, 4)$.

We expect the line to bisect the first and third quadrants, but it doesn't. We need to adjust the selections for Xmin, Xmax, Ymin, and Ymax so that a **square screen** results. On most graphing utilities, this is accomplished by setting the ratio of x to y at $3 : 2$.*
In other words,

$$2(X \max - X \min) = 3(Y \max - Y \min)$$

EXAMPLE 1 — Examples of Viewing Rectangles That Result in Square Screens

(a) X min $= -3$
X max $= 3$
X scl $= 1$
Y min $= -2$
Y max $= 2$
Y scl $= 1$

(b) X min $= -6$
X max $= 6$
X scl $= 1$
Y min $= -4$
Y max $= 4$
Y scl $= 1$

(c) X min $= -6$
X max $= 6$
X scl $= 2$
Y min $= -4$
Y max $= 4$
Y scl $= 1$

FIGURE 11

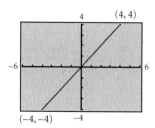

Figure 11 shows the graph of the line $y = x$ on a square screen using the viewing rectangle given in Example 1(b). Notice that the line now bisects the first and third quadrants. Compare this illustration to Figure 10.

*Some graphing utilities have a built-in function that automatically squares the screen. For example, the TI-85 has a ZSQR function that does this. Some graphing utilities require a ratio other than 3:2 to square the screen. For example, the HP 48G requires the ratio of x to y to be 2:1 for a square screen. Consult your manual.

EXERCISE A.3

In Problems 1–8, determine which of the given viewing rectangles result in a square screen.

1. X min $= -3$
 X max $= 3$
 X scl $= 2$
 Y min $= -2$
 Y max $= 2$
 Y scl $= 2$

2. X min $= -5$
 X max $= 5$
 X scl $= 1$
 Y min $= -4$
 Y max $= 4$
 Y scl $= 1$

3. X min $= 0$
 X max $= 9$
 X scl $= 3$
 Y min $= -2$
 Y max $= 4$
 Y scl $= 2$

4. X min $= -6$
 X max $= 6$
 X scl $= 1$
 Y min $= -4$
 Y max $= 4$
 Y scl $= 2$

5. X min $= -6$
 X max $= 6$
 X scl $= 1$
 Y min $= -2$
 Y max $= 2$
 Y scl $= 0.5$

6. X min $= -6$
 X max $= 6$
 X scl $= 2$
 Y min $= -4$
 Y max $= 4$
 Y scl $= 2$

7. X min $= 0$
 X max $= 9$
 X scl $= 1$
 Y min $= -2$
 Y max $= 4$
 Y scl $= 1$

8. X min $= -6$
 X max $= 6$
 X scl $= 2$
 Y min $= -4$
 Y max $= 4$
 Y scl $= 2$

9. If Xmin $= -4$, Xmax $= 8$, and Xscl $= 1$, how should Ymin, Ymax, and Yscl be selected so that the viewing rectangle contains the point $(4, 8)$ and the screen is square?

10. If Xmin $= -6$, Xmax $= 12$, and Xscl $= 2$, how should Ymin, Ymax, and Yscl be selected so that the viewing rectangle contains the point $(4, 8)$ and the screen is square?

A.4 Using a Graphing Utility to Locate Intercepts and Check for Symmetry

Value and Zero (or Root)

Most graphing utilities have an eVALUEate feature that, given a value of x, determines the value of y for an equation. We can use this feature to evaluate an equation at $x = 0$ to determine the y-intercept. Most graphing utilities also have a ZERO (or ROOT) feature that can be used to determine the x-intercept(s) of an equation.

EXAMPLE 1 Finding Intercepts Using a Graphing Utility

Use a graphing utility to find the intercepts of the equation $y = x^3 - 8$.

SOLUTION Figure 12(a) shows the graph of $y = x^3 - 8$.

FIGURE 12

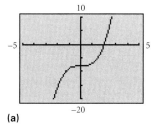
(a)

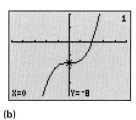
(b)

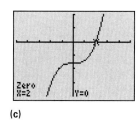
(c)

The eVALUEate feature of a TI-83 Plus graphing calculator accepts as input a value of x and determines the value of y. If we let $x = 0$, we find $y = -8$ so that the y-intercept is $(0, -8)$. See Figure 12(b).

The ZERO feature of a TI-83 Plus is used to find the x-intercept(s). See Figure 12(c). The x-intercept is $(2, 0)$.

TRACE

Most graphing utilities allow you to move from point to point along the graph, displaying on the screen the coordinates of each point. This feature is called TRACE.

EXAMPLE 2 Using TRACE to Locate Intercepts

Graph the equation $y = x^3 - 8$. Use TRACE to locate the intercepts.

SOLUTION Figure 13 shows the graph of $y = x^3 - 8$.

FIGURE 13

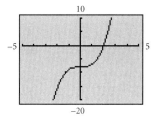

Activate the TRACE feature. As you move the cursor along the graph, you will see the coordinates of each point displayed. When the cursor is on the y-axis, we find that the y-intercept is −8. See Figure 14.

FIGURE 14

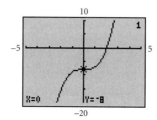

Continue moving the cursor along the graph. Just before you get to the x-axis, the display will look like the one in Figure 15(a). (Due to differences in graphing utilities, your display may be slightly different from the one shown here.)

FIGURE 15

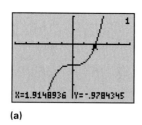 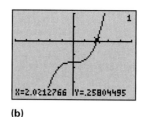

(a)   (b)

In Figure 15(a), the negative value of the y-coordinate indicates that we are still below the x-axis. The next position of the cursor is shown in Figure 15(b). The positive value of the y-coordinate indicates that we are now above the x-axis. This means that between these two points the x-axis was crossed. The x-intercept lies between 1.9148936 and 2.0212766. ▶

EXAMPLE 3   **Graphing the Equation $y = \dfrac{1}{x}$**

FIGURE 16

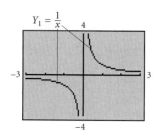

Graph the equation: $y = \dfrac{1}{x}$

with the viewing window set as

$$X\min = -3 \qquad Y\min = -4$$
$$X\max = 3 \qquad Y\max = 4$$
$$X\text{ scl} = 1 \qquad Y\text{ scl} = 1$$

Use TRACE to infer information about the intercepts and symmetry.

SOLUTION   Figure 16 illustrates the graph. We infer from the graph that there are no intercepts; we may also infer that symmetry with respect to the origin is a possibility. The TRACE feature on a graphing utility can provide further evidence of symmetry with respect to the origin. Using TRACE, we observe that for any ordered pair (x, y) the ordered pair $(-x, -y)$ is also a point on the graph. For example, the points $(0.95744681, 1.0444444)$ and $(-0.95744681, -1.0444444)$ both lie on the graph. ▶

EXERCISE A.4

In Problems 1–6, use ZERO (or ROOT) to approximate the smaller of the two x-intercepts of each equation. Express the answer rounded to two decimal places.

1. $y = x^2 + 4x + 2$
2. $y = x^2 + 4x - 3$
3. $y = 2x^2 + 4x + 1$
4. $y = 3x^2 + 5x + 1$
5. $y = 2x^2 - 3x - 1$
6. $y = 2x^2 - 4x - 1$

*In Problems 7–14, use ZERO (or ROOT) to approximate the **positive** x-intercepts of each equation. Express each answer rounded to two decimal places.*

7. $y = x^3 + 3.2x^2 - 16.83x - 5.31$
8. $y = x^3 + 3.2x^2 - 7.25x - 6.3$
9. $y = x^4 - 1.4x^3 - 33.71x^2 + 23.94x + 292.41$
10. $y = x^4 + 1.2x^3 - 7.46x^2 - 4.692x + 15.2881$
11. $y = \pi x^3 - (8.88\pi + 1)x^2 - (42.066\pi - 8.88)x + 42.066$
12. $y = \pi x^3 - (5.63\pi + 2)x^2 - (108.392\pi - 11.26)x + 216.784$
13. $y = x^3 + 19.5x^2 - 1021x + 1000.5$
14. $y = x^3 + 14.2x^2 - 4.8x - 12.4$

In Problems 15–18, the graph of an equation is given.
(a) List the intercepts of the graph.
(b) Based on the graph, tell whether the graph is symmetric with respect to the x-axis, y-axis, and/or origin.

15.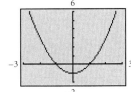

16.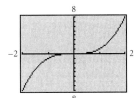

17.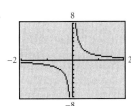

18.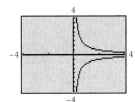

A.5 Using a Graphing Utility to Solve Equations

For many equations, there are no algebraic techniques that lead to a solution. For such equations, a graphing utility can often be used to investigate possible solutions. When a graphing utility is used to solve an equation, usually *approximate* solutions are obtained. Unless otherwise stated, we shall follow the practice of giving approximate solutions *rounded to two decimal places*.

The ZERO (or ROOT) feature of a graphing utility can be used to find the solutions of an equation when one side of the equation is 0. In using this feature to solve equations, we make use of the fact that the *x*-intercepts (or zeros) of the graph of an equation are found by letting $y = 0$ and solving the equation for *x*.

EXAMPLE 1 Using ZERO (or ROOT) to Approximate Solutions of an Equation

Find the solution(s) of the equation $x^2 - 6x + 7 = 0$. Round answers to two decimal places.

SOLUTION The solutions of the equation $x^2 - 6x + 7 = 0$ are the same as the x-intercepts of the graph of $Y_1 = x^2 - 6x + 7$. We begin by graphing the equation. See Figure 17(a).

FIGURE 17

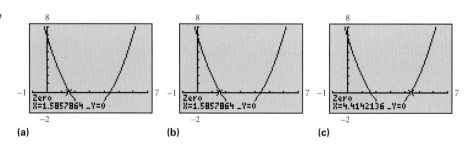

(a)   (b)   (c)

From the graph there appear to be two x-intercepts (solutions to the equation): one between 1 and 2, the other between 4 and 5.

Using the ZERO (or ROOT) feature of our graphing utility, we determine that the x-intercepts, and so the solutions to the equation, are $x = 1.59$ and $x = 4.41$, rounded to two decimal places. See Figures 17(b) and (c). ▶

A second method for solving equations using a graphing utility involves the INTERSECT feature of the graphing utility. This feature is used most effectively when one side of the equation is not 0.

EXAMPLE 2 Using INTERSECT to Approximate Solutions of an Equation

Find the solution(s) to the equation $3(x - 2) = 5(x - 1)$.

SOLUTION We begin by graphing each side of the equation as follows: graph $Y_1 = 3(x - 2)$ and $Y_2 = 5(x - 1)$. See Figure 18(a).

FIGURE 18

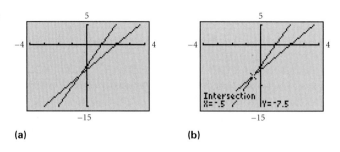

(a)   (b)

At the point of intersection of the graphs, the value of the y-coordinate is the same. We conclude that the x-coordinate of the point of intersection represents the solution to the equation. Do you see why? The INTERSECT feature on a graphing utility determines the point of intersection of the graphs. Using this feature, we find that the graphs intersect at $(-0.5, -7.5)$. See Figure 18(b). The solution of the equation is therefore $x = -0.5$. ▶

CHECK: We can verify our solution by evaluating each side of the equation with -0.5 STOred in x. See Figure 19. Since the left side of the equation equals the right side of the equation, the solution checks.

FIGURE 19

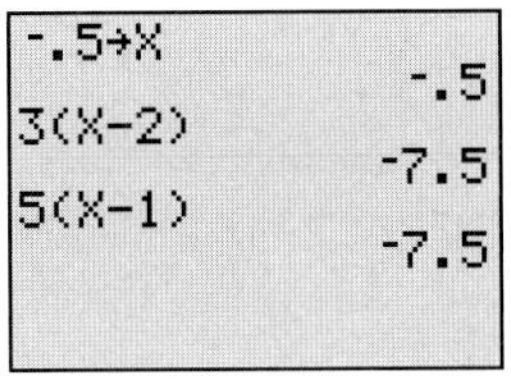

```
-.5→X
              -.5
3(X-2)
              -7.5
5(X-1)
              -7.5
```

SUMMARY The steps to follow for approximating solutions of equations are given next.

> **Steps for Approximating Solutions of Equations Using ZERO (or ROOT)**
>
> **STEP 1:** Write the equation in the form {expression in x} $= 0$.
> **STEP 2:** Graph $Y_1 =$ {expression in x}.
> **STEP 3:** Use ZERO (or ROOT) to determine each x-intercept of the graph.
>
> **Steps for Approximating Solutions of Equations Using INTERSECT**
>
> **STEP 1:** Graph $Y_1 =$ {expression in x on left side of equation}.
>   Graph $Y_2 =$ {expression in x on right side of equation}.
> **STEP 2:** Use INTERSECT to determine each x-coordinate of the point(s) of intersection, if any.

EXAMPLE 3 **Solving a Radical Equation**

Find the real solutions of the equation $\sqrt[3]{2x - 4} - 2 = 0$.

SOLUTION Figure 20 shows the graph of the equation $Y_1 = \sqrt[3]{2x - 4} - 2$. From the graph, we see one x-intercept near 6. Using ZERO (or ROOT), we find that the x-intercept is $(6, 0)$. The only solution is $x = 6$.

FIGURE 20

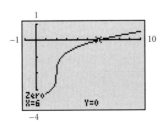

Answers to Odd-Numbered Problems

CHAPTER 0 Review

Exercise 0.1 (p. 15)

1. (a) 2 and 5 are natural numbers.
   (b) $-6, 2,$ and 5 are integers.
   (c) $-6, \frac{1}{2}, -1.333\ldots, 2,$ and 5 are rational numbers.
   (d) π is an irrational number.
   (e) All the numbers are real numbers.

3. (a) 1 is a natural number.
   (b) 0 and 1 are integers.
   (c) All the numbers are rational numbers.
   (d) There are no irrational numbers in the set C.
   (e) All the numbers are real numbers.

5. (a) There are no natural numbers in the set E.
   (b) There are no integers in the set E.
   (c) There are no rational numbers in the set E.
   (d) All the numbers are irrational.
   (e) All the numbers are real numbers.

7. (a) 18.953 (b) 18.952
9. (a) 28.653 (b) 28.653
11. (a) 0.063 (b) 0.062
13. (a) 9.999 (b) 9.998
15. (a) 0.429 (b) 0.428
17. (a) 34.733 (b) 34.733
19. $3 + 2 = 5$
21. $x + 2 = 3 \cdot 4$
23. $3y = 1 + 2$
25. $x - 2 = 6$
27. $\frac{x}{2} = 6$
29. 7
31. 6
33. 1
35. $\frac{13}{3}$
37. -11
39. 11
41. -4
43. 1
45. 6
47. $\frac{2}{7}$
49. $\frac{4}{45}$
51. $\frac{23}{20}$
53. $\frac{79}{30}$
55. $\frac{13}{36}$
57. $-\frac{16}{45}$
59. $\frac{1}{60}$
61. $\frac{15}{22}$
63. $6x + 24$
65. $x^2 - 4x$
67. $x^2 + 6x + 8$
69. $x^2 - x - 2$
71. $x^2 - 10x + 16$
73. $x^2 - 4$
75. Answers will vary.
77. Answers will vary.
79. Subtraction is not commutative. Examples will vary.
81. Division is not commutative. Examples will vary.
83. This is true by the symmetric property of real numbers.
85. All real numbers are either rational or irrational; no real number is both.
87. $0.9999\ldots = 1$

Exercise 0.2 (p. 26)

1.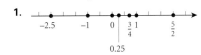

3. $>$
5. $>$
7. $>$
9. $=$
11. $<$
13. $x > 0$
15. $x < 2$
17. $x \leq 1$
19.
21.
23. 1
25. 2
27. 6
29. 4
31. -28
33. $\frac{4}{5}$
35. 0
37. 1
39. 5
41. 1
43. 22
45. 2
47. (c) $x = 0$
49. (a) $x = 3$
51. none
53. (b) $x = 1$, (c) $x = 0$, (d) $x = -1$
55. $\{x \mid x \neq 5\}$
57. $\{x \mid x \neq -4\}$
59. $C = 0°$
61. $C = 25°$
63. 16
65. $\frac{1}{16}$
67. $\frac{1}{9}$
69. 9
71. 5
73. 4
75. $64x^6$
77. $\frac{x^4}{y^2}$
79. $\frac{x}{y}$
81. $-\frac{8x^3 z}{9y}$
83. $\frac{16x^2}{9y^2}$
85. -4
87. 5
89. 4

Answers to Odd-Numbered Problems

91. 2 **93.** $\sqrt{5}$ **95.** $\frac{1}{2}$ **97.** 10; 0
99. 81 **101.** 304,006.671 **103.** 0.004
105. 481.890 **107.** 0.000
109. 4.542×10^2 **111.** 1.3×10^{-2}
113. 3.2155×10^4 **115.** 4.23×10^{-4}
117. 61,500 **119.** 0.001214
121. 110,000,000 **123.** 0.081
125. $A = l \cdot w$; $\{l|l > 0\}$, $\{w|w > 0\}$, $\{A|A > 0\}$
127. $C = \pi \cdot d$; $\{d|d > 0\}$, $\{C|C > 0\}$
129. $A = \frac{\sqrt{3}}{4}x^2$; $\{x|x > 0\}$, $\{A|A > 0\}$
131. $V = \frac{4}{3}\pi r^3$; $\{r|r > 0\}$, $\{V|V > 0\}$
133. $V = x^3$; $\{x|x > 0\}$, $\{V|V > 0\}$
135. (a) It costs $6,000 to produce 1000 watches.
(b) It costs $8,000 to produce 2000 watches.
137. (a) $|113 - 115| = |-2| = 2 \le 5$
(b) $|109 - 115| = |-6| = 6 > 5$
139. (a) Yes, $|2.999 - 3| = |-0.001| = 0.001 \le 0.01$.
(b) No, $|2.89 - 3| = |-0.11| = 0.11 > 0.01$.
141. No, $\frac{1}{3}$ is larger by 0.000333....
143. No; answers will vary.
145. Answers will vary.

Exercise 0.3 (p. 39)

1. $10x^5 + 3x^3 - 10x^2 + 6$ **3.** $2ax + a^2$
5. $2x^2 + 17x + 8$ **7.** $x^4 - x^2 + 2x - 1$
9. $6x^2 + 2$ **11.** $(x - 6)(x + 6)$
13. $(1 - 2x)(1 + 2x)$ **15.** $(x + 2)(x + 5)$
17. prime **19.** prime
21. $(x + 3)(5 - x)$ **23.** $3(x - 6)(x + 2)$
25. $y^2(y + 5)(y + 6)$ **27.** $(2x + 3)^2$
29. $(3x + 1)(x + 1)$ **31.** $(x - 3)(x + 3)(x^2 + 9)$
33. $(x - 1)^2(x^2 + x + 1)^2$ **35.** $x^5(x - 1)(x + 1)$
37. $(4x + 1)(5 - 4x)$ **39.** $(2y - 3)(2y - 5)$
41. $(x^2 + 1)(3x + 1)(1 - 3x)$ **43.** $(x - 6)(x + 3)$
45. $(x + 2)(x - 3)$ **47.** $3x(x - 2)^3(5x - 4)$
49. $(x - 1)(x + 1)(x + 2)$
51. $(x - 1)(x + 1)(x^2 - x + 1)$

53. $\dfrac{3(x - 3)}{5x}$ **55.** $\dfrac{x(2x - 1)}{x + 4}$

57. $\dfrac{5x}{(x - 6)(x - 1)(x + 4)}$ **59.** $\dfrac{2(x + 4)}{(x - 2)(x + 2)(x + 3)}$

61. $\dfrac{x^3 - 2x^2 + 4x + 3}{x^2(x - 1)(x + 1)}$ **63.** $-\dfrac{1}{x(x + h)}$

65. $2(3x + 4)(9x + 13)$ **67.** $2x(3x + 5)$
69. $5(x + 3)(x + 1)(x - 2)^2$ **71.** $3(4x - 1)(4x - 3)$

73. $6(3x - 5)(5x - 4)(2x + 1)^2$ **75.** $\dfrac{19}{(3x - 5)^2}$

77. $\dfrac{(x - 1)(x + 1)}{(x^2 + 1)^2}$ **79.** $\dfrac{x(3x + 2)}{(3x + 1)^2}$

81. $\dfrac{(x + 3)(1 - 3x)}{(x^2 + 1)^2}$

Exercise 0.4 (p. 49)

1. $\{7\}$ **3.** $\{-3\}$ **5.** $\{4\}$ **7.** $\{\frac{5}{4}\}$
9. $\{-1\}$ **11.** $\{-18\}$ **13.** $\{-3\}$ **15.** $\{-16\}$
17. $\{0.5\}$ **19.** $\{2\}$ **21.** $\{2\}$ **23.** $\{3\}$
25. $\{0, 9\}$ **27.** $\{0, 9\}$ **29.** $\{21\}$ **31.** $\{-2, 2\}$
33. $\{6\}$ **35.** $\{-3, 3\}$ **37.** $\{-4, 1\}$ **39.** $\{-1, \frac{3}{2}\}$
41. $\{-4, 4\}$ **43.** $\{2\}$
45. The equation has no solution. **47.** $\{-2, 2\}$ **49.** $\{-1, 3\}$
51. $\{-2, -1, 0, 1\}$ **53.** $\{0, 4\}$ **55.** $\{-6, 2\}$
57. $\{-\frac{1}{2}, 3\}$ **59.** $\{3, 4\}$ **61.** $\{\frac{3}{2}\}$ **63.** $\{-\frac{2}{3}, \frac{3}{2}\}$
65. $\{-\frac{3}{4}, 2\}$ **67.** $\{-5, 5\}$ **69.** $\{-1, 3\}$ **71.** $\{-3, 0\}$
73. 16 **75.** $\frac{1}{16}$ **77.** $\frac{1}{9}$ **79.** $\{-7, 3\}$
81. $\{-\frac{1}{4}, \frac{3}{4}\}$ **83.** $\left\{\dfrac{-1 - \sqrt{7}}{6}, \dfrac{-1 + \sqrt{7}}{6}\right\}$
85. $\{2 - \sqrt{2}, 2 + \sqrt{2}\}$ **87.** $\left\{\dfrac{5 - \sqrt{29}}{2}, \dfrac{5 + \sqrt{29}}{2}\right\}$
89. $\{1, \frac{3}{2}\}$ **91.** The equation has no real solution.
93. $\left\{\dfrac{-1 - \sqrt{5}}{4}, \dfrac{-1 + \sqrt{5}}{4}\right\}$
95. $\left\{-\dfrac{\sqrt{3} - \sqrt{15}}{2}, -\dfrac{\sqrt{3} + \sqrt{15}}{2}\right\}$
97. The equation has no real solution.
99. The equation has a repeated real solution.
101. The equation has two unequal real solutions.

103. $\left\{\dfrac{b + c}{a}\right\}$ **105.** $\left\{\dfrac{abc}{a + b}\right\}$ **107.** $\{a^2\}$

109. $\left\{\dfrac{R_1 R_2}{R_1 + R_2}\right\}$ **111.** $\left\{\dfrac{mv^2}{F}\right\}$ **113.** $\left\{\dfrac{S - a}{S}\right\}$

115. The solution set to the quadratic equation
$ax^2 + bx + c = 0, a \neq 0$ is
$\left\{\dfrac{-b - \sqrt{b^2 - 4ac}}{2a}, \dfrac{-b + \sqrt{b^2 - 4ac}}{2a}\right\}$. Adding the two roots, we obtain

$\dfrac{-b - \sqrt{b^2 - 4ac}}{2a} + \dfrac{-b + \sqrt{b^2 - 4ac}}{2a} =$

$\dfrac{-b - b - \sqrt{b^2 - 4ac} + \sqrt{b^2 - 4ac}}{2a} = -\dfrac{2b}{2a} = -\dfrac{b}{a}.$

117. $k = -\dfrac{1}{2}$ or $k = \dfrac{1}{2}$

119. Because $b^2 - 4ac = (-b)^2 - 4ac \geq 0$, both equations $ax^2 + bx + c = 0$ and $ax^2 - bx + c = 0$ have real solutions. The solutions to the first equation are

$x = \dfrac{-b - \sqrt{b^2 - 4ac}}{2a}$ and $x = \dfrac{-b + \sqrt{b^2 - 4ac}}{2a}$,

and the solutions to the second equation are

$x = \dfrac{b - \sqrt{b^2 - 4ac}}{2a} = -\left(\dfrac{-b + \sqrt{b^2 - 4ac}}{2a}\right)$ and

$x = \dfrac{b + \sqrt{b^2 - 4ac}}{2a} = -\left(\dfrac{b + \sqrt{b^2 - 4ac}}{2a}\right)$.

121. The equations in (b) are equivalent, because $\sqrt{9} = 3$. In (a), -3 is a solution of $x^2 = 9$ but not of $x = 3$. In (c), 1 is a solution of $(x-1)(x-2) = (x-1)^2$ but not of $x - 2 = x - 1$.

123. Answers will vary.   **125.** Answers will vary.

127. Answers will vary.

Exercise 0.5 (p. 61)

1. $(0, 2), 0 \leq x \leq 2$ **3.** $(-1, 2), -1 < x < 2$

5. $[0, 3), 0 \leq x < 3$

7. $[0, 4)$

9. $[4, 6)$

11. $[4, \infty)$

13. $(-\infty, -4)$

15. $2 \leq x \leq 5$

17. $-3 < x < 2$

19. $x \geq 4$

21. $x < -3$

23. (a) $6 < 8$
(b) $-2 < 0$
(c) $9 < 15$
(d) $-6 > -10$

25. (a) $7 > 0$
(b) $-1 > -8$
(c) $12 > -9$
(d) $-8 < 6$

27. (a) $2x + 4 < 5$
(b) $2x - 4 < -3$
(c) $6x + 3 < 6$
(d) $-4x - 2 > -4$

29. $<$ **31.** $>$ **33.** $\geq$ **35.** $<$

37. $\leq$ **39.** $>$ **41.** $\geq$

43. $(-\infty, 4) = \{x | x < 4\}$

45. $[-1, \infty) = \{x | x \geq -1\}$

47. $(3, \infty) = \{x | x > 3\}$

49. $[2, \infty) = \{x | x \geq 2\}$

51. $(-7, \infty) = \{x | x > -7\}$

53. $\left(-\infty, \dfrac{2}{3}\right] = \left\{x | x \leq \dfrac{2}{3}\right\}$

55. $(-\infty, -20) = \{x | x < -20\}$

57. $\left[\dfrac{4}{3}, \infty\right) = \left\{x | x \geq \dfrac{4}{3}\right\}$

59. $[3, 5] = \{x | 3 \leq x \leq 5\}$

61. $[\frac{2}{3}, 3] = \{x | \frac{2}{3} \le x \le 3\}$

63. $(-\frac{11}{2}, \frac{1}{2}) = \{x | -\frac{11}{2} < x < \frac{1}{2}\}$

65. $(-6, 0) = \{x | -6 < x < 0\}$

67. $(-\infty, -\frac{1}{2}) = \{x | x < -\frac{1}{2}\}$

69. $(\frac{10}{3}, \infty) = \{x | x > \frac{10}{3}\}$

71. $(-1, 3) = \{x | -1 < x < 3\}$

73. $(-3, 3) = \{x | -3 < x < 3\}$

75. $(-\infty, -4) \cup (3, \infty) = \{x | x < -4 \text{ or } x > 3\}$

77. $(-\infty, 3) \cup (4, \infty) = \{x | x < 3 \text{ or } x > 4\}$

79. $\{\}$

81. $(1, \infty) = \{x | x > 1\}$

83. $(-\infty, 1) \cup (2, 3) = \{x | x < 1 \text{ or } 2 < x < 3\}$

85. $(-2, 0) \cup (4, \infty) = \{x | -2 < x < 0 \text{ or } x > 4\}$

87. $(-1, 0) \cup (1, \infty) = \{x | -1 < x < 0 \text{ or } x > 1\}$

89. $(1, \infty) = \{x | x > 1\}$

91. $(-\infty, -1) \cup (1, \infty) = \{x | x < -1 \text{ or } x > 1\}$

93. $(-\infty, -1) \cup (0, 1) = \{x | x < -1 \text{ or } 0 < x < 1\}$

95. $(-1, 1) \cup [2, \infty) = \{x | -1 < x < 1 \text{ or } x \ge 2\}$

97. $(-\infty, 2) = \{x | x < 2\}$

99. $(-\infty, -3) \cup (-1, 1) \cup (2, \infty)$
$= \{x | x < -3 \text{ or } -1 < x < 1 \text{ or } x > 2\}$

101. The solution is $74 \le x < 124$, but assuming that the highest possible test score is 100, the range of possible exam scores that will enable you to earn a B is from 74 to 100.

103. The range of possible commissions is \$45,00 to \$95,00. The commission varies from 5% of the selling price to 8.6% of the selling price.

105. The amount withheld varies from \$81.35 to \$131.35.

107. Usage ranged from 657.41 kilowatt-hours to 2500.91 kilowatt-hours.

109. The dealer's cost range from \$7457.63 to \$7857.14.

111. Answers will vary.   **113.** Answers will vary.

Exercise 0.6 (p. 70)

1. 3   **3.** -2 **5.** $2\sqrt{2}$

7. $-2x\sqrt[3]{x}$ **9.** $x^3 y^2$ **11.** $x^2 y$

13. $6\sqrt{x}$ **15.** $6x\sqrt{x}$ **17.** $15\sqrt[3]{3}$

19. $12\sqrt{3}$ **21.** $2\sqrt{3}$ **23.** $x - 2\sqrt{x} + 1$

25. $-5\sqrt{2}$ **27.** $(2x - 1)\sqrt[3]{2x}$ **29.** $\dfrac{\sqrt{2}}{2}$

31. $-\dfrac{\sqrt{15}}{5}$ **33.** $\dfrac{5\sqrt{3} + \sqrt{6}}{23}$ **35.** $\dfrac{8\sqrt{5} - 19}{41}$

37. $\dfrac{5\sqrt[3]{4}}{2}$ **39.** $\dfrac{2x + h - 2\sqrt{x^2 + xh}}{h}$

41. $\{\frac{9}{2}\}$ **43.** $\{3\}$ **45.** 4

47. -3 **49.** 64   **51.** $\frac{1}{27}$

53. $\dfrac{27\sqrt{2}}{32}$ **55.** $\dfrac{27\sqrt{2}}{32}$ **57.** $x^{7/12}$

59. xy^2 **61.** $x^{4/3} y^{5/3}$ **63.** $\dfrac{8x^{3/2}}{y^{1/4}}$

65. $\dfrac{3x + 2}{(x + 1)^{1/2}} = \dfrac{3x + 2}{\sqrt{x + 1}}$

67. $\dfrac{x(3x^2 + 2)}{(x^2 + 1)^{1/2}} = \dfrac{x(3x^2 + 2)}{\sqrt{x^2 + 1}}$

69. $\dfrac{22x + 5}{10(4x^2 - 17x - 15)^{1/2}} = \dfrac{22x + 5}{10\sqrt{4x^2 - 17x - 15}}$

71. $\dfrac{x + 2}{2(x + 1)^{3/2}}$ **73.** $\dfrac{4 - x}{(x + 4)^{3/2}}$

75. $\dfrac{1}{x^2(x^2 - 1)^{1/2}} = \dfrac{1}{x^2\sqrt{x^2 - 1}}$

77. $\dfrac{1 - 3x^2}{2x^{1/2}(x^2 + 1)^2} = \dfrac{1 - 3x^2}{2\sqrt{x}(x^2 + 1)^2}$

79. $\dfrac{(5x + 2)(x + 1)^{1/2}}{2}$

81. $2x^{1/2}(x + 1)(3x - 4)$

83. $(x^2 + 4)^{1/3}(11x^2 + 12)$

85. $(2x + 3)^{1/2}(3x + 5)^{1/3}(17x + 27)$

87. $\dfrac{3(x + 2)}{2x^{1/2}}$

89. $\dfrac{2(2 - x)(2 + x)}{(8 - x^2)^{1/2}} = \dfrac{2(2 - x)(2 + x)}{\sqrt{8 - x^2}}$

Exercise 0.7 (p. 74)

1. $c = 13$ **3.** $c = 26$ **5.** $c = 25$

7. This is a right triangle. The hypotenuse is the side of length 5.

9. This is not a right triangle.

11. This is a right triangle. The hypotenuse is the side of length 25.

13. This is not a right triangle.   **15.** $A = 8$ inches²

17. $A = 4$ inches²   **19.** $A = 25\pi$ m², $C = 10$ m

21. $V = 224$ ft³, $S = 232$ ft²   **23.** $V = \dfrac{256\pi}{3}$ cm³, $S = 64\pi$ cm²

25. $V = 648\pi$ inches³, $S = 306\pi$ inches²

27. The area is π units².   **29.** The area is 2π units².

31. The wheel travels 64π inches after four revolutions.

33. The area is 64 ft².

35. The area of the window is $24 + 2\pi$ ft² ≈ 30.28 ft². $16 + 2\pi \approx 22.28$ ft of wood frame are needed to enclose the window.

37. You can see a distance of 28,920 ft, which is about 5.478 miles.

39. A person can see 64,667 ft or 12.248 miles from the deck. A person can see 79,200 ft or 15.0 miles from the bridge.

41. The areas of the rectangular pools vary from 0 ft² to 62,500 ft². The shape of the rectangle of largest area is a square with side length 250 ft. The area of the circular pool is $\dfrac{250\,000}{\pi}$ ft² $\approx 79,577$ ft². The best choice for a pool of largest area would be the circular pool.

Exercise 0.8 (p. 80)

1. (a) Quadrant II   (b) Positive x-axis   (c) Quadrant III
   (d) Quadrant I   (e) Negative y-axis   (f) Quadrant IV

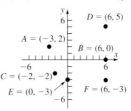

3. The points will be on a vertical line that is 2 units to the right of the y-axis.

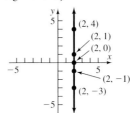

5. $\sqrt{5}$ **7.** $\sqrt{10}$ **9.** $2\sqrt{17}$
11. $\sqrt{85}$ **13.** $\sqrt{53}$ **15.** $\sqrt{6.89} \approx 2.625$
17. $\sqrt{a^2 + b^2}$

19. $d(A, B) = \sqrt{13}$
$d(B, C) = \sqrt{13}$
$d(A, C) = \sqrt{26}$
$(\sqrt{13})^2 + (\sqrt{13})^2 = (\sqrt{26})^2$
Area $= \dfrac{13}{2}$ square units

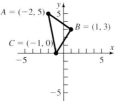

21. $d(A, B) = \sqrt{130}$
$d(B, C) = \sqrt{26}$
$d(A, C) = 2\sqrt{26}$
$(\sqrt{26})^2 + (2\sqrt{26})^2 = (\sqrt{130})^2$
Area $= 26$ square units

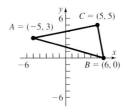

23. $d(A, B) = 4$
$d(B, C) = \sqrt{41}$
$d(A, C) = 5$
$4^2 + 5^2 = (\sqrt{41})^2$
Area = 10 square units

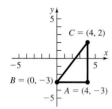

25. $(2, 2); (2, -4)$ **27.** $(0, 0); (8, 0)$

29. $4\sqrt{10}$ **31.** $2\sqrt{65}$

33. $90\sqrt{2} \approx 127.28$ ft

35. (a) First base $(90, 0)$; second base $(90, 90)$; third base $(0, 90)$
(b) $5\sqrt{2161}$ ft ≈ 232.4 ft
(c) $30\sqrt{149}$ ft ≈ 366.2 ft

37. $d = 50\,t$

Exercise 0.9 (p. 92)

1.

x	0	-2	2	-2	4	-4
y	4	0	8	0	12	-4

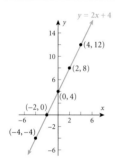

3.

x	0	3	2	-2	4	-4
y	-6	0	-2	-10	2	-14

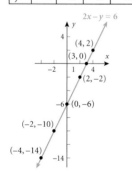

5. (a) Vertical line: $x = 2$
(b) Horizontal line: $y = -3$

7. (a) Vertical line: $x = -4$
(b) Horizontal line: $y = 1$

9. $m = \frac{1}{2}$;   A slope of $\frac{1}{2}$ means that for every 2 unit change in x, y changes 1 unit.

11. $m = -1$;   A slope of -1 means that for every 1 unit change in x, y changes by (-1) units.

13. $m = 3$;   A slope of 3 means that for every 1 unit change in x, y will change 3 units.

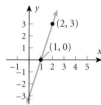

15. $m = -\frac{1}{2}$;   A slope of $-\frac{1}{2}$ means that for every 2 unit change in x, y will change (-1) unit.

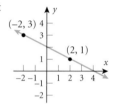

17. $m = 0$;   A slope of 0 means that regardless how x changes, y remains constant.

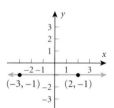

19. The slope is not defined.

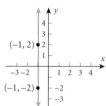

21.   **23.**

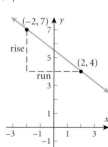

25. 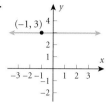 **27.** 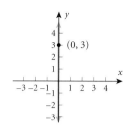 **59.** Slope is not defined; there is no y-intercept.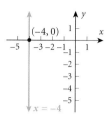

29. $x - 2y = 0$
31. $x + y = 2$
33. $2x - y = -9$
35. $2x + 3y = -1$
37. $x - 2y = -5$
39. $2x + y = 3$
41. $3x - y = -12$
43. $4x - 5y = 0$
45. $x - 2y = 2$
47. $x = 1$
49. $y = 4$

51. slope: $m = 2$; y-intercept: $(0, 3)$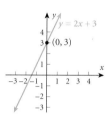

53. slope: $m = 2$; y-intercept: $(0, -2)$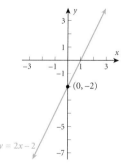

55. slope: $m = \frac{2}{3}$; y-intercept: $(0, -2)$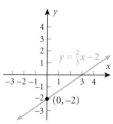

57. slope: $m = -1$; y-intercept: $(0, 1)$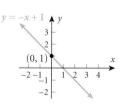

61. slope: $m = 0$; y-intercept: $(0, 5)$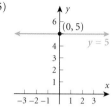

63. slope: $m = 1$; y-intercept: $(0, 0)$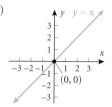

65. slope: $m = \frac{3}{2}$; y-intercept: $(0, 0)$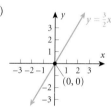

67. $y = -3$ **69.** $C = 0.122x$

71. (a) $C = 0.08275x + 7.58, 0 \leq x \leq 400$
(b)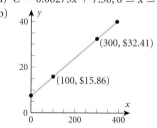
(c) The monthly charge for using 100 KWH is $15.86.
(d) The monthly charge for using 300 KWH is $32.41.
(e) The slope indicates that for every extra KWH used (up to 400 KWH), the electric bill increases by 8.275 cents.

73. $w = 4h - 129$ **75.** $C = 0.53x + 1{,}070{,}000$

77. (a) $C = \frac{5}{9}(F - 32)$ (b) $C = 20°$

79. (a) $A = -\frac{1}{75}t + 53.007$ (b) $A = 52.74$ billion gallons
(c) The slope tells us that the reservoir loses 1 billion gallons of water every 75 days.
(d) $A = 52.194$ billion gallons
(e) In 10.89 years the reservoir will be empty.
(f) Answers will vary.

81. Window: X min $= -10$; X max $= 10$
    Y min $= -10$; Y max $= 10$
    x-intercept: $(1.67, 0)$; y-intercept: $(0, 2.50)$

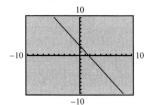

83. Window: X min $= -10$; X max $= 10$
    Y min $= -10$; Y max $= 10$
    x-intercept: $(2.52, 0)$; y-intercept: $(0, -3.53)$

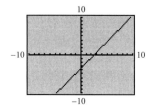

85. Window: X min $= -10$; X max $= 10$
    Y min $= -10$; Y max $= 10$
    x-intercept: $(2.83, 0)$; y-intercept: $(0, 2.56)$

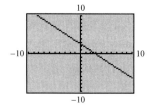

87. Window: X min $= -10$; X max $= 10$;
    Y min $= -10$; Y max $= 10$
    x-intercept: $(0.78, 0)$; y-intercept: $(0, -1.41)$

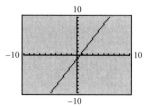

89. (b) 91. (d)

93. $y = x + 2$ or $x - y = -2$

95. $y = -\frac{1}{3}x + 1$ or $x + 3y = 3$

97. (b), (c), (e), (g)

99. $y = 0$

101. Answers will vary.

103. No; answers will vary; No; answers will vary.

105. The lines are identical.

107. Two lines can have the same y-intercept and the same x-intercept but different slopes only if their y-intercept is the point $(0, 0)$.

CHAPTER 1 Functions and Their Graphs

Exercise 1.1 (p. 107)

1. (a) $(3, -4)$ (b) $(-3, 4)$ (c) $(-3, -4)$

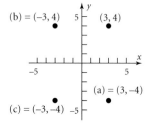

3. (a) $(-2, -1)$ (b) $(2, 1)$ (c) $(2, -1)$

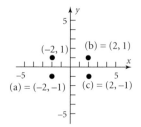

Exercise 1.1

5. (a) $(1, -1)$ (b) $(-1, 1)$ (c) $(-1, -1)$

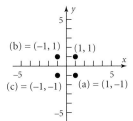

7. (a) $(-3, 4)$ (b) $(3, -4)$ (c) $(3, 4)$

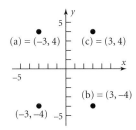

9. (a) $(0, 3)$ (b) $(0, -3)$ (c) $(0, 3)$

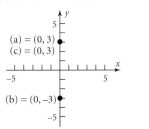

11. (a) The x-intercepts are $(-1, 0)$ and $(1, 0)$. There is no y-intercept.
(b) The graph is symmetric with respect to the x-axis, y-axis, and the origin.

13. (a) The x-intercepts are $\left(-\frac{\pi}{2}, 0\right)$ and $\left(\frac{\pi}{2}, 0\right)$. The y-intercept is $(0, 1)$.
(b) The graph is symmetric with respect to the y-axis.

15. (a) The x-intercept and the y-intercept are both $(0, 0)$.
(b) The graph is symmetric with respect to the x-axis.

17. (a) The x-intercept is $(1, 0)$. There is no y-intercept.
(b) The graph is not symmetric with respect to either axis or to the origin.

19. (a) The x-intercepts are $(-1, 0)$ and $(1, 0)$. The y-intercept is $(0, -1)$.
(b) The graph is symmetric with respect to the y-axis.

21. (a) There are no intercepts.
(b) The graph is symmetric with respect to the origin.

23. The point $(0, 0)$ is on the graph. The points $(1, 1)$ and $(-1, 0)$ are not on the graph.

25. The point $(0, 3)$ is on the graph. The points $(3, 0)$ and $(-3, 0)$ are not on the graph.

27. The points $(0, 2)$ and $(\sqrt{2}, \sqrt{2})$ are on the graph. The point $(-2, 2)$ is not on the graph.

29. The x-intercept and the y-intercept are $(0, 0)$. The graph is symmetric with respect to the y-axis.

31. The x-intercept and the y-intercept are $(0, 0)$. The graph is symmetric with respect to the origin.

33. The x-intercepts are $(-3, 0)$ and $(3, 0)$. The y-intercept is $(0, 9)$. The graph is symmetric with respect to the y-axis.

35. The x-intercepts are $(-2, 0)$ and $(2, 0)$. The y-intercepts are $(0, -3)$ and $(0, 3)$. The graph is symmetric with respect to the x-axis, the y-axis, and the origin.

37. The x-intercept is $(3, 0)$. The y-intercept is $(0, -27)$. The graph is not symmetric with respect to either axis or to the origin.

39. The x-intercepts are $(-1, 0)$ and $(4, 0)$. The y-intercept is $(0, -4)$. The graph is not symmetric with respect to either axis or to the origin.

41. The x-intercept and the y-intercept are both $(0, 0)$. The graph is symmetric with respect to the origin.

43. The x-intercept and the y-intercept are both $(0, 0)$. The graph is symmetric with respect to the y-axis.

45.

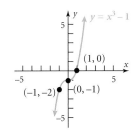

47.

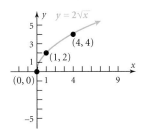

49. $a = -1$ **51.** $b = -\frac{2}{3}a + 2$

53. (a)

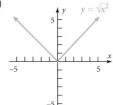

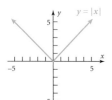
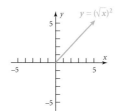

The graphs of $y = \sqrt{x^2}$ and $y = |x|$ are the same.

55. Let (x, y) be a point on the graph of the equation. Assume that the graph is symmetric with respect to both axes. Because of symmetry with respect to the y-axis, the point $(-x, y)$ is on the graph. Similarly, because of symmetry with respect to the x-axis, the point $(-x, -y)$ is also on the graph. Thus, the graph is symmetric with respect to the origin.

Assume that the graph is symmetric with respect to the x-axis and to the origin. Because of symmetry with respect to the x-axis, $(x, -y)$ is on the graph. Because of symmetry with respect to the origin, $(-x, y)$ is also on the graph. Thus, the graph is symmetric with respect to the y-axis.

Assume that the graph is symmetric with respect to the y-axis and to the origin. Because of symmetry with respect to the y-axis, $(-x, y)$ is on the graph. Because of symmetry with respect to the origin, $(x, -y)$ is also on the graph. Thus, the graph is symmetric with respect to the x-axis.

Exercise 1.2 (p. 118)

1. (a) $f(0) = -4$ (b) $f(1) = 1$
(c) $f(-1) = -3$ (d) $f(-x) = 3x^2 - 2x - 4$
(e) $-f(x) = -3x^2 - 2x + 4$
(f) $f(x+1) = 3x^2 + 8x + 1$
(g) $f(2x) = 12x^2 + 4x - 4$
(h) $f(x+h) = 3x^2 + 6xh + 2x + 3h^2 + 2h - 4$

3. (a) $f(0) = 0$ (b) $f(1) = \frac{1}{2}$
(c) $f(-1) = -\frac{1}{2}$ (d) $f(-x) = -\frac{x}{x^2+1}$
(e) $-f(x) = -\frac{x}{x^2+1}$ (f) $f(x+1) = \frac{x+1}{x^2+2x+2}$

(g) $f(2x) = \dfrac{2x}{4x^2+1}$

(h) $f(x+h) = \dfrac{x+h}{x^2+2hx+h^2+1}$

5. (a) $f(0) = 4$ (b) $f(1) = 5$
(c) $f(-1) = 5$ (d) $f(-x) = |x| + 4$
(e) $-f(x) = -|x| - 4$ (f) $f(x+1) = |x+1| + 4$
(g) $f(2x) = 2|x| + 4$ (h) $f(x+h) = |x+h| + 4$

7. (a) $f(0) = -\frac{1}{5}$ (b) $f(1) = -\frac{3}{2}$
(c) $f(-1) = \frac{1}{8}$ (d) $f(-x) = \dfrac{2x-1}{3x+5}$
(e) $-f(x) = \dfrac{2x+1}{5-3x}$ (f) $f(x+1) = \dfrac{2x+3}{3x-2}$
(g) $f(2x) = \dfrac{4x+1}{6x-5}$ (h) $f(x+h) = \dfrac{2x+2h+1}{3x+3h-5}$

9. 4   **11.** $2x + h - 1$
13. $3x^2 + 3xh + h^2$ **15.** $4x^3 + 6hx^2 + 4h^2x + h^3$
17. function   **19.** function
21. not a function   **23.** not a function
25. function   **27.** not a function
29. all real numbers   **31.** all real numbers
33. $\{x | x \neq -4, x \neq 4\}$ **35.** $\{x | x \neq 0\}$
37. $\{x | x \geq 4\}$ or the interval $[4, \infty)$
39. $\{x | x > 9\}$ or the interval $(9, \infty)$
41. $\{x | x > 1\}$ or the interval $(1, \infty)$
43. $A = -\frac{7}{2}$
45. $A = -4$, f is undefined at $x = -2$
47. $A = 8$, f is undefined at $x = 3$
49. $G(x) = 10x$ **51.** $R(x) = -\frac{1}{5}x^2 + 100x$
53. $R(x) = -\frac{1}{20}x^2 + 5x$
55. 28,027 thousand acres of wheat will be planted in 2010.
57. The expected Mathematics SAT score would be 456 in 2010.
59. (a) The cost per passenger is $222.
(b) The cost per passenger is $225.
(c) The cost per passenger is $220.
(d) The cost per passenger is $230.
61. (a) Yes   (b) No   (c) No   (d) No
63. Answers will vary.

Exercise 1.3 (p. 132)

1. This is not the graph of a function.
3. This is the graph of a function.
   (a) The domain is $\{x|-\pi \leq x \leq \pi\}$ or the interval $[-\pi,\pi]$. The range is $\{y|-1 \leq y \leq 1\}$ or the interval $[-1, 1]$.
   (b) The x-intercepts are $\left(-\frac{\pi}{2}, 0\right)$ and $\left(\frac{\pi}{2}, 0\right)$. The y-intercept is $(0, 1)$.
   (c) The graph is symmetric with respect to the y-axis.
5. This is not the graph of a function.
7. This is the graph of a function.
   (a) The domain is $\{x|x > 0\}$ or the interval $(0, \infty)$. The range is all real numbers or the interval $(-\infty, \infty)$.
   (b) The x-intercept is $(1, 0)$. There is no y-intercept.
   (c) This graph does not have symmetry with respect to the x-axis, y-axis, or the origin.
9. This is the graph of a function.
   (a) The domain is all real numbers or the interval $(-\infty,\infty)$. The range is $\{y|y \leq 2\}$ or the interval $(-\infty, 2]$.
   (b) The x-intercepts are $(-3, 0)$ and $(3, 0)$. The y-intercept is $(0, 2)$.
   (c) The graph is symmetric with respect to the y-axis.
11. This is the graph of a function.
    (a) The domain is all real numbers or the interval $(-\infty,\infty)$. The range is $\{y|y \geq -3\}$ or the interval $[-3,\infty)$.
    (b) The x-intercepts are $(1, 0)$ and $(3, 0)$. The y-intercept is $(0, 9)$.
    (c) This graph does not have symmetry with respect to the x-axis, y-axis, or the origin.
13. (a) $f(0) = 3, f(-6) = -3$ (b) $f(6) = 0, f(11) = 1$
    (c) $f(3)$ is positive. (d) $f(-4)$ is negative.
    (e) $x = -3, x = 6,$ and $x = 10$
    (f) $f(x) > 0$ on the intervals $(-3, 6)$ and $(10, 11]$.
    (g) The domain of f is $\{x|-6 \leq x \leq 11\}$ or the interval $[-6, 11]$.
    (h) The range of f is $\{y|-3 \leq y \leq 4\}$ or the interval $[-3, 4]$.
    (i) The x-intercepts are $(-3, 0), (6, 0),$ and $(10, 0)$.
    (j) The y-intercept is $(0, 3)$.
    (k) The line $y = \frac{1}{2}$ intersects the graph 3 times.
    (l) The line $x = 5$ intersects the graph once.
    (m) $f(x) = 3$ when $x = 0$ and $x = 4$.
    (n) $f(x) = -2$ when $x = -5$ and $x = 8$.

15. (a) Yes
    (b) $f(-2) = 9$; the point $(-2, 9)$ is on the graph of f.
    (c) $x = 0$ or $x = \frac{1}{2}$; the points $(0, -1)$ and $\left(\frac{1}{2}, -1\right)$ are on the graph of f.
    (d) The domain of f is all real numbers or the interval $(-\infty, \infty)$.
    (e) The x-intercepts are $\left(-\frac{1}{2}, 0\right)$ and $(1, 0)$.
    (f) The y-intercept is $(0, -1)$
17. (a) No
    (b) $f(4) = -3$; the point $(4, -3)$ is on the graph of f.
    (c) $x = 14$; the point $(14, 2)$ is on the graph of f.
    (d) The domain is the set $\{x \,|\, x \neq 6\}$.
    (e) The x-intercept is $(-2, 0)$.
    (f) The y-intercept is $\left(0, -\frac{1}{3}\right)$.
19. (a) Yes
    (b) $f(2) = \frac{8}{17}$; the point $(2, \frac{8}{17})$ is on the graph of f.
    (c) $x = -1$ or $x = 1$; the points $(-1, 1)$ and $(1, 1)$ are on the graph of f.
    (d) The domain is the set of all real numbers or the interval $(-\infty, \infty)$.
    (e) The x-intercept is $(0, 0)$.
    (f) The y-intercept is $(0, 0)$.
21. Yes   23. No
25. f is increasing on the intervals $(-8, -2), (0, 2),$ and $(5, \infty)$ or for $-8 < x < -2, 0 < x < 2$ and $x > 5$.
27. There is a local maximum at $x = 2$. The local maximum is $f(2) = 10$.
29. f has local maxima at $x = -2$ and $x = 2$. The local maxima are $f(-2) = 6,$ and $f(2) = 10$.
31. (a) The x-intercepts are $(-2, 0)$ and $(2, 0)$. The y-intercept is $(0, 3)$.
    (b) The domain is $\{x|-4 \leq x \leq 4\}$or the interval $[-4, 4]$. The range is $\{y|0 \leq y \leq 3\}$ or $[0, 3]$.
    (c) The function is increasing on $(-2, 0)$ and $(2, 4)$ or for $-2 < x < 0$ and $2 < x < 4$. The function is decreasing on $(-4, -2)$ and $(0, 2)$ or for $-4 < x < -2$ and $0 < x < 2$.
    (d) The function is even.
33. (a) The y-intercept is $(0, 1)$. There is no x-intercept.
    (b) The domain is the set of all real numbers. The range is set of positive numbers or $\{y|y > 0\}$ or the interval $(0, \infty)$.
    (c) The function is increasing on $(-\infty, \infty)$ or for all x-values.
    (d) The function is neither even nor odd.

35. (a) The x-intercepts are $(-\pi, 0), (0, 0)$ and $(\pi, 0)$. The y-intercept is $(0, 0)$.
(b) The domain is $[-\pi, \pi]$. The range is $[-1, 1]$.
(c) The function is increasing on $(-\frac{\pi}{2}, \frac{\pi}{2})$ or for $-\frac{\pi}{2} < x < \frac{\pi}{2}$. The function is decreasing on $(-\pi, -\frac{\pi}{2})$ and $(\frac{\pi}{2}, \pi)$ or for $-\pi < x < -\frac{\pi}{2}$ and $\frac{\pi}{2} < x < \pi$.
(d) The function is odd.

37. (a) The x-intercepts are $(\frac{1}{2}, 0)$ and $(\frac{5}{2}, 0)$. The y-intercept is $(0, \frac{1}{2})$.
(b) The domain is $\{x | -3 \le x \le 3\}$ or the interval $[-3, 3]$. The range is $\{y | -1 \le y \le 2\}$ or the inerval $[-1, 2]$.
(c) The function is increasing on $(2, 3)$ or for $2 < x < 3$. The function is decreasing on $(-1, 1)$ or for $-1 < x < 1$. The function is constant on $(-3, -1)$ and $(1, 2)$ or for $-3 < x < 1$ and $1 < x < 2$.
(d) The function is neither even nor odd.

39. (a) The function has a local maximum of 3 at $x = 0$.
(b) The function has local minima of 0 at $x = -2$ and $x = 2$.

41. (a) The function has a local maximum of 1 at $x = \frac{\pi}{2}$.
(b) The function has a local minimum of -1 at $x = -\frac{\pi}{2}$.

43. (a) $\frac{\Delta y}{\Delta x} = -4$ (b) $\frac{\Delta y}{\Delta x} = -8$
(c) $\frac{\Delta y}{\Delta x} = -10$

45. (a) $\frac{\Delta y}{\Delta x} = 5$ (b) $\frac{\Delta y}{\Delta x} = 5 = m_{sec}$
(c) $y = 5x$

47. (a) $\frac{\Delta y}{\Delta x} = -3$ (b) $\frac{\Delta y}{\Delta x} = -3 = m_{sec}$
(c) $y = -3x + 1$

49. (a) $\frac{\Delta y}{\Delta x} = x - 1$ (b) $\frac{\Delta y}{\Delta x} = 1 = m_{sec}$
(c) $y = x - 2$

51. (a) $\frac{\Delta y}{\Delta x} = x^2 + x$ (b) $\frac{\Delta y}{\Delta x} = 6 = m_{sec}$
(c) $y = 6x - 6$

53. (a) $\frac{\Delta y}{\Delta x} = -\frac{1}{1 + x}$ (b) $\frac{\Delta y}{\Delta x} = -\frac{1}{3} = m_{sec}$
(c) $y = -\frac{1}{3}x + \frac{4}{3}$

55. (a) $\frac{\Delta y}{\Delta x} = \frac{\sqrt{x} - 1}{x - 1}$ (b) $\frac{\Delta y}{\Delta x} = \sqrt{2} - 1 = m_{sec}$
(c) $y = (\sqrt{2} - 1)x + 2 - \sqrt{2}$

57. odd **59.** even **61.** odd
63. neither even nor odd **65.** even **67.** odd

69.

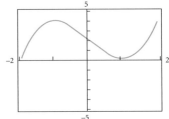

The function has a local minimum of 0 at $x = 1$ and a local maximum of 4 at $x = -1$. The function is increasing on $(-2, -1)$ and $(1, 2)$ and is decreasing on $(-1, 1)$.

71.

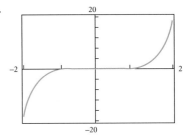

The function has a local minimum of -0.19 at $x = 0.77$ and a local maximum of 0.19 at $x = -0.77$. The function is increasing on $(-2, -0.77)$ and $(0.77, 2)$ and is decreasing on $(-0.77, 0.77)$.

73.

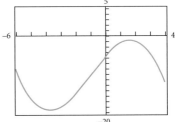

The function has a local minimum of -18.89 at $x = -3.77$ and a local maximum of -1.91 at $x = 1.77$. The function is increasing on $(-3.77, 1.77)$ and it is decreasing on $(-6, -3.77)$ and $(1.77, 4)$.

75.

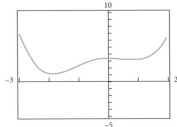

The function has a local minimum of 0.95 at $x = -1.87$, a local maximum of 3 at $x = 0$, and a local minimum of 2.65 at $x = 0.97$. The function is increasing on $(-1.87, 0)$ and $(0.97, 2)$ and is decreasing on $(-3, -1.87)$ and $(0, 0.97)$.

77. (a) $\dfrac{\Delta y}{\Delta x} = 1$ (b) $\dfrac{\Delta y}{\Delta x} = 0.5$ (c) $\dfrac{\Delta y}{\Delta x} = 0.1$
(d) $\dfrac{\Delta y}{\Delta x} = 0.01$ (e) $\dfrac{\Delta y}{\Delta x} = 0.001$

(f)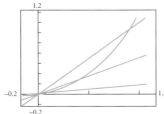

(g) The secant lines are approaching the tangent line to the graph of f at $x = 0$.

(h) The slopes of the secant lines are approach the slope of the tangent line to the graph of f at $x = 0$, which is 0.

79. (a) 81.07 ft. (b) 129.59 ft.
(c) 26.63 ft. (d) The golf ball was hit 528.13 feet.

(e)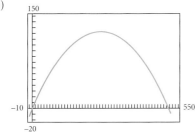

(f) The ball is at a height of 90 feet when it has traveled 115.07 feet and 413.05 feet.

(g)

X	Y1
0	0
25	23.817
50	45.266
75	64.349
100	81.065
125	95.414
150	107.4

X = 0

(h) The ball travels about 275 feet before reaching its maximum height. The maximum height of the ball is 132 feet.

(i) The ball travels 264 feet before reaching its maximum height.

81. (a) $V(x) = x(24 - 2x)^2 = 4x^3 - 96x^2 + 576x$
(b) The volume is 972 in³.
(c) The volume is 160 in³.

(d)

The volume V is largest when $x = 4$ in.

83. (a)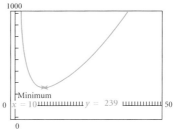

(b) Producing 10 riding lawn mowers minimizes average cost.

(c) The minimum average cost is $239.

85. Reasons will vary. (a) II (b) V (c) IV (d) III (e) I

87. Answers will vary. **89.** Answers will vary.

91. Answers will vary. **93.** Answers will vary.

Exercise 1.4 (p. 145)

1. C. **3.** E. **5.** B. **7.** F.

9.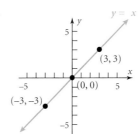

11.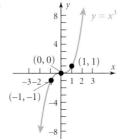

13.

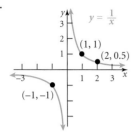

15.

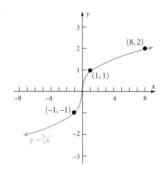

17. (a) $f(-2) = 4$ (b) $f(0) = 2$ (c) $f(2) = 5$

19. (a) $f(1.2) = 2$ (b) $f(1.6) = 3$ (c) $f(-1.8) = -4$.

21. (a) The domain is all real numbers, or the interval $(-\infty, \infty)$.
(b) There is no x-intercept. The y-intercept is $(0, 1)$.
(c)

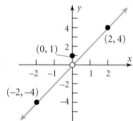

(d) The range is $\{y | y \neq 0\}$ or the intervals $(-\infty, 0)$ and $(0, \infty)$.

23. (a) The domain is all real numbers.
(b) There is no x-intercept. The y-intercept is $(0, 3)$.
(c)

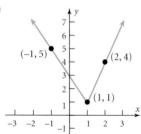

(d) The range is $\{y | y \geq 1\}$ or the interval $[1, \infty)$.

25. (a) The domain is the set of real numbers greater than or equal to -2, $\{x | x \geq -2\}$ or $[-2, \infty)$.
(b) The x-intercept is $(2, 0)$. The y-intercept is $(0, 3)$.
(c)

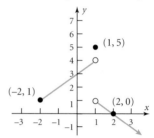

(d)

27. (a) The domain is set of all real numbers or the interval $(-\infty, \infty)$.
(b) The x-intercepts are $(-1, 0)$ and $(0, 0)$. The y-intercept is $(0, 0)$.
(c)

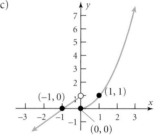

(d) The range is all real numbers or the interval $(-\infty, \infty)$.

29. (a) The domain is $\{x | x \geq -2\}$ or the interval $[-2, \infty)$.
(b) There is no x-intercept. The y-intercept is $(0, 1)$.
(c)

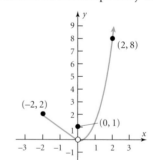

(d) The range is $\{y | y > 0\}$ or the interval $(0, \infty)$.

31. (a) The domain is all real numbers.
(b) The x-intercepts lie in the interval [0, 1). The y-intercept is (0, 0).
(c)

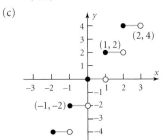

(d) The range is the set of even integers.

33. $f(x) = \begin{cases} -x & \text{if } -1 \le x \le 0 \\ \frac{1}{2}x & \text{if } 0 < x \le 2 \end{cases}$

35. $f(x) = \begin{cases} -x & \text{if } x \le 0 \\ 2 - x & \text{if } 0 < x \le 2 \end{cases}$

37. (a) $39.99 (b) $43.74 (c) $40.24

39. (a) $59.33 (b) $396.04

(c) $C(x) = \begin{cases} 0.99755x + 9.45 & \text{if } 0 \le x \le 50 \\ 0.74825x + 21.915 & \text{if } x > 50 \end{cases}$

(d)

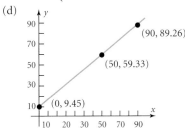

41. (a) 10°C. (b) 3.98°C. (c) −2.67°C.
(d) −3.70°C.
(e) For wind speeds under 1.79 m/sec, the wind chill factor is simply the air temperature.
(f) For wind speeds above 20 m/sec, the wind chill factor is a function of the air temperature.

43. $y = \begin{cases} 0.10x & \text{if } 0 < x \le 7000 \\ 700 + 0.15(x - 7000) & \text{if } 7000 < x \le 28,400 \\ 3910 + 0.25(x - 28,400) & \text{if } 28,400 < x \le 68,800 \\ 34,926 + 0.33(x - 143,500) & \text{if } 143,500 < x \le 311,950 \\ 90,514.5 + 0.35(x - 311,950) & \text{if } x > 311,950 \end{cases}$

45.

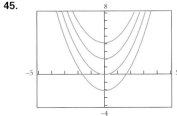

If $a > 0$, the graph of $y = x^2 + a$ is a vertical shift of the graph of $y = x^2$ upward by a units. If $a < 0$, the graph of $y = x^2 + a$ is a vertical shift of the graph of $y = x^2$ downward by a units. The graph of $y = x^2 - 4$ is a vertical shift of the graph of $y = x^2$ downward by 4 units. The graph of $y = x^2 + 5$ is a vertical shift of the graph of $y = x^2$ upward by 5 units.

47.

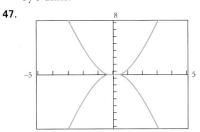

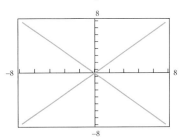

The graph of $y = -x^2$ is the reflection of $y = x^2$ in the x-axis. The graph of $y = -|x|$ is the reflection of $y = |x|$ in the x-axis.

49.

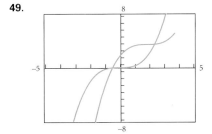

The graph of $y = (x - h)^3 + k$ is obtained by shifting the graph of $y = x^3$ h units to the right and k units upward.

51.

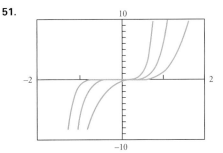

Each of the graphs increases in height as x increases. The graphs become increasingly flatter at $x = 0$ as the exponent increases.

Exercise 1.5 (p. 154)

1. B. **3.** H. **5.** A. **7.** F.

9. $y = (x - 4)^3$ **11.** $y = x^3 + 4$

13. $y = -x^3$ **15.** $y = -\sqrt{-x} - 2$

17. $y = -\sqrt{x + 3} + 2$ **19.** (c) $(3, 0)$

21.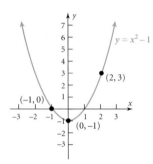

23.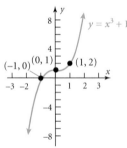

25.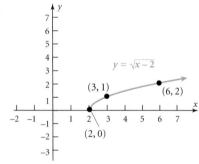

27.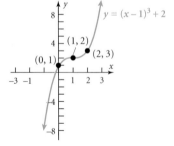

29.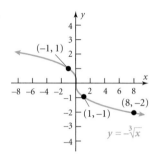

31.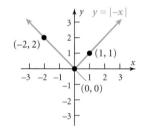

33.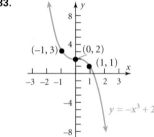

35.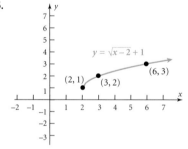

37.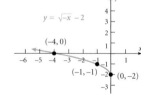

Exercise 1.5 AN-17

39. (a)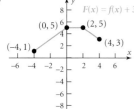

(b)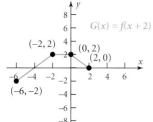

(c)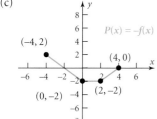

(d)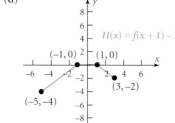

(e)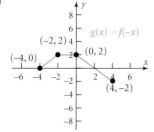

41. (a)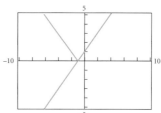

(b)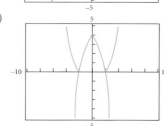

(c)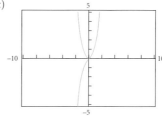

(d) The graph of $y = |f(x)|$ is obtained from the graph of $y = f(x)$ by reflecting the portions of the graph of $y = f(x)$ below the x-axis across the x-axis.

43. (a)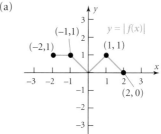

(b)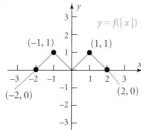

CHAPTER 1 Review

True – False Items (p.158)

1. False   2. False   3. False
4. False   5. True   6. False

Fill in the Blanks (p.158)

1. independent, dependent   2. vertical
3. 5, −3   4. $a = -2$ 5. $(-5, 0), (-2, 0), (2, 0)$

Review Exercises (p.158)

1.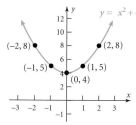

3. The x-intercept and the y-intercept are both $(0, 0)$. The graph is symmetric with respect to the x-axis.

5. The x-intercepts are $(-4, 0)$ and $(4, 0)$. The y-intercepts are $(0, -2)$ and $(0, 2)$. The graph is symmetric with respect to the x-axis, the y-axis, and the origin.

7. There is no x-intercept. The y-intercept is $(0, 1)$. The graph is symmetric with respect to the y-axis.

9. The x-intercepts are $(-1, 0)$ and $(0, 0)$. The y-intercepts are $(0, 0)$ and $(0, -2)$. The graph is not symmetric with respect to either axis nor the origin.

11. (a) $f(2) = 2$ (b) $f(-2) = -2$
    (c) $f(-x) = -\dfrac{3x}{x^2 - 1}$ (d) $-f(x) = -\dfrac{3x}{x^2 - 1}$
    (e) $f(x - 2) = \dfrac{3x - 6}{x^2 - 4x + 3}$ (f) $f(2x) = \dfrac{6x}{4x^2 - 1}$

13. (a) $f(2) = 0$ (b) $f(-2) = 0$
    (c) $f(-x) = \sqrt{x^2 - 4}$ (d) $-f(x) = -\sqrt{x^2 - 4}$
    (e) $f(x - 2) = \sqrt{x^2 - 4x}$ (f) $f(2x) = 2\sqrt{x^2 - 1}$

15. (a) $f(2) = 0$ (b) $f(-2) = 0$
    (c) $f(-x) = \dfrac{x^2 - 4}{x^2}$ (d) $-f(x) = -\dfrac{x^2 - 4}{x^2}$
    (e) $f(x - 2) = \dfrac{x^2 - 4x}{x^2 - 4x + 4}$ (f) $f(2x) = \dfrac{x^2 - 1}{x^2}$

17. $\{x | x \neq -3, x \neq 3\}$

19. $\{x | x \leq 2\}$ or the interval $(-\infty, 2]$ 21. $\{x | x \neq 0\}$

23. $\{x | x \neq -3, x \neq 1\}$ 25. $-4x - 2h + 1$

27. (a) Domain is $\{y | -4 \leq x \leq 3\}$ or the interval $[-4, 3]$, range is $\{y | -3 \leq y \leq 3\}$ or the interval $[-3, 3]$.
    (b) The x-intercept and the y-intercept are both $(0, 0)$.
    (c) $f(-2) = -1$ (d) $x = -4$
    (e) The interval $(0, 3]$ or $\{x | 0 < x \leq 3\}$

29. (a) Domain is the interval $(-\infty, \infty)$, range is the interval $(-\infty, 1)$.
    (b) The function f is increasing on the intervals $(-\infty, -1)$ and $(3, 4)$ and is decreasing on the intervals $(-1, 3)$ and $(4, \infty)$.
    (c) The function has a local maximum of 1 at $x = -1$ and a local maximum of 0 at $x = 4$. The function has a local minimum of -3 at $x = 3$.
    (d) The graph is not symmetric with respect to either axis nor the origin.
    (e) The function is neither even nor odd.
    (f) The x-intercepts are $(-2, 0)$, $(0, 0)$, and $(4, 0)$. The y-intercept is $(0, 0)$.

31. odd   33. even   35. neither even nor odd

37. odd

39.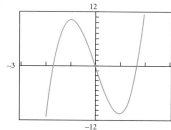

The function has a local maximum of 4.043 at $x = -0.913$ and a local minimum of -2.043 at $x = 0.913$. The function is increasing on the intervals $(-3, -0.913)$ and $(0.913, 3)$ and is decreasing on the interval $(-0.913, 0.913)$.

41.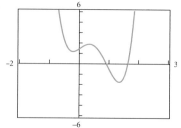

The function has local minima $f(-0.336) = 0.543$ and $f(1.798) = -3.565$, and a local maximum $f(0.414) = 1.532$. The function is increasing on the intervals $(-0.336, 0.414)$ and $(1.798, 3)$ and is decreasing on the intervals $(-2, -0.336)$ and $(0.414, 1.798)$.

43. (a) $\dfrac{\Delta y}{\Delta x} = 23$ (b) $\dfrac{\Delta y}{\Delta x} = 7$
(c) $\dfrac{\Delta y}{\Delta x} = 47$

45. $\dfrac{\Delta y}{\Delta x} = -5$ **47.** $\dfrac{\Delta y}{\Delta x} = -4x - 5$

49. Graphs (b), (c), (d), and (e) are graphs of functions.

51.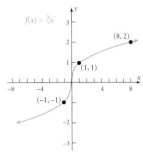

53. (a) The domain is $\{x|x > -2\}$ or the interval $(-2, \infty)$.
(b) The x-intercept and y-intercept are both $(0, 0)$.
(c)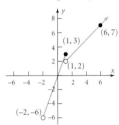

(d) The range is $\{y|y > -6\}$ or the interval $(-6, \infty)$.

55. (a) The domain is $\{x|x \geq -4\}$ or the interval $[-4, \infty)$.
(b) The y-intercept is $(0, 1)$. There is no x-intercept.
(c)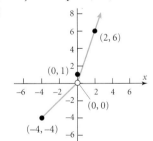

(d) The range is the intervals $[-4, 0)$ and $(0, \infty)$.

57.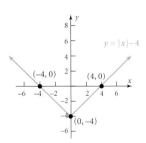

The x-intercepts are $(-4, 0)$ and $(4, 0)$. The y-intercept is $(0, -4)$. The domain is all real numbers or the interval $(-\infty, \infty)$, and the range is $\{y|y \geq -4\}$ or the interval $[-4, \infty)$.

59.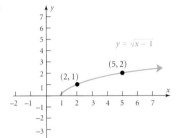

The x-intercept is $(1, 0)$. There is no y-intercept. The domain is $\{x|x \geq 1\}$ or $[1, \infty)$, and the range is $\{y|y \geq 0\}$ or the interval $[0, \infty)$.

61.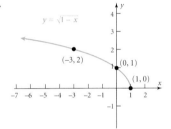

The x-intercept is $(1, 0)$. The y-intercept is $(0, 1)$. The domain is $\{x|x \leq 1\}$ or the interval $(-\infty, 1]$, and the range is $\{y|y \geq 0\}$ or the interval $[0, \infty)$.

63.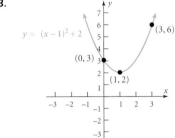

There is no x-intercept. The y-intercept is $(0, 3)$. The domain is $(-\infty, \infty)$, and the range is $\{y|y \geq 2\}$ or $[2, \infty)$.

65. (a)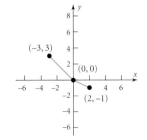

AN-20 Answers to Odd-Numbered Problems

(b)
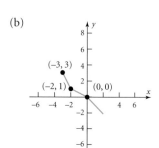

(c)
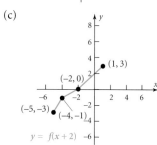

(d)
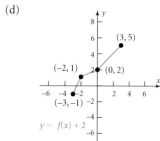

67. $f(x) = -2x + 3$ **69.** $A = 11$

71. $V(r) = 2\pi r^3$

73. (a) $R(x) = -\frac{1}{6}x^2 + 100x;\ 0 \le x \le 600$

(b) The revenue is $13,333.33.

75. (a) $R(x) = -\frac{1}{5}x^2 + 20x,\ 0 \le x \le 100$

(b) The revenue is $255.

77. (a) The total cost C in dollars is given by
$C(r) = 0.12\pi r^2 + \frac{40}{r}$.

(b) The cost is $16.03.   (c) The cost is $29.13.

(d)
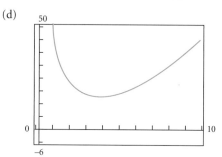

The cost is least when the radius is 3.758 cm.

Mathematical Questions from Professional Exams (p. 162)

1. (d) $0 \le y < 4$ **2.** (d) 87   **3.** (c) $[-1, 0] \cup [1, \infty)$

CHAPTER 2   Classes of Functions

Exercise 2.1 (p. 174)

1. C.   **3.** F.   **5.** G.   **7.** H.

9.
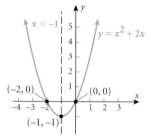

The graph opens upward. The vertex is located at the point $(-1, -1)$. The axis of symmetry is the line $x = -1$. The y-intercept is $(0, 0)$. The x-intercepts are $(-2, 0)$ and $(0, 0)$. The domain is the interval $(-\infty, \infty)$, and the range is the interval $[-1, \infty)$. The function is increasing on the interval $(-1, \infty)$, and is decreasing on the interval $(-\infty, 1)$.

11.
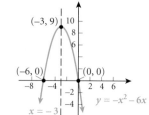

The graph opens downward. The vertex is located at the point $(-3, 9)$. The axis of symmetry is the line $x = -3$. The y-intercept is $(0, 0)$. The x-intercepts are $(-6, 0)$ and $(0, 0)$. The domain is the interval $(-\infty, \infty)$, and the range is the interval $(-\infty, 9]$. The function is increasing on the interval $(-\infty, -3)$, and is decreasing on the interval $(-3, \infty)$.

13.

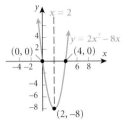

The graph opens upward. The vertex is located at the point $(2, -8)$. The axis of symmetry is the line $x = 2$. The y-intercept is $(0, 0)$. The x-intercepts are $(0, 0)$ and $(4, 0)$. The domain is the interval $(-\infty, \infty)$, and the range is the interval $[-8, \infty)$. The function is increasing on the interval $(2, \infty)$, and is decreasing on the interval $(-\infty, 2)$.

15.

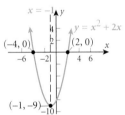

The graph opens upward. The vertex is located at the point $(-1, -9)$. The axis of symmetry is the line $x = -1$. The y-intercept is $(0, -8)$. The x-intercepts are $(-4, 0)$ and $(2, 0)$. The domain is the interval $(-\infty, \infty)$, and the range is the interval $[-9, \infty)$. The function is increasing on the interval $(-1, \infty)$, and is decreasing on the interval $(-\infty, -1)$.

17.

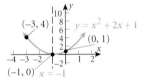

The graph opens upward. The vertex is located at the point $(-1, 0)$. The axis of symmetry is the line $x = -1$. The y-intercept is $(0, 1)$. The x-intercept is $(-1, 0)$. The domain is the interval $(-\infty, \infty)$, and the range is the interval $[0, \infty)$. The function is increasing on the interval $(-1, \infty)$, and is decreasing on the interval $(-\infty, -1)$.

19.

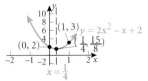

The graph opens upward. The vertex is located at the point $(\frac{1}{4}, \frac{15}{8})$. The axis of symmetry is the line $x = \frac{1}{4}$. The y-intercept is $(0, 2)$. There is no x-intercept. The domain is the interval $(-\infty, \infty)$, and the range is the interval $[\frac{15}{8}, \infty)$. The function is increasing on the interval $(\frac{1}{4}, \infty)$, and is decreasing on the interval $(-\infty, \frac{1}{4})$.

21.

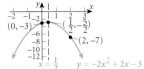

The graph opens downward. The vertex is located at the point $(\frac{1}{2}, -\frac{5}{2})$. The axis of symmetry is the line $x = \frac{1}{2}$. The y-intercept is $(0, -3)$. There is no x-intercept. The domain is the interval $(-\infty, \infty)$, and the range is the interval $(-\infty, -\frac{5}{2}]$. The function is increasing on the interval $(-\infty, \frac{1}{2})$, and is decreasing on the interval $(\frac{1}{2}, \infty)$.

23.

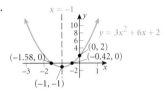

The graph opens upward. The vertex is located at the point $(-1, -1)$. The axis of symmetry is the line $x = -1$. The y-intercept is $(0, 2)$. The x-intercepts are $(\frac{-\sqrt{3} - 3}{3}, 0)$ $\approx (-1.58, 0)$ and $(\frac{\sqrt{3} - 3}{3}, 0) \approx (-0.42, 0)$. The domain is the interval $(-\infty, \infty)$, and the range is the interval $[-1, \infty)$. The function is increasing on the interval $(-1, \infty)$, and is decreasing on the interval $(-\infty, -1)$.

25.

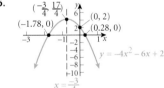

The graphs opens downward. The vertex is located at the point $(-\frac{3}{4}, \frac{17}{4})$. The axis of symmetry is the line $x = -\frac{3}{4}$. The y-intercept is $(0, 2)$. The x-intercepts are $(\frac{-\sqrt{17} - 3}{4}, 0)$ $\approx (-1.78, 0)$ and $(\frac{\sqrt{17} - 3}{4}, 0) \approx (0.28, 0)$. The domain is the interval $(-\infty, \infty)$, and the range is the interval $(-\infty, \frac{17}{4}]$. The function is increasing on the interval $(-\infty, -\frac{3}{4})$, and is decreasing on the interval $(-\frac{3}{4}, \infty)$.

27. The function has a minimum value of $f(-3) = -18$.

29. The function has a minimum value of $f(-3) = -21$.

31. The function has a maximum value of $f(5) = 21$.

33. The function has a maximum value of $f(2) = 13$.

35. (a) $f(x) = (x + 3)(x - 1)$, $f(x) = 2(x + 3)(x - 1)$, $f(x) = -2(x + 3)(x - 1)$, $f(x) = 5(x + 3)(x - 1)$

(b) The x-intercepts are unaffected. The y-coordinate of the y-intercept changes from that of $y = (x + 3)(x - 1)$ by a factor of a.

(c) The value of a has no effect on the axis of symmetry.

(d) The x-coordinate of the vertex is the same for each function. The y-coordinate of the vertex changes from that of $y = (x + 3)(x - 1)$ by a factor of a.

(e) The x-coordinate of the vertex is equal to the x-coordinate of the midpoint of the x-intercepts.

37. A unit price of $500 should be established to maximize revenue. The maximum revenue is $1,000,000$.

39. (a) $R(x) = -\frac{1}{6}x^2 + 100x$

(b) The revenue is $13,333.33$.

(c) A quantity of 300 units maximizes revenue. The maximum revenue is $15,000$.

(d) The company should charge $50.

41. (a) $R(x) = -\frac{1}{5}x^2 + 20x$

(b) The revenue is $255.

(c) A quantity of 50 units maximizes revenue. The maximum revenue is $500.

(d) The company should charge $10.

43. (a) $A(x) = 200x - x^2$

(b) $x = 100$

(c) The maximum area is 10,000 square yards.

45. The largest area is 2,000,000 m².

47. (a) The projectile is 39.0625 feet horizontally from the base of the cliff when it achieves is maximum height.

(b) The maximum height is 219.53125 feet above the water.

(c) The projectile strikes the water 170.024 feet from the base of the cliff.

(d)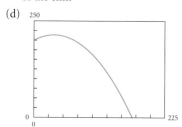

(e) The projectile is 135.698 feet from the cliff.

49. A depth of 3 inches will provide the maximum cross-sectional area.

51. 100 meters by $\frac{100}{\pi}$ meters $\approx$ 100 meters by 31.8 meters

53. (a) There are the most hunters at the income level of $56,584. At this income level, there are about 3685 hunters.

(b)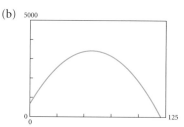

The number of hunters is increasing between the $20,000 and $40,000 income levels.

55. (a) The number of 23-year old male murder victims is 1795.

(b) The number of male murder victims at age 28 years is about 1456.

(c)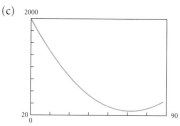

(d) The number of male murder victims decreases with age until age 70 and then begins to increase.

57. $x = \frac{a}{2}$

59. Area $= \frac{38}{3}$ units²

61. Area $= \frac{248}{3}$ units²

63. $A = 25$ units²

65. Answers will vary.

67.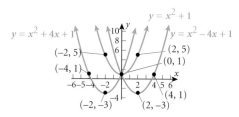

Each parabola opens upward, has its vertex at $(-\frac{b}{2}, \frac{4-b^2}{4})$, and has the line $x = -\frac{b}{2}$ as its axis of symmetry.

69. Write the formula for a quadratic function as $f(x) = a\left(ax + \frac{b}{2a}b\right)^2 + \frac{4ac - b^2}{4a}$. If $a > 0$, then $a\left(ax + \frac{b}{2a}b\right)^2$ is zero if $x = -\frac{b}{2a}$ and is positive otherwise. Hence the function has a minimum at $x = -\frac{b}{2a}$, and the graph opens upward.

If $a < 0$, then $\left(ax + \frac{b}{2a}b\right)^2$ is zero if $x = -\frac{b}{2a}$ and is negative otherwise. Hence the function has a maximum at $x = -\frac{b}{2a}$, and the graph opens downward.

Exercise 2.2 (p. 185)

1. Answers will vary. Possible answers include $(-1, -1)$, $(0, 0)$ and $(1, 1)$.
3. origin
5.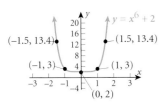
7.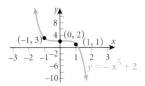
9.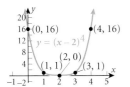
11. This is a polynomial function of degree 3.
13. This is a polynomial function of degree 2.
15. This is not a polynomial functon because the term $\frac{1}{x}$ has the monomial x in the denominator.
17. This is not a polynomial function because the exponent of the term $x^{3/2}$ is not a nonnegative integer.
19. This is a polynomial function of degree 4.
21. This is a polynomial function of degree 4.
23. $y = 3x^4$
25. $y = -2x^5$
27. $y = 5x^3$
29. $\{x | x \neq 3\}$
31. $\{x | x \neq 2, x \neq -4\}$
33. $\{x | x \neq -\frac{1}{2}, x \neq 3\}$
35. $\{x | x \neq 2\}$
37. $(-\infty, \infty)$
39. $\{x | x \neq -3, x \neq 3\}$
41. (a) The percentage of union membership in the labor force in 2000 was 13.2%.
    (b) $u(75) = 18.02$. The percentage of union membership in the labor force in 2005 will be 18.02%.

Exercise 2.3 (p. 197)

1. (a) 11.2116 (b) 11.5873 (c) 11.6639
   (d) 11.6648
3. (a) 8.8152 (b) 8.8214 (c) 8.8244
   (d) 8.8250
5. (a) 21.2166 (b) 22.2167 (c) 22.4404
   (d) 22.4592
7. 3.3201
9. 0.4274
11. This is not an exponential function.
13. This is an exponential function with base $a = 4$.
15. This is an exponential function with base $a = 2$.
17. This is not an exponential function.
19. B.   21. D.   23. A.   25. E.
27.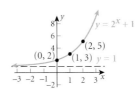

Domain $= (-\infty, \infty)$, Range $= (1, \infty)$; the horizontal asymptote is $y = 1$.

29.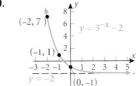

Domain $= (-\infty, \infty)$, Range $= (-2, \infty)$; the horizontal asymptote is $y = -2$.

31.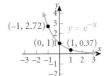

Domain $= (-\infty, \infty)$, Range $= (0, \infty)$; the horizontal asymptote is $y = 0$.

33.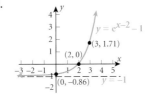

Domain $= (-\infty, \infty)$, Range $= (-1, \infty)$; the horizontal asymptote is $y = -1$.

35. $\{\frac{1}{2}\}$ 37. $\{-\sqrt{2}, 0, \sqrt{2}\}$ 39. $\{\frac{3 - \sqrt{6}}{3}, \frac{3 + \sqrt{6}}{3}\}$

41. {0} **43.** {4} **45.** {$\frac{3}{2}$}

47. {1, 2} **49.** $\frac{1}{49}$ **51.** $\frac{1}{4}$

53. $y = 3^x$ **55.** $y = -6^x$

57. (a) 74.1% of light will pass through 10 panes.
(b) 47.2% of light will pass through 25 panes.

59. (a) There will be 44.35 watts after 30 days.
(b) There will be 11.61 watts after one year.

61. There will be 3.35 mg of the drug in the bloodstream after 1 hour. There will be 0.45 mg of the drug in the bloodstream after 6 hours.

63. (a) The probability that a car will arrive within 10 minutes of 12:00 PM is 0.632.
(b) The probability that a car will arrive within 40 minutes of 12:00 PM is 0.982.
(c) $F(t)$ approaches 1 as t becomes unbounded in the positive direction.
(d)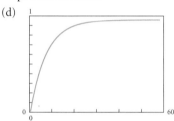
(e) About 6.931 minutes are needed for the probability to reach 50%.

65. (a) The probability that 15 cars will arrive between 5:00 PM and 6:00 PM is 5.2%.
(b) The probability that 20 cars will arrive between 5:00 PM and 6:00 PM is 8.9%.

67. (a) A 3-year old Civic DX Sedan costs $12,123.27.
(b) A 9-year old Civic DX Sedan costs $6442.80.

69. (a) 5.4 amperes, 7.6 amperes, 10.4 amperes
(b)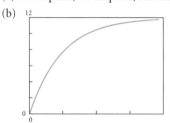
(c) The maximum current is 12 amperes.
(d) 3.3 amperes, 5.3 amperes, 9.4 amperes

(e)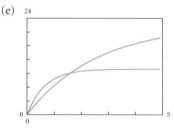

(f) The maximum current is 24 amperes.

71.

n	$2 + \frac{1}{2!} + \frac{1}{3!} + \cdots + \frac{1}{n!}$	Difference
4	2.7083333333	0.0099484951
6	2.7180555556	0.0002262729
8	2.7182787698	0.0000030586
10	2.7182818011	0.0000000273

73. $\dfrac{f(x+h) - f(x)}{h} = \dfrac{a^{x+h} - a^x}{h}$

$= \dfrac{a^x a^h - a^x}{h}$

$= a^x \cdot \dfrac{a^h - 1}{h}$ b

75. $f(-x) = a^{-x}$
$= (a^x)^{-1}$
$= \dfrac{1}{a^x}$
$= \dfrac{1}{f(x)}$

77. (a) The relative humidity is 71%.
(b) The relative humidity is 73%.
(c) The relative humidity is 100%.

79. (a) $\sinh(-x) = \dfrac{e^{-x} - e^{-(-x)}}{2}$

$= \dfrac{e^{-x} - e^x}{2}$

$= -\dfrac{e^x - e^{-x}}{2}$

$= -\sinh x$

(b)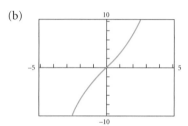

81. It took 59 minutes to fill half of the container.

83. There is no power function that increases more rapidly than an exponential function whose base is greater than 1. Explanations will vary.

85. The graphs are identical because $y = a^{-x}$ and $y = \left(\frac{1}{a}\right)^x$ represent the same function.

Exercise 2.4 (p. 210)

1. $\log_3 9 = 2$ **3.** $\log_a 1.6 = 2$ **5.** $\log_{1.1} M = 2$

7. $\log_2 7.2 = x$ **9.** $\log_x \pi = \sqrt{2}$ **11.** $\ln 8 = x$

13. $2^3 = 8$ **15.** $a^6 = 3$ **17.** $3^x = 2$

19. $2^{1.3} = M$ **21.** $(\sqrt{2})^x = \pi$ **23.** $e^x = 4$

25. 0   **27.** 2   **29.** -4

31. $\frac{1}{2}$ **33.** 4   **35.** $\frac{1}{2}$

37. $\{x | x > 3\}$ **39.** $\{x | x \neq 0\}$ **41.** $\{x | x > 0\}$

43. $\{x | x \geq 1\}$ **45.** 0.511   **47.** 30.099

49. $a = \sqrt{2}$

51.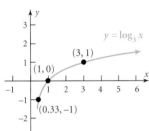

53.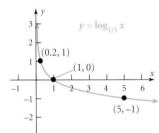

55. B.   **57.** D.   **59.** A.

61. E.

63.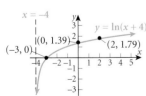

Domain $(-4, \infty)$, Range $= (-\infty, \infty)$; the vertical asymptote is $x = -4$.

65.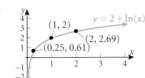

Domain $= (0, \infty)$, Range $= (-\infty, \infty)$; the vertical asymptote is $x = 0$.

67.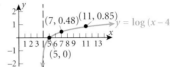

Domain $= (4, \infty)$, Range $= (-\infty, \infty)$; the vertical asymptote is $x = 4$.

69.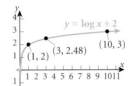

Domain $= (0, \infty)$
Range $= (-\infty, \infty)$; the vertical asymptote is $x = 0$.

71. $\{9\}$ **73.** $\frac{7}{2}$ **75.** $\{2\}$

77. $\{5\}$ **79.** $\{3\}$ **81.** $\{2\}$

83. $e^{\frac{\ln 10}{3}}$ f   **85.** $e^{\frac{\ln 8 - 5}{2}}$ f   **87.** $-2\sqrt{2}, 2\sqrt{2}$

89. $\{-1\}$

91. (a) 1   (b) 2   (c) 3
(d) The pH increases as the hydrogen ion concentration decreases.
(e) $[\text{H}^+] = 0.000316$
(f) $[\text{H}^+] = 3.981 \times 10^{-8}$

93. (a) The aircraft is 5.965 km above sea level.
(b) The height of the mountain is 0.900 km above sea level.

95. (a) It will take 6.931 minutes for the probability to reach 50%.
(b) It will take 16.094 minutes for the probability to reach 80%.
(c) No, the probability cannot equal 100% because the exponential term $e^{-0.1t}$ is never equal to zero.

97. The time between injections is about 2 hours, 17 minutes and 27 seconds.

99. It takes 0.269 seconds to achieve a current of 0.5 ampere and 0.896 seconds to achieve a current of 1.0 ampere.

101. The population will be 309,123,927 people.

103. 50 decibels   **105.** 110 decibels

107. The magnitude of the earthquake was 8.1 on the Richter scale.

109. (a) $k = \frac{50}{3} \ln\left(\frac{10}{3}\right) \approx 20.066$
(b) The risk is 90.9%.
(c) A blood alcohol concentration of 0.175 corresponds to a risk of 100%.
(d) A driver should be arrested with a blood alcohol concentration of 0.08 or greater.
(e) Answers will vary.

111. Explanations will vary.

Exercise 2.5 (p. 221)

1. 71   **3.** −4   **5.** 7
7. 1   **9.** 1   **11.** 2
13. $\frac{5}{4}$ **15.** 4   **17.** $a + b$
19. $b - a$ **21.** $3a$ **23.** $\frac{a+b}{5}$
25. $2 \log_5 5 - \log_5 x$ **27.** $3 \log_2 z$
29. $\ln e + \ln x$ **31.** $\ln x + x \ln e$ **33.** $2 \log u + 3 \log a\, v$
35. $2 \ln x + \frac{1}{2} \ln(1 - x)$
37. $3 \log_2 x - \log_2(x - 3)$
39. $\log x + \log(x + 2) - 3 \log(x + 3)$
41. $\frac{1}{3} \ln(x - 2) + \frac{1}{3} \ln(x + 1) - \frac{2}{3} \ln(x + 4)$
43. $\ln 5 + \ln x + \frac{1}{2} \ln(1 + 3x) - 3 \ln(x - 4)$
45. $\log_5 u^3 v^4$ **47.** $\log_3 x^{7/2}$
49. $\log_4 a \dfrac{x^2 - 1}{(x + 1)^5} b$
51. $\ln a \dfrac{1}{(x - 1)^2} b$
53. $\log_2 x(3x - 2)^4$ **55.** $\log_a a \dfrac{25 x^6}{\sqrt{2x + 3}} b$
57. $\log_2 a \dfrac{(x + 1)^2}{x + 2x - 3} b$ **59.** 2.771
61. −3.880   **63.** 5.615   **65.** 0.874

67.
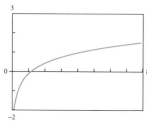

69.
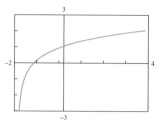

71.
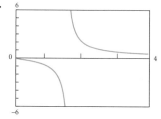

73. $y = Cx$ **75.** $y = C(x^2 + x)$
77. $y = Ce^{3x}$ **79.** $y = Ce^{-4x} + 3$
81. $y = a\dfrac{C\sqrt{2x + 1}}{(x + 4)^{1/3}} b^{1/3}$ **83.** 3   **85.** 1

87. $\log_a(x + \sqrt{x^2 - 1}) + \log_a(x - \sqrt{x^2 - 1})$
$= \log_a[(x + \sqrt{x^2 - 1})(x - \sqrt{x^2 - 1})]$
$= \log_a[x^2 - (x^2 - 1)]$
$= \log_a(1)$
$= 0$

89. $\ln(1 + e^{2x}) = \ln(e^{2x}(e^{-2x} + 1))$
$= \ln e^{2x} + \ln(1 + e^{-2x})$
$= 2x + \ln(1 + e^{-2x})$

91. $-f(x) = -\log_a x$
$= -\dfrac{\ln x}{\ln a}$
$= \dfrac{\ln x}{-\ln a}$
$= \dfrac{\ln x}{\ln a^{-1}}$
$= \dfrac{\ln x}{\ln(1/a)}$
$= \log_{1/a} x$

93. $f a \dfrac{1}{x} b = \log_a a \dfrac{1}{x} b$
$= \log_a x^{-1}$
$= -\log_a x$
$= -f(x)$

95. $\log_a\left(\dfrac{M}{N}\right) = \log_a(MN^{-1})$

$\qquad = \log_a M + \log_a N^{-1}$

$\qquad = \log_a M - \log_a N$

97. The functions are not equivalent. Explanations will vary.

Exercise 2.6 (p. 226)

1. $1127.50   **3.** $580.92   **5.** $98.02

7. $466.20

9. The amount is $1020.20, and the interest is $20.20.

11. (a) $4434.60   (b) $3933.14

13. A 23.1% interest rate is required.

15. It will take approximately 11 years for the investment to triple.

17. $913.93 is needed to get $1000 in 1 year. $835.27 is needed to get $1000 in 2 years.

19. They should invest $35,476.82.

21. A 22.0% interest rate is required.

23. (a) The Rule of 70 approximation is 70 years, which is greater than the actual solution of 69.3147 years by about 0.685 year.
(b) The Rule of 70 approximation is 14 years, which is greater than the actual solution of 13.8629 years by about 0.137 year.
(c) The Rule of 70 approximation is 7 years, which is greater than the actual solution of 6.9315 years by about 0.069 year.

CHAPTER 2   Review

True – False Items (p. 230)

1. True   **2.** False   **3.** True
4. True   **5.** False   **6.** False
7. False   **8.** True   **9.** False

Fill in the Blanks (p. 230)

1. parabola   **2.** axis of symmentry
3. $-\dfrac{b}{2a}$ **4.** $(0, 1), (1, a), \left(-1, \dfrac{1}{a}\right)$
5. 1   **6.** 4   **7.** $(0, \infty)$
8. $(1, 0), (a, 1), \left(\dfrac{1}{a}, -1\right)$ **9.** 1
10. $r \log_a M$

Review Exercises (p. 230)

1.

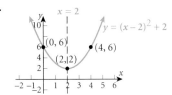

The graph opens upward. The vertex is $(2, 2)$. The axis of symmetry is the line $x = 2$. The y-intercept is $(0, 6)$. There is no x-intercept.

3.

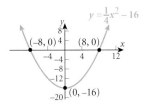

The graph opens up. The vertex is $(0, -16)$. The axis of symmetry is the line $x = 0$. The y-intercept is $(0, -16)$. The x-intercepts are $(-8, 0)$ and $(8, 0)$.

5.

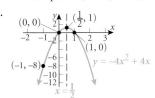

The graph opens downward. The vertex is $\left(\dfrac{1}{2}, 1\right)$. The axis of symmetry is the line $x = \dfrac{1}{2}$. The y-intercept is $(0, 0)$. The x-intercepts are $(0, 0)$ and $(1, 0)$.

7.

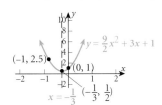

The graph opens upward. The vertex is $\left(-\dfrac{1}{3}, \dfrac{1}{2}\right)$. The axis of symmetry is the line $x = -\dfrac{1}{3}$. The y-intercept is $(0, 1)$. There is no x-intercept.

9.

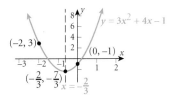

The graph opens upward. The vertex is $\left(-\dfrac{2}{3}, -\dfrac{7}{3}\right)$. The axis of symmetry is the line $x = -\dfrac{2}{3}$. The y-intercept is $(0, -1)$. The x-intercepts are $\left(\dfrac{-2 - \sqrt{7}}{3}, 0\right)$ and $\left(\dfrac{-2 + \sqrt{7}}{3}, 0\right)$ or approximately $(-1.55, 0)$ and $(0.22, 0)$.

11. Minimum value = 1   **13.** Maximum value = 12

15. Maximum value = 16

17. Answers will vary. Possibilities include $(-1, 1), (0, 0),$ and $(1, 1)$.

19.
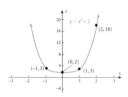

21.
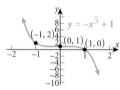

23. This is a polynomial function of degree 5.

25. This is not a polynomial function because the exponent in the term $5x^{1/2}$ is not a nonnegative integer.

27. $y = -2x^4$

29. The domain is $\{x|x \neq -3, x \neq 3\}$.

31. The domain is $\{x|x \neq -2\}$.

33. (a) 81   (b) 2   (c) $\frac{1}{9}$ (d) -3

35. $\log_5 z = 2$ **37.** $5^{13} = u$ **39.** $\{x|x > \frac{2}{3}\}$

41. $\{x|x < \frac{2}{3}\}$ **43.** -3 **45.** 4

47. 2   **49.** $\sqrt{2}$ **51.** 0.4

53. $\log_3 u + 2\log_3 v - \log_3 w$ **55.** $2\log x + \frac{1}{3}\log(x^3 + 1)$

57. $\ln x + \frac{1}{3}\ln(x^2 + 1) - \ln(x - 3)$ **59.** $\log_4 x^{25/4}$

61. $\ln a \dfrac{1}{(x+1)^2} b$ **63.** $\ln a \dfrac{4x^3}{\sqrt{x^2 + x - 6}} b$

65. 2.124

67.
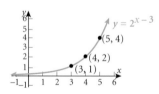

Domain = $(-\infty, \infty)$
Range = $(0, \infty)$; the x-axis is an asymptote.

69.
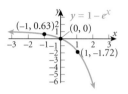

Domain = $(-\infty, \infty)$, Range = $(-\infty, 1)$; the line $y = 1$ is an asymptote.

71.
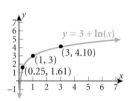

Domain = $(0, \infty)$, Range = $(-\infty, \infty)$; the y-axis is an asymptote.

73. $\frac{1}{4}$ **75.** $\{\frac{-1 - \sqrt{3}}{2}, \frac{-1 + \sqrt{3}}{2}\}$ **77.** $\frac{1}{4}$

79. $\{4\}$ **81.** $\{11\}$ **83.** $125.23

85. $923.12

87. It will take almost 11.6 years to double.

89. The Piper Cub is 3229.5 meters above sea level.

91. 50 feet by 50 feet

93. 25 feet by $\frac{50}{\pi}$ feet ≈ 25 feet by 15.92 feet

95. (a) The limiting magnitude is 11.77.
(b) A diameter of 9.56 inches is required.

97. (a) The annual interest rate was 10.436%.
(b) The actual value will be $32,249.24.

99. (a) 63 clubs should be manufactured.
(b) The marginal cost is $151.90.

Mathematical Questions from Professional Exams (p. 235)

1. (e) $2b + \log_6 27$ **2.** (e) I, II, and III

3. (c) $\ln(y + \sqrt{y^2 + 1})$ **4.** (e) 125

5. (b) 1   **6.** (a) 2

7. (a) 3

CHAPTER 3 The Limit of a Function

Exercise 3.1 (p. 241)

1.

x	0.9	0.99	0.999
$f(x) = 2x$	1.8	1.98	1.998
x	1.1	1.01	1.001
$f(x) = 2x$	2.2	2.02	2.002

$\lim\limits_{x \to 1} f(x) = 2$

3.

x	-0.1	-0.01	-0.001
$f(x) = x^2 + x$	2.01	2.0001	2.000001
x	0.1	0.01	0.001
$f(x) = x^2 + x$	2.01	2.0001	2.000001

$\lim\limits_{x \to 0} f(x) = 2$

5.

x	-2.1	-2.01	-2.001
$f(x) = \dfrac{x^2 - 4}{x + 2}$	-4.1	-4.01	-4.001
x	-1.9	-1.99	-1.999
$f(x) = \dfrac{x^2 - 4}{x + 2}$	-3.9	-3.99	-3.999

$\lim\limits_{x \to -2} f(x) = -4$

7.

x	-1.1	-1.01	-1.001
$f(x) = \dfrac{x^3 + 1}{x + 1}$	3.31	3.0301	3.003001
x	-0.9	-0.99	-0.999
$f(x) = \dfrac{x^3 + 1}{x + 1}$	2.71	2.9701	2.997001

$\lim\limits_{x \to -1} f(x) = 3$

9. 32 **11.** 1 **13.** 4

15. 2 **17.** 3 **19.** 4

21. The limit does not exist.

23.

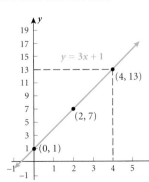

$\lim\limits_{x \to 4} f(x) = 13$

25.

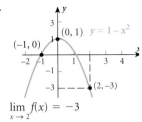

$\lim\limits_{x \to 2} f(x) = -3$

27.

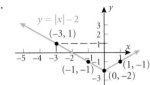

$\lim\limits_{x \to -3} f(x) = 1$

29.

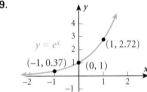

$\lim\limits_{x \to 0} f(x) = 1$

31.

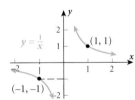

$\lim\limits_{x \to -1} f(x) = -1$

33.

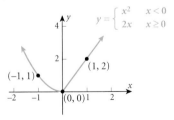

$\lim\limits_{x \to 0} f(x) = 0$

35.

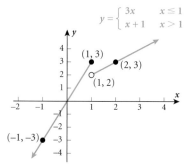

$$y = \begin{cases} 3x & x \le 1 \\ x+1 & x > 1 \end{cases}$$

The limit does not exist.

37.

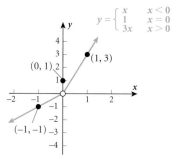

$$y = \begin{cases} x & x < 0 \\ 1 & x = 0 \\ 3x & x > 0 \end{cases}$$

$\lim_{x \to 0} f(x) = 0$

39.

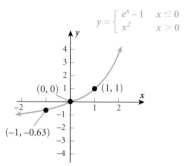

$$y = \begin{cases} e^x - 1 & x \le 0 \\ x^2 & x > 0 \end{cases}$$

$\lim_{x \to 0} f(x) = 0$

41. 0.67 **43.** 1.6 **45.** 0

Exercise 3.2 (p. 250)

1. 5 **3.** 4 **5.** 8
7. 8 **9.** −1 **11.** 8
13. 3 **15.** −1 **17.** 32
19. 2 **21.** $\frac{7}{6}$ **23.** 3
25. 0 **27.** $\frac{2}{3}$ **29.** $\frac{8}{5}$
31. 0 **33.** 5 **35.** 6
37. −1 **39.** 0 **41.** −1

43. $\frac{1}{4}$ **45.** 10 **47.** 8
49. $\frac{4}{5}$ **51.** 10

Exercise 3.3 (p. 256)

1. $[-8, -3) \cup (-3, 4) \cup (4, 6]$ **3.** $(-8, 0), (-5, 0)$
5. $f(-8) = 0, f(-4) = 2$ **7.** 3
9. 2 **11.** 1
13. Yes, $\lim_{x \to 4} f(x) = 0$ **15.** No. **17.** Yes.
19. No. **21.** 5 **23.** 7
25. 1 **27.** 4 **29.** $-\frac{2}{3}$
31. $\frac{3}{2}$ **33.** Continuous **35.** Continuous
37. Discontinuous **39.** Discontinuous **41.** Discontinuous
43. Continuous **45.** Discontinuous **47.** Continuous
49. f is continuous on the interval $(-\infty, \infty)$.
51. f is continuous on the interval $(-\infty, \infty)$.
53. f is continuous on the interval $(0, \infty)$. f is not discontinuous for any numbers in its domain.
55. f is continuous on the interval $(-\infty, \infty)$.
57. f is continuous for all numbers in the set $\{x | x \ne -2, x \ne 2\}$. f is discontinuous at $x = -2$ and $x = 2$.
59. f is continuous on the intervals $(0, 1)$ and $(1, \infty)$. f is discontinuous at $x = 1$.
61. f is continuous for all number in the set $\{x | x \ne 0\}$. f is discontinuous at $x = 0$.
63. (a) $\lim_{x \to 350^-} C(x) = 39.99$
(b) $\lim_{x \to 350^+} C(x) = 39.99$
(c) The function $C(x)$ is continuous at $x = 350$.
(d) Answers will vary.
65. (a) $W(v) = $

$$\begin{cases} 10 & 0 \le v < 1.79 \\ 33 - \dfrac{23(10.45 + 10\sqrt{v} - v)}{22.04} & 1.79 \le v \le 20 \\ -3.7034 & v > 20 \end{cases}$$

(b) $\lim_{v \to 0^+} W(v) = 10$
(c) $\lim_{v \to 1.79^-} W(v) = 10$
(d) $\lim_{v \to 1.79^+} W(v) = \dfrac{26{,}407}{1102} - \dfrac{575\sqrt{179}}{551} \approx 10.00095$
(e) $W(1.79) = \dfrac{26{,}407}{1102} - \dfrac{575\sqrt{179}}{551} \approx 10.00095$
(f) W is not continuous at $v = 1.79$ since $10 \ne 10.00095$.
(g) $\lim_{v \to 1.79^-} W(v) = 10.00, \lim_{v \to 1.79^+} W(v) = 10.00$, $W(1.79) = 10.00$; W is continuous at $v = 1.79$.

(h) Answers will vary.

(i) $\lim_{v \to 20^-} W(v) = \dfrac{94{,}697}{2204} - \dfrac{11{,}500\sqrt{5}}{551} \approx -3.70332$

(j) $\lim_{v \to 20^+} W(v) = -3.7034$

(k) $W(20) = \dfrac{94{,}697}{2204} - \dfrac{11{,}500\sqrt{5}}{551} \approx -3.70332$

(l) W is not continuous at $v = 20$ since $-3.70332 \ne -3.7034$.

(m) $\lim_{v \to 20^-} W(v) = -3.70,\ \lim_{v \to 20^+} W(v) = -3.70$, $W(20) = -3.70$; W is continuous at $v = 20$.

(n) Answers will vary.

Exercise 3.4 (p. 265)

1. 1
3. 2
5. 3
7. 0
9. ∞
11. $-\infty$
13. ∞
15. ∞
17. ∞
19. ∞

21. The horizontal asymptote is $y = 3$. The vertical asymptote is $x = 0$.

23. The horizontal asymptote is $y = 2$. The vertical asymptote is $x = 1$.

25. The horizontal asymptote is $y = 1$. The vertical asymptotes are $x = -2$ and $x = 2$.

27. (a) $\{x | x \ne 6\}$
    (b) $[0, \infty)$
    (c) The x-intercepts are $(-4, 0)$, and $(0, 0)$. The y-intercept is $(0, 0)$.
    (d) $f(-2) = 2$ (e) $x = 4$ or $x = 8$
    (f) f is discontinuous at $x = 6$.
    (g) $x = 6$ (h) $y = 4$
    (i) There is a local maximum of 2 at $x = -2$.
    (j) There are local minima of 0 at $x = -4$ and $x = 0$ and a local minimum of 4 at $x = 8$.
    (k) f is increasing on $(-4, -2) \cup (0, 6) \cup (4, \infty)$
    (l) f is decreasing on $(-\infty, -4) \cup (-2, 0) \cup (6, 8)$
    (m) 4 (n) ∞
    (o) ∞ (p) ∞

29. $\lim_{x \to -1^-} R(x) = -\infty$, and $\lim_{x \to -1^+} R(x) = \infty$, so the graph of R will have a vertical asymptote at $x = -1$. $\lim_{x \to 1} R(x) = \tfrac{1}{2}$, but R is not defined at $x = 1$, so the graph of R will have a hole at $(1, \tfrac{1}{2})$.

31. $\lim_{x \to -1} R(x) = \tfrac{1}{2}$, but R is not defined at $x = -1$, so the graph of R will have a hole at $(-1, \tfrac{1}{2})$. $\lim_{x \to 1^-} R(x) = -\infty$ and $\lim_{x \to 1^+} R(x) = \infty$, so the graph of R will have a vertical asymptote at $x = 1$.

33. $R(x)$ is undefined at $x = 1$, where a hole appears, and at $x = -2$, where a vertical asymptote occurs.

35. $R(x)$ is undefined at $x = 2$, where a hole appears, and at $x = -3$, where a vertical asymptote occurs.

37. $R(x)$ is undefined at $x = -1$, where a hole appears.

39. (a) $C(x) = 79{,}000 + 10x$
    (b) $\{x | x \ge 0\}$
    (c) $\overline{C}(x) = \dfrac{79{,}000}{x} + 10$
    (d) $\{x | x > 0\}$
    (e) $\lim_{x \to 0^+} \overline{C}(x) = \infty$; the average cost of producing a few calculators is very high due to the fixed costs. Therefore, the average cost of producing no calculators is unbounded.
    (f) $\lim_{x \to \infty} \overline{C}(x) = 10$; the more calculators that are produced, the closer the average cost gets to $10 per calculator.

41. (a) $\lim_{x \to 100^-} C(x) = \infty$
    (b) No, it is not possible to remove 100% of the pollutant. Explanations will vary.

43. Answers will vary. One possibility is graphed below.

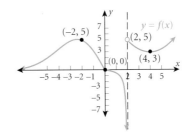

CHAPTER 3 Review

True – False Items (p. 269)

1. True
2. False
3. True
4. True
5. True
6. True
7. True

Fill in the Blanks (p. 269)

1. $\lim_{x \to c} f(x) = N$
2. equals
3. not exist
4. continuous
5. $\neq$
6. equals
7. $y = 2$, horizontal

Review Exercises (p. 269)

1. 12
3.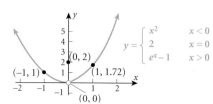
$\lim_{x \to 0} f(x) = 0$
5. 9
7. 25
9. 4
11. 0
13. 64
15. -16
17. $\frac{1}{3}$
19. $\frac{6}{7}$
21. 0
23. $\frac{3}{2}$
25. $\frac{28}{11}$
27. $\frac{5}{3}$
29. $-\infty$
31. $-\infty$
33. ∞
35. Continuous
37. Discontinuous
39. Discontinuous
41. Continuous
43. The line $y = 0$ is a horizontal asymptote. The lines $x = -1$ and $x = 1$ are vertical asymptotes.
45. The $y = 5$ is a horizontal asymptote. The line $x = -2$ is a vertical asymptote.
47. (a) $(-\infty, 2) \cup (2, 5) \cup (5, \infty)$
    (b) $(-\infty, \infty)$
    (c) $(-2, 0), (0, 0), (1, 0), (6, 0)$
    (d) $(0, 0)$
    (e) $f(-6) = 2, f(-4) = 1$
    (f) $f(-2) = 0, f(6) = 0$
    (g) $\lim_{x \to -4^-} f(x) = 4, \lim_{x \to -4^+} f(x) = -2$
    (h) $\lim_{x \to -2^-} f(x) = -2, \lim_{x \to -2^+} f(x) = 2$
    (i) $\lim_{x \to 5^-} f(x) = 2, \lim_{x \to -5^+} f(x) = 2$
    (j) $\lim_{x \to 0} f(x)$ does not exist.
    (k) $\lim_{x \to 2} f(x)$ does not exist.
    (l) No (m) No (n) No
    (o) No (p) Yes (q) No
    (r) $(-6, -4) \cup (-2, 0) \cup (6, \infty)$
    (s) $(-\infty, -6) \cup (0, 2) \cup (2, 5) \cup (5, 6)$
    (t) $\lim_{x \to -\infty} f(x) = \infty, \lim_{x \to \infty} f(x) = 2$
    (u) There is no local maximum. There is a local minimum of 2 at $x = -6$, a local minimum of 0 at $x = 0$, and a local minimum of 0 at $x = 6$.
    (v) The line $y = 2$ is a horizontal asymptote. The line $x = 2$ is a vertical asymptote.
49. -11
51. $-\frac{1}{4}$
53. The graph has a hole at $x = -4$ because $\lim_{x \to -4} R(x) = -\frac{1}{8}$, but R is not defined at $x = -4$. The graph has a vertical asymptote at $x = 4$ because $\lim_{x \to 4^-} R(x) = -\infty$ and $\lim_{x \to 4^+} R(x) = \infty$.
55. $R(x)$ is undefined at $x = 2$ and $x = 9$. A hole appears at $x = 2$, and a vertical asymptote appears at $x = 9$.
57. Answers will vary. One possibility is graphed below.

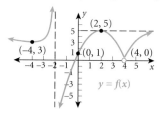

59. (a) $\lim_{x \to \infty} S(x) = \frac{4000}{7} \approx 571.4$
    (b) For larger and larger advertising expenditures, the sales level will eventually level off at 571 units.

Mathematical Questions from Professional Exams (p. 272)

1. (b) $\frac{5}{6}$
2. (c) III
3. (d) $\frac{\sqrt{2}}{4}$
4. (e) $x \neq -1$

CHAPTER 4   The Derivative of a Function

Exercise 4.1 (p. 283)

1. $m_{\tan} = 3, y = 3x + 5$

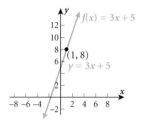

3. $m_{\tan} = -2, y = -2x + 1$

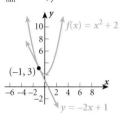

5. $m_{\tan} = 12, y = 12x - 12$

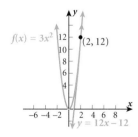

7. $m_{\tan} = 5, y = 5x - 2$

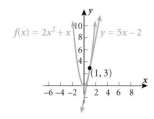

9. $m_{\tan} = -4, y = -4x + 2$

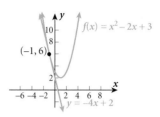

11. $m_{\tan} = 1, y = x + 1$

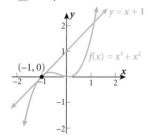

13. $f'(3) = -4$ **15.** $f'(0) = 0$ **17.** $f'(1) = 7$
19. $f'(0) = 4$ **21.** $f'(1) = 3$ **23.** $f'(1) = -1$
25. $f'(x) = 2$ **27.** $f'(x) = -2$ **29.** $f'(x) = 2x$
31. $f'(x) = 6x - 2$ **33.** $f'(x) = 3x^2$ **35.** $f'(x) = m$

37. (a) The average rate of change is 3.
    (b) The instantaneous rate of change at $x = 1$ is 3.

39. (a) The average rate of change is 12.
    (b) The instantaneous rate of change at $x = 1$ is 6.

41. (a) The average rate of change is 6.
    (b) The instantaneous rate of change at $x = 1$ is 4.

43. (a) The average rate of change is 7.
    (b) The instantaneous rate of change at $x = 1$ is 3.

45. $f'(-2) = 60$ **47.** $f'(8) = -\frac{3527}{4107} \approx -0.859$

49. $f'(0) = 1$ **51.** $f'(1) = 3e \approx 8.155$

53. $f'(1) = 0$ **55.** No.

57. The pilot should release the bomb at the point (2, 4).

59. (a) The average rate of change in sales is 74 tickets per day.
    (b) The average rate of change in sales is 94 tickets per day.
    (c) The average rate of change in sales is 110 tickets per day.
    (d) The instantaneous rate of change of sales on day 5 is 90 tickets per day.
    (e) The instantaneous rate of change of sales on day 10 is 130 tickets per day.

AN-34 Answers to Odd-Numbered Problems

61. (a) The farmer is willing to supply 4500 crates for $10 per crate.
(b) The farmer is willing to supply 7800 crates for $13 per crate.
(c) The average rate of change in supply is 1100 crates per dollar.
(d) The instantaneous rate of change is 950 crates per dollar.
(e) Answers will vary.

63. (a) $R'(x) = 8 - 2x$
(b) $C'(x) = 2$
(c) The break-even points are $x = 1$ and $x = 5$.
(d) $x = 3$
(e)

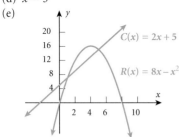

65. (a) $R(x) = -10x^2 + 2000x$
(b) $R'(x) = -20x + 2000$
(c) $R'(100) = \$0$
(d) The average rate of change in revenue is -10 dollars per ton.

67. (a) $R(x) = 90x - 0.02x^2$
(b) $R'(x) = 90x - 0.04x$
(c) $C'(x) = 10$
(d) The break-even points are $x = 0$ and $x = 4000$.
(e) The marginal revenue equals marginal cost at a production level of 2000 units.

69. The instantaneous rate of change of the volume with respect to the radius is $18\pi \approx 56.55$ cubic feet per foot.

Exercise 4.2 (p. 292)

1. $f'(x) = 0$ **3.** $f'(x) = 3x^2$ **5.** $f'(x) = 12x$

7. $f'(t) = t^3$ **9.** $f'(t) = 2x + 1$

11. $f'(x) = 3x^2 - 2x$ **13.** $f'(t) = 4t - 1$

15. $f'(x) = 4x^7 + 3$ **17.** $f'(x) = \frac{5}{3}x^4$

19. $f'(x) = 2ax + b$

21. $\frac{d}{dx}(-6x^2 + x + 4) = -12x + 1$

23. $\frac{d}{dt}(-16t^2 + 80t) = -32t + 80$

25. $\frac{dA}{dr} = 2\pi r$ **27.** $\frac{dV}{dr} = 4\pi r^2$

29. $f'(-3) = -24$ **31.** $f'(4) = 15$ **33.** $f'(3) = -4$

35. $f'(1) = 1$ **37.** $f'\left(-\frac{b}{2a}\right) = 0$ **39.** $\frac{dy}{dx} = 4$

41. $\frac{dy}{dx} = 8$ **43.** $\frac{dy}{dx} = -7$ **45.** $\frac{dy}{dx} = 1$

47. $\frac{dy}{dx} = 10$ **49.** $m_{\tan} = 3, y = 3x - 1$

51. $x = 2$ **53.** $x = -1, x = 1$

55. There are no such values of x.

57. $\left(-\frac{1}{3}, -\frac{1}{3}\right), \left(\frac{1}{3}, \frac{1}{3}\right)$ **59.** $y = -4x + 1, y = 4x - 7$

61. (a) The average cost is $45.00.
(b) $C'(x) = 0.4x + 3$
(c) The marginal cost of production level of 100 pairs of eyeglasses is $43.00.
(d) We can interpret $C'(100)$ to be the cost of producing the 101st pair of eyeglasses.

63. (a) $V'(R) = 4kR^3$
(b) $V'(0.3) = 0.108k$ cm^3/cm
(c) $V'(0.4) = 0.256k$ cm^3/cm
(d) The amount of blood flowing through the artery increases by about $0.0175k$ cm^3.

65. (a) The daily cost of producing 40 microwave ovens is $3920.
(b) The marginal cost function is $C'(x) = 50 - 0.1x$.
(c) $C'(40) = 46$. The marginal cost of producing 40 microwave ovens may be interpreted as the cost to produce the 41st microwave oven.
(d) The cost of producing 41 microwave ovens is approximately $3966.
(e) The actual cost of producing 41 microwave ovens is $3965.95. The actual cost is $0.05 less than the estimated cost.
(f) The actual cost of producing the 41st microwave oven is $45.95.
(g) The average cost function is $\overline{C}(x) = \frac{2000}{x} + 50 - 0.05x$.
(h) The average cost of producing 41 microwave ovens is $96.73.
(i) The average cost of producing 41 microwave ovens is $50.78 greater than the actual cost of producing the 41st microwave oven, and $50.73 greater than the estimated cost of producing the 41st microwave oven. Explanations will vary.

67. (a) The marginal price of beans in 1995 was $-\$2.431$.
(b) The marginal price of beans in 2002 was $-\$9.634$.
(c) Answers will vary.

69. $V'(2) = 16\pi \approx 50.27$ cubic feet per foot

71. $A'(t) = 3a_3 t^2 + 2a_2 t + a_1$

73. Let $f(x) = x^n$.

$$f'(x) = \lim_{h \to 0} \frac{f(x+h) - f(x)}{h} = \lim_{h \to 0} \frac{(x+h)^n - x^n}{h}$$

$$= \lim_{h \to 0} \frac{x^n + nx^{n-1}h + \frac{n(n-1)}{2}x^{n-2}h^2 + \cdots + h^n - x^n}{h}$$

$$= \lim_{h \to 0} \frac{nx^{n-1}h + h^2(\frac{n(n-1)}{2})x^{n-2} + \text{terms involving } x \text{ and } h}{h}$$
$$= \lim_{h \to 0} nx^{n-1} + h(\frac{n(n-1)}{2})x^{n-2} + \text{terms involving } x \text{ and } h$$
$$= nx^{n-1}$$

Exercise 4.3 (p. 301)

1. $f'(x) = 16x - 2$
3. $f'(t) = 4t^3 - 6t$
5. $f'(x) = 18x^2 - 20x + 3$
7. $f'(x) = 24x^7 + 40x^4 + 9x^2$
9. $f'(x) = \dfrac{1}{(1+x)^2}$
11. $f'(x) = -\dfrac{11}{(2x-1)^2}$
13. $f'(x) = \dfrac{x^2 - 8x}{(x-4)^2}$
15. $f'(x) = -\dfrac{6x^2 + 6x - 8}{(3x^2 + 4)^2}$
17. $f'(t) = \dfrac{4}{t^3}$
19. $f'(x) = -\dfrac{1}{x^2} - \dfrac{2}{x^3}$
21. $m_{\tan} = 3, y = 3x - 1$
23. $m_{\tan} = \frac{5}{4}, y = \frac{5}{4}x - \frac{3}{4}$
25. $x = -\frac{2}{3}, x = 1$
27. $x = -2, x = 0$
29. $y' = 9x^2 - 4x$
31. $y' = 32x - \dfrac{6}{x^3}$
33. $y' = \dfrac{1}{(3x+5)^2}$
35. $y' = -\dfrac{8x}{(x^2 - 4)^2}$
37. $y' = \dfrac{12x^2 + 12x + 23}{(2x+1)^2}$
39. $y' = -\dfrac{4x^2(x^2 + 12)}{(x^2 + 4)^2}$

41.

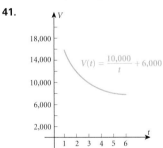

(a) The average rate of change is $-\$1000$ per year.
(b) $V'(t) = -\dfrac{10,000}{t^2}$ dollars per year.
(c) The instantaneous rate of change after two years is $-\$2500$ per year.
(d) The instantaneous rate of change after five years is $-\$400$ per year.
(e) Answers will vary.

43. (a) $R(x) = 10x + 40$ (b) $R'(x) = 10$
    (c) $R'(4) = 10$ (d) $R'(6) = 10$

45. (a) $D'(p) = \dfrac{-1,000,000 - 200,000p}{(p^2 + 10p + 50)^2}$
    (b) $D'(5) = -128, D'(10) = -48,$
    $D'(15) = -\dfrac{6400}{289} \approx -22.15$
    (c) Answers will vary.

47. (a) The population is growing at a rate of 38.820 bacteria per hour.
    (b) The population is growing at a rate of 35.503 bacteria per hour.
    (c) The population is growing at a rate of 30.637 bacteria per hour.
    (d) The population is growing at a rate of 24.970 bacteria per hour.

49. The rate of change of the intensity is -2 units per meter.

51. (a) $C'(x) = \dfrac{1}{10} - \dfrac{36,000}{x^2}$
    (b) The marginal cost is $-\$0.044$ dollars per mph.
    (c) The marginal cost is $-\$0.019$ dollars per mph.
    (d) The marginal cost is $-\$0.078$ dollars per mph.

53. (a) $S'(r) = \dfrac{d}{dr}\left[\dfrac{ar}{g-r}\right]$
    $= \dfrac{a(g-r) - (-1)(ar)}{(g-r)^2}$
    $= \dfrac{ag}{(g-r)^2}$
    (b) Answers will vary.

Exercise 4.4 (p. 307)

1. $f'(x) = 8(2x - 3)^3$
3. $f'(x) = 6x(x^2 + 4)^2$
5. $f'(x) = 12x(3x^2 + 4)$
7. $f'(x) = (4x + 1)(x + 1)^2$
9. $f'(x) = 8x(6x + 1)(2x + 1)^3$
11. $f'(x) = 3x^2(2x - 1)(x - 1)^2$
13. $f'(x) = -\dfrac{6}{(1 - 3x)^3}$
15. $f'(x) = -\dfrac{8x}{(x^2 + 4)^2}$
17. $f'(x) = \dfrac{24x}{(x^2 - 9)^4}$
19. $f'(x) = \dfrac{3x^2}{(x+1)^4}$
21. $f'(x) = \dfrac{(4x - 2)(2x + 1)^3}{3x^3}$
23. $f'(x) = \dfrac{(x^2 + 1)^2(5x^2 - 1)}{x^2}$
25. $f'(x) = \dfrac{3(x^2 - 1)(x^2 + 1)^2}{x^4}$
27. $f'(x) = \dfrac{6x(1 - x^2)}{(x^2 + 1)^3}$

29. (a) The car is depreciating at a rate of $\$7733.33$ per year.
    (b) The car is depreciating at a rate of $\$4793.39$ per year.
    (c) The car is depreciating at a rate of $\$3017.69$ per year.
    (d) The car is depreciating at a rate of $\$1972.79$ per year.

31. (a) $p'(x) = -\dfrac{2000}{(x + 20)^2}$ (b) $R(x) = \dfrac{10,000x}{5x + 100} - 5x$
    (c) $R'(x) = \dfrac{40,000}{(x + 20)^2} - 5$

(d) $R'(10) = 39.44$ dollars per pound, $R'(40) = 6.11$ dollars per pound
(e) Answers will vary.

33. (a) The average rate of change in mass is -3.5 grams per hour.
(b) $M'(0) = -7$ grams per hour
(c) Answers will vary.

Exercise 4.5 (p. 317)

1. $f'(x) = 3x^2 - e^x$

3. $f'(x) = e^x(x^2 + 2x)$

5. $f'(x) = \dfrac{e^x(x-2)}{x^3}$

7. $f'(x) = \dfrac{8x - 4x^2}{e^x}$

9. $y = (x^3 + 1)^5$, $\dfrac{dy}{dx} = 15x^2(x^3 + 1)^4$

11. $y = \dfrac{x^2 + 1}{x^2 + 2}$, $\dfrac{dy}{dx} = \dfrac{2x}{(x+2)^2}$

13. $y = a\left(\dfrac{x+1}{x}\right)^2 b$, $\dfrac{dy}{dx} = -\dfrac{2(x+1)}{x^3}$

15. $y = \dfrac{(1-x^6)^5}{x^{30}}$, $\dfrac{dy}{dx} = \dfrac{-30(1-x^6)^4}{x^{31}}$

17. $y = e^{3x}$, $\dfrac{dy}{dx} = 3e^{3x}$

19. $y = e^{x^3}$, $\dfrac{dy}{dx} = 3x^2 e^{x^3}$

21. $y' = 6x^2(x^3 + 1) = 6x^5 + 6x^2$

23. $f'(x) = 5e^{5x}$

25. $f'(x) = -16xe^{-x^2}$

27. $f'(x) = e^{x^2}(2x^3 + 2x)$

29. $f'(x) = 15e^{3x}$

31. $f'(x) = \dfrac{2x - x^2}{e^x}$

33. $f'(x) = \dfrac{e^{2x}(2x - 1)}{x^2}$

35. $f'(x) = 2x - \dfrac{3}{x}$

37. $f'(x) = x + 2x \ln x$

39. $f'(x) = \dfrac{3}{x}$

41. $f'(x) = \ln(x^2 + 1) + \dfrac{2x^2}{x^2 + 1}$

43. $f'(x) = 1 + \dfrac{8}{x}$

45. $f'(x) = \dfrac{24(\ln x)^2}{x}$

47. $f'(x) = \dfrac{1}{x \ln 3}$

49. $f'(x) = \dfrac{x + 2x \ln x}{\ln 2}$

51. $f'(x) = (\ln 3)3^x$

53. $f'(x) = 2^x(2x + x^2 \ln 2)$

55. $y = 3x + 1$

57. $y = x - 1$

59. $y = 3x - 1$

61. $y = x - 1$

63. $y = x + 1$

65. (a) The reaction rate is 1.1 per unit.
(b) The reaction rate is 0.55 per unit.
(c) Answers will vary.

67. The rate of change of the pressure with respect to the height is -1.130 kilograms per square meter per meter at a height of 500 meters and is -1.103 kilograms per square meter per meter at a height of 700 meters.

69. (a) $A'(t) = 18.9e^{-0.21t}$ percent of the market per year
(b) $A'(5) = 6.614$ percent of the market per year. In the sixth year, DVD players will penetrate approximately an additional 6.614 percent of the market.
(c) $A'(10) = 2.314$ percent of the market per year. In the eleventh year, DVD players will penetrate approximately an additional 2.314 percent of the market.
(d) $A'(30) = 0.035$ percent of the market per year. In the thirty-first year, DVD players will penetrate approximately an additional 0.035 percent of the market.

71. (a) $S'(x) = \dfrac{400,000}{x}$ dollars in sales per thousands of dollars of advertising cost
(b) $S'(10) = 40,000$ dollars in sales per thousands of dollars of advertising cost
(c) $S'(20) = 20,000$ dollars in sales per thousands of dollars of advertising cost
(d) Answers will vary.

73. (a) 1000 t-shirts can be sold at $40.41.
(b) 5000 t-shirts can be sold at $34.27.
(c) $p'(1000) = -\$0.0036$. This means that another t-shirt will be demanded if the price were reduced by $0.0036.
(d) $p'(5000) = -\$0.00078$. This means that another t-shirt will be demanded if the price were reduced by $0.00078.
(e) $R(x) = 50x - 4x \ln\left(\dfrac{x}{100} + 1\right)$
(f) $R'(1000) = \$36.77$. The revenue received for selling the 1001st t-shirt is $36.77.
(g) $R'(5000) = \$30.35$. The revenue received for selling the 5001st t-shirt is $30.35.
(h) $P(x) = 46x - 4x \ln\left(\dfrac{x}{100} + 1\right)$
(i) $P(1000) = \$36,408.42$
(j) $P(5000) = \$151,363.49$
(k) For $x = 3,631,550$, the profit is the greatest.
(l) A price of $8.00 should be charged to maximize profit.

75. (a) $p'(t) = \dfrac{0.026}{t}$ dollars per year
(b) $p'(5) = 0.0052$ dollars per year
(c) $p'(10) = 0.0026$ dollars per year
(d) Answers will vary.

77. Let $y = \ln u$ and $u = g(x)$. Thus $y(u(x)) = \ln(g(x))$.

Thus $\dfrac{d}{dx} \ln g(x) = \dfrac{d}{dx}(y(u(x))) = \dfrac{dy}{du}\dfrac{du}{dx} = \dfrac{1}{u(x)}\dfrac{d}{dx}g(x) = \dfrac{\frac{d}{dx}g(x)}{g(x)}$.

Exercise 4.6 (p. 326)

1. $f'(x) = 2, f''(x) = 0$

3. $f'(x) = 6x + 1, f''(x) = 6$

5. $f'(x) = -12x^3 + 4x, f''(x) = -36x^2 + 4$

7. $f'(x) = -\dfrac{1}{x^2}, f''(x) = \dfrac{2}{x^3}$

9. $f'(x) = 1 - \dfrac{1}{x^2}, f''(x) = \dfrac{2}{x^3}$

11. $f'(x) = \dfrac{1}{(x+1)^2}, f''(x) = -\dfrac{2}{(x+1)^3}$

13. $f'(x) = e^x, f''(x) = e^x$

15. $f'(x) = 6x(x^2+4)^2, f''(x) = 6(x^2+4)(5x^2+4)$

17. $f'(x) = \dfrac{1}{x}, f''(x) = -\dfrac{1}{x^2}$

19. $f'(x) = e^x(x+1), f''(x) = e^x(x+2)$

21. $f'(x) = 2e^{2x}, f''(x) = 4e^{2x}$

23. $f'(x) = -\dfrac{1}{x(\ln x)^2}, f''(x) = \dfrac{2 + \ln(x)}{x^2(\ln x)^3}$

25. (a) The domain is the interval $(-\infty, \infty)$.
    (b) $f'(x) = 2x$
    (c) The domain is the interval $(-\infty, \infty)$.
    (d) $x = 0$
    (e) The derivative $f'(x)$ exists for all values of x.
    (f) $f''(x) = 2$
    (g) The domain is the interval $(-\infty, \infty)$.

27. (a) The domain is the interval $(-\infty, \infty)$.
    (b) $f'(x) = 3x^2 - 18x + 27$
    (c) The domain is the interval $(-\infty, \infty)$.
    (d) $x = 3$
    (e) The derivative $f'(x)$ exists for all values of x.
    (f) $f''(x) = 6x - 18$
    (g) The domain is the interval $(-\infty, \infty)$.

29. (a) The domain is the interval $(-\infty, \infty)$.
    (b) $f'(x) = 12x^3 - 36x^2$
    (c) The domain is the interval $(-\infty, \infty)$.
    (d) $x = 0, x = 3$
    (e) The derivative $f'(x)$ exists for all values of x.
    (f) $f''(x) = 36x^2 - 72x$
    (g) The domain is the interval $(-\infty, \infty)$.

31. (a) The domain is the set $\{x | x \neq -2, x \neq 2\}$.
    (b) $f'(x) = -\dfrac{x^2 + 4}{(x^2 - 4)^2}$
    (c) The domain is the set $\{x | x \neq -2, x \neq 2\}$.
    (d) The derivative $f'(x)$ is never zero.
    (e) The derivative $f'(x)$ does not exist for $x = -2$ and $x = 2$.
    (f) $f''(x) = \dfrac{2x(x^2 + 12)}{(x^2 - 4)^3}$
    (g) The domain is the set $\{x | x \neq -2, x \neq 2\}$.

33. $f^{(4)}(x) = 0$ 35. $f^{(20)}(x) = 0$ 37. $f^{(8)}(x) = 5040$

39. $v = 32t + 20, a = 32$ 41. $v = 9.8t + 4, a = 9.8$

43. $f^{(n)}(x) = e^x$

45. $f^{(n)}(x) = \dfrac{(-1)^{n+1}(n-1)!}{x^n}$

47. $f'(x) = 1 + \ln x, f^{(n)}(x) = \dfrac{(-1)^n(n-2)!}{x^{n-1}}$ for $n > 1$

49. $f^{(n)}(x) = n! \, 2^n$

51. $f^{(n)}(x) = a^n e^{ax}$

53. $f^{(n)}(x) = \dfrac{(-1)^{n+1}(n-1)!}{x^n}$

55. $y'' - 4y = 0$

57. $f''(x) = x^2 g''(x) + 4xg'(x) + 2g(x)$

59. (a) The velocity is 16 feet per second.
    (b) The ball will reach its maximum height 2.5 seconds after it is thrown.
    (c) The maximum height of the ball is 106 feet.
    (d) The acceleration is -32 feet per second per second.
    (e) The ball is in the for air 5.074 seconds.
    (f) The velocity of the ball is -82.365 feet per second upon impact.
    (g) The total distance traveled by the ball is 206 feet.

61. $v(1) = 3$ meters per second, $a(t) = 6t - 12$ meters per second.

63. (a) It takes 4.24 seconds for the rock to hit the ground.
    (b) The average velocity is -20.8 meters per second.
    (c) The average velocity is -14.7 meters per second.
    (d) The velocity is -41.6 meters per second when the rock hits the ground.

Exercise 4.7 (p. 333)

1. $\dfrac{dy}{dx} = -\dfrac{x}{y}$ 3. $\dfrac{dy}{dx} = -\dfrac{2y}{x}$ 5. $\dfrac{dy}{dx} = \dfrac{2x - y}{x - 2y}$

7. $\dfrac{dy}{dx} = \dfrac{2x + 4y}{1 - 4x - 2y}$ 9. $\dfrac{dy}{dx} = -\dfrac{2x}{y^2}$ 11. $\dfrac{dy}{dx} = \dfrac{x - 6x^2}{3y^2}$

13. $\dfrac{dy}{dx} = \dfrac{y^3}{x^3}$ 15. $\dfrac{dy}{dx} = -\dfrac{y^2}{x^2}$

17. $\dfrac{dy}{dx} = \dfrac{ye^x - 2x}{2y - e^x}$ 19. $\dfrac{dy}{dx} = \dfrac{6x^2y^2e^x + y^3 - x^2y}{xy^2 - x^3}$

21. $\dfrac{dy}{dx} = \dfrac{2x^2 - y^2}{2xy \ln x}$ 23. $\dfrac{dy}{dx} = \dfrac{-3x - 6y}{6x + 8y}$

25. $\dfrac{dy}{dx} = \dfrac{3x^2 - 6xy + 3y^2 - 4x^3 - 4xy^2}{3x^2 - 6xy + 3y^2 + 4x^2y + 4y^3}$

27. $\dfrac{dy}{dx} = \dfrac{xy^2 - 3x^5 - 3x^2y^3}{3x^3y^2 + 3y^5 - x^2y}$ 29. $\dfrac{dy}{dx} = \dfrac{2xe^{x^2+y^2}}{1 - 2ye^{x^2+y^2}}$

31. $y' = -\dfrac{x}{y}, y'' = -\dfrac{x^2 + y^2}{y^3}$

33. $y' = -\dfrac{2xy + y}{x^2 + x}, y'' = \dfrac{6x^2y + 6xy + 2y}{x^4 + 2x^3 + x^2}$

35. $m_{\tan} = -\dfrac{1}{2}, y = -\dfrac{1}{2}x + \dfrac{5}{2}$

37. There is no tangent line at $(0, 0)$.

39. $(0, -2)$ and $(0, 2)$

41. $(0, -4)$ and $(0, 4)$

43. (a) $m_{\tan} = \dfrac{dy}{dx} = -\dfrac{y+1}{x+4y}$ (b) $y = -\dfrac{1}{3}x + \dfrac{5}{3}$
    (c) $(2, 1)$

45. $\dfrac{dV}{dp} = \dfrac{V^3(V-b)^2}{2a(V-b)^2 - cV^3}$

47. (a) $\dfrac{dN}{dt} = \dfrac{430{,}163t^4 + 1{,}720{,}649t^2 + 1{,}720{,}658}{430{,}163t^5 + 1{,}720{,}655t^3 + 1{,}720{,}658t}$

(b) $N'(2) = \dfrac{2{,}580{,}977}{5{,}161{,}962} \approx 0.500$

$N'(4) = \dfrac{23{,}228{,}795}{92{,}915{,}244} \approx 0.250$

(c) Answers will vary.

Exercise 4.8 (p. 338)

1. $f'(x) = \dfrac{4}{3}x^{1/3}$ **3.** $f'(x) = \dfrac{2}{3x^{1/3}}$ **5.** $f'(x) = -\dfrac{1}{2x^{3/2}}$

7. $f'(x) = 3(2x+3)^{1/2}$ **9.** $f'(x) = 3x(x^2+4)^{1/2}$

11. $f'(x) = \dfrac{x}{\sqrt{2x+3}}$ **13.** $f'(x) = \dfrac{9x}{\sqrt{9x^2+1}}$

15. $f'(x) = 5x^{2/3} - \dfrac{2}{x^3}$ **17.** $f'(x) = \dfrac{7}{3}x^{4/3} - \dfrac{4}{3x^{2/3}}$

19. $f'(x) = -\dfrac{4}{(x^2-4)^{3/2}}$ **21.** $f'(x) = \dfrac{e^{x/2}}{2}$

23. $f'(x) = \dfrac{1}{2x\sqrt{\ln x}}$ **25.** $f'(x) = \dfrac{e^{\sqrt[3]{x}}}{3x^{2/3}}$

27. $f'(x) = \dfrac{1}{3x(\ln x)^{2/3}}$ **29.** $f'(x) = e^x\left(\sqrt{x} + \dfrac{1}{2\sqrt{x}}\right)$

31. $f'(x) = \dfrac{e^{2x}(2x^2+x+2)}{\sqrt{x^2+1}}$ **33.** $\dfrac{dy}{dx} = -\dfrac{\sqrt{y}}{\sqrt{x}}$

35. $\dfrac{dy}{dx} = \dfrac{\sqrt{x^2+y^2} - x}{y}$ **37.** $\dfrac{dy}{dx} = \dfrac{y^{2/3}}{x^{2/3}}$

39. $\dfrac{dy}{dx} = -\dfrac{\sqrt{y}e^{\sqrt{x} - \sqrt{y}}}{\sqrt{x}}$

41. (a) The domain is the interval $[0, \infty)$.

(b) $f'(x) = \dfrac{1}{2\sqrt{x}}$

(c) The domain is the interval $[0, \infty)$.
(d) The derivative $f'(x)$ is nonzero on its domain.
(e) $x = 0$
(f) $f''(x) = -\dfrac{1}{4x^{3/2}}$
(g) The domain is the interval $(0, \infty)$.

43. (a) The domain is the interval $(-\infty, \infty)$.

(b) $f'(x) = \dfrac{2}{3x^{1/3}}$

(c) The domain is the set $\{x | x \neq 0\}$.
(d) The derivative $f'(x)$ is nonzero on its domain.
(e) $x = 0$

(f) $f''(x) = -\dfrac{2}{9x^{4/3}}$

(g) The domain is the set $\{x | x \neq 0\}$.

45. (a) The domain is the interval $(-\infty, \infty)$.

(b) $f'(x) = \dfrac{2}{3x^{1/3}} + \dfrac{2}{3x^{2/3}}$

(c) The domain is the set $\{x | x \neq 0\}$.
(d) $x = -1$
(e) $x = 0$

(f) $f''(x) = -\dfrac{2}{9x^{4/3}} - \dfrac{4}{9x^{5/3}}$

(g) The domain is the set $\{x | x \neq 0\}$.

47. (a) The domain is the interval $(-\infty, \infty)$.

(b) $f'(x) = \dfrac{4x}{3(x^2-1)^{1/3}}$

(c) The domain is the set $\{x | x \neq -1, x \neq 1\}$.
(d) $x = 0$
(e) $x = -1, x = 1$

(f) $f''(x) = \dfrac{4(x^2-3)}{9(x^2-1)^{4/3}}$

(g) The domain is the set $\{x | x \neq -1, x \neq 1\}$.

49. (a) The domain is the interval $[-1, 1]$.

(b) $f'(x) = \dfrac{1 - 2x^2}{\sqrt{1 - x^2}}$

(c) The domain is the interval $(-1, 1)$.

(d) $x = -\dfrac{\sqrt{2}}{2}, x = \dfrac{\sqrt{2}}{2}$

(e) $x = -1, x = 1$

(f) $f''(x) = \dfrac{x(2x^2-3)}{(1-x^2)^{3/2}}$

(g) The domain is the interval $(-1, 1)$.

51. (a) $N'(t) = \dfrac{500}{(1 + 0.1t)^{3/2}}$

(b) Student enrollment will be increasing at the rate of 176.78 students per year.

53. $\dfrac{dz}{dx} = \dfrac{d}{dx}(x^{0.5}y^{0.4})$

$0 = 0.5x^{-0.5}y^{0.4} + 0.4x^{0.5}y^{-0.6}\dfrac{dy}{dx}$

$\dfrac{dy}{dx} = -\dfrac{0.5x^{-0.5}y^{0.4}}{0.4x^{0.5}y^{-0.6}}$

$\dfrac{dy}{dx} = -\dfrac{5y}{4x}$

55. (a) $A'(t) = \dfrac{3(t^{\frac{1}{4}} + 3)^2}{4t^{3/4}}$ units per year

(b) The instantaneous rate of change of pollution is 2.34 units per year.

57. After 1 second, the child's velocity is 1.5 feet per second. The child strikes the ground with a velocity of 3 feet per second.

CHAPTER 4 Review

True – False Items (p. 341)

1. True   2. False   3. True
4. True   5. False   6. False
7. True

Fill in the Blanks (p. 341)

1. tangent   2. marginal cost
3. chain rule   4. velocity
5. 0   6. implicit

Review Exercises (p. 342)

1. $f'(2) = 2$ 3. $f'(2) = 4$ 5. $f'(1) = 0$
7. $f'(0) = 3$ 9. $f'(x) = 4$ 11. $f'(x) = 4x$
13. $f'(x) = 5x^4$ 15. $f'(x) = x^3$
17. $f'(x) = 4x - 3$ 19. $f'(x) = 14x$
21. $f'(x) = 15(x^2 - 6x + 6)$
23. $f'(x) = 24(16x^3 + 3x^2 - 5x + 1)$
25. $f'(x) = -\dfrac{16}{(5x-3)^2}$ 27. $f'(x) = -\dfrac{24}{x^{13}}$
29. $f'(x) = -\dfrac{3}{x^2} - \dfrac{8}{x^3}$ 31. $f'(x) = \dfrac{17}{(x+5)^2}$
33. $f'(x) = 10x^4(3x - 1)(3x - 2)^4$
35. $f'(x) = 7(x + 1)^3(5x + 1)$
37. $f'(x) = -\dfrac{2(x+1)}{(3x+2)^3}$ 39. $f'(x) = -\dfrac{42x^2}{(x^3+4)^3}$
41. $f'(x) = \dfrac{3(3x^2 - 4)(3x^2 + 4)^2}{x^4}$
43. $f'(x) = 3e^x + 2x$ 45. $f'(x) = 3e^{3x+1}$
47. $f'(x) = e^x(2x^2 + 11x + 7)$ 49. $f'(x) = -\dfrac{x}{e^x}$
51. $f'(x) = \dfrac{2e^{2x}(x-1)}{9x^3}$ 53. $f'(x) = \dfrac{1}{x}$
55. $f'(x) = x + 2x \ln x$ 57. $f'(x) = \dfrac{6x^2}{2x^3 + 1}$
59. $f'(x) = (\ln 2)2^x + 2x$ 61. $f'(x) = 1 + \dfrac{1}{x \ln 10}$
63. $f'(x) = \dfrac{1}{2\sqrt{x}}$ 65. $f'(x) = 5x^{2/3}$
67. $f'(x) = \dfrac{2x-3}{2\sqrt{x^2-3x}}$ 69. $f'(x) = \dfrac{x+9}{2(x+5)^{3/2}}$
71. $f'(x) = \dfrac{e^{\frac{x}{2}}(x+3)}{2}$ 73. $f'(x) = \dfrac{2 + \ln x}{2\sqrt{x}}$
75. $f'(x) = 3x^2, f''(x) = 6x$
77. $f'(x) = -3e^{-3x}, f''(x) = 9e^{-3x}$
79. $f'(x) = \dfrac{1}{(2x+1)^2}, f''(x) = -\dfrac{4}{(2x+1)^3}$
81. $\dfrac{dy}{dx} = \dfrac{10-y}{x+6y}$ 83. $\dfrac{dy}{dx} = \dfrac{8x - e^y}{xe^y}$
85. $m_{\tan} = -1, y = -x - 9$ 87. $m_{\tan} = 1, y = x + 1$
89. (a) The average rate of change is 7.
    (b) The instantaneous rate of change at 2 is $f'(2) = 15$.
91. (a) 2.5 seconds elapse before the stone hits the water.
    (b) The average velocity is -40 feet per second.
    (c) The instantaneous velocity is -80 feet per second.
93. (a) The ball reaches its maximum height 4 seconds after it is thrown.
    (b) The maximum height that the ball reaches is 262 feet.
    (c) The total distance the ball travels is 518 feet.
    (d) The velocity of the ball at any time t is $v(t) = 128 - 32t$ feet per second.
    (e) The velocity of the ball is zero at $t = 4$ seconds, which is the time at which the ball's velocity changes from upward motion to downward motion.
    (f) The ball is in the air for 8.0 seconds.
    (g) The velocity of ball is -129.5 feet per second when it hits the ground.
    (h) The acceleration at any time t is -32 feet per second per second.
    (i) The velocity of ball is 64 feet per second when it has been in the air for 2 seconds and is -64 feet per second when it has been in the air for 6 seconds.
    (j) Answers will vary.
95. (a) $R(x) = -0.50x^2 + 75x$
    (b) $R'(x) = -1.00x + 75$
    (c) $C'(x) = 15$
    (d) The break even points are $x = 10$ and $x = 110$.
    (e) The marginal revenue and the marginal cost are equal when $x = 60$ units are produced.

Mathematical Questions from Professional Exams (p. 345)

1. (e) $2x^3 e^{x^2} + 2xe^{x^2}$ 2. (d) $x \neq -3$
3. (d) $2bxe^{c^2 + x^2}$ 4. (a) $2xe^{x^2}$
5. (e) 15   6. (c) $t = 2$

CHAPTER 5 Applications: Graphing Functions; Optimization

Exercise 5.1 (p. 353)

1. There is a horizontal tangent at $(2, 0)$.
3. There is a horizontal tangent at $(4, 16)$.
5. There is a horizontal tangent at $(2, 9)$.
7. There is a vertical tangent at $(0, 1)$.
9. There are horizontal tangents at $(-1, -1)$ and $(1, 3)$.
11. There is a vertical tangent at $(0, -2)$.
13. There are horizontal tangents at $(0, 0)$ and $(8, -8192)$.
15. There is a horizontal tangent at $(0, -1)$.
17. There is a horizontal tangent at $(-1, -1)$, and there is a vertical tangent at $(0, 0)$.
19. There is a horizontal tangent at $(4, -12\sqrt[3]{2})$, and there is a vertical tangent at $(0, 0)$.
21. There are horizontal tangents at $(-2, -12\sqrt[3]{4})$ and $(2, -12\sqrt[3]{4})$, and there is a vertical tangent at $(0, 0)$.
23. There is a horizontal tangent at $\left(-4, -\frac{\sqrt[3]{2}}{3}\right)$, and there is a vertical tangent at $(0, 0)$.
25. There is a horizontal tangent at $\left(-\frac{1}{2}, \frac{\sqrt[3]{4}}{3}\right)$, and a vertical tangent at $(0, 0)$.
27. (a) Yes (b) No (c) vertical tangent at $(0, 0)$
29. (a) No (b) No (c) f is not continuous at $x = 1$.
31. (a) Yes (b) No (c) no tangent at $(0, 0)$
    (d) $f(x) = \begin{cases} 3x & \text{if } x < 0 \\ x^2 & \text{if } x \geq 0 \end{cases}$

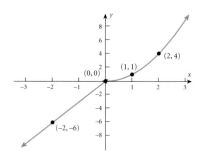

33. (a) No (b) No (c) f is not continuous at $x = 2$.

(d)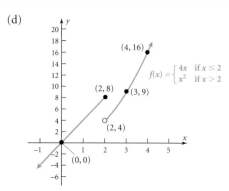

$f(x) = \begin{cases} 4x & \text{if } x \leq 2 \\ x^2 & \text{if } x > 2 \end{cases}$

35. (a) Yes (b) Yes, $f'(0) = 0$ (c) $f'(0)$ does exist.
(d)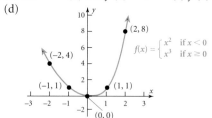

$f(x) = \begin{cases} x^2 & \text{if } x < 0 \\ x^3 & \text{if } x \geq 0 \end{cases}$

Exercise 5.2 (p. 363)

1. The domain is the interval $[x_1, x_9]$.
3. The graph is increasing on the intervals (x_1, x_4), (x_5, x_7), and (x_8, x_9).
5. $x = x_4$, $x = x_6$, $x = x_7$, and $x = x_8$
7. f has a local maximum at (x_4, y_4) and at (x_7, y_7).
9. Step 1: The domain is the interval $(-\infty, \infty)$.
   Step 2: The x-intercept is $(1, 0)$, and the y-intercept is $(0, -2)$.
   Step 3: The graph is increasing on the interval $(-\infty, 1)$ and is decreasing on the interval $(1, \infty)$.
   Step 4: There is a local maximum at $(1, 0)$.
   Step 5: The tangent line is horizontal at $(1, 0)$.
   Step 6: The end behavior is $y = -2x^2$.

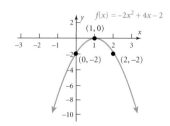

11. Step 1: The domain is the interval $(-\infty, \infty)$.
Step 2: The x-intercept is $(3, 0)$, and the y-intercept is $(0, -27)$.
Step 3: The graph is increasing on the interval $(-\infty, \infty)$.
Step 4: There are no local extrema.
Step 5: The tangent line is horizontal at $(3, 0)$.
Step 6: The end behavior is $y = x^3$.

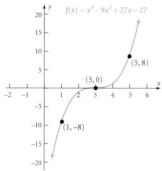

13. Step 1: The domain is the interval $(-\infty, \infty)$.
Step 2: The x-intercept and the y-intercept are both $(0, 0)$.
Step 3: The graph is increasing on the intervals $(-\infty, 2)$ and $(3, \infty)$ and is decreasing on the interval $(2, 3)$.
Step 4: There is a local maximum at $(2, 28)$, and there is a local minimum at $(3, 27)$.
Step 5: The tangent line is horizontal at the points $(2, 28)$ and $(3, 27)$.
Step 6: The end behavior is $y = 2x^3$.

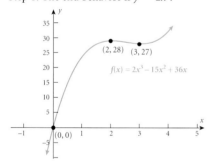

15. Step 1: The domain is the interval $(-\infty, \infty)$.
Step 2: The y-intercept is $(0, -1)$.
Step 3: The graph is increasing on the interval $(-1, 1)$ and is decreasing on the intervals $(-\infty, -1)$ and $(1, \infty)$.
Step 4: There is a local minimum at $(-1, -3)$, and there is a local maximum at $(1, 1)$.
Step 5: The tangent line is horizontal at the points $(-1, -3)$ and $(1, 1)$.
Step 6: The end behavior is $y = -x^3$.

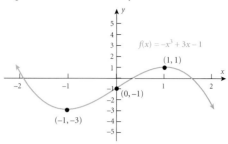

17. Step 1: The domain is the interval $(-\infty, \infty)$.
Step 2: The y-intercept is $(0, 2)$.
Step 3: The graph is increasing on the interval $(3, \infty)$ and is decreasing on the interval $(-\infty, 3)$.
Step 4: There is a local minimum at $(3, -79)$.
Step 5: The tangent line is horizontal at the points $(0, 2)$ and $(3, -79)$.
Step 6: The end behavior is $y = 3x^4$.

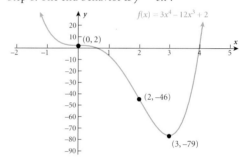

19. Step 1: The domain is the interval $(-\infty, \infty)$.
Step 2: The y-intercept is $(0, 1)$.
Step 3: The graph is increasing on the intervals $(-\infty, -1)$ and $(1, \infty)$, and is decreasing on the interval $(-1, 1)$.
Step 4: There is a local maximum at $(-1, 5)$, and there is a local minimum at $(1, -3)$.
Step 5: The tangent line is horizontal at $(-1, 5)$ and $(1, -3)$.
Step 6: The end behavior is $y = x^5$.

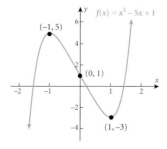

21. Step 1: The domain is the interval $(-\infty, \infty)$.
    Step 2: The y-intercept is $(0, 1)$.
    Step 3: The graph is increasing on the intervals $(-\infty, -2)$ and $(2, \infty)$ and is decreasing on $(-2, 2)$.
    Step 4: There is a local maximum at $(-2, 65)$, and there is a local minimum at $(2, -63)$.
    Step 5: The tangent line is horizontal at $(-2, 65)$, $(0, 1)$, and $(2, -63)$.
    Step 6: The end behavior is $y = 3x^5$.

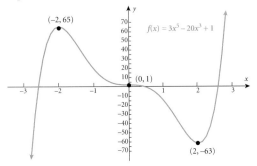

23. Step 1: The domain is the interval $(-\infty, \infty)$.
    Step 2: The x-intercepts are $(0, 0)$ and $(0, -8)$, and the y-intercept is $(0, 0)$.
    Step 3: The graph is increasing on the interval $(-1, \infty)$, and is decreasing on the interval $(-\infty, -1)$.
    Step 4: There is a local minimum at $(-1, -1)$.
    Step 5: The tangent line is vertical at $(0, 0)$.
    Step 6: The end behavior is $y = x^{2/3}$.

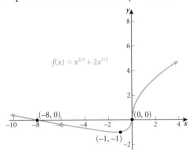

25. Step 1: The domain is the interval $(-\infty, \infty)$.
    Step 2: The x-intercepts are $(-1, 0)$ and $(1, 0)$, and the y-intercept is $(0, 1)$.
    Step 3: The graph is increasing on the intervals $(-1, 0)$ and $(1, \infty)$ and is decreasing on the intervals $(-\infty, -1)$ and $(0, 1)$.
    Step 4: There is a local maximum at $(0, 1)$, and there are local minima at $(-1, 0)$ and $(1, 0)$.
    Step 5: The tangent line is horizontal at $(0, 1)$ and is vertical at $(-1, 0)$ and $(1, 0)$.

Step 6: The end behavior is $y = x^{4/3}$.

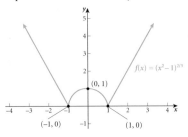

27. Step 1: The domain is the set $\{x \mid x \neq -4, x \neq 4\}$.
    Step 2: The y-intercept is $(0, -\frac{1}{2})$.
    Step 3: The graph is increasing on the intervals $(-\infty, -4)$ and $(-4, 0)$ and is decreasing on the intervals $(0, 4)$ and $(4, \infty)$.
    Step 4: There is a local maximum at $(0, -\frac{1}{2})$.
    Step 5: The tangent line is horizontal at $(0, -\frac{1}{2})$.
    Step 6: The end behavior is the horizontal asymptote $y = 0$. The lines $x = -4$ and $x = 4$ are vertical asymptotes.

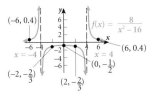

29. Step 1: The domain is the set $\{x \mid x \neq -3, x \neq 3\}$.
    Step 2: The x-intercept and the y-intercept are both $(0, 0)$.
    Step 3: The graph is decreasing on the intervals $(-\infty, -3)$, $(-3, 3)$, and $(3, \infty)$.
    Step 4: There are no local extrema.
    Step 5: There are no horizontal tangents or vertical tangents.
    Step 6: The end behavior is the horizontal asymptote $y = 0$. The lines $x = -3$ and $x = 3$ are vertical asymptotes.

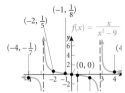

31. Step 1: The domain is the set $\{x \mid x \neq -2, x \neq 2\}$.
    Step 2: The x-intercept and y-intercept are both $(0, 0)$.
    Step 3: The graph is increasing on the intervals $(-\infty, -2)$ and $(2, 0)$, and is decreasing on the intervals $(0, 2)$ and $(2, \infty)$.
    Step 4: There is a local maximum at $(0, 0)$.
    Step 5: There is a horizontal tangent at $(0, 0)$.

Step 6: The end behavior is the horizontal asymptote $y = 1$. The lines $x = -2$ and $x = 2$ are vertical asymptotes.

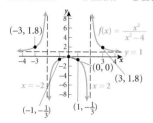

33. Step 1: The domain is the set $\{x | x > 0\}$.
    Step 2: The x-intercept is $(1, 0)$.
    Step 3: The graph is increasing on the interval $(0.37, \infty)$ and is decreasing on the interval $(0, 0.37)$.
    Step 4: There is a local minimum at $(0.37, -0.37)$.
    Step 5: There is a horizontal tangent at $(0.37, -0.37)$.
    Step 6: The end behavior is $y = x \ln x$. $\lim\limits_{x \to 0+} = 0$ and $\lim\limits_{x \to \infty} = \infty$

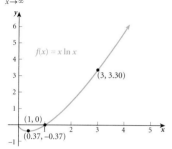

35. $S'(x) = 8x + 50 > 0$ for $1 \leq x \leq 10$. Therefore S is an increasing function.

37. (a) The graph of R is increasing on $(0, 2000)$ and is decreasing on $(2000, \infty)$.
    (b) 2000 trucks need to be sold.
    (c) The maximum revenue is $20,000.
    (d)

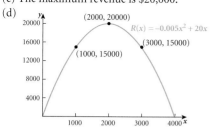

39. (a) The function is increasing on the interval $\left(0, \frac{681}{106}\right) \approx (0, 6.42)$.
    (b) The acreage of wheat planted from 2004 to 2008 will be decreasing.

41. (a) The yield will be increasing for amounts of nitrogen between 0 units and $\frac{20{,}164}{19{,}321} \approx 1.044$ units.
    (b) The yield will be decreasing for amounts of nitrogen greater than $\frac{20{,}164}{19{,}321} \approx 1.044$ units.

43. $c = \frac{1}{2}$

45. $c = 0$

47. $c = \frac{3}{2}$

49. $c = \frac{2\sqrt[3]{9}}{3}$

Exercise 5.3 (p. 378)

1. The domain is $[x_1, x_4] \cup (x_4, x_7]$

3. The graph of f is increasing on the intervals (x_1, x_3), $(0, x_4)$, and (x_4, x_6).

5. $x = 0$ and $x = x_6$

7. (x_3, y_3), (x_6, y_6)

9. The graph of f is concave up on the intervals (x_1, x_3) and (x_3, x_4).

11. The line $x = 4$ is a vertical asymptote.

13. The graph is concave down on $(-\infty, 2)$ and is concave up on $(2, \infty)$. The point $(2, -15)$ is the only inflection point.

15. The graph is concave down on $(0, 1)$ and is concave up on $(-\infty, 0)$ and $(1, \infty)$. The points $(0, -1)$ and $(1, 4)$ are the inflection points.

17. The graph is concave down on $(-\infty, 1)$ and is concave up on $(1, \infty)$. The point $(1, 1)$ is the only inflection point.

19. The graph is concave down on $(-\infty, -1)$ and $(0, 1)$ and is concave up on $(-1, 0)$ and $(1, \infty)$. The points $(-1, 7)$, $(0, 10)$, and $(1, 13)$ are the inflection points.

21. The graph is concave down on $(-\infty, 1)$ and is concave up on $(1, \infty)$. The point $(1, -5)$ is the only inflection point.

23. The graph is concave down on $(0, \infty)$ and is concave up on $(-\infty, 0)$. The point $(0, 2)$ is the only inflection point.

25. The graph is concave down on $(-\infty, -2)$ and is concave up on $(-2, 0)$ and $(0, \infty)$. The point $(-2, -12\sqrt[3]{4})$ is the inflection point.

27. The graph is concave up on $(-\infty, 0)$ and $(0, \infty)$. There are no inflection points.

29. Step 1: The domain is the interval $(-\infty, \infty)$.
    Step 2: The y-intercept is $(0, 1)$.
    Step 3: The graph is increasing on the intervals $(-\infty, 0)$ and $(4, \infty)$ and is decreasing on the interval $(0, 4)$.
    Step 4: There is a local maximum at $(0, 1)$, and there is a local minimum at $(4, -31)$.
    Step 5: The tangent line is horizontal at $(0, 1)$ and $(4, -31)$.
    Step 6: The end behavior is $y = x^3$.

Step 7: The graph is concave up on the interval $(2, \infty)$ and is concave down on the interval $(-\infty, 2)$. The point $(2, -15)$ is the only inflection point.

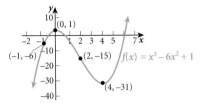

31. Step 1: The domain is the interval $(-\infty, \infty)$.
    Step 2: The x-intercepts are $(-1, 0)$ and $(1, 0)$, and the y-intercept is $(0, 1)$.
    Step 3: The graph is increasing on the intervals $(-1, 0)$ and $(1, \infty)$ and is decreasing on the intervals $(-\infty, -1)$ and $(0, 1)$.
    Step 4: There is a local maximum at $(0, 1)$, and there are local minima at $(-1, 0)$ and $(1, 0)$.
    Step 5: The tangent line is horizontal at $(-1, 0)$, $(0, 1)$, and $(1, 0)$.
    Step 6: The end behavior is $y = x^4$.
    Step 7: The graph is concave up on the intervals $(-\infty, -\frac{\sqrt{3}}{3})$ and $(\frac{\sqrt{3}}{3}, \infty)$ and is concave down on the interval $(-\frac{\sqrt{3}}{3}, \frac{\sqrt{3}}{3})$. The points $(-\frac{\sqrt{3}}{3}, \frac{4}{9})$ and $(\frac{\sqrt{3}}{3}, \frac{4}{9})$ are the inflection points.

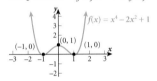

33. Step 1: The domain is the interval $(-\infty, \infty)$.
    Step 2: The x-intercepts are $(10, 0)$ and $(0, 0)$, and the y-intercept is $(0, 0)$.
    Step 3: The graph is increasing on the intervals $(-\infty, 0)$ and $(8, \infty)$ and is decreasing on $(0, 8)$.
    Step 4: There is a local maximum at $(0, 0)$, and there is a local minimum at $(8, -8192)$.
    Step 5: The tangent line is horizontal at $(0, 0)$ and $(8, -8192)$.
    Step 6: The end behavior is $y = x^5$.
    Step 7: The graph is concave up on the interval $(6, \infty)$ and is concave down on the interval $(-\infty, 6)$. The point $(6, -5184)$ is the only inflection point.

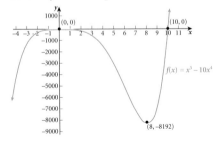

35. Step 1: The domain is the interval $(-\infty, \infty)$.
    Step 2: The x-intercepts are $(3, 0)$ and $(0, 0)$, and the y-intercept is $(0, 0)$.
    Step 3: The graph is increasing on the interval $(\frac{5}{2}, \infty)$ and is decreasing on the intervals $(-\infty, 0)$ and $(0, \frac{5}{2})$.
    Step 4: There is a local minimum at $\left(\frac{5}{2}, -\frac{3125}{64}\right)$.
    Step 5: The tangent line is horizontal at $(0, 0)$ and $\left(\frac{5}{2}, -\frac{3125}{64}\right)$.
    Step 6: The end behavior is $y = x^6$.
    Step 7: The graph is concave up on the intervals $(-\infty, 0)$, and $(2, \infty)$ and is concave down on the interval $(0, 2)$. The points $(0, 0)$ and $(2, -32)$ are the inflection points.

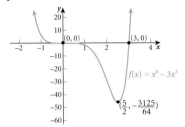

37. Step 1: The domain is the interval $(-\infty, \infty)$.
    Step 2: The x-intercepts are $(4, 0)$ and $(0, 0)$, and the y-intercept is $(0, 0)$.
    Step 3: The graph is increasing on the interval $(3, \infty)$ and is decreasing on the intervals $(-\infty, 0)$ and $(0, 3)$.
    Step 4: There is a local minimum at $(3, -81)$.
    Step 5: The tangent line is horizontal at $(0, 0)$ and $(3, -81)$.
    Step 6: The end behavior is $y = 3x^4$.
    Step 7: The graph is concave up on the intervals $(-\infty, 0)$ and $(2, \infty)$, and is concave down on the interval $(0, 2)$. The points $(0, 0)$ and $(2, -48)$ are the inflection points.

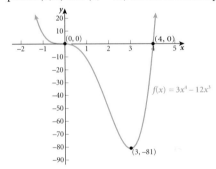

39. Step 1: The domain is the interval $(-\infty, \infty)$.
    Step 2: The y-intercept is $(0, 4)$.
    Step 3: The graph is increasing on the intervals $(-\infty, 0)$ and $(\sqrt[3]{4}, \infty)$ and is decreasing on the interval $(0, \sqrt[3]{4})$.
    Step 4: There is a local maximum at $(0, 4)$, and there is a local minimum at $(\sqrt[3]{4}, 4 - 12\sqrt[3]{2})$.

Step 5: The tangent line is horizontal at (0, 4) and $(\sqrt[3]{4}, 4-12\sqrt[3]{2})$.
Step 6: The end behavior is $y = x^5$.
Step 7: The graph is concave up on the interval $(1, \infty)$ and is concave down on the interval $(-\infty, 1)$. The point $(1, -5)$ is the only inflection point.

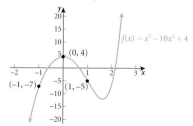

41. Step 1: The domain is the interval $(-\infty, \infty)$.
Step 2: The x-intercepts are $(10, 0)$ and $(0, 0)$, and the y-intercept is $(0, 0)$.
Step 3: The graph is increasing on the intervals $(-\infty, 0)$ and $(4, \infty)$ and is decreasing on the interval $(0, 4)$.
Step 4: There is a local maximum at $(0, 0)$, and there is a local minimum at $(4, -6\sqrt[3]{16})$.
Step 5: The tangent line is horisontal at $(4, -6\sqrt[3]{16})$ and is vertical at $(0, 0)$.
Step 6: The end behavior is $y = x^{5/3}$.
Step 7: The graph is concave up on the intervals $(-2, 0)$, $(0, \infty)$ and is concave down on the interval $(-\infty, -2)$. The point $(-2, -12\sqrt[3]{4})$ is the only inflection point.

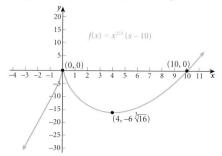

43. Step 1: The domain is the interval $(-\infty, \infty)$.
Step 2: The x-intercepts are $(-4, 0), (4, 0)$, and $(0, 0)$, and the y-intercept is $(0, 0)$.
Step 3: The graph is increasing on the intervals $(-2, 0)$ and $(2, \infty)$ and is decreasing on the intervals $(-\infty, -2)$ and $(0, 2)$.
Step 4: There is a local maximum at $(0, 0)$, and there are local minima at $(-2, -12\sqrt[3]{4})$ and $(2, -12\sqrt[3]{4})$.
Step 5: The tangent line is horizontal at $(-2, -12\sqrt[3]{4})$ and $(2, -12\sqrt[3]{4})$ and is vertical at $(0, 0)$.
Step 6: The end behavior is $y = x^{8/3}$.

Step 7: The graph is concave up on the intervals $(-\infty, 0)$ and $(0, \infty)$. There is no inflection point.

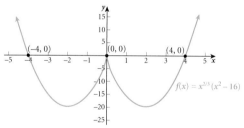

45. Step 1: The domain is the interval $(-\infty, \infty)$.
Step 2: The x-intercept and the y-intercept are both $(0, 0)$.
Step 3: The graph is increasing on the interval $(-1, \infty)$ and is decreasing on the interval $(-\infty, -1)$.
Step 4: There is a local minimum at $(-1, -\frac{1}{e})$.
Step 5: The tangent line is horizontal at $(-1, -\frac{1}{e})$.
Step 6: The end behavior is $y = xe^x$. $\lim_{x \to \infty} xe^x = \infty$ and. $\lim_{x \to -\infty} xe^x = 0$. The line $y = 0$ is a horizontal asymptote.
Step 7: The graph is concave up on the interval $(-2, \infty)$ and is concave down on the interval $(-\infty, -2)$. The point $\left(-2, -\frac{2}{e^2}\right)$ is the only inflection point.

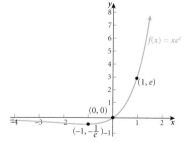

47. There is a local maximum at $(-1, 4)$ and a local minimum at $(1, 0)$.

49. There is a local minimum at $(-1, -4)$.

51. There is a local maximum at $(0, 2)$ and a local minimum at $(4, -254)$.

53. There is a local maximum at $(-1, -2)$ and a local minimum at $(1, 2)$.

55. Answers will vary.   57. Answers will vary.

59. $a = -3, b = 9$

61. (a) $\overline{C}(x) = 2x + \frac{50}{x}$
(b) The minimum average cost per item is $20.
(c) $C'(x) = 4x$

(d)
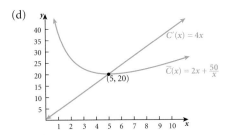

Point of intersection is (5, 20).

(e) The minimum average costs occurs at the production level for which the average cost equals the marginal cost.

63. (a) $\overline{C}(x) = \frac{500}{x} + 10 + \frac{x}{500}$

(b) The minimum average cost per item is $12.

(c) $C'(x) = 10 + \frac{x}{250}$

(d)
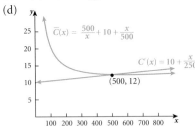

Point of intersection is (500, 12).

(e) The minimum average costs occurs at the production level for which the average cost equals the marginal cost.

65. (a) The domain of N is the interval $[0, \infty)$.

(b) The N-intercept is $(0, 1)$. There is no t-intercept.

(c) N is increasing on the interval $(0, \infty)$.

(d) N is concave up on the interval $(0, \ln(49, 999))$ and is concave down on the interval $(\ln(49, 999), \infty)$.

(e) The inflection point is $(\ln(49, 999), 25, 000))$.

(f)
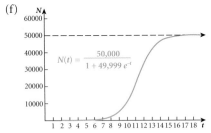

(g) The rumor is spreading at its greatest rate at $t = \ln(49, 999) \approx 10.82$ units of time since the rumor began to spread.

67. (a) $p'(t) = \frac{72,000 e^t}{(e^t + 9)^2}$

(b) The maximum growth rate occurs at $t = \ln(9) \approx 2.20$ days.

(c) The equilibrium population is 8000.

(d)
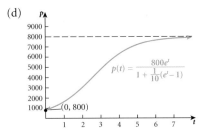

69. (a) The sales rate is a maximum at $x = \ln(50) \approx 3.91$ months.

(b)
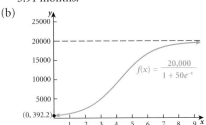

Exercise 5.4 (p. 391)

1. The absolute maximum is 15, and the absolute minimum is -1.

3. The absolute maximum is 1, and the absolute minimum is -39.

5. The absolute maximum is 16, and the absolute minimum is -4.

7. The absolute maximum is 1, and the absolute minimum is 0.

9. The absolute maximum is 1, and the absolute minimum is 0.

11. The absolute maximum is 4, and the absolute minimum is 2.

13. The absolute maximum is $\frac{1}{2}$, and the absolute minimum is $-\frac{1}{2}$.

15. The absolute maximum is 0, and the absolute minimum is $-\frac{1}{2}$.

17. The absolute maximum is $98\sqrt[3]{2}$, and the absolute minimum is 0.

19. The absolute maximum is $\frac{\sqrt[3]{12}}{9}$, and the absolute minimum is $-\sqrt[3]{2}$.

21. The absolute maximum is $10e^{10}$, and the absolute minimum is $-\frac{1}{e}$.

23. The absolute maximum is $\frac{1}{e}$, and the absolute minimum is 0.

25. The dimensions are 2 cm × 8 cm × 8 cm.

27. The dimensions are $20\sqrt[3]{2}$ cm × $20\sqrt[3]{2}$ cm × $10\sqrt[3]{2}$ cm ≈ 25.2 cm × 25.2 cm × 12.6 cm.

29. The radius is $\frac{4\sqrt{25\pi^2}}{\pi} \approx 7.99$ cm, and the height is $\frac{10\sqrt{25\pi^2}}{\pi} \approx 19.96$ cm.

31. The company should connect the telephone line 1.98 km from the box.

33. The most economical speed is 40 miles per hour.

35. The dimensions are 7 inches by 14 inches.

37.

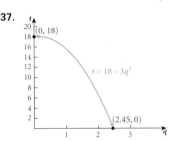

Demand for the product decreases as the tax rate increases. The optimal tax rate is 12%, and the revenue generated by this tax rate is 16.97 monetary units.

39. Let r be the radius, h be the height, and S be the surface area of the cylinder of volume V. We know that $V = \pi r^2 h$, and hence $h = \frac{V}{\pi r^2}$. The formula for S is
$S = 2\pi r^2 + 2\pi r h = 2\pi r^2 + \frac{2V}{r}$. Now $S'(r) = 4\pi r - \frac{2V}{r^2}$.

The only critical number of S is $r_c = \left(\frac{V}{2\pi}\right)^{1/3}$.
Furthermore, $S'(r) < 0$ if $0 < r < r_c$, and $S'(r) > 0$ if $r > r_c$. Therefore, by the First Derivative Test, the surface area is least when $r = r_c = \left(\frac{V}{2\pi}\right)^{1/3}$. The height of the cylinder when $r = r_c$ is $h = \frac{V}{\pi r_c^2} = \left(\frac{V}{\pi}\right)\left(\frac{2\pi}{V}\right)^{2/3}$
$= 2\left(\frac{V}{2\pi}\right)^{1/3} = 2r_c$, as desired.

41. The concentration is greatest two hours after the injection.

Exercise 5.5 (p. 396)

1. (a) $x = 4000 - 100p$ (b) $E(p) = \frac{p}{p - 40}$
(c) $E(5) = -0.143$, The demand decreases by approximately 1.43%.
(d) $E(15) = -0.6$, The demand decreases by approximately 6%.
(e) $E(20) = -1$, The demand decreases by approximately 10%.

3. (a) $x = 10{,}000 - 200p$ (b) $E(p) = \frac{p}{p - 50}$

(c) $E(10) = -0.25$, The demand decreases by approximately 1.25%.
(d) $E(25) = -1$, The demand decreases by approximately 5%.
(e) $E(35) = -2.333$, The demand decreases by approximately 11.665%.

5. $E(p) = \frac{p}{p - 200}$, $E(50) = -0.333$, The demand is inelastic.

7. $E(p) = -\frac{p}{p + 4}$, $E(10) = -0.714$, The demand is inelastic.

9. $E(p) = \frac{2p^2}{p^2 - 1000}$, $E(10) = -0.222$, The demand is inelastic.

11. $E(p) = \frac{p}{2p - 200}$, $E(10) = -0.056$, The demand is inelastic.

13. $E(p) = \frac{3p}{p - 4}$, $E(2) = -3$, The demand is elastic.

15. $E(p) = \frac{3\sqrt{p}}{6\sqrt{p} - 40}$, $E(4) = -0.214$, The demand is inelastic.

17. $E(4) = -1.23$ **19.** $E(5) = -1.18$

21. $E(5) = -39$ **23.** $E(2) = -0.125$

25. $E(100) = -3$

27. (a) The demand is elastic.
(b) The revenue will decrease.

29. (a) The demand is inelastic.
(b) The revenue will decrease.

31. (a) $x = 3000 - \frac{200}{3}p$
(b) $E(18) = -0.667$
(c) The demand will decrease by approximately 3.33%.
(d) The revenue will increase.

Exercise 5.6 (p. 404)

1. $\frac{dx}{dt} = -3$ **3.** $\frac{dx}{dt} = -\frac{8}{9}$

5. $\frac{dV}{dt} = 40$ **7.** $\frac{dV}{dt} = 5\pi^2$

9. The volume is increasing at a rate of 900 cubic centimeters per second.

11. The length of the leg of side length y is decreasing at a rate of $-\frac{8\sqrt{2009}}{2009} \approx 0.178$ centimeters per minute.

13. The surface area is shrinking at a rate of -0.75 square meters per minute.

15. The water level is rising at a rate of $\frac{1}{30}$ meter per minute.

17. The area of the spill is increasing at a rate of 316.67 square feet per minute.

19. (a) The rate of change in daily cost is $500 per day.
(b) The rate of change in daily revenues is $1480 per day.
(c) Revenue is increasing.
(d) $P(x) = -\frac{x^2}{10{,}000} + 10x - 5000$
(e) The rate of change in daily profit is $980 per day.

21. Revenues are increasing at a rate of $260,000 per year.

Exercise 5.7 (p. 411)

1. $dy = (3x^2 - 2)dx$ **3.** $dy = \frac{-x^2 + 2x - 6}{(x^2 + 2x - 8)^2}dx$

5. $\frac{dy}{dx} = -\frac{y}{x}, \frac{dx}{dy} = -\frac{x}{y}$ **7.** $\frac{dy}{dx} = -\frac{x}{y}, \frac{dx}{dy} = -\frac{y}{x}$

9. $\frac{dy}{dx} = \frac{2xy - x^2}{y^2 - x^2}, \frac{dx}{dy} = \frac{y^2 - x^2}{2xy - y^2}$

11. $d(\sqrt{x-2}) = \frac{1}{2\sqrt{x-2}}dx$

13. $d(x^3 - x - 4) = (3x^2 - 1)dx$

15. $y = 2x - 3$

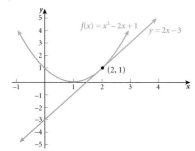

17. $y = \frac{1}{4}x + 1$

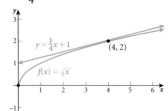

19. $y = x + 1$

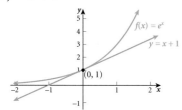

21. (a) The change is approximately 0.006.
(b) The change is approximately 0.00125.

23. The approximate increase in surface area is $2\pi \approx$ 6.28 square centimeters.

25. The approximate increase in volume is $3.6\pi \approx$ 11.31 cubic meters.

27. The percentage error is 6%.

29. The approximate loss is $0.8\pi \approx 25.13$ cubic centimeters.

31. The estimated height is 30 meters, and the percentage error of this estimate is 0.9%.

33. The clock will lose 70.28 minutes each day.

CHAPTER 5   Review

True – False Items (p. 414)

1. False   **2.** True   **3.** False
4. False   **5.** True   **6.** False

Fill in the Blanks (p. 414)

1. decreasing   **2.** decreasing, increasing
3. concave up   **4.** horizontal
5. concavity   **6.** $f'(x)\, dx$
7. linear approximation

Review Exercises (p. 414)

1. The graph has no horizontal or vertical tangent lines.

3. The graph has a horizontal tangent line at $\left(2, \frac{\sqrt[3]{2}}{6}\right)$ and a vertical tangent line at $(0, 0)$.

5. (a) Yes   (b) No   (c) vertical tangent line
7. (a) Yes   (b) No   (c) no tangent line

9. (a) The graph is increasing on the intervals $(-\infty, -2)$ and $(2, \infty)$ and is decreasing on the interval $(-2, 2)$.
(b) There is a local maximum at $\left(-2, \frac{48}{5}\right)$ and a local minimum at $\left(2, -\frac{48}{5}\right)$.

11. (a) The graph is increasing on the intervals $(-\infty, -2\sqrt{2})$ and $(-2\sqrt{2}, 0)$ and is decreasing on the intervals $(0, 2\sqrt{2})$ and $(2\sqrt{2}, \infty)$.
(b) The graph has a local maximum at $(0, 0)$.

13. (a) The graph is decreasing on $(-\infty, \infty)$.
(b) The graph has no local maxima or minima.

15. (a) The domain is the interval $(-\infty, \infty)$.
(b) The x-intercept is $(1, 0)$, and the y-intercept is $(0, -1)$.
(c) The graph is increasing on the interval $(-\infty, \infty)$.
(d) The graph has no local maxima or minima.

(e) The graph has a horizontal tangent at $(1, 0)$.
(f) The end behavior is $y = x^3$.
(g) The graph is concave down on the interval $(-\infty, 1)$ and is concave up on the interval $(1, \infty)$. The point $(1, 0)$ is the only inflection point.

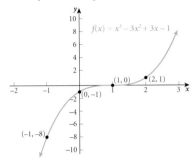

17. (a) The domain is the interval $(-\infty, \infty)$.
(b) The x-intercepts are $(-\sqrt[4]{5}, 0)$, $(0, 0)$, and $(-\sqrt[4]{5}, 0)$, and the y-intercept is $(0, 0)$.
(c) The graph is increasing on the intervals $(-\infty, -1)$ and $(1, \infty)$ and is decreasing on the interval $(-1, 1)$.
(d) The graph has a local maximum at $(-1, 4)$ and a local minimum at $(1, -4)$.
(e) The tangent lines to the graph are horizontal at $(-1, 4)$ and $(1, -4)$.
(f) The end behavior is $y = x^5$.
(g) The graph is concave down on the interval $(-\infty, 0)$ and is concave up on the interval $(0, \infty)$. The point $(0, 0)$ is the only inflection point.

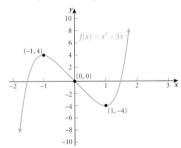

19. (a) The domain is the interval $(-\infty, \infty)$.
(b) The x-intercepts are $(-4, 0)$ and $(0, 0)$. The y-intercept is $(0, 0)$.
(c) The graph is increasing on the interval $(-1, \infty)$ and is decreasing on the interval $(-\infty, -1)$.
(d) The graph has a local minimum at $(-1, -3)$.
(e) The graph has a horizontal tangent at $(-1, -3)$ and a vertical tangent at $(0, 0)$.
(f) The end behavior is $y = x^{4/3}$.
(g) The graph is concave up on the intervals $(-\infty, 0)$ and $(2, \infty)$ and is concave down on the interval $(0, 2)$.

The points $(0, 0)$ and $(2, 6\sqrt[3]{2})$ are the inflection points.

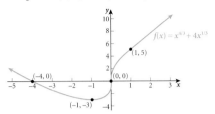

21. (a) The domain is the interval $(-\infty, \infty)$.
(b) The x-intercept and the y-intercept are both $(0, 0)$.
(c) The graph is increasing on the interval $(-1, 1)$ and is decreasing on the intervals $(-\infty, -1)$ and $(1, \infty)$.
(d) The graph has a local maximum at $(1, 1)$ and a local minimum at $(-1, -1)$.
(e) The graph has horizontal tangent lines at $(-1, -1)$ and $(1, 1)$.
(f) The end behavior is $y = 0$, which is a horizontal asymptote.
(g) The graph is concave down on the intervals $(-\infty, -\sqrt{3})$ and $(0, \sqrt{3})$ and is concave up on the intervals $(-\sqrt{3}, 0)$ and $(\sqrt{3}, \infty)$. The points $\left(-\sqrt{3}, -\frac{\sqrt{3}}{2}\right)$, $(0, 0)$, and $\left(\sqrt{3}, \frac{\sqrt{3}}{2}\right)$ are the inflection points.

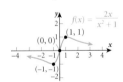

23. There is a local maximum at $\left(-\frac{1}{2}, 1\right)$ and a local minimum at $\left(\frac{1}{2}, -1\right)$.

25. There is a local maximum at $(0, 0)$ and local minima at $(-1, -1)$ and $(1, -1)$.

27. There is a local minimum at $\left(-1, -\frac{1}{e}\right)$.

29. The absolute maximum is 8, and the absolute minimum is -1.

31. The absolute maximum is 9, and the absolute minimum is 0.

33. The absolute maximum is 8, and the absolute minimum is -3.

35. $E(p) = \dfrac{2p^2}{p^2 - 500}$, The demand is elastic at $p = 20$.

37. $E(p) = \dfrac{p^2}{p^2 - 500}$, The demand is inelastic at $p = 10$.

39. (a) The demand function is elastic.
(b) Revenue will decrease if the price is raised.

41. $dy = (12x^3 - 6x^2 + 1)dx$ **43.** $dy = -\dfrac{5}{(x+1)^2}dx$

45. $y = 6x - 18$

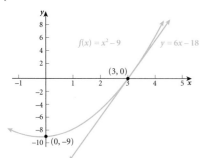

47. $\dfrac{dx}{dt} = -3$ **49.** $\dfrac{dy}{dt} = -\dfrac{9}{29}$

51. The surface area is increasing at a rate of $\dfrac{20}{3}$ square meters per minute.

53. (a) $\overline{C}(x) = 5x + \dfrac{1125}{x}$

(b) The minimum average cost is $150.

(c) $C'(x) = 10x$.

(d)

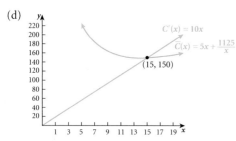

Intersection point is (15,150).

(e) The minimum average cost occurs at the production level where the average cost equals the marginal cost.

55. 98 units need to be sold.

57. The radius is $\dfrac{5\sqrt[3]{2\pi^2}}{\pi} \approx 4.30$ centimeters, and the height is $\dfrac{10\sqrt[3]{2\pi^2}}{\pi} \approx 8.60$ centimeters.

59. The decrease in area is approximately $40\pi \approx 125.7$ square millimeters.

61. (a) The demand decreases by approximately 1800 pounds.
(b) The demand decreases by approximately 9000 pounds.

63. (a) The concentration increases by approximately 0.0071 units.
(b) The concentration decreases by approximately 0.0208 units.

65. (b)

Mathematical Questions from Professional Exams (p. 418)

1. (b) $6X^2 + 8X + 3$ **2.** (d) $4X^2 + 6X + 2 + \dfrac{10}{X}$

3. (a) The average cost function multiplied by X.

4. (c) Substitute the solution(s) in the second derivative equation, and a positive solution indicates a minimum.

5. (a) $y = x^3 e^{x^2}$ **6.** (b) Dec, 96

7. (d) 8,645 per year   **8.** (d) 1, −2

9. (c) (0, 2)   **10.** (c) $\dfrac{1}{3}$

11. (c) $\dfrac{7e^3}{4}$ **12.** (b) $-\dfrac{1}{2}$

13. (a) $\dfrac{1}{2\pi}$ **14.** (b) 0

15. (d) 54   **16.** (b) $-\dfrac{16}{5}$

CHAPTER 6 The Integral of a Function and Applications

Exercise 6.1 (p. 430)

1. $F(x) = \dfrac{x^4}{4} + K$

3. $F(x) = x^2 + 3x + K$

5. $F(x) = 4\ln|x| + K$

7. $F(x) = \dfrac{2\sqrt{3}}{3} X^{3/2} + K$

9. $3x + K$

11. $\dfrac{x^2}{2} + K$

13. $\dfrac{3x^{4/3}}{4} + K$

15. $-\dfrac{1}{x} + K$

17. $2x^{1/2} + K$

19. $\dfrac{x^4}{2} + \dfrac{5x^2}{2} + K$

21. $\dfrac{x^3}{3} + 2e^x + K$

23. $\dfrac{x^4}{4} - \dfrac{2x^3}{3} + \dfrac{x^2}{2} - x + K$

25. $x - \ln|x| + K$

27. $2e^x - 3\ln|x| + K$

29. $3x + 2\sqrt{x} + K$

31. $\dfrac{x^2}{2} - 2x + K$

33. $\dfrac{x^3}{3} - \dfrac{x^2}{2} + K$

35. $\dfrac{3x^5}{5} + 2\ln|x| + K$

37. $4x + e^x + K$

39. $R(x) = 600x$

41. $R(x) = 10x^2 + 5x$

43. $C(x) = 7x^2 - 2800x + 4300$, The cost is a minimum when $x = 200$.

45. $C(x) = 10x^2 - 8000x + 500$, The cost is a minimum when $x = 400$.

47. (a) $C(x) = 9000 + 1000x - 10x^2 + \frac{x^3}{3}$
(b) $R(x) = 3400x$
(c) $P(x) = -9000 + 2400x + 10x^2 - \frac{x^3}{3}$
(d) A sales volume of 60 units yields maximum profit.
(e) The profit is $99,000.
(f)

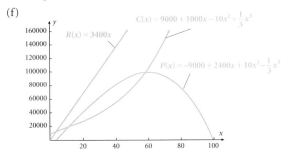

49. There will be 1,142,462 inmates in 2008.

51. The population will be 20,055 people in 10 months.

53. There will be 22,700 voting citizens in 3 years.

55. $2 - 2\ln 2 \approx 0.614$ milligrams were produced.

57. The reservoir will be empty after 500.28 days.

Exercise 6.2 (p. 436)

1. $\frac{(2x+1)^6}{12} + K$ **3.** $\frac{e^{2x-3}}{2} + K$

5. $\frac{1}{6-4x} + K$ **7.** $\frac{(x^2+4)^3}{6} + K$

9. $\frac{e^{x^2+1}}{3} + K$ **11.** $\frac{1}{2}(e^x - e^{-x}) + K$

13. $\frac{(x^3+2)^7}{21} + K$ **15.** $\frac{3(1+x^2)^{2/3}}{4} + K$

17. $\frac{2(x+3)^{5/2}}{5} - 2(x+3)^{3/2} + K$

19. $\ln(e^x + 1) + K$ **21.** $2e^{\sqrt{x}} + K$

23. $\frac{3(x^{1/3}-1)^7}{7} + K$ **25.** $-\frac{1}{2x^2+4x+6} + K$

27. $-\frac{2}{3(1+\sqrt{x})^3} + K$ **29.** $\frac{\ln|2x+3|}{2} + K$

31. $\frac{\ln(4x^2+1)}{8} + K$ **33.** $\frac{\ln(x^2+2x+2)}{2} + K$

35. The value of the car after two years is $19,414.55. The value of the car after four years is $16,624.02.

37. (a) $B(t) = 68.6e^{0.025t}$
(b) The budget will exceed $100 billion when $t = 15.08$, which is during January, 2016.

39. (a) $N(t) = 2000\,e^{0.01t} - 1600$
(b) The work force will reach 800 employees in 18.23 years.

41. Let $u = ax + b$. Then $du = a\,dx$, so $\frac{1}{a}du = dx$.
Substituting, we have
$\int (ax+b)^n\,dx = \int u^n \cdot \frac{1}{a}du$
$= \frac{1}{a}\int u^n\,du$
$= \frac{1}{a}\left[\frac{1}{n+1}u^{n+1}\right] + K$
$= \frac{(ax+b)^{n+1}}{a(n+1)} + K$

Exercise 6.3 (p. 441)

1. $\frac{xe^{4x}}{4} - \frac{e^{4x}}{16} + K$ **3.** $\frac{xe^{2x}}{2} - \frac{e^{2x}}{4} + K$

5. $-(x^2+2x+2)e^{-x} + K$ **7.** $\frac{2x^{3/2}\ln x}{3} - \frac{4x^{3/2}}{9} + K$

9. $x(\ln x)^2 - 2x\ln x + 2x + K$

11. $\frac{x^3 \ln 3x}{3} - \frac{x^3}{9} + K$ **13.** $\frac{x^3(\ln x)^2}{3} - \frac{2x^3 \ln x}{9} + \frac{2x^3}{27} + K$

15. $\frac{(\ln x)^2}{2} + K$

17. The population is 5389 ants after four days and is 6012 ants after one week.

19. The car is worth $14,061.64 after 2 years.
The car is worth $9,640.02 after 4 years.

Exercise 6.4 (p. 449)

1. $\frac{7}{2}$ **3.** e **5.** $\frac{2}{3}$

7. $-\frac{1}{15}$ **9.** $\frac{5}{3}$ **11.** $\frac{5}{24}$

13. $3\ln 2 + \frac{95}{3}$ **15.** $\frac{20}{3}$ **17.** 0

19. $\frac{8}{3}$ **21.** $\frac{e^2-3}{2}$ **23.** $1 - \frac{1}{e}$

25. $\ln 2$ **27.** $\frac{2\ln 2}{3}$ **29.** $\frac{5e^6}{4} - \frac{e^2}{4}$

31. $\frac{4}{9e^3} - \frac{7}{9e^6}$ **33.** $5\ln 5 - 4$ **35.** 0

37. 0 **39.** 2 **41.** 64

43. 36 **45.** 12

47. The cost increase is 567,000.

49. (a) The total number of labor-hours needed is 15,921.
(b) The total number of labor-hours needed is 35,089.
(c) The quantity k represents the difficulty of learning a new skill. The closer k is to 0, the longer the time required to master the new skill.

51. The projected deficit is $197 billion.

53. The total sales during the first year were $13,450.01.

55. The total number of labor-hours needed is 3660.

57. (a) $\frac{2}{3}$
(b) $\frac{16}{15}$

Exercise 6.5 (p. 462)

1. 56

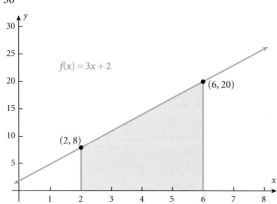

3. $\frac{8}{3}$

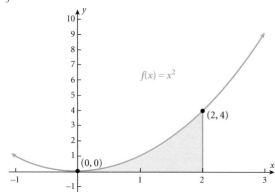

5. $\frac{8}{3}$

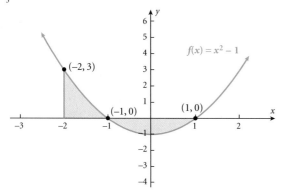

7. $\frac{51}{4}$

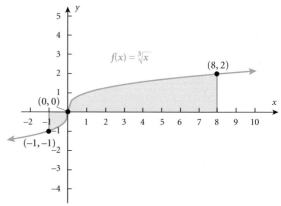

9. $e - 1$

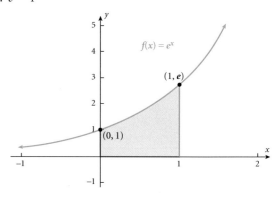

11. $\frac{1}{2}$

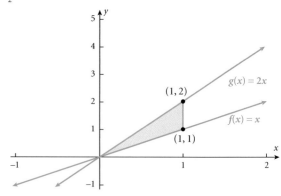

Exercise 6.5 AN-53

13. $\frac{1}{6}$

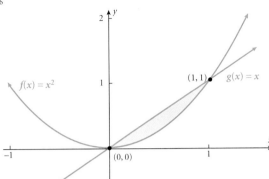

15. $\frac{1}{6}$

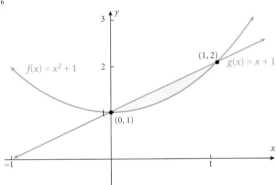

17. $\frac{5}{12}$

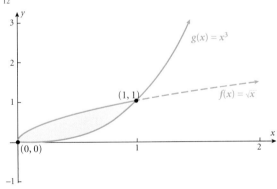

19. $\frac{4}{15}$

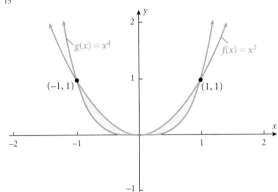

21. $\frac{8}{3}$

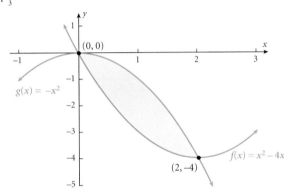

23. $\frac{9}{2}$

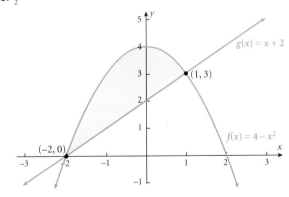

25. 8

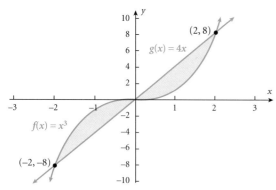

27. $\frac{1}{3}$

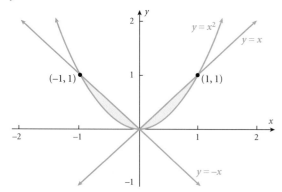

29. (a) The integral represents area below the line $y = 3x + 1$, above the the x-axis, and between the vertical lines $x = 0$ and $x = 4$.

(b)
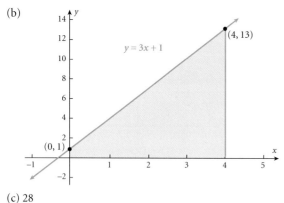

(c) 28

31. (a) The integral represents area below the curve $y = x^2 - 1$, above the the x-axis, and between the vertical lines $x = 2$ and $x = 5$.

(b)
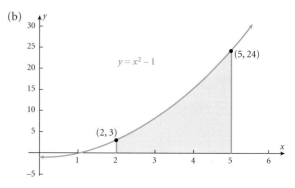

(c) 36

33. (a) The integral represents area below the curve $y = e^x$, above the the x-axis, and between the vertical lines $x = 0$ and $x = 2$.

(b)
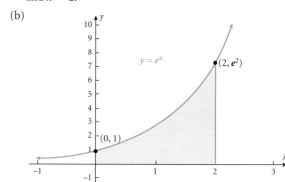

(c) $e^2 - 1$

35. The consumer's surplus is $\$\frac{40}{9} \approx \$4.44$, and the producer's surplus is $\$\frac{32}{9} \approx \$3.56$.

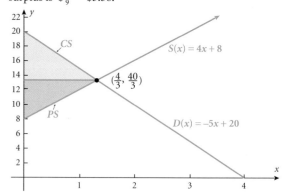

37. The operation should continue for 16 years. The profit that can be generated during this period is $85.33 million.

39. (a) $c = \frac{\sqrt{3}}{3}$

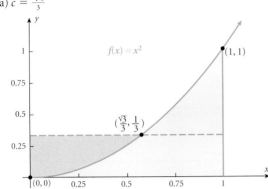

(b) $c = 2$

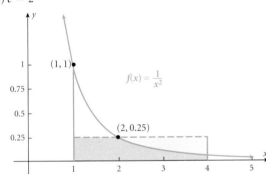

41. (a) x^2

(b) $\sqrt{x^2 - 2}$

(c) $\sqrt{x^x + 2x}$

Exercise 6.6 (p. 468)

1. $A \approx 3$

3. $A \approx 56$

5. (a)

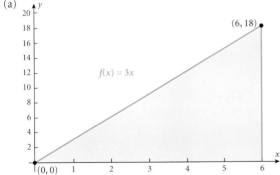

(b) $A \approx 36$
(c) $A \approx 72$
(d) $A \approx 45$
(e) $A \approx 63$
(f) $A = 54$

7. (a)

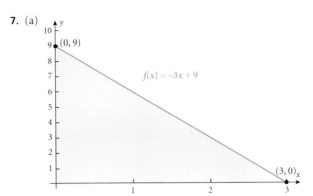

(b) $A \approx 18$
(c) $A \approx 9$
(d) $A \approx \frac{63}{4}$
(e) $A \approx \frac{45}{4}$
(f) $A = \frac{27}{2}$

9. (a)

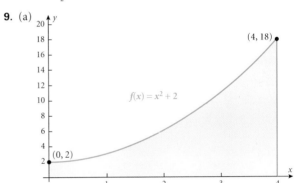

(b) $A \approx 22$
(c) $A \approx \frac{51}{2}$
(d) $A = \int_0^4 (x^2 + 2)dx$
(e) $A = \frac{88}{3}$

11. (a)

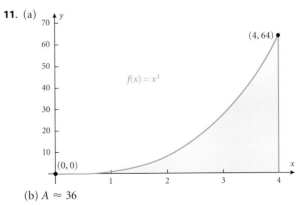

(b) $A \approx 36$

(c) $A \approx 49$

(d) $A = \int_0^4 x^3 \, dx$

(e) $A = 64$

13. (a)

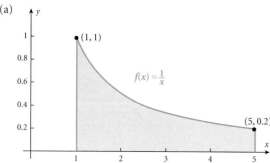

(b) $A \approx \frac{25}{12}$

(c) $A \approx \frac{4609}{2520} \approx 1.829$

(d) $A = \int_1^5 \frac{1}{x} \, dx$

(e) $A = \ln 5$

15. (a)

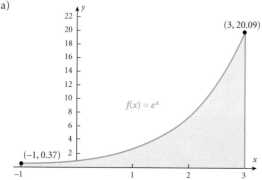

(b) $A \approx 11.475$

(c) $A \approx 15.197$

(d) $A = \int_{-1}^3 e^x \, dx$

(e) $A = e^3 - \frac{1}{e}$

17. 1.46

19. 38.29

21. (a)

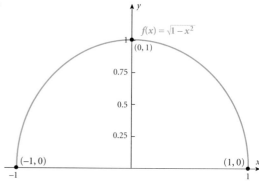

(b) $A \approx 1.42$

(c) $A \approx 1.52$

(d) $A = \int_{-1}^1 \sqrt{1 - x^2} \, dx$

(e) $A = 1.57$

(f) The area is $\frac{\pi}{2}$.

Exercise 6.7 (p. 476)

1. $y = \frac{x^3}{3} - x$

3. $y = \frac{x^3}{3} - \frac{x^2}{2} - \frac{3}{2}$

5. $y = \frac{x^4}{4} - \frac{x^2}{2} + 2x + 3$

7. $y = e^x + 3$

9. $y = \frac{x^2}{2} + x + \ln x - \frac{3}{2}$

11. There are 12,975 bacteria after one hour. There are 147,789 bacteria after 90 minutes. It will take 113.58 minutes to reach 1,000,000 bacteria.

13. There will be 7.68 grams of radium

15. The tree died 9727 years ago.

17. The population size is 6944 mosquitoes.

19. There are 55,418 bacteria.

21. (a) $P(t) = 10,000 \cdot 5^{t/10}$ bacteria

(b) There were 250,000 bacteria.

(c) There were 20,000 bacteria after $t_1 = \frac{10 \ln 2}{\ln 5} \approx 4.31$ minutes.

23. The half-life is 4,620,981 years.

25. (a) $p(x) = 300 \left(\frac{1}{2}\right)^{x/200}$

(b) A price of $106.07 should be charged.

(c) A price of $89.19 should be charged.

CHAPTER 6 Review

True – False Items (p. 478)
1. True 2. False 3. False
3. False 4. False 5. False
6. True 7. True 8. True
9. False 10. True

Fill in the Blanks (p. 479)
1. $F'(x) = f(x)$ 2. $\int f(x)dx$
3. integration by parts
4. lower, upper limits, integration 5. 0
6. $F(b) - F(a)$ 7. $\int_0^2 \sqrt{x^2 + 1}\ dx$

Review Exercises (p. 479)
1. $F(x) = x^6 + K$ 3. $F(x) = \frac{x^4}{4} + \frac{x^2}{2} + K$
5. $F(x) = 2\sqrt{x} + K$ 7. $7x + K$
9. $\frac{5x^4}{4} + 2x + K$ 11. $\frac{x^5}{5} - x^3 + 6x + K$
13. $3\ln|x| + K$ 15. $\ln|x^2 - 1| + K$
17. $\frac{e^{3x}}{3} + K$ 19. $\frac{(x^3 + 3x)^6}{18} + K$
21. $\frac{2x^3}{3} - 3x^2 + K$ 23. $e^{3x^2 + x} + K$
25. $\frac{2(x-5)^{5/2}}{5} + \frac{10(x-5)^{3/2}}{3} + K$ 27. $\frac{xe^{4x}}{4} - \frac{e^{4x}}{16} + K$
29. $-\frac{\ln(2x)}{x} - \frac{1}{x} + K$ 31. $R(x) = \frac{5x^2}{2} + 2x$
33. $C(x) = \frac{5x^2}{2} + 120{,}000x + 7500$; the cost is minimum at a production level of zero.
35. (a) $R(x) = 500x - 0.005x^2$
   (b) 50,000 televisions need to be sold.
   (c) The maximum revenue is $12,500,000.
   (d) The increase in revenue is $625,000.
37. $-\frac{9}{2}$ 39. $\frac{304}{3}$ 41. $e + \frac{1}{e} - 2$
43. $\frac{1}{7}$ 45. $\frac{e^6 - e^{-6}}{3}$ 47. $3 - \frac{4}{e}$
49. 1 51. -10

53. 15

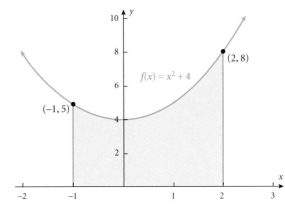

55. $e - \frac{1}{2}$

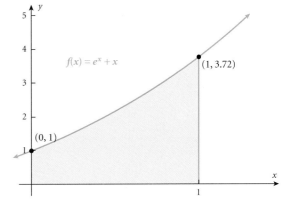

57. $\frac{31}{6}$

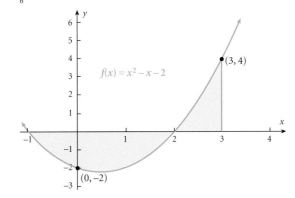

59. $\frac{125}{6}$

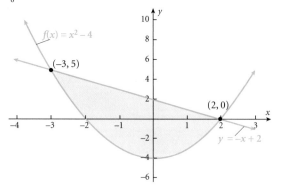

61. 8

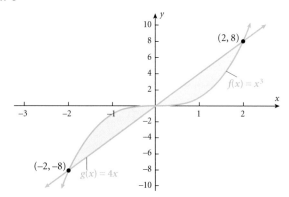

63. The monthly profit stays the same. The increase in profit is 0.

65. (a) 1 time unit
(b) 6 monetary units

67. (a)

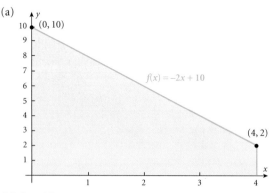

(b) $A \approx 28$
(c) $A \approx 20$

(d) $A \approx 26$
(e) $A \approx 22$
(f) $A = \int_0^4 (-2x + 10)dx$
(g) $A = 24$

69. 90.38

71. $y = \frac{x^3}{3} + \frac{5x^2}{2} - 10x + 1$

73. $y = \frac{e^{2x}}{2} - \frac{x^2}{2} + \frac{5}{2}$

75. $y = e^{10x}$

77. There will be 23,689 bacteria.

79. The burial ground is 7403 years old.

81. (a) $E(t) = \frac{t^3}{150} + \frac{t^2}{2} + 5$
(b) The economy will total $18.33 million.

83. Margo should allow for 5432 labor-hours.

85. (a) The market price is $10, and the demand level is 100 units.
(b) The consumer's surplus is $100, and the producer's surplus is $250.
(c)

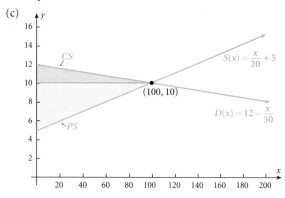

87. (a) $p(x) = 800 \left(\frac{3}{4}\right)^{x/80}$
(b) A price of $389.71 should be charged.
(c) A price of $325.58 should be charged.

Mathematical Questions from Professional Exams (p. 484)

1. (c) 1

2. (d) $-\frac{1}{4}$

3. (c) $x^2 e^x + 2xe^x$

4. (e) None of the above

5. (c) 62,208

6. (c) $(\ln x)^2 + C$

CHAPTER 7 Other Applications and Extensions of the Integral

Exercise 7.1 (p. 489)

1. This integral is improper because the upper limit of integration is infinite.
3. This integral is improper because the integrand is discontinuous at the lower limit of integration.
5. This integral is improper because the integrand is discontinuous at the lower limit of integration.
7. $\frac{e^{-4}}{4}$
9. The improper integral has no value.
11. $-\frac{5}{4}$
13. The improper integral has no value.
15. Area $= 2$ square units
17. The capital value of the apartment is $102,480.
19. (a) The area is equal to the integral, $\int_0^\infty r(t)\,dt$, which is the total reaction.
    (b) This total reaction is $\frac{1}{2}$ units.
21. (a) The improper integral has no value.
    (b) $\frac{3(\sqrt[3]{18} - \sqrt[3]{2})}{2}$

Exercise 7.2 (p. 492)

1. $\frac{1}{3}$
3. $\frac{2}{3}$
5. 9
7. -26
9. $e - 1$
11. The average population would be $8.22 \cdot 10^9$ people.
13. The average temperature is 37.5 degrees Celsius.
15. The average speed is 12 meters per second.
17. The average annual revenue is $207.32 billion.
19. The average rainfall is 0.188118 inches.

Exercise 7.3 (p. 501)

1. $f(x) = \frac{1}{2} > 0$ on $[0, 2]$. $\int_0^2 f(x)\,dx = 1$.
3. $f(x) = 2x \geq 0$ on $[0, 1]$. $\int_0^1 f(x)\,dx = 1$.
5. $f(x) = \frac{3}{250}(10x - x^2) \geq 0$ on $[0, 5]$. $\int_0^5 f(x)\,dx = 1$.
7. $f(x) = \frac{1}{x} > 0$ on $[1, e]$. $\int_1^e f(x)\,dx = 1$.

9. $k = \frac{1}{3}$
11. $k = \frac{1}{2}$
13. $k = \frac{3}{250}$
15. $k = \frac{1}{\ln 2}$
17. $E(x) = 1$
19. $E(x) = \frac{2}{3}$
21. $E(x) = \frac{25}{8}$
23. $E(x) = e - 1$
25. The probability is $\frac{2}{5}$.
27. The probability is $\frac{1}{e^3} \approx 0.0498$.
29. The probability is 0.865.
31. (a) The probability is 0.0737.
    (b) The probability is 0.2865.
33. (a) The probability is 0.5276.
    (b) The probability is 0.0490.
    (c) The probability is 0.3867.
    (d) Answers will vary.
35. The probability is 0.2865.
37. (a) $f(x) = \frac{1}{10}e^{-x/10}$
    (b) The probability is 0.1954.
    (c) The probability is 0.2231.
    (d) The probability is 0.7769.
39. The expected waiting time is 9 minutes.
41. The contractor can be expected to be off by 2.29% on average.
43. (a) The probability is 0.242.
    (b) The probability is 0.516.
45. (a) Let x be the cost of a new car, in thousands of dollars.
$$f(x) = \begin{cases} \frac{1}{35}x - \frac{2}{7} & \text{if } 10 \leq x \leq 17 \\ -\frac{1}{15}x + \frac{4}{3} & \text{if } 17 < x \leq 20 \end{cases}$$
    (b) The probability is $\frac{5}{14} \approx 0.3571$.
    (c)

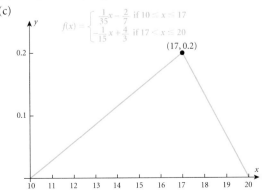

(d) The expected price is about $16,000.
(e) The expected price is $15,666.67.
(f) Answers will vary.

47. (a) Answers will vary.
(b) $Pr(0.6 \leq X < 0.9) = 0.3$.

49. Let $f(x) = \frac{1}{b-a}$ on $[a, b]$. Then

$$\sigma^2 = \int_a^b x^2 \cdot \frac{1}{b-a} dx - [E(x)]^2$$

$$= \int_a^b x^2 \cdot \frac{1}{b-a} dx - \left[\int_a^b x \cdot \frac{1}{b-a} dx\right]^2$$

$$= \frac{x^3}{3(b-a)}\Big|_a^b - \frac{x^2}{2(b-a)}\Big|_a^b{}^2$$

$$= \frac{b^2 + ab + a^2}{3} - \frac{b^2 + 2ab + a^2}{4}$$

$$= \frac{4b^2 + 4ab + 4a^2 - 3b^2 - 6ab - 3a^2}{12}$$

$$= \frac{(b-a)^2}{12}.$$

CHAPTER 7 Review

True – False Items (p. 506)
1. False   **2.** True   **3.** False

Fill in the Blanks (p. 506)
1. average value
2. random variable
3. probability density
4. $\lim_{b \to \infty} \int_2^b f(x)\, dx$
5. $\lim_{b \to 2^-} \int_0^b f(x)\, dx$

Review Exercises (p. 506)
1. 1   **3.** 6
5. The improper integral has no value.
7. Area = 1 square unit.   **9.** 5
11. $\frac{64}{3}$ **13.** 4
15. (a) $f(x) = \frac{8}{9}x \geq 0$ on $[0, \frac{3}{2}]$. $\int_0^{3/2} f(x)\, dx = 1$.
(b) $E(x) = 1$

17. (a) $f(x) = 12x^3(1 - x^2) \geq 0$ on $[0, 1]$.
$\int_0^1 f(x)\, dx = 1$.
(b) $E(x) = \frac{24}{35}$

19. The average yearly sales is 1255 units.

21. The average price is $14.47.

23. (a) The probability is $\frac{1}{4}$.
(b) The probability is $\frac{5}{12}$.
(c) $E(X) = 4$

25. (a) $f(x) = \frac{3}{635,840}(x^2 - 28x + 196) \geq 0$ on $[20, 100]$.
$\int_{20}^{100} f(x)\, dx = 1$.
(b) The probability is $\frac{217}{7948} \approx 0.0273$.
(c) The probability is $\frac{607}{3974} \approx 0.153$.
(d) The man's expected age of death is 78.52 years.

27. (a) $f(x) = \frac{1}{2}x > 0$ on $[0, 2]$. $\int_0^2 f(x)\, dx = 1$.
(b) The probability is $\frac{1}{4} = 0.25$.
(c) The probability is $\frac{5}{16} = 0.3125$.
(d) The probability is $\frac{7}{16} = 0.4375$.
(e) $E(x) = \frac{4}{3}$.

29. (a) The probability is $\frac{1}{5} = 0.2$.
(b) The probability is $\frac{1}{3} \approx 0.3333$.
(c) The expected wait time is 7.5 minutes.

31. The probability is 0.0821.

33. (a) The probability is 0.1055.
(b) The probability is 0.3189.

35. The probability is 0.5134.

Mathematical Questions from Professional Exams (p. 509)
1. (d) $-\frac{1}{4}$ **2.** (e) 0   **3.** (b) $\frac{1}{4}$
4. (b) $\frac{7}{16}$ **5.** (d) $\frac{1}{4}$

CHAPTER 8 — Calculus of Functions of Two or More Variables

Exercise 8.1 (p. 515)

1.
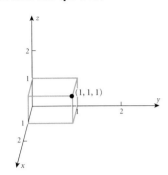

3.
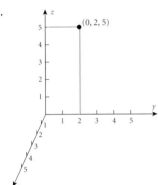

5.
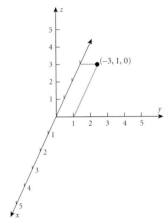

7. $(0, 0, 3), (0, 1, 0), (0, 1, 3), (2, 0, 0), (2, 0, 3), (2, 1, 0)$

9. $(1, 2, 5), (1, 4, 3), (1, 4, 5), (3, 2, 3), (3, 2, 5), (3, 4, 3)$

11. $(-1, 0, 5), (-1, 2, 2), (-1, 2, 5), (4, 0, 2), (4, 0, 5), (4, 2, 2)$

13. The plane through the point $(0, 3, 0)$ that is parallel to the xz-plane

15. The yz-plane

17. The plane through the point $(0, 0, 5)$ that is parallel to the xy-plane

19. $\sqrt{17}$ units   **21.** $\sqrt{57}$ units   **23.** $\sqrt{26}$ units

25. $(x - 3)^2 + (y - 1)^2 + (z - 1)^2 = 1$

27. $(x + 1)^2 + (y - 1)^2 + (z - 2)^2 = 9$

29. Center $= (-1, 1, 0)$, radius $= 2$

31. Center $= (-2, -2, -1)$, radius $= 3$

33. Center $= (2, 0, -1)$, radius $= 2$

35. $x^2 + (y - 3)^2 + (z - 6)^2 = 17$

Exercise 8.2 (p. 520)

1. $f(2, 1) = 5$ **3.** $f(2, 1) = \sqrt{2}$ **5.** $f(2, 1) = \frac{1}{5}$

7. $f(2, 1) = 3$ **9.** $f(2, 1) = 0$

11. (a) $f(1, 0) = 3$ (b) $f(0, 1) = 2$ (c) $f(2, 1) = 10$
(d) $f(x + \Delta x, y) = 3x + 3\Delta x + 2y + xy + \Delta xy$
(e) $f(x, y + \Delta y) = 3x + 2y + 2\Delta y + xy + x\Delta y$

13. (a) $f(0, 0) = 0$ (b) $f(0, 1) = 0$
(c) $f(a^2, t^2) = at + a^2$
(d) $f(x + \Delta x, y) = \sqrt{xy + \Delta xy} + x + \Delta x$
(e) $f(x, y + \Delta y) = \sqrt{xy + x\Delta y} + x$

15. (a) $f(1, 2, 3) = 14$ (b) $f(0, 1, 2) = 2$
(c) $f(-1, -2, -3) = -14$

17. The domain is the set $\{(x, y) \mid x \geq 0 \text{ and } y \geq 0\}$. This set is the first quadrant together with its border.

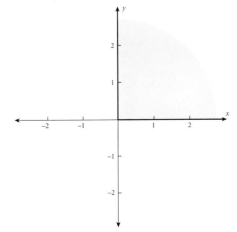

19. The domain is the set $\{(x, y) \mid x^2 + y^2 \leq 9\}$. This set is the circle of radius 3 centered at the origin and its interior.

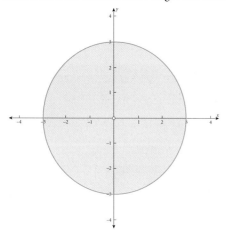

21. The domain is the set $\{(x, y) \mid x > 0 \text{ and } y > 0 \text{ and } y \neq 1\}$. This set is the first quadrant.

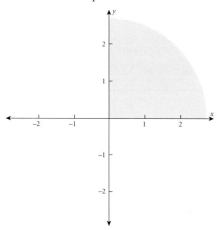

23. The domain is the set $\{(x, y) \mid x^2 + y^2 \neq 4\}$. This set is the union of the regions inside and outside of the circle of radius 2 centered at the origin.

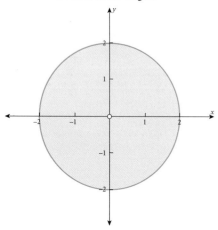

25. The domain is the set $\{(x, y) \mid (x, y) \neq (0, 0)\}$. This set is the entire xy-plane except for the origin.

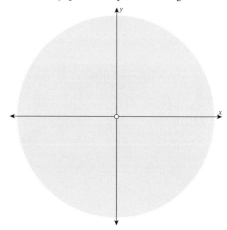

27. The domain is the set $\{(x, y, z) \mid x^2 + y^2 + z^2 \geq 16\}$. This set is the region on and outside of the sphere of radius 4 centered at the origin.

29. The domain is the set $\{(x, y, z) \mid x^2 + y^2 + z^2 \neq 0\}$. This set is the set of all points in space except for the origin $(0, 0, 0)$.

31. (a) $f(x + \Delta x, y) = 3x + 3\Delta x + 4y$
(b) $f(x, y + \Delta y) = 3x + 4y + 4\Delta y$
(c) $\frac{f(x + \Delta x, y) - f(x, y)}{\Delta x} = 3$
(d) $\lim_{\Delta x \to 0} \frac{f(x + \Delta x, y) - f(x, y)}{\Delta x} = 3$

33. $C(r, h) = 600\pi r^2 + 1000\pi rh$ dollars

35. (a) 6.75   (b) 18   (c) 2   (d) 1.5

37. The total monthly bill is $159.99.

39. (a) The heat index is 105.2°F.
(b) The relative humidity is 43%.
(c) The relative humidity is 55%.

Exercise 8.3 (p. 529)

1. $f_x(x, y) = 3$, $f_y(x, y) = -2 + 9y^2$,
$f_x(2, -1) = 3$, $f_y(-2, 3) = 79$

3. $f_x(x, y) = 2x - 2y$, $f_y(x, y) = 2y - 2x$,
$f_x(2, -1) = 6$, $f_y(-2, 3) = 10$

5. $f_x(x, y) = \frac{x}{\sqrt{x^2 + y^2}}$, $f_y(x, y) = \frac{y}{\sqrt{x^2 + y^2}}$,
$f_x(2, -1) = \frac{2\sqrt{5}}{5}$, $f_y(-2, 3) = \frac{3\sqrt{13}}{13}$

7. $f_x(x, y) = -24x - 2y$, $f_y(x, y) = -2x + 3y^2 + 2y$,
$f_{xx}(x, y) = -24$, $f_{yy}(x, y) = 6y + 2$,
$f_{yx}(x, y) = -2$, $f_{xy}(x, y) = -2$

9. $f_x(x, y) = ye^x + e^y + 1$, $f_y(x, y) = e^x + xe^y$, $f_{xx}(x, y) = ye^x$, $f_{yy}(x, y) = xe^y$, $f_{yx}(x, y) = e^x + e^y$, $f_{xy}(x, y) = e^x + e^y$

11. $f_x(x, y) = \frac{1}{y}$, $f_y(x, y) = -\frac{x}{y^2}$, $f_{xx}(x, y) = 0$, $f_{yy}(x,y) = \frac{2x}{y^3}$, $f_{yx}(x, y) = -\frac{1}{y^2}$, $f_{xy}(x, y) = -\frac{1}{y^2}$

13. $f_x(x, y) = \frac{2x}{x^2 + y^2}$, $f_y(x, y) = \frac{2y}{x^2 + y^2}$, $f_{xx}(x, y) = \frac{2y^2 - 2x^2}{(x^2 + y^2)^2}$, $f_{yy}(x, y) = \frac{2x^2 - 2y^2}{(x^2 + y^2)^2}$, $f_{yx}(x, y) = -\frac{4xy}{(x^2 + y^2)^2}$, $f_{xy}(x, y) = -\frac{4xy}{(x^2 + y^2)^2}$

15. $f_x(x, y) = \frac{-2y + 10}{x^2 y}$, $f_y(x, y) = \frac{x - 10}{xy^2}$, $f_{xx}(x, y) = \frac{8y + 20}{x^3 y}$, $f_{yy}(x, y) = \frac{20 - 2x}{xy^3}$, $f_{yx}(x, y) = \frac{10}{x^2 y^2}$, $f_{xy}(x, y) = \frac{10}{x^2 y^2}$

17. $f_{xy} = 0 = f_{yx}$

19. $f_{xy} = 24x^3 y + 14x = f_{yx}$

21. $f_{xy} = -\frac{2}{x^3} = f_{yx}$

23. $f_x(x, y, z) = 2xy - 3yz$, $f_y(x, y, z) = x^2 - 3xz$, $f_z(x, y, z) = 3z^2 - 3xy$

25. $f_x(x, y, z) = e^y$, $f_y(x, y, z) = xe^y + e^z$, $f_z(x, y, z) = ye^z$

27. $f_x(x, y, z) = \ln(yz) + \frac{y}{x}$, $f_y(x, y, z) = \ln(xz) + \frac{x}{y}$, $f_z(x, y, z) = \frac{x + y}{z}$

29. $f_x(x, y, z) = \frac{2x}{x^2 + y^2 + z^2}$, $f_y(x, y, z) = \frac{2y}{x^2 + y^2 + z^2}$, $f_z(x, y, z) = \frac{2z}{x^2 + y^2 + z^2}$

31. The slope is 20.

33. The slope is $-\frac{2\sqrt{11}}{11}$.

35. The slope is 1.

37. The slope is 1.

39. We have that $\frac{\partial z}{\partial x} = 2x$ and $\frac{\partial z}{\partial y} = 8y$. Now
$x\frac{\partial z}{\partial x} + y\frac{\partial z}{\partial y} = x(2x) + y(8y)$
$= 2(x^2 + 4y^2)$
$= 2z$.

41. We have that $\frac{\partial z}{\partial x} = \frac{x}{x^2 + y^2}$ and $\frac{\partial z}{\partial y} = \frac{y}{x^2 + y^2}$.
Then $\frac{\partial^2 z}{\partial x^2} = \frac{y^2 - x^2}{(x^2 + y^2)^2}$ and $\frac{\partial^2 z}{\partial y^2} = \frac{x^2 - y^2}{(x^2 + y^2)^2}$. Now
$\frac{\partial^2 z}{\partial x^2} + \frac{\partial^2 z}{\partial y^2} = \frac{y^2 - x^2}{(x^2 + y^2)^2} + \frac{x^2 - y^2}{(x^2 + y^2)^2}$
$= \frac{y^2 - x^2 + x^2 - y^2}{(x^2 + y^2)^2}$
$= 0$.

43. (a) $\frac{\partial z}{\partial x} = -20$, $\frac{\partial z}{\partial y} = -50$

(b) If the average price per pound of margarine remains fixed and the average price per pound of butter is increased by \$1, $\frac{\partial z}{\partial x}$ is the change in demand for butter. If the average price per pound of butter remains fixed and the average price per pound of margarine is increased by \$1, $\frac{\partial z}{\partial y}$ is the change in demand for margarine.

45. (a) $\frac{\partial A}{\partial N} = \frac{9}{I}$, $\frac{\partial A}{\partial I} = -\frac{9N}{I^2}$

(b) $\frac{\partial A}{\partial N}(78, 217) = 0.041$, $\frac{\partial A}{\partial I}(78, 217) = -0.015$.

(c) If he pitched 217 innings and he gave up 79 earned runs, his earned run average would increase by 0.041. If he pitched 218 innings and gave up 78 earned runs, his earned run average would decrease by 0.015.

47. (a) $\frac{\partial H}{\partial t} = 2.04901523 - 0.2247554 \, | \, r$
$- 0.01367566t$
$+ 0.00245748tr$
$+ 0.00085282r^2$
$- 0.00000398tr^2$

(b) $\frac{\partial H}{\partial t}$ is the change in the heat index with respect to temperature, given a fixed humidity.

(c) $\frac{\partial H}{\partial r} = 10.141333127 - 0.2247554 \, | \, t$
$- 0.10963434r$
$+ 0.00122874t^2$
$+ 0.00170564tr$
$- 0.00000398t^2 r$

(d) $\frac{\partial H}{\partial r}$ is the change in the heat index with respect to humidity, given a fixed temperature.

49. No. Explanations will vary.

Exercise 8.4 (p. 536)

1. $(-1, 0)$, $(0, 0)$, $(1, 0)$

3. $(-1, -1)$, $(0, 0)$, $(1, 1)$

5. $(0, 0)$

7. The point $(0, 0)$ is a local minimum.

9. The point $\left(\frac{3}{2}, 0\right)$ is a local minimum.

11. The point $(-2, 4)$ is a saddle point.

13. The point $(2, -1)$ is a local minimum.

15. The point $(4, -2)$ is a local minimum.

17. The point $(0, 0)$ is a saddle point.

19. The point $(0, 0)$ is a saddle point, and the point $(2, 2)$ is a local minimum.

21. The point $(0, 0)$ is a critical point that is neither a saddle point nor a local extremum, and the point $\left(3, -\frac{9}{2}\right)$ is a saddle point.

23. The funtion has no critical points.

25. The maximize profits, the quantities sold should be $x \approx \frac{20{,}000}{7} \approx 2857$ units and $y = \frac{2000}{7} \approx 286$ units. The corresponding prices are $P = \frac{64{,}000}{7} \approx \$9{,}143$ and $q = \frac{54{,}000}{7} \approx \$7{,}714$. The maximum profit is $P\left(\frac{20}{7}, \frac{2}{7}\right) = \frac{128{,}000}{7} \approx \$18{,}286$.

27. The manufacturer should produce 15,250 tons of grade A and 4100 tons of grade B to maximize profit.

29. For a fixed amount of the first drug, an amount of $\frac{2b}{3}$ of the second drug maximizes the reaction. For a fixed amount of the second drug, an amount of $\frac{2a}{3}$ of the first drug maximizes the reaction. If the amounts of both drugs are variable, $\frac{2a}{3}$ units of the first drug and $\frac{2b}{3}$ units of the second drug maximize the reaction.

31. There are no such values of x and t that will maximize y.

33. (a) The dimensions are $43\frac{1}{3}$ inches by $21\frac{2}{3}$ inches by $21\frac{2}{3}$ inches.

(b) For the cylinder of maximum volume, the radius is $\frac{130}{3\pi} \approx 13.79$ inches and the height is $43\frac{1}{3}$ inches.

35. $x = 50$ tons, $y = 1$ ton per week

Exercise 8.5 (p. 545)

1. The maximum value is 15.

3. The minimum value is $\frac{1}{2}$.

5. The maximum value is 528.

7. The minimum value is 612.

9. The maximum value is 16,000.

11. The minimum value is $\frac{233}{12}$.

13. The two numbers are both 50.

15. The three numbers are all $\frac{5\sqrt{3}}{3}$.

17. To minimize cost, the factory should produce 18 units of type x and 36 units of type y.

19. The dimensions are $20\frac{2}{3}$ inches by $20\frac{2}{3}$ inches by $20\frac{2}{3}$.

21. (a) $178\frac{4}{7}$ units of capital and 750 units of labor will maximize the total production.

(b) The maximum number of units of production is 529.14 units.

23. The dimensions are $\sqrt[3]{175}$ feet by $\sqrt[3]{175}$ feet by $\sqrt[3]{175}$ feet, which is approximately 5.593 feet by 5.593 feet by 5.593 feet.

25. The dimensions are $\sqrt[3]{12}$ feet by $\sqrt[3]{12}$ feet by $\frac{3\sqrt[3]{12}}{2}$ feet, which is approximately 2.29 feet by 2.29 feet by 3.43 feet.

Exercise 8.6 (p. 551)

1. $2y^3 + \frac{8}{3}$ **3.** $18x^2 + 4x$ **5.** $3y + \frac{5}{2}$

7. $8x - 22$ **9.** $\frac{1}{3\sqrt[3]{1+y^2}}$ **11.** $e^y(e^2 - 1)$

13. $(e^4 - 1) e^{-4y}$ **15.** $\frac{2}{\sqrt{y+6}}$ **17.** 8

19. $\frac{35}{3}$ **21.** $\frac{17}{6}$ **23.** 22

25. 12 **27.** 24 **29.** 21

31. Volume $= \frac{35}{2}$ cubic units

CHAPTER 8 Review

True – False Items (p. 553)

1. True **2.** False **3.** False **4.** False

Fill in the Blanks (p. 553)

1. surface **2.** $2 - \sqrt{2}$ **3.** $x = x_0$ **4.** saddle point

Review Exercises (p. 553)

1. 3 units **3.** $\sqrt{69}$ units **5.** 5 units

7. The radius is 3 units.

9. $(x + 6)^2 + (y - 3)^2 + (z - 1)^2 = 4$

11. The center is the point $(1, -3, -8)$, and the radius is 5.

13. (a) $(x - 1)^2 + (y + 4)^2 + (z - 3)^2 = 36$

(b) The center is the point $(1, -4, 3)$, and the radius is 6.

15. (a) $f(1, -3) = 11$ (b) $f(4, -2) = -8$

17. (a) $f(1, -3) = -\frac{1}{2}$ (b) $f(4, -2) = 0$

19. The domain is the entire xy-plane.

21. The domain is the set $\{(x, y) \mid y > x^2 + 4\}$, which is the set of points above the parabola $y = x^2 + 4$.

23. The domain is the set $\{(x, y) \mid (x + 2)^2 + y^2 \geq 9\}$, which is the set of points on or outside of the circle of radius 3 centered at the point $(-2, 0)$.

25. $f_x(x, y) = 2xy + 4$, $f_y(x, y) = x^2$, $f_{xx}(x, y) = 2y$, $f_{xy}(x, y) = 2x$, $f_{yx}(x, y) = 2x$, $f_{yy}(x, y) = 0$

27. $f_x(x, y) = y^2 e^x + \ln y$, $f_y(x, y) = 2ye^x + \frac{x}{y}$, $f_{xx}(x, y) = y^2 e^x$, $f_{xy}(x, y) = 2ye^x + \frac{1}{y}$, $f_{yx}(x, y) = 2ye^x + \frac{1}{y}$, $f_{yy}(x, y) = 2e^x - \frac{x}{y^2}$

29. $f_x(x, y) = \frac{x}{\sqrt{x^2 + y^2}}$, $f_y(x, y) = \frac{y}{\sqrt{x^2 + y^2}}$, $f_{xx}(x, y) = \frac{y^2}{(x^2 + y^2)^{3/2}}$, $f_{xy}(x, y) = -\frac{xy}{(x^2 + y^2)^{3/2}}$, $f_{yx}(x, y) = -\frac{xy}{(x^2 + y^2)^{3/2}}$, $f_{yy}(x, y) = \frac{x^2}{(x^2 + y^2)^{3/2}}$

31. $f_x(x, y) = e^x a \ln(5x + 2y) + \frac{5}{5x + 2y}b$, $f_y(x, y) = \frac{2e^x}{5x + 2y}$,

$f_{xx}(x, y) = e^x a \ln(5x + 2y) + \frac{50x + 20y - 25}{(5x + 2y)^2}b$,

$f_{xy}(x, y) = \frac{e^x(10x + 4y - 10)}{(5x + 2y)^2}$,

$f_{yx}(x, y) = \frac{e^x(10x + 4y - 10)}{(5x + 2y)^2}$,

$f_{yy}(x, y) = -\frac{4e^x}{(5x + 2y)^2}$

33. $f_x(x, y, z) = 3e^y + ye^z - 24xy$,
$f_y(x, y, z) = 3xe^y + xe^z - 12x^2$,
$f_z(x, y, z) = xye^z$

35. The slope is 12. **37.** The slope is 1.

39. (a) The only critical point is $(-4, -2)$.
(b) The point $(-4, -2)$ is a local maximum.

41. (a) The only critical point is $(1, 2)$.
(b) The point $(1, 2)$ is a local maximum.

43. (a) The only critical point is $a0, \frac{9}{2}b$.
(b) The point $a0, \frac{9}{2}b$ is a local minimum.

45. The maximum value is $\frac{5900}{19}$.

47. The minimum value is $\frac{16}{5}$.

49. $-\frac{40y}{3}$ **51.** $24x^2 + 8$ **53.** 51

55. 448   **57.** 32   **59.** $\frac{27}{2}$

61. The volume is 672 cubic units.

63. (a) $\frac{\partial z}{\partial K} = 20(\frac{L}{K})^{3/4}$, $\frac{\partial z}{\partial L} = 60(\frac{K}{L})^{1/4}$
(b) $\frac{\partial z}{\partial K} = 1.257$, $\frac{\partial z}{\partial L} = 150.892$.

(c) The factory should increase the use of labor. Explanations will vary.

65. $C_x(x, y) = 40$, $C_y(x, y) = 45$, If the number of deluxe vacuum cleaners produced remains fixed, increasing the production of standard vacuum cleaners by one will increase cost by $40. If the number of standard vacuum cleaners produced remains fixed, increasing the production of deluxe vacuum cleaners by one will increase cost by $45.

67. (a) $R(x, y) = -6x^2 + 3xy - 8y^2 + 350x + 400y$
(b) $R_x(x, y) = -12x + 3y + 350$,
$R_y(x, y) = 3x - 16y + 400$, If the demand for deluxe vacuum cleaners produced remains fixed, an increase of one in the demand for standard vacuum cleaners will change revenue by R_x dollars. If the demand for standard vacuum cleaners produced remains fixed, an increase of one in the demand for deluxe vacuum cleaners will change revenue by R_y dollars.

69. (a) $P(x, y) = -6x^2 + 3xy - 8y^2 + 310x + 355y - 1050$
(b) $P_x(50, 30) = -160$, $P_y(50, 30) = 70$, If the demand for deluxe vacuum cleaners produced remains at 30 vacuum cleaners, increasing the demand of standard vacuum cleaners from 50 to 51 will decrease profit by $160. If the demand for standard vacuum cleaners produced remains at 50 vacuum cleaners, increasing the demand of deluxe vacuum cleaners from 30 to 31 will increase profit by $70.

71. (a) 4000 units of brand x at a price of $4,000 and 5000 units of brand y at a price of $11,000 will maximize profit.
(b) The maximum profit is a loss of $159,000.

73. (a) $15,300 should be allocated to capital, and $35,700 should be allocated to labor.
(b) The maximum number of units is 3409 units.

APPENDIX A Graphing Utilities

Exercise Appendix A.1 (p. 557)

1. $(-1, 2)$ quadrant II   **3.** $(3, 1)$ quadrant I

5. X min $= -6$
X max $= 6$
X scl $= 2$
Y min $= -4$
Y max $= 4$
Y scl $= 2$

7. X min $= -6$
X max $= 6$
X scl $= 2$
Y min $= -1$
Y max $= 3$
Y scl $= 1$

9. X min $= 3$
X max $= 9$
X scl $= 1$
Y min $= 2$
Y max $= 10$
Y scl $= 2$

11. X min $= -12$
X max $= 6$
X scl $= 1$
Y min $= -4$
Y max $= 8$
Y scl $= 1$

13. X min $= -30$
X max $= 50$
X scl $= 10$
Y min $= -100$
Y max $= 50$
Y scl $= 10$

15. X min $= -10$
X max $= 110$
X scl $= 10$
Y min $= -20$
Y max $= 180$
Y scl $= 20$

Exercise Appendix A.2 (p. 561)

1. (a) 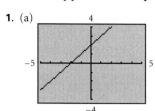 (b) 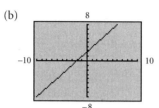 (c) 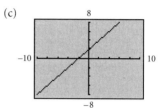 (d)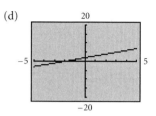

3. (a) 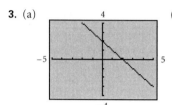 (b) 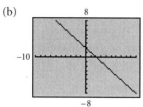 (c) 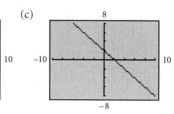 (d)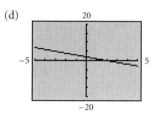

5. (a) 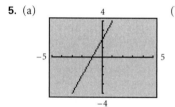 (b) 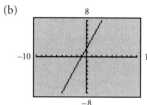 (c) 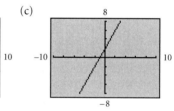 (d)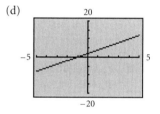

7. (a) 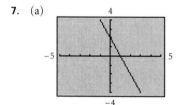 (b) 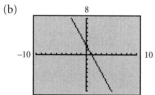 (c) 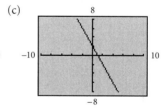 (d)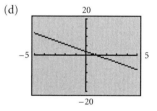

9. (a) 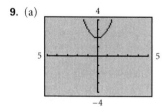 (b) 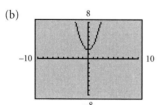 (c) 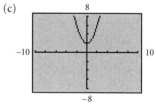 (d)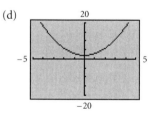

11. (a) 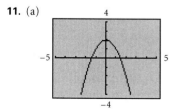 (b) 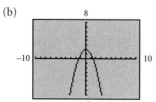 (c) 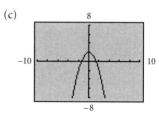 (d)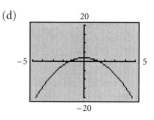

13. (a) 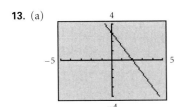 (b) 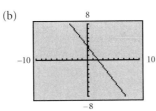 (c) 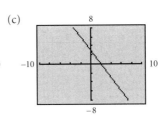 (d)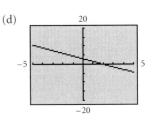

15. (a) 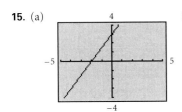 (b) 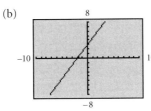 (c) 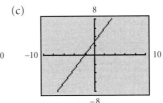 (d)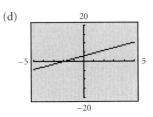

17. 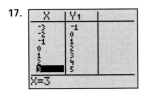 **19.** 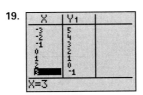 **21.** 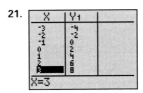 **23.**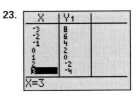

25. **27.**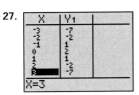

29. **31.**

Exercise Appendix A.3 (p. 562)

1. Yes **3.** Yes **5.** No **7.** Yes

9. $Y \min = 1$
$Y \max = 9$
$Y \text{scl} = 1$

Exercise Appendix A.4 (p. 565)

1. The smaller of the two x-intercepts is -3.41.

3. The smaller of the two x-intercepts is -1.71.

5. The smaller of the two x-intercepts is -0.28.

7. The positive x-intercept is 3.

9. The positive x-intercept is 4.5.

11. The positive x-intercepts are 0.32 and 12.3.

13. The positive x-intercepts are 1 and 23.

15. (a) The x-intercepts are -1 and 1. The y-intercept is -1.

(b) The graph is symmetric with respects to the y-axis.

17. (a) The graph has no intercepts.

(b) The graph is symmetric with respects to the origin.

Photo Credits

Chapter 0
Page 1: PhotoDisc, Inc./Getty Images. Page 63: ThinkStock LLC//Index Stock.

Chapter 1
Page 97: Digital Vision. Page 115: Simon Fraser/SPLPhoto Researchers Inc. Page 118: Corbis Digital Stock. Page 135: Monika Graff/The Image Works. Page 144: Lester Lefkowitz/Stone/Getty Images.

Chapter 2
Page 163: John Henley/Corbis Images. Page 201: Corbis Digital Stock. Page 209: Corbis Digital Stock. Page 227: PhotoDisc, Inc./Getty Images.

Chapter 3
Page 236: Syracuse Newspapers/Dick Blume/The Image Works. Page 257: Corbis Digital Stock. Page 264: PhotoDisc Red/Getty Images. Page 271: PhotoDisc Green/Getty Images.

Chapter 4
Page 273: PhotoDisc, Inc./Getty Images. Page 284: ©AP/Wide World Photos. Page 318: Corbis Digital Stock. Page 339: Corbis Digital Stock.

Chapter 5
Page 380: Digital Vision/Getty Images. Page 392 (left): Francesco Reginato/The Image Bank/Getty Images. Page 392 (right): Image State. Page 419: HIRB//Index Stock.

Chapter 6
Page 420: ©AP/Wide World Photos. Page 447: Comstock Images/Getty Images. Page 459: PhotoDisc, Inc./Getty Images. Page 473: SciMAT/Photo Researchers, Inc.

Chapter 7
Page 485: Kim Kulish/Corbis Images. Page 490: PhotoDisc, Inc./Getty Images. Page 499: PhotoDisc, Inc./Getty Images. Page 500: Monika Graff/The Image Works.

Chapter 8
Page 511: Stone/Getty Images. Page 521: Brendan Byrne/Digital Vision. Page 541: Rich La Salle/Index Stock.

Index

... (ellipsis), 2
$|a|$ (absolute value), 18
$\int$ (integral sign), 423
$+$ (addition), 6
$\emptyset$ (null set), 2
$\approx$ (approximately equal to), 5
/ (cancellation mark), 12
/ (division), 6
dx, 406
dy, 406
$=$ (equal sign), 6
$f(x)$, 110
$>$ (greater than), 17
$\geq$ (greater than or equal to), 17
∞ (infinity), 52
$<$ (less than), 17
$\leq$ (less than or equal to), 17
· (multiplication), 6
$\sqrt{\ }$ (square root), 23
( ) (parentheses), 7
π (pi), 3
$-$ (subtraction), 6

A

Abscissa (x-coordinate), 77
Absolute maximum/minimum, 381–382
   in applied problems, 384–390
   finding, 382–383
   necessary condition for a function to have an, 382
   test for, 382
Absolute value, 18
Absolute value function, 138–139, 142
Acceleration, 324
Addition, 6
Addition property of inequalities, 53–54
Additive identity, 10
Additive inverse, 10
Additive inverse property, 10
Additive property of area, 451
Algebraic expressions, 19
Antiderivative(s), 421–422
   of $f(x) = x^n$, 423
Approximations, 4–5

Area:
   additive property of, 451
   applications involving, 457–460
   as a definite integral, 461
   enclosed by two graphs, 454–457
   under a graph, 451–452
   maximizing, 385–386
   nonnegative property of, 451
Area problem, 250–251
Argument (of a function), 111
Arithmetic calculators, 6
Associative property, 9
Asymptotes:
   horizontal, 260–261
   vertical, 261–262
Average cost function, 302
Average rate of change, 130
   limit of, 250
   of y with respect to x, 85
Average value of a function, 490–491
   defined, 491
   geometric interpretation of, 491–492
Average velocity, 321–322
Axis of symmetry, 167

B

Base, 21
Binomials, 30
   cubes of, 32
   squares of, 31
Boyle's law, 334

C

Calculators, 5–6
Cancellation marks, 12
Cancellation properties, 12
Carbon dating, 474
Cartesian (rectangular) coordinates, 76–77
Center of a sphere, 514
Chain Rule, 310–311
Change, average rate of, 130
   limit of, 250
Change-of-base formula, 220

Circle:
   area of a, 74
   circumference of a, 20, 74
Closed intervals, 52
Cobb, Charles W., 530
Cobb-Douglas Model, 530
Coefficient(s), 30
   leading, 30
Common logarithms (log), 219
Common logarithmic function, 207
Common multiples, 14
Commutative property, 9
Complete graph, 99
Completely factored polynomials, 33
Completing the square, 45–46
Composite functions, 310
Composition, 310
Compound interest (compounding), 224–225
   continuous compounding, 224
Concave up/down, 365–366
Concavity, test for, 366
Connected mode, 143
Constant(s):
   defined, 19
   derivative of a function times a, 288–289
Constant function(s), 127–128, 140
   derivative of the, 286–287
   limit of the, 243–244
Consumer's surplus (CS), 458
Continuity and differentiability, 351–352
Continuous compounding, 224
Continuous function(s), 254–255
Continuously compounded interest, law of, 472
Continuous probability functions, 495
Continuous random variable, 494
   expected value for a, 500
Coordinates, 16
   of the point, 76
   rectangular, 76–77
Coordinate axes, 77, 512

I-1

Cost:
   of increasing production by one
      unit, 281
   marginal, 280, 428
   maximizing, 386–388
Cost function, 116
   average, 302
Counting (natural) numbers, 2
Critical points, 532
Cubes:
   of binomials, 32
   difference of two, 32
   perfect, 32
   sum of two, 32
Cube function, 111
Cube root, 64
Cube root function, 137–138, 141
Cubic function, 141
Cylinder, surface area of a right circular, 74

D

Decay, law of uninhibited, 472
Decimals, 3
   as approximations, 5
   changing scientific notation to, 25–26
   converting, to scientific notation, 24
   real numbers as, 4
Decreasing functions, 127–128
Definite integral(s):
   applications of, 446–448
   approximating, 463–465
   area as a, 461
   defined, 442
   notation for, 443
   properties of, 444, 445
Degree:
   of a monomial, 30
   of a polynomial function, 181, 182
   of a power function, 178–180
Demand, elasticity of, *see* Elasticity of demand
Demand equation, 116, 475
Demand level, 457–458
Denominator:
   defined, 3
   rationalizing the, 66
Density functions, *see* Probability density functions
Dependent variable, 110, 516
Derivative(s). *See also* Partial derivative(s)
   of the constant function, 286–287
   of a constant times a function, 288–289

Derivative(s) (*continued*)
   defined, 276
   of a difference, 290
   of exponential functions, 308–309, 311–312, 315
   of $f(x) = a^x$, 315
   of $f(x) = e^{g(x)}$, 311–312
   of $f(x) = \ln g(x)$, 313
   of $f(x) = \ln x$, 312–313
   of $f(x) = \log_a x$, 314
   of $f(x) = x^{p/q}$, 336
   finding, from differentials, 408–409
   formula for, 278
   geometric interpretation of, 408
   higher-order, 320
   Leibniz notation for, 286–287
   of logarithmic functions, 312–314
   of a polynomial function, 290
   of power functions, 287–288, 298
   prime notation for, 286
   of a product, 295–296
   of a quotient, 296–297
   second, 320
   steps for finding, 277
   of a sum, 289
   using the Chain Rule to find, 310–311
   using the Power Rule to find a, 304–305
Difference, 6, 11
   derivative of a, 290
   limit of a, 245
   of two cubes, 32
   of two squares, 31
Difference quotient (of a function), 112
Differentiability and continuity, 351–352
Differentiable functions, 278
Differential(s), 405–406
   defined, 406
   finding derivatives from, 408–409
   formulas for, 406–407
Differential equation(s), 470
   applications of, 471–476
   boundary conditions with, 470
   general solution of, 470
   particular solution of, 470
   solution of, 470
Differentiation, 278
   implicit, 328
Digits, set of, 2
Diminishing returns, 375
Dirichlet, Lejeune, 98
Discontinuous functions, 254
Discrete probability functions, 495
Discrete random variable, 494

Discriminant (of a quadratic equation), 48
Distance:
   between points in space, 513
   between points on a plane, 77–78
   on real number line, 18
Distance formula, 78, 513
Distribution(s):
   exponential, 498–499
   uniform, 498
Distributive property, 9
Division, 6
Division properties, 11–12
Domain:
   of a function, 109, 114–115, 123
   of a function of two variables, 516, 517
   of a function of three variables, 519–520
   of the variable, 20
Dot mode, 143
Double integrals, 547, 549–550
   defined, 549
   finding volume by using, 550–551
Double root, 44
Douglas, Paul H., 530

E

e, 193–194
Elasticity, 393–394
Elasticity of demand:
   defined, 394
   determination of, 394–395
   and revenue, 395–396
Elements, in a set, 2
Ellipsis, 2
Empty set, 2
End behavior:
   and limit at infinity of a function, 260
   of a polynomial function, 183–184
Endpoints (of an interval), 52
Equality, 2
Equal sign, 6
Equation(s). *See also* Graphs of equations; *specific equations,* e.g. Demand equation
   equivalent, 41
   finding intercepts from an, 102–103
   as functions, 113
   general, 82
   linear, 82–85
   logarithmic, 207
   in one variable, 40
   quadratic, 43
   radical, 67

Equation(s) (*continued*)
   roots of, 40
   satisfying an, 40, 98
   sides of an, 40, 98
   solution set of, 40
   solving, 40
   solving, with graphic utility, 565–567
   in two variables, 82–85, 98
Equilibrium point, 457
Equivalent equations, 41
Equivalent inequalities, 54–55
Euler, Leonhard, 98
eVALUATE (graphing utilities), 563
Even functions, 125
Eventually diminishing marginal productivity, law of, 527
Expected value, 500
Explicit form of a function, 113
Exponents, 20–21
   Laws of, 22, 187
   and logarithms, 203–204
   negative, 22
   rational, 68
Exponential density function (exponential distribution), 498–499
Exponential equations, 195
Exponential function(s):
   defined, 187
   derivatives of, 308–309, 311–312, 315
   e, 193–194
   graphing, 189–195
   indefinite integrals of, 426
   properties of, where $0 < a < 1$, 193
   properties of, where $a > 1$, 191
Exponential law, 472
Expressions:
   algebraic, 19
   finding the value of, 7–8
   rational, 34

F
Factors, 6, 33
Factoring, 32–33
   by grouping, 34
   over the integers, 33
   solving quadratic equations by, 43–44
Fermat, Pierre de, 202
First Derivative Test, 356
First-order partial derivatives, 527–528
Function(s), 98. *See also* Graphs of functions; *specific functions,* e.g. Quadratic function(s)
   antiderivatives of a, 421–423
   argument of a, 111

Function(s) (*continued*)
   average rate of change of a, 130
   average value of a, 490–492
   composite, 310
   constant, 127–128, 140, 243–244
   continuous vs. discontinuous, 254–255
   decreasing, 127–128
   defined, 109
   defined/not defined at x, 114
   derivative of a, 276
   determining whether an equation is a, 113
   difference quotient of a, 112
   differentiable, 278
   domain of a, 109, 114–115
   even vs. odd, 125
   implicit vs. explicit forms of a, 113
   increasing, 127–128
   increasing and decreasing, 354–355
   inflection point of, 368–369
   linear, 121, 139–140
   linear approximations of, 408–409
   local maximum/minimum of a, 128
   notation for, 110–111
   piecewise-defined, 143
   polynomial, 181–183
   power, 178–180
   probability, 495
   range of a, 109, 114
   rational, 184–185
   step, 143
   symmetry of, 125
   value of, at x, 109, 110
   vertical-line test for, 121–122
   zeroes of a, 125
Function keys, 6
Function(s) of three variables, 519–520
   Lagrange multipliers with, 543–544
   partial derivatives of, 528
Function(s) of two variables. *See also* Partial derivative(s)
   defined, 516
   domain of a, 516, 517
   graphing, 517–519
   range of a, 516
   surface of, 517
Fundamental Theorem of Calculus, 468

G
General equations, 82
General solution (of a differential equation), 470
Gini coefficient, 483

Graphs of equations:
   complete graph, 99
   creating, by plotting points, 99–101
   creating, with graphing utility, 558–561
   finding intercepts from, 102
   linear equation, 82–85
   symmetric, 103–104
   two variables, equation in, 98
Graphs of functions, 121
   absolute value function, 138–139, 142
   area enclosed by two graphs, 454–457
   area under a graph, 451–452
   concave up/down, 365–366
   constant functions, 140
   cube root function, 137–138, 141
   cubic function, 141
   decreasing functions, 127–128
   even vs. odd functions, 126
   exponential functions, 189–195
   greatest integer function, 142–143
   identity function, 140
   increasing and decreasing functions, 356–363
   increasing functions, 127–128
   inflection point on, 368–369
   intercepts of, 125
   linear functions, 139–140
   logarithmic functions, 205–207
   polynomial functions, 181–184
   power functions, 178–181
   reciprocal function, 141
   and reflections, 152
   and shifts, 148–151
   square function, 140
   square root function, 137, 141
   steps for creating, 356, 369
   two variables, functions of, 517–519
   and vertical-line test, 121–122
   viewing domain of, 123
   viewing range of, 123
Graphs of inequalities, 18
Graphs of quadratic functions, 165–171
   axis of symmetry of, 167
   creating, from vertex, axis, and intercepts, 168–170
   vertex of, 167
   x-intercepts of, 168
Graphing utility(-ies):
   calculators, 6
   checking for symmetry with, 564
   defined, 556
   finding intercepts using, 563–564
   graphing equations with, 558–561

Graphing utility(-ies) (*continued*)
  solving equations with, 565–567
  square screen on, 562
  viewing rectangle/window of, 556–557
Greatest integer function, 142–143
Growth, law of uninhibited, 472

H
Half-life, 474
Half-open (half-closed) intervals, 52
Higher-order derivatives, 320
Higher-order partial derivatives, 527–528
Horizontal asymptotes, 260–261
Horizontal line, equation for, 89
Horizontal shifts:
  combining vertical and, 151
  to the left, 150–151
  to the right, 150
Horizontal tangent lines, 348–349
Hyperbolic cosine function, 202
Hyperbolic sine function, 202
Hypoteneuse, 71

I
Identity(-ies), 41
  additive, 10
  multiplicative, 10
Identity function, 140
  limit of the, 243–244
Identity properties, 10
Image, 109
Implicit differentiation, 328
  finding first and second derivatives with, 331–332
  steps for, 330
Implicit form of a function, 113
Improper integral(s), 486–489
Increasing and decreasing functions, 354–355
  First Derivative Test for, 356
  graphing, 356–363
  test for, 355
Increasing functions, 127–128
Indefinite integral(s), 423–424.
       See also Integration
Independent variable, 110, 516
Index, 64
Inequality(-ies):
  defined, 17
  equivalent, 54–55
  graphing, 18
  in one variable, 54
  polynomial, 59–60
  properties of, 53–54
  quadratic, 57–59

Inequality (*continued*)
  rational, 60–61
  sides of, 17
  solutions of an, 54
  solving, 54–55
  strict vs. nonstrict, 17
  writing, 51–52
Inequality symbols, 17
  procedures reversing direction of, 55
Infinite limits, 259–260
  one-sided, 260
Infinity, 52
  limits at, 258–259
Inflection point, 368–369
Instantaneous rate of change, 279
Instantaneous velocity, 322
Integers, 3
  factoring over the, 33
Integral(s):
  definite, 442
  double, 547, 550–551
  improper, 486–489
  indefinite, 423–424
  iterated, 548–549
Integrand, 424
  discontinuous, 488
Integration, 424
  of exponential functions, 426
  formulas for, 424
  of logarithmic functions, 426
  lower/upper limits of, 442
  by substitution, 432–436
Integration by parts, 437–438
  formula for, 438
  hints for using, 439
Intercepts, 82–83, 101–102
  finding, from equation, 102–103
  finding, from graph, 102
  finding, with graphing utility, 563–564
  of the graph of a function, 125
Interest, law of continuously compounded, 472
INTERSECT feature (graphing utilities), 566
Intervals, 52
Inverse:
  additive, 10
  multiplicative, 10
Irrational numbers, 3
Iterated integrals, 548–549

L
Labor, marginal cost of, 526
Lagrange multiplier(s):
  defined, 539

Lagrange multiplier(s) (*continued*)
  with functions of three variables, 543–544
  method of, 539–541
Law of continuously compounded interest, 472
Law of eventually diminishing marginal productivity, 527
Laws of Exponents, 22, 187
Law of uninhibited growth/decay, 472
LCM, *see* Least common multiple
Leading coefficients, 30
Learning curves, 446–447
Least common multiple (LCM), 14
  of polynomials, 37
Left endpoint, 52
Left limit, 252
Legs (of a right triangle), 71
Leibniz, Gottfried Wilhelm von, 98, 274
Leibniz notation, 286–287
Like terms, 30
Limit(s), 237–238
  of the constant function, 243–244
  of a difference, 245
  finding, by graphing, 240–241
  finding, by using a table, 238–240
  of the identity function, 243–244
  infinite, 259–260
  at infinity, 258–259
  left vs. right, 252
  one-sided, 251–253
  of a polynomial function, 247
  of a power function, 246–247
  of a power or root, 247–248
  of a product, 245–246
  of a quotient, 248
  of a rational function, 248–249
  of a sum, 244–245
Line(s), 82
  finding equation of, given two points, 89–90
  graphing, given slope and one point, 87–88
  horizontal, 89
  point-slope form of equation of a, 88
  secant, 131
  slope-intercept form of equation of a, 90
  slope of a, 85
  tangent, 275
  vertical, 84
Linear approximations, 408–409
Linear equations, 82–85
Linear functions, 121, 139–140
ln (natural logarithmic function), 206

Local maxima/minima, 128
   First Derivative Test for locating, 356
   of functions of two variables, 531–533
   Second Derivative Test for locating, 373–374
Logarithms:
   and exponents, 203–204
   properties of, 215–217, 219
Logarithmic equations, 207
Logarithmic function(s):
   change-of-base formula with, 220
   common logarithmic function, 207
   defined, 203
   derivatives of, 312–314
   domain of a, 204–205
   graphs of, 205–207
   indefinite integrals of, 426
   natural logarithmic function, 206
   properties of, 210
log (common logarithms), 219
Logistic (saturation) curves, 377–378
Lorenz curve, 482
Loudness, 213
Lower limit of integration, 442
Lowest terms:
   rational functions, 185
   reduced to the, 34

M
Magnitude:
   of an earthquake, 214
   of a telescope, 233
Marginal analysis, 280
Marginal cost, 280
   finding cost from, 428
   of labor, 526
   of raw materials, 526
Marginal product, 527
Marginal productivity, 527
Marginal revenue, 282, 306–307, 332–333
   finding revenue from, 428
Market price, 457–458
Maxima, local, 128
   First Derivative Test for locating, 356
   of a function of two variables, 531–533
   Second Derivative Test for locating, 373–374
Maximum, absolute, 381–382
   in applied problems, 384–386, 388–390
   finding, 382–383
   necessary condition for a function to have an, 381–382
   test for, 382

Maximum value of a quadratic function, 170
Mean Value Theorem, 365
Method of Lagrange multipliers, 539–541
Minima, local, 128
   First Derivative Test for locating, 356
   of a function of two variables, 531–533
   Second Derivative Test for locating, 373–374
Minimum, absolute:
   in applied problems, 386–388
   finding, 382–383
   necessary condition for a function to have an, 381–382
   test for, 382
Minimum value of a quadratic function, 170
Mixed numbers, 6
Mixed partials, 528
Monomials, 30
Multiplication, 6
   by zero, 11
Multiplication properties (of inequalities), 54
Multiplicative identity, 10
Multiplicative inverse, 10
Multiplicative inverse property, 10

N
Natural (counting) numbers, 2
Natural logarithmic function (ln), 206
Negative exponents, 22
Negative real numbers, 17
Newton, Sir Isaac, 274, 321
Nonnegative property (of inequalities), 53
Nonstrict inequalities, 17
Norm of the partition, 465
nth roots, 64–65
Null set, 2
Numbers:
   classification of, 2–4
   mixed, 6
Numerator, 3

O
Odd functions, 125
One-sided infinite limits, 260
One-sided limits, 251–253
Open intervals, 52
Operations, 6
   order of, 7–8
Ordered pairs, 77
Ordered triples, 512
Order of operations, 7–8
Ordinate (y-coordinate), 77

Origin, 16, 76
   in space, 512
   symmetry with respect to, 104

P
Parentheses, 7
Partial derivative(s), 522–523
   applications of, 526–527
   defined, 523
   finding, 523–524
   of functions of three variables, 528
   geometric interpretation of, 524–525
   higher-order, 527–528
   mixed partials, 528
Partial integrals, 547–548
Particular solution (of a differential equation), 470
Perfect cubes, 32
Perfect roots, 65
Perfect squares, 31
Pi, 3
Piecewise-defined functions, 143
   continuity of, 255
Pixels, 556
Plane(s):
   distance between points on a, 77–78
   xy, 77, 512–513
   xz, 512–513
   yz, 512–513
Plotting, 77
Point(s):
   coordinate of the, 76
   critical, 532
   distance between, in space, 513
   distance between, on a plane, 77–78
   graphing an equation by plotting, 99–101
   inflection, 368–369
   plotting a, 77
   saddle, 532, 533
   symmetry of, 104
Point-slope form of equation of a line, 88
Polynomial(s), 30
   completely factored, 33
   factoring, 32–33
   prime, 33
   standard form for, 31
   zero, 31
Polynomial function(s):
   defined, 181
   degree of a, 181, 182
   derivative of a, 290
   end behavior of, 183–184
   graphs of, 182–184
   limit of a, 247

Polynomial inequalities, solving, 59–60
"Pooling the sample," 344
Positive real numbers, 17
Powers, *see* Exponents
Power function(s):
    defined, 178
    derivatives of, 287–288, 298
    limit of a, 246–248
    properties of, where degree is even integer, 179
    properties of, where degree is odd integer, 180
Power Rule, 304, 337
    using, to find a derivative, 304–305
Present value, 225
Price-demand equations, 475
Price function, 116
Prime notation, 286
Prime polynomials, 33
Principal nth root of a number a, 64
Principal square root, 23
Principle of substitution, 8
Probability density functions, 495–497
    defined, 496–497
    exponential density function, 498–499
    uniform density function, 498
Probability functions, 495
Producer's surplus *(PS)*, 458
Product(s), 6
    derivative of a, 295–296
    limit of a, 245–246
    special, 31–32
Production function(s), 339, 527
Productivity, marginal, 527
Profit function, 116, 390
Pythagorean Theorem, 71–72
Pythagorean triples, 76

Q

Quadrants, 77
Quadratic equation(s), 43
    discriminant of a, 48
    maximum/minimum values of a, 170
    repeated solution to, 44
    solving, by completing the square, 45–46
    solving, by factoring, 43–44
    solving, by Square Root Method, 44–45
    solving, with quadratic formula, 48–49
    standard form for, 43
Quadratic formula, 46–48
Quadratic function(s):
    defined, 164
    graphing, 165–171

Quadratic inequalities, solving, 57–59
Quadratic models, 170
Quotient(s), 6, 11
    arithmetic of, 13
    derivative of a, 296–297
    limit of a, 248

R

Radicals, 64
    properties of, 65
Radical equations, 67
Radical sign, 23
Radicand, 64
Radius of a sphere, 514
Raising to a power, 21
Random variable(s), 493–494
    continuous, 494, 500
    defined, 494
    discrete, 494
Range:
    of a function, 109, 114, 123
    of a function of two variables, 516
Rate of change:
    average, 85, 130, 250
    instantaneous, 279
    related rates, 398
Rational exponents, 68
Rational expressions, 34
Rational function(s), 184–185
    limit of a, 248–249
Rational inequalities, solving, 60–61
Rationalizing the denominator, 66
Rational numbers, 3
Raw materials, marginal cost of, 526
Real numbers:
    defined, 4
    properties of, 8–13
Real number line, 16–17
Reciprocal, 10
Reciprocal function, 141
Reciprocal property of inequalities, 54
Rectangle:
    area of a, 73
    perimeter of a, 73
Rectangular box:
    surface area of a, 74
    volume of a, 74
Rectangular (Cartesian) coordinates, 76–77
Reduced to the lowest terms, 34
Reflection:
    about the x-axis, 152
    about the y-axis, 152

Reflexive property, 8
Related rates, 398
Related rate problems, 398
    steps for solving, 400
Repeated solution (to quadratic equation), 44
Revenue:
    and elasticity of demand, 395–396
    marginal, 282, 306–307, 332–333
    maximizing, 388–389
    from sale of one additional unit, 282
Richter scale, 214
Riemann sums, 466
Right angle, 71
Right endpoint, 52
Right-handed systems, 512
Right-hand rule, 512
Right limit, 252
Right triangles, 71
Rolle's Theorem, 364
ROOT feature (graphing utilities), 563–566
Root(s):
    of an equation, 40
    limit of a, 247–248
    of multiplicity 2, 44
    perfect, 65
Roster method, 2
Rounding, 5
Rules of signs, 12

S

S^2 (variance), 505
Saddle points, 532, 533
Satisfying an equation, 40, 98
Saturation (logistic) curves, 377–378
Scale (of number line), 16
Scientific calculators, 6
Scientific notation, 24
Secant line, 131
Second derivative, 320
Second Derivative Test, 373–374
Second-order partial derivatives, 527–528
Set(s):
    classifying numbers in a, 2–4
    defined, 2
    elements in, 2
    empty (null), 2
    notation for, 2
Set-builder notation, 2
Shifts:
    horizontal, 150–151
    vertical, 148–149

Sides:
  of an equation, 40, 98
  of an inequality, 17
Signs, rules of, 12
Simplified rational expressions, 34
Slope:
  of a line, 85
  of the secant line, 131
  undefined, 85
Slope-intercept form of equation of a line, 90
Solution(s):
  of an equation, 40
  of an inequality, 54
  of a differential equation, 470
Solution set, 40
Solving an equation, 40
Space:
  distance between two points in, 513
  origin in, 512
  x-axis in, 512
  y-axis in, 512
Special products, 31–32
Sphere, 514
  surface area of a, 74
  volume of a, 74
Squares, 23
  of binomials, 31
  difference of two, 31
  perfect, 31
Square function, 140
Square root function, 137, 141
Square Root Method, 44–45
Square root(s), 23, 64
  principal, 23
Square screen (on graphing utility), 562
Standard equation of a sphere, 514
Standard form:
  for polynomials, 31
  for quadratic equations, 43
Statements, writing, 6
Step functions, 143
Strict inequalities, 17
Subscripts, 30n.
Substitution, principle of, 8
Substitution method, 432–436
Subtraction, 6
Sum, 6
  derivative of a, 289
  limit of a, 244–245
  of two cubes, 32
Surface (of a function of two variables), 517
Symmetric property, 8

Symmetry:
  axis of, 167
  checking for, with graphic utility, 564
  of functions, 125
  with respect to the origin, 104
  with respect to x-axis, 103–104
  with respect to y-axis, 104
  tests for, 105

T
Tangent line(s), 275
  horizontal, 348–349
  vertical, 349–350
The Tangent Problem, 274–275
Terms, 31
Thurstone, L. L., 339
Time. *See also* Rate of change
  maximizing profit over, 459–460
  total sales over, 448
Total reaction, 490
Total sales over time, 448
TRACE feature (graphing utilities), 563–564
Transformations, 148
Transitive property, 8
Triangle(s):
  area of a, 73
  right, 71
Trinomials, 30, 32
Truncation, 5

U
Uniform density function (uniform distribution), 498
Uninhibited growth, law of, 472
Upper limit of integration, 442

V
Van der Waals equation, 334
Variable(s):
  defined, 19
  dependent vs. independent, 110, 516
  domain of the, 20
  equations in one, 40
  equations in two, 82–85, 98
  functions of two, 516–519
  functions of three, 519–520
  random, 494
Variance, 505
Velocity:
  average, 321–322
  instantaneous, 322

Vertex (of the graph of a quadratic equation), 167
Vertical asymptotes, 261–262
Vertical line, equation for, 84
Vertical-line test, 121–122
Vertical shifts:
  combining horizontal and, 151
  down, 149
  up, 148–149
Vertical tangent lines, 349–350
Viewing rectangle/window (of graphing utility), 556–557
Volume:
  finding, by using double integrals, 550–551
  maximizing, 384–385

W
Weber-Fechner Law, 318
Whole numbers, 2

X
x-axis, 76
  reflection about the, 152
  in space, 512
  symmetry with respect to, 103–104
x-coordinate (abscissa), 77
x-intercept(s), 82, 101–102
  of the graph of a function, 125
xy-plane, 77, 512–513
xz-plane, 512–513

Y
y-axis, 76
  reflection about the, 152
  in space, 512
  symmetry with respect to, 104
y-coordinate (ordinate), 77
y-intercept(s), 82, 102
  of the graph of a function, 125
yz-plane, 512–513

Z
z-axis, 512
Zero(s), 2, 17
  division by, 12
  of a function, 125
  multiplication by, 11
ZERO feature (graphing utilities), 563–566
Zero-level earthquake, 214
Zero polynomial, 31
Zero-product property, 13